Experimental Methods in the Physical Sciences

Volume 47

Optical Radiometry for Ocean Climate Measurements

Experimental Methods in the Physical Sciences

Thomas Lucatorto, Albert C. Parr and Kenneth Baldwin
Editors in Chief

Experimental Methods in the Physical Sciences

Volume 47

Optical Radiometry for Ocean Climate Measurements

Edited by

Giuseppe Zibordi
Institute for Environment and Sustainability
Joint Research Centre
Ispra, Italy

Craig J. Donlon
European Space Agency/ESTEC
Noordwijk
The Netherlands

Albert C. Parr
Space Dynamics Laboratory, Utah State
University, Logan, UT, USA

AMSTERDAM • BOSTON • HEIDELBERG • LONDON
NEW YORK • OXFORD • PARIS • SAN DIEGO
SAN FRANCISCO • SINGAPORE • SYDNEY • TOKYO

Academic Press is an imprint of Elsevier

Academic Press is an imprint of Elsevier
225 Wyman Street, Waltham, MA 02451, USA
525 B Street, Suite 1800, San Diego, CA 92101-4495, USA
32 Jamestown Road, London NW1 7BY, UK
The Boulevard, Langford Lane, Kidlington, Oxford OX5 1GB, UK

Copyright © 2014 Elsevier Inc. All rights reserved.

Except as follows:
The European Union retains the copyright to the chapters — 3, 3.1, 4, 4.1, 5, 5.1, 6 and 6.1
The following chapters are in public domain — 2, 2.1 and 4.2
James A. Yoder retains the copyright for his contribution

No part of this publication may be reproduced or transmitted in any form or by any means, electronic or mechanical, including photocopying, recording, or any information storage and retrieval system, without permission in writing from the publisher. Details on how to seek permission, further information about the Publisher's permissions policies and our arrangements with organizations such as the Copyright Clearance Center and the Copyright Licensing Agency, can be found at our website: www..elsevier.com/permissions.

This book and the individual contributions contained in it are protected under copyright by the Publisher (other than as may be noted herein).

Notices
Knowledge and best practice in this field are constantly changing. As new research and experience broaden our understanding, changes in research methods, professional practices, or medical treatment may become necessary.

Practitioners and researchers must always rely on their own experience and knowledge in evaluating and using any information, methods, compounds, or experiments described herein. In using such information or methods they should be mindful of their own safety and the safety of others, including parties for whom they have a professional responsibility.

To the fullest extent of the law, neither the Publisher nor the authors, contributors, or editors, assume any liability for any injury and/or damage to persons or property as a matter of products liability, negligence or otherwise, or from any use or operation of any methods, products, instructions, or ideas contained in the material herein.

ISBN: 978-0-12-417011-7
ISSN: 1079-4042

For information on all Academic press publications visit our web site at http://store.elsevier.com/

Contents

List of Contributors — xv
Volumes in Series — xvii
Foreword — xxi
Preface — xxiii

1. **Introduction to Optical Radiometry and Ocean Climate Measurements from Space**
   James A. Yoder and B. Carol Johnson

   1.1. **Ocean Climate and Satellite Optical Radiometry**
   James A. Yoder, Kenneth S. Casey and Mark D. Dowell

1. Introduction	3
1.1 Characteristics of a Climate-Observing System	4
2. Global Climate Observing System Requirements for ECVs and CDRs	6
2.1 Ocean Color Radiometry	7
2.2 Sea Surface Temperature	8
3. From Essential Climate Variables to Climate Data Records	10
4. Conclusion	11
References	12

   1.2. **Principles of Optical Radiometry and Measurement Uncertainty**
   B. Carol Johnson, Howard Yoon, Joseph P. Rice and Albert C. Parr

1. Basics of Radiometry	14
1.1 Introduction	14
1.2 Radiance	17
1.3 Irradiance	21
1.4 Reflectance	23
1.5 Distance and Aperture Areas in Radiometry	28
2. Radiometric Standards and Scale Realizations	30
2.1 Sources	30
2.2 Radiometers	38

	3. The Measurement Equation	42
	3.1 Background and a Review of the Concepts	42
	3.2 Measurement Equation Examples	46
	3.3 Uncertainty in Ocean Color Measurements	57
	4. Summary	61
	Acknowledgments	62
	References	62

2. Satellite Radiometry

Charles R. McClain and Peter J. Minnett

2.1. Satellite Ocean Color Sensor Design Concepts and Performance Requirements

Charles R. McClain, Gerhard Meister and Bryan Monosmith

1. Introduction	74
2. Ocean Color Measurement Fundamentals and Related Science Objectives	75
3. Evolution of Science Objectives and Sensor Requirements	80
4. Performance Parameters and Specifications	84
4.1 Spectral Coverage and Dynamic Range	84
4.2 Coverage and Spatial Resolution	86
4.3 Radiometric Uncertainty	87
4.4 SNR and Quantization	89
4.5 Polarization	90
4.6 Additional Characterization Requirements	91
4.7 On-Board Calibration Systems	92
5. Sensor Engineering	93
5.1 Basic Sensor Designs: Whiskbroom and Pushbroom	95
5.2 Design Fundamentals and Radiometric Equations	96
5.3 Performance Considerations	99
5.4 Sensor Implementation	104
6. Summary	107
Acronyms	108
Symbols and Dimensions	109
7. Appendix. Historical Sensors	109
7.1 CZCS and OCTS	110
7.2 SeaWiFS	111
7.3 MODIS	113
7.4 MERIS	115
References	116

2.2. On Orbit Calibration of Ocean Color Reflective Solar Bands

Robert E. Eplee, Jr and Sean W. Bailey

1.	Introduction	121
2.	Solar Calibration	124
	2.1 SD Degradation	125
	2.2 SD Radiometric Response Trends	126
	2.3 SNR on Orbit	128
	2.4 Uncertainties in the Solar Calibration Data	128
3.	Lunar Calibrations	128
	3.1 ROLO Photometric Model of the Moon	129
	3.2 Lunar Radiometric Response Trending	130
	3.3 Uncertainties in Lunar Calibration	131
	3.4 Lunar Calibration Intercomparisons	133
4.	Spectral Calibration of Grating Instruments	135
5.	Vicarious Calibration	137
	5.1 NIR/SWIR Band Calibration	139
	5.2 Visible Band Calibration	140
	5.3 Alternative Approaches	142
6.	On-orbit Calibration Uncertainties	142
	6.1 Accuracy	143
	6.2 Long-term Stability of the TOA Radiances	143
	6.3 Precision of the TOA Radiances	144
	6.4 Combined Uncertainty Assessment	144
7.	Comparison of Uncertainties Across Instruments	145
8.	Summary of On-orbit Calibration	149
	References	150

2.3. Thermal Infrared Satellite Radiometers: Design and Prelaunch Characterization

David L. Smith

1.	Introduction	154
2.	Radiometer Design Principles	155
	2.1 Performance Model	159
	2.2 Signal to Noise	160
3.	Remote Sensing Systems	161
	3.1 Along Track Scanning Radiometers (ATSR)	161
	3.2 Sea and Land Surface Temperature Radiometer (SLSTR)	164
	3.3 Advanced Very High Resolution Radiometer (AVHRR)	165
	3.4 MOderate Resolution Imaging Spectroradiometer (MODIS)	166
	3.5 Visible Infrared Imaging Suite (VIIRS)	167
	3.6 Spinning Enhanced Visible and Infrared Imager (SEVIRI)	171

	4.	Calibration Model	172
		4.1 Radiometric Noise	174
		4.2 Nonlinearity	174
		4.3 Offset Variations	176
	5.	On-Board Calibration	176
		5.1 Calibration Sources	178
	6.	Pre-launch Characterization and Calibration	182
		6.1 Blackbody Calibration	182
		6.2 Instrument Radiometric Calibration	184
	7.	Conclusions	197
		References	198

2.4. Postlaunch Calibration and Stability: Thermal Infrared Satellite Radiometers

Peter J. Minnett and David L. Smith

1.	Introduction	201
2.	On-Board Calibration	203
	2.1 (A)ATSR Radiometric Calibration	204
	2.2 AVHRR Calibration	209
	2.3 MODIS and VIIRS Radiometric Calibration	213
	2.4 MODIS Spectroradiometric Calibration Assembly for On-Orbit Stability	214
	2.5 MODIS Mirror Response versus Scan Angle	216
3.	Comparisons with Reference Satellite Sensors	218
	3.1 Spatial Comparisons	219
	3.2 Temporal Comparisons	220
	3.3 Simultaneous Nadir Overpasses	222
	3.4 Instruments on the Same Satellite	223
4.	Validating Geophysical Retrievals	225
	4.1 Cloud Screening	229
	4.2 Atmospheric Correction Algorithm	230
	4.3 Geophysical Validation	232
	4.4 Ship-Board Radiometers	236
5.	Discussion	237
6.	Conclusions	239
	References	239

3. In Situ Optical Radiometry

Craig J. Donlon and Giuseppe Zibordi

3.1. In situ Optical Radiometry in the Visible and Near Infrared

Giuseppe Zibordi and Kenneth J. Voss

1.	Introduction and History	248
2.	Field Radiometer Systems	249

	2.1	General Classification: Multispectral and Hyperspectral	249
	2.2	Irradiance Sensors	250
	2.3	Basic Radiance Sensors	252
3.	**System Calibration**		254
	3.1	Linearity Response	255
	3.2	Temperature Response	255
	3.3	Polarization Sensitivity	256
	3.4	Stray Light Perturbations	257
	3.5	Spectral Response	257
	3.6	Angular Response of Irradiance Sensors	258
	3.7	Rolloff of Imaging Systems	260
	3.8	Immersion Effects	260
	3.9	Absolute Response	263
4.	**Measurement Methods**		264
	4.1	In Water Systems	265
	4.2	Above Water Systems	267
	4.3	Radiometric Data Products	268
5.	**Errors and Uncertainty Estimates**		273
	5.1	Calibration Specific Sources of Uncertainties	274
	5.2	Instrument Specific Sources of Uncertainties	276
	5.3	Methods and Field Specific Sources of Uncertainties	277
	5.4	Examples of Uncertainty Budget for Radiometric Products	282
6.	**Applications**		285
	6.1	Sky and Sea Radiance Distribution	285
	6.2	In-water Light Field Polarization	287
	6.3	Bio-Optical Models	289
	6.4	Validation of Satellite Radiometric Products	291
	6.5	In situ Data and System Vicarious Calibration	293
7.	**Summary and Outlook**		294
	References		295

3.2. Ship-Borne Thermal Infrared Radiometer Systems

Craig J. Donlon, Peter J. Minnett, Andrew Jessup, Ian Barton, William Emery, Simon Hook, Werenfrid Wimmer, Timothy J. Nightingale, and Christopher Zappa

1.	**Introduction and Background**		306
2.	**TIR Measurement Theory**		311
	2.1	General Considerations	311
	2.2	SST_{skin} Ship-Borne Radiometer Measurement Challenges	317
	2.3	Practical Measurement of SST_{skin} from a Ship-Borne Radiometer	320
3.	**TIR Field Radiometer Design**		321
	3.1	TIR Detectors	328
	3.2	TIR Radiometer Spectral Definition	336
	3.3	Beam Shaping and Steering	341
	3.4	Thermal Control System	350

		3.5	An Environmental System to Protect and Thermally Stabilize the Radiometer	351

	3.6	Instrument Control and Data Acquisition	353
	3.7	A Calibration System	354
	3.8	Summary	361
	3.9	Additional Comments	363
4.	Examples of FRM Ship-Borne TIR Radiometer Design and Deployments		363
	4.1	The DAR-011 Filter Radiometer	363
	4.2	The SISTeR Filter Radiometer	364
	4.3	NASA JPL NNR	368
	4.4	The Calibrated Infrared In situ Measurement System	371
	4.5	ISAR—Quasi Operational Ocean Field Radiometers	375
	4.6	Use of Unmanned Airborne Vehicles BESST Radiometer	380
	4.7	Spectroradiometers	382
	4.8	Derivation of Air Temperature Using a Spectroradiometer	387
	4.9	TIR Cameras	389
5.	Future Directions		393
6.	Conclusions		395
	Acknowledgments		395
	References		395

4. Theoretical Investigations

Barbara Bulgarelli, Menghua Wang and Christopher J. Merchant

4.1. Simulation of In Situ Visible Radiometric Measurements

Barbara Bulgarelli and Davide D'Alimonte

1.	Overview		407
2.	The RTE and Its Solution Methods		408
	2.1	The Radiative Transfer Equation	408
	2.2	Deterministic Solutions of the RTE	410
	2.3	Monte Carlo Solutions of the RTE	410
3.	Simulations of In Situ Radiometric Measurement Perturbations		413
	3.1	Overstructure Perturbations	414
	3.2	Perturbations Induced by Sea-Surface Waves	429
4.	Summary and Remarks		441
	References		442

4.2. Simulation of Satellite Visible, Near-Infrared, and Shortwave-Infrared Measurements

Menghua Wang

1.	Introduction	452
2.	Ocean—Atmospheric System	455
3.	Simulations	457

		3.1 Ocean Radiance Contributions	457
		3.2 The TOA Atmospheric Path Radiance Contributions	464
		3.3 Atmospheric Diffuse Transmittance	470
		3.4 Simulated and Satellite-Measured TOA Radiances	471
	4.	Summary	478
		Disclaimer	479
		References	479

4.3. Simulation and Inversion of Satellite Thermal Measurements

Christopher J. Merchant and Owen Embury

1.	Introduction	489
2.	Radiative Transfer Simulation for Thermal Remote Sensing	490
3.	Propagation of Thermal Radiation through Clear Sky	493
4.	Simulation of Interaction with Aerosol and Cloud	500
5.	Simulation of Surface Emission and Reflection	502
6.	Use of Simulations in Thermal Image Classification (Cloud Detection)	504
7.	Use of Simulations in Geophysical Inversion (Retrieval)	509
8.	Use of Simulations in Uncertainty Estimation	516
9.	Conclusion	521
	References	523

5. In Situ Measurement Strategies

Giuseppe Zibordi and Craig J. Donlon

5.1. Requirements and Strategies for In situ Radiometry in Support of Satellite Ocean Color

Giuseppe Zibordi and Kenneth J. Voss

1.	Introduction		532
2.	Overview of Past and Current Field-Related Radiometric Activities		533
	2.1	Field Measurements	533
	2.2	Intercomparisons	538
	2.3	Data Repositories	542
3.	Requirements and Strategies for Future Satellite Ocean-Color Missions		543
	3.1	Field Measurements for System Vicarious Calibration	544
	3.2	Field Measurements for the Validation of Satellite Data Products	546
	3.3	Field Measurements for Bio-Optical Modeling	547
	3.4	Protocols Revision and Consolidation	547
	3.5	Calibration and Characterization of Field Radiometers	547
	3.6	Data Reduction, Quality Control, and (re)Processing	548

		3.7	Accuracy Tailored to Applications	549
		3.8	Archival and Access	549
		3.9	Intercomparisons to Secure Accuracy and Best Practice	549
		3.10	Standardization and Networking	550
		3.11	Development and Implementation	551
	4.	Summary and Way Forward		551
		References		552

5.2. **Strategies for the Laboratory and Field Deployment of Ship-Borne Fiducial Reference Thermal Infrared Radiometers in Support of Satellite-Derived Sea Surface Temperature Climate Data Records**
Craig J. Donlon, Peter J. Minnett, Nigel Fox, and Werenfrid Wimmer

	1.	Introduction		558
	2.	Fiducial Reference Measurements for SST CDRs and Uncertainty Budgets		559
		2.1	FRM TIR Ship-Borne Radiometer Network	562
		2.2	The Importance of Uncertainty Budgets	563
	3.	Laboratory Intercalibration Experiments for FRM Ship-Borne Radiometers		585
	4.	Ship-Borne Radiometer Field Intercomparison Exercises		590
	5.	Protocols to Maintain the SI Traceability of FRM Ship-Borne TIR Radiometers for Satellite SST Validation		595
		5.1	Definition of Measurement Methodology	595
		5.2	Definition of Laboratory Calibration and Verification Methodology and Procedures	595
		5.3	Predeployment Calibration Verification	596
		5.4	Postdeployment Calibration Verification	596
		5.5	Uncertainty Budgets	596
		5.6	Improving Traceability of Calibration and Verification Measurements	596
		5.7	Accessibility to Documentation	597
		5.8	Archiving of Data	597
		5.9	Periodic Consolidation and Update of Calibration and Verification Procedures	598
	6.	Summary and Future Perspectives		598
		Acknowledgments		598
		References		599

6. **Assessment of Satellite Products for Climate Applications**
Frédéric Mélin and Gary K. Corlett

6.1. Assessment of Satellite Ocean Colour Radiometry and Derived Geophysical Products
Frédéric Mélin and Bryan A. Franz

1.	Introduction	609
2.	Validation of Satellite Products	610
	2.1 Validation Protocol	610
	2.2 Validation Metrics	612
	2.3 Analysis of Validation Results	614
	2.4 Model-Based Approaches to Uncertainty Analysis and Error Propagation	618
3.	Comparison of Cross-Mission Data Products	621
	3.1 Band Shift Correction	622
	3.2 Point-by-Point Comparison	624
	3.3 Analysis of Time Series	626
	3.4 Climate Signal Analysis	628
4.	Conclusions	631
	Acknowledgments	632
	References	632

6.2. Assessment of Long-Term Satellite Derived Sea Surface Temperature Records
Gary K. Corlett, Christopher J. Merchant, Peter J. Minnett and Craig J. Donlon

1.	Introduction	639
2.	Background	640
	2.1 Assessment of Top of Atmosphere Brightness Temperatures	641
	2.2 Validation Uncertainty Budget	643
	2.3 Reference Data Sources	647
3.	Assessment of Long-Term SST Datasets	649
	3.1 Example 1: Long-Term SST Data Record Assessment	652
	3.2 Example 2: Long-Term Component Assessment	654
	3.3 Quantitative Metrics	657
	3.4 Demonstrating Traceability to SI	659
	3.5 Stability	663
	3.6 Validation of Uncertainties	669
4.	Summary and Recommendations	673
	References	674

Index   679

List of Contributors

Sean W. Bailey, Ocean Biology Processing Group, NASA Goddard Space Flight Center, Greenbelt, MD, USA; FutureTech Corporation, Greenbelt, MD, USA

Ian Barton, CSIRO Marine and Atmospheric Research, Hobart, Tasmania, Australia

Barbara Bulgarelli, European Commission, Joint Research Centre, Ispra, Italy

Kenneth S. Casey, NOAA Oceanographic Data Center, Silver Spring, MD, USA

Gary K. Corlett, Department of Physics and Astronomy, University of Leicester, Leicester, UK

Davide D'Alimonte, Centre for Marine and Environmental Research, University of Algarve, Faro, Portugal

Craig J. Donlon, European Space Agency/ESTEC, Noordwijk, The Netherlands

Mark D. Dowell, European Commission, Joint Research Centre, Ispra, Varese, Italy

Owen Embury, Department of Meteorology, University of Reading, Reading, UK

William Emery, Aerospace Engineering Sciences Department, University of Colorado, Boulder, CO, USA

Robert E. Eplee, Jr, Ocean Biology Processing Group, NASA Goddard Space Flight Center, Greenbelt, MD, USA; Science Applications International Corporation, Beltsville, MD, USA

Nigel Fox, National Physical Laboratory (NPL), Teddington, Middlesex, UK

Bryan A. Franz, NASA, Goddard Space Flight Center, Greenbelt, MD, USA

Simon Hook, NASA Jet Propulsion Laboratory, California Institute of Technology, Pasadena, CA, USA

Andrew Jessup, Applied Physics Laboratory, University of Washington, Seattle, WA, USA

B. Carol Johnson, Sensor Science Division, National Institute of Standards and Technology, Gaithersburg, MD, USA

Charles R. McClain, NASA Goddard Space Flight Center, Greenbelt, MD, USA

Gerhard Meister, NASA Goddard Space Flight Center, Greenbelt, MD, USA

Frédéric Mélin, European Commission, Joint Research Centre, Ispra, Italy

Christopher J. Merchant, Department of Meteorology, University of Reading, Reading, UK

Peter J. Minnett, Meteorology & Physical Oceanography, Rosenstiel School of Marine and Atmospheric Science, University of Miami, Miami, FL, USA

Bryan Monosmith, NASA Goddard Space Flight Center, Greenbelt, MD, USA

Timothy J. Nightingale, RAL Space STFC Rutherford Appleton Laboratory, Harwell, Oxford, Didcot, UK

Albert C. Parr, Sensor Science Division, National Institute of Standards and Technology, Gaithersburg, MD, USA; Space Dynamics Laboratory, Utah State University, Logan, UT, USA

Joseph P. Rice, Sensor Science Division, National Institute of Standards and Technology, Gaithersburg, MD, USA

David L. Smith, RAL Space, Science and Technologies Facilities Council, Harwell Oxford, Oxford, UK

Kenneth J. Voss, Physics Department, University of Miami, Coral Gables, FL, USA

Menghua Wang, NOAA Center for Satellite Applications and Research, College Park, Maryland, USA

Werenfrid Wimmer, Ocean and Earth Science, University of Southampton, European Way, Southampton, UK

James A. Yoder, Woods Hole Oceanographic Institution, Woods Hole, MA, USA

Howard Yoon, Sensor Science Division, National Institute of Standards and Technology, Gaithersburg, MD, USA

Christopher Zappa, Ocean and Climate Physics Division, Lamont-Doherty Earth Observatory of Columbia University, Palisades, NY, USA

Giuseppe Zibordi, European Commission, Joint Research Centre, Ispra, Italy

Volumes in Series

Experimental Methods in the Physical Sciences (Formerly Methods of Experimental Physics)

Volume 1. Classical Methods
Edited by Immanuel Estermann

Volume 2. Electronic Methods, Second Edition (in two parts)
Edited by E. Bleuler and R. O. Haxby

Volume 3. Molecular Physics, Second Edition (in two parts)
Edited by Dudley Williams

Volume 4. Atomic and Electron Physics - Part A: Atomic Sources and Detectors; Part B: Free Atoms
Edited by Vernon W. Hughes and Howard L. Schultz

Volume 5. Nuclear Physics (in two parts)
Edited by Luke C. L. Yuan and Chien-Shiung Wu

Volume 6. Solid State Physics - Part A: Preparation, Structure, Mechanical and Thermal Properties; Part B: Electrical, Magnetic and Optical Properties
Edited by K. Lark-Horovitz and Vivian A. Johnson

Volume 7. Atomic and Electron Physics - Atomic Interactions (in two parts)
Edited by Benjamin Bederson and Wade L. Fite

Volume 8. Problems and Solutions for Students
Edited by L. Marton and W. F. Hornyak

Volume 9. Plasma Physics (in two parts)
Edited by Hans R. Griem and Ralph H. Lovberg

Volume 10. Physical Principles of Far-Infrared Radiation
Edited by L. C. Robinson

Volume 11. Solid State Physics
Edited by R. V. Coleman

Volume 12. Astrophysics - Part A: Optical and Infrared Astronomy
Edited by N. Carleton
Part B: Radio Telescopes; Part C: Radio Observations
Edited by M. L. Meeks

Volume 13. Spectroscopy (in two parts)
Edited by Dudley Williams

Volume 14. Vacuum Physics and Technology
Edited by G. L. Weissler and R. W. Carlson

Volume 15. Quantum Electronics (in two parts)
Edited by C. L. Tang

Volume 16. Polymers - Part A: Molecular Structure and Dynamics; Part B: Crystal Structure and Morphology; Part C: Physical Properties
Edited by R. A. Fava

Volume 17. Accelerators in Atomic Physics
Edited by P. Richard

Volume 18. Fluid Dynamics (in two parts)
Edited by R. J. Emrich

Volume 19. Ultrasonics
Edited by Peter D. Edmonds

Volume 20. Biophysics
Edited by Gerald Ehrenstein and Harold Lecar

Volume 21. Solid State Physics: Nuclear Methods
Edited by J. N. Mundy, S. J. Rothman, M. J. Fluss, and L. C. Smedskjaer

Volume 22. Solid State Physics: Surfaces
Edited by Robert L. Park and Max G. Lagally

Volume 23. Neutron Scattering (in three parts)
Edited by K. Skold and D. L. Price

Volume 24. Geophysics - Part A: Laboratory Measurements; Part B: Field Measurements

Edited by C. G. Sammis and T. L. Henyey

Volume 25. Geometrical and Instrumental Optics
Edited by Daniel Malacara

Volume 26. Physical Optics and Light Measurements
Edited by Daniel Malacara

Volume 27. Scanning Tunneling Microscopy
Edited by Joseph Stroscio and William Kaiser

Volume 28. Statistical Methods for Physical Science
Edited by John L. Stanford and Stephen B. Vardaman

Volume 29. Atomic, Molecular, and Optical Physics - Part A: Charged Particles; Part B: Atoms and Molecules; Part C: Electromagnetic Radiation
Edited by F. B. Dunning and Randall G. Hulet

Volume 30. Laser Ablation and Desorption
Edited by John C. Miller and Richard F. Haglund, Jr.

Volume 31. Vacuum Ultraviolet Spectroscopy I
Edited by J. A. R. Samson and D. L. Ederer

Volume 32. Vacuum Ultraviolet Spectroscopy II
Edited by J. A. R. Samson and D. L. Ederer

Volume 33. Cumulative Author lndex and Tables of Contents, Volumes 1-32

Volume 34. Cumulative Subject lndex

Volume 35. Methods in the Physics of Porous Media
Edited by Po-zen Wong

Volume 36. Magnetic Imaging and its Applications to Materials
Edited by Marc De Graef and Yimei Zhu

Volume 37. Characterization of Amorphous and Crystalline Rough Surface: Principles and Applications
Edited by Yi Ping Zhao, Gwo-Ching Wang, and Toh-Ming Lu

Volume 38. Advances in Surface Science
Edited by Hari Singh Nalwa

Volume 39. Modern Acoustical Techniques for the Measurement of Mechanical Properties
Edited by Moises Levy, Henry E. Bass, and Richard Stern

Volume 40. Cavity-Enhanced Spectroscopies
Edited by Roger D. van Zee and J. Patrick Looney

Volume 41. Optical Radiometry
Edited by A. C. Parr, R. U. Datla, and J. L. Gardner

Volume 42. Radiometric Temperature Measurements. I. Fundamentals
Edited by Z. M. Zhang, B. K. Tsai, and G. Machin

Volume 43. Radiometric Temperature Measurements. II. Applications
Edited by Z. M. Zhang, B. K. Tsai, and G. Machin

Volume 44. Neutron Scattering — Fundamentals
Edited by Felix Fernandez-Alonso, and David L. Price

Volume 45. Single-Photon Generation and Detection
Edited by Alan Migdall, Sergey Polyakov, Jingyun Fan, and Joshua Bienfang

Volume 46. Spectrophotometry: Accurate Measurement of Optical Properties of Materials
Edited by Thomas A. Germer, Joanne C. Zwinkels, and Benjamin K. Tsai

Volume 47. Optical Radiometry for Ocean Climate Measurements
Edited by Giuseppe Zibordi, Craig J. Donlon, and Albert C. Parr

Foreword

The view of the Earth from space has become an icon of our time. First seen through the spectacular photographs taken by the Apollo astronauts, it showed us the Earth, which had seemed limitless to our ancestors, to be small and fragile, a vulnerable oasis for life in the vast vacuum of space. If no other benefit had ever come from the space age, those pictures alone would have justified the effort to leave the Earth, for they changed our view of the planet forever.

But those photographs, it turned out, were just the beginning of what can be learned by looking down on the Earth from space. Only from the vantage point in orbit above the planet can we really get the whole picture—seeing far enough to give a truly global view, but also with sufficient detail to get down to the local scale. Since the time of the early satellites, the number and sophistication of remote sensing measurements has grown hugely, so that we now have a nearly continuous view of the Earth from space that is highly resolved in area, time, and wavelength. Terabytes of data now flood down from our satellites, documenting the view of Earth from space in unprecedented detail. If only we can make sense of it all, it offers the chance to understand our home planet as never before, allowing us to see how every locality fits into the whole picture. For the oceans in particular, this is a transformative view, because over large areas they are only rarely visited by people or instruments to make in situ observations. Much of our uncertainty over prediction of seasonal and longer term changes originates in this ignorance of the oceans, which are the main storage for heat in the climate system and the site of half the world's biological productivity.

This book describes the latest knowledge and techniques in visible and infrared radiometry from satellites. These regions of the electromagnetic spectrum can be used to give important information about several aspects of the oceans: the infrared observations can be used to measure sea surface temperature, which is a fundamental variable needed for climate and weather prediction studies. Visible measurements characterize ocean color, from which we can derive estimates of chlorophyll and other pigments to enable characterization of the plankton community. The plankton are in turn the base of the ocean food chain and play important roles in the Earth's carbon cycle, both in the rapid changes occurring today as a result of human activities—climate change and ocean acidification—and over the longer term for in maintaining a habitable planet.

As the contributions here illustrate, making sense of the flood of data from satellites is no easy task: it requires meticulous attention to detail. The sensors must be continually calibrated and the data validated, so that long-term records, constructed over time from successive instruments, can be relied on to be free from drift. This is of critical importance for studies of climate change, where any long-term change in temperature must be carefully separated from instrumental effects. To achieve this kind of reliability requires continuous and extended free exchange and cooperation between all those involved—from the designers and engineers who build the sensors, those interpreting the data, and researchers making in situ observations who provide the ground truth. However, there is a rich return on this effort for our civilization as a whole, for from it we can understand our home planet as never before.

Andrew Watson
University of Exeter 27th July 2014

Preface

Climate change science relies on the combined use of models and measurements to advance understanding of climate fluctuations and trends, and ultimately to formulate predictions. Gathering measurements for climate change investigations requires well-characterized observing systems and the implementation of strategies to detect decadal variations that are much smaller than those occurring at daily or interannual scales. This requirement imposes the collection of uninterrupted time series of highly accurate measurements traceable to accepted international standards that collectively constitute the evidence baseline for climate research.

Satellite systems provide a quasi-synoptic global sampling dimension of climate data measured using a variety of instruments operated over the Earth's surface. Like any observing system devoted to the generation of climate-quality data records, space-based instruments supporting climate change investigations need to deliver continuous highly accurate measurements with defined uncertainties. This imposes lifetime calibration and validation processes for each component of the end-to-end observing system and for the derived data products.

During the last few decades, several space missions have been designed to support ocean climate studies through measurement of physical, biological, and chemical variables. Among the various remote sensing technologies, optical sensors operating in the visible, near-infrared, and thermal infrared spectrum are well suited to measure variables such as sea surface temperature and water leaving radiance at timescales varying from hours to days and geographical-scales from tens of meters to kilometers. While the sea surface temperature has relevance for the heat, gas, and momentum coupling between the atmosphere and ocean, reconstruction of patterns associated with dynamical processes such as surface currents, eddies, and upwelling, the water-leaving radiance in the visible spectral region is fundamental for the quantification of optically significant seawater constituents, including phytoplankton biomass, that play a major role in the Earth's carbon cycle.

Optical remote sensing technologies used to generate climate-quality data records share the need for thorough prelaunch characterization and absolute calibration of the satellite radiometer. These activities are then followed by the postlaunch monitoring of the radiometer stability over the mission lifetime, the continuous assessment of data product quality, and finally, successive reanalysis and reprocessing of all data in conjunction with better understanding of error sources. The postlaunch activities largely rely on in situ reference measurements for the development and assessment of the algorithms and methods applied to determine each climate variable, and successively for the continuous validation of derived satellite products. Furthermore, reference measurements are required to homogenize climate data records obtained from multiple or successive satellite instruments. Because of this, advances in remote sensing optical technology demand progress to deliver in situ reference instrumentation, measurement methods, and field strategies. Such progress embraces the design of increasingly precise and stable field optical radiometers, the improvement of laboratory techniques for their characterization and absolute calibration, the assessment of measurement methods and field intercomparison strategies, and finally, advances in the creation and handling of data repositories.

This book, through a number of contributions from various authors, presents the state of the art for optical remote sensing and shows how it can be applied for the generation of marine climate-quality data products. The various chapters are grouped into six thematic parts each introduced by a brief overview. The different parts include: (1) requirements for the generation of climate data records from satellite ocean measurements and the basic radiometry principles addressing terminology, standards, measurement equation, and uncertainties; (2) satellite visible and thermal infrared radiometry embracing instrument design, characterization, and pre- and postlaunch calibration; (3) in situ visible and thermal infrared reference radiometry including overviews on basic principles, technology, and measurement methods required to support satellite missions devoted to climate change investigations; (4) computer model simulations as fundamental tools to support interpretation and analysis of both in situ and satellite radiometric measurements; (5) strategies for in situ reference radiometry to satisfy mission requirements for the generation of climate data records; and finally, (6) methods for the assessment of satellite data products.

The expectation of the editors is that this book will become a working tool, as either a reference text or as background literature for discussions, for students and scientists interested in ocean climate studies and satellite radiometry.

Giuseppe Zibordi
Craig J. Donlon
Albert C. Parr

Chapter 1

Introduction to Optical Radiometry and Ocean Climate Measurements from Space

James A. Yoder,[1,]* B. Carol Johnson[2]
[1] *Woods Hole Oceanographic Institution, Woods Hole, MA, USA;* [2] *Sensor Science Division, National Institute of Standards and Technology, Gaithersburg, MD, USA*
*Corresponding author: Email: jyoder@whoi.edu

This first part of Optical Radiometry for Ocean Climate Measurements is intended to give the reader an introductory background in radiometry and lay the groundwork that establishes the context of the measurements that are necessary to understanding larger climate issues.

Chapter 1.1 introduces the terminology and requirements of oceanic radiometric measurements, which include both spectral measurements to determine a wide range of biogeochemical constituents of the upper ocean as well as methodologies to determine the sea surface temperature. The authors review the recommendations of the Intergovernmental Panel on Climate Change as well as the requirements established by the Global Climate Observing System. Some of these requirements are listed in Table 2 of Chapter 1.1. This table gives the data quality required for a determination of the sea surface temperature to an accuracy of 0.1 K over 100 km spatial scales with a stability of 0.03 K. These stringent requirements, and those of determining water-leaving radiance to an accuracy of 5% and a stability of 0.5%, place exacting demands upon the calibration and reliability of the instrumentation performing the measurements. Additionally, the need to intercompare and utilize data from different systems and different national efforts requires commonly agreed upon standards and calibration procedures.

Chapter 1.2 is an introduction to radiometric nomenclature, methods, and procedures pertinent to the remote sensing community performing optical radiometry for determining ocean properties. The meanings of the terms radiance and irradiance are shown and methods of measuring these radiometric quantities are outlined in some detail. This includes reviewing some of the basic instrumentation and their characteristics and calibration needs.

The development and use of appropriate reference standards to calibrate instruments is discussed for various radiometer systems. Examples of systems, such as filter radiometers that are commonly applied by the ocean remote sensing community, are used to demonstrate the uncertainty issues in characterizing and calibrating instruments for radiometric applications. The uses of common types of laboratory calibration sources such as blackbodies and calibrated lamps are discussed with a view to integrating their performance into the instrument uncertainty statements.

The concept of the measurement equation is emphasized for its utility in the propagation of uncertainty statements for the entire radiometer system. The use of a measurement equation forms the basis for the international standards for uncertainty reporting given in the Guide to the expression of uncertainty in measurement adopted by the international metrology community centered at the Bureau International des Poids and Mesures in Sèvres, France (see Reference 85 in Chapter 1.2).

Chapter 1.1

Ocean Climate and Satellite Optical Radiometry

James A. Yoder,[1,*] Kenneth S. Casey,[2] Mark D. Dowell[3]
[1] Woods Hole Oceanographic Institution, Woods Hole, MA, USA; [2] NOAA Oceanographic Data Center, Silver Spring, MD, USA; [3] European Commission, Joint Research Centre, Ispra, Varese, Italy
*Corresponding author: Email: jyoder@whoi.edu

Chapter Outline

1. Introduction 3
 1.1 Characteristics of a Climate-Observing System 4
2. Global Climate Observing System Requirements for ECVs and CDRs 6
 2.1 Ocean Color Radiometry 7
 2.2 Sea Surface Temperature 8
3. From Essential Climate Variables to Climate Data Records 10
4. Conclusion 11
References 11

1. INTRODUCTION

The following two statements are from a summary report of the recent Intergovernmental Panel on Climate Change (IPCC) [1].

> Warming of the climate system is unequivocal, and since the 1950s, many of the observed changes are unprecedented over decades to millennia. The atmosphere and ocean have warmed, the amounts of snow and ice have diminished, sea level has risen, and the concentrations of greenhouse gases have increased.

> Ocean warming dominates the increase in energy stored in the climate system, accounting for more than 90% of the energy accumulated between 1971 and 2010 (high confidence). It is virtually certain that the upper ocean (0–700 m) warmed from 1971 to 2010, and it likely warmed between the 1870s and 1971.

These recent statements of the IPCC point to the importance of determining both the rates of changes to the climate system, as well as the impacts of those changes, including on marine ecosystems. Both tasks require global observing systems of which space-borne radiometers are crucial components.

Earth observation satellites measuring in the visible and infrared spectral domain provide a global perspective for many measurements required to determine the role of the ocean in the global climate system, as well as the effects on the ocean of a changing climate. The focus of this book is on measurements of sea surface temperature (SST) and on ocean color radiometery (OCR). SST is directly related to warming of the ocean and to the oceans role in the hydrologic cycle in general. OCR provides measures of biogeochemical constituents of the upper ocean, including phytoplankton pigments (e.g., Chl a), colored organic matter, particulate carbon and estimates of phytoplankton size and taxonomic composition. Variability or trends in these constituents can be related to changes in ocean productivity and to the taxonomic structure of the organisms responsible for primary production. These changes also have implications for higher tropic levels, including fisheries. Measurements from satellites provide a regional to global scale perspective not possible from in situ and airborne measurements that are more limited in spatial and temporal coverage. Comparing satellite measurements with in situ observations, however, is essential to establish the credibility of satellite-based measurements to be used as essential climate variables (ECVs) leading to climate data records (CDRs) (see Table 1 for definitions).

1.1 Characteristics of a Climate-Observing System

The discussions in the present monograph, on the use of satellite visible and infrared radiometry for studies of ocean climate should be seen in the broader context of the development of a climate observing system, based on determined requirements, which should be adopted systematically. In order to characterize climate and climate change, data need to be accurate and homogeneous over long time scales. The relevant signals for the detection of climate change can easily be lost in the noise of a changing observing system. This enforces the need for continuity in an observing system, where observations can be tied to an invariant reference. Such a system needs to be maintained over at least several decades and preferably indefinitely.

Climate-monitoring principles, requirements, and guidelines for the creation of CDRs have been formulated to increase awareness of the specific observational and procedural needs for establishing a successful approach to climate monitoring. In this respect the task of climate monitoring has specific requirements that go beyond one-time research missions. For instance, it is important that the design of an observing system for climate monitoring, including satellite and in situ systems, takes account of all required observations and legacy instruments, and that it guarantees effective continuity in measurements. At the very least, appropriate transfer standards must be provided to enable robust linkage to an invariant, International System of Units (SI) reference system at an appropriate level of accuracy. The provision of such an observing system requires a global strategy in which agencies agree to collaborate to fulfill such a generic continuity requirement. It is simply too

TABLE 1 Basic Terminology for Data Records Relating to Climate

An understanding of the terminology used when talking about climate-related data records is important. This box therefore lists established definitions, with respect to data records in general and satellite data records in particular:

An *essential climate variable (ECV)* is a geophysical variable that is associated with climate variation and change as well as the impact of climate change onto Earth. GCOS has defined a set of ECVs for three spheres, atmospheric, terrestrial, and oceanic [2].

A *climate data record (CDR)* is a series of observations over time that measures variables believed to be associated with climate variation and change. These changes may be small and occur over long time periods (seasonal, interannual, and decadal to centennial) compared to the short-term changes that are monitored for weather forecasting. Thus, a CDR is a time series of a climate variable that tries to account for systematic errors and noise in the measurements [3].

Stability [4] may be thought of as the extent to which the accuracy remains constant with time. Over time periods of interest for climate, the relevant component of total uncertainty is expected to be its systematic component as measured over the averaging period. Stability is therefore measured by the maximum excursion of the difference between a true value and the short-term average measured value of a variable under identical conditions over a decade. The smaller the maximum excursion, the greater the stability of the data set.

The term *fundamental climate data record (FCDR)* denotes a well-characterized, long-term data record, usually involving a series of instruments, with potentially changing measurement approaches, but with overlaps and calibrations sufficient to allow the generation of products that are accurate and stable, in both space and time, to support climate applications [3]. FCDRs are typically calibrated radiances, backscatter of active instruments, or radio occultation bending angles. FCDRs also include the ancillary data used to calibrate them. The term FCDR has been adopted by GCOS and can be considered as an international consensus definition.

The term *thematic climate data record (TCDR)* denotes the counterpart of the FCDR in geophysical space [3]. It is closely connected to the ECVs but strictly covers one geophysical variable, whereas an ECV can encompass several variables. For instance, the ECV cloud property includes at least five different geophysical variables, each of them constitutes a TCDR. The term TCDR has been taken up by many space agencies and can be considered as de facto standard.

GCOS, Global Climate Observing System.

large a task for a single agency or even a single country to implement effectively. Although most space agencies accept the climate-monitoring principles, there is still only limited coordination of the long-term commitment to collect climate observations.

Well-calibrated and stable satellite measurements can be used for climate monitoring, studies of trends and variability, climate impacts, and verification of climate models. The following sections specifically address the fundamental requirements for OCR and SST ECVs, and additionally summarize ongoing effort for the creation of CDRs.

2. GLOBAL CLIMATE OBSERVING SYSTEM REQUIREMENTS FOR ECVs AND CDRs

For space-based observations the most relevant and comprehensive set of specific user requirements is provided by GCOS within their supplement *Systematic Observation Requirements for Satellite-Based Products for Climate* [5] to the *GCOS Implementation Plan* [6]. The GCOS requirements are given for those ECVs for which the feasibility of satellite measurements has been demonstrated. The requirements are based on expert opinion and are updated every five or six years. This subset of ECVs is intended to reflect the most important climate variables needed to monitor the complete climate system.

GCOS has also developed a set of climate-monitoring principles that set out a general guideline to achieve observations with the required quality [7]. The monitoring principles address the key satellite-specific operational issues. This includes the availability of high quality in situ data for calibration and validation of the satellite instruments. Many international collaborative initiatives as well as individual agency programs have provided concrete responses to these requirements via their mission plans and data products. In some cases these responses have been made in a coordinated manner at the international scale, e.g., the Committee on Earth Observing Satellites (CEOS) response to the first GCOS Implementation Plan.

The broad GCOS requirements have specific consequences for visible and thermal radiometry, both in the context of domain-specific requirements for these observations, as well as in initial activities to address coordination and implementation of the necessary system. Table 2 summarizes the GCOS requirements for OCR and SST ECVs.

The rationales for the SST requirements are based in part on determinations of the interannual and longer-term temperature variability. Assuming a global surface temperature change signal of 0.1 K/decade, a global average temperature time series should be stable to much better than 0.1 K/decade in order to distinguish the signal from the instability of the time series. To detect such slow and small, yet significant changes it is prudent to aim for a target stability of at least 0.03 K/decade and ideally 0.01 K/decade. The 30% for chlorophyll is intended for the concentration range $0.01-10$ mg/m^3 in waters in which chlorophyll-a dominates the bio-optical properties (Case-1 waters). Four kilometers horizontal resolution and a daily observing cycle are required at the global scale. Current achievable performance for accuracy is $5-15\%$ for water-leaving radiances (for the blue and green wavelengths), and $30-70\%$ for chlorophyll-a in the concentration range $0.01-10$ mg/m^3 in open ocean waters. Errors are considerably higher for coastal waters and regional seas.

The following subsections provide brief overviews of the measurements required for OCR and SST ECVs, with details provided in the following chapters of this book.

TABLE 2 GCOS Requirements for Satellite SST and OCR

Variable/ Parameter	Horizontal Resolution	Temporal Resolution	Accuracy	Stability
SST	10 km	Daily	0.1 K over 100 km scales	Less than 0.03 K over 100 km scales
Water-leaving radiance[a]	4 km	Daily	5%	0.5%
Chlorophyll-a concentration	30 km	Weekly averages	30%	3%

Terms Defined in Table 1. SST, sea surface temperature; OCR, ocean color radiometry; GCOS, Global Climate Observing System.
[a]5% requirement is for the blue and green wavelength bands.
From Ref. [5].

2.1 Ocean Color Radiometery

A system of measurements and models is required to measure and calculate ECVs from satellite ocean color radiometers (SOCRs). In brief, the measurements include top of the atmosphere (TOA) spectral radiance in narrow (10–20 nm) bands at visible and near-infrared (VNIR) wavelengths when the satellite sensor is over the ocean. In addition, in situ spectral radiance and other measurements are required for calibration of the sensor and validation of derived products. Models used for processing imagery include the radiative transfer equations to calculate the effects of absorption and scattering properties of the atmosphere. This is a critical step since the radiance reaching the top of the atmosphere is dominated by atmospheric properties, rather than by the water-leaving radiance (L_w) at the key wavelengths used to calculate biogeochemical products. Bio-optical models (algorithms) are applied to L_w to calculate phytoplankton chlorophyll-a (Chl a), particulate organic carbon, and other in water constituents related to ocean biogeochemistry. It is a major challenge to derive products from this system of measurements and models that meet the GCOS requirements for ECVs. It is even more of a challenge to sustain a calibrated and consistent time series across multiple satellite sensors at the accuracies specified by GCOS (see Table 2), even for the basic measurement of L_w.

SeaWiFS achieved an overall uncertainty level of its calibration gains of 0.3% for TOA radiance measurements by minimizing three primary independent sources: calibration trends with time, uncertainty in the primary in situ calibration source (MOBY buoy [8]), and uncertainty estimating the sensor gain corrections through system vicarious [9]. Overall uncertainty of less than 0.5% for the blue-green spectral bands for oligotrophic and mesotrophic waters has now become a de facto target for TOA radiances. This target requires key sensor design features for high stability, capability to avoid sensor saturation over bright targets (the ocean is very dark at VNIR wavelength bands compared to land and clouds) and to minimize polarization sensitivity. In addition, preflight characterization of the sensor for temperature effects, stray light, band-to-band spatial registration, signal-to-noise ratios, relative spectral and out-of-band response, and other characteristics is required [10]. Mission operations also have to contend with the impact of specular reflection from the sea surface (sun glint) by either tilting the sensor away from the sun or by masking pixels significantly affected by glint. Ensuring stability or having the ability to quantify changes in sensor drift is essential. Gain changes are known to occur following launch and with time in the harsh environment of space and have to be monitored throughout any given mission (see subsequent chapters for details).

One of the important lessons learned from SOCR missions to date is the importance of periodic reprocessing throughout a mission to incorporate changes to the gain factors, and to include new developments in atmospheric and bio-optical algorithms. In fact, one cannot envision how a mission can measure ECVs, if it does not include reprocessing within mission operations. As the relation between L_w and derived products depend in part upon the taxonomic and size composition of marine phytoplankton, algorithms used to process measurements from a global SOCR mission need to be validated within different water types in the global ocean. This is a difficult, if not impossible task, for a single space agency to accomplish. The International Ocean Colour Coordinating Group (IOCCG) and more recently CEOS's Ocean Color Radiometry Virtual Constellation (OCR-VC) have both encouraged international cooperation for calibration/validation measurements to help meet the challenge of collecting global in situ data for validation. The data are most useful if collected using standard measurement protocols which have been defined and are used by many investigators. It is also most useful if the data are archived and easily accessible. SeaBASS (http://seabass.gsfc.nasa.gov/seabam/) is an example of a publicly shared archive of in situ oceanographic and atmospheric data maintained by the NASA Ocean Biology Processing Group (OBPG).

2.2 Sea Surface Temperature

For SST, a major scientific challenge is introduced by the substantial differences between the "SST" as measured by different space-borne and in situ instruments. These differences must be accounted for when attempting to combine or compare measurements from multiple platforms. Even for series of nearly identical instruments, challenges still arise in connecting them consistently without introducing biases into the overall time series. Calculating, documenting, and understanding the uncertainty of any given data set also remains a large scientific challenge and being able to meaningfully compare those uncertainties between data sets is an even larger issue.

Scientific challenges associated with SST products are being addressed through coordinated international science teams like Group for High

Resolution Sea Surface Temperature (GHRSST). GHRSST brings together the world's experts in SST to exchange knowledge, share updates, and tackle the scientific issues at hand. One of the earliest scientific issues addressed by GHRSST, for example, was to clearly define the different kinds of "SST" and propagate the use of those definitions throughout the SST data producer community and data standards organizations. Techniques and methods for blending those different kinds of SST and quantifying their uncertainties are also shared among the members of the GHRSST science team and documented for the broader community.

Pragmatic computational and data bandwidth challenges are also being addressed and in a growing number of cases are being overcome through cloud-computing pilot projects. Activities to enable cloud-based access to large volume data sets and to associate computing cycles with those data sets in cloud environments are beginning to yield successes. In some cases, these successes are taking the form of community-based cloud deployments, where individuals and organizations bring the algorithms into a shared data environment as opposed to downloading hundreds of terabytes to their local computing platforms. European Space Agency (ESA)'s Felyx project is one such example, and provides a free, open-source software platform to enable the analysis of large volumes of environmental data. Organizations are also tackling issues of data reproducibility and citation through a range of activities. The NOAA Climate Data Record program (CDRP), for example, has established archive requirements, a maturity matrix, workflow diagram requirements, and standards for software documentation and deployment to ensure CDRs can be consistently reproduced. In addition, data citation has been enabled through the use of digital object identifiers (DOIs). In the same manner that journal articles can be given a citable DOI, the Natural Environment Research Council in the UK, NOAA in the US, and other organizations are now "minting" DOIs for published data sets. These DOIs provide long-term, stable pointers to the data as they were used in scientific publications and other applications, and can be used to provide the appropriate credit to researchers for publishing their data.

Throughout these and related efforts, the need for careful, well-calibrated, accurate in situ measurements is being emphasized and shortcomings in the in situ networks have been identified. For example, the GHRSST community has worked with the Joint Technical Commission for Oceanography and Marine Meteorology Data Buoy Cooperation Panel to lobby for improvements to the network of drifting surface buoys, whose SST measurements are often lacking in both SST and positional accuracy. Similarly, calls for non-pumped near-surface sampling by the Argo profiler network have been supported. New requirements for in situ radiometers, calibrated to reference blackbodies, have also been identified and documented by the GHRSST community. Efforts to rescue historical in situ data have been highlighted as more important than ever, to support the consistent development and assessment of accuracy and stability of satellite-based climate records of time. Successful examples include the International Comprehensive Ocean Atmosphere Data Set, now managed by NOAA's National Climatic Data Center and focusing on surface marine and meteorological observations. Another example is the World Ocean Database [11], a consistent, quality-controlled set of in situ data at the surface and at depth created by NOAA's National Oceanographic Data Center.

3. FROM ESSENTIAL CLIMATE VARIABLES TO CLIMATE DATA RECORDS

The OCR and SST communities (both scientific and space agencies) have taken significant steps for initiating mechanisms and activities to address the above-described broad requirements from the climate-observing system for their specific domains. CDRs are beginning to meet the needs of climate change studies and societal application of ocean climate information, with the main constraint being the comparatively short length of satellite records compared to ocean climate phenomena. These activities include both competitively selected grants programs, as well as operational commitments by mission-oriented agencies to the long-term development, production, and distribution of these space-based and related in situ CDRs. Examples of successful competitive programs include the ESA Climate Change Initiative and the NOAA CDRP. NASA's commitment to operational reprocessing of ocean color and SST through the long-term activities of the OBPG, NOAA's maintenance of the AVHRR Pathfinder SST project (which dates to a 1990 NOAA-NASA Cooperative Agreement) through the base activities of the National Oceanographic Data Center [12,13] and the NOAA CDRP, and EUMETSAT's sustained support of the Ocean and Sea Ice Satellite Application Facility, are just a few examples of ongoing, sustained activities with significant contribution to the provision of CDRs for SST and OCR. Collaborative efforts, like the international contributions to the IOCCG, the GHRSST [14], and the CEOS Virtual Constellations for SST and Ocean Colour Radiometry (SST-VC and OCR-VC, respectively) provide venues for the coordinated development and production of CDRs around the world. With a focus on data sharing and provision of fit-for-purpose products, the Virtual Constellations complement the more scientifically oriented science teams. Participants in the Virtual Constellation discuss, coordinate, and deliver sets of related products through globally shared data management and discovery systems, optimized on-orbit constellations, and application-oriented projects, thereby lowering barriers to data access and use.

From an organizational perspective, data sharing arrangements present particular challenges, and for satellite-based data sets, data sharing typically involves high-level national policies. Yet, sharing of data is essential to produce SST and OCR global ECVs. Procedural and pragmatic challenges must also be addressed. For all climate-related data sets, reproducibility is a key challenge as the underlying data sets and ancillary information often go through numerous processing steps and are often conducted at different institutions. Encouraging individual researchers to overcome issues of data sharing and documentation by enabling data citation is a work-in-progress, but when successful will provide additional incentives to the free and open exchange of the knowledge and data needed to create high-quality CDRs. Modern sensors, with more channels, finer space-time resolutions, and longer periods of records create substantial computing and network bandwidth challenges, even when policies are supportive of reprocessing and exchanging data sets.

Turning measurements of ECVs into CDRs is the final challenge. The ocean shows considerable variability at interannual and decadal time scales owing to El Nino - Southern Osciallation (ENSO), Pacific Decadal Oscillation, North Atlantic Oscillation, and other phenomena. Thus, regional to global scale CDRs require decadal and longer time series implying that ECVs derived from TOA radiances must involve multiple sensors flown by multiple space agencies. In general, no two sensors have the same characteristics and that adds considerable complexity for generating CDRs.

The following chapters cover many of the issues raised in this Introduction with appropriate detail for a technical readership. Techniques discussed in this volume for processing satellite data, in situ data, and making appropriate comparisons between the two for calibration and validation took decades to bring to the point at which producing ECVs is possible. Challenges still remain, although the reader of the following chapters will find that much progress has been made. That progress is a credit to the hard work and dedication of scientists throughout the international VNIR and IR radiometry community many of whom are authors of the chapters in this volume.

4. CONCLUSION

ECVs from satellite radiometry are being used to produce CDRs that are essential for determining changes to global upper ocean temperature as well as changes to important biogeochemical parameters. Many of the challenges have been overcome to produce CDRs and the GCOS requirements can be met for SST, although not yet for water-leaving radiance. CEOS, GHRSST, and IOCCG are promoting international cooperation to share satellite and in situ data and techniques. The stage is set to produce CDRs from satellite radiometers and for sustaining these long and calibrated time series into the future.

REFERENCES

[1] Climate Change. The physical science basis. Summary for policymakers. Working Group I Contribution to the Fifth Assessment Report of the Intergovernmental Panel on Climate Change (IPCC) (2013).

[2] GCOS-82, The Second Report on the Adequacy of the Global Observing Systems for Climate in Support of the UNFCCC, World Metrological Organization, Intergovernmental Oceanographic Commission, April 2003. GCOS-82, WMO/TD 1143.

[3] National Research Council, Climate Data Records from Environmental Satellites: Interim Report, National Academies Press, Washington, D.C, 2004, ISBN 0-309-09168-3, 150 pp.

[4] G. Ohring, B. Wielicki, R. Spencer, B. Emery, R. Datla (Eds.), Satellite Instrument Calibration for Measuring Global Climate Change, National Institute of Standards and Technology, 2004. NISTIR-7047, March 2004. Available at: http://www.nist.gov/pml/div685/pub/upload/nistir7047.pdf.

[5] GCOS-154, Systematic Observation Requirements for Satellite-based Products for Climate, 2011 Update, World Meteorological Organization, GenevaPublisher, December 2011, 139 pp.

[6] GCOS-138, Implementation Plan for the Global Observing System for Climate in Support of the UNFCCC, World Meteorological Organization, GenevaPublisher, August 2010 (2010 Update), GCOS-138 (GOOS-176, GTOS-84, WMO/TD-No. 1523), http://www.wmo.int/pages/prog/gcos/Publications/gcos-138.pdf ('IP-10').

[7] GCOS-143, Guideline for the generation of datasets and products meeting GCOS requirements, in: An Update of the "Guideline for the Generation of Satellite-based Datasets and Products Meeting GCOS Requirements" (GCOS-128, WMO/TD-no. 1488), Including In Situ Datasets and Amendments, World Meteorological Organization, GenevaPublisher, May 2010, p. 12. Available at: http://www.wmo.int/pages/prog/gcos/Publications/gcos-143.pdf.

[8] D.K. Clark, M. Feinholz, M.B. Yarbrough, B.C. Johnson, S.W. Brown, Y.S. Kim, R.A. Barnes, Overview of the radiometric calibration of MOBY, Proc. SPIE 4483 (2002) 64–76.

[9] IOCCG, Mission requirements for future ocean-colour sensors, in: C.R. McClain, G. Meister (Eds.), Reports of the International Ocean-colour Coordinating Group, No. 13, IOCCG, Dartmouth, Canada, 2012.

[10] National Research Council, Assessing Requirements for Sustained Ocean Color Research and Operations, National Academies Press, Washington, D.C, 2011, ISBN 978-0-309-21044-7, 126 pp.

[11] T.P. Boyer, J.I. Antonov, O.K. Baranova, C. Coleman, H.E. Garcia, A. Grodsky, D.R. Johnson, R.A. Locarnini, A.V. Mishonov, T.D. O'Brien, C.R. Paver, J.R. Reagan, D. Seidov, I.V. Smolyar, M.M. Zweng, in: S. Levitus (Ed.), A. Mishonov, (Technical Ed.), World Ocean Database 2013. NOAA Atlas NESDIS 72, National Oceanic and Atmospheric Administration, Washington, D.C, 2013, p. 209.

[12] K.S. Casey, T.B. Brandon, P. Cornillon, R. Evans, The past, present and future of the AVHRR pathfinder SST program, in: V. Barale, J.F.R. Gower, L. Alberotanza (Eds.), Oceanography from Space: Revisited, Springer, 2010. http://dx.doi.org/10.1007/978-90-481-8681-5_16.

[13] Kenneth S. Casey, Robert H. Evans, Warner Baringer, Katherine A. Kilpatrick, Guillermo P. Podesta, Susan Walsh, Elizabeth Williams, Tess B. Brandon, Deirdre A. Byrne, Gregg Foti, Yuanjie Li, Sheri A. Phillips, Dexin Zhang, Yongsheng Zhang, AVHRR Pathfinder Version 5.2 Level 3 Collated (L3C) Global 4km Sea Surface Temperature, National Oceanographic Data Center, NOAA. Dataset, 2011. http://dx.doi.org/10.7289/V5WD3XHB (accessed on 18.12.13).

[14] C.J. Donlon, K.S. Casey, I.S. Robinson, C.L. Gentemann, R.W. Reynolds, I. Barton, O. Arino, J. Stark, N. Rayner, P. Le Borgne, D. Poulter, J. Vazquez-Cuervo, E. Armstrong, H. Beggs, D. Llewelly-Jones, P.J. Minnett, C.J. Merchant, R. Evans, The GODAE high resolution sea surface temperature pilot project, Oceanography 22 (3) (2009) 34–45.

Chapter 1.2

Principles of Optical Radiometry and Measurement Uncertainty

B. Carol Johnson,[1,*] Howard Yoon,[1] Joseph P. Rice,[1] Albert C. Parr[1,2]

[1] Sensor Science Division, National Institute of Standards and Technology, Gaithersburg, MD, USA; [2] Space Dynamics Laboratory, Utah State University, Logan, UT, USA
*Corresponding author: Email: carol.johnson@nist.gov

Chapter Outline

1. Basics of Radiometry   14
    1.1 Introduction   14
    1.2 Radiance   17
    1.3 Irradiance   21
    1.4 Reflectance   23
    1.5 Distance and Aperture Areas in Radiometry   28
2. Radiometric Standards and Scale Realizations   30
    2.1 Sources   30
       2.1.1 Blackbody Sources   30
       2.1.2 Lamp Sources   32
       2.1.3 Integrating Spheres   33
       2.1.4 Diffuse Reflectance Standards   36
    2.2 Radiometers   38
       2.2.1 Electrical Substitution Radiometers   39
       2.2.2 Radiance and Irradiance Responsivity   41
3. The Measurement Equation   42
    3.1 Background and a Review of the Concepts   42
    3.2 Measurement Equation Examples   46
       3.2.1 Filter Radiometer for Validation or Measurement of Spectral Radiance   46
       3.2.2 Filter Radiometers and HTBB for Realization of Spectral Irradiance   50
    3.3 Uncertainty in Ocean Color Measurements   57
       3.3.1 Correlations   58
       3.3.2 Comparisons and Reproducibility   60
4. Summary   61
Acknowledgments   62
References   62

1. BASICS OF RADIOMETRY

1.1 Introduction

Remote sensing of ocean properties involves the measurement of electromagnetic radiation being reflected or emitted by the ocean in selected wavelength regions or wavelength bands appropriate to the property under consideration. The Sun is the source of the reflected radiation and the primary driver for heating of the oceans. Remote sensing is usually done with Earth orbiting satellite systems but in some instances, aircraft and balloons are used [1,2]. The focus of this book is the wavelength region from the ultraviolet (UV) through the thermal infrared (TIR) region of the spectrum. In this region, the interaction of light and matter involves outer electrons and vibrations or rotations of molecules, and is often referred to as the optical region; the commonly named regions are listed in Table 1. At wavelengths shorter than UV-C there is the deep UV, soft X-ray, and even more energetic regions including the gamma ray region. The other end of the spectrum, at wavelengths longer than 1 mm, are the microwave and radio frequency regions of the electromagnetic spectrum. These regions have substantially different measurement methodologies and purposes and are beyond the focus of this book.

A device to measure optical radiation is called a radiometer and if it has spectral resolving capabilities it is referred to as a spectroradiometer. The radiometric terminology in this chapter conforms to the definitions accepted by the International Commission on Illumination and by the International Organization for Standardization [3,4]. Table 2 summarizes a few of the commonly used radiometric quantities. If the quantities in Table 2 are

TABLE 1 Commonly Named Wavelength Regions of Optical Radiation Taken from the *International Lighting Vocabulary* [3]

Region	Wavelength Interval
UV-C	100–280 nm
UV-B	280–315 nm
UV-A	315–400 nm
Visible	380–780 nm
IR-A	780–1400 nm
IR-B	1.4–3.0 μm
IR-C	3.0–1.0 mm

TABLE 2 Radiometric Quantities and Commonly Used Symbols to Describe Them

Radiometric Quantity	Symbol	Units (mks)
Radiant energy	Q	J
Radiant flux (power)	Φ	W
Irradiance	E	$W\,m^{-2}$
Radiance	L	$W\,m^{-2}\,sr^{-1}$
Radiant intensity	I	$W\,sr^{-1}$
Bidirectional reflectance distribution function, BRDF	f_r	sr^{-1}
Reflectance	ρ	None
Reflectance factor	R	None

spectrally dependent, the same symbol is used with a shown functional dependence on the variable or a subscript of the variable. For example, the spectral radiance is indicated as $L(\lambda)$ or L_λ. Typically, the quantity is dependent on other variables including geometric and spatial variables, such as angles of incidence or position as well as temperature or other environmental quantities. The denominators of spectral units have an additional unit of wavelength, usually measured in nanometers.

The first group of radiometric quantities listed in Table 2 can be visualized by reference to Figure 1(a), which is adapted from [5]. An emitting surface designated by dA_1 acts as a source of radiation that impinges upon a receiving surface designated by dA_2. For the purposes of this discussion we assume dA_1 emits uniformly in all directions and the two surfaces are both centered on and perpendicular to the centerline. A source of radiation that has the same radiance when viewed from any angle is called a Lambertian emitter. The radiant *intensity* is defined as the *power* (or *flux*) from a point on the surface that is emitted into the solid angle shown by the cone of light starting at the origin of dA_1 and intersecting dA_2. This defines the radiant intensity as the power per steradian. More practically, the radiant intensity is the amount of power per steradian passing through a surface subtending a given solid angle. The term intensity in optics is often used in differing ways, which can cause confusion [6–8].

The quantity of *radiance* of the source area dA_1 is important for remote sensing. From Figure 1(a), the radiance is defined as the amount of optical power emitted from dA_1 within the angular space defined by the truncated cone shown with its vertex centered in dA_1 in Figure 1. Radiance is the optical

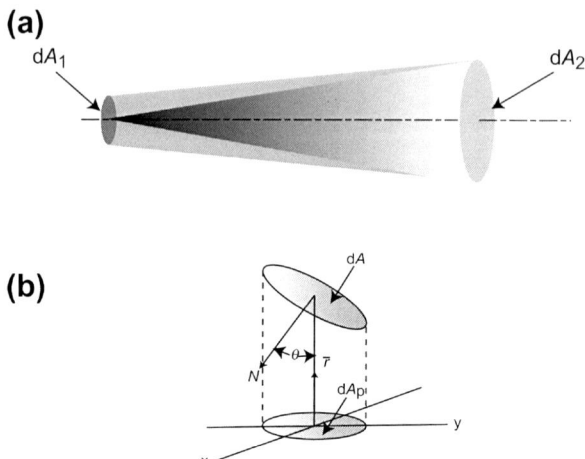

FIGURE 1 (a) Schematic of a source of radiation at dA_1 that illuminates a receiving area dA_2. (b) Schematic of the geometry of the projected area dA_p from an area dA inclined at an angle θ. *Reproduced from Figure 1.2 in Ref. [5].*

power per area of the source per steradian of solid angle. The radiance can depend on direction and hence upon the direction of the propagation.

Assuming the total optical power passing through the surface dA_2 is evenly distributed over the element of area dA_2, the *irradiance* is defined by dividing the radiant flux on the surface by the area dA_2. If the distribution of optical power on A_2 is not uniform then appropriate integration of the power over the surface would be necessary to determine the total irradiance. Irradiance is the power per unit area in some region of space and is also a fundamental quantity in remote sensing studies. The total power of an optical detector senses is the integral of the irradiance from all directions over the area of the receiver. In a general sense, irradiance is a quantity used to understand detectors and radiance is a property used to understand or describe sources.

Also stated in Table 2 are quantities associated with surfaces. *Reflectance* is defined as the ratio of the incident to reflected flux, while the *bidirectional reflectance distribution function* (BRDF) is an inherent property of the surface, with units of inverse steradians. *Reflectance factors* are determined by consideration of the flux reflected from an idealized surface.

The quantity, $dA \cos \theta$, appears often in radiometry and is the projected area dA_p. This concept can be understood from the geometry shown in Figure 1(b) where an area dA, whose normal vector $\overline{N}$ is oriented at an angle θ with respect to a plane defined by coordinates x and y and which has a normal direction shown by $\bar{r}$. The inclination produces a projected area $dA_p = dA \cos \theta$ in the x–y plane and represents the area of dA as viewed from the x–y plane. This concept is useful in describing the amount of flux passing through a plane due to some external source such as an emitting surface element dA.

Geometrically, the projected area is the effective area the source or detector makes with an observer due to angles of observation.

The radiometric quantities listed in Table 2 can be functions of the wavelength, λ, or the frequency, ν, or the wave number, σ. These are the symbols commonly used for these quantities but the reader needs to ensure what symbols an individual author may use as there is not a universal consensus on usage. These quantities are related to each other, the speed of light, c, and the index of refraction, n, by

$$\lambda = \frac{c}{n\nu} = \frac{1}{\sigma}. \tag{1}$$

The index of refraction is a function of the medium, wavelength, and environmental factors such as temperature and pressure.

Radiance, irradiance, and reflectance are the usual quantities measured for remote sensing purposes and will be discussed in more detail in the following sections. This material is adapted from previous publications [5,9], but many other sources of this material exist, such as, in particular, the NIST Self-Study Manual on Optical Radiation Measurements [10] or texts such as the description of in-water radiative transfer by Mobley [11] or those that cover the solar-reflected (SR), TIR, and longer wavelengths in the study of oceans from space [2]. The optical area of apertures and separation distance will also be discussed. There are other quantities developed from these basic concepts that will be introduced as needed later in this book.

1.2 Radiance

Radiometric detectors measure optical power; incorporated into a radiometer, the output can be made traceable to the appropriate International System of Units (SI) units by use of radiometric reference standards. The relationship between optical power, radiance, and irradiance can be shown by reference to Figure 2, which is adapted from [5]. In Figure 2, x_1 and y_1 describe a coordinate system centered on a source that emits radiation from a differential element of area dA_1 positioned on a source whose area is A_1. This source is characterized by its radiance. In order to visualize these quantities it is useful to describe a bundle radiation from dA_1 that is incident on a second surface in a differential element dA_2 of a larger area A_2 centered on coordinate system x_2 and y_2. This area A_2 could, for example, be the entrance pupil of a radiometer system. The lines connecting surface elements dA_1 and dA_2 in Figure 2 indicate some, but certainly not all, of the possible paths rays of light traverse between the surfaces.

The radiation incident on A_2 can be characterized in terms of the radiance, but often the irradiance is a more useful quantity for characterizing the radiation incident upon a surface. A_1 and A_2 are shown in Figure 2 as being circular for schematic reasons and ease of discussion, but they can be of any shape that describes a source of radiation and a surface of interest through

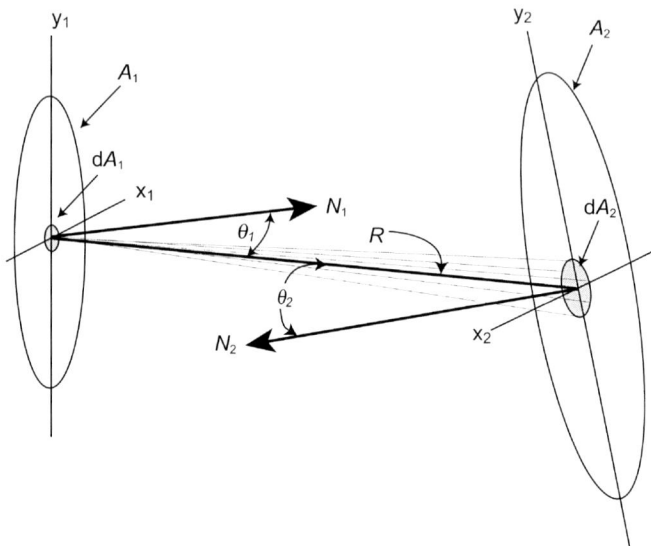

FIGURE 2 Schematic of a generalized configuration of optical radiation passing between two surfaces A_1 and A_2 that may be inclined with respect to each other. *Reproduced from Figure 1.3 in Ref. [5].*

which the radiation propagates. In a typical example for a radiometric measurement, A_2 is the entrance pupil of a detector system and A_1 describes the aperture of the source providing the radiation. In Figure 2, R is a line whose length is the distance between the origins of the two differential surface area elements, and $\overline{N}_1$ and $\overline{N}_2$ are the normal vectors to the surfaces at angles θ_1 and θ_2 with respect to R. The coordinate systems are centered on the apertures for convenience.

If L_1 is the radiance of the source at dA_1, the amount of flux $d^2\Phi_1$ in the beam that leaves the element of area dA_1 and that passes through element of area dA_2 is,

$$d^2\Phi_1 = \frac{L_1 dA_1 \cos\theta_1 dA_2 \cos\theta_2}{R^2}. \qquad (2)$$

This equation defines radiance and underscores its fundamental properties in describing the propagation of fluxes of optical radiation. The angular terms correctly describe the effective or projected areas that the apertures present to each other. Equation (2) relates the optical power passing through a region of space to a property of the source, the radiance L, and purely geometric considerations. In general, the radiance is a function of the coordinates defining dA_1 as well as the angles that define the direction of propagation of the light leaving surface dA_1, and thus evaluating Eqn (2), in real circumstances, can be difficult. In many cases simplifying assumptions can be made. It is important, however, to start with this complex definition of radiance in order to understand the implications of the approximations that are made to evaluate the flux in configurations used in practical radiometric measurements. Terms on the right-hand side of Eqn (2) can be grouped and the expression written in a different manner. Using the following definition,

$$d\omega_{12} = \frac{dA_2 \cos \theta_2}{R^2}, \qquad (3)$$

where $d\omega_{12}$ is the solid angle subtended at dA_1 by area dA_2, we can rearrange Eqn (2) in the following form:

$$L_1 = \frac{d^2\Phi_1}{d\omega_{12} dA_1 \cos \theta_1}. \qquad (4)$$

This equation defines the radiance in terms of the optical flux, the geometry of the source, and the solid angle containing the flux. We see in Table 2 that radiance has units of $W\ m^{-2}\ sr^{-1}$ and from Eqn (4) that radiance is the optical power of a source per unit area emitted into a solid angle defined by the region of space into which the flux is directed.

In the absence of any dissipative mechanisms in the space between A_1 and A_2, the flux in the beam leaving dA_1 within the solid angle shown is equal to that which passes through dA_2. One can create an equation much like that of Eqn (2) to describe the relationship between the radiance and flux at dA_2 and show that the radiance is a conserved quantity in the beam [10,12,13]. This can be seen if we introduce a radiance L_2 that represents the radiance at surface element dA_2 and realize that Eqn (2) can express the flux $d^2\Phi_2$ at that surface by permuting the variable subscripts, $1 \leftrightarrow 2$, and generate the same equation as Eqn (2) due to the symmetry in the variables. This argument, which assumes that the radiance is not varying over the elements of area dA_1 and dA_2, leads to Eqn (5) which states the conservation of radiance in a beam which is propagating in a nondissipative medium whose index of refraction is unity:

$$L_1 = L_2 = \text{constant}. \qquad (5)$$

This fundamental relationship underscores the importance of the concept of radiance in radiometry. This relationship undergoes a slight modification if the index of refraction of the medium in which the optical radiation is traveling differs from unity. Application of Snell's law to the angles defining the angular quantities in the definitions of radiance leads to a factor of n^2 normalizing the radiance. Using n_1 and n_2 to represent the index of refraction of the medium at the surfaces dA_1 and dA_2 respectively, we can write Eqn (6), which defines the conserved quantity in circumstances when the index of refraction is different from unity [14]:

$$\frac{L_1}{n_1^2} = \frac{L_2}{n_2^2} = \text{constant}. \qquad (6)$$

The quantity L/n^2 is sometimes referred to as the *reduced radiance* and becomes the generalized conserved quantity in the presence of media with varying indices of refraction. This relationship depends on the conservation of flux in a beam. Factors due to scattering or absorption because of index variation or other factors are not accounted for and would have to be dealt with appropriately.

The invariance of radiance in Eqn (6) holds for total radiance or for spectral radiances with units of per frequency or per wave number since those quantities do not have additional dependencies on the refractive index of the medium. Equation (6) must be modified if the spectral radiance is per wavelength because these quantities depend on the index of refraction of the media. The relationship of spectral radiance per wavelength to radiance is given by

$$L_{\lambda_1} = \frac{dL_1(\lambda_1)}{d\lambda_1}, \qquad (7)$$

where the wavelength is dependent on the refractive index of the medium,

$$\lambda_1 = \frac{\lambda_0}{n_1}, \qquad (8)$$

and where λ_1 is the wavelength in the medium with refractive index n_1, and λ_0 is the wavelength in vacuum. The elementary spectral interval is given by

$$d\lambda_1 = \frac{d\lambda_0}{n_1}. \qquad (9)$$

From Eqn (6), the differential element of radiance is also constant so that

$$\frac{dL_1}{n_1^2} = \frac{dL_2}{n_2^2}. \qquad (10)$$

Putting Eqns (7) and (9), into Eqn (10) we obtain,

$$\frac{L(\lambda_1)}{n_1^3} = \frac{L(\lambda_2)}{n_2^3}. \qquad (11)$$

A discussion of this issue may be found in the NIST Self Study Manual [10].

Using the concept of projected area, $dA_{1p} = \cos\theta_1 \, dA_1$, Eqn (4) can be written in as

$$L_1 = \frac{d^2\Phi_1}{d\omega_{12} dA_{1p}}. \qquad (12)$$

The reader is referred to the literature for details of the mathematical limiting procedures to arrive at the mathematically correct differential form of radiance [10,14,15]. An alternative and sometimes useful variant of the development of

Principles of Optical Radiometry and Measurement Uncertainty

the projected area is the concept of projected differential solid angle $d\Omega_{12} = \cos\theta_1 d\omega_{12}$. In this notation Eqn (12) becomes

$$L_1 = \frac{d^2\Phi_1}{d\Omega_{12}dA_1}. \tag{13}$$

The projected solid angle Ω_{12} is

$$\Omega_{12} = \int_\omega d\Omega_{12} = \int_\omega \cos\theta_1 d\omega_{12}. \tag{14}$$

1.3 Irradiance

Equation (2) allows the calculation of the radiant flux Φ_1 passing through the surface A_2 which originates from source surface A_1. The flux can be expressed with an integral relation over the source surface A_1 and the receiving surface A_2. Then

$$\Phi_1 = \int_{A_1}\int_{A_2} \frac{L_1 dA_1 \cos\theta_1 dA_2 \cos\theta_2}{R^2}. \tag{15}$$

In Eqn (15) Φ_1 and L_1 represent the *total radiant flux* and the *total radiance*, that is, integrated over wavelength. Since it is true in general that these quantities are a function of wavelength as well as the spatial coordinates of the emitting surface, Eqn (15) can be modified to describe the *spectral flux* leaving the surface that is a result of the *spectral radiance* of the surface by substituting the terminology $\Phi_1(\lambda)$ and $L_1(\lambda)$ for Φ_1 and L_1.

In general the integral in Eqn (15) is difficult to perform. Additionally, in Figure 2, if the apertures are large, the angles θ_1 and θ_2 and the distances R between the differential areas all have complicated relationships across the apertures that make the solution of Eqn (15) difficult. The propagation of optical flux represented by Eqn (15) is related to problems in other areas of physics such as heat transfer, and some of the techniques developed for those problems can be utilized for calculations of radiometric flux transfer [16,17].

Although Eqn (15) is in general difficult to solve exactly, in many situations in radiometry sources can be considered uniform and Lambertian. If this is the case Eqn (15) becomes,

$$\Phi_1 = L_1 \int_{A_1}\int_{A_2} \frac{dA_1 \cos\theta_1 dA_2 \cos\theta_2}{R^2} \equiv L_1 \int_{A_1} dA_1 \int_{F_{12}} \pi dF_{12} \tag{16}$$
$$= L_1 A_1 \pi F_{12} = L_1 T_{12},$$

where F_{12} only depends on the geometry and is called the configuration factor, see Appendix 3 in Chapter 4 of [10]. Other terms for the configuration factor are view, shape or exchange factor. Then, for uniform and Lambertian sources, this geometric term can be looked up in the literature [16,17] or evaluated using numerical techniques. The quantity $T_{12} = A_1 \pi F_{12}$ is called the *throughput* of the particular optical arrangement. The units of throughput are area times steradian. In the absence of dissipative effects the invariance of the radiance implies the invariance of the throughput of the system. Using the same arguments as above, one can readily show the reciprocity relations for radiometric systems, $A_1 F_{12} = A_2 F_{21}$ or $T_{12} = T_{21}$, which indicates the reversibility of the propagation of the optical beam. If the radiance and hence flux are wavelength dependent the spectral terms need to be substituted in Eqn (16).

With a uniform and Lambertian source, we can simplify Eqn (13) (or Eqn (15) using Eqns (3) and (14)) to find

$$\Phi_1 \cong L_1 \int_{A_1} dA_1 \int_{\Omega_{12}} d\Omega_{12} = L_1 A_1 \Omega_{12}, \quad (17)$$

where the approximation is that the separation distance between the areas dA_1 and dA_2 is so large that it is practically the same value across both finite areas. In other words, the projected solid angle of the beam is not strongly dependent of the position of dA_1 on A_1 or dA_2 on A_2. Note we have made no restrictions on the tilts or shapes of the aperture areas, only that the source is uniform and Lambertian. When this large R approximation is valid, then $\Omega_{12} \cong A_2/R^2$ and

$$\Phi_2 = \Phi_1 \cong L_1 \frac{A_1 A_2}{R^2}, \quad E_2 = \frac{\Phi_2}{A_2} = \frac{L_1 A_1}{R^2} = L_1 \Omega_{21}. \quad (18)$$

The irradiance E_2 at the surface A_2 is simply the product of L_1 times the projected solid angle $\Omega_{21} \cong A_1/R^2$. This is a fundamental relationship, sometimes termed the *flux transfer* equation. Equation (18) also makes it clear that for detectors calibrated for optical power, knowledge of the aperture area gives the irradiance. A third important consequence of Eqn (18) is the $1/R^2$ law. Considering the irradiance at two distances R_a and R_b from the radiance source L_1 of area A_1, Eqn (18) states that $E_{2b}/E_{2a} = (R_b/R_a)^2$, e.g., the irradiance falls off as the inverse of the separation distance squared.

Next we apply this treatment to the case where there are no restrictions on the sizes of the two apertures or the separation distances, but to a specific aperture configuration. Here, the apertures are circular, the centers are on a common centerline, and the apertures are oriented such that their aperture planes are normal to the centerline. In Figure 2 this corresponds to $\overline{N}_1$ and $\overline{N}_2$ lying along R. Letting r_1 be the radius of A_1 and r_2 be the radius of A_2, the configuration factor determined by the integral in Eqn (16) becomes [17,18],

$$F_{12} = \frac{1}{2}\left[\left(\frac{r_1^2 + R^2 + r_2^2}{r_1^2}\right) - \left[\left(\frac{r_1^2 + R^2 + r_2^2}{r_1^2}\right)^2 - 4\frac{r_2^2}{r_1^2}\right]^{1/2}\right]. \quad (19)$$

From Eqn (16) and writing $A_1 = \pi r_1^2$ with our assumption that the radiance is uniform and Lambertian over the surface, we can write an exact expression for the optical power at the second aperture due to the radiance from A_1 as,

$$\Phi_1 = L_1 \frac{\pi^2}{2}\left[(r_1^2 + r_2^2 + R^2) - \left[(r_1^2 + r_2^2 + R^2)^2 - 4r_1^2 r_2^2\right]^{1/2}\right]. \qquad (20)$$

In many cases R is sufficiently larger than either of the radii of the apertures and the expression can be simplified through a Taylor series expansion. Then

$$\Phi_1 = \frac{L_1 \pi r_1^2 \pi r_2^2}{(r_1^2 + r_2^2 + R^2)}\left(1 + \frac{r_1^2 r_2^2}{(r_1^2 + r_2^2 + R^2)^2} + \text{higher terms}\right)$$

$$= \frac{L_1 A_1 A_2}{D^2}\left[1 + \delta + 2\delta^2 + 5\delta^3 + \ldots\right] \qquad (21)$$

where we have $D^2 = r_1^2 + r_2^2 + R^2$,

and $\delta = \dfrac{r_1^2 r_2^2}{D^4}$.

This result gives a correction to the approximations used in Eqn (18) when it is evaluated for the case of two circular apertures, centered and perpendicular to common centerline. The correction is both a multiplicative and additive bias to the simpler expression, and it is straightforward, given a target uncertainty component, to select between these two forms of the flux transfer equation for circular, aligned apertures. As in Eqn (18), Eqn (21) relates flux, radiance, and irradiance. Note that the $1/R^2$ dependence for irradiance still holds true.

The reader is referred to the extensive literature on optical systems to find approximations used for the configuration factor to calculate the throughput for various specific optical arrangements, including those with lenses and other beam forming and steering devices [10,12,13,19,20]. If a source is not Lambertian and uniform, extensive measurements and characterizations are necessary to understand the relationship between the flux, radiance, or irradiance of the source.

1.4 Reflectance

The interactions of light with matter are complex and depend on the optical properties of the medium. Absorption, emission, elastic and inelastic scattering, fluorescence, and Fresnel reflection at a boundary associated with a change in index of refraction are all processes encountered in radiometry and ocean optics. In this section, we discuss reflectance, which is a general term for the ratio of reflected to incident flux. It is important for both the SR and TIR spectral regions. In general, reflectance must be described in terms of spectral, directional, geometrical, spatial, temporal, and polarization variables. Careful attention is required to define the terms and concepts in order to accurately convey the various underlying dependencies, as the notation may not be consistent across disciplines, see, for example [9,21]. In addition, authors sometimes fail to describe fully the physical quantity in question. Adding to the potential confusion is that unlike radiance or irradiance, the dimensional analysis of mathematical expressions is not likely to offer the necessary insights. We begin by introducing the terminology, following Nicodemus et al., see Ref. [9], followed by brief comments applicable to the SR region in ocean optics.

The result of reflection is a function of the incident radiance distribution, the reflectance properties of the real (or imagined, in the case of in-water radiometry) surface, the incident and viewing directions, and the associated solid angles. In what follows, we will make simplifying assumptions for both the incident radiance distribution and the surface's scattering properties, and state results for particular geometries. As in Sections 1.2 and 1.3, we will distinguish between differential and finite geometries. Consider an opaque material with a flat reflecting surface that has spatially uniform and azimuthally isotropic scattering properties over an elemental area dA. It is assumed that there are no nonlinear interactions with the incident flux and the surface. There is an incident beam of flux from a particular direction (subset "i" for incident) contained within a differential solid angle $d\omega_i$; within this solid angle we assume the radiance is constant and independent of direction. Note we are not constraining the radiance to be the same for all incident directions. The reflected radiance is viewed from a different direction (subscript "r" for reflected). It can be shown that for the area dA, there is an invariant quantity, the BRDF, that describes its reflection properties. The BRDF, or $f_r(\theta_i, \phi_i; \theta_r, \phi_r)$, is

$$\text{BRDF} = f_r(\theta_i, \phi_i; \theta_r, \phi_r) = \frac{dL_r(\theta_i, \phi_i; \theta_r, \phi_r; E_i)}{dE_i(\theta_i, \phi_i)}, \tag{22}$$

where the polar angles θ are measured from the normal to the surface and the azimuthal angles ϕ are measured from some reference plane on the surface [22]. For convenience, in Eqn (22) and below, we are ignoring the spectral dependence in the quantities. The BRDF is the ratio of the differential reflected radiance, $dL_r(\theta_i, \phi_i; \theta_r, \phi_r; E_i)$, to the differential incident irradiance, $dE_i(\theta_i, \phi_i) = L_i(\theta_i, \phi_i)\cos\theta_i \, d\omega_i$, and has units of inverse steradians. If the surface does not polarize the incident flux and there are no magnetic fields present, then Helmholtz reciprocity holds and,

$$f_r(\theta_1, \phi_1; \theta_2, \phi_2) = f_r(\theta_2, \phi_2; \theta_1, \phi_1). \tag{23}$$

The reflectance ρ is defined as the ratio of reflected to incident flux,

$$d\rho = \frac{d\Phi_r}{d\Phi_i} \tag{24}$$

and can be expressed in terms of the BRDF of a surface for a variety of geometries [9,22]. Reflectance is a dimensionless quantity. For the situation described above, the bidirectional reflectance, a differential quantity, is

$$d\rho(\theta_i, \phi_i; \theta_r, \phi_r) = f_r(\theta_i, \phi_i; \theta_r, \phi_r) d\Omega_r, \tag{25}$$

where $d\Omega_r = \cos\theta_r d\omega_r$ is the projected solid angle for the reflected beam. To derive expressions for ρ and particular geometries we must integrate over solid angle, and here we introduce three cases: over $d\Omega_r$, over $d\Omega_i$, or over both $d\Omega_r$ and $d\Omega_i$. Nicodemus et al. [9] introduce the term conical to indicate this averaging, hence we have *directional-conical*, *conical-directional*, and *biconical*. Clearly, the directional-conical situation is independent of the incident radiance distribution because we have restricted the radiance to be isotropic over this very narrow range of incident angles. This directional-conical reflectance follows directly from integration of Eqn (25):

$$\rho(\theta_i, \phi_i; \omega_r) = \int_{\omega_r} f_r(\theta_i, \phi_i; \theta_r, \phi_r) d\Omega_r. \tag{26}$$

It is also apparent that for the other two cases, the reflectance will depend on the incident radiance distribution. The general result for the conical-directional reflectance is

$$d\rho(\omega_i; \theta_r, \phi_r) = \frac{d\Omega_r \int_{\omega_i} L_i(\theta_i, \phi_i) f_r(\theta_i, \phi_i; \theta_r, \phi_r) d\Omega_i}{\int_{\omega_i} L_i(\theta_i, \phi_i) d\Omega_i}. \tag{27}$$

The conical-directional reflectance remains a differential quantity because we have yet to integrate over the viewing angles. A common situation is for ω_r or ω_i to include the entire hemisphere in which case, with the limits of integration corresponding to the hemisphere above the sample, Eqn (26) is the *directional-hemispherical reflectance* and Eqn (27) is the *hemispherical-directional reflectance*. Now, for the special case, $L_i(\theta_i, \phi_i) = L_i$, a constant for all incident directions contained within ω_i, then the incident radiance factors out of Eqn (27) and we have for the conical-directional reflectance

$$d\rho(\omega_i; \theta_r, \phi_r) = \frac{d\Omega_r}{\Omega_i} \int_{\omega_i} f_r(\theta_i, \phi_i; \theta_r, \phi_r) d\Omega_i. \tag{28}$$

The general result for the biconical reflectance is

$$\rho(\omega_i; \omega_r) = \frac{\int_{\omega_i} \int_{\omega_r} L_i(\theta_i, \phi_i) f_r(\theta_i, \phi_i; \theta_r, \phi_r) d\Omega_r d\Omega_i}{\int_{\omega_i} L_i(\theta_i, \phi_i) d\Omega_i}, \tag{29}$$

and with $L_i(\theta_i,\phi_i) = L_i$, over the directions contained within ω_i, the biconical reflectance reduces to

$$\rho(\omega_i;\omega_r) = \frac{1}{\Omega_i} \int_{\omega_i} \int_{\omega_r} f_r(\theta_i,\phi_i;\theta_r,\phi_r) d\Omega_r d\Omega_i. \tag{30}$$

One can idealize the surface as well as the incident radiance distribution. Any surface that reflects such that the reflected radiance is the same at all viewing angles, independent of the incident angle, is termed perfectly diffuse. Note this characteristic of the reflected radiance makes this a Lambertian source, and the terms diffuse and Lambertian surface are used interchangeably. If the BRDF of this surface is independent of position, the source is uniform. In general, we assume the BRDF does not depend on polarization, and that the surface exhibits no fluorescence or inelastic scattering. If there is no absorption, the surface is termed ideal as well as being diffuse and uniform. To state that a surface is perfectly diffuse, but possibly absorbing, is equivalent to stating that the *diffuse BRDF*, $f_{r,d}$, is a constant. Otherwise, from Eqn (22) the reflected radiance would depend on incident and view angles. Then from Eqn (22), $f_{r,d}$ is equal to the reflected radiance divided by the incident irradiance,

$$f_{r,d} = \frac{L_r}{E_i}. \tag{31}$$

If the surface is perfectly diffuse, then using Eqn (26) the directional-hemispherical reflectance becomes

$$\rho_d(\theta_i,\phi_i;2\pi) = f_{r,d} \int_{2\pi} d\Omega_r = \pi f_{r,d}. \tag{32}$$

The diffuse directional-hemispherical reflectance $\rho_d(\theta_i,\phi_i;2\pi)$ is the fraction of the incident flux that is reflected equally into all angles of the hemisphere above the surface. Combining Eqns (31) and (32), we have

$$L_r = E_i \frac{\rho_d(\theta_i,\phi_i;2\pi)}{\pi}. \tag{33}$$

Note the connection between Eqns (18) and (33): in both cases the irradiance is proportional to the radiance times a solid angle. For an ideal (lossless) surface we have $\rho_d = 1$, $f_{r,d} = 1/\pi$, and $L_r = (E_i/\pi)$.

From Eqns (26) and (32), the diffuse directional-hemispherical reflectance is independent of the incident radiance distribution and holds for any value of ω_i: $\rho_d(d\omega_i;2\pi) = \rho_d(\omega_i;2\pi) = \rho_d(2\pi;2\pi)$. However, there are other implications to having a perfectly diffuse and uniform surface. For an arbitrary incident radiance distribution, Eqn (27) and (29) give the conical-directional and biconical reflectances. In either case, when the BRDF is a constant, the factors involving the incident radiance distribution cancel, and we find that the diffuse conical-directional, diffuse directional-conical, and diffuse biconical reflectance are equivalent:

$$\rho_d(\theta_i, \phi_i; \omega_r) = \rho_d(\omega_i; \theta_r, \phi_r) = \rho_d(\omega_i; \omega_r) = f_{r,d}\Omega_r. \tag{34}$$

Note the diffuse directional-hemispherical reflectance, $\rho_d(\theta_i,\phi_i;2\pi) = f_{r,d}\,\pi$ is related to the sample absorption when the sample is opaque, because absorption is $1 - \rho_d(\theta_i,\phi_i;2\pi)$.

A final concept, the *reflectance factor*, is also encountered in the literature. A reflectance factor is defined as the ratio of the actual reflected flux to the flux that would be reflected by a perfectly diffuse, uniform, and ideal (non-absorbing) sample. Generally the symbol R is used to denote reflectance factors, and this quantity is dimensionless. For example, the expression for the *bidirectional reflectance factor* is

$$R(\theta_i, \phi_i; \theta_r, \phi_r) = \pi f_r(\theta_i, \phi_i; \theta_r, \phi_r), \tag{35}$$

and for the biconical reflectance factor

$$R(\omega_i; \omega_r) = \frac{\pi \int\int_{\omega_i\,\omega_r} L_i(\theta_i, \phi_i) f_r(\theta_i, \phi_i; \theta_r, \phi_r) d\Omega_r d\Omega_i}{\int\int_{\omega_i\,\omega_r} L_i(\theta_i, \phi_i) d\Omega_r d\Omega_i}, \tag{36}$$

which reduces to

$$R(\omega_i; \omega_r) = \frac{\pi}{\Omega_r \Omega_i} \int\int_{\omega_i\,\omega_r} f_r(\theta_i, \phi_i; \theta_r, \phi_r) d\Omega_r d\Omega_i \tag{37}$$

for isotropic incident radiance within the full solid angle of incidence ω_i. Note the reflectance factors are not simply π times the corresponding reflectance, for example, the *directional-hemispherical reflectance* and the *directional-hemispherical reflectance factor* are identical. Additional examples are given in [9].

We conclude this section with a brief mention of reflectance terminology specific to ocean color remote sensing. In remote sensing of the Earth's oceans from satellites in the SR spectral region, the spectral radiance exiting the surface in direction (θ_v,ϕ_v) is due to backscatter by seawater, denoted the water-leaving spectral radiance, or $L_w(\theta_v,\phi_v;\lambda)$. It is a fundamental quantity of interest. When normalized by the incident (downwelling) spectral irradiance at the surface, $E_s(\lambda)$, the result is termed the remote sensing reflectance, $R_{rs}(\theta_v,\phi_v;\lambda)$,

$$R_{rs}(\theta_v, \phi_v; \lambda) = \frac{L_w(\theta_v, \phi_v; \lambda)}{E_s(\lambda)}, \tag{38}$$

with units of inverse steradians [11]. Here we use the subscript "v" to denote a viewing angle to emphasize that Fresnel reflection from the water surface is not included in the definition of $R_{rs}(\theta_v,\phi_v;\lambda)$. The downwelling spectral irradiance includes the direct (sun) and diffuse (sky, clouds) contributions. The remote sensing reflectance is important because it is a function of the optical properties of the seawater, namely the ratio of backscatter to absorption coefficients, leading to biooptical algorithms that relate optical fluxes to the constituents of the seawater. It can also be specified for an imaginary surface within the water column. Note that the $R_{rs}(\theta_v,\phi_v;\lambda)$ has the same form as BRDF, but it is *not* a BRDF—for Eqn (22), the incoming irradiance is directional, while in Eqn (38) the irradiance corresponds to the entire hemisphere.

A second important quantity in ocean optics is termed the spectral irradiance reflectance, $R_{Irr}(z;\lambda)$, measured at some depth z in the water column:

$$R_{Irr}(z;\lambda) = \frac{E_u(z;\lambda)}{E_d(z;\lambda)}, \tag{39}$$

where $E_{u,d}(z;\lambda)$ are the upwelling and downwelling plane irradiances [11]. As with $R_{rs}(\theta_v,\phi_v;\lambda)$, the irradiance reflectance is a function of the optical properties of the medium. Finally, the BRDF of seawater must be measured—one application is for matching in situ values of the water-leaving spectral radiance to satellite observations in order to correct for differences in illumination and viewing geometry, see Ref. [23].

1.5 Distance and Aperture Areas in Radiometry

The realization of absolute spectral radiance and irradiance scales requires knowledge of the optical area of apertures and determinations of the distances between them. Absolute BRDF also depends on these quantities. Distance is straightforward; methods of measurement include digital inside micrometers and digital linear rulers with values that are traceable to wavelength standards. In the flux transfer method, Eqn (21), the $1/R^2$ dependence gives a sensitivity factor of two, $u(\Phi)/\Phi = 2u(R)/R$ (see Section 3.1), so at 50 cm an uncertainty of 250 μm would contribute 0.1% to the uncertainty in the flux. An example expanded uncertainty ($k=2$) for a calibrated inside micrometer at 50 cm is 18 μm, commensurate with the 0.1% goal. In some applications, the physical location of the aperture or radiometric reference plane is not available, such as a diffuse transmitting irradiance collector. Then measurements at different distances and the $1/R^2$ scaling law are used to determine the radiometric separation distance. An example is the use of FEL lamps at distances other than 50 cm, see Section 2.2.2 below; another example the distance determination between two apertures using a linear encoder [24]. With care, the use of commercial electronic rulers results in distance uncertainties in the tens of micrometers, which is adequate to support the uncertainty requirements for flux (or irradiance) at the distances in question.

Achieving the desired uncertainty goals for the optical area of radiometric apertures has stimulated the development of custom facilities dedicated to this application. In the simplest cases of irradiance measurement, for an unfiltered detector calibrated for spectral flux responsivity, the irradiance responsivity is equal to the area of the flux-limiting aperture times the flux responsivity, see Sections 2.2.2 and 3.2.2, for examples. The measurement of the Sun's irradiance at the top of the Earth's atmosphere over the entire SR region, termed the total solar irradiance (TSI) provides a remote sensing example. The precision of these flight instruments is more than adequate to discern the 0.12% change in TSI over an 11-year solar cycle but establishment of the absolute TSI at the desired uncertainty of <0.01% remains a goal [25]. One component in the TSI uncertainty budget is due to aperture area.

A traditional method of determining area of an optical aperture is to translate a ball-ended stylus from one edge to another at several locations around the circumference and determine the diameter and then the area. This mechanical method requires a flat reference surface on the inner diameter of the aperture, termed the land. Although designed to be thin, e.g., 0.1 mm, along the optical axis, the flat surface acts to reflect and scatter light, so the resulting optical area is dependent on the range of incident angles in the incident beam of flux and the collection geometry of the radiometer. The land can also be distorted from the mechanical pressure during measurement, depending on the aperture material and design. Knife-edged apertures with much sharper edges, e.g., tens of nanometers, can be manufactured using diamond turning, but they cannot be measured mechanically without destroying the edge. This led to the development of noncontact methods of aperture area determination, with the edge detected optically.

The noncontact method at NIST is implemented using an interferometrically controlled x-y stage that translates the aperture under a microscope that is coupled to a charge-coupled device camera [26,27]. The aperture is illuminated from below and the camera's z position is automated to move relative to the aperture to determine the best focus for points around the circumference. An edge-detection algorithm locates the x,y coordinates of the edge relative to the stage location. The results are fit to a circle and corrected to a 20 °C reference temperature using the material's linear expansion coefficient. Monte Carlo-based resampling of the data is used to estimate the uncertainties in the mean radius and center coordinates of the circle. For a quality, knife-edge, diamond-turned aperture 5.26 mm in diameter, relative standard uncertainty in the area was reported to be 0.0025% [27]. A comparison of apertures with heritage to TSI instruments revealed that the uncertainty in aperture area is a function of the quality of the edge, the flatness of the aperture, and the ability to mount it in the plane with the moving stages [28]. A final point regarding apertures, whether they are flux limiting, field-of-view limiting, or nonlimiting baffles, is the edge bevel—the slanted surface on the inner diameter connecting the main body of the aperture structure to the optical edge.

Breault gives an excellent discussion on this topic [29], which, if neglected, will lead to unwanted scattered light in the optical system.

2. RADIOMETRIC STANDARDS AND SCALE REALIZATIONS

In this section we give examples of radiometric source and detector standards, including descriptions of how the associated radiometric scales are realized. The methods of scale realization result in the ability at a laboratory, typically a National Measurement Institute (NMI) to assign values of spectral (ir)radiance, spectral (ir)radiance responsivity, spectral BRDF (and related reflectance quantities), or aperture area to particular artifacts. These artifacts can then be used in different laboratories, including field work, to realize the associated radiometric scale specific to the user's purposes, thus disseminating the scale by assigning values to their source, radiometer, reflectance standard, or aperture. In this way, the metrological traceability of the user's results is established. The formal definition of metrological traceability is

property of a measurement result whereby the result can be related to a reference through a documented unbroken chain of calibrations, each contributing to the measurement uncertainty [4].

Metrological traceability is critical for climate change research because inherent to the principle is the concept of establishing a calibration hierarchy, which includes estimates of measurement uncertainty. The measurement of absolute changes required for climate change research is best served when the metrological traceability is to a measurement unit of the SI.

2.1 Sources

2.1.1 Blackbody Sources

Blackbody sources are used for spectral radiance calibration of sea surface temperature radiometers. They are used indirectly for other ocean radiometric applications, such as ocean color, in the establishment of spectral radiance scales in the visible through the infrared spectral range. A typical blackbody source consists of a black-coated metallic cavity and a means to measure the temperature of the cavity. Given the spectral emittance of the cavity and the temperature, the emitted spectral radiance is computed from the Planck radiation formula [30,31]:

$$L(\lambda; T_{BB}) = \varepsilon \frac{c_{1L}}{n^2 \lambda^5} \frac{1}{\exp(c_2/(n\lambda T_{BB})) - 1}, \qquad (40)$$

where T_{BB} is the thermodynamic temperature of the blackbody, c_{1L} is the first radiation constant for spectral radiance, and c_2 is the second radiation constant. The values for c_{1L} and c_2 can be found in [32]. For the in-air case, n is near unity, but see the work by Ciddor for proper evaluation [33]. The emittance ε

of the blackbody, which is a correction factor for its nonideal nature, is also generally wavelength dependent, and is a function of the blackbody cavity geometry, the reflectance of the inner walls, and temperature gradients in the cavity walls [30]. In Eqn (40), the spectral radiance is per unit wavelength.

Blackbody designs depend on the temperature range, coating material, and environmental conditions. In the TIR emissive spectral range, relevant for sea surface temperature measurements, blackbody temperatures within a few tens of degrees of room temperature will suffice. In the SR spectral range, much higher blackbody temperatures are required in order to adequately represent the relative spectral distribution of reflected solar radiation. Aperture diameters typically used in the TIR can be relatively large, in the range of several centimeters, providing an extended area scene to the radiometer being calibrated. However, in the SR range, much smaller aperture diameters (few millimeters) are common to facilitate maintaining the much higher temperatures (700–3000 K) in a practical manner. Often blackbody sources are used in an ambient air environment, except those used to calibrate satellite sensors, which are typically used in a vacuum environment.

For radiometric calibration of sea surface radiometers and other thermal infrared spectroradiometers, a temperature range of 15–50 °C is common [34]. NIST uses a water bath blackbody (WBBB) to cover this range [35]. It has a 10.8-cm-diameter cylindrical-conical cavity coated on the inside with a specular black paint to provide an emittance of 0.9997. The cavity is immersed in the side of a stirred bath of water such that its outer surface is surrounded by water. Some sea surface temperature radiometry groups use calibration sources having a very similar design to the NIST WBBB, except with antifreeze or other fluids instead of water to achieve slightly lower temperatures [34]. Other sea surface temperature groups have developed lower-cost, higher-uncertainty variants whereby the rear surface of the cavity and thermometer are immersed in an insulated cooler containing ice water [36]. Diffuse black coatings are also used, depending on the blackbody cavity shape.

For radiometric calibration in the SR spectral range, a laboratory blackbody operating at temperatures that simulate the solar spectrum requires temperatures near 6000 K, which is too high to be practical. In practice, a high-temperature blackbody (HTBB) is one operating at temperature near 3000 K, with operation at 3200 K with compromised lifetimes [37]. Temperature measurement is one of the main issues for HTBB sources. Fixed-point blackbodies are commonly used, whereby the entire outer surface of the cavity is immersed in pure metal or metal–carbon eutectics operated at the liquid–solid critical temperature, and the inner surface is the radiating surface. Since the transition temperatures of the pure metals used are defined by the ITS-90, the blackbody radiance temperature is known if the material is not contaminated by impurities and other experimental conditions are met. As an example, NIST uses a gold-point blackbody source at a temperature of 1337.33 K to define the radiance scale [38]. A new, independent method to determine the blackbody temperature, described in [39], uses filter radiometers, previously calibrated on a detector-based responsivity scale with values traceable to the electrical watt and the meter rather than the kelvin, to directly measure the radiance in a known bandwidth and deduce the temperature from Planck's equation [40,41]. This method is used for the blackbody temperature determination described in Section 3.2.2 below.

2.1.2 Lamp Sources

Tungsten ribbon filaments, or strip lamps are important sources in the radiometric standards community, serving as convenient proxies for fixed-point blackbodies. They can also be calibrated as standards of spectral radiance or as standards of radiance temperature [15,38,42,43] The lamp envelope can be evacuated or filled with an inert gas. Remarkable long-term stability is achieved with vacuum strip lamps operated as standards of spectral radiance at 655 nm. Routine use of tungsten strip lamps for radiance scale realizations is not common; the lamps are difficult to acquire and the small filament requires an area of 0.6×0.8 mm to be imaged. Polarization and alignment issues must be considered. However, they are encountered and do provide a means to validate spectral radiance scale realizations at a user facility [44].

As opposed to strip lamps, 1000 W type FEL irradiance standard lamps are common for routine use. In the past, NIST assigned values of spectral irradiance these FEL irradiance standard lamps using the gold-point blackbody as the primary standard. More recently, these values are derived from a detector-based spectral responsivity scale that is discussed below [39,45]. FEL lamps are suited to operation as spectral irradiance sources. The double coil nature of the tungsten filament creates small cavities that enhance the emittance from that of bare tungsten. The lamps can be very stable and they are portable and compact. NIST calibrates the FEL lamp for spectral irradiance at 50 cm from the front of the lamp posts over a region 1 cm^2 in area. If they are used for irradiance calibration of a radiometer at exactly the same distance and receiving aperture as the NIST calibrations, the irradiance distribution need not be considered. If the receiving area of the irradiance radiometer is larger than this, then the source nonuniformity will contribute to the uncertainty. Treating the FEL lamp as a point source, we expect the irradiance on a plane to fall off as $\cos^3 \theta$, so to achieve uniformity over a large receiving area, the distance must be increased. The filaments of FEL lamps are larger in the vertical dimension compared to the horizontal dimension, leading to asymmetry in the sensitivity to tilts. FEL lamps issued by NIST have been screened and do not exhibit more than a 1% change in spectral irradiance from a $\pm 1.0°$ rotation in any direction [46]. For the best results during operations not identical to the NIST conditions, the irradiance uniformity must be mapped. Results with three lamps indicate that the geometric and radiometric centers of the irradiance uniformity are not coaligned [47], due to the fact that the filament can be tilted with respect to the mechanical reference coordinate system.

To determine the irradiance at distances other than 50 cm, the $1/R^2$ behavior of the irradiance from a point source is utilized. However, for FEL lamps, the radiometric center is not known—the 50 cm separation is measured from the lamp posts, a mechanical reference. An offset, χ, has to be determined from measurements at different distances and fits to the $1/R^2$ model. If the offset is not known, one approximation is that the lamp filament is centered over the posts, in which case $\chi = 3.175$ mm. Yoon and colleagues estimate from studies of three lamps that use of this approximation would result in a distance uncertainty of about 1 mm [47].

As mentioned in Section 2.1.1, in the SR spectral region the relative spectral distributions of tungsten strip lamps, FEL lamps, or an HTBB are a poor match to natural sources. Figure 3 illustrates this effect. In Figure 3(a), the spectral irradiance of an FEL lamp at $R = 50$ cm and an HTBB at 2950 K are plotted on the left ordinate and the spectral radiance of the solar-illuminated Earth, for the Earth as a diffuse reflector with a $\rho_d(\theta_i,\phi_i;2\pi) = 0.8$ on the right ordinate. The exoatmospheric solar irradiance is designated $F_0(\lambda)$. Reflectance values in this range correspond to clouds or bright regions such as deserts. In Figure 3(b), the two standards are plotted as before, but on the right ordinate the solar-illuminated Earth for $\rho_d(\theta_i,\phi_i;2\pi) = 0.05$ and the upwelling spectral radiance for low-chlorophyll open ocean at 1 m depth, $L_u(z = 1$ m$)$ is plotted. For either natural target, the spectral distribution of the standards has a blue to red ratio that is much smaller than that required to match the spectral distributions that are to be measured. However, the relative spectral distributions of the HTBB at 2950 K and the FEL lamp are a good match; this is discussed in Section 3.2.2 below. Figure 3 illustrates that the systematic effect of spectral stray light is not extremely critical when comparing the two standards under the conditions stated, but can be extremely important when an instrument is calibrated with a blue-poor, red-rich standard and then used to measure a blue-rich, red-poor natural target. Spectral stray light is when the instrument is set to measure at some wavelength but due to finite out-of-band transmittance in filters or scatter in grating instruments, flux at other wavelengths is sensed and therefore measured at the set wavelength. In order to improve the spectral match between calibration and unknown sources, research and development into alternatives to the incandescent lamps is ongoing. Examples of alternative sources include Xe arc lamps, metal halide arc lamps, light-emitting diodes, novel laser sources that produce hot plasmas, and spectrally programmable broadband light sources.

2.1.3 Integrating Spheres

Lamp-illuminated integrating spheres are sources of broadband spectral radiance for absolute responsivity calibration of radiometers in the SR spectral range. Examples are the NIST Portable Radiance source that is used for

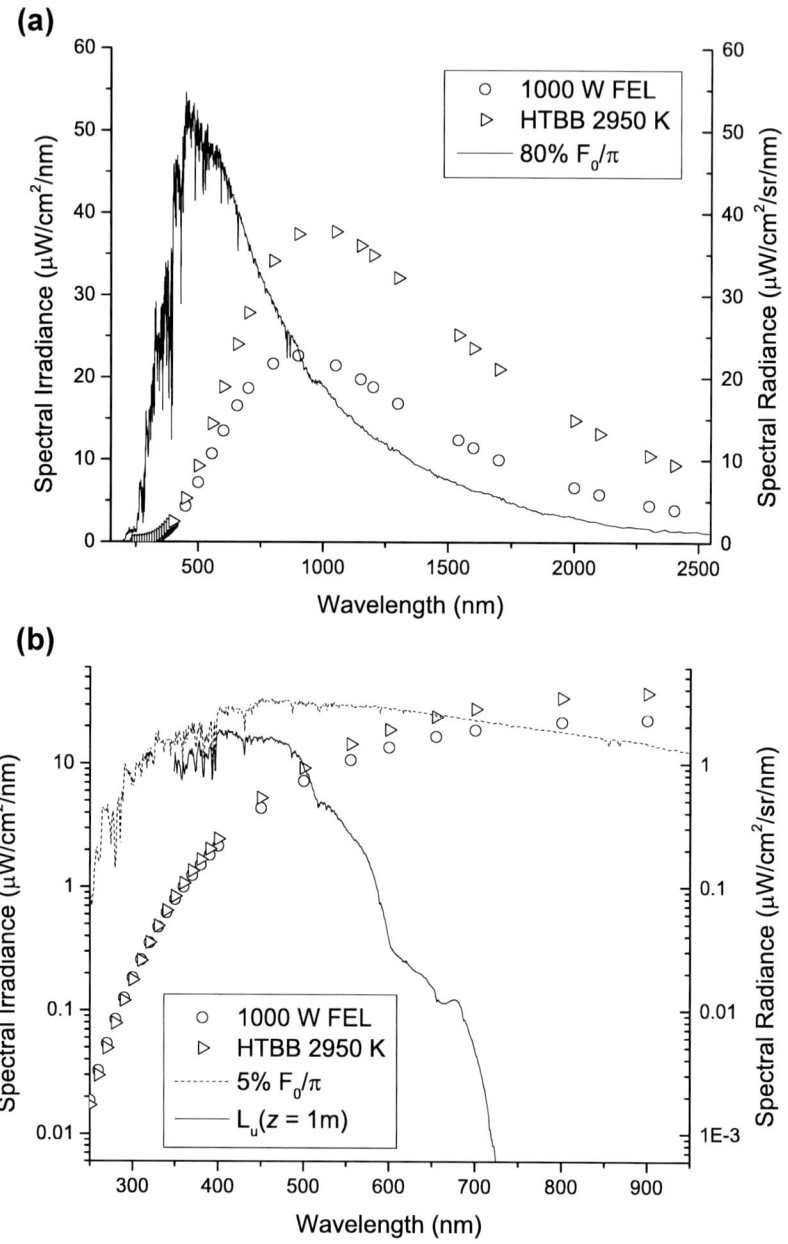

FIGURE 3 (a) Left ordinate: the spectral irradiance of a 1000 W FEL lamp at $R = 50$ cm (open circles) and an HTBB at 2950 K with a 10-mm-diameter aperture at $R = 43.4$ cm (open right triangles); right ordinate: the spectral radiance of the Earth, illuminated by the Sun and modeled as an 80% diffuse reflector (solid line). (b) Left ordinate: as in Figure 3(a); right ordinate: the spectral radiance of the Earth modeled as a 5% diffuse reflector (dashed line) and the upwelling spectral radiance at 1 m depth at the MOBY site (solid line) [48].

validation activities in ocean color [49], or much larger spheres used to calibrate satellite sensors, see for example [50]. Because the inner wall reflects diffusely, the radiance in the exit aperture is unpolarized and has good spatial uniformity. The uniformity is dependent on the specific configuration, with the results improving as the ratio of the open aperture areas to the overall sphere wall area decreases [51]. Generally, tungsten quartz halogen lamps are used, either for internal or external illumination. As with FEL lamps, the radiance depends on lamp current. The spectral radiance may also change with time from environmental contamination. The short- and long-term stability of the spectral radiance should be recorded during operation. Example devices include monitor detectors mounted in the sphere wall, shunt resistors in series with the lamp current, and lamp sockets wired to provide the voltage drop across the lamp. In addition to this integrated monitoring of the sphere output, the spectral radiance of integrating sphere sources routinely used to calibrate ocean color radiometers is verified using external radiometers specifically designed for that purpose [50,52—54].

Laser-illuminated integrating spheres are also used as radiometric sources for calibration and characterization of radiometers. Here a flux-stabilized laser illuminates the integrating sphere, either through a small port or with fiber-optical coupling. A second, larger exit aperture serves as the radiance source. The result is a spatially uniform, diffuse, and monochromatic radiance source with a wide dynamic range in terms of flux level. At NIST, the facility for Spectral Irradiance and Radiance responsivity Calibrations using Uniform Sources (SIRCUS) incorporates a variety of tunable lasers and Spectralon® [55] and diffuse gold spheres to cover the SR and a portion of the TIR spectral regions [24]. In some cases the lasers can be automated to cover selected spectral regions under computer control. Care is taken to reduce the effects of laser speckle, and a monitor detector in the sphere wall serves as a transfer standard in the flux transfer procedure that realizes radiance in the exit aperture or irradiance at some distance from the exit aperture. The wavelength is measured using an interferometer-based wavemeter. The flux transfer method utilizes broadband radiometers calibrated for flux responsivity to a detector-based scale, discussed in Section 2.2.2 below, and apertures measured on the NIST aperture facility discussed in Section 1.5 above.

For any type of integrating sphere, including those used as part of the radiometer's input optics, as well as for diffuse reflectance standards (see Section 2.1.4 below), the goal is to utilize uniform, perfectly diffuse, and ideal reflectors. Of course, no real material is perfectly diffuse, uniform, and non-absorbing, but some are adequate to qualify as diffuse reflectance standards. For the SR region, polytetrafluoroethylene (PTFE) can be used in the pressed powder form [56—59] or purchased commercially in solid forms such as Spectralon® from Labsphere, Inc. [60]. White diffuse opal glass, diffuse white paints, barium sulfate coatings, and bead-blasted aluminum are other examples of diffuse reflectance surfaces. In the TIR spectral range, diffuse gold coatings are used in integrating spheres [61].

2.1.4 Diffuse Reflectance Standards

A direct application of the reflectance concepts introduced in Section 1.4 is to utilize very diffuse and uniform reflectance standards in the SR spectral region to realize a scale of spectral radiance when combined with an FEL lamp standard of spectral irradiance. The spectral radiance values are traceable to spectral irradiance and the BRDF, reflectance, or reflectance factor of the diffuse reflectance standard. Examples of this approach include the preflight calibration of ocean color satellite sensors, e.g., the Moderate Resolution Imaging Spectroradiometer (MODIS) [50] and laboratory calibrations of field radiometers used for in situ validation [62]. The typical measurement geometry for this "lamp/plaque" method of realizing spectral radiance is normal incidence for the illumination and 45° from normal for the viewing. Assuming that there is no dependence on azimuth, the incident beam is collimated, and Ω_r is large enough to produce measurable signal but small enough to qualify as the bidirectional geometry, we can write the measurement equation, now including the spectral dependence, either in terms of BRDF or bidirectional reflectance factor:

$$L_r(0°;45°;\lambda) = f_r(0°;45°;\lambda)E_i(0°;\lambda) = \frac{R(0°;45°;\lambda)}{\pi}E_i(0°;\lambda). \quad (41)$$

Lamp-to-plaque distances greater than 50 cm are sometimes utilized in order to have uniform irradiance over the entire area of the plaque. At normal incidence, the irradiance at the plaque surface is given by

$$E(0°;\lambda;d) = E(0°;\lambda;d=50\text{ cm})\frac{(50+\chi)^2}{(d+\chi)^2}, \quad (42)$$

where d and χ are in centimeters. Combining equations,

$$L_r(0°;45°;\lambda) = E_i(0°;\lambda;d=50\text{ cm})\frac{R(0°;45°;\lambda)}{\pi}\frac{(50+\chi)^2}{(d+\chi)^2}, \quad (43)$$

where $R(0°;45°;\lambda)$ is the bidirectional reflectance factor for the 0°/45° geometry.

If the sample is perfectly diffuse and uniform, comparison of Eqns (32) and (33) to Eqn (41) predicts that the diffuse directional-hemispherical reflectance, $\rho_d(\theta_i,\phi_i;2\pi)$, should be identical to the $R(0°;45°;\lambda)$ reflectance factor. To reduce the source radiance to match values in the SR natural light fields, it can be desirable to have $\rho_d(\theta_i,\phi_i;2\pi)$ less than unity and as spectrally flat as possible; one method is to add carbon to the sample or the paint. Gray-sintered PTFE is commercially available.

The BRDF of a sample can be determined using incident monochromatic radiation and a broadband detector. In the NIST Spectral Tri-function

Automated Reference Reflectometer (STARR) facility the sample is mounted in a goniometer and the same detector, by rotating about the sample, measures the incident and reflected flux [22]. Alternatively, a broadband source of irradiance and a spectroradiometer can be used in an analogous but inverted configuration from the STARR facility [63]. A third method utilizes calibrated spectral radiance and spectral irradiance standards [64]. Absolute measurements of directional-hemispherical reflectance are accomplished using an integrating sphere coupled to a detector, with the sample mounted in the sphere wall and the ability to rotate the sphere collector so that the incident and reflected fluxes are measured [58,61]. If the directional-hemispherical reflectance is determined for normal incidence, any specular component to the reflectance is not measured as the reflected flux exits the entrance aperture of the integrating sphere collector. Most measurement requests correspond to the specular component being included, and therefore angles of incidence near, but not equal to, normal incidence are used.

Diffuse reflectance standards can be used to establish scales of spectral reflectance, with a spectroradiometer performing as a reflectometer. This technique is employed in establishment of scales of spectral reflectance on samples at user facilities using a diffuse reflectance standard calibrated by a standards laboratory, see, for example, [65]. In SR remote sensing, the Sun is the source. Measurements from orbit determine the spectral radiance or reflectance of the Earth using an on-board diffuse reflectance standards, see, for example, [66,67]. Similarly, the reflectance of natural targets, e.g., desert playas or the ocean, can be determined by reference to diffuse reflectance standards [68–70].

In Section 1.4, we discussed reflectance standards for extreme cases—completely arbitrary or perfectly diffuse incident radiance distributions or scattering surfaces. Actual diffuse reflectance standards are not perfectly diffuse, uniform, or nonabsorbing, and they depend weakly on polarization. First, the BRDF (or alternatively, the bidirectional reflectance factor) is not a constant as stated in Eqn (31). As part of a National Atmospheric and Space Administration (NASA) Earth Observing System (EOS)-sponsored round-robin, pressed PTFE, sintered PTFE, Spectralon®, and vacuum-deposited aluminum were measured by four laboratories at four incident angles (0, 30, 45, and 60°) for a range of reflected angles and wavelengths [71]. This study found weak spectral dependence in the BRDF in the ocean color region but strong dependence, 20% or more, on θ_r for a given θ_i. The BRDF for these samples was fairly symmetric in θ_r for $\theta_i = 0°$, but asymmetry in θ_r was present at the other three angles of incidence. The BRDF of all samples increased in the specular direction for $\theta_i = 45$ and $60°$. These measurements were in plane, so no information was gained on the degree of azimuthal isotropicity. Other investigations of angular dependencies include the SeaWiFS Intercalibration Round-Robin Experiments (SIRREXs), where a 4% change in $R(0°;\theta_r)$ was found for Spectralon® for $\theta_r = 7-40°$ at 400 and 632.8 nm [72].

Using the Sun as a source, Jackson, Clarke, and Moran, found over 20% variation in $R(0°;\theta_r)$ for Spectralon® for θ_r up to 80° [73]. Butler and Georgiev found complicated behavior in the BRDF of gray Spectralon®, and report detailed results for a set of gray samples in Ref. [74]. The facility at NASA Goddard Space Flight Center has the ability for out-of-plane measurements [75].

Second, it follows that the diffuse directional-hemispherical reflectance, $\rho_d(\theta_i,\phi_i;2\pi)$, is not a constant, see Eqn (32); it depends on angle of incidence and wavelength. See Ref. [58] for results for pressed PTFE, where up to 2% difference was observed at 250 nm and 75° angle of incidence. Third, for the perfectly diffuse surface, we derived from Eqns (26), (32), and (35) $R(\theta_i, \phi_i; \theta_r, \phi_r) = \rho_d(\theta'_i, \phi'_i; 2\pi)$, e.g., $R(0°;45°) = \rho(6°;2\pi)$ for the lamp/plaque radiance scale realization if the plaque is perfectly diffuse. If the calibration laboratory reports only the directional-hemispherical reflectance to the customer (6° for NIST and 8° for Labsphere, Inc.), who then uses this equality to evaluate the bidirectional reflectance factor for scale realizations or stability monitoring, biases will be introduced into the result. Comparison of measurements of $R(0°;45°)$ with measured values of $\rho(6°;2\pi)$ for pressed PTFE gave $R(0°;45°)/\rho(6°;2\pi)$ values between 1.02 and 1.025 for the spectral interval from 400 to 1600 nm [57,64]. Studies at SIRREX-4 found $R(0°;45°)/\rho(8°;2\pi) = 1.028$ for a single Spectralon® plaque at 400 and 632.8 nm [72].

The illumination and viewing conditions are in general very different between the standards laboratory and the natural environment of ocean optics. A critical factor is temporal change in the incident radiation, which is controlled or monitored in the laboratory but constantly varying in the natural environment. Simultaneous measurements at all wavelengths for all quantities of interest, which would reduce the measurement uncertainty, are often not possible, and some form of normalization must be performed to mitigate this effect. The incident radiance distribution and the scattering properties of the oceans are very different than the diffuse reflectance standards used in some protocols to transfer the reflectance scales [69]. For best results, the diffuse reflectance standard must be calibrated for the geometry of the application, with full consideration of the uniformity of the incident irradiance distribution and the field-of-view of the spectroradiometer.

2.2 Radiometers

Radiometers function as transducers, converting radiant flux into electrical outputs depending on the nature of the detector and associated electronics. In solid-state devices the absorption of photons by silicon or another semiconductor results in an electrical current generation, which can be calibrated to be proportional to the optical power. In radiometers dependent on thermal effects, the optical power is determined by knowledge of the physics of the thermal effects in the detector [14,40,76,77]. The authors in these references offer complete discussions of the various types of optical detectors suitable for the wavelength region of interest to the optical remote sensing community. Radiometers for specific applications in remote sensing will be discussed as needed in other chapters of this volume. Here we describe detector-based radiometry and how these values are disseminated.

2.2.1 Electrical Substitution Radiometers

The starting point of detector-based spectral responsivity scales used for calibration of transfer radiometers, such as filter radiometers and spectrographs used in ocean radiometry, is an instrument known as an electrical substitution radiometer (ESR). ESRs are used primarily at the national standards laboratories such as NIST and the National Physical Laboratory, United Kingdom (NPL, UK) to provide radiant flux responsivity scales traceable to the watt through electrical standards [78]. The principle of operation of an ESR is that electrical power and radiant power have the same heating effect on the cavity. Figure 4 shows a schematic of an ESR that consists of a receiving cone thermally connected to a heat sink. When the shutter is opened, the radiant flux to be measured enters the receiving cone and is absorbed. The receiving cone has an absorptive material on the inside surface and is designed to maximize the absorbance. The radiant flux heats the cone to a temperature T that depends on the conductance G of the thermal link between the receiving cone and the heat sink. The heat sink is controlled at temperature T_0. Typically, the temperature of the receiver cone is also controlled by an external servo loop that supplies electrical power to the heater of resistance R. When the shutter is closed, the servo loop increases the electrical current through the

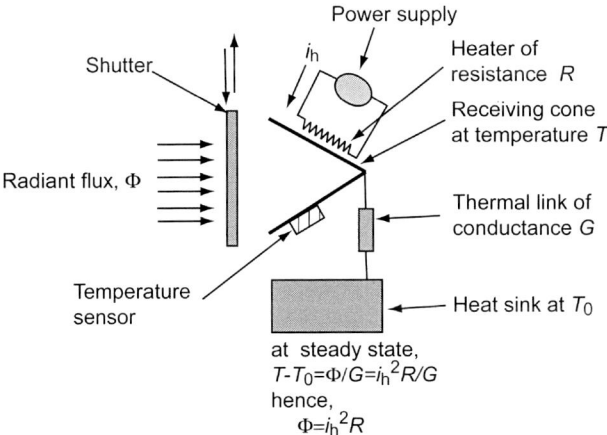

FIGURE 4 Schematic diagram of an electrical substitution radiometer. *Reproduced from Figure 1.1 in Ref. [5].*

heater by an amount i_h in order to maintain the temperature of the cone at T. Except for a number of small effects discussed below, the radiant flux Φ is equal to the electrical power $i_h^2 R$, so the radiant flux can be measured by measuring i_h and R.

Using an ESR, the measurement of radiant flux can be made traceable to electrical standards for current (or voltage) and resistance. That is, in contrast to blackbody sources, the traceability of radiant power to SI units as measured by an ESR is independent of temperature standards. (Note that although temperature is monitored while the shutter is opened and closed, an absolute temperature scale is not required.) Thus, ESRs provide an independent method of establishing radiometric standards—a feature that can be used to verify temperature-based blackbody source scales. Also, ESRs work just as well in the SR spectral range as they do in the thermal infrared if the surface is spectrally flat in absorptance.

A number of small effects are corrected for when using ESRs. Most obvious is the absorptance of the receiver cone. As with the emittance of blackbody cavities, the absorptance of the receiver cone can be determined from measurements of the reflectance. Typical cavity reflectance values are in the range 5×10^{-6}–10^{-4}. The reflectance can be modeled from coating reflectance measurements and cone geometry. It is usually dominated by the diffuse reflectance from the first surface encountered by incident light. As a verification of the model, cavity reflectance of the coated cavity can also be measured, using reflected laser light collected by an integrating sphere [78,79].

The corrections from several other small effects are minimized by operating ESRs in vacuum and at cryogenic temperatures, typically between 1.4 and 9 K. The vacuum eliminates convective thermal coupling between the receiver cone and its surroundings, which would otherwise cause thermal drift. Also, since the radiative coupling to the surroundings is reduced as T^4, the operation under cryogenic temperatures virtually eliminates thermal drift caused by such coupling. Superconducting heater leads are used, such that there is no joule heating of the leads which would otherwise provide a systematic error. At cryogenic temperatures, the thermal diffusivity of copper is high, so the spatial uniformity of the temperature of the copper receiver cone can be maintained, providing better equivalence of the spatial profile of the heating between radiant power and electrical power. The residual nonequivalence is modeled and corrected. Also, as compared with ambient temperatures, the lower specific heat of copper at cryogenic temperatures provides for faster measurements at a given sensitivity level. Typically, the polarized laser beam enters the cryostat through a window oriented at Brewster's angle, in which case the window transmittance is near unity. The exact value for a given window is measured and a correction is made. The uncertainty budget for cryogenic ESRs includes the uncertainty for the corrections from all of these small effects. The total combined uncertainty for radiant power measurements is typically about 0.01% ($k = 1$) [80].

2.2.2 Radiance and Irradiance Responsivity

Radiant flux scales established by cryogenic ESRs at standards labs using flux-stabilized lasers are immediately transferred to stable detectors. In the visible to near-IR spectral range, these are typically silicon photodiodes. Usually photodiodes in a "trap" arrangement are used, consisting of three or more silicon photodiodes arranged optically in series and with the photocurrents summed. To determine the absolute radiant flux responsivity at each laser wavelength, the photocurrent response measurement at each laser wavelength is divided by the corresponding laser power measured by the ESR.

As introduced in Section 2.1.3, in order to form a detector-based radiance responsivity scale, the irradiance from a laser-illuminated integrating sphere is measured using a trap detector that is calibrated for spectral flux responsivity using the ESR and fitted with a precision aperture of known aperture area. If the light from the sphere uniformly illuminates this irradiance trap detector by overfilling the aperture, and if the beam defined by the aperture underfills the individual detectors in the trap, the irradiance of the beam is determined from the ratio of radiant flux measurement to the aperture area as in Eqn (18). In order to accurately apply the radiant flux spectral responsivity value determined from the direct laser measurements against the ESR, the individual detectors in the trap must have sufficient spatial uniformity. If the aperture area of the sphere and the distance between the sphere aperture and the trap detector aperture are also known, the flux transfer method gives the radiance value of the sphere. These (ir)radiance scales can then be immediately applied to any other radiometers, such as filter radiometers or spectrographs used for ocean radiometry, that view the laser-illuminated sphere source in either radiance or irradiance mode. Absolute (ir)radiance responsivity of filter radiometers or spectrographs can then be measured by repeating this process as the wavelength of the laser is tuned across the spectral range [24,81,82]. Relative (ir)radiance responsivity characterizations can also be performed, resulting in the necessary data to correct for spectral stray light in spectroradiometers, see, for example, [83,84].

In practice, tuning a laser across a wide range as described above may be very time consuming. One alternative is to use a scanning monochromator to filter the output of a broadband source, which then illuminates the device under test (a filter radiometer or a spectroradiometer). A broadband, calibrated detector also measures the output, so the relative spectral responsivity of the device under test can be determined. Then the laser-illuminated sphere method is used at a limited number of tie-point wavelengths to determine the absolute scale of the spectral responsivity. Another alternative for determining the absolute scale is to utilize the relative spectral responsivity determined from the scanning monochromator and broadband detector and measure a calibrated source of spectral radiance—a blackbody or a lamp-illuminated integrating sphere, see Section 3.2.1. Whether the source is the output of the scanning monochromator or the laser- or lamp-illuminated integrating sphere, it should fill the entrance pupil of the filter radiometer or spectrograph in order that it is calibrated in the same way it is used in the field. The laser-illuminated sphere approach has an advantage over the monochromator approach in this regard, as the direct output of the monochromator is not usually large enough or Lambertian enough to fill the pupil, and diffusing it with an integrating sphere or plaque usually leads to signal-to-noise limitations.

Radiometers are critical tools in optical remote sensing and this brief introduction on their use is not sufficient for the construction or calibration of instruments. The reader is encouraged to consult the specific literature to gain a detailed understanding of radiometer construction and calibration. The book, *Optical Radiometry*, edited by Parr, Datla, and Gardner contains discussions of most types of radiometers and their calibration [40]. Hengstberger's book, Absolute Radiometry, discusses the construction and use of ESRs and methods of calibration [77]. Books by Boyd, Budde, and Rieke discuss various types of solid-state detectors for various wavelength regions and discuss their integration into appropriate measurement electronics [14,85,86].

3. THE MEASUREMENT EQUATION

3.1 Background and a Review of the Concepts

The internationally agreed upon methods for reporting measurement uncertainties are outlined in the Guide to the expression of uncertainty in measurement (GUM) [87]. The GUM, as it is commonly referred to, gives the framework for constructing uncertainty statements and outlines appropriate statistical techniques to arrive at estimates of uncertainty in the measured quantities. An excellent summary of the GUM as applied to radiometry can be found in [88]. In the GUM, the quantity to be determined or measured is called the measurand and is denoted by Y. The measurand is related to a set of input quantities $X_1, X_2 \ldots X_N$ by a functional relationship,

$$Y = f(X_1, X_2 \ldots X_N), \qquad (44)$$

where the function f is defined by the physics of the measurement problem. This equation is referred to as the measurement equation as it relates the input quantities to the desired measurand.

An actual measurement in general does not determine the measurand but it is a methodology to make an estimate of the measurand using the best estimates of the input quantities X_i from the measurement equation. The GUM uses lower case variables to distinguish these best estimates from the physical quantities represented in Eqn (44). Thus $x_1, x_2, \ldots x_N$, are variables that are the best estimates of $X_1, X_2, \ldots X_N$, and y is the best estimate of Y. We then have as an experimental measurement equation

$$y = f(x_1, x_2 \ldots x_N). \qquad (45)$$

The variables x_1, x_2,...x_N, are experimentally determined and will have uncertainties associated with them that give rise to an uncertainty in the estimate for the measurand y. The quantities, x_1, x_2,...x_N, may themselves have measurement equations representing separate determinations. We will explore this issue later in some detail for an actual measurement problem.

The writers of the GUM had specific intent, namely to outline procedures for estimating uncertainty that are universal, internally consistent, and transferable. In the GUM, there is no difference between an uncertainty component arising from a random effect and one arising from a correction (or correction factor) to a known bias (also termed a systematic effect). Emphasis is placed on realistic determinations of uncertainty, not simply on making conservative estimates. The result is that the GUM is framed in specific language that may be at odds or unfamiliar with the reader's experience, so here we give a brief overview.

The GUM makes a distinction between quantities that can be realized compared to those that can only be idealized and recommends avoiding reference to idealized concepts. Estimates of uncertainty are realized, while the concepts of error, true value, and accuracy remain forever idealized because they involve knowledge of the *true value*, which remains unknown in spite of the level of effort associated with the measurement. Formerly, *error* is the result of a measurement minus the true value, and *accuracy* is the closeness of the agreement between the measurement and the true value. Note accuracy is a qualitative concept. *Precision* is sometimes confused for accuracy, but they are not related—precision is simply a measure of the closeness of agreement of repeat measurements under the specified conditions [4].

Given that errors contribute to the lack of agreement of the measurement result and the true value, the GUM recognizes two error sources: random error and systematic error. Unpredictable, nondeterministic variations in some influencing quantity result in random effects, while deterministic variations result in systematic effects. Random effects are mitigated by increasing the number of measurements while maintaining similar measurement conditions; systematic effects are mitigated by performing characterization experiments that result in corrections, e.g., offsets, or correction factors, e.g., multiplicative factors. Measurement equations and corresponding uncertainty estimates can be associated with the systematic corrections or correction factors.

The concepts of repeatability and reproducibility add depth to the notions of precision as well as random and systematic effects. The repeatability condition of measurement is a

> condition of measurement, out of a set of conditions that includes the same measurement procedure, same operators, same measuring system, same operating conditions and same location, and replicate measurements on the same or similar objects over a short period of time [4].

so a practical definition of measurement precision, or repeatability, is the result of realizing a repeatability condition of measurement. In the laboratory, the radiometric measurement precision is often determined by the stability of the sources and radiometers, and with care it will be a negligible component to the uncertainty budget. In the natural environment, the time scale over which the same conditions hold will most likely constrain the achievable measurement precision. In contrast to repeatability, the reproducibility condition of measurement is a

> condition of measurement, out of a set of conditions that includes different locations, operators, measuring systems, and replicate measurements on the same or similar objects [4].

In radiometry, it is quite easy to be unaware of systematic effects. In this case, the reported uncertainty will be inadequate as a measure of the likelihood that the measurement result is representative of the true value. Intercomparison and validation activities are examples of implementing a reproducibility condition of measurement. This practice is strongly advised, as it may reveal unidentified differences that can be understood in terms of systematic effects.

If the uncertainty values are determined by statistical methods, they are called Type A uncertainties. Uncertainty values determined by all other means are called Type B uncertainties. Type A uncertainties are characterized in terms of the estimated variances—the uncertainty is the positive square root of the variance, or standard deviation; Type B uncertainties are to be considered approximations to these standard deviations. This terminology is not to be confused with random and systematic effects; for either effect, the uncertainty can be Type A or Type B. Type A and B uncertainties are combined as equivalent quantities to form the combined standard uncertainty, lower case u in the GUM, see Eqn (47) below. The result we are after is to express the quantity Y as

$$Y = y \pm U, \quad \text{or} \quad y - U \leq Y \leq y + U, \qquad (46)$$

where, in the GUM, upper case U represents the expanded uncertainty in Y such that the full interval includes a large fraction of the values that can be reasonably attributed to Y (see Chapter 6 in [87]). The interval expressed in Eqn (46) is associated with a specified coverage probability, or level of confidence, termed p. The expanded uncertainty U is simply a coverage factor k times the standard uncertainty, $U = ku$. Coverage factor $k = 1$ gives the standard uncertainty and $p = 68.27\%$. For formal calibration reports, setting instrument specifications, or as input to comparison (validation) exercises, the level of confidence should be higher—so typical expanded uncertainties are determined at $k = 2$ ($p = 95.45\%$) or $k = 3$ ($p = 99.73\%$). Uncertainty budgets should state the evaluation method for each component (Type A or B) as well as the coverage factor. NIST recommends $k = 2$ for expanded uncertainties.

The combined standard uncertainty in y, $u_c(y)$, can be expressed by a Taylor series for the expression in Eqn (45). The GUM, in Section 5.2, gives for the combined standard uncertainty,

$$u_c^2(y) = \sum_{i=1}^{N} \left(\frac{\partial f}{\partial x_i}\right)^2 u^2(x_i) + 2 \sum_{i=1}^{N-1} \sum_{j=i+1}^{N} \frac{\partial f}{\partial x_i} \frac{\partial f}{\partial x_j} u(x_i, x_j). \quad (47)$$

The quantities $u(x_i)$ are the uncertainties assigned to the individual variables x_i that are used to calculate y. The partial derivatives are referred to as sensitivity coefficients as they relate the rate of change in y with respect to the variable x_i. Values for the sensitivity coefficients can also be determined empirically, see Section 5.1 in the GUM. The second term in Eqn (47) is the second-order term in the Taylor expansion and needs to be considered if there are correlations between the random variables x_i and x_j. The term $u(x_i, x_j)$ is the estimated covariance associated with the variables x_i and x_j, and is often expressed in terms of the estimated correlation coefficient $r(x_i, x_j)$, where,

$$r(x_i, x_j) = \frac{u(x_i, x_j)}{u(x_i)u(x_j)}. \quad (48)$$

The GUM, in Chapter 5, suggests a method of evaluation of the correlation coefficient and which can be written in the following manner,

$$r(x_i, x_j) = \frac{1}{(N-1)} \frac{1}{s(x_i)s(x_j)} \sum_{k=1}^{N} (x_{i,k} - \bar{x}_i)(x_{j,k} - \bar{x}_j), \quad (49)$$

where the $s(x_i)$ and $s(x_j)$ are the experimental standard deviations, see Eqn (50) below.

While the issue of correlations can be important, we first concentrate our discussion on obtaining the appropriate measurement equation and using it to determine the evaluation of the first-order terms in Eqn (47) and thereby establishing an estimate of the combined standard uncertainty in the measurand y. The experimenter will need to look at all the variables and their relationships to decide if they are correlated and if the additional terms are necessary for their purposes, and we provide an example in Section 3.3.1 below.

If the determination of the measurand X_i is made by a series of measurements x_i whose variability is random, the best estimate of the measurand is the average value of the x_i's and the uncertainty can be estimated in terms of the standard deviation, $s(x_i)$, of the measurements. In our notation, for a series of N measurements of the variable x_i, the estimated average value, $\bar{x}_i$, and the experimental estimated variance $s^2(x_i)$ are given by,

$$\bar{x}_i = \frac{1}{N} \sum_{k=1}^{N} x_{i,k} \quad \text{and} \quad s^2(x_i) = \frac{1}{N-1} \sum_{k=1}^{N} (x_{i,k} - \bar{x}_i)^2 \quad (50)$$

The experimental variance of the mean is $s^2(\bar{x}_i) = s^2(x_i)/N$. The GUM recommends the square root of the experimental variance of the mean as the best estimate of the uncertainty for these statistically determined quantities. See Section 4.2 of the GUM for a discussion of this topic. In the expansion given in Eqn (47) for the total uncertainty there will be uncertainties for which statistical techniques are not relevant. An example might be the uncertainty of measurement given by a manufacturer of an instrument used in the measurement. Other examples of these Type B uncertainties will show up in examples throughout this volume. The NIST Statistical Engineering Division has developed a number of resources to aid researchers in the area of measurement uncertainty. The NIST "Uncertainty Machine" is a software application available online, see Ref. [89].

3.2 Measurement Equation Examples

To acquaint the reader with some of the details of constructing and using measurement equations we will derive measurement equations for a filter radiometer for spectral radiance, followed by an example of realizing a spectral irradiance scale using a blackbody source.

3.2.1 Filter Radiometer for Validation or Measurement of Spectral Radiance

A common radiometer system in use in remote sensing is shown in Figure 5. The system consists of a detector denoted "D," a wavelength filter denoted "F," a detector aperture of area A_D, and a source of optical radiation to be measured. In Figure 5 the source is an extended source that illuminates a source aperture of area A_S. For simplicity the apertures will be circular of radii r_S and r_D for the source and detector. The source aperture and detector aperture are a distance d apart and are collinear. When the source and detector aperture

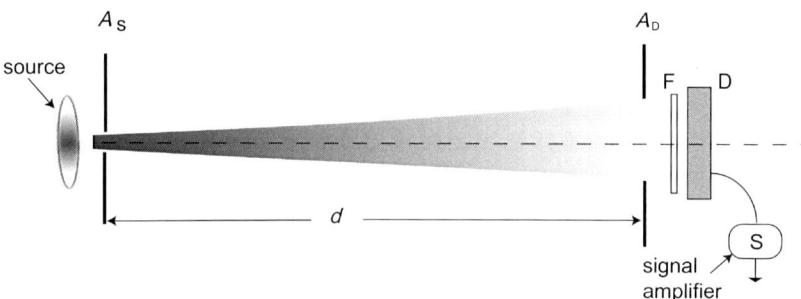

FIGURE 5 This simple filter detector system consists of source, a defining aperture A_S separated from a detector plane aperture A_D by a distance d. The radiant flux is selected in wavelength by filter F and illuminates detector D, which generates a response that is amplified to produce output signal S. *Reproduced from Figure 1.4 in Ref. [5].*

are mechanically integrated with the filter and detector into one system, the result is a filter radiometer for radiance measurements. The aperture pair forms a type of foreoptic known as a Gershun tube. The output signal from the detector is symbolized by S. S can be either an electrical current or voltage that is recorded digitally. The output signal S is related to the incoming flux at the detector by the spectral flux responsivity $R_\phi(\lambda)$ defined by the following equation,

$$S(\lambda) = R_\phi(\lambda)\Phi(\lambda). \tag{51}$$

In the configuration shown in Figure 5, the objective is to generate an expression for the radiance in the source aperture in terms of the detector output. This can be accomplished by using the results of Eqn (21). Using $\tau(\lambda)$ as the spectral transmittance of the filter, we can express the signal from the detector for this Gershun tube radiometer as,

$$S = \frac{A_S A_D (1+\delta)}{(r_S^2 + r_D^2 + d^2)} \int_\lambda L(\lambda)\tau(\lambda)R_\phi(\lambda)\mathrm{d}\lambda = \int_\lambda L(\lambda)R_L(\lambda)\mathrm{d}\lambda \tag{52}$$

where $R_L(\lambda) = \dfrac{A_S A_D (1+\delta)}{(r_S^2 + r_D^2 + d^2)}\tau(\lambda)R_\phi(\lambda).$

The radiance responsivity $R_L(\lambda)$ has units of output signal per unit radiance. Figure 6 gives an example of the $R_L(\lambda)$ for one channel of the Visible Transfer Radiometer (VXR) constructed by NIST to monitor and calibrate a variety of sources [81]. A lens and focusing mirrors are used in the VXR foreoptics, rather than a Gershun tube as illustrated in Figure 5, so $R_L(\lambda)$ has to be determined experimentally. The output of the VXR electronics is a voltage and hence the radiance responsivity, which was measured using the NIST SIRCUS facility, is given in units of volts per unit radiance.

If both the radiance responsivity $R_L(\lambda)$ and the source spectral radiance $L(\lambda)$ are known from ancillary calibration measurements, Eqn (52) can be solved to determine a predicted value for the signal for the source under study. This predicted value is compared to the measured value to validate or verify the source/detector radiometric system. Alternatively, if $R_L(\lambda)$ is known but only the relative spectral shape of $L(\lambda)$ is available, then Eqn (52) can be numerically solved given the measured signal S to obtain the absolute source spectral radiance. This concept is discussed below in the context of the HTBB and results in the determination of the temperature and the spectral radiance from Eqn (40). In other cases the source may be a standard lamp where the general spectral output distribution is known but the absolute level of the irradiance or radiance is unknown.

Parr and Johnson demonstrated a useful procedure for evaluation of Eqn (52) in cases where narrowband filters with very low transmittance outside of the bandpass, or out-of-band spectral region, are used [90].

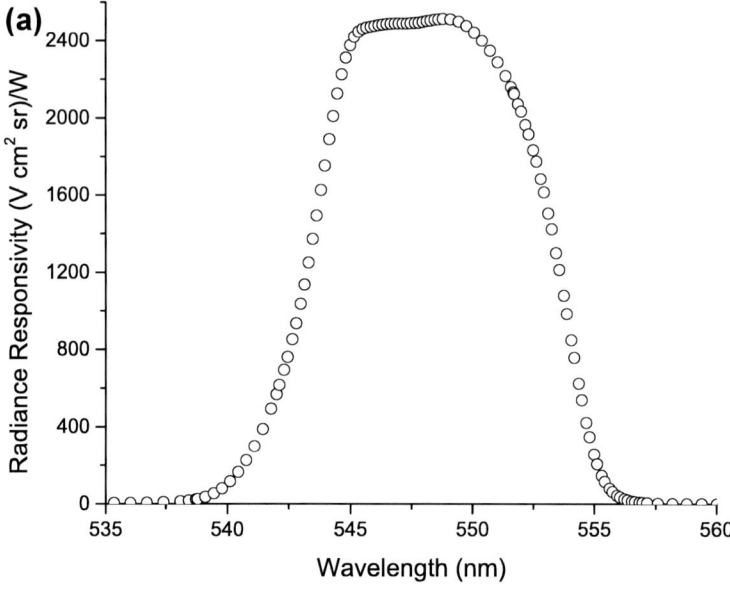
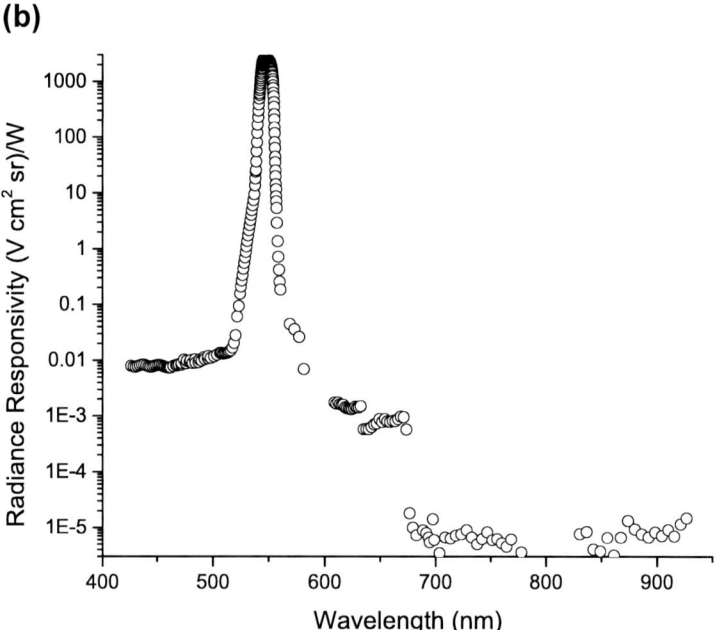

FIGURE 6 The radiance responsivity of the 550-nm channel of the VXR as determined on the SIRCUS facility: (a) linear ordinate; (b) logarithmic ordinate.

Practically this requires that the responsivity outside the main peak in Figure 6 should be down by at least five orders of magnitude, and this is confirmed by the SIRCUS measurements, see Figure 6(b). Another way to assess the performance is to assume a shape for the source distribution (a blackbody, for example) and evaluate the ratio of Eqn (52) for the in-band to the out-of-band spectral region. The out-of-band contribution should be much less than 1% of the integral or the approximation presented in [90] will yield spurious results.

The method uses the product of the radiance responsivity and spectral radiance as a distribution to determine a mean value of the wavelength and an effective width. It is a self-consistent approach utilizing Eqn (52) to determine spectral radiances with narrowband filter radiometers with negligible spectral out-of-band when the spectral radiance distribution is smooth. The mean wavelength λ_m defined as,

$$\lambda_m \equiv \frac{\int \lambda L(\lambda) R_L(\lambda) d\lambda}{\int L(\lambda) R_L(\lambda) d\lambda}. \tag{53}$$

Next we approximate the integral in Eqn (52) as the product of three terms—an effective spectral width, $\Delta\lambda$, the spectral radiance evaluated at the mean wavelength, and the radiance responsivity evaluated at the mean wavelength. This gives an expression for the effective spectral width:

$$\Delta\lambda L(\lambda_m) R_L(\lambda_m) \equiv \int_\lambda L(\lambda) R_L(\lambda) d\lambda \quad \text{or}$$

$$\Delta\lambda = \frac{\int L(\lambda) R_L \lambda) d\lambda}{L(\lambda_m) R_L(\lambda_m)}. \tag{54}$$

The mean wavelength and effective width depend on the ratios of the radiance responsivity and the spectral radiance and hence these parameters are independent of their absolute values. Thus, the relative radiance responsivity can be substituted for $R_L(\lambda)$ and a reasonable estimate of the spectral shape of the spectral radiance distribution can be substituted for $L(\lambda)$.

We can then write the signal shown in Eqn (52) as,

$$S = \Delta\lambda L(\lambda_m) R_L(\lambda_m), \tag{55}$$

or, solving for spectral radiance we can define a spectral radiance responsivity F for each channel of the filter radiometer,

$$L(\lambda_m) = \frac{S}{\Delta\lambda R_L(\lambda_m)} = \frac{S}{F} \quad \text{where} \quad F = \Delta\lambda R_L(\lambda_m). \tag{56}$$

Note that in this last step, there are two ways to derive F: (1) from measurements of a calibrated broadband spectral radiance standard and (2) from absolute values of the radiance responsivity. In other words, in the one case the radiance responsivity is relative but the spectral radiance is absolute, and vice versa for the other case.

In addition to determining the spectral radiance of an unknown source according to Eqn (56), if a radiometer with these characteristics is used to measure sources of similar spectral distributions, the spectral radiance can be transferred from a calibrated source to an unknown source. Finally, if the long-term stability of the filter radiometer is established, it can be used to monitor the output of radiance sources used in remote sensing calibrations [53].

In this section, several measurement equations have been presented, and all can be subjected to a detailed uncertainty analysis. For the purpose of validating spectral radiance values, Eqn (52), with $R_L(\lambda)$ known from calculation or measurement, results in a value for the predicted signal that would be compared to a net measured signal for the same source. For the purpose of determining spectral radiances, Eqn (56) is valid under the approximations discussed above. In either case, the measurement equations involve integrals. One approach for computing the combined uncertainty would be to approximate the integrals as sums and then apply Eqn (47); this approach was taken with the SeaWiFS Transfer Radiometer, the predecessor of the VXR, see [91]. A Monte Carlo approach could also be used.

3.2.2 Filter Radiometers and HTBB for Realization of Spectral Irradiance

A technique that is used to generate standard irradiance scales in NMIs utilizes calibrated filter radiometers in irradiance mode and blackbody sources. We will follow the treatment given by Yoon and Gibson in their discussion of the irradiance scale realization at NIST [46]. In the NIST laboratory, three filter radiometers and a spectroradiometer are mounted side by side on the detector table. The FEL lamps and the blackbody reside on the source table, which translates horizontally under computer control so that the blackbody or FEL lamps can be aligned to the filter radiometers or the spectroradiometer. Figure 3 shows the spectral irradiance values from a typical FEL standard lamp and the HTBB at 2950 K.

There are four major steps in the calibrations of lamps for customers. First, the temperature of the blackbody is determined using the detector-based filter radiometers and the radiometric flux transfer procedure. The blackbody, with its known temperature, is then used as a source of spectral irradiance, as computed from the Planck function (Eqn (40)), source area, and distance to the receiving aperture using Eqn (18) or Eqn (21). The receiving aperture is the entrance pupil to the spectroradiometer, and the result is the determination of the spectral irradiance responsivity of the spectroradiometer. The spectroradiometer is then used to calibrate check standard (CS) and working standard (WS) FEL lamps. The WS lamps are used to calibrate FEL lamps-under-test (LUTs) that are issued to NIST customers. Use of the WS lamps to hold the

Principles of Optical Radiometry and Measurement Uncertainty

spectral irradiance scale preserves the HTBB, which has a limited lifetime. The WS lamps also provide a mechanism to monitor the temporal stability of the irradiance responsivity of the spectroradiometer.

3.2.2.1 Irradiance Measurement

The idea is to utilize a calibrated filter radiometer and the flux transfer method outlined in Section 1.3 to determine the spectral radiance and hence the blackbody temperature. The experimental configuration is similar to Figure 5, with the source the HTBB fitted with an aperture and the filter radiometer consisting of an aperture, filter, detector, and electronics. Explicitly allowing for tilt of the apertures, and following Eqns (21) and (40) and Figure 5, the spectral flux in watts per nanometer is

$$\Phi(\lambda) = \frac{L(\lambda; T_{BB}) A_{FR} \cos(\alpha) A_{HTBB} \cos(\beta)}{D^2}(1+\delta), \quad (57)$$

where the HTBB standard source has an aperture of radius r_{BB}, an area A_{HTBB}, and is a distance d_{FR} from the aperture of the filter radiometer that has radius r_{FR} and area A_{FR}. From Eqn (21), $D^2 = d_{FR}^2 + r_{FR}^2 + r_{BB}^2$ and $\delta = r_{FR}^2 r_{BB}^2 / D^4$. The angles α and β, which are minimized to a few milliradians using alignment lasers, are due to rotational misalignment of the apertures.

Equation (52) showed that for a source of known spectral radiance and a radiometer calibrated for radiance responsivity, the expected signal can be determined from numerical integration. The filter radiometers are calibrated in the NIST Spectral Comparator Facility [92,93], so we have the spectral flux responsivity, $\tau(\lambda) R_\phi(\lambda)$ of the filtered detector in amperes per watt. Using Eqn (52) and allowing for the conversion to voltage using a transimpedance amplifier with gain G we have for the signal S from the calibrated filter radiometer

$$S = G \frac{A_{FR} \cos(\alpha) A_{HTBB} \cos(\beta)}{D^2}(1+\delta) \int \tau(\lambda) R_\phi(\lambda) L(\lambda; T_{BB}) d\lambda. \quad (58)$$

Equation (58) is solved iteratively, with numerical integration, to arrive at a value for the temperature T_{BB} that gives the best agreement with the measured signal.

With the temperature of the HTBB determined it can be used as a source of spectral radiance or, through Eqn (18), a source of spectral irradiance at an aperture plane. The spectral irradiance is known from

$$E_{HTBB}(\lambda, T_{BB}) = \frac{A_{HTBB}}{D'^2}(1+\delta') L(\lambda; T_{BB}), \quad (59)$$

where we take the misalignment angles to be zero. We use D' and δ' to indicate that the separation d_{ISR} between the blackbody aperture and the precision aperture of the integrating sphere receiver (ISR) of radius r_{ISR} are not the same as the filter radiometer–HTBB configuration in Eqns (57) and (58). The ISR is the foreoptic to a scanning double monochromator, or spectroradiometer, which is used as a transfer radiometer from the HTBB to the irradiance standard lamps.

With the blackbody irradiance at the ISR aperture plane known, the remaining three steps take the form of ratios. First, the irradiance responsivity of the spectroradiometer at each wavelength setting, $F(\lambda)$, is determined from the signals measured when observing the HTBB, $S_{HTBB}(\lambda)$, and its known irradiance at the aperture plane of the ISR, see Eqn (59):

$$F(\lambda) = \frac{S_{HTBB}(\lambda)}{G_{HTBB} E_{HTBB}(\lambda; T_{BB})}. \tag{60}$$

The gain factor G_{HTBB} converts $F(\lambda)$ to current per unit of spectral irradiance, or $(A\,cm^2\,nm)\,W^{-1}$, anticipating that the gain of the amplifier is different for the HTBB and the FEL lamps. Next, a set of WS FEL lamps are measured, giving $S_{WS}(\lambda; t_0)$ at gain G_{WS} where time t_0 refers to the date of the irradiance scale realization using the HTBB. The assigned spectral irradiance for the WS lamp is

$$E_{WS}(\lambda) = \frac{S_{WS}(\lambda; t_0)}{G_{WS} F(\lambda)} = \frac{S_{WS}(\lambda; t_0)}{S_{HTBB}(\lambda)} \frac{G_{HTBB}}{G_{WS}} E_{HTBB}(\lambda; T_{BB}). \tag{61}$$

Last, the routine calibrations of the LUTs, which occur at time t, are found by comparison using the spectroradiometer and the WS lamps. The spectral irradiance $E_{LUT}(\lambda)$ is

$$\begin{aligned} E_{LUT}(\lambda) &= \frac{S_{LUT}(\lambda)}{S_{WS}(\lambda; t)} \frac{G_{WS}}{G_{LUT}} E_{HTBB}(\lambda; T_{BB}) \\ &= \frac{S_{LUT}(\lambda)}{S_{WS}(\lambda; t)} \frac{S_{WS}(\lambda; t_0)}{S_{HTBB}(\lambda)} \frac{G_{HTBB}}{G_{LUT}} E_{HTBB}(\lambda; T_{BB}), \end{aligned} \tag{62}$$

where we have taken the gain used during the WS lamp measurement to be the same at t_0 and t. From Eqn (62), we see that the spectral irradiance of the LUT is given in terms of ratios of amplifier gain and signal measurements, and the blackbody temperature and fundamental constants via Eqn (59). The gain ratios as well as the linearity within a gain setting must be determined by instrument characterization. The bandpass of the spectroradiometer is assumed to be narrow compared to any spectral features present in the FEL lamp flux, so these measurement equations do not have the integration over wavelength as did the filter radiometer example of Section 3.2.1 and Eqn (58).

3.2.2.2 Uncertainty Evaluations

In this section we describe the procedures used to estimate the uncertainties for the spectral irradiance values assigned to the LUT lamps, see Table 3. The table is organized so that the first section deals with the uncertainty in the

TABLE 3 The Uncertainty Budget, Expressed in Percent, for the Determination of the Spectral Irradiance of the LUT Lamps with the Type (A or B) Indicated and where $u(T_{BB}) = 0.43$ K at 2950 K With a Temporal Stability of 0.5 K/h and a Wavelength Uncertainty of the Spectroradiometer of 0.05 nm

Item	Source of Uncertainty (%)\Wavelength (nm)	250	350	450	555	654.6	900	1600	2000	2300	2400
1	HTBB temperature (B)	0.28	0.20	0.16	0.13	0.11	0.08	0.05	0.04	0.03	0.03
2	HTBB spectral emittance (B)	0.05	0.05	0.05	0.05	0.05	0.05	0.05	0.05	0.05	0.05
3	HTBB spatial uniformity (B)	0.05	0.05	0.05	0.05	0.05	0.05	0.05	0.05	0.05	0.05
4	HTBB temporal stability (B)	0.03	0.02	0.02	0.02	0.01	0.01	0.01	0.00	0.00	0.00
5	Geometric factors, flux transfer (B)	0.05	0.05	0.05	0.05	0.05	0.05	0.05	0.05	0.05	0.05
6	Stability of irradiance responsivity (B)	0.30	0.30	0.15	0.15	0.15	0.15	0.15	0.15	0.15	0.50
7	Wavelength uncertainty (B)	0.29	0.13	0.06	0.04	0.02	0.00	0.01	0.01	0.01	0.01
8	WS Lamp to spectroradiometer transfer (B)	0.05	0.05	0.05	0.05	0.05	0.05	0.05	0.05	0.05	0.05
9	WS Lamp current stability (B)	0.03	0.02	0.02	0.02	0.02	0.01	0.01	0.01	0.01	0.01
	Combined standard uncertainty, $u(E_{WS})/E_{WS}$ $k = 1$	**0.52**	**0.40**	**0.25**	**0.23**	**0.21**	**0.20**	**0.19**	**0.18**	**0.18**	**0.51**
10	WS to LUT transfer (A)	0.25	0.15	0.10	0.10	0.10	0.10	0.10	0.15	0.15	0.20
11	Long-term stability of the WS lamps (B)	0.66	0.47	0.36	0.30	0.25	0.18	0.10	0.08	0.07	0.07
	Combined standard uncertainty, $u(E_{LUT})/E_{LUT}$ $k = 1$	**0.87**	**0.63**	**0.45**	**0.39**	**0.34**	**0.29**	**0.24**	**0.25**	**0.25**	**0.55**
	Combined expanded uncertainty, $U(E_{LUT})/E_{LUT}$ $k = 2$	**1.74**	**1.26**	**0.91**	**0.77**	**0.69**	**0.57**	**0.47**	**0.50**	**0.50**	**1.11**

HTBB, high-temperature blackbody; WS, working standard; LUT, lamps-under-test.

WS lamp, including the HTBB and its temperature determination. The results of this section are the standard uncertainties, expressed in percent, of the spectral irradiance of the WS lamps, see Eqn (61). The remainder of the table deals with the transfer to the LUT and the expression of the combined standard uncertainty as combined expanded uncertainty at $k = 2$.

The measurement equation for a customer's LUT lamp is Eqn (62), which is in the form of products and quotients. From the first-order term in Eqn (47), the combined standard uncertainty for a measurement equation in this form is most easily expressed in terms of relative uncertainty, $u(x_i)/x_i$ (for additive forms of the measurement equation, expressions involving $u(x_i)$ will result). Applying Eqn (47) to Eqn (62) gives

$$\frac{u^2(E_{\text{LUT}})}{E_{\text{LUT}}^2} = \frac{u^2(S_{\text{LUT}})}{S_{\text{LUT}}^2} + \frac{u^2(S_{\text{WS}}(t))}{S_{\text{WS}}^2(t)} + \frac{u^2(S_{\text{WS}}(t_0))}{S_{\text{WS}}^2(t_0)} + \frac{u^2(S_{\text{HTBB}})}{S_{\text{HTBB}}^2} + \frac{u^2(G_{\text{HTBB}})}{G_{\text{HTBB}}^2} + \frac{u^2(G_{\text{LUT}})}{G_{\text{LUT}}^2} + \frac{u^2(E_{\text{HTBB}})}{E_{\text{HTBB}}^2}. \tag{63}$$

In Eqn (63), we eliminate the spectral dependence in the variables for reasons of clarity.

The relative standard uncertainty components in the spectral irradiance $E_{\text{HTBB}}(\lambda;T_{\text{BB}})$ resulting from the HTBB as the primary standard are listed as Items 1–5 in Table 3 and discussed in detail by Yoon and colleagues [39,46]. The measurement equations for T_{BB} are Eqns (40) and (58), and for $E_{\text{HTBB}}(\lambda;T_{\text{BB}})$, the measurement equation is Eqn (59). From Eqns (40) and (58) we see that $u(T_{\text{BB}})$ depends on S, $R_\phi(\lambda)$, r_{BB}, r_{FR}, d_{FR}, G, α, β, and the HTBB emittance ε. The values of $u(T_{\text{BB}})$ that are due to uncertainty in the radiation constants c_{1L} and c_2 and $n(\lambda)$ are negligible. The uncertainties $u(R_\phi(\lambda))$ dominate $u(T_{\text{BB}})$ [39,46]. The filters in the filter radiometers are broad and values for $R_\phi(\lambda)$ were determined using the Visible to Near-Infrared Spectral Comparator Facility [92]. The relative uncertainty in the HTBB spectral radiance is directly proportional to the relative uncertainty in the spectral flux responsivity. The relative sensitivity coefficient for spectral radiance with blackbody temperature is $c_2/(\lambda T_{\text{BB}})$ from the Wien approximation to Eqn (40), and this contributes to the spectral dependence in Item 1 in Table 3. The components related to the filter radiometer signals S, amplifier gain G, and the aperture areas and distances indicated in Eqns (58) and (59) each contribute no more than 0.07% in the relative standard uncertainty of $E_{\text{HTBB}}(\lambda;T_{\text{BB}})$ for Item 1 of Table 3: see Table 1 in [39]. The angular alignment uncertainty components associated with Eqns (58) and (59) are included in Item 5 in Table 3.

Values for $u(\varepsilon)$, Item 2 of Table 3, were estimated by comparing values of $L(\lambda;T_{\text{BB}})$ for the HTBB determined using the filter radiometers to values determined by comparison to a variable temperature blackbody (VTBB) over a range of wavelengths. The emittance of the VTBB is estimated to be >0.999

from theoretical modeling and experimental characterization, and its temperature is determined by reference to a blackbody operated at the freezing temperature of gold [38]. The comparison agreed to within 0.5% in spectral radiance: see Figure 3 in [46]. This comparison also allowed an assessment of the spatial uniformity of the HTBB, included as Item 3 in Table 3. The temporal stability of the HTBB, Item 4 in Table 3, was assessed using three filter radiometers to monitor the radiant flux over a 30 min interval.

There is uncertainty in the signals of the output electronics of the spectroradiometer. These components are included in Items 8 and 10 of Table 3. Item 8 includes $u(S_{WS}(t_0))$ and $u(S_{HTBB})$, and Item 10 includes $u(S_{WS}(t))$ and $u(S_{LUT})$. The signals are the result of averaging a number of individual random measurements and Eqn (50) applies. This results in a Type A uncertainty which in part may be due to short-term instability in the system. The Type B contributions to Item 8 are related to the alignment repeatability.

There is uncertainty in the amplifier gain in the spectroradiometer electronics. This uncertainty component contains estimates of the linearity of the system response. In practice, the same gain is used with the WS and the LUT lamps, so the uncertainty only impacts the second and third step in the procedure, see Eqn (61), and it is incorporated into Item 8 of Table 3. The gains were characterized using a calibrated current source [94].

Item 6 in Table 3, the stability of the irradiance responsivity of the spectroradiometer, is important because its determination, see Eqn (60), occurs infrequently. The short-term stability of $F(\lambda)$ was evaluated by using a single FEL lamp and assessing the measurement repeatability with three independent calibrations. The long-term stability of $F(\lambda)$ is held on the CS and WS lamps. Item 7 in Table 3 quantifies the effect of error in the wavelength calibration of the spectroradiometer on the assigned spectral irradiance values. Through the Wien approximation of Eqn (40), the wavelength uncertainty of the spectroradiometer has a relative sensitivity coefficient of $c_2/(\lambda T_{BB})-5$, resulting in the spectral dependences indicated.

Item 9 accounts for the uncertainty and stability of the current driving the FEL lamp. Biases in the FEL lamp current affect the spectral irradiance as a function of wavelength. This is a case where the sensitivity coefficient could be evaluated empirically, by varying the lamp current and noting the relative change is spectral irradiance. However, there are published expressions; Early gives

$$\frac{u(E)}{E} = \left(\frac{654.6}{\lambda}\right) 0.0006 \, u(I), \quad (64)$$

with wavelength in nanometers and the uncertainty in current in milliamps [95]. The uncertainty in the current is evaluated from the measured voltages of the calibrated shunt resistor that is in series with the FEL lamp.

Item 11 accounts for potential drift in the WS lamps between the time of the scale realization and the time they are used to assign spectral irradiance values to the LUT lamps. All FEL lamps calibrated at NIST are evaluated for their temporal stability; drift of more than 0.5% at 650 nm over a 24 h interval disqualifies the lamp for calibration. From these types of data, which indicate the temporal drift is inversely proportional to wavelength, a Type B uncertainty for the long-term stability of the WS lamps was estimated. This component of the spectral irradiance uncertainty of a LUT lamp is important at wavelengths below 1600 nm.

Other potential sources of uncertainty that have not been discussed in the context of the FEL lamps are spectral stray light, spatial stray light, and polarization sensitivity. The presence of spectral stray light can cause very large biases when the radiometric reference standard and the unknown source have very different spectral distributions, as we saw previously for the case of ocean color radiometry. Figure 3 illustrates the similarity of the spectral distributions of the HTBB and the WS or LUT lamps, which combined with the use of a double monochromator with stray light rejection of 10^{-5}, indicate that spectral stray light is not a significant component in the $u(E_{LUT})$ budget. Spatial stray light refers to sensitivity to flux that is outside the source area imaged by radiance radiometers. It is mitigated by maintaining the same source area during radiance measurements for the standard and the unknown source. Irradiance radiometers may have a poor cosine response; in principle bias can be mitigated by maintaining the same portion of the irradiance collector's field of view for the calibration and the unknown source, but, in practice, this is not often the case and full characterization of the cosine response is necessary. As for polarization, typically the response of grating spectroradiometers depends on the polarization of the incoming flux, but in the example described here, the use of the ISR removes this sensitivity. The HTBB is unpolarized, so any polarization sensitivity in the filter radiometer does not impact the measurement result.

There are additional uncertainty components in Table 3 that are not explicitly stated in, nor immediately apparent from, the measurement equations. It is not unusual to identify such sources of uncertainty upon reflection of the entire measurement system; one might even go so far as to designate these components as "environmental," which may be a helpful concept for field measurements. This is an important issue in uncertainty evaluation—the analyst must have a good understanding of the experimental procedure and its inherent limitations in order to include possible sources of systematic error.

For example, great care was taken to optimize the entire system for signal-to-noise ratio and temporal stability. Multiple, cooled, and temperature-stabilized detectors, each optimized for a particular wavelength region, are utilized. The monochromator is equipped with multiple gratings and order sorting filters, an absolute encoder with 2^{14} pulses per revolution of the sine bar, and the instrument is temperature-stabilized near ambient. The monochromator, detector chamber, and foreoptics are purged with either nitrogen gas boil off or dry air in order to maintain the long-term spectral irradiance responsivity. The current to the FEL lamps is actively controlled with 16 bit resolution, using calibrated shunt resistors with temperature coefficients of $3 \times 10^{-6}/°C$. The four independent lamp stations, along with the three independent filter radiometers for the T_{BB} determination, provide a reproducibility condition of measurement.

For dissemination of the NIST spectral irradiance scale, the calibrated LUTs are issued to customers along with a calibration report. Included in the report are a description of the activities performed and the spectral irradiance values and their uncertainties, which are documented in a table similar to Table 3 discussed above. In Table 3, the combined expanded uncertainties expressed in percent at $k = 2$ for the LUT range from 1.74% at 250 nm to about 0.5% from 1600 nm to 2300 nm, and these values are typical for issued lamps, see [39,45].

It is emphasized that the user must develop their own uncertainty budget for their particular application. The uncertainty for their reference standard, $u(E_{LUT})$, would be one component in their uncertainty budget, and it would be a Type B. Critical factors in the user's setup include the lamp current, the lamp distance, the wavelength calibration of the instrumentation, and the alignment of the optical systems. A thorough discussion of uncertainties in spectral irradiance and use of FEL lamps is found in Ref. [95]. Many users transfer the spectral irradiance values from the reference lamp to their own lamps using a spectroradiometer; these lamps then serve as secondary standards. This preserves the lifetime of the reference lamp and makes best use of resources at the expense of increased uncertainty due to the transfer process.

3.3 Uncertainty in Ocean Color Measurements

Uncertainty estimates for in situ and satellite ocean color measurements are an ongoing area of research. A full review is outside the scope of this chapter, but we do cite a few representative cases and then address the issue of correlation through an example problem. Uncertainties associated with in-water radiometry using moored systems have been described for the Bouée pour l'acquisition de Séries Optiques à Long Terme (BOUSSOLE) facility [96] and the Marine Optical Buoy [97]. Uncertainties for in-water profiling instruments are presented in [98] as part of an in situ comparison exercise. Uncertainties associated with in-air radiometry using modified sun photometers, the AERONET-OC network, are described in [99,100]. The measurement uncertainties for reflectance in the SR bands of MODIS, a satellite sensor, are presented in [101]. These results apply to the satellite measurements and do not include the uncertainty associated with the atmospheric correction algorithms. There are numerous examples of sensitivity analyses in the literature, and these studies can be utilized to perform uncertainty estimates. In many cases in remote sensing, Monte Carlo techniques are appropriate. For example, recently Lenhard utilized Monte Carlo methods to estimate measurement uncertainty for an imaging spectroradiometer [102].

3.3.1 Correlations

In ocean color radiometry, measurements of spectral radiance and irradiance are often made over short time intervals using paired instruments, for example, independent filter irradiance and radiance radiometers with the filters at the same wavelengths or spectroradiometers with the ability to switch between irradiance and radiance foreoptics. Often the desired product is in the form of radiance to irradiance ratios—examples are the remote sensing reflectance, $R_{rs}(\theta_v, \phi_v; \lambda)$, Eqn (38), as well as the normalized water-leaving spectral radiance $L_{WN}(\theta_v, \phi_v; \lambda)$,

$$L_{WN}(\theta_v, \phi_v; \lambda) = R_{rs}(\theta_v, \phi_v; \lambda) F_0(\lambda), \tag{65}$$

and the $Q(0^-; \theta, \phi; \lambda)$ quantity

$$Q(0^-; \theta, \phi; \lambda) = \frac{E_{u;}(0^-; \lambda)}{L_u(0^-; \theta, \phi; \lambda)}, \tag{66}$$

(see, for example, Chapter 4 in Ref. [103] for these definitions). In Eqn (65), $F_0(\lambda)$ is the exoatmospheric solar irradiance, corrected to a standard Earth–Sun distance. In Eqn (66), the upwelling fluxes are just below the air–sea interface—the 0^- notation—and the angles and wavelengths are in the seawater medium.

If the irradiance sensor for these measurements is calibrated with a reference standard FEL lamp, and the radiance sensor is calibrated with the same FEL lamp using a diffuse standard of spectral reflectance, it might be tempting to state that the uncertainties in the products $R_{rs}(\lambda)$, $L_{WN}(\lambda)$, or $Q(\lambda)$ determined using these calibrated radiometers are independent of the uncertainty in the spectral irradiance values of the FEL lamp, since the spectral irradiance of the standard lamp factors out of the final products. This is not the case. The GUM recognizes that the two calibration factors are correlated, in this case through the physical measurement standard (the FEL lamp), and possibly through the measurement system, e.g., the lamp current or distance to the receiver aperture, which is the diffuse reflectance standard for the spectral radiance realization. The GUM recommends an approach in Section F.1.2, which we illustrate here using simplified notation.

The calibration factors for spectral irradiance $C_{FEL}(\lambda)$ and radiance $C_{L/P}(\lambda)$ are

$$\begin{aligned} C_{FEL}(\lambda) &= \frac{E_{FEL}(\lambda)}{S_{FEL}(\lambda)} \frac{(50+\chi)^2}{(d_E+\chi)^2} \quad \text{and} \\ C_{L/P}(\lambda) &= \frac{E_{FEL}(\lambda)}{S_{L/P}(\lambda)} \frac{R(0°; 45°; \lambda)}{\pi} \frac{(50+\chi)^2}{(d_{L/P}+\chi)^2}, \end{aligned} \tag{67}$$

where we allow for different distances d_E and $d_{L/P}$ in the two measurements and where all distances are in centimeters. The measured net signals are denoted $S_{FEL}(\lambda)$ and $S_{L/P}(\lambda)$ for the irradiance and radiance radiometers with the respective reference sources. The only quantity common to the equations defining the two calibration factors is $E_{FEL}(\lambda)$. Therefore, from Eqn F.2 in the GUM, the estimated covariance is

$$u(C_{FEL}, C_{L/P}) = \frac{\partial C_{FEL}}{\partial E_{FEL}} \frac{\partial C_{L/P}}{\partial E_{FEL}} u^2(E_{FEL}) = C_{FEL} C_{L/P} \frac{u^2(E_{FEL})}{E_{FEL}^2}. \tag{68}$$

In the field, the two instruments measure net signals due to the downwelling surface spectral irradiance, $S_s(\lambda)$ and the water-leaving radiance, $S_w(\lambda)$, which are interpreted in physical units using the calibration factors

$$E_s(\lambda) = C_{FEL}(\lambda) S_s(\lambda) \tag{69}$$

and

$$L_w(\lambda) = C_{L/P}(\lambda) S_w(\lambda) \xi(\lambda), \tag{70}$$

where $\xi(\lambda)$ is the product of various factors according to the in-air or in-water measurement method of determining $L_w(\lambda)$. Then our measurement equation is

$$R_{RS}(\lambda) = \frac{L_w(\lambda)}{E_s(\lambda)} = \frac{C_{L/P}(\lambda) S_w(\lambda) \xi(\lambda)}{C_{FEL}(\lambda) S_s(\lambda)}. \tag{71}$$

Now we apply Eqn (47) to determine the uncertainty in the remote sensing reflectance. Using Eqns (71) and (68), the result is

$$\frac{u^2(R_{RS})}{R_{RS}^2} = \frac{u^2(C_{L/P})}{C_{L/P}^2} + \frac{u^2(S_w)}{S_w^2} + \frac{u^2(\xi)}{\xi^2} + \frac{u^2(C_{FEL})}{C_{FEL}^2} + \frac{u^2(S_s)}{S_s^2} - 2\frac{u^2(E_{FEL})}{E_{FEL}^2}, \tag{72}$$

where the last term is from the correlation between C_{FEL} and $C_{L/P}$. If independent measurement standards had been used, this negative term would have been replaced by two positive terms representing the relative standard uncertainty in the spectral irradiance and spectral radiance standard values. As we expected, we find that use of the same FEL lamp reduces the overall uncertainty. In evaluating Eqn (72), the terms representing the calibration factors C_{FEL} and $C_{L/P}$ are evaluated using the first term in Eqn (47), see Eqn F.1 in the GUM. Additional correlations, e.g., through the lamp current, would be handled in a similar fashion after appropriate modification of the calibration factor measurement equations to express this dependency.

Correlation coefficients can be estimated directly from data sets, see Eqn (49). In a study of AERONET-OC results at various sites, Gergely and Zibordi, following the GUM, included correlations in the measurement equation in their uncertainty analyses [99]. The results showed that as applied to Acqua Alta Oceanographic Tower, the estimated uncertainties in $L_{WN}(\lambda)$ decreased by 1−3 percentage points, e.g., from 8.3% to 6.2% at 412 nm.

3.3.2 Comparisons and Reproducibility

Instrument characterization, whether for a source or a radiometer, is the process of establishing and defining the measurement equations (or numerical models) that will allow for proper interpretation, e.g., removal of bias from systematic effects, of the results for all possible measurement scenarios that will be encountered. Radiometric calibration of the instruments is the process of establishing the relationship between the instrument output and physical quantities. Validation and verification of results are procedures implemented under the reproducibility condition of measurement. Validation of results is especially important as new procedures or protocols are introduced to a particular measurement objective. Verification is similar to validation, except it is performed to assess the long-term performance of the measurement. For example, if the FEL lamp type were to be discontinued, a new lamp type would have to be substituted. One could imagine validation of its performance would involve development followed by robust and intensive comparisons to the FEL-based scale, and that verification some years later might involve random substitution of a few preserved and previously calibrated WS FEL lamps into the new calibration procedure.

The question arises as to how to interpret the results of such validation or verification comparisons? Two types of reproducibility comparisons are possible: (1) measurements at different times, locations, etc., of a stable radiometric artifact; and (2) simultaneous measurements of the same object by different instruments. Examples of the nonsimultaneous case include the circulation of diffuse reflectance standards in the EOS BRDF round-robin [71] and the circulation of radiometers at SIRREX-6 [104]. There are numerous examples of the simultaneous case, involving all possible combinations of in situ and satellite ocean color results, see, for example, the work presented in references [96,98,105,106]. In the context of the simultaneous comparisons, pairs of results are termed "matchups" in the ocean color community. Referring back to Eqn (46), every comparison participant would report their result and their combined expanded uncertainty. Given the object under study is the same, and if there are no unidentified sources of bias, we would expect the intervals defined by Eqn (46) to overlap if the selected level of confidence was reasonable, e.g., $k = 2$ or $p = 95.45\%$. We say the results agree within their uncertainties. However, we cannot also conclude that: (1) the uncertainty estimates for any or all participants are accurate or (2) a mean or weighted mean of the results is the true value; or, most egregiously in terms of the GUM, that (3) the observed difference is a *quantitative measure* of each participant's uncertainty or of the overall uncertainty in the procedure. Examples of all three types of misinterpretations of comparison results can be found in the ocean color literature. It is important to realize that assigning meaning to the average result of comparisons is valid only when the two (or more) measurements are from the same sample sets and hence amenable to statistical treatments, and this is not known a priori. If the number of independent measurements is large, then comparisons of the standard deviations might be a way to assess the similarity of the underlying distributions. Alternatively, the manner in which the independent samples are distributed with respect to the comparison mean might be a useful question to address.

There is much to learn from performing intercomparisons under reproducibility conditions of measurement (or the closest thing achievable in field exercises). Statistical indicators based on the ratio of the results, various ways to express the average differences, linear fits to the matchup data sets, or investigations of correlation are utilized by the ocean color community. Outliers, indicating possible unidentified systematic bias, then become apparent, and point to the need for further characterization efforts. Investigation of the results of matchups for different ocean color products may indicate where certain procedures are underperforming. To mitigate the impact of bias due to differences in the underlying radiometric scales, to the extent possible all intercomparison exercises should include a "laboratory" component, where all instruments observe a common source that has established stability and is not subject to the varying conditions found in the natural environment; examples are found in the work of Zibordi [106] and Voss [98]. The results of matchup intercomparisons in ocean color are influenced by many environmental factors, and it is necessary to quality control the matchup data set. Typical quality control factors involving in situ data include solar zenith angle, aerosol optical thickness, wind speed, cloud coverage, thresholds on the variability of the results, anomalous instrument housekeeping data, evidence of instrument instability, internal reproducibility, comparisons to model results, and so on. These procedures, which result in selected data from much larger data sets, need to be exercised with care so as not to introduce bias into the final result.

4. SUMMARY

In this chapter, we have outlined the basic concepts in optical radiometry, with an emphasis on this field as encountered at a radiometric standards laboratory. The quantities of radiance, irradiance, reflectance, and dimensional metrology for optical radiometry were described. Examples of radiometers and sources relevant to radiometric standards encountered by the ocean optics user community were introduced. Some of the methods used at NIST to realize and disseminate the basic radiometric quantities were discussed. The reader is referred to the text by Mobley [11] for complete description of the in-water light fields and associated radiative transfer relevant to ocean color. The text by Robinson covers the topic of satellite oceanography for the SR and TIR spectral regions, as well as passive and active techniques at longer wavelengths [2].

The fundamental notions in developing a measurement equation were presented, along with examples used by the remote sensing community.

The reader is strongly encouraged to pursue this introduction by reference to the literature on this topic. The GUM, is a good starting point and it is readily accessible on the BIPM Web site [107]. The BIPM Web site also has links to the various NMIs of the member states, so it can identify the availability of standard lamps and other artifacts needed by instrumentation scientists. There are a number of good books on measurement equations and uncertainty analyses at various levels of sophistication. A classic is the 1964 volume by John Mandel, which is noted for introducing topics in a manner readily accessible to most scientists and engineers [108]. Two more modern books on the topic of statistics and measurement uncertainty are a traditional text by Stanford and Vardeman [109] and a treatise that introduces Bayesian techniques and ideas by Gregory [110].

ACKNOWLEDGMENTS

Albert C. Parr was supported by Space Dynamics Laboratory, Logan, Utah, under a Joint NIST/Utah State University Program in Optical Sensor Calibration.

REFERENCES

[1] A.P. Cracknell, L.W.B. Hayes, Introduction to Remote Sensing, Taylor & Francis, New York, 1991.
[2] I.S. Robinson, Measuring the Ocean from Space: The Principles and Methods of Satellite Oceanography, Springer-Verlag, Berlin, Germany, 2004.
[3] International Lighting Vocabulary, Commision Internationale de L'Eclairage (CIE), Vienna, Austria, 1987.
[4] International Vocabulary of Metrology − Basic and General Concepts and Associated Terms (VIM), third ed., International Organization for Standardization, Geneva, Switzerland, 2012.
[5] R.U. Datla, A.C. Parr, Introduction to optical radiometry, in: A.C. Parr, R.U. Datla, J.L. Gardner (Eds.), Optical Radiometry, Elsevier, Amsterdam, The Netherlands, 2005, pp. 1−34.
[6] S.F. Johnston, A History of Light and Colour Measurement, Institute of Physics, Philadelphia, Pennsylvania, 2001.
[7] J.M. Palmer, Getting intense on intensity, Metrologia 30 (1993) 371−372.
[8] S. Perkowitz, Empire of Light, first ed., Joseph Henry Press, Washington, D.C., 1996.
[9] F.E. Nicodemus, J.C. Richmond, J.J. Hsia, I.W. Ginsberg, T. Limperis, Geometrical Considerations and Nomenclature for Reflectance, Washington, D.C. 20402, 1977.
[10] F.E. Nicodemus, Self-study Manual on Optical Radiation Measurements, U.S. Government Printing Office, Washington, D.C. 20402−9325, 1976. Available at: http://www.nist.gov/pml/div685/pub/studymanual.cfm.
[11] C.D. Mobley, Light and Water: Radiative Transfer in Natural Waters, Academic Press, San Diego, California, 1994.
[12] F. Grum, R. Becherer, Radiometry, first ed., Academic Press, San Diego, California, 1979.
[13] C.L. Wyatt, Radiometric Calibration: Theory and Methods, Academic Press, Orlando, Florida, 1978.

[14] R.W. Boyd, Radiometry and the Detection of Optical Radiation, John Wiley & Sons, New York, 1983.
[15] H.J. Kostkowski, Reliable Spectroradiometry, first ed., Spectroradiometry Consulting Co, LaPlata, Maryland, 1997.
[16] F.P. Incropera, D.P. DeWitt, Fundamentals of Heat and Mass Transfer, fourth ed., John Wiley & Sons, New York, 1996.
[17] R. Siegel, J.R. Howell, Thermal Radiation Heat Transfer, Hemisphere Publishing Corporation, Washington, D.C, 1992.
[18] A.C. Parr, A National Measurement System for Radiometry, Photometry, and Pyrometry Based upon Absolute Detectors, U.S. Government Printing Office, Washington, D.C. 20402−9325, 1996.
[19] W.R. McCluney, Introduction to Radiometry and Photometry, Artech House, Boston, Massachusetts, 1994.
[20] C.L. Wyatt, Radiometric System Design, Macmillian Publishing Co., New York, 1987.
[21] C.D. Mobley, C. Mazel, Informal Notes on Reflectance, Sequoia Scientific, Inc., Redmond, Washington, 2000 (unpublished report).
[22] P.Y. Barnes, E.A. Early, A.C. Parr, Spectral Reflectance, U.S. Government Printing Office, Washington, D.C. 1998.
[23] A.C.R. Gleason, K.J. Voss, H.R. Gordon, M. Twardowski, J. Sullivan, C. Trees, A. Weidemann, J.-F. Berthon, D.K. Clark, Z.-P. Lee, Detailed validation of the bidirectional effect in various case I and case II waters, Opt. Express 20 (2012) 7630−7645.
[24] S.W. Brown, G.P. Eppeldauer, K.R. Lykke, Facility for spectral irradiance and radiance responsivity calibrations using uniform sources, Appl. Opt. 45 (2006) 8218−8237.
[25] G. Kopp, J.L. Lean, A new, lower value of total solar irradiance: evidence and climate significance, Geophys. Res. Lett. 38 (2011) L01706.
[26] J.B. Fowler, R.S. Durvasula, A.C. Parr, High-accuracy aperture-area measurement facilities at the National Institute of Standards and Technology, Metrologia 35 (1998) 497−500.
[27] J.B. Fowler, M. Litorja, Geometric area measurements of circular apertures for radiometry at NIST, Metrologia 40 (2003) S9−S12.
[28] B.C. Johnson, M. Litorja, J.B. Fowler, E.L. Shirley, R.A. Barnes, J.J. Butler, Results of aperture area comparisons for exo-atmospheric total solar irradiance measurements, Appl. Opt. 52 (2013) 7963−7980.
[29] R.P. Breault, Control of stray light, in: M. Bass (Ed.), Handbook of Optics Volume I Fundamentals, Techniques, and Design, McGraw Hill, Inc., New York, 1995 pp. 38.31−38.35.
[30] D.P. DeWitt, G.D. Nutter, Theory and Practice of Radiation Thermometry, John Wiley & Sons, New York, 1988.
[31] M. Planck, The Theory of Heat Radiation, Tomash/American Institute of Physics, New York, 1989.
[32] Fundamental Constants Data Center. 2014. http://www.nist.gov/pml/div684/fcdc/.
[33] P.E. Ciddor, Refractive index of air: new equations for the visible and near infrared, Appl. Opt. 35 (1996) 1566−1573.
[34] J.P. Rice, J.J. Butler, B.C. Johnson, P.J. Minnett, K.A. Maillet, T. Nightingale, S.J. Hook, A. Abtahi, C.J. Donlon, I.J. Barton, The Miami2001 infrared radiometer calibration and intercomparison. Part I: laboratory characterization of blackbody targets, J. Atmos. Oceanic Technol. 21 (2004) 258−267.
[35] J.B. Fowler, A third generation water bath based blackbody source, J. Res. NIST 100 (1995) 591−599.

[36] C.J. Donlon, T. Nightingale, L. Fiedler, G. Fisher, D. Baldwin, S. Robinson, The calibration and intercalibration of sea-going infrared radiometer systems using a low cost blackbody cavity, J. Atmos. Oceanic Technol. 16 (1999) 1183–1197.
[37] V.I. Sapritsky, Black-body radiometry, Metrologia 32 (1995) 411–417.
[38] J.H. Walker, R.D. Saunders, A.T. Hattenburg, Spectral Radiance Calibrations, U.S. Government Printing Office, Washington, D.C. 1987.
[39] H.W. Yoon, C.E. Gibson, P.Y. Barnes, Realization of the National Institute of Standards and Technology detector-based spectral irradiance scale, Appl. Opt. 41 (2002) 5879–5890.
[40] A.C. Parr, R.U. Datla, J.L. Gardner, Optical Radiometry, Elsevier, Amsterdam, The Netherlands, 2005.
[41] Z.M. Zhang, B.K. Tsai, G. Machin, Radiation Temperature Measurements I. Fundamentals, Academic Press (Elsevier), Amsterdam, The Netherlands, 2010.
[42] J.C. De Vos, A new determination of the emissivity of tungsten ribbon, Physica 20 (1954) 690–714.
[43] C.E. Gibson, B.K. Tsai, A.C. Parr, Radiance Temperature Calibrations, U.S. Department of Commerce, Washington, D.C. 20402–9325, 1998.
[44] H.W. Yoon, B.C. Johnson, D. Kelch, S.F. Biggar, P.R. Spyak, A 400 nm to 2500 nm absolute spectral radiance comparison using filter radiometers, Metrologia 35 (1998) 563–568.
[45] H.W. Yoon, C.E. Gibson, P.Y. Barnes, The realization of the NIST detector-based spectral irradiance scale, Metrologia 40 (2003) S172–S176.
[46] H.W. Yoon, C.E. Gibson, Spectral Irradiance Calibrations, U.S. Department of Commerce, Washington, D.C. 20402–9325, 2011.
[47] H.W. Yoon, G.D. Graham, R.D. Saunders, Y. Zong, E.L. Shirley, The distance dependences and spatial uniformities of spectral irradiance standard lamps, Proc. SPIE 8510 (2012) 85100D.
[48] S.J. Flora, 2014. (Personal communication) http://moby.mlml.calstate.edu/.
[49] S.W. Brown, B.C. Johnson, Development of a portable integrating sphere source for the earth observing system's calibration validation program, Int. J. Remote Sens. 24 (2003) 215–224.
[50] J.J. Butler, S.W. Brown, R.D. Saunders, B.C. Johnson, S.F. Biggar, E.F. Zalewski, B.L. Markham, P.N. Gracey, J.B. Young, Radiometric measurement comparison on the integrating sphere source used to calibrate the Moderate Resolution Imaging Spectroradiometer (MODIS) and the Landsat 7 Enhanced thematic Mapper Plus (ETM$^+$), J. Res. NIST 108 (2003) 199–228.
[51] D.G. Goebel, Generalized integrating-sphere theory, Appl. Opt. 6 (1967) 125–128.
[52] J.J. Butler, R.A. Barnes, The use of transfer radiometers in validating the visible to shortwave infrared calibrations of radiance sources used by instruments in NASA's earth observing system, Metrologia 40 (2003) S70–S77.
[53] D.K. Clark, M.E. Feinholz, M.A. Yarbrough, B.C. Johnson, S.W. Brown, Y.S. Kim, R.A. Barnes, Overview of the the radiometric calibration of MOBY, Proc. SPIE 4483 (2002) 64–76.
[54] B.C. Johnson, F. Sakuma, J.J. Butler, S.F. Biggar, J.W. Cooper, J. Ishida, K. Suzuki, Radiometric measurement comparison using the Ocean Color Temperature Scanner (OCTS) visible and near infrared integrating sphere, J. Res. NIST 102 (1997) 627–646.
[55] Certain commercial equipment, instruments, or materials are identified in this chapter to foster understanding. Such identification does not imply recommendation or endorsement

by the National Institute of Standards and Technology (NIST), nor does it imply that the materials or equipment identified are necessarily the best available for the purpose.

[56] P.Y. Barnes, J.J. Hsia, 45/0 Reflectance Factors of Pressed Polytetrafluoroethylene (PTFE) Powder, U.S. Government Printing Office, Washington, D.C. 20402−9325, 1995.

[57] M.E. Nadal, P.Y. Barnes, Near infrared 45deg/0deg reflectance factor of pressed polytetrafluoroethylene (PTFE) powder, J. Res. NIST 104 (1999) 185−188.

[58] V.R. Weidner, J.J. Hsia, Reflection properties of pressed polytetrafluorothylene powder, J. Opt. Soc.Am. 71 (1981) 856−861.

[59] V.R. Weidner, J.J. Hsia, B. Adams, Laboratory intercomparison study of pressed polytetrafluoroethylene powder reflectance standards, Appl. Opt. 24 (1985) 2225−2230.

[60] G.T. Georgiev, J.J. Butler, Long-term calibration monitoring of Spectralon diffusers BRDF in the air-ultraviolet, Appl. Opt. 46 (2007) 7892−7899.

[61] L.M. Hanssen, S. Kaplan, Infrared diffuse reflectance instrumentation and standards at NIST, Anal. Chim. ACTA 380 (1999) 289−302.

[62] J.L. Mueller, B.C. Johnson, C.L. Cromer, S.B. Hooker, J.T. McLean, S.F. Biggar, The Third SeaWiFS Intercalibration Round-robin Experiment (SIRREX-3), 19−30 September 1994, NASA Goddard Space Flight Center, Greenbelt, Maryland, 1996.

[63] D. Hunerhoff, U. Grusemann, A. Hope, New robot-based gonioreflectometer for measuring spectral diffuse reflection, Metrologia 43 (2006) S11−S16.

[64] H.W. Yoon, D.W. Allen, G.P. Eppeldauer, B.K. Tsai, The extension of the NIST BRDF scale from 1100 nm to 2500 nm, Proc. SPIE 7452 (2009) 745204.

[65] C. Wells, S.F. Pellicori, M. Pavlov, Polarimetry and scatterometry using a Wollaston polarimeter, Proc. SPIE 2265 (1994) 105−112.

[66] R.A. Barnes, E.F. Zalewski, Reflectance-based calibration of SeaWiFS. I. Calibration coefficients, Appl. Opt. 42 (2003) 1629−1647.

[67] C.J. Bruegge, A.E. Stiegman, R.A. Rainen, A.W. Springsteen, Use of Spectralon as a diffuse reflectance standard for in-flight calibration of earth-orbiting sensors, Opt. Eng. 32 (1993) 805−814.

[68] R.D. Jackson, M.S. Moran, P.N. Slater, S.F. Biggar, Field calibration of reference reflectance panels, Rem. Sens. Environ. 22 (1987) 145−158.

[69] C.D. Mobley, Estimation of the remote-sensing reflectance from above-surface measurements, Appl. Opt. 38 (1999) 7442−7455.

[70] K.J. Thome, K. Arai, S. Tsuchida, S.F. Biggar, Vicarious calibration of ASTER via the reflectance-based approach, IEEE Trans. Geo. Remote Sens. 46 (2008) 3285−3295.

[71] E.A. Early, P.Y. Barnes, B.C. Johnson, J.J. Butler, C.J. Bruegge, S.F. Biggar, P.R. Spyak, M. Pavlov, Bidirectional reflectance round-robin in support of the earth observing system program, J. Atmos. Oceanic Technol. 17 (2000) 1077−1091.

[72] B.C. Johnson, S.S. Bruce, E.A. Early, J.M. Houston, T.R. O'Brian, A. Thompson, S.B. Hooker, J.L. Mueller, The Fourth SeaWiFS Intercalibration Round−Robin Experiment (SIRREX-4), May 1995, NASA Goddard Space Flight Center, Greenbelt, Maryland, 1996.

[73] J.D. Jackson, T.R. Clarke, M.S. Moran, Bidirectional calibration results for 11 Spectralon and 16 BaSO$_4$ reference reflectance panels, Rem. Sens. Environ. 40 (1992) 231−239.

[74] G.T. Georgiev, J.J. Butler, BRDF study of gray-scale Spectralon, Proc. SPIE 7081 (2008) 708107.

[75] G.T. Georgiev, J.J. Butler, Laboratory-based bidirectional reflectance distribution functions of radiometric tarps, Appl. Opt. 47 (2008) 3313−3323.

[76] E.L. Derniak, D.G. Crowe, Optical Radiation Detectors, first ed., John Wiley & Sons, New York, 1984.
[77] F. Hengstberger, Absolute Radiometry, Academic Press, Boston, Massachusetts, 1989.
[78] N.P. Fox, J.P. Rice, Absolute radiometers, in: A.C. Parr, R.U. Datla, J.L. Gardner (Eds.), Optical Radiometry, Elsevier Academic Press, Amsterdam, The Netherlands, 2005, pp. 35–96.
[79] N.P. Fox, P.R. Haycocks, J.E. Martin, I. Ul-haq, A mechanically cooled portable cryogenic radiometer, Metrologia (1995) 581–584.
[80] R. Goebel, M. Stock, R. Kohler, Report on the international comparison of cryogenic radiometers based on transfer detectors, in: Rapport BIPM-2000/9, September 2000, BIPM, Paris, France, 2000.
[81] B.C. Johnson, S.W. Brown, G.P. Eppeldauer, K.R. Lykke, System-level calibration of a transfer radiometer used to validate EOS radiance scales, Int. J. Remote Sens. 24 (2003) 339–356.
[82] J. McIntire, D. Moyer, J.K. McCarthy, S.W. Brown, K.R. Lykke, F. DeLuccia, X. Xiong, J.J. Butler, B. Guenther, Results from solar reflective band end-to-end testing for VIIRS F1 sensor using T-SIRCUS, Proc. SPIE 8153 (2011) 81530I.
[83] M.E. Feinholz, S.J. Flora, S.W. Brown, Y. Zong, K.R. Lykke, M.A. Yarbrough, B.C. Johnson, D.K. Clark, Stray light correction algorithm for multichannel hyperspectral spectrographs, Appl. Opt. 51 (2012) 3631–3641.
[84] M.E. Feinholz, S.J. Flora, M.A. Yarbrough, K.R. Lykke, S.W. Brown, B.C. Johnson, Stray light correction of the MOBY optical system, J. Atmos. Oceanic Technol. 26 (2009) 57–73.
[85] W. Budde, Optical Radiation Measurements: Physical Detectors of Optical Radiation, Academic Press, New York, 1983.
[86] G.H. Rieke, Detection of Light: From the Ultraviolet to the Submillimeter, Cambridge University Press, Cambridge, U.K, 1994.
[87] Evaluation of Measurement Data – Guide to the Expression of Uncertainty in Measurement, first ed., International Organization for Standardization, Geneva, Switzerland, 2008.
[88] J.L. Gardner, Uncertainty estimates in radiometry, in: A.C. Parr, R.U. Datla, J.L. Gardner (Eds.), Optical Radiometry, Elsevier, Amsterdam, The Netherlands, 2005, pp. 291–325.
[89] T. Lafarge, A. Possolo, Uncertainty Machine V. 1.0, 2013. http://www.nist.gov/itl/sed/gsg/uncertainty.cfm.
[90] A.C. Parr, B.C. Johnson, The use of filtered radiometers for radiance measurements, J. Res. NIST 116 (2011) 751–760.
[91] B.C. Johnson, J.B. Fowler, C.L. Cromer, The SeaWiFS Transfer Radiometer (SXR), NASA Goddard Space Flight Center, Greenbelt, Maryland, 1998.
[92] T.C. Larason, S.S. Bruce, C.L. Cromer, The NIST high accuracy scale for absolute spectral response from 406 nm to 920 nm, J. Res. NIST 101 (1996) 133–140.
[93] T.C. Larason, S.S. Bruce, A.C. Parr, Spectroradiometric Detector Measurements: Part I – Ultraviolet Detectors and Part II – Visible to Near Infrared Detectors, U.S. Department of Commerce, Washington, D.C. 20402–9325, 1998.
[94] G.P. Eppeldauer, H.W. Yoon, D.G. Jarrett, T.C. Larason, Development of an in situ calibration method for current-to-voltage converters for high-accuracy SI-traceable low dc current measurements, Metrologia 50 (2013) 509–517.
[95] E.A. Early, A. Thompson, J. DeLuisi, P. Disterhoft, D. Wardle, E. Wu, W. Mou, Y. Sun, T. Lucas, T. Mestechkina, L. Harrison, J. Berndt, D.S. Hayes, The 1995 North American

interagency intercomparison of ultraviolet monitoring spectroradiometers, J. Res. NIST 103 (1998) 15—62.

[96] D. Antoine, F. d'Ortenzio, S.B. Hooker, G. Becu, B. Gentili, D. Tailliez, A.J. Scott, Assessment of uncertainty in the ocean reflectance determined by three satellite ocean color sensors (MERIS, SeaWiFS, and MODIS-A) at an offshore site in the Mediterranean sea (BOUSSOLE project), J. Geophys. Res. 113 (2008) C07013.

[97] S.W. Brown, S.J. Flora, M.E. Feinholz, M.A. Yarbrough, T. Houlihan, D. Peters, Y.S. Kim, J.L. Mueller, B.C. Johnson, D.K. Clark, The Marine Optical Buoy (MOBY) radiometric calibration and uncertainty budget for ocean color satellite sensor vicarious calibration, Proc. SPIE 6744 (2007) 67441M.

[98] K.J. Voss, S. McLean, M. Lewis, B.C. Johnson, S.J. Flora, M.E. Feinholz, M.A. Yarbrough, C. Trees, M. Twardowski, D.K. Clark, An example crossover experiment for testing new vicarious calibration techniques for satellite ocean color radiometry, J. Atmos. Oceanic Technol. 27 (2010) 1747—1759.

[99] M. Gergely, G. Zibordi, Assessment of AERONET-OC Lwn uncertainties, Metrologia 51 (2014) 40—47.

[100] G. Zibordi, B. Holben, I. Slutsker, D. Giles, D. D'Alimonte, F. Melin, J.-F. Berthon, D. Vandemark, H. Feng, G. Schuster, B.E. Fabbri, S. Kaitala, J. Seppala, AERONET-OC: a network for the validation of ocean color primary products, J. Atmos. Oceanic Technol. 26 (2009) 1634—1651.

[101] J.A. Esposito, X. Xiong, A. Wu, J. Sun, W.L. Barnes, MODIS reflective solar bands uncertainty analysis, Proc. SPIE 5542 (2004) 448—458.

[102] K. Lenhard, Determination of combined measurement uncertainty via Monte Carlo analysis for the imaging spectrometer ROSIS, Appl. Opt. 51 (2012) 4065—4072.

[103] A. Morel, J.L. Mueller, Normalized water-leaving radiance and remote sensing reflectance: bidirectional reflectance and other factors, in: J.L. Mueller, G.S. Fargion, C.R. McClain (Eds.), Ocean Optics Protocols for Satellite Ocean Color Sensor Validation, Revision 4, Radiometric Measurements and Data Analysis Protocols, Volume III, NASA Goddard Space Flight Center, Greenbelt, Maryland, 2003, pp. 32—59.

[104] T. Riley, S.W. Bailey, The Sixth SeaWiFS/SIMBIOS Intercalibration Round-robin Experiment (SIRREX-6) August — December 1997, NASA Goddard Space Flight Center, Greenbelt, Maryland, 1998.

[105] G. Zibordi, F. Melin, J.-F. Berthon, Comparison of SeaWiFS, MODIS, and MERIS radiometric products at a coastal site, Geophys. Res. Lett. 33 (2006). L06617.

[106] G. Zibordi, K. Ruddick, I. Ansko, G. Moore, S. Kratzer, J. Icely, A. Reinart, In situ determination of the remote sensing reflectance: an inter-comparison, Ocean Sci. 8 (2012) 567—586.

[107] JCGM/WG1, Guide to Uncertainty in Measurement, 2008. http://www.bipm.org/en/publications/guides/gum.html.

[108] J. Mandel, The Statistical Analysis of Experimental Data, John Wiley & Sons, New York, 1964.

[109] J.L. Stanford, S.B. Vardeman, Statistical Methods for the Physical Sciences, Academic Press, New York, 1994.

[110] P.C. Gregory, Bayesian Logical Data Analysis for the Physical Sciences, Cambridge Univesity Press, Cambridge, U.K., 2005.

Chapter 2

Satellite Radiometry

Charles R. McClain,[1,]* Peter Minnett[2]
[1] *NASA Goddard Space Flight Center, Greenbelt, MD, USA;* [2] *Meteorology & Physical Oceanography, Rosenstiel School of Marine and Atmospheric Sciences, University of Miami, Miami, FL, USA*
*Corresponding author: Email: charles.r.mcclain@nasa.gov

In 1978, the first satellite sensors specifically designed for quantitative measurements of ocean color (the Nimbus-7 Coastal Zone Color Scanner, CZCS) and deriving quantitative estimates of sea surface temperature (SST; the TIROS-N Advanced Very High Resolution Radiometer, AVHRR) were launched and this laid the foundations for a series of subsequent international global missions, the data from which are now being used for tracking long-term trends in not only the ocean, but also the atmosphere and land. The AVHRR time series has been incorporated into the operational NOAA satellite series and the EUMETSAT MetOp series. Additional satellite radiometers in the infrared and microwave have enhanced the AVHRR time series, bringing improved or complementary SST capabilities with different spectral bands, measurement geometry, or less sensitivity to the presence of clouds. An uninterrupted ocean color time series did not begin until 1997 with the Sea-viewing Wide Field-of-view Sensor (SeaWiFS). The ADEOS-1 Ocean Color and Temperature Sensor (OCTS) was launched in 1996, and there was a 3-month gap between the OCTS and SeaWiFS time series. In each case, advances in sensor, calibration metrology, ground system, and data processing technologies have improved data coverage, quality, and accessibility to the point where high quality data products are available online within hours of the satellite overpass. The scientific synergy between chlorophyll-a (a proxy for phytoplankton biomass) and SST is well established as SST often indicates areas of upwelling and enhanced nutrients and several sensors have included both ocean color and thermal infrared bands or wavelengths, e.g., OCTS, and the Moderate Resolution Imaging Spectroradiometer (MODIS). Figure 1 provides an example of global chlorophyll and SST data from SeaWiFS and AVHRR, respectively, during the 1997–1998 El Niño/La Niña.

Chapter 2 deals with satellite sensor technology and engineering, measurement requirements, and calibration and validation methods (on-board and in situ). Chapters 2.1 and 2.2, "Visible satellite radiometers: design and pre-launch characterization" (McClain, Monosmith, and Meister) and "Post-launch

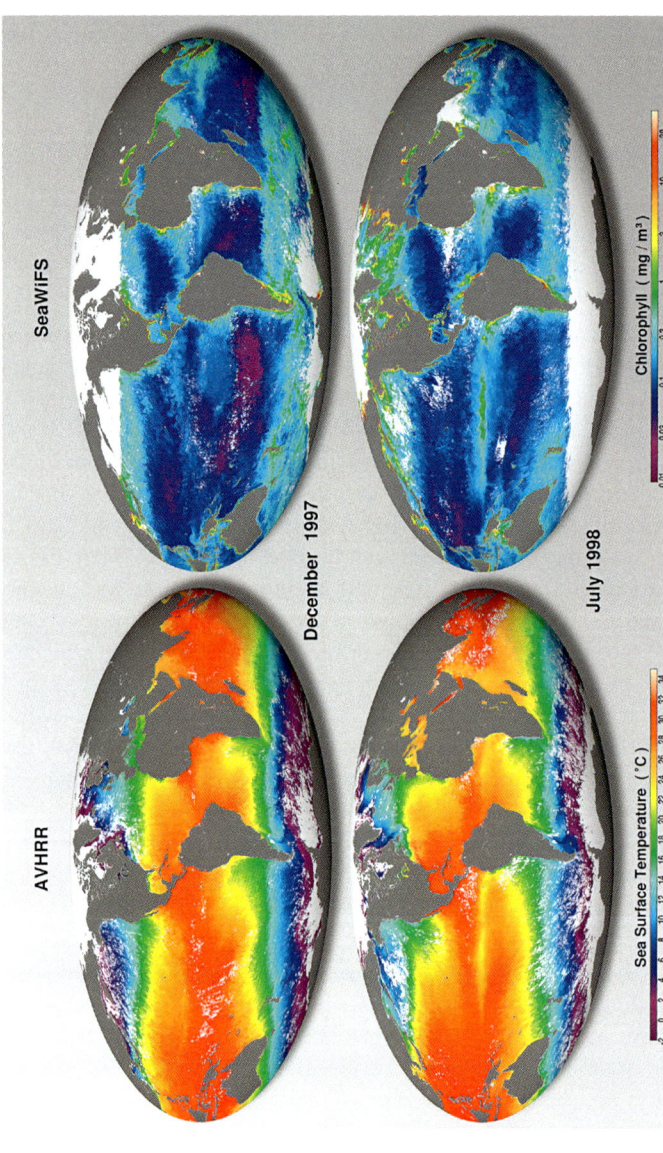

FIGURE 1 Global monthly composites of chlorophyll-a Sea-viewing Wide Field-of-view Sensor (SeaWiFS) and sea surface temperature (SST) (Advanced Very High Resolution Radiometer (AVHRR)) during the warm (December, 1997) and cold (July, 1998) phases of the 1997–1998 El Niño and La Niña, respectively. The lack of high latitude ocean color data in the wintertime high latitudes is due to low solar elevations which do not affect SST retrievals in the thermal IR.

calibration and stability check of visible satellite radiometers" (Eplee and Bailey), focus on ocean color. McClain et al. discuss the basic measurement concepts, the evolution of science requirements, sensor calibration and characterization specifications and requirements, engineering considerations, and a review of "heritage" or historical ocean color sensor designs and performance, i.e., CZCS, OCTS, SeaWiFS, MODIS, and the Medium Resolution Imaging Spectrometer (MERIS). With respect to the evolution of science requirements, the CZCS ocean products were pigment concentration and diffuse attenuation, i.e., only two geophysical quantities. The product suite envisioned for future missions includes a diverse set of biological, chemical, and optical properties. Thus, Chapter 2.1 discusses how sensor requirements such as spectral coverage, signal-to-noise ratios (SNR), and polarization sensitivity impact the design, the different engineering disciplines involved in fabricating a flight instrument, and the process of qualifying an instrument for flight. Eplee and Bailey provide an overview of on-orbit sensor sensitivity or stability monitoring including methods such as solar diffusers, stability monitors, and periodic lunar observations as well as in situ or vicarious methods for adjusting the total sensor—atmosphere system calibration. These chapters focus on low earth orbit (LEO) missions because of the authors' experience and the fact that the Korean Geostationary Ocean Color Imager (GOCI) mission is the only dedicated geostationary ocean color mission launched to date. Other such missions are in the design phases. The science, technology, and engineering are similar to LEO sensors, but there are important differences such as how sampling and signal-to-noise requirements are achieved.

The third and fourth chapters of this part of the book, "Thermal infrared satellite radiometers: design and pre-launch characterization" (Smith) and "Post-launch Calibration and Stability: Thermal Infrared Satellite Radiometers" (Minnett and Smith) address the requirements and procedures for ensuring the accuracy of the radiance and temperature measurements of infrared radiometers on spacecraft. The demands for accurate SSTs for climate research and monitoring impose very stringent accuracy requirements: 0.1 K accuracy and 0.04 K decade^{-1} stability. It is not only challenging to achieve these accuracies, it is also challenging to demonstrate whether they have been achieved. The prelaunch calibration and characterization is an important aspect of being able to determine the fundamental accuracy of the satellite measurement. The prelaunch calibration and characterization can be done at the component level, at which the performance of individual parts of the radiometers is determined, and at the system level, where the entire instrument is tested, ideally in a large thermal-vacuum chamber that facilitates the simulation of on-orbit conditions. Of course once the satellite is on-orbit, direct checking of the performance of the radiometer in ways comparable with the prelaunch calibration is impossible, and confirmation of the integrity of the on-orbit calibration procedure has to be done using indirect methods. Minnett and Smith describe how this can be achieved by monitoring house-keeping data, adjusting the operating parameters of the instrument, and assessing uncertainty characteristics of the derived SST.

Both ocean color and SST satellite sensors are used for a variety of applications ranging from global observations for numerical weather and ocean forecasting and climate research to coastal zone management applications with near-real time latency requirements, e.g., harmful algal bloom and oil spill detection. For global missions, 1- or 2-day coverage, not considering cloud cover, at 1 km spatial resolution has been considered adequate. Continuing developments in detector technology and design result in improving spatial resolution at the surface without loss of signal to noise. For, example Visible Infrared Imaging Radiometer Suite (VIIRS) on the Suomi National Polar-orbiting Partnership satellite (S-NPP), launched on October 28, 2011, has a native spatial resolution of 750 m and a novel approach, called "pixel aggregation," to reduce the effect of the growth in the field of view in the instrument scan away from nadir. The S-NPP VIIRS is first in a new series of visible and infrared radiometers to be flown on the Joint Polar Satellite System into the next decade which should ensure the continuation of some key ocean time series even though VIIRS has fewer ocean color spectral bands than heritage sensors like SeaWiFS, MODIS, and MERIS. However, for near-shore and estuarine studies, more frequent (several times daily) and higher spatial resolution (250 m or finer) data are required. To achieve both high resolution and broad-swath imagery in a single instrument can be challenging if not impossible, depending on the SNR and other requirements. Certainly, at low latitudes, LEO orbit cannot provide more than a single observation per day for ocean color or two observations per satellite of SST (day and night).

Other challenges come into play if the sensor is to serve multiple science communities and if reflective and emissive thermal infrared (IR) data are to be collected. These collective requirements put constraints on the design that usually compromise the performance for a particular science discipline set of requirements. For example, ocean color measurements should avoid sun glint requiring sensor tilting or skewed viewing away from nadir, while the terrestrial community prefers nadir views. Also, ocean color sensors should have a depolarizer to minimize uncertainties in the sensor polarization sensitivity and the atmospheric correction. Depolarizers have a spectral dependence and may interfere with the thermal IR measurements at longer wavelengths. The trade comes down to mission cost (multiple sensors vs. a single sensor; spacecraft impacts such as size and viewing geometries) and complexity.

Chapter 2.1

Satellite Ocean Color Sensor Design Concepts and Performance Requirements

Charles R. McClain,* Gerhard Meister, Bryan Monosmith
NASA Goddard Space Flight Center, Greenbelt, MD, USA
Corresponding author: Email: charles.r.mcclain@nasa.gov

Chapter Outline

1. Introduction ... 74
2. Ocean Color Measurement Fundamentals and Related Science Objectives ... 75
3. Evolution of Science Objectives and Sensor Requirements ... 80
4. Performance Parameters and Specifications ... 84
   - 4.1 Spectral Coverage and Dynamic Range ... 84
   - 4.2 Coverage and Spatial Resolution ... 86
   - 4.3 Radiometric Uncertainty ... 87
     - 4.3.1 Prelaunch Absolute Radiance-Based Radiometric Calibration ... 88
     - 4.3.2 Prelaunch Absolute Reflectance-Based Radiometric Calibration ... 88
     - 4.3.3 Relative Radiometric Calibration ... 89
   - 4.4 SNR and Quantization ... 89
   - 4.5 Polarization ... 90
   - 4.6 Additional Characterization Requirements ... 91
   - 4.7 On-Board Calibration Systems ... 92
5. Sensor Engineering ... 93
   - 5.1 Basic Sensor Designs: Whiskbroom and Pushbroom ... 95
   - 5.2 Design Fundamentals and Radiometric Equations ... 96
   - 5.3 Performance Considerations ... 99
     - 5.3.1 Dynamic Range and Sensitivity ... 99
     - 5.3.2 Noise ... 100
     - 5.3.3 End-Of-Life Performance ... 103
   - 5.4 Sensor Implementation ... 104
     - 5.4.1 Design Controls and Margins ... 104
     - 5.4.2 Electronic Parts Selection ... 105
     - 5.4.3 Materials Selection and Control ... 105

5.4.4 Life Test and Component Screening	105	6. **Summary** **Acronyms** **Symbols and Dimensions**	**107** **108** **109**
5.4.5 Process Controls	106	7. **Appendix. Historical Sensors**	**109**
5.4.6 Environmental Test and Performance Verification	106	7.1 CZCS and OCTS 7.2 SeaWiFS 7.3 MODIS	110 111 113
5.4.7 Reviews and Schedule	107	7.4 MERIS **References**	115 **116**

1. INTRODUCTION

In late 1978, the National Aeronautics and Space Administration (NASA) launched the Nimbus-7 satellite with the Coastal Zone Color Scanner (CZCS) and several other sensors, all of which provided major advances in Earth remote sensing. The inspiration for the CZCS is usually attributed to a article in *Science* by Clarke et al. [1] who demonstrated that large changes in open ocean spectral reflectance are correlated to chlorophyll-a concentrations. Chlorophyll-a is the primary photosynthetic pigment in green plants (marine and terrestrial) and is used in estimating primary production, i.e., the amount of carbon fixed into organic matter during photosynthesis. Thus, accurate estimates of global and regional primary production are key to studies of the earth's carbon cycle. Because the investigators used an airborne radiometer, they were able to demonstrate the increased radiance contribution of the atmosphere with altitude that would be a major issue for spaceborne measurements.

Since 1978, there has been much progress in satellite ocean color remote sensing such that the technique is well established and is used for climate change science and routine operational environmental monitoring. Also, the science objectives and accompanying methodologies have expanded and evolved through a succession of global missions, e.g., the Ocean Color and Temperature Sensor (OCTS), the Sea-viewing Wide Field-of-view Sensor (SeaWiFS), the Moderate Resolution Imaging Spectroradiometer (MODIS), the Medium Resolution Imaging Spectrometer (MERIS), and the Global Imager (GLI). With each advance in science objectives, new and more stringent requirements for sensor capabilities (e.g., spectral coverage) and performance (e.g., signal-to-noise ratio, SNR) are established. The CZCS had four bands for chlorophyll and aerosol corrections. The Ocean Color Imager (OCI) recommended for the NASA Pre-Aerosol, Cloud, and ocean Ecosystems (PACE) mission includes 5 nm hyperspectral coverage from 350 to 800 nm with three additional discrete near infrared (NIR) and shortwave infrared (SWIR) ocean aerosol correction bands. Also, to avoid drift in sensor sensitivity from being interpreted as environmental change, climate change research requires rigorous monitoring of sensor stability. For SeaWiFS, monthly lunar imaging accurately tracked stability at an accuracy of $\sim 0.1\%$ that allowed the data to be used for climate studies [2]. It is now acknowledged by the international community that future missions and sensor designs need to accommodate lunar calibrations. An overview of ocean color remote sensing and a review of the progress made in ocean color remote sensing and the variety of research applications derived from global satellite ocean color data are provided in Refs [3] and [4], respectively.

The purpose of this chapter is to discuss the design options for ocean color satellite radiometers, performance and testing criteria, and sensor components (optics, detectors, electronics, etc.) that must be integrated into an instrument concept. These ultimately dictate the quality and quantity of data that can be delivered as a trade against mission cost. Historically, science and sensor technology have advanced in a "leap-frog" manner in that sensor design requirements for a mission are defined many years before a sensor is launched and by the end of the mission, perhaps 15–20 years later, science applications and requirements are well beyond the capabilities of the sensor. Section 3 provides a summary of historical mission science objectives and sensor requirements. This progression is expected to continue in the future as long as sensor costs can be constrained to affordable levels and still allow the incorporation of new technologies without incurring unacceptable risk to mission success. The IOCCG[1] Report Number 13 [5] discusses future ocean biology mission Level-1 requirements in depth.

2. OCEAN COLOR MEASUREMENT FUNDAMENTALS AND RELATED SCIENCE OBJECTIVES

The basis of ocean color remote sensing lies primarily in the selective absorption of key pigments found in phytoplankton and other biogenic substances like colored dissolved organic matter (CDOM), but also in the scattering properties of some species like coccolithophores and particulates. Generally, as pigment concentrations increase, the ocean reflectance spectral slope "rotates" from negative to positive, i.e., from blue to red, as absorption suppresses the blue and scattering elevates the red (more pigment is associated with more particles). Water is highly transmissive in the blue, but highly absorbing in the red so that the ocean water-leaving radiance is derived from increasingly shallower depths with increasing wavelength. According to Pope and Fry [6], the greatest transmission is between 400 and 450 nm with the maximum being at 418 nm. One important point to make is that the chlorophyll-a and -b in vivo absorption peaks (440 and 470 nm, respectively [7]) coincide with the extraterrestrial solar spectrum peak around 450 nm as well as the maximal water transmission. Given that chlorophyll-a concentrations range from ~ 0.02 mg l^{-1} to over 200 mg l^{-1} (more than four orders of magnitude), the dynamic ranges of downwelling irradiance and, therefore, water-leaving radiance are greatest in the blue which is optimal for remote sensing of chlorophyll. Also,

---

1. Purpose and current membership of the International Ocean-Colour Coordinating Group (IOCCG) is provided at www.ioccg.org.

high water absorption in the NIR and SWIR means ocean reflectance is small and allows for estimation of top-of-the-atmosphere (TOA) aerosol radiance which must be subtracted along with atmospheric molecular scattering (Rayleigh radiance) in the estimation of the ocean reflectances at shorter wavelengths [8–10]. In Wang [9] and subsequent papers, MODIS SWIR bands at 1260, 1640, and 2130 were used for aerosol corrections over turbid waters having finite NIR ocean reflectance even though the SNRs of these bands are significantly lower than what would be desired, i.e., the bands were not designed for this purpose. The European Space Agency's Ocean and Land Colour Instrument (OLCI) will have a band at 1020 nm for this purpose.

Of course, there are complications. One is that CDOM absorption exponentially increases in the visible and ultraviolet (UV). At 440 nm, both chlorophyll-a and CDOM are highly absorbing. To separate the two constituents requires measurements at lower wavelengths, e.g., 360 nm. Historically, including UV bands below 410 nm in satellite ocean color sensors has proven to be a challenge for a number of reasons. SeaWiFS, MODIS, MERIS, and other sensors included bands around 410 nm, but with limited success for this application. To date, only one sensor, the GLI on Advanced Earth Observing Satellite-2 (ADEOS-2), included an ocean color band below 410 nm, i.e., a band at 380 nm (the follow-on sensor, the Second Generation Global Imager or SGLI also has a 380 nm band). The issues include sensor optical throughput and the rapidly decreasing solar irradiance in the UV that limit SNR, as well as relatively greater Rayleigh scattering atmospheric contributions.

Aside from chlorophyll-a and CDOM, other pigments with different absorption spectra may be useful in identifying the presence of key classes of phytoplankton or functional groups [11] and [12]. The identification of these pigments requires additional spectral bands than those of historical multispectral sensors like SeaWiFS, MODIS, and MERIS. For instance, bands at 495, 545, and 625 nm have been recommended for *Trichodesmium*, 655 nm for chlorophyll-b, 470 nm for carotenoids, and 620 nm for phycocyanin [5]. The approach applied in [12] is based on derivative analyses which require a continuous spectrum over the UV–visible domain, i.e., hyperspectral data. Also, the research community is moving to spectral inversion algorithms to estimate derived products [13], the accuracy of which improves with the number and range of the input wavelengths. The distinction between multispectral and hyperspectral is essentially that multispectral implies discrete bands at specific wavelengths while hyperspectral implies a continuous spectrum at a designated resolution, e.g., 5 nm.

Historically, multispectral ocean color sensors placed bands within atmospheric "windows" which are outside major gas (particularly O_2, O_3, NO_2, and water vapor) absorption bands when possible. However, gases like O_3 and NO_2 have absorption bands in the ocean color critical visible which are too broad to avoid and require explicit corrections relying on other ancillary data sources for the global distributions of these gas concentrations. NO_2 absorbs in the UV and blue portions of the spectrum making corrections in bands between 340 and 490 nm essential, especially in coastal areas where pollution is high and water-leaving radiances are small [14]. O_3 has significant absorption in the green portion of the spectrum, around 555 nm (e.g., SeaWiFS) in particular, making accurate corrections necessary because of the sensitivity of bio-optical band ratio algorithms that use 555 nm in the denominator. However, O_3 absorption is nearly zero between 340 and 400 nm, but its absorption does increase rapidly at wavelengths below 340 nm. O_2 has a strong absorption band, the A-band, at 758—770 nm. The SeaWiFS 765 nm band straddled the A-band requiring a correction [15,16]. There are reasons for making A-band measurements that could be beneficial to ocean color atmospheric corrections such as estimation of aerosol plume heights [17], although, in their study, this application required aerosol optical depths >0.3 which exceeds the value normally allowed for valid ocean color retrievals. Finally, water vapor has strong absorption bands around 820, 940, 1125, 1375, and 1875 nm that broaden with wavelength. Water vapor also has a minor absorption band around 720 nm. There have also been recommendations for a thin cirrus cloud flag or correction using a 1380 nm band [18,19], but the necessity of the correction is not unanimous [20,21]. For continuous hyperspectral data spanning the UV—NIR as is being proposed for the NASA PACE mission, corrections for all these absorbing gases will be necessary. The hyperspectral data may allow inversion techniques to be used to estimate gas concentrations simultaneously, but this remains to be demonstrated.

While there are bio-optical signatures below 340 nm, atmospheric ozone effectively blocks any upwelling ocean radiance, at least at detectable levels for a satellite sensor. Other major problems in the UV are the rapid drop in the solar spectrum and increased atmospheric Rayleigh scattering (Rayleigh radiance is proportional to λ^{-4}) making the UV atmospheric radiance especially large compared to the relatively small ocean signals. For comparison, the Rayleigh radiance at 350 nm is about 16 times that at 700 nm. An additional consideration is that the Rayleigh radiance is highly polarized, the degree of polarization (DOP) being determined by the solar geometry with respect to the viewing geometry (Figure 1). In the visible domain, ocean upwelling radiances are no more than about 15% of the TOA radiance with the remaining 85% being largely Rayleigh radiance for clear ocean scenes. Therefore, if a sensor has a 5% sensitivity to polarization that is not characterized or corrected for and the scene or pixel has a Rayleigh radiance equaling 85% of the total radiance with a DOP of 70%, the estimated uncertainty or error is 0.05 * 0.85 * 0.70 or about 3% of the TOA radiance which translates to a ~30% or greater error in water-leaving radiance estimates. For this reason, ocean color sensors need to be designed to minimize polarization sensitivity in order to minimize uncertainties in the on-orbit calibration and atmospheric

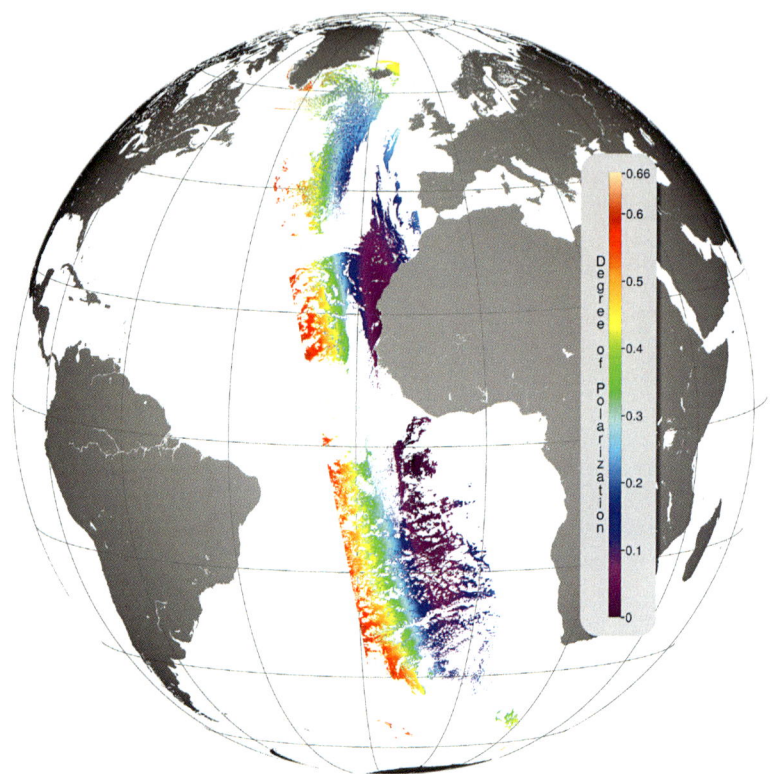

FIGURE 1 The degree of polarization (DOP) at 412 nm computed for a MODIS Aqua orbit on March 22, 2003. DOP = $(I_p - I_s)/(I_p + I_s)$ where I_p and I_s are the intensities of the parallel and perpendicular components of the polarized light, respectively. The range of values in the figure is 0−0.662.

corrections. Because of the sensitivity of bio-optical algorithms for quantities like chlorophyll-a, accurate removal of atmospheric radiance (Rayleigh and aerosol) and high signal-to-noise performance are required. Also, another reason for measurements in the UV is the potential for identifying and correcting for absorbing aerosols at low optical thicknesses (e.g., less than 0.3 in the blue), a problem that has not been solved for heritage ocean color sensors.

Finally, a critical consideration in sensor design is the orbit and the temporal coverage desired. In the past, ocean color missions have been in low earth sun-synchronous orbits meaning that the satellite orbits the earth in essentially a fixed plane with the earth rotating under it such that the satellite passes over head at about the same local time each orbit. The overpass time for ocean color missions has been between 10:00 am and 2:00 pm in order to optimize solar illumination. Low solar zenith angles (high solar elevation) increase sunglint, so sensors like the CZCS and SeaWiFS had a tilt capability to minimize glint contamination. Orbital altitudes for low earth orbits (LEO) have ranged from 705 km for SeaWiFS and MODIS to 955 km for the CZCS. The altitude affects the orbital period, the swath width for a given sensor view or scan angle range, and the sensor instantaneous field of view (IFOV) for a specified ground resolution. The orbital velocity and period can be calculated as.

$$v = (Gm_e/r)^{1/2} \tag{1}$$

$$T = 2\pi r^{3/2}(Gm_e)^{-1/2} \tag{2}$$

where G is the universal gravitational constant ($6.67 * 10^{-11}$ m^{3}Kg^{-1}s^{-2}), m_e is the earth's mass ($5.98 * 10^{24}$ kg), and r is the sum of the earth's radius ($6.37 * 10^6$ m) + the orbital altitude. For a 650 km orbit, $v \approx 7.5$ km/s and $T \approx 97$ min. The orbital velocity dictates many aspects of the design, e.g., the scan rate for sensors like SeaWiFS and MODIS. In the SeaWiFS design, the telescope rotated at ~6 Hz to achieve the required 1.1 km ground resolution at nadir (with some overlap). At that rate of rotation, a time-delay-integration (TDI) scheme using four detectors was implemented to meet the SNR requirements, i.e., each detector sees the ground pixel at slightly different times and the signals are summed because a single detector would not accumulate an adequate number of photons over the sample integration or dwell time for each ground pixel to achieve the desired SNR. The benefit of LEO sun-synchronous orbits is that the entire global ocean can be routinely observed subject to cloud cover. The frequency of global coverage depends on the sensor swath width and orbital altitude. For example, a sensor having an FOV (also called field of regard) of ±60° with a 20° tilt at 650 km altitude, views the entire globe daily with no gaps between swaths, even at the equator.

Besides LEO, there are geostationary orbits where the spacecraft rotates with the earth so that the surface area viewed remains constant. Geostationary orbit altitudes are ~36,000 km with the spacecraft usually positioned on the equator. Variations of geostationary orbits that allow the spacecraft to move north and south of the equator in a periodic fashion, e.g., seasonally, are possible. The advantage of geostationary orbits is frequent views daily, depending on the rapidity in which the sensor can collect the data, the area to be sampled, and the spacecraft transmission and ground station receiving data rates. Another advantage is that the sensor can "stare" at a scene for much longer than LEO, thereby improving SNR by offsetting the "distance-squared" decrease in photons received from a ground pixel. Staring can require jitter control to avoid ground resolution degradation and adds complexity and cost to the sensor. To date, the only geostationary ocean biology mission is the Korean Geostationary Ocean Color Imager (GOCI), although a follow-on mission has been approved. The IOCCG Report Number 12 [22] provides a detailed description of the science and sensor design considerations for a geostationary ocean color mission.

3. EVOLUTION OF SCIENCE OBJECTIVES AND SENSOR REQUIREMENTS

Sensor design and performance requirements are necessarily linked to the science objectives of the mission. Normally, a science traceability matrix (STM) is defined which provides (1) the scientific questions to be addressed, (2) the approaches to answering the questions using the satellite sensor data, complementary field data, modeling, etc., (3) the satellite geophysical data products, and (4) other mission requirements and activities that must be supported to ensure mission success. In outlining these, the sensor measurement requirements (e.g., spectral bands and SNR; [23]) must be specified as well. An STM for future ocean color missions is outlined in [5]. From a historical perspective, the science objectives have evolved dramatically from those of the CZCS. Table 1 provides a brief (and simplified) summary of how mission science objectives have expanded over time with the corresponding impacts on sensor design and complexity. Overall, the objectives have evolved from simply demonstrating that a useful pigment product could be estimated from space to measuring a variety of phytoplankton pigments, dissolved and particulate constituents, phytoplankton functional groups and physiological properties, and more.

Not only has the number of research products increased, each with spectral coverage requirements, but over time, algorithms have incorporated more spectral information, all of which expand the spectral coverage requirements. The CZCS band ratio algorithm [25] correlated ratios of 443/550 and 490/550 to pigment concentration (chlorophyll-a + phaeophytin) with a switch to the latter when the 443 nm water-leaving radiance dropped below a threshold value. O'Reilly et al. [26] used the sum of three band ratios to avoid discrete algorithm switching which generally produces discontinuities in the pigment distributions. One aspect of product development is the substantial lag between algorithm formulation and postlaunch product verification. Product verification requires substantial numbers of field samples for match-up comparisons with satellite estimates. Typically, only about 10–15% of the possible match-up samples pass quality control criteria, e.g., cloud cover [27]. Semianalytical models as discussed in [28] invert ocean reflectance spectra to estimate inherent optical properties (IOPs; absorption and scattering coefficients) and provide estimates of chlorophyll-a, but the inversion fidelity increases with the number of spectral reflectance wavelengths. Thus, as the research community moves to more sophisticated and accurate algorithms based on semianalytical models, spectral requirements are increasing as well as ocean reflectance spectral accuracy because these models are more sensitive to error than band-ratio algorithms, i.e., additional emphasis on sensor performance and calibration accuracy.

Overall, this progress has been the result of the research community constantly pushing beyond each sensor and mission's original science

TABLE 1 Chronological Sequence of Research Ocean Color Missions Illustrating the Evolution of Science Applications and Sensor Complexity (e.g., Number of Spectral Bands)

Sensor	Primary Ocean Science Objectives	Primary Bio-optical Data Products (Prelaunch)	Bio-optical Spectral Bands (nm)	Comments
CZCS (1978)	Feasibility of measuring chlorophyll-a (Chl-a) from space	Total pigment (Chl-a + phaeophytin), K(490)	3 bands: 443, 490, 550	• 670 nm band used for aerosol corrections. • 8-bit digitization • Noise equivalent radiance (NEΔL) at 443 nm: 0.21 Wm^{-2} sr^{-1} μm^{-1}
OCTS (1996) SeaWiFS (1997)	• Global net primary productivity (NPP) • Global Chl-a time series	Chl-a, K(490)	6 bands: 412, 443, 490, 510/520, 555, 670	• Bands at 765 and 865 nm included for aerosol corrections. • Solar diffusers added for on-orbit stability monitoring. • SeaWiFS monthly lunar maneuvers of stability monitoring • 10-bit digitization • NEΔL at 443 nm: 0.11 and 0.077 Wm^{-2} sr^{-1} μm^{-1}, respectively • NPP, CDOM, IOPs, calcite, particulate organic carbon (POC), photosynthetically available radiation (PAR) added after launch. • 412 nm band inadequate for accurate separation of Chl-a and CDOM limiting accuracy of PP estimates based on Chl-a.

Continued

TABLE 1 Chronological Sequence of Research Ocean Color Missions Illustrating the Evolution of Science Applications and Sensor Complexity (e.g., Number of Spectral Bands)—cont'd

Sensor	Primary Ocean Science Objectives	Primary Bio-optical Data Products (Prelaunch)	Bio-optical Spectral Bands (nm)	Comments
MODIS (2000 & 2002)	• Global NPP • Global Chl-a time series • Fluorescence line height (FLH) applications	Chl-a, K(490), CDOM, FLH, calcite, IOPs	7 bands: 412, 443, 488, 531, 547, 668, 678	• 12-bit digitization • NEΔL at 443 nm: 0.032 Wm^{-2} sr^{-1} μm^{-1} • FLH found useful indicator of Fe limitation, but of limited use for Chl-a. • Spacecraft roll maneuvers approved postlaunch for lunar calibration. • Saturation of red and NIR bands over bright areas problematic for lunar calibration.
MERIS (2002)	• Global NPP • Global Chl-a time series • Coastal water quality evaluation (e.g., turbidity, red tides)	Chl-a, K(490), CDOM, FLH, total suspended matter (TSM)	9 bands: 412, 443, 490, 510, 560, 620, 665, 681, 709	• High resolution (300 m) for coastal applications • NEΔL at 443 nm: 0.025 Wm^{-2} sr^{-1} μm^{-1} • Additional bands for suspended sediment and red tides.

GLI (2002)	• Global NPP • Global Chl-a time series • Water quality (e.g., turbidity, red tides)	Chl-a, K(490), CDOM, FLH, TSM	13 bands: 380, 400, 412, 443, 460, 490, 520, 545, 565, 625, 666, 680, 710	• 380 and 400 nm added for improved separation of Chl-a and CDOM. • NEΔL at 443 nm: 0.054 Wm^{-2} sr^{-1} μm^{-1}
OCI (as discussed in IOCCG, 2012; see Table 2 in Section 4 below for details)	• Global NPP • Global Chl-a time series • Water quality • Marine carbon budget • Ecosystem structure (e.g., phytoplankton functional groups) • Particle properties (e.g., size distribution) • Phytoplankton physiological properties	Chl-a, K(490), CDOM, FLH, calcite, IOPs, NPP, POC, PAR, TSM, phytoplankton functional groups, particle size distributions, C:Chl ratio, fluorescent yield	19 bands: 350, 360, 385, 412, 425, 443, 460, 475, 490, 510, 532, 555, 583, 617, 640, 655, 665, 678, 710	• NASA PACE science definition team recommended hyperspectral coverage with 5 nm resolution from 350 to 800 nm, two SWIR (1260 and 1640 nm) bands for ocean color aerosol corrections over highly turbid water, a 900 nm band for water vapor corrections, no sensor saturation over clouds (high L_{max}) in any band, monthly full moon lunar calibrations and 14-bit digitization.

"Primary ocean science objectives" are not all-inclusive and have been paraphrased somewhat from what is stated in various mission documents. Similarly, "primary bio-optical data products" have been redefined or standardized. In both instances the purpose is to show commonality between missions and to reflect international efforts to converge on common products. Also, in most cases, the product suite was expanded in the post-launch phase as new algorithms were developed and data sets were shared and reprocessed by different groups. The visible infrared imaging radiometer suite (VIIRS), SGLI, and OLCI are not included for the sake of space, but all three have fewer bio-optical bands (5, 6, and 11, respectively) than GLI (13) and provide products similar to previous sensors with commensurate capabilities. OLCI will be the first ocean color sensor with a 1024 nm band for aerosol corrections over turbid water. The CZCS NEΔL at 443 nm values are taken from IOCCG report number 1 [24].

objectives after launch, thereby laying the groundwork for the next mission. Column 5 of Table 1 includes some of the additional products developed in the postlaunch phase of the missions, most of which were incorporated into the product suites of other subsequent missions.

4. PERFORMANCE PARAMETERS AND SPECIFICATIONS

The performance specifications laid out in this chapter follow the suggestions presented in [5], which are the consensus as agreed upon by representatives from the following space agencies (in alphabetical order): Center national d'etudes spatiales (CNES), European Space Agency (ESA), Japan Aerospace Exploration Agency (JAXA), Korean Aerospace Research Institute (KARI), NASA, National Oceanic and Atmospheric Administration (NOAA). The report was also reviewed by the IOCCG which has representation from essentially all space agencies with an active interest in ocean color research. The specifications are also very similar to the sensor requirements for an advanced ocean color radiometer developed by the NASA Goddard Space Flight Center Ocean Ecology Laboratory [23]. This section includes specific recommendations for the verification of the requirements.

4.1 Spectral Coverage and Dynamic Range

An overview of the wavelengths needed to address the ocean color science issues discussed in [5] is given in Table 2. Usually, it is not required to match the exact wavelengths of Table 2. However, for all bands, the center wavelength should be known to within ~ 0.1 nm because processing algorithms are tuned to the band centers and relative spectral response (RSR) functions. The NASA PACE Science Definition Team requirements for the OCI are outlined in [29].

Table 2 also provides the typical radiances (L_{typ}), the nominal bandwidth (the bandwidth used for SNR calculation), as well as the *minimum* required SNR. L_{typ} is generally specified as the most frequent clear sky radiance over the open ocean. The L_{typ} at the wavelengths common to the SeaWiFS and MODIS sensors were derived from on-orbit data (MODIS values were scaled to the SeaWiFS values). The L_{typ} of the remaining bands were calculated using the Thuillier solar irradiance (F_0) values [30] and interpolations or extrapolations of the L_{typ}/F_0 ratios of the SeaWiFS/MODIS bands. The maximum radiance L_{max} is provided in Table 2 as well to help define the dynamic range. It was calculated using an albedo of 1.1 and $0°$ incidence angle to simulate the brightest case of a white cloud for an orbit with an equator overpass time of around noon. The SNRs in Table 2 are comparable to those of SeaWiFS. Sensors like MODIS had much higher SNRs ($\sim$factor of 2 or more at the listed L_{typ}'s which should be the goal of future sensors).

The RSR needs to be measured for each band and each sensor element (e.g., mirror, camera, and detector). The out-of-band (OOB) response should

TABLE 2 Multispectral Band Centers, Bandwidths, Typical TOA Clear Sky Ocean Radiances (L_{typ}), Saturation Radiances (L_{max}), and *Minimum* SNRs at L_{typ}

λ	Δλ	L_{typ}	L_{max}	L_{min}	L_{high}	SNR (min)
350	15	74.6	356			300
360	15	72.2	376			1000
385	15	61.1	381			1000
412	15	78.6	602	50	125	1000
425	15	69.5	585			1000
443	15	70.2	664	42	101	1000
460	15	68.3	724			1000
475	15	61.9	722			1000
490	15	53.1	686	32	78	1000
510	15	45.8	663	28	66	1000
532	15	39.2	651			1000
555	15	33.9	643	19	52	1000
583	15	28.1	624			1000
617	15	21.9	582			1000
640	10	19.0	564			1000
655	15	16.7	535			1000
665	10	16.0	536	10	38	1000
678	10	14.5	519			1400
710	15	11.9	489			1000
748	10	9.3	447			600
765	40	8.3	430	3.8	19	600
820	15	5.9	393			600
865	40	4.5	333	2.2	16	600
1245	20	0.88	158	0.2	5	250
1640	40	0.29	82	0.08	2	180
2135	50	0.08	22	0.02	0.8	100

Radiance units are W/m² μm str. SNR is to be measured at L_{typ}. L_{min} and L_{high} are TOA radiance ranges for valid ocean color retrievals derived from a SeaWiFS global one-day data set for the respective SeaWiFS bands after removing the 0.5% highest and 0.5% lowest radiances. These values need to be derived for the remaining bands in the future. Adjustments may be necessary for sensors with different solar and viewing geometries. This table is taken from IOCCG report number 13 [5].

be less than 1% of the total response (where OOB region is defined as those wavelengths where RSR < 0.01; in-band region are wavelengths RSR ≥ 0.01). The characterization is typically achieved by shining light of well-defined wavelength and small bandwidth (e.g., <1 nm) into the sensor. The spectral sampling resolution is ideally related to the response: the larger the response, the finer the sampling. The spectral sampling range needs to be broad enough to capture all significant energy contributions. In the case of a silicon-based detector, this could be 340–1000 nm, for example. For the OOB measurements, the light intensity is increased because of the low expected response. For the in-band measurements, the light intensity is decreased to avoid saturation. The center wavelength λ_c can be calculated from the RSR measurements with the full-width-half-maximum value and should be known with an accuracy of <0.5 nm.

The RSR should be characterized for every sensor element or at least for a representative subset. Variations of the center wavelength for different sensor elements should be less than 0.5 nm. For cross-track scanning sensors, it is generally sufficient to characterize the RSR at one view angle such as nadir, especially if an instrument model has shown that the dependence of the RSR on scan angle is negligible. The RSR should be characterized, as much as possible, involving the complete optical path.

Depending on the instrument design, an on-orbit spectral calibration approach may be required. It is generally accepted that such an approach is not required for filter-based instruments such as SeaWiFS and MODIS. For MODIS, it was demonstrated using an on-board spectral calibration device that the on-orbit spectral change was negligible [31]. However, for instruments such as MERIS an on-orbit spectral calibration approach is required because the dispersion from a grating is very sensitive to alignment changes which may occur, e.g., during launch. MERIS used a doped solar diffuser as well as absorption lines (solar and atmospheric) to determine its wavelength calibration [32].

4.2 Coverage and Spatial Resolution

At large sensor and solar zenith angles, the radiances contributed from the atmosphere become very large relative to the water-leaving radiances, which limits the useful solar and sensor zenith angle range for ocean color products [33]. For SeaWiFS and MODIS, 60° is the maximum sensor zenith angle that is used for level-3 (L3, spatially and/or temporally averaged or binned) data. For SeaWiFS, this translates to a maximum scan angle that is used of about 45° (because of the SeaWiFS tilt). Because MODIS is not tilted, its maximum scan angle used for L3 data binning is about 50° (less than 60° because of the Earth curvature). Another drawback to wide swaths and LEOs is the range of solar and sensor zenith angles which requires an accurate ocean bidirectional reflectance function (BRDF) correction. Experience from

SeaWiFS, MODIS, and MERIS show that reasonably accurate ocean color products can be derived for solar zenith angles $\leq \sim 70-75°$ and sensor zenith angles $\leq \sim 60°$ [5]. For global ocean color applications, a spatial resolution of 1 km at nadir has proven to be sufficient. For coastal and estuarine waters, a higher spatial resolution of 50–300 m is desirable. Global coverage is improved with sensor tilting to minimize sunglint. According to Gregg and Patt [34], a tilted sensor can obtain 20% more coverage than an untilted sensor for a noontime orbit. Such a mechanism should be considered for any ocean color sensor.

For most science questions it is not sufficient to have a measurement at one point in time, but the measurements are required over a certain period of time (e.g., to study the seasonal variation of an ocean color product). Cloud coverage strongly reduces the number of valid retrievals, such that in many areas of the world (e.g., equatorial regions) with a revisit time of every other day there are locations with no valid ocean observations even over a week's time. Other examples are the arctic and Antarctic regions, where the revisit time is even higher due to the convergence of LEO orbits at the poles [35].

4.3 Radiometric Uncertainty

The IOCCG Report Number 10 [33] states that a goal of 0.5% for the accuracy of the TOA radiance at 443 nm is required to achieve a water-leaving radiance accuracy of 5% (at 443 nm) and an accuracy of the chlorophyll product of $\sim 30\%$ (see also [36]). Ideally, the required uncertainties should be defined for each science question. The ocean color community has accepted the method of vicarious calibration [37]. In practice, this means the initial prelaunch calibration is adjusted by the vicarious calibration, and the focus of the calibration effort shifts to the trending of the radiometric gains and the characterization of artifacts like spectral response changes, polarization, etc.

The accuracy goal of about 0.5% for the TOA signal is very challenging. Assuming error sources are uncorrelated, the total error is estimated by taking the square root of the sum of the squares of all individual uncertainty components (such as polarization, linearity, straylight, etc.). This requires the uncertainty of each individual component to be much smaller than 0.5%, preferably less than 0.2%.

There are two separate phases of the radiometer characterization: prelaunch and on-orbit. The prelaunch characterization is very extensive and includes as many aspects of the instrument as possible, whereas the on-orbit characterization is usually restricted to the measurement of the radiometric gain and the SNR, and possibly trending of the spectral responsivity and polarization. The testing protocols and procedures should be mature and vetted with the science community well before the start of the prelaunch characterization phase, in particular.

4.3.1 Prelaunch Absolute Radiance-Based Radiometric Calibration

The absolute radiometric calibration of the instrument is achieved by letting the sensor measure a calibrated light source. The radiance level of the light source should be SI (International System of Units) traceable to standards from national metrology institutes such as the National Institute of Standards and Technology in the United States. Spherical integrating spheres (SIS) are a popular light source, because their spectral output can be easily traced to standards, and they can achieve a high level of spatial uniformity at their exit aperture. Note that for nonscanning instruments such as MERIS, calibration of the complete FOV of the sensor can only be covered using an SIS by scanning the sensor's FOV across the aperture of the SIS, increasing significantly the uncertainty. The spheres are often illuminated by light from tungsten lamps, and a large number of lamps (placed at different positions in the sphere), in conjunction with the scattering inside the sphere (which is coated on the inside with a diffuse, highly reflective material) assures a high degree of spatial uniformity of the light output. The actual non-uniformity of both the output aperture and the back of the sphere need to be characterized (in the sensor's geometric configuration—pupil location and FOV) to reduce the errors. The multiple scattering inside the sphere leads to a very low DOP of the radiance exiting the SIS, the goal should be a DOP of less than 0.2%. After the light output of the SIS has been calibrated, it needs to be monitored (e.g., by sensors internal to the sphere) to ensure that the SIS radiance does not change from the time of the sphere calibration to the time of the radiometer calibration.

It may seem unnecessary to define a prelaunch radiance uncertainty requirement for sensors like MODIS or MERIS, whose ocean color products do not use the prelaunch gain. However, many of the prelaunch characterization tests (e.g., straylight, saturation, etc.) require an instrument gain to calculate the radiance, and therefore such a requirement is justified. The requirement for SeaWiFS and MODIS of 5% was relatively high, and modern technology can achieve better accuracies.

4.3.2 Prelaunch Absolute Reflectance-Based Radiometric Calibration

The reflectance calibration of an instrument applies to instruments that use a solar diffuser as their main on-orbit calibration source. The BRDF of the solar diffuser needs to be determined. As defined by Nicodemus et al. [38], the BRDF describes the absolute reflectance of a surface, as well as the dependence of the reflectance on incidence and view angles. These measurements need to be made so that all combinations of angles that are expected on-orbit are bracketed, with an angular resolution of better than 5°. The absolute uncertainty for the reflectance measurements should be better than 1%, and the relative uncertainty at different angles with respect to each other should be ~0.2%. If a device like a solar diffuser screen is used to avoid sensor saturation (e.g., MODIS), the characterization measurements should be done with the screen in place to determine the combined effect. An analysis of the MODIS on-orbit calibration measurements revealed a significant detector dependency of the vignetting (reduction in brightness) function [39] that was not measured prelaunch.

4.3.3 Relative Radiometric Calibration

The two previous sections described uncertainty goals for the absolute calibration. The calibration requirements of different sensor elements relative to each other (e.g., half-angle mirror sides for SeaWiFS, detectors or cameras for MERIS) need to be even tighter. The reason is that very small relative calibration inaccuracies for adjacent sensor elements are easily identifiable in images of ocean color products as stripes, which reduce the confidence of the user community in the overall product quality and is detrimental to the detection of spatial features in the level-2 (L2, derived products like ocean reflectance and chlorophyll-a) data. A SIS can provide a spatially homogeneous light field that can be used for relative calibration measurements. The gains of detector elements should be calibrated with an uncertainty relative to each other of ~0.2%.

4.4 SNR and Quantization

The minimum SNR requirements are given in Table 2. They are the result of studies for the Aerosol, Cloud, Ecosystems (ACE; a NASA decadal survey mission in formulation) mission that were adopted by the PACE SDT. For the bands from 360 to 710 nm, the SNR requirements were derived from simulations using a semianalytical ocean color model [40], varying the spectral marine remote sensing reflectance and assessing the impact on biogeochemical variables. The 350 nm band is primarily for absorbing aerosol detection, so the SNR requirement (300) is lower than for other bands. The value of 1400 for the 678 nm band was derived from an analysis of MODIS retrievals of the fluorescence line height, which is a very small signal. The NIR and SWIR values were derived from a study of the sensitivity of the reflectance inversion bio-optical model to noise in atmospheric correction algorithms [8,9].

A 14-bit resolution is sufficient for most ocean color applications even when bright cloud radiance levels are included in the dynamic range. The requirements for quantization depend strongly on the radiance level and the sensitivity of the ocean reflectance to a particular ocean constituent: a very high degree of quantization is required at radiances typical of ocean scenes, but at higher radiance levels (e.g., over clouds and over land) a reduced degree of quantization is acceptable. This was achieved in the SeaWiFS instrument with a bilinear gain (see the SeaWiFS description in the appendix). Generally, ocean color sensors have multiple gain modes where the gain is set via command (e.g., the CZCS) or using automatic gain switching (e.g., the Visible Infrared Imaging Radiometer Suite, VIIRS). However, different gain modes add considerable complexity to the sensor design, characterization and on-orbit calibration, and are generally not recommended now that 14-bit flight qualified analog-to-digital converters (ADC) are available. The main reason is that many on-orbit calibration or validation methods (e.g., lunar measurements or deep convective cloud analysis) operate at radiance levels higher than the typical clear sky ocean radiances. For bilinear gains or different gain modes, results obtained from these methods need additional analysis before they can be applied to the lower radiance levels, increasing the total uncertainty.

The instrument SNR is calculated using the noise of a single detector element when viewing a constant light source. The SNR must be determined for each band at L_{typ} (see Table 2). A SIS with a spatially homogenous output is often used for this test. Obviously, an excellent (and well-characterized) short-term temporal stability of the SIS light output is crucial for this test. Additionally, the SNR should be determined at various light levels within the dynamic range. This is often done in conjunction with the dynamic range test, and leads to a reduction in schedule and cost associated with sensor characterization.

4.5 Polarization

Circular polarization of the TOA signal is very low [41] and, therefore, does not need to be considered during sensor characterization. The degree of *linear* polarization of the TOA signal over the ocean can be up to 70% (44; Figure 1). This is not a problem for a sensor without polarization sensitivity. On the other hand, a sensor like MODIS/Aqua, with a polarization sensitivity of up to 5.4%, may produce radiance errors of up to 2.7% if the TOA signal is 50% polarized. Sensors like MERIS and SeaWiFS used polarization scramblers to reduce the instrument polarization sensitivity to low levels (SeaWiFS: about 0.3% or less, MERIS: less than 0.1% in the blue, $\sim 0.2\%$ in the NIR) and carry the residual polarization sensitivity as an uncertainty without modifying the measured radiances. Sensors with significant polarization sensitivity like MODIS need a correction to the TOA measured radiances using the sensor prelaunch polarization characterization data and radiative transfer model [41]. An incorrect polarization correction can lead to large regional and seasonal biases [42]. Thus, it is important to accurately characterize instrument polarization sensitivity.

One proven polarization characterization method is to use a SIS with low DOP, and to place a linear polarizer sheet (with well characterized polarization characteristics) between the SIS and the sensor. This method was used to characterize the polarization sensitivity of VIIRS. The polarizer sheet must be rotated 180° (or preferably 360°, to confirm that the results 0—180° agree with the results from 180—360°), taking measurements with the sensor at intervals of about every 15°. These measurements must be obtained such that all scan angles (or the desired FOV) are covered. In many cases, this requires repeating the measurement sequence with different orientations of the sensor relative to the SIS. The overall goal should be to characterize the sensor polarization sensitivity with an uncertainty of about 0.2% [29].

4.6 Additional Characterization Requirements

Straylight refers to optical processes within the sensor, such as ghosts and optical scatter, and should be reduced as much as possible. Therefore, straylight must be a consideration early in the design process as it can seriously degrade data quality and straylight sources can be very difficult to isolate during testing. However, straylight is part of any optical sensor and can be minimized using baffling, special black paints, antireflection coatings on optics, etc. In the vicinity of strong radiance gradients, straylight effects often exceed the accuracy goal of 0.5%. Straylight can be particularly prominent in the vicinity of bright objects like clouds adjacent to relatively dark ocean areas and can seriously reduce global ocean coverage. As an example, in the case of MODIS Aqua, the masking of pixels due to straylight from clouds leads to a data loss of about 50% of all L2 ocean pixels for a given day [43]. If properly characterized prior to launch, straylight corrections can be made (e.g., SeaWiFS [44]) to recover some of the data.

Due to space limitations, only the sensor requirements most relevant to ocean color products have been discussed above. As for most Earth remote sensing sensors, the following items need to be characterized as well:

1. Linearity of the counts to radiance conversion
2. Temperature dependence
3. Dark current (offset) characterization
4. Spectral registration (or band coregistration, i.e., overlap of the footprint of different bands)
5. Pointing accuracy and knowledge (for geolocation purposes)
6. Modulation Transfer Function
7. IFOV

Additionally, every sensor needs comprehensive instrument models, e.g., throughput models for SNR estimation and ray trace models for component specification, straylight avoidance and alignment. Component, e.g., mirrors, lens, dichroics, gratings, detectors, and depolarizers, characteristics need to be tested and verified. Such models are essential in predicting performance in the design phase, in evaluating system performance during the characterization phase, and diagnosing problems on-orbit.

4.7 On-Board Calibration Systems

For space-based ocean color remote sensing, four different calibration approaches have been used historically:

- Lamps (e.g., CZCS, MODIS)
- Lunar observations (e.g., SeaWiFS, MODIS)
- Solar diffuser (e.g., MERIS, MODIS)
- Earth observations (e.g., MODIS)

Due to the high predictability of the lunar irradiance, the moon is an excellent calibration source. The main limitation of the moon is its small size relative to the instrument FOV. Lunar calibrations are described in a separate chapter (2.2).

On-orbit calibration with light bulbs has been only moderately successful (at best) in the past, because the brightness variation of a lamp over time is often larger than the ocean color radiometric stability requirements. Monitoring lamp output with photodiodes is necessary, but adds complexity. Lamp sources should only be considered for specific calibration subtasks (like spectral calibration, linearity, short term monitoring), not for absolute calibration or long-term trending [36].

Solar diffusers are a well-established approach for on-orbit calibration. The most common type is a reflective solar diffuser (e.g., MERIS, MODIS). Transmissive solar diffusers (e.g., GOCI) have been used, but much less frequently. For some instruments, they cover the full FOV (e.g., MERIS). The most commonly used material is space grade Spectralon. The main challenge regarding solar diffusers is to determine the reflectance change on-orbit. There are two main approaches to overcome this challenge:

a. The use of two solar diffusers, one of which is exposed to sunlight very infrequently (e.g., only every 3 months) to limit its reflectance degradation. The other diffuser is used for the more frequent calibration measurements. The ratio of the ocean color sensor measurements of the two solar diffusers is used to determine the reflectance degradation of the more frequently used solar diffuser. Additionally, by calculating the degradation as a function of exposure time for the more frequently used solar diffuser, the expected degradation of the less frequently solar diffuser can be calculated. This degradation can then be used in a correction algorithm. Note that for MERIS, the degradation of the less frequently used solar diffuser was negligible (less than 0.2% over the first 7 years; [45]).
b. The use of a solar diffuser stability monitor (SDSM). The SDSM on MODIS is a ratioing radiometer that successively views the solar diffuser and the sun. A screen is needed in the optical path between the SDSM detector and the sun, because the sun is so much brighter than the light reflected off the solar diffuser. Characterizing the vignetting function of this screen has been a challenge for the MODIS instruments [46]. An

additional potential problem is that the SDSM necessarily views the solar diffuser at a different angle than the MODIS instrument, and is therefore not able to capture any change in the relative BRDF of the solar diffuser. This is only a minor concern for small changes in solar diffuser reflectance, but the MODIS/Terra solar diffuser reflectance as measured by the SDSM has declined by about 50%. A similar degradation is expected for the solar diffuser used for VIIRS on the Suomi National Polar-orbiting Partnership project (US) mission. Limiting the solar exposure of the solar diffuser reduces the degradation of its reflectance and, therefore, this should be a design goal. MODIS achieves this goal by employing a door (unfortunately, this door has stopped working properly for MODIS/Terra) whereas MERIS moves the solar diffuser into a protected area. The solar diffuser on VIIRS is only protected by a screen, not a door, so it receives solar radiation every orbit. Also, the diffuser faces the velocity vector which increases its degradation. Therefore, its solar diffuser reflectance has degraded much faster than for MODIS or MERIS [47].

The MODIS ocean bands have a limited dynamic range. Therefore, it was necessary to reduce the illumination of the solar diffuser for the calibration of the MODIS ocean bands. This was achieved by a screen that transmits about 8% of the incoming light from the sun (via pinholes). The characterization of the vignetting function of this screen did not accurately capture the MODIS detector to detector differences seen on-orbit [39]. If possible, this source of radiometric uncertainty should be eliminated by choosing a dynamic range for the sensor that does not require a solar diffuser screen.

Using earth view data (e.g., ocean observations) is a common approach for ocean color sensors to adjust the absolute calibration of the sensor by one constant factor per spectral band ("vicarious calibration" [37]). For the case of MODIS/Terra, the standard calibration methods did not produce reasonable ocean color products, even after vicarious calibration. Because of the serious degradation of the MODIS/Terra mirror (e.g., reflectivity and polarization attributes), the SeaWiFS time series of global ocean products were used to provide time-dependent corrections to the MODIS/Terra standard calibration by modifying the scan angle dependence of the radiometric gains and the polarization sensitivity tables [48]. Although the approach was rather effective, it relies on the existence of reliable concurrent ocean color products from another global sensor, and therefore it should not be considered for sensors that claim to derive independent climate data records.

5. SENSOR ENGINEERING

The usual approach to defining a satellite sensor and scoping a mission (cost, facilities, etc.) is to formulate an STM as discussed in Section 4. In this section, some of the sensor engineering considerations are presented. Ocean

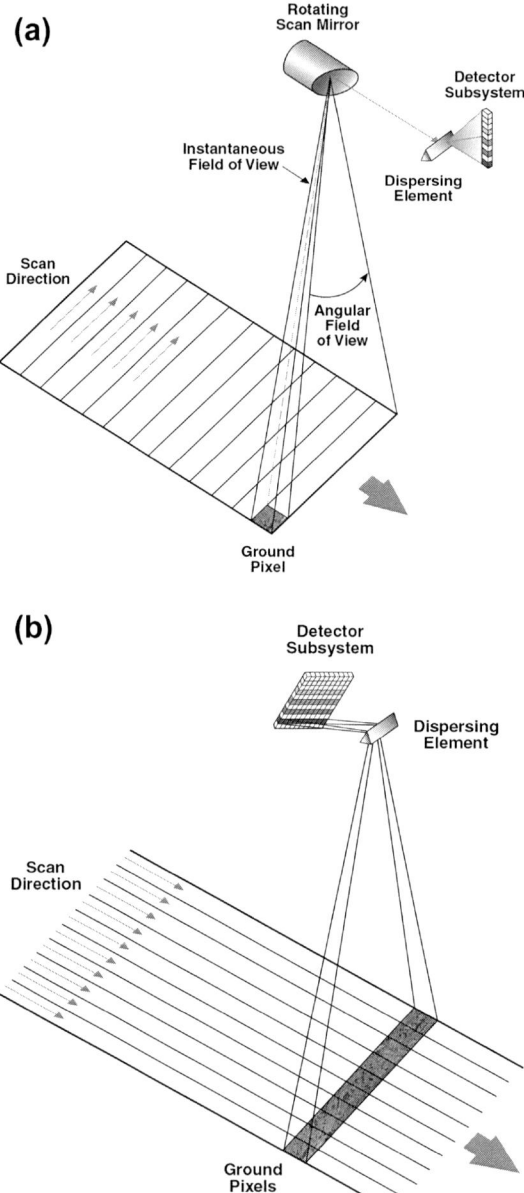

FIGURE 2 (a) Conceptual whiskbroom design and (b) conceptual pushbroom design. The large arrow on the right of each is the spacecraft ground track direction of motion. In panels (a) and (b), "dispersing element" can be a system of dichroic beam splitters and bandpass filters as in most multispectral instruments to date like SeaWiFS and MODIS or a prism or grating (e.g., MERIS). Whiskbroom designs include SeaWiFS and MODIS. In the case of SeaWiFS, each rotation of the telescope produced a single ground "swath" in the cross-track direction. Because MODIS had

10 detectors on each ocean color band focal plane aligned in the along track direction, a single rotation of the mirror produced 10 ground swaths in the cross-track direction. In (b), the 2D detector system has one dimension for spatial sampling corresponding to the along track line of ground pixels in the "scan direction" which is actually the satellite track direction as there is no mechanical scan. The other detector subsystem dimension is spectral. For a pushbroom design, the width of the swath can be increased by adding cameras or increasing the number of detectors in the spatial dimension.

color sensor design and fabrication requires the expertise of a broad range of engineering disciplines including the following: optical, mechanical, electro-mechanical, electrical, detector systems, thermal, contamination, quality assurance, calibration and characterization metrology, system integration and testing, and software development. Also, knowledge of the behavior and compatibility of all materials in a space environment is critical, e.g., outgassing and solder joints. All disciplines must work collaboratively because of the interdependencies of various design requirements and constraints. For instance, the optical, mechanical, and electro-mechanical design teams need to collectively ensure all optical elements (mirrors, lens, dichroics, spectrographs, detectors, filters, depolarizers, baffles, mounts, etc.) can be fit into place without any interference with the optical path from the sensor entrance aperture to the detectors and allow space within the instrument to insert and accurately align components. Another example is the interface between those providing the detectors and those designing the electrical system (e.g., detector taps and formats, read-out integrated circuits, ADCs). An important consideration is avoidance of electrical cross-talk between closely packaged circuits. Overall, the design team's goal is to minimize sensor size, weight, and power requirements while achieving science performance requirements. Page limitations for this chapter do not allow for a detailed or comprehensive description of all aspects of sensor design, so brief overviews of some design fundamentals and an overview of one particularly important performance parameter, SNR, are highlighted.

5.1 Basic Sensor Designs: Whiskbroom and Pushbroom

There are a variety of sensor designs that have been flown (see the appendix for some examples) or are being proposed for future missions. In general, they fall into two categories, whiskbroom and pushbroom, each having advantages and disadvantages. Figure 2 provides a representation of each. Whiskbroom sensors use a scanning mechanism that rotates a mirror (e.g., CZCS, MODIS) or telescope assembly (e.g., SeaWiFS, VIIRS) perpendicular to the orbit track at a rate that matches the spacecraft velocity such that there are no gaps between the scans. The sample rate in the scan direction is determined by the IFOV of the sensor that, in turn, is determined by the altitude and specified ground pixel size at nadir or the subsatellite point along the ground track for tilted sensors. Scan mechanisms usually rotate a full 360° resulting in much of the scan being outside the desired ground swath, e.g., roughly 70% of the MODIS scan is not used. This has implications on the SNR as that lost sampling time (or integration time per ground pixel τ) limits the number of photons collected for each IFOV. $\tau = \text{IFOV}/(2\pi * \text{revolutions per second})$. For example, SeaWiFS at 705 km altitude and a ground resolution of 1.1 km at nadir had an IFOV of ~0.09° and scanned 360° in 0.167 s (telescope rotation rate of 6 Hz) so the time per IFOV was about 4.2×10^{-5} s (this does not take into account the SeaWiFS TDI scheme using four detectors which increases final signal). For narrow swath sensors having high spatial resolution like the Landsat Thematic Mapper, scan mechanisms that sweep back and forth over the swath or FOV have been implemented to avoid this problem. Pushbroom designs use an array of detectors aligned in the cross-track direction, thereby avoiding a scan mechanism. The sampling rate is determined by the IFOV and the spacecraft velocity so as to achieve contiguous data in the along track direction. However, despite the substantial increase in sampling time over whiskbroom designs, other system parameters, such as detector "well depth" (maximum number of photoelectrons a detector can hold) and saturation, limit the photon count and place constraints on other design parameters such as aperture size. A limitation for pushbroom designs is the number of such subsystems that must be incorporated to achieve the desired swath for the specified spatial resolution, For example, MERIS uses five optical subsystems (cameras) and detector arrays, but has a ground swath half that of MODIS (1150 vs 2330 km). Other disadvantages include the number of detectors that must be calibrated and only partial illumination of the detector array during a lunar calibration. An advantage of pushbroom designs is the spatial resolution does not degrade with scan angle, i.e., no cosine effect, except for the increase due to the Earth's curvature.

Finally, there is the issue of optimizing the design for a specific science application, e.g., ocean color, or accommodating multiple sets of science requirements. SeaWiFS was designed specifically for ocean color and included the depolarizer, tilt mechanism, and a limited set of spectral bands. MODIS, MERIS, and GLI were multidiscipline sensors requiring additional spectral bands, but no depolarizer (incompatible with thermal IR bands) or tilt mechanism that compromised ocean data quality (especially for MODIS/Terra) and coverage. To date, SeaWiFS provided the highest quality time series and proved to be an excellent design, although it too had deficiencies due to certain performance specifications being too lax, e.g., OOB spectral response.

5.2 Design Fundamentals and Radiometric Equations

From a systems analysis perspective, the optical system of an orbital sensor can be represented by a simple lens and detector element as in Figure 3. The detector pixel and ground pixel are in the same ratio as the effective

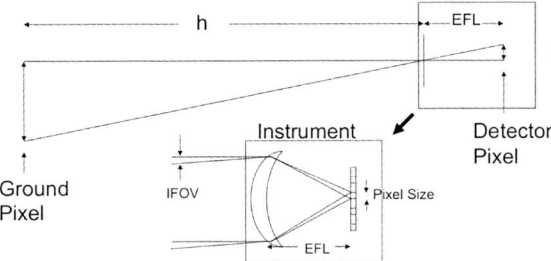

FIGURE 3 Simplified instrument optics represented by a lens aperture and effective focal length (EFL). The angle defined by detector size and EFL is geometrically similar to the angle defined by the altitude and ground pixel, and the value of the angle is IFOV.

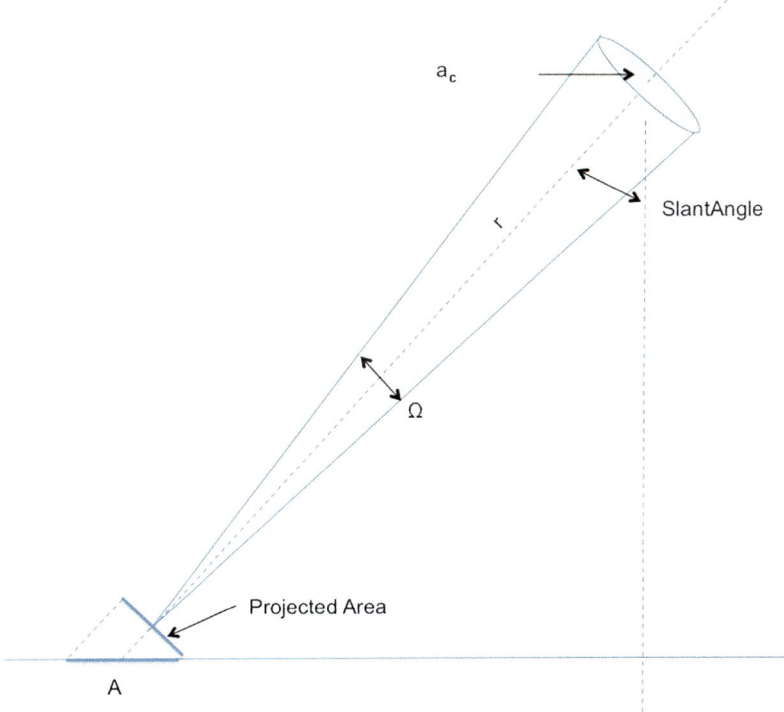

FIGURE 4 Aperture area geometry used to calculate detected power. A_c is the "clear" aperture of the instrument and is the diameter of the circular area through which light enters or is collected by the sensor.

focal length (EFL) and altitude. From Figures 4 and 5, the IFOV is the angle that encompasses the ground pixel from the satellite altitude and is calculated given the desired along-track resolution (d), tilt or slant angle (θ), and altitude (h).

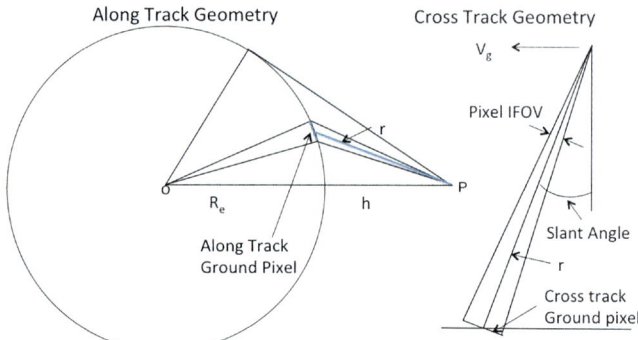

FIGURE 5 Orbit geometry terms used in the text. The cross-track ground pixel has been rotated 90° for ease of viewing.

$$\text{IFOV} = \sin^{-1}\left[\frac{d \cdot \sin\left(\cos^{-1}\left(\frac{(R_e+h)\sin\theta}{R_e}\right)\right)}{r}\right] \quad (3)$$

or IFOV = 2 * $\tan^{-1}(a_c/2r)$. Note that for typical tilt angles of 20°, the slant range (r) is somewhat larger than $h * \tan^{-1}\theta$ due to the curvature of the earth (not shown in Figures 4 and 5 for simplicity). The corresponding cross-track resolution is simply r * IFOV. Both along- and cross-track resolutions are for ground pixels on the suborbital track.

To understand how SNR influences the optical design, the geometry of the measurement must be explained (Figures 4 and 5). The solid angle of the aperture from a distance of r, is the ratio of the area of the aperture to the area of the half sphere of radius r, times the number of steradians in a hemisphere i.e., 2π [49]. Thus,

$$\Omega = \frac{\pi}{4}\left(\frac{a_c}{r}\right)^2 \quad (4)$$

Therefore, the observed power is

$$P = LA\beta\Omega \quad (5)$$

where L is the TOA radiance observed at the sensor, e.g., W/(m² * μm * sr) and β is the bandwidth in μm. Finally, the power actually reaching the detector surface (P_d) is simply P * E, where E the optical efficiency of the instrument.

The electronic signal output by a detector is proportional to the number of photoelectrons generated by the absorption of incident photons. The power incident multiplied by the time the power is applied, or the time before the photoelectrons are transferred out of the detector, is the energy deposited. The maximum number of photoelectrons produced is the energy deposited divided by the energy per photon. Finally, in a manner similar to the optical efficiency, the detector produces photoelectrons with an efficiency referred to as the quantum efficiency (QE). Accounting for these factors, the final expression for the number of photoelectrons that produce the electronic signal is

$$\varepsilon = P_d \tau QE \left(\frac{\lambda}{hc}\right) \qquad (6)$$

where h is Plank's constant, c is the speed of light, and τ is the dwell or integration time, i.e., the time during which photons from an IFOV are collected. The signal output from the detectors is directly proportional to ε, so ε is key to estimating SNR.

5.3 Performance Considerations

Sensor data quality depends upon a host of design factors as discussed in §4. This discussion centers on intrinsic system attributes of concern to all designs, i.e., dynamic range and sensitivity, noise, and sensor degradation due to component deterioration.

5.3.1 Dynamic Range and Sensitivity

As discussed earlier, radiance from the ocean is a small fraction of the TOA radiance measured by a satellite sensor. The range of L_{max}/L_{typ} in Table 2 is 4.75 (350 nm) to 275 (2135 nm). This huge range of light into the sensor puts a dynamic range burden on the engineering team. The problem of dynamic range is coupled to sensor sensitivity.

There are a number of strategies to deal with the large dynamic range. The obvious one is to ignore the large signal, design for clear ocean radiances, and simply let the sensor saturate when clouds are in the IFOV as was the case for some of the MODIS ocean bands. When the cloud radiance exceeds the sensor dynamic range, either the detector or the analog front-end electronics are allowed to saturate. This is rarely a satisfactory solution, as saturation by either usually leads to unacceptably long recovery time of the system as was the case with the CZCS [50]. This "bright target recovery" problem (it has many names) tends to cause uncorrectable distortions in the time decay of the output from saturation conditions. When charge-coupled device (CCD) elements or pixels saturate, charge leaks to adjacent elements and this is called "blooming" and is irreparable. One strategy to deal with a large dynamic range is to prevent the sensor from saturation by some sort of large signal overflow drain in the detector. Such structures can be designed into detectors, but frequently with an unacceptable cost in performance. The impact on performance is detector dependent, but lower efficiency and signal response nonlinearity are common issues.

Yet another strategy, and the preferable one, is to make sure neither the detector, nor the electronics, saturate. Accommodating the especially large dynamic range requirement in the red portion of the spectrum, e.g., $L_{max}/L_{typ} = 74$ (865 nm), can have a deleterious effect on sensitivity for an

ADC with 12 bits or less, and accommodating the large dynamic range can result in a signal increment per ADC count that is larger than the required sensitivity, i.e., the noise equivalent radiance (NE$\Delta L = L_{typ}$/SNR) is greater than the water-leaving radiance resolution required by the bio-optical algorithms for accurate estimation of ocean properties like chlorophyll-a. This bit resolution can be the ultimate limit on sensitivity in a sensor system where the total system noise is dominated by digitization noise (particularly true of early sensors like the CZCS). The total subsystem noise component at the digitization stage further limits sensitivity because much of the signal increment per digital count will be due to additional noise from the digitizer itself. Many of the currently orbiting sensors have 12-bit ADCs as discussed in the appendix. In recent years, a selection of 14-bit ADCs has become available for space flight. This increase in number of ADC bits promises a new generation of sensors where the total system noise, across the spectrum, is dominated by the intrinsic noise from photon counting rather than noise from the ADC, which is the desired regime for the sensor designer.

If the design choices result in loss of sensitivity in the red, given the limited digitization range and noise contributions, a solution that has been used on existing sensors is to incorporate a bilinear gain as in SeaWiFS or an automatic electronically switchable gain as in VIIRS, effectively boosting the low signal gain, and lowering the large signal gain. This effectively provides low sensitivity at large cloud radiances where the gain is low, and higher sensitivity at clear ocean radiances where the gain in high.

5.3.2 Noise

There are many types of noise that must be considered by the system designer. Some, like Johnson noise due to thermal effects in circuits, exists everywhere in the low signal analog electronics chain, irrespective of device or applied voltage. Others, like quantization noise and shot noise (related to the particle nature of light), are specific to a particular type of electronic component. Quantization noise is intrinsic to ADCs, while shot noise is counting noise, and arises as a consequence of the statistics of the Poisson distribution. The Poisson distribution describes the probabilistic nature of counting photons.

For SNR estimation, the noise contributions arise from three major conceptual subsystems: the detector, the analog front-in amplifiers, and finally the ADC. In truth, there is radiometric uncertainty in the calibration process that contributes to the overall SNR as estimated in prelaunch testing. Estimation of on-orbit uncertainty involves additional sources as well [51,52], but those sources are not considered here. The total system noise contribution from these three sources is the root mean square (RMS) of the individual contributions, assuming the sources are uncorrelated. A more thorough treatment of electronic noise and its physical origins can be found in reference texts dedicated to the subject [53]. One of the tradeoffs in defining mission science requirements is spatial resolution versus SNR. Aggregation of ground pixels increases SNR approximately as the square root of the number of samples averaged. For example, aggregating sixteen 250-m pixels increases the 1-km aggregated sample SNR by only four, or stated another way, a 1-km pixel would have four times the SNR of sixteen aggregated 250-m pixels. Measurements or models of the individual noise contributions allow the system designer to focus on reducing, where possible, the dominant contribution in the system root sum square (RSS) noise and can result in a dramatic increase in sensor performance.

5.3.2.1 Detector Noise

There are many different types of detectors suitable for satellite ocean color sensors, the most common being silicon diodes and arrays (e.g., CCDs) for the UV-NIR and HgCdTe and InGaAs for the SWIR [54]. Some detectors and detector arrays have integral amplifiers (usually silicon) built into the device, or in the case of nonsilicon detector arrays, these amplifiers reside on the readout integrated circuit, or ROIC, and this ROIC is electrically contacted, or bump bonded, to the detector. The contact material is usually indium because it is both somewhat physically compliant as well as electrically conductive.

In discussing detector noise, the intrinsic detector noise as well as the radiation noise that manifests itself in the detector is considered. The latter includes shot noise resulting from light being composed of discrete photons and the blackbody background radiation that can become a serious problem in the thermal region of the spectrum. Within the UV−NIR spectral range, detectors generally operate in one of two modes, photovoltaic or photoconductive. A photoconductive detector can be viewed conceptually as a variable resistance device where the resistance is a function of incident radiation having energy greater than the band-gap of the material. A photoconductive detector is essentially a *p-n* junction diode. Incident radiation with energy exceeding the band gap of the material creates electron hole pairs, increasing the carrier population. Typically, the devices operate with reverse (voltage) bias, resulting in significant increase in reverse current when the detector absorbs the incident radiation.

The total noise at the detector is specific to the type of detector (photoconductive or photovoltaic) as well as the bias voltage, the material, and a number of other factors. A comprehensive discussion of detector noise can be found in [54]. In some cases the noise can be modeled or predicted, as for many shot noise limited *p-n* devices, but in general measurements are necessary.

5.3.2.2 Read or Preamplifier Noise

Read noise is the electrical noise generated in the front-end analog electronics after the detector, though as already mentioned, it may originate on the same physical device as the actual detector. At the initial amplifier stage the signal is at its lowest, and any noise arising from the amplifier stage itself is increased along with the signal. For this reason the noise at the analog front-end is usually dominated by the noise of the first high gain amplifier, or set of amplifiers. It is for this reason that engineers sometimes refer to the device at this first stage of amplification as the low noise amplifier, since it is here that high gain is secondary to low noise behavior when mated with detectors having low signal level outputs. Electronic noise superimposed on signal data at this early stage of amplification is bad because both the noise as well and the signal are amplified by further gain stages.

5.3.2.3 Digitization Noise

ADCs used for satellite ocean color sensors are generally successive approximation digitizers, with as many bits as possible to maximize dynamic range and maintain sensitivity, though other types of digital converters exist. An idealized ADC, where the noise is dependent only on the number of bits, does not exist since all ADCs are actually hybrid devices with an analog front-end followed by the digitization stage providing the output digital word. This analog front-end contributes to digitizer output noise, as does the following digitizing stage.

The noise performance of an ADC is summarized in the manufacturer's specification sheet, which will also specify the test conditions and the circuit configuration. Unfortunately, most commercial ADC applications relate to analog signal sampling and reconstruction, and the test parameters in the specification sheet reflect this fact. Sensor designers are interested in DC performance more than AC signal reconstruction. Satellite sensors, irrespective of the design specifics, accumulate a ground pixel signal over the integration time and digitize this essentially DC signal. The signal is usually stable, or nearly so, during the conversion. Under these conditions digitizer performance may be better than indicated by the specification sheet.

A more realistic test of the candidate ADC would be to sample a stable DC signal at the frequency desired and examine the histogram of output count values. The RMS of count values distribution is called input-referred noise or code transition noise [55] and is a good measure of how the ADC will perform in the sensor, assuming the test set-up and lay out reflects the conditions in which the converter will be used. Code-transition noise is rarely, if ever, found on the specification sheet for the device and should always be measured under realistic conditions, e.g., temperature range.

5.3.2.4 Total System Noise Reduction

Measurement of the noise associated with each component subsystem in a relevant sensor configuration is most beneficial because it highlights the component or components that dominate overall system noise. Since the individual noise terms are squared then added, a term significantly larger than others will dominate the RMS sum. This shows the system designer where to concentrate efforts to affect performance improvements. Even good estimates at an early stage can have a large payoff.

As discussed above in the context of aggregation of ground pixels, averaging can have a significant effect on reducing system noise and increasing SNR, but does come at a price compared to having larger ground pixels. However, depending on the specific design and component capabilities, signal averaging can be accomplished elsewhere in the signal chain. A common technique is to oversample at the digitization stage. Digitizing twice within the integration period and averaging the resulting digital counts will reduce the noise contribution by the square root of 2.

5.3.2.5 SNR and Noise Equivalent Radiance

SNR and NEΔL are determined from a base radiance, e.g., L_{typ}, and total system noise that is the RSS sum of all uncorrelated sensor noise sources. For early sensors, such as CZCS with 8-bit digitization, the digitization noise dominated and it was valid to consider sensor sensitivities with reference to the digitization granularity ($L_{max}/2^n$ where n is the number of bits). This is not true for sensors with digitizers having 12 bits and greater. These sensors are primarily signal limited, that is to say Poisson or shot limited.

5.3.3 End-Of-Life Performance

Sometimes overlooked in the early system design stage, lack of attention to end-of-life (EOL) performance can result in a sensor with stellar performance in its early years becoming a sensor with severely degraded performance in its latter years. Most of the issues affecting EOL are known to those organizations that routinely build flight sensors, but the comprehensiveness of EOL mitigation measures is inevitably a budget issue.

Most causes of optical sensor degradation fall into two categories: contamination and radiation damage. Radiation includes alpha and beta particles and gamma rays. Mitigating these via materials control coupled with a materials test program early in the design phase is the best way to minimize potential long term sensor degradation and is the responsibility of specialists in contamination and radiation damage. Contamination affects an optical system throughput by way of both particulates and volatile organics. Particulates are generally a concern for thermal IR systems with wavelength bands beyond 3000 nm, whereas volatile organic controls are generally more important at wavelengths shorter than 500 nm, depending on the thickness of the contaminant. However, all satellite instruments are fabricated in clean rooms. There are a number of clean room classifications. One of the more commonly used schemes, ISO 14,644−1, has nine classes categorized by the particle size and number/unit volume. Class 4 is typically used for spacecraft and sensors. Volatile carbon-based organics are sensitive to UV radiation, which is energetic enough to break organic bonds. The exact chemical mechanisms are varied and contaminant specific, but the net result is loss of transmission or reflectivity in the blue region of the spectrum.

Many transmissive optical and detector materials are susceptible to radiation damage. The mechanism is energetic defect creation in the material and can occur in both crystalline and amorphous materials. Solid-state physicists actually refer to categories of these damage sites as "color centers," because the material can visually appear to take on a color tint if enough of these damage centers are created. Lens and optical fiber materials must be carefully chosen or screened to avoid materials with damage susceptibility. Detector devices are often adversely affected by damage centers created by energetic radiation. The exact mechanisms are varied, reflecting the number of materials used as detectors, and the subject is complex. Detector damage can be mitigated by cooling and shielding which have cost and design implications. What can be said is that defect creation usually changes the electrical properties of the material, and detectors are both electrical as well as optical devices. Any flight detector must be evaluated for radiation damage effects prior to use on a space sensor. The radiation environment is determined by the altitude. The South Atlantic Anomaly is a location where the inner Van Allen belt is closest to the Earth allowing a higher flux of energetic particles. For example, Poivey et al. [56] discuss the frequency and orbital distribution of "single event upsets" for the Orbview-2 (SeaWiFS) solid state recorders.

5.4 Sensor Implementation

A space flight mission includes a number of elements, to include the ground communications, mission operations, launch (including the appropriately sized launch vehicle), data processing, space segments (e.g., the sensors and spacecraft or bus), and mission science. The science sensor is but one element of the space segment and the total mission cost cap as well as technical resource constraints in other elements may result in design compromises that limit sensor performance. This is an important point because it is easy to conceive, for example, of a sensor with data rate that is not commensurate with the capacity of the data link between spacecraft and ground station. The message here is that science requirements must be considered in a mission context, not just in the context of the sensor. Building space flight sensors and spacecraft that must survive on orbit and perform to specifications for years, without maintenance, is a challenging engineering endeavor, and the importance of quality assurance throughout the entire build process is a distinguishing feature of space flight. Quality assurance, in the context of the sensor, is the totality of the effort devoted to ensure science mission success and includes the following elements:

5.4.1 Design Controls and Margins

The design process is governed by discipline specific rules. The purpose of these rules is to prevent failure or undesirable performance degradation. These rules dictate required margins such as mechanical strength, electrical current and voltage capacity, software processing and storage, and component temperature sensitivities to name but a few. In most cases the engineering organization maintains these rules in configuration-managed documentation.

5.4.2 Electronic Parts Selection

Parts engineering is a specialty discipline in space flight. The design engineer must stay current with changing technology in their field, both design tools as well as component technology. Knowing which specific electronic components are suitable for space flight is outside their area of expertise. The flight parts engineer's job is to monitor component vendor processes and flight part screening, keep current with bulletins regarding part restrictions and warnings, and understand parts performance in general to allow them to suggest substitutions in the electronic engineer's design, for example, when the engineer's preferred part does not exist as a flight screened or qualified version.

5.4.3 Materials Selection and Control

The materials engineer has oversight and approval authority for all materials used. The concern is primarily twofold, contamination and corrosion, although issues of material properties such as brittleness, toxicity and others are concerns. Many dissimilar metals will chemically corrode at the metallic junction over time, leading to parts failure. This includes solder joints. Many materials, especially oils, greases, plastics, and other organics will outgas in the vacuum of space, depositing films on optics and thermal control surfaces. Outgassing greases may leave a mechanism without proper lubricant, leading to increased wear and possible component failure, e.g., bearings.

5.4.4 Life Test and Component Screening

Mechanical, optical, and electronic components that are subject to degradation or failure, and for which there is little or no flight heritage or screening data, must be verified as suitable for the mission. For an electronic part, this may involve radiation testing and thermal cycling followed by electrically stressing the part to voltages, currents, or clocking speeds beyond the design limits. This screening is only valid for the lot produced during a production run where processes and materials are documented and remain constant. Parts used for flight must be from the same lot as those tested. A mechanism (e.g., scan motors and momentum compensators), or individual components of the mechanism (e.g., bearings), must be placed in a relevant thermal and vacuum environment, with the approved lubricant, and life tested. In some circumstances using rotation rates or higher duty cycles than needed for the mission may allow for an accelerated life test. In other cases the frictional properties of the lubricant will change with higher rotation rates, precluding this form of accelerated test.

5.4.5 Process Controls

Soldering and electrostatic discharge (ESD) are two examples of processes requiring special training. Soldering and ESD handling are governed by strict documented procedures, and the technicians doing this work are trained and certified, and their work and workplaces are subject to periodic quality assurance monitoring.

Plating and coating processes, whether done for optical, thermal, or other surface properties reasons, must adhere to strictly documented processes and procedures. Witness samples are usually produced along with the processed flight part to ensure, by test, that the surface modification meets standards of uniformity, surface adherence and corrosion resistance.

Printed circuit boards (PCBs) are potential failure points for electronic subsystems and quality can never be taken for granted. A PCB with dense, fine features and a large number of board layers must survive the launch vibration environment without creating mechanically weak, failure prone traces. Thermal stress caused by varying operating and survival temperatures on orbit can also cause mechanical failures. PCB fabrication is a complex process utilizing many steps and involves the use of many chemicals. Because of the many materials and chemicals used in the process, there is the possibility of process-induced corrosion. PCB quality is verified by way of PCB coupon testing. Testing involves both environmental stresses and subsequent destructive sectioning and microscopic analysis of the stressed boards.

5.4.6 Environmental Test and Performance Verification

Environmental tests are conducted on individual sensors and the fully configured spacecraft. Environmental testing at the sensor level is fundamentally a quality or workmanship battery of tests. The specific tests are designed to reveal flaws in the fabrication of the sensor or its subsystems and usually require special facilities such as large thermal-vacuum chambers that are large enough for large multi-instrument platforms such as Aqua (US), Envisat (ESA), and ADEOS (Japan). Environmental tests are not performance tests, though various levels of, or subsets of, complete performance tests are performed between or during some environmental tests. Environmental tests done with the sensor unpowered include simulation of the launch vibrational environment and sometimes load tests on the structure. The acoustic and shock tests to simulate launch conditions are also done at some level, either at the subsystem level or the space segment level with the sensor integrated to the spacecraft. Electromagnetic interference testing may also be required.

Other tests are performed with the sensor powered and operating. The thermal balance test is done with the sensor operating at select temperatures in vacuum and is designed to verify the accuracy of the thermal model for the sensor. Electromagnetic emissions and susceptibility tests are designed to ensure that the sensor neither emits interfering radiation to other elements such as other instruments, nor is susceptible to defined levels of allowable emission from other elements. The thermal cycling test characterizes the behavior of the sensor at a variety of temperatures and places thermal stresses on the components. The temperature extremes also include sensor nonoperating survival temperatures as when the satellite goes into a "safe haven" status. Some subset of performance testing is generally performed during thermal cycling.

Various performance tests and calibrations can be performed either in vacuum or not, as appropriate. The usual approach is to perform a comprehensive performance test before environmental testing and again after environmental testing, with a limited set of performance tests done during or between environmental tests. These limited performance tests are carefully chosen to reveal anomalous behavior and to identify when in the test process the anomalous behavior occurred.

5.4.7 Reviews and Schedule

A typical satellite sensor build and test schedule is about 5 years. This assumes a specific design that conceptually, e.g., modeled, has been developed which meets the sensor performance specifications. There are a number of formal reviews held at certain milestones in the sensor development where issues may be raised. Each issue must be addressed in detail and cleared by the review panel before the instrument development can proceed. The sensor development reviews are part of a larger set of mission reviews which cover all aspects of the mission, e.g., sensor, spacecraft, launch vehicle, and ground system (including the data processing system). At NASA, the reviews have titles like the mission confirmation review, system definition review, preliminary design review, critical design review, and launch readiness review.

6. SUMMARY

Since the 1970s when the CZCS was conceived as a proof-of-concept experiment to determine if basic biological and optical properties, i.e., near surface pigment concentration and diffuse attenuation, could be estimated from space, science objectives have advanced considerably as planning for missions like PACE proceed and as outlined in [5]. Beyond hyperspectral sensors, even more advanced concepts can be envisioned such as inclusion of polarization bands as demonstrated by Loisel et al. [57] using the POLarization and Directionality of Earth Reflectance (POLDER) sensor, for example. The SGLI has polarization bands as well, but neither POLDER nor the SGLI polarization bands were designed specifically for ocean biogeochemistry applications. As the science objectives evolve and become more exacting in terms of the number of parameters or derived products and the accuracy and range of values to be quantified, sensor technology and design engineering concepts are constantly challenged to meet the associated performance requirements. Thus, science and engineering must move forward hand-in-hand as scientists work closely with the design engineers to ensure that requirements are thoroughly documented and concisely understood by the engineers. These two groups can have very different perspectives and approaches. Advances in technology, e.g., detector systems, and optical and electronic components, required to meet future measurement requirements must be identified and funded well in advance of a flight project to ensure the technology is proven and qualified for flight prior to when a mission budget and schedule is defined. Otherwise, the mission can be at risk of cost overruns and launch delays or even cancellation. Finally, some ocean color missions are for the purpose of technology development and can accept more risk than missions providing climate research quality data to both research and operational users, e.g., MODIS. Also, these climate research missions require a comprehensive calibration and validation program as well as a robust and flexible processing system designed to provide data to operational users at short latency times while accommodating frequent data quality and algorithm tests and mission reprocessings.

ACRONYMS

ADC Analog to digital converter
CCD Charge-coupled device
CDOM Colored dissolved organic matter
Chl Chlorophyll
CZCS Coastal zone color scanner
DOP Degree of polarization
FLH Fluorescence line height
FOV Field of view
FWHM Full width half maximum
GLI Global imager
GOCI Geostationary ocean color imager
HICO Hyperspectral imager for coastal ocean
IFOV Instantaneous field of view
IOCCG International ocean colour coordinating group
IOP Inherent optical property
K(490) Diffuse attenuation at 490 nm
MERIS Medium resolution imaging spectrometer
MODIS Moderate resolution imaging spectroradiometer
NEΔL Noise equivalent radiance
NIR Near-infrared radiation
NPP Net primary production
OCI Ocean color imager
OCTS Ocean color and temperature scanner
OLCI Ocean and land colour instrument
OOB Out-of-band
PAR Photosynthetically available radiation
POC Particulate organic carbon

POLDER POLarization and directionality of earth reflectance
RSR Relative spectral response
SeaWiFS Sea-viewing wide field-of-view sensor
SGLI Second generation global imager
SNR Signal to noise ratio
STM Science traceability matrix
SWIR Shortwave infrared radiation
TDI Time-delay-integration
TOA Top of atmosphere
TSM Total suspended matter
VIIRS Visible infrared imaging radiometer suite

SYMBOLS AND DIMENSIONS

a_c Aperture (clear) diameter (millimeter (mm), centimeter (cm), meter (m))
Ω Aperture solid angle (steradians (sr))
A Area ground (square kilometers (km^2))
β Bandwidth (nanometers (nm))
m_e Earth mass (kilograms (kg))
R_e Earth radius (kilometers (km))
v_g Ground velocity (km s^{-1})
τ Integration time (seconds (s))
E Optical throughput (dimensionless)
h Orbit altitude (km)
T Orbital period (minutes (min))
ε Photoelectrons (dimensionless)
P Power (watts (W))
P_d Power (detector) (W)
QE Quantum Efficiency (dimensionless)
L Radiance (W m^{-2} μm^{-1} sr^{-1})
r Slant range (km)
V Spacecraft velocity (km/s)
θ Tilt or slant angle (degrees)

7. APPENDIX. HISTORICAL SENSORS

The sections below discuss the designs of the CZCS (US, 1978–1986), the OCTS (Japan, 1996–1997), SeaWiFS (US, 1997–2010), MODIS (US, 2000-present), and MERIS (ESA, 2002–2012). This suite of instruments includes both whiskbroom designs with various unique features (CZCS, OCTS, SeaWiFS, MODIS) and a pushbroom design (MERIS). VIIRS is a whiskbroom design which incorporates a scanning telescope like SeaWiFS and a focal plane similar to MODIS, so it is not discussed here even though it too has some unique features like aggregation zones and electronic gain switching.

7.1 CZCS and OCTS

The CZCS was a grating spectrometer design. The fore optics consisted of a rotating mirror that could be tilted in 2° increments up to ±10° (a 10° tilt results in a 20° viewing angle) to avoid sun glint. The CZCS also had another innovative element, the polarization scrambler. This component was inserted because the Rayleigh molecular scattering and surface Fresnel reflections are highly polarized, thus requiring the full Stokes parameters and sensor Mueller matrix for the atmospheric correction if no depolarization was incorporated. The sensor had six bands at 443 nm (chlorophyll-a absorption peak), 520 nm (near the spectral location least sensitive to chlorophyll-a, the "hinge point"), 550 nm (measures increased water-leaving radiance as particulate concentrations and backscatter increase), 670 nm (a secondary chlorophyll-a absorption peak), 750 nm (cloud detection), and 11.5 µm (sea surface temperature). The four visible bands had nominal bandpasses of about 20 nm. The Nimbus-7 orbit was sun-synchronous at local noon and descending (altitude = 955 km). Earth data were collected between scan angles ±39.36° with a spatial resolution of ∼800 m at nadir and a swath of 1566 km.

The sensor had four commandable gain settings (visible bands only) to compensate for the range of expected illumination conditions and, as it turns out, decreased sensitivity over time. This was necessitated by the 8-bit digitization in order to maintain the desired quantization. The SNRs ranged from about 400 (520 nm) to 140 (670 nm) for typical open ocean clear sky TOA radiances.

The sensor also had internal lamps for on-orbit calibration stability tracking, but these proved to be too unstable to be useful. The final post–mission calibration was based on global analyses of the time series using "clear-water" radiances to set the "vicarious" gain factors. Indeed, over the lifetime of the sensor, the 443 nm band sensitivity decreased by about 40% [58]. This degradation was presumably due to contamination of the scan mirror.

Being a proof-of-concept mission, some components of the system worked well and others did not. The gain on the 750 nm band was coarse so it was used only for cloud detection. Therefore, the 670 nm band was used for aerosol corrections where it was assumed that the water-leaving radiances at 670 nm were zero [25]. Ironically, this made the CZCS least reliable for measurements in turbid coastal waters.

The system polarization sensitivity was reduced by inclusion of a dual-wedge depolarizer and by positioning the folding mirror such that it compensated for the scan mirror polarization. All mirrors (scan mirror, two telescope mirrors, threefold mirrors, and the collimating mirror) had protective silver coatings. A dichroic located after the scan and two telescope mirrors separated the visible and infrared light and the depolarizer was positioned further down the optical train after the first fold mirror and the collimating mirror. Prelaunch testing showed a maximum polarization sensitivity at 443 nm of about 3% for a 10° tilt (most data was collected at a 10° mirror tilt).

Having the depolarizer located in the aft optics increases the polarization uncertainty. Assuming a DOP of 60% and a Rayleigh component of 80% of the total radiance, the effect is roughly 1.4%. If the polarization properties of the system components stay constant, there is no issue, i.e., the Mueller matrix is known. If component reflectances and transmissions change on orbit and are sensitive to polarization, then having the depolarization wedges near the tail end of the optical path means that the system's actual polarization sensitivity is unknown, i.e., the Mueller matrix has changed.

The CZCS preamplifiers on the detectors tended to "ring" off bright targets. This electronic overshoot often persisted for tens of downscan pixels [50] and depended on how bright the up scan pixels were. No completely satisfactory algorithm for masking contaminated pixels was ever developed.

Both the CZCS and the Ocean Color and Temperature Sensor (OCTS) used a 45° "barrel roll" mirror. In this configuration the sensor aft optics were positioned either forward or aft of the mirror assembly (along the spacecraft velocity direction), and the incoming light was reflected from the Earth-viewing direction along that axis. The tilt mechanism rotated the mirror assembly within the instrument. This had the effect of changing the pixel spacing, and the total scan width, as a function of tilt angle. For example, on OCTS (± 40 ° scan), the scan angle per pixel was 0.83 mrad at tilt $-20°$ (aft), 0.72 mrad at tilt 0, and 0.58 mrad at tilt $+20°$. Since data are collected primarily at $\pm$ 20° degrees tilt, this resulted in a large difference in spatial resolution and coverage north and south of the tilt change (subsolar point).

On OCTS, the 45° mirror, combined with the MODIS-like focal plane design (a large 2-D array of detectors), also had the effect of rotating the effective focal plane footprint on the ground as the mirror scanned from one side to the other. As a result, the individual bands were only co-registered near nadir. As the scan angle increased from nadir, the rotation of the viewed area caused the individual bands to separate in the along-track direction. At the largest scan angles, a given location on the Earth required five consecutive scans to be viewed by all of the bands. This required substantial resampling of the bands to achieve approximate coregistration, and this process increased the noise level in the resampled data.

Both the CZCS and OCTS incorporated internal calibration lamps and OCTS also included a solar calibration capability. The OCTS digitization was 10 bits. Unlike the CZCS, OCTS did not have a depolarizer. The ADEOS-1 orbit was sun-synchronous at local 10:30 AM and descending (altitude = 800 km) and the OCTS swath was 1400 km.

7.2 SeaWiFS

SeaWiFS was a NASA data buy from Orbital Sciences Corporation who subcontracted the sensor to Hughes Santa Barbara Research Center (SBRC). The SBRC sensor design was a huge departure from the CZCS. Rather than a scan mirror, a rotating telescope with a half-angle mirror was used. The half-angle mirror rotates in the same direction and at half the speed of the telescope, thereby maintaining a constant light path into the aft optical subsystem containing four focal planes. As a result, both sides of the half-angle mirror are in the optical path on alternating scans and slight differences in mirror reflectivity are present in the imagery, but this effect was accurately removed via the on-orbit calibration procedures. This design helped minimize polarization and protected the fore optics from contamination. VIIRS also uses a rotating telescope, but (presumably) because of the finer spatial resolution a longer focal length was required resulting in two additional telescope folding mirrors.

The SeaWiFS SNR values are 2–3 times higher than CZCS in the blue and green bands and about 6 times higher at 670 nm for the same radiances. To achieve this, each spectral band has four detectors, the signals from which are summed in a TDI scheme, i.e., each detector sees a ground pixel at a slightly different time. This requires the synchronization of the scan mechanisms and the detector read-out electronics. This feature also eliminated striping that is problematic in other designs such as MODIS and VIIRS.

Another strength of the SeaWiFS detector array or focal plane design is the bilinear gain that prevents bright pixels from saturating any band. This design was implemented to allow for a straylight correction. The original copy of SeaWiFS failed to meet straylight specifications and a number of design adjustments were made to ameliorate the problem, e.g., putting a wedge angle on the front surface of the depolarizer to "collapse" these reflections onto that of the main reflection off the mirror-coated back side [44]. A bilinear gain without electronic switching was implemented by setting the saturation of one of the four detectors at a high maximum radiance producing a "knee" in the total response as the other three detectors saturate at a lower value. Additional measures not implemented because of cost and schedule constraints included higher quality mirrors and the addition of "septums" between the detectors that would have reduced straylight even more.

The SeaWiFS sensor has eight bands in the visible (412, 443, 490, 510, 555, and 670 nm) and near-infrared (765 and 876 nm). The 412 nm band was added to improve separation of chlorophyll-a and CDOM. The 490 band was added to provide better sensitivity for chlorophyll-a estimation in coastal waters where 443 nm water-leaving radiances are small. The two NIR bands are for aerosol corrections in open ocean waters. The visible bandpasses are roughly 20 nm and the NIR bandpasses are 40 nm. The 765 nm band straddles the O_2 A-band absorption feature and requires a correction for this effect [15,16]. Also, SeaWiFS has significant OOB contamination, particularly at 555, 765 and 865 nm, due to poorly specified filter requirements [59] requiring additional corrections [60]. Improved filters to reduce OOB response should have been incorporated when the straylight issues were addressed. The SeaWiFS OOB does complicate the processing and makes comparisons with other sensors more difficult (including those used for in situ validation). The OOB was substantially higher than that of MODIS.

Like the CZCS, SeaWiFS also incorporated four commandable electronic gains and a polarization scrambler. The polarization scrambler was located behind the primary mirror (second optical component) and the sensor polarization sensitivity is estimated to be about 0.25%. Rather than internal lamps for on-orbit calibration, it had a solar diffuser with a solar diffuser cover of the same material. More importantly, the mission allowed for a monthly spacecraft pitch maneuver to scan the moon at a constant phase angle ($\sim 7°$). The solar diffuser cover was never activated to expose the solar diffuser. The diffuser cover time series provided a record for estimating changes in the SeaWiFS SNRs [51], but was not used for correcting the sensor calibration over time. Along with the daily solar calibrations, the electronic gains of each band were checked with calibration pulses. The lunar calibration established the long-term stability of the sensor at a very high accuracy [2].

The SeaWiFS orbit was initially sun-synchronous at noon, but the node drifted past 2:00 pm over the ensuing 12 years on orbit. SeaWiFS Local Area Coverage (LAC) had a spatial resolution of 1.1 km at nadir and a swath of about 2800 km ($\pm 58.3°$ scan). The SeaWiFS Global Area Coverage (GAC) subsampled the data (every fourth line and pixel for a data volume reduction of 16) and truncates the scan to $\pm 45°$ scan angles resulting in a 1500 km swath (SeaSTAR altitude = 705 km). LAC data was broadcast real time and GAC was stored on-board and downlinked to specific ground stations. The sensor tilt positions included $\pm 20°$ and $0°$, although the $0°$ position was only used for the solar and lunar calibrations. Unlike the CZCS, the whole sensor was tilted.

The SeaWiFS subsampling allows small clouds to escape detection in the GAC processing in which case straylight is uncorrected (straylight is scattered light within the instrument that contaminates measurements in adjacent pixels), thereby elevating the total radiance values. The prelaunch characterization data provided enough information for a straylight correction algorithm to be derived. This correction works well in the LAC data processing and for correcting the effects of large bright targets in the GAC.

SeaWiFS data is truncated from 12 to 10 bits on the data recorder resulting in coarser digitization, especially in the NIR bands where the SNRs are relatively low. Noise can cause jitter in the aerosol model selection amplifying the variability in visible water-leaving radiance values via the aerosol correction. Undetected clouds in the GAC data, digitization truncation, and low NIR band SNR values are thought to be the primary reasons for speckling in the SeaWiFS derived products [61].

7.3 MODIS

The design for MODIS was targeted to serve a number of research communities and, therefore, had a broader set of design requirements resulting in a much more complex sensor than CZCS and SeaWiFS. It incorporated 36 bands with wavelengths between 412 nm and 12 µm, including bands with different spatial resolutions (1000, 500, and 250 m). Like SeaWiFS, it was built at SBRC, but about the only thing the two sensors have in common is that they both are filter radiometers, i.e., filters over the detectors for spectral separation rather than dispersive optics like gratings or prisms. Also, the MODIS data is recorded at 12 bits and provides global 1 km ocean color data (no sub-sampling). The MODIS scan is ±55° about nadir resulting in a 2330 km swath for Aqua (1:30 pm, ascending) and Terra (10:30 am, descending) orbital altitude of 705 km.

The MODIS design uses a large rotating mirror similar to that of the CZCS and OCTS, but with no tilt. Unlike CZCS and OCTS, the mirror is not tilted relative to the nadir view, i.e., it is parallel to the local Earth tangent plane when viewing nadir. This is because the receiving optics are to the side of the scan mirror (cross-scan direction) in line with the orbit track (orthogonal to the scan). Because MODIS does not tilt, sunglint contamination is more serious than for CZCS and SeaWiFS even though the MODIS orbits are 10:30 am and 1:30 pm (the orbits have been maintained at these times) rather than noon. Having the mirror exposed does subject it to contamination, but this is tracked using the solar diffuser and solar diffuser stability monitor which provide a much more robust calibration than the SeaWiFS diffuser, but was an expensive addition to MODIS. To date, MODIS (Terra and Aqua) have experienced degradations as high as 50% (412 nm) for the ocean color bands after 12 and 10 years on orbit, respectively. The degradations are significantly different for the two mirror sides of MODIS/Terra (data is collected using both sides of the scan mirror).

MODIS can view the moon at high phase angles and spacecraft roll maneuvers are executed monthly to provide a time series at ∼56° phase angle (a partial moon). One problem with the MODIS lunar calibration is that the ocean color bands (667–869 nm) on the NIR focal plane saturate. Also, all ocean color bands saturate over clouds and those between 490 and 869 nm saturate over other bright targets such as deserts. Avoiding saturation over bright targets while maintaining high SNR and low NEΔL is one of the primary sensor engineering challenges as science objectives become more demanding.

The four MODIS focal planes (Visible, NIR, SWIR/MWIR, and LWIR) have 7–10 bands with 10–40 detectors per band. The MODIS ocean color bands are 412, 443, 531, 547, 667, 678, 748, and 869 nm. The 678 nm band is for chlorophyll-a fluorescence measurements that CZCS and SeaWiFS did not have. The 10 detectors sample 10 adjacent pixels along track allowing for a much slower scan rate (more dwell time) providing higher SNR (∼1.5–3 times higher than SeaWiFS; average of ∼2.1 times). This is a very different strategy to achieve SNR than the SeaWiFS TDI scheme. The downside is the accurate calibration of the 10 detectors in each band. Slight differences leads to striping in the imagery.

MODIS does not have a polarization scrambler and had a prelaunch polarization sensitivity of as high as 5.4% at 412 nm. Methods for accounting for this in the atmospheric correction have been developed [41,42], but uncertainties in the characterization and changes on orbit remain problematic, especially when other sources of error, e.g., response versus scan uncertainty (RVS), are convolved together. Indeed, for MODIS/Terra, the RVS and polarization sensitivity has changed dramatically over time, changes that cannot be accurately estimated using the on-board calibration capabilities such as the solar diffuser. A methodology for correcting these artifacts using concurrent SeaWiFS observations has been demonstrated [48].

While not designed for ocean color applications, the MODIS 1240, 1640, and 2130 nm SWIR bands (500 m) have applications for aerosol corrections over turbid water where the NIR surface reflectance is nonzero. Water absorption is orders of magnitude higher in the SWIR. The SNR values for these bands are low [62], but can be used to some degree of success [63], particularly at higher solar zenith angles (brighter illuminations).

7.4 MERIS

MERIS was an earth-observing spectrometer onboard ESA's ENVISAT satellite (altitude = 800 km, 10:00 am, descending). Remarkably, MERIS did not show significant performance degradation during its 10 years on-orbit.

The primary objective of MERIS was ocean color applications, but land and atmosphere products are an important part of the MERIS product suite as well. MERIS measured (12-bit digitization) the TOA radiances in 15 discrete bands with center wavelengths from 412 to 900 nm, with bandwidths from 3.75 to 20 nm. MERIS operated as a pushbroom scanner with five distinct cameras, pointing at five different angles in the cross-track direction, resulting in a swath width of 1150 km (FOV = 68°). This resulted in global coverage every 3 days. Each camera had its own CCD, with an imaging area of 520 lines for the spectral dimension and 740 columns in the spatial (cross-track) dimension for each CCD. Gratings are used for spectral dispersion. All MERIS bands can measure at a spatial resolution of 300 m (selected acquisitions only), but in the standard mode, 4 × 4 pixels are averaged to obtain an image with 1.2 km pixel size (global data set).

The calibration of MERIS was based on three solar diffusers: a white diffuser viewed frequently (diffuser-1, every 15 days), another white diffuser viewed rarely (diffuser-2, every 3 months), and a diffuser doped with Erbium. The doped diffuser was used for the spectral calibration (every 3 months), the other two to monitor (and correct) the radiometric sensitivity degradation of the instrument. The degradation of diffuser-2 was kept to a minimum by minimizing its exposure to solar radiation. The unavoidable small degradation due to the solar exposure during the rare diffuser-2 calibration events was modeled based on the degradation measured for diffuser-1 and the different solar exposure times for the two solar diffusers.

The MERIS instrument did not have a tilt capability, which leads to a relatively large loss of coverage due to glint contamination (MERIS equator crossing time was 10:00 am, so glint occurs in the eastern part of the scan) because the MERIS swath is narrow compared to MODIS for instance. The swath of the MERIS follow-on sensor, OLCI, will be shifted to the west to reduce glint contamination (this is accomplished by skewing the camera fields of view to the west side of nadir). This will increase the maximum scan angle for the western part of the scan of OLCI. Due to the pushbroom design, pixel growth for high scan angles is minimal relative to MODIS and SeaWiFS.

Each camera is an independent optical system, each with its own polarization scrambler, grating, filters (inverse filter to improve NIR performance and avoid saturation in the visible and a second-order filter to remove the second-order grating reflection), and CCD (thinned/backside illuminated for greater quantum efficiency). The transition region in the image from one camera to the next has been a challenge regarding calibration consistency, in many cases vertical lines appear in the ocean color products at the camera boundaries. The SNRs achieved vary by spectral band from 575 to 1060 for typical ocean radiances (300 m resolution [5]). The MERIS dynamic range includes typical cloud radiances without having to use different gain states.

REFERENCES

[1] G.L. Clarke, G.C. Ewing, C.J. Lorenzen, Spectra of backscattered light from sea obtained from aircraft as a measure of chlorophyll concentration, Science 16 (1970) 1119−1121.
[2] R.E. Eplee Jr., G. Meister, F.S. Patt, B.A. Franz, S.W. Bailey, C.R. McClain, The on-orbit calibration of SeaWiFS, Appl. Opt. 51 (36) (2013a) 8702−8730.
[3] C.R. McClain, Satellite remote sensing: ocean color, in: Encyclopedia of Ocean Sciences, Elsevier Ltd., London, 2009a, pp. 4403−4416.
[4] C.R. McClain, A decade of satellite ocean color observations, Annu. Rev. Marine Sci. 1 (2009b) 19−42.
[5] IOCCG, Mission requirements for future ocean-colour sensors, in: C.R. McClain, G. Meister (Eds.), Reports of the International Ocean-Colour Coordinating Group, Number 13, IOCCG, Dartmouth, Canada, 2012, 98 pp.
[6] R.M. Pope, E.S. Fry, Absorption spectrum (380-700 nm) of pure water. II. Integrating cavity measurements, Appl. Opt. 36 (33) (1997) 8710−8723.
[7] R. Bidigare, M.E. Ondrusek, J.H. Morrow, D.A. Kiefer, In vivo absorption properties of algal pigments, Proc. SPIE, Ocean Opt. X 1302 (1990) 290−302, http://dx.doi.org/10.1117/12.21451.
[8] H.R. Gordon, M. Wang, Retrieval of water-leaving radiance and aerosol optical thickness over the oceans with SeaWiFS: a preliminary algorithm, Appl. Opt. 33 (3) (1994) 443−452.
[9] M. Wang, Remote sensing of the ocean contributions from ultraviolet to near-infrared using the shortwave infrared bands: simulations, Appl. Opt. 46 (2007) 1535−1547.
[10] S.W. Bailey, B.A. Franz, P.J. Werdell, Estimation of near-infrared water-leaving reflectance for satellite ocean color data processing, Opt. Express 18 (7) (2010) 7521−7527.

[11] S. Alvain, C. Moulin, Y. Dandonneau, F.M. Bréon, Remote sensing of phytoplankton groups in case 1 waters from global SeaWiFS imagery, Deep-sea Res. 52 (2005) 1989–2004.
[12] Z.-P. Lee, K. Carder, R. Arnone, M.-X. He, Determination of primary spectral bands for remote sensing of aquatic environments, Sensors 7 (2007) 3428–3441.
[13] P.J. Werdell, et al., Generalized ocean color inversion model for retrieving Marine inherent optical properties, Appl. Opt. 52 (10) (2013) 2019–2037.
[14] Z. Ahmad, C.R. McClain, J.R. Herman, B.A. Franz, E.J. Kwaitkowska, W.D. Robinson, E.J. Bucsela, M. Tzortziou, Atmospheric correction of NO_2 absorption in retrieving water-leaving reflectances from the SeaWiFS and MODIS measurements, Appl. Opt. 46 (26) (2007) 6504–6512.
[15] K. Ding, H.R. Gordon, Analysis of the influence of O_2 A-band absorption on atmospheric correction of ocean-color imagery, Appl. Opt. 34 (12) (1995) 2068–2080.
[16] M. Wang, Validation study of the SeaWiFS oxygen a-band absorption correction: comparing the retrieved cloud optical thicknesses from SeaWiFS measurements, Appl. Opt. 38 (6) (1999) 937–944.
[17] P. Dubuisson, R. Frouin, D. Dessailly, L. Duforêt, J.-F. Léon, K. Voss, D. Antoine, Estimating the altitude of aerosol plumes over the ocean from reflectance ratio measurements in the O_2 A-band, Remote Sens. Environ. 113 (2009) 1899–1911.
[18] B.-C. Gao, Y.J. Kaufman, W. Han, W.J. Wiscombe, Correction of thin cirrus path radiances in the 0.4–1.0 μm Spectral range using the sensitive 1.375 μm Cirrus detecting channel, J. Geophys. Res. 103 (D24) (1998) 32,169–32,176.
[19] B.-C. Gao, P. Yang, W. Han, R.-R. Li, W.J. Wiscombe, An algorithm using visible and 1.38-μm channels to retrieve cirrus cloud reflectances from aircraft and satellite data, IEEE Trans. Geosci. Remote Sens. 40 (8) (2002) 1659–1668.
[20] H.R. Gordon, T. Zhang, F. He, K. Ding, Effects of stratospheric aerosols and thin cirrus clouds on the atmospheric correction of ocean color imagery: simulations, Appl. Opt. 36 (3) (1997a) 682–697.
[21] G. Meister, B.A. Franz, C.R. McClain, Influence of Thin Cirrus Clouds on Ocean Color Products, in: Ocean Remote Sensing: Methods and Applications, vol. 7459, SPIE, San Diego, 2009, 12 pp. http://dx.doi.org/10.1117/12.827272.
[22] IOCCG, Ocean-colour observations from a geostationary orbit, in: D. Antoine (Ed.), Reports of the International Ocean-colour Coordinating Group, Number 12, IOCCG, Dartmouth, Canada, 2012b, 103 pp.
[23] G. Meister, C. McClain, Z. Ahmad, S.W. Bailey, R.A. Barnes, S. Brown, G.E. Eplee, B. Franz, A. Holmes, W.B. Monosmith, F.S. Patt, R.P. Stumpf, K.R. Turpie, P.J. Werdell, Requirements for an Advanced Ocean Radiometer, NASA T/M-2011-215883, NASA Goddard Space Flight Center, Greenbelt, Maryland, 2011, 37 pp.
[24] IOCCG, Minimum requirements for an operational ocean-Colour sensor for the open ocean, in: A. Morel (Ed.), Reports of the International Ocean-colour Coordinating Group, Number 1, IOCCG, Dartmouth, Canada, 1998, 46 pp.
[25] H.R. Gordon, D.K. Clark, J.W. Brown, O.B. Brown, R.H. Evans, W.W. Broenkow, Phytoplankton pigment concentrations in the Middle Atlantic Bight: comparison of ship determinations and CZCS estimates, Appl. Opt. 22 (1) (1983) 20–36.
[26] J.E. O'Reilly, S. Maritorena, B.G. Mitchell, D.A. Siegel, K.L. Carder, S.A. Garver, M. Kahru, C. McClain, Ocean color chlorophyll algorithms for SeaWiFS, J. Geophys. Res. 103 (C11) (1998) 24937–24953.
[27] S.W. Bailey, P.J. Werdell, A multi-sensor approach for the on-orbit validation of ocean color satellite data products, Remote Sens. Environ. 106 (2006) 12–23.

[28] IOCCG, Remote sensing of inherent optical properties: fundamentals, tests of algorithms and applications, in: Z.P. Lee (Ed.), Reports of the International Ocean-colour Coordinating Group, Number 6, IOCCG, Dartmouth, Canada, 2006, 126 pp.
[29] PACE, Pre-aerosol, Clouds, and Ocean Ecosystem, 2012 (PACE) Mission Science Definition Team Report, 274 pp. http://decadal.gsfc.nasa.gov/PACE.html
[30] G. Thuiller, M. Herse, D. Labs, T. Foujols, W. Peetermans, D. Gillotay, P.C. Simon, H. Mandel, The solar spectral irradiance from 200 to 2400 nm as measured by the SOLSPE and EURECA Missions, Solar Phys. 214 (2003) 1–22.
[31] X. Xiaoxiong, N. Che, W.L. Barnes, Terra MODIS on-orbit spectral characterization and performance, IEEE Trans. Geosci. Remote Sens. 44 (8) (2006) 2198–2206.
[32] S. Delwart, R. Preusker, L. Bourg, R. Santer, D. Ramon, J. Fischer, MERIS in-flight spectral calibration, Int. J. Remote Sens. 28 (3–4) (2007) 479–496.
[33] IOCCG, Atmospheric correction for remotely-sensed ocean-colour products, in: M. Wang (Ed.), Reports of International Ocean-color Coordinating Group, Number 10, IOCCG, Dartmouth, Canada, 2010, 78 pp.
[34] W.W. Gregg, F.P. Patt, Assessment of tilt capability for spaceborne global ocean color sensors, IEEE Trans. Geosci. Remote Sens. 32 (4) (1994) 866–877.
[35] E.J. Kwiatkowska, C.R. McClain, Capabilities for extracting phytoplankton diurnal variability using ocean color data from SeaWiFS, MODIS-Terra, and MODIS-Aqua, Int. J. Remote Sens 30 (24) (2009) 6441–6459.
[36] IOCCG, In-flight calibration of ocean-colour sensors, in: R. Frouin, (Ed.), Reports of International Ocean-Color Coordinating Group, Number 15, Dartmouth, Canada, 106 pp., 2013.
[37] B.A. Franz, S.W. Bailey, P.J. Werdell, C.R. McClain, Sensor-independent approach to the vicarious calibration of satellite ocean color radiometry, Appl. Opt. 46 (22) (2007) 5068–5082.
[38] F.E. Nicodemus, J.C. Richmonds, J.J. Hsia, I.W. Ginsberg, T. Lamperis, Geometric Considerations and Nomenclature for Reflectance, vol. 160, US Department of Commerce, National Bureau of Standards, Monogram, 1977, 67 pp.
[39] G. Meister, J. Sun, R. Eplee, F. Patt, X. Xiong, C. McClain, Sun beta angle residuals in solar diffuser measurements of the MODIS ocean bands, Earth Observing Systems XIII, SPIE, Vol. 7081 (2008) 12. http://dx.doi.org/10.1117/12.796291.
[40] S. Maritorena, D.A. Siegel, A. Peterson, Optimization of a semi-analytical ocean color model for global scale applications, Appl. Opt. 41 (15) (2002) 2705–2714.
[41] H.R. Gordon, T. Du, T. Zhang, Atmospheric correction of ocean color sensors: analysis of the effects of residual instrument polarization sensitivity, Appl. Opt. 36 (1997b) 6938–6948.
[42] G. Meister, E.J. Kwiatkowska, B.A. Franz, F.S. Patt, G.C. Feldmam, C.R. McClain, Moderate-resolution imaging spectroradiometer ocean color polarization correction, Appl. Opt. 44 (26) (2005) 5524–5535.
[43] G. Meister, C.R. McClain, Point-spread function of the ocean color bands of the moderate resolution imaging spectrometer on aqua, Appl. Opt. 49 (32) (2010) 6276–6285.
[44] R.A. Barnes, A.W. Holmes, W.E. Esaias, in: S.B. Hooker, E.R. Firestone, J.G. Acker (Eds.), Stray; Light in the SeaWiFS Radiometer, NASA Tech. Memo. 104566, vol. 31, NASA Goddard Space Flight Center, Greenbelt, Maryland, 1995, 76 pp.
[45] S. Delwart, L. Bourg, Radiometric calibration of MERIS, Proc. SPIE. 7474 (2009) 12. http://dx.doi.org/10.1117/12.567989.
[46] J. Sun, X. Xiong, W.L. Barnes, MODIS solar diffuser stability monitor sun view modeling, IEEE Trans. Geosci. Remote Sens. 43 (8) (2005) 1845–1854.

[47] R.E. Eplee, K.R. Turpie, G. Meister, F.S. Patt, G.F. Fireman, B.A. Franz, C.R. McClain Jr., A Synthesis of VIIRS Solar and Lunar Calibrations, in: Observing Systems XVIII, vol. 8866, Proc. SPIE, San Diego, 2013b, 21 pp. http://dx.doi.org/10.1117/12.2024069.
[48] E.J. Kwiatkowska, B.A. Franz, G. Meister, C.R. McClain, X. Xiong, Cross-calibration of ocean color bands from moderate resolution imaging spectroradiometer on terra platform, Appl. Opt. 47 (36) (2008) 6796–6810.
[49] J. Wertz, W. Larson, Space Mission Design and Analysis, Kluwer Academic Publishers, Dordrecht, Netherlands, 1991, 811 pp.
[50] J.L. Mueller, Nimbus-7 CZCS: electronic overshoot due to cloud reflectance, Appl. Opt. 27 (3) (1988) 438–440.
[51] R.E. Eplee, F.S. Patt, R.A. Barnes, C.R. McClain, SeaWiFS long-term solar diffuser reflectance and sensor signal-to-noise analyses, Appl. Opt. 46 (5) (2007) 762–773.
[52] C. Hu, L. Feng, Z. Lee, C. Davis, A. Mannino, C. McClain, B. Franz, Dynamic range and sensitivity requirements of satellite ocean color sensors: learning from the past, Appl. Opt. 51 (25) (2012) 6045–6062.
[53] G. Vasilescu, Electronic Noise and Interfering Signals, Springer-Verlag, Berlin, 2005, 709 pp.
[54] A. Rogalski, Infrared Detectors, second ed., CRC Press, Boca Raton, Florida, 2011, 876 pp.
[55] W. Kester, ADC noise: the good, the bad and the ugly. Is no noise good noise? Analog Dialog 40 (2) (2006) 1–5.
[56] C. Poivey, J.L. Barth, K.A. LaBel, G. Gee, H. Safren, In-flight observations of long-term single-event effect (SEE) performance on Orbview-2 solid state recorders (SSR), in: Radiation Effects Data Workshop 2003, IEEE, vol. 102(107), 2003, pp. 21–25. http://dx.doi.org/10.1109/REDW, 1281357, 2003.
[57] H. Loisel, L. Duforet, D. Dessailly, M. Chami, P. Dubuisson, Investigation of the variations in water leaving polarized reflectance from the POLDER satellite data over two biogeochemical contrasted oceanic areas, Opt. Exp. 16 (17) (2008) 12905–12918.
[58] R.H. Evans, H.R. Gordon, CZCS "System calibration": a retrospective examination, J. Geophys. Res. 99 (C4) (1994) 7293–7307.
[59] R.A. Barnes, A.W. Holmes, W.L. Barnes, W.E. Esaias, C.R. McClain, in: S.B. Hooker, E.R. Firestone, J.G. Acker (Eds.), SeaWiFS Prelaunch Radiometric Calibration and Spectral Characterization, NASA Tech. Memo. 104566, vol. 23, NASA Goddard Space Flight Center, Greenbelt, Maryland, 1994, 55 pp.
[60] H.R. Gordon, Remote sensing of ocean color: a methodology of dealing with broad spectral bands and significant out-of-band response, Appl. Opt. 34 (1995) 8363–8374.
[61] C. Hu, K.L. Carder, F.E. Muller-Karger, How precise are SeaWiFS ocean color estimates? implications of digitization-noise errors, Remote Sens. Environ. 76 (2) (2001) 239–249.
[62] P.J. Werdell, B.A. Franz, S.W. Bailey, Evaluation of shortwave infrared atmospheric correction for ocean color remote sensing in Chesapeake Bay, Remote Sens. Environ. 114 (2010) 2238–2247.
[63] M. Wang, W. Shi, Estimation of ocean contribution at MODIS near-infrared wavelengths along the East Coast of the U.S.: two case studies, Geophys. Res. Lett. 32 (2005) L13606. http://dx.doi.org/10.1029/2005GL022917, 5 pp.

Chapter 2.2

On Orbit Calibration of Ocean Color Reflective Solar Bands

Robert E. Eplee, Jr [1,2,*] **Sean W. Bailey** [1,3]
[1] *Ocean Biology Processing Group, NASA Goddard Space Flight Center, Greenbelt, MD, USA;*
[2] *Science Applications International Corporation, Beltsville, MD, USA;* [3] *FutureTech Corporation, Greenbelt, MD, USA*
Corresponding author: E-mail: robert.e.eplee@nasa.gov

Chapter Outline

1. Introduction	121
2. Solar Calibration	124
2.1 SD Degradation	125
2.2 SD Radiometric Response Trends	126
2.3 SNR on Orbit	128
2.4 Uncertainties in the Solar Calibration Data	128
3. Lunar Calibrations	128
3.1 ROLO Photometric Model of the Moon	129
3.2 Lunar Radiometric Response Trending	130
3.3 Uncertainties in Lunar Calibration	131
3.4 Lunar Calibration Intercomparisons	133
4. Spectral Calibration of Grating Instruments	135
5. Vicarious Calibration	137
5.1 NIR/SWIR Band Calibration	139
5.2 Visible Band Calibration	140
5.3 Alternative Approaches	142
6. On-orbit Calibration Uncertainties	142
6.1 Accuracy	143
6.2 Long-term Stability of the TOA Radiances	143
6.3 Precision of the TOA Radiances	144
6.4 Combined Uncertainty Assessment	144
7. Comparison of Uncertainties Across Instruments	145
8. Summary of On-orbit Calibration	149
References	150

1. INTRODUCTION

One goal of climate change research is to discern small secular trends in geophysical processes that have comparatively large daily, seasonal, annual, or longer-scale periodic signals. Accordingly, a climate data record (CDR) for ocean color data is defined to be: a time series of measurements of sufficient length, consistency, and continuity that will allow the determinations of climate variability and change [1]. This endeavor requires that the remote sensing data be collected from satellite instruments with long-term radiometric stability, where the radiometric uncertainty in the data is less than the magnitude of the possible climate change signal. For ocean color data, the radiometric requirements are 5% absolute uncertainty and 1% relative uncertainty (uncertainty over time) for the water-leaving radiances [2]. Because ocean surface reflectances are low, approximately 90% of the top-of-the-atmosphere (TOA) signal observed by ocean color satellite instruments arises from scattering of sunlight by gases and aerosols within the atmosphere. The ocean color atmospheric correction algorithm must remove the atmospheric signal to yield water-leaving radiances. Due to amplification of errors in the sensor calibration by the atmospheric correction, the 1% uncertainty over time requirement on water-leaving radiances translates into a 0.1% long-term stability requirement for the TOA radiances [3]. This stability goal assumes incorporation of radiometric corrections for short-term orbital and seasonal response variations. Uncertainties in the sensor calibration and atmospheric correction algorithm necessitate a vicarious calibration of the sensor/atmospheric correction algorithm system to meet the accuracy requirements for the TOA radiances [4].

The long-term radiometric stability requirements for ocean color sensors require the implementation of robust on-orbit calibration programs for the reflective solar bands of these instruments. The primary components of these calibration programs are solar calibrations, through observations of sunlight reflected by solar diffusers (SDs), and lunar calibrations. SD observations provide frequent calibration opportunities (once per orbit to once per day or less frequent observations) while lunar observations provide monthly calibration opportunities. Spectral instruments such as MEdium Resolution Imaging Spectrometer (MERIS) also require on on-orbit spectral calibration through observation of solar or atmospheric spectral absorption lines or of spectral absorption lines from doped diffusers. The ocean color instruments for which robust on-orbit calibration program have been developed are Sea-viewing Wide Field-of-view Sensor (SeaWiFS) (1997–2010), Terra MODerate resolution Imaging Spectroradiometer (MODIS) (1999–Present), Aqua MODIS (2002–Present), MERIS (2002–2012), and Suomi National Polar-orbiting Partnership (NPP) Visible Infrared Imaging Radiometer Suite (VIIRS) (2011–Present). The reflective solar bands of SeaWiFS, MODIS, MERIS, and VIIRS are shown in Table 1. An overview of the solar, lunar, and vicarious calibration programs will be presented in the following pages.

Solar calibration is the first method for monitoring the on-orbit radiometric performance of the reflective solar bands. The change in radiometric sensitivity of the ocean color instrument over time is computed from the SD observation time series, corrected for changes in the diffuser bidirectional reflectance distribution function (BRDF) by observations of the solar diffuser

TABLE 1 Reflective Solar Bands of SeaWiFS, MODIS, MERIS, and VIIRS

SeaWiFS	λ (nm)	MODIS	λ (nm)	MERIS	λ (nm)	VIIRS	λ (nm)
Band 1	412	Band 8	412	Band 1	412.5	M1	412
Band 2	443	Band 9	443	Band 2	442.5	M2	445
		Band 3	469				
Band 3	490	Band 10	488	Band 3	490	M3	488
Band 4	510	Band 11	531	Band 4	510		
Band 5	555	Band 12	551	Band 5	560	M4	555
		Band 4	555				
		Band 1	645	Band 6	620	I1	640
Band 6	670	Band 13	667	Band 7	665	M5	672
		Band 14	678	Band 8	681.25		
				Band 9	705		
				Band 10	753.75		
Band 7	765	Band 15	748	Band 11	760	M6	746
Band 8	865	Band 2	858	Band 13	865	M7	865
		Band 16	869			I2	865
				Band 14	885		
				Band 15	900		

stability monitor (SDSM). This methodology yields a calibration of the instrument on a per-band, per-detector, and per-mirror side basis. While the SeaWiFS SD is an aluminum plate coated with YB-71 paint [5], the SDs for MODIS [5–7] and VIIRS [8] are Spectralon panels placed behind attenuation screens. The SDSM for MODIS and VIIRS are 9- and 8-channel radiometers with wavelengths corresponding to the RSB bands with a reference channel at 935 nm.

Lunar calibration is the second method for monitoring the radiometric response of the reflective solar bands on orbit. Once a month, the spacecraft is either pitched across the Moon at a phase angle of 7° (SeaWiFS) [5] or rolled to observe the Moon through the Space View at a nominal phase angle of 55° (MODIS) [9] or 51° (VIIRS) [10]. The resulting lunar images are processed into disk-integrated lunar irradiances, which are then compared with the United States Geological Surver (USGS) Robotic Lunar Observatory (ROLO) photometric model of the moon [11] to yield a time series of instrument measurements to model prediction ratios, which represent the radiometric response of the instrument over time. This methodology yields a calibration of the instrument on a per-band, per-mirror-side basis. Changes in the lunar observing geometry over the course of a year limit the number of lunar observations available to spacecraft rolls (MODIS or VIIRS) to 8–9 per year. Lunar calibrations using the ROLO model allow the on-orbit calibration of different ocean color instruments to be compared without the necessity of simultaneous observations or contemporaneous missions.

Spectral calibration of grating instruments is performed by observing spectral absorption lines from Earth or instrument targets [12a,12b]. The atmospheric oxygen absorption line at 760 nm provides a spectral reference for bright desert scenes with clear skies. Solar Fraunhofer lines at 393, 485, 588, 655, 855, and 867 nm provide spectral references for observations of SD. Labsphere provides Spectralon diffusers doped with rare earth elements for spectral references at 378–381 nm, 521–523 nm, and 1009–1013 nm (erbium oxide) and at 447–449 nm and 453–456 nm (holmium oxide) [13]. Spectral calibration requires the development of an instrument spectral model to convert the spectral observations into a wavelength calibration for the instrument.

Unlike SeaWiFS and MODIS, VIIRS views the SD and the moon at the same incidence angle on the mirror [10]. This design allows the VIIRS solar-derived and lunar-derived radiometric responses to be compared directly. Thus, two independent calibrations of the VIIRS reflective solar bands are afforded by the solar and lunar calibration time series.

Once a stable TOA instrument radiometric performance has been established for an ocean color instrument, the generation of ocean color CDRs requires that the TOA radiances be vicariously calibrated to yield accurate water-leaving radiances or remote sensing reflectances. The vicarious calibration mitigates biases in the instrument calibration, in the calibration of the in situ radiometer, and the atmospheric correction algorithm. The evolving strategies for vicariously calibrating ocean color instruments is discussed in Section 5.

One requirement for producing CDRs from ocean color data is a quantification of the uncertainties in the calibrated TOA radiances. The error budgets for water-leaving radiances or remote-sensing reflectances begin with the uncertainties in the TOA radiances. The on-orbit calibration data from which these uncertainties can be characterized include solar calibrations, lunar calibrations, spectral calibrations, and vicarious calibrations. The uncertainties in the TOA radiances can be addressed in terms of accuracy (biases in the measurements), stability (repeatability of the measurements over time), and precision (scatter in the measurements).

2. SOLAR CALIBRATION

Satellite instruments view sunlight reflecting off of SDs as the spacecraft pass over the earth's pole, moving from shadow into sunlight. SD observations provide the monitor for the radiometric response of the instrument, while the ratio of SDSM diffuser observations to SDSM solar observations provide the monitor for the diffuser BRDF. The SD observations also provide a measure of the signal-to-noise ratio (SNR) for instruments on orbit, though the SNR is computed at diffuser radiance s rather than typical radiances. The VIIRS solar calibrations [10] will be used as a case study, though a similar set of observations and analyses are performed for the MODIS instruments.

Geometric corrections that are applied to the solar calibration data include instrument-sun distances, transmission functions for the SD and SDSM attenuation screens, and the diffuser BRDF. The prelaunch characterizations of the screen transmission functions and diffuser BRDF were performed over a limited range of observations. These functions have been evaluated on orbit through the spacecraft yaw maneuvers, with resulting updates to the combined screen transmission functions and diffuser BRDF. Uncertainties in these functions are still a major factor in the uncertainty in the solar observations.

2.1 SD Degradation

The SDSM time series are used to monitor the changes in the diffuser BRDF. The SDSM H-factor is the ratio of the SD measurements to the direct solar measurements [10]:

$$h(\lambda, t) = \frac{dn_{\text{sol}}(\lambda, t)}{dn_{\text{sun}}(\lambda, t)} \frac{\cos(\phi(t)) \tau_{\text{sdsm}}(\lambda)}{\tau_{\text{sds}}(\lambda) \text{BRDF}(\lambda, t_0) \Omega_{\text{sdsm}}} \quad (1)$$

where:

λ = wavelength;
t = time;
ϕ = incidence angle of the Sun on the diffuser;
dn_{sol} = dark-subtracted counts of the diffuser observations;
dn_{sun} = dark-subtracted counts of the solar observations;
τ_{sdsm} = SDSM screen transmission function;
τ_{sds} = SD screen transmission function;
BRDF = SD BRDF; and
Ω_{sdsm} = cone angle of the SDSM view of the SD.

The time-dependent BRDF correction becomes:

$$H(\lambda, t) = \frac{h(\lambda, t)}{h(\lambda, t_0)} \quad (2)$$

The SDSM trends over time are subject to measurement noise and instrument artifacts. The diffuser BRDF is assumed to be invariant with time at 935 nm, so the SDSM reference channel at 935 nm is used to normalize the SDSM trends to correct for these artifacts.

2.2 SD Radiometric Response Trends

The radiometric gain over time (the F-factor) is computed as the ratio of the predicted solar irradiance on the diffuser (L_{pred}) to the observed solar irradiance reflected by the diffuser (L_{obs}) [10]:

$$F(\lambda, t) = \text{RVS}(\theta_{\text{sd}}, \lambda) \cos(\phi(t)) \frac{L_{\text{pred}}(\lambda, t)}{L_{\text{obs}}(\lambda, t)} \quad (3)$$

where:

θ_{sd} = incidence angle of the Sun on the half-angle mirror; and
RVS = response-vs-scan angle function.

The predicted solar irradiance is:

$$L_{\text{pred}}(\lambda, t) = \frac{E_{\text{sun}}(\lambda)}{R_{\text{se}}^2(t)} \tau_{\text{sds}}(\lambda) \text{BRDF}(\lambda, t_0) H(\lambda, t) \quad (4)$$

where:

E_{sun} = solar irradiance; and
R_{se} = earth–sun distance.

The observed solar irradiance is:

$$L_{\text{obs}}(\lambda, t) = c_0(\lambda, t) + c_1(\lambda, t) dn_{\text{sd}}(\lambda, t) + c_2(\lambda, t) dn_{\text{sd}}^2(\lambda, t) \quad (5)$$

where:

c_i = counts-to-radiance conversion coefficients; and
dn_{sd} = dark-subtracted counts of the diffuser observations.

In terms of instrument measurements, the F-factor becomes:

$$F(\lambda, t) = \frac{E_{\text{sun}}(\lambda)}{R_{\text{se}}^2(t)} \frac{\text{RVS}(\theta_{\text{sds}}, t) \cos(\phi(t)) \tau_{\text{sds}}(\lambda) \text{BRDF}(\lambda, t_0)}{c_0(\lambda, t) + c_1(\lambda, t) dn_{\text{sd}}(\lambda, t) + c_2(\lambda, t) dn_{\text{sd}}^2(\lambda, t)} H(\lambda, t) \quad (6)$$

The inverse of the F-factor time series is the radiometric response over time. Figure 1 shows the SDSM-derived H-factor time series and the inverse of the F-factor time series (the radiometric response) for VIIRS.

Several approaches are possible for incorporating the F-factor time series into the calibration of Earth data. The measured F-factors can be used directly to calibrate the data on a per-orbit basis. A methodology of smoothing the F-factors can be adopted, as with the autocalibration approach that National Atmospheric and Oceanic Administration (NOAA) has adopted for their VIIRS data processing. The National Aeronautics and Space Administration (NASA) MODIS Characterization Support Team (MCST) uses solar calibrations as the primary radiometric correction for MODIS data processing. The NASA Ocean Biology Processing Group (OBPG) constructs lookup

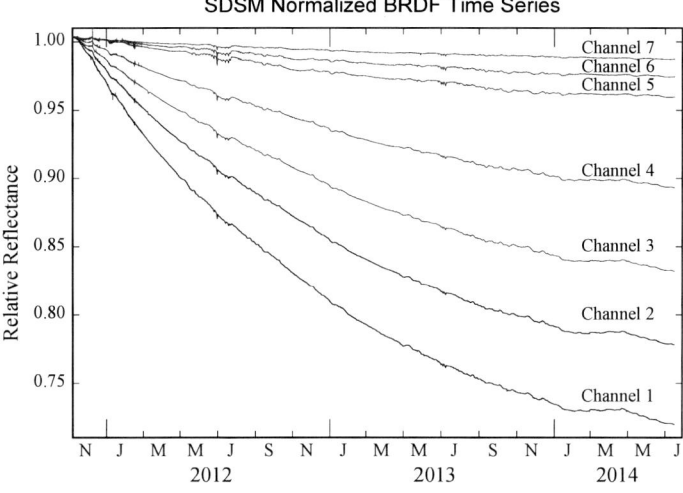

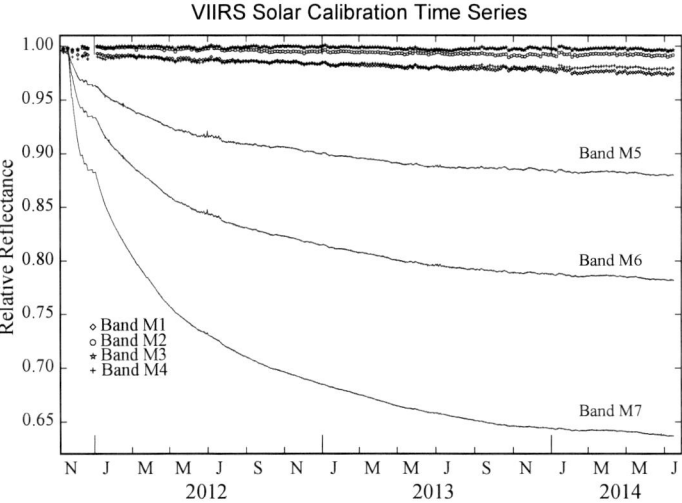

FIGURE 1 VIIRS solar calibration time series.

tables for VIIRS from fits to the F-factor time series, with an F-factor computed for each day over the time range of the lookup table. These F-factor fits provide the radiometric correction for the OBPG VIIRS data processing. These fits are typically exponential plus linear functions of time or simultaneous exponential functions of time [10]:

$$f(\lambda, t) = A_0(\lambda) - A_1(\lambda)\left[1 - e^{-A_2(\lambda)(t-t_0)}\right] - A_3(\lambda)(t - t_0) \quad (7)$$

$$f(\lambda, t) = A_0(\lambda) - A_1(\lambda)\left[1 - e^{-A_2(\lambda)(t-t_0)}\right] - A_3(\lambda)\left[1 - e^{-A_4(\lambda)(t-t_0)}\right] \quad (8)$$

where A_i is the fit coefficients. The goal of the various F-factor implementations is to provide for a methodology of calibrating earth data that can be maintained and updated with minimal effort.

2.3 SNR on Orbit

The solar calibration data provide a stable radiance source that allows computation of the instrument SNR on orbit [5,14]:

$$\text{SNR}(\lambda, t) = \frac{<L_{\text{obs}}(\lambda, t)>}{\sigma(<L_{\text{obs}}(\lambda, t)>)} \quad (9)$$

The difficulty in evaluating these on-orbit SNRs is that the on-orbit SNRs are computed at the diffuser radiance, while the prelaunch SNRs are computed at typical radiances.

2.4 Uncertainties in the Solar Calibration Data

Uncertainties in the solar calibration data include both uncertainties in the prelaunch characterization of the instrument and uncertainties in the on-orbit calibration. The predominant prelaunch uncertainties are in the characterization of the diffuser BRDF, the SD screen transmission, and the SDSM screen transmission. The predominant on-orbit uncertainties are in the characterization of the combined diffuser BRDF/SD transmission function and in the SDSM screen transmission. Additional uncertainties in the on-orbit data include periodic residuals in the geometric corrections and any inhomogeneous degradation in the diffuser BRDF. As the length of the solar calibration times series increases to several years, the periodic residuals in the time series can possibly be identified and corrected, thus reducing the uncertainties in the data.

3. LUNAR CALIBRATIONS

Several lunar calibration strategies have been developed for the on-orbit calibration of ocean color instruments, each of which use nominally monthly observations. For SeaWiFS, the spacecraft is pitched across the Moon to observe the Moon through the Earth View at a phase angle of approximately 7°, thus maximizing the illuminated lunar disk, while avoiding the opposition effect [5]. SeaWiFS has a single detector in the along-track direction for each band, yielding elongated, oversampled images of the moon. The observations require an oversampling correction that is derived from the ratio of the apparent to actually size of the lunar image.

For MODIS [9] and VIIRS [10], the spacecraft is rolled to observe the moon through the Space View at a target phase angle (+55° for Terra MODIS, −55° for Aqua MODIS, −51° for VIIRS). Occasionally, the Moon moves through the space view as the spacecraft/Moon geometries coincide, though the phase angle is larger than the target phase angle. MODIS and VIIRS have multiple detectors in the along-track direction, yielding full disk images of the moon from multiple detectors with no oversampling. VIIRS on-orbit pixel aggregation in the along-scan direction for single-gain bands does require an oversampling correction to account for the aggregation.

3.1 ROLO Photometric Model of the Moon

The ROLO model is used to normalize the lunar calibration time series for variations in observing geometry: instrument/Moon distances, Sun/Moon distances, phase and libration angles [11,15,16]. The ROLO model predicts the reflectances of the moon based on the phase and libration angles of the observation, converts the lunar reflectances to irradiances using the solar reference irradiances, uses instrument RSR's to convert the lunar irradiances to the instrument band passes, then uses the time of the observation and the position of the spacecraft to normalize the lunar irradiances to the values seen by the instrument.

The ROLO model requires disk-integrated lunar irradiances as input:

$$L_T(\lambda, t) = K_{os}(\lambda) \sum_{\text{pixels}} K_c(\lambda) dn_{\text{moon}}(\lambda, t) \tag{10}$$

where:

K_{os} = oversampling correction (if required);
K_c = counts-to-radiance conversion factors; and
dn_{moon} = dark-subtracted counts of the lunar pixels.

The disk-integrated lunar irradiances are computed from the lunar radiances and the instantaneous field of view (IFOV) of the instrument:

$$E_{\text{inst}}(\lambda, t) = \text{IFOV}_{\text{along-scan}} \text{IFOV}_{\text{along-track}} L_T(\lambda, t) \tag{11}$$

The radiometric output of the ROLO model is the ratio of the irradiance measured by the instrument to the irradiance predicted by the model (E_{rolo}):

$$P(\lambda, t) = \frac{E_{\text{inst}}(\lambda, t)}{E_{\text{rolo}}(\lambda)} - 1 \tag{12}$$

The ROLO model computes an effective albedo for the moon based on the phase and libration angles of the observations. The model uses the solar irradiance to convert the albedo into irradiance, then uses an Apollo soil reference spectrum and the instrument relative spectral response function to convert the irradiance to the instrument bandpass. Finally, the model uses the sun–moon ($R_{\text{sun-moon}}$) and instrument-moon ($R_{\text{inst-moon}}$) distances to normalize the irradiance to the geometry of the observation. Thus, the model lunar irradiances is [5]:

$$E_{\text{rolo}}(\lambda, r, t) = \frac{E_{\text{sun}}(\lambda) A_{\text{moon}}(\lambda, r, t)}{\Omega_{\text{moon}}(r) K_r(r)} \quad (13)$$

where:

AU = astronomical unit;
MLD = mean lunar distance (earth-moon distance); and
Ω_{moon} = solid angle of the Moon = 6.4177E-05 sr.

$$K_r(r) = \left(\frac{R_{\text{sun-moon}}(r)}{\text{AU}}\right)^2 \left(\frac{R_{\text{inst-moon}}(r)}{\text{MLD}}\right)^2 \quad (14)$$

where:

AU = astronomical unit; and
MLD = mean lunar distance (earth–moon distance).

3.2 Lunar Radiometric Response Trending

The lunar calibration time series used to monitor the instrument radiometric performance over time takes the form:

$$\frac{E_{\text{inst}}(\lambda, t)}{E_{\text{rolo}}(\lambda, r, t)} = P(\lambda, r, t) + 1 \quad (15)$$

Fits to these lunar time series are used in the on-orbit instrument calibration. These fits are typically exponential plus linear functions of time or simultaneous exponential functions of time, the same fits as used in the solar calibration analysis [5]:

$$f(\lambda, t) = A_0(\lambda) - A_1(\lambda)\left[1 - e^{-A_2(\lambda)(t-t_0)}\right] - A_3(\lambda)(t - t_0) \quad (16)$$

$$f(\lambda, t) = A_0(\lambda) - A_1(\lambda)\left[1 - e^{-A_2(\lambda)(t-t_0)}\right] - A_3(\lambda)\left[1 - e^{-A_4(\lambda)(t-t_0)}\right] \quad (17)$$

Several approaches are possible for incorporating the fits to the lunar calibration time series into the calibration of Earth data. The OBPG used the fits to the lunar data as the radiometric correction for SeaWiFS data processing. The SeaWiFS lunar calibration item series are shown in Figure 2. The MCST uses lunar calibrations to track the MODIS RVS trending. The OBPG uses the fits to the lunar calibration time series to validate the radiometric corrections from the solar-derived F-factors.

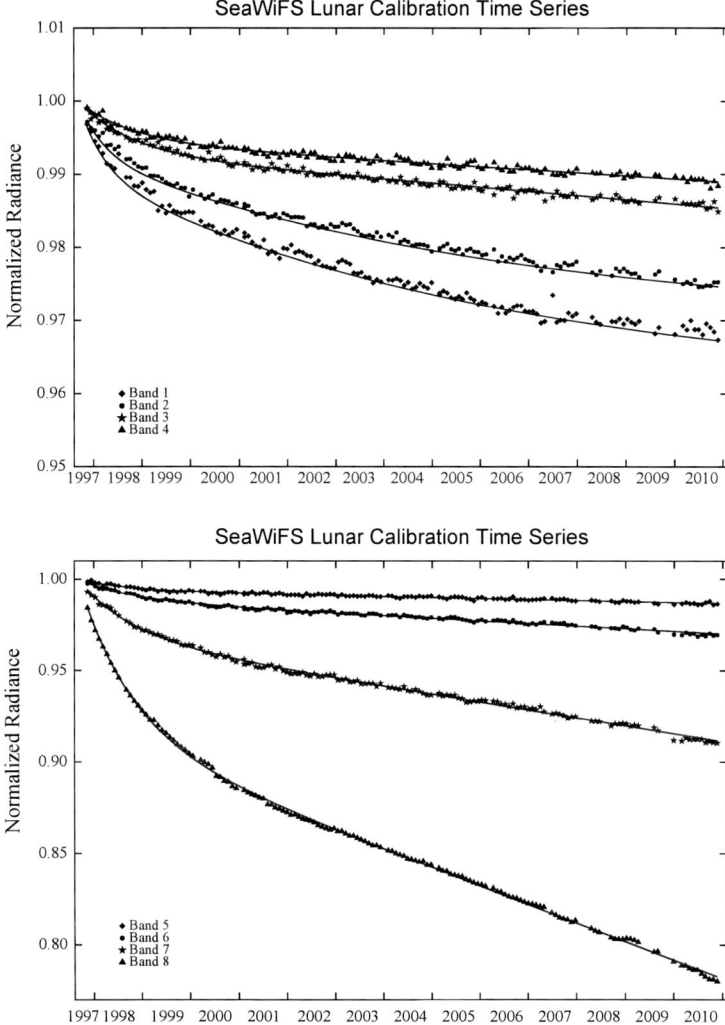

FIGURE 2 SeaWiFS lunar calibration time series.

3.3 Uncertainties in Lunar Calibration

Since lunar observations are obtained on a monthly basis, several years of lunar observations are typically required to obtain sufficient lunar observations to discern the long-term radiometric trends in the lunar data. Nevertheless, the scatter in the lunar observations represents a first take on the uncertainties in the lunar calibrations. The magnitude of the scatter in the lunar observations (the output of the ROLO model) is comparable for SeaWiFS, MODIS, and VIIRS. A residual analysis allows further investigation of uncertainties in the lunar data. As noted above, time series fits to the lunar data should model the radiometric performance of the instrument.

The residuals for SeaWiFS are highly correlated between bands, and their primary source being errors in the oversampling correction of the lunar images [5]. Due to the high correlation of the residuals, a coherent noise correction and be computed for and applied to the SeaWiFS lunar time series. The coherent noise of 0.5% can be reduced to a corrected noise of 0.1%.

The residuals for MODIS are not as correlated as those for SeaWiFS [17]. The noise source is presumably a systematic observational error. The coherent noise of 0.5% can be reduced to a corrected noise of 0.2%

The residuals for VIIRS are periodic, increase with increasing wavelength, and are correlated with the subspacecraft libration effect [10]. The ROLO model derives corrections for the libration of the lunar surface at the subsolar and subspacecraft points. The subsolar libration corrections are wavelength dependent, while the subspacecraft corrections are independent of wavelength. To assess possible residual libration correlations in the lunar time series, the OBPG fit the lunar time series with exponential plus linear or double exponential fits of time plus additional lunar functions of the longitude and latitude of the subspacecraft point. The fits have the form of:

$$f(\lambda,t) = A_0(\lambda) - A_1(\lambda)\left[1 - e^{-A_2(\lambda)(t-t_0)}\right] - A_3(\lambda)(t-t_0) + A_4(\lambda)\theta + A_5(\lambda)\phi \tag{18}$$

$$f(\lambda,t) = A_0(\lambda) - A_1(\lambda)\left[1 - e^{-A_2(\lambda)(t-t_0)}\right] - A_3(\lambda)\left[1 - e^{-A_4(\lambda)(t-t_0)}\right] \\ + A_5(\lambda)\theta + A_6(\lambda)\phi \tag{19}$$

where:

θ = subspacecraft libration longitude; and
ϕ = subspacecraft libration latitude.

While the ROLO model provides the primary, wavelength-independent subspacecraft point libration correction for the lunar time series, the wavelength-dependent libration effect requires this second-order empirical correction. The size of the residual libration effect varies with wavelength, from 0.7% at 412 nm to 0.5% at 555 nm to 0.2% at 865 nm. Figure 3 shows the VIIRS lunar calibration time series with the radiometric plus libration fits for bands M1–M4 and M5–M7.

The two MODIS instruments observe the Moon at comparable phase angles to VIIRS. However, the observational noise in the MODIS lunar data sets is larger than in the VIIRS data, so these residual libration effects are just discernable in the MODIS lunar data. The SeaWiFS lunar data obtained at 7° phase angle have significantly higher disk-integrated irradiances than do the MODIS and VIIRS observations, so the residual libration effect is not discernable in the SeaWiFS lunar data.

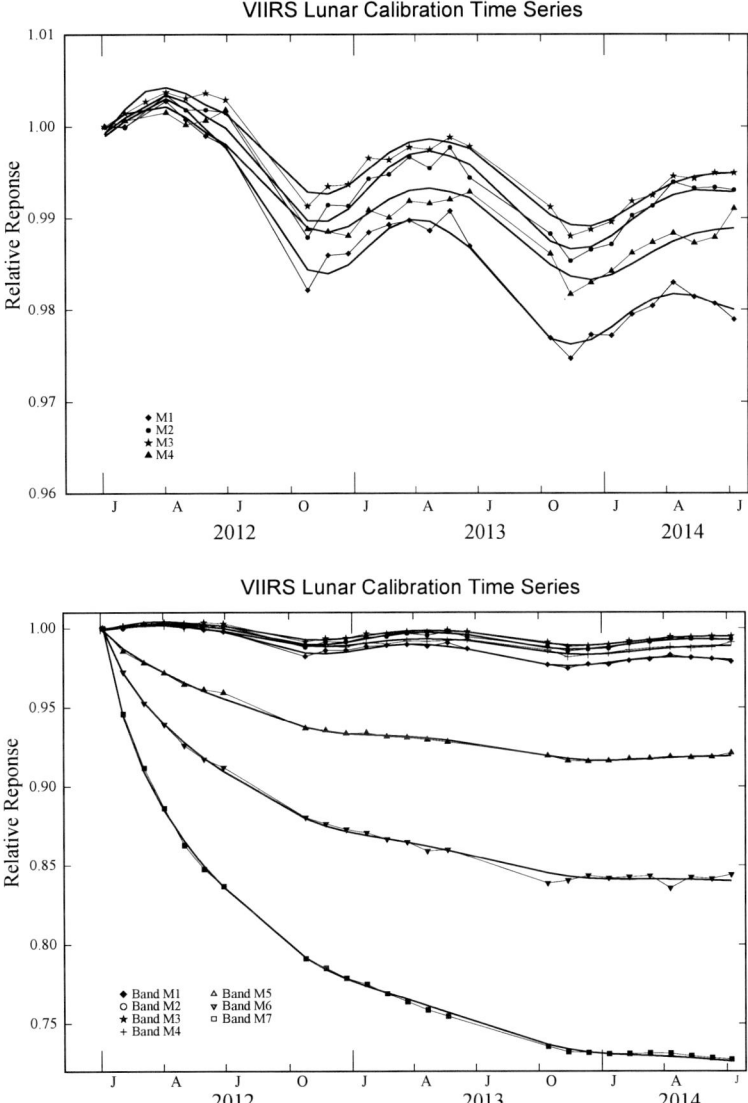

FIGURE 3 VIIRS lunar calibration time series.

3.4 Lunar Calibration Intercomparisons

Observations of the Moon provide a unique way of intercomparing the radiometric performance of two or more ocean color instruments on orbit [17]. The on-orbit radiometric corrections are applied to the lunar data, thus allowing comparisons to be made with stable lunar irradiances. Use of the

ROLO model for the intercomparison removes the requirement of contemporaneous instrument operational lifetimes, aiding in the production of consistently calibrated data sets from these instruments. This intercomparison technique has been applied to SeaWiFS, Terra MODIS, Aqua MODIS, and NPP VIIRS observations. The intercomparison includes all eight SeaWiFS bands, the MODIS visible and near-infrared bands that do not saturate on the Moon, and VIIRS bands M1–M7 [5,17].

The intercomparison over wavelength using the ROLO model allows the on-orbit calibration biases between the instruments to be determined. The data sets used for this analysis are the mission-long average of the ROLO residuals for the primary lunar calibration data set for each instrument:

1. The monthly observations for SeaWiFS at $-7°$ and $+7°$ phase angle.
2. The scheduled observations for Terra MODIS at $+55°$ phase angle.
3. The scheduled observations for Aqua MODIS at $-55°$ phase angle.
4. The scheduled observations for VIIRS at $-51°$ phase angle.

The results of the intercomparisons are shown in Figure 4 and the relative biases between the instruments are derived for Table 2. When corrected for the residual libration effects, the uncertainties in the VIIRS lunar time series are comparable to those achieved for SeaWiFS after its coherent noise correction.

The full lunar data sets for SeaWiFS, MODIS, and VIIRS lunar observations include the primary observations plus the SeaWiFS high phase angle observations and the MODIS and VIIRS unscheduled observations.

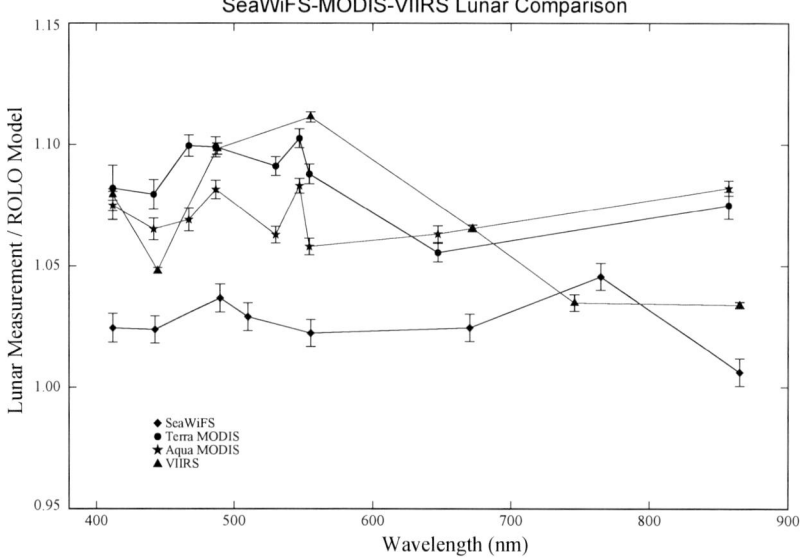

FIGURE 4 Lunar cross calibration.

TABLE 2 Instrument Relative Calibration Biases

Instrument Comparison	Calibration Bias (%)	Uncertainty (%)
SeaWiFS/Terra MODIS	3−8	1.4
SeaWiFS/Aqua MODIS	3−8	1.3
Terra MODIS/Aqua MODIS	1−3	1.3
SeaWiFS/VIIRS	1−9	1.3
Terra MODIS/VIIRS	1−4	1.3
Aqua MODIS/VIIRS	1−6	1.2

Comparison of the full data sets allows the observational scatter in the data sets to be compared over a phase angle range of −85° to +85°. The intercomparison results at 412 nm are shown in Figure 5. The different data sets are represented by different colored dots in the plots. The second plot in the figure shows the mean residuals for a number of phase angle ranges, which allow the significance of the variations as a function of phase angle to be examined. The magnitude of the observed scatter in the lunar residuals is consistent for all four instruments.

4. SPECTRAL CALIBRATION OF GRATING INSTRUMENTS

Spectral calibration on orbit is necessary for grating instruments. The grating dispersion can vary with temperature and may change over time. Two methods of spectral calibration have been developed for MERIS [12a,12b]. The first uses spectral line absorption features, namely the O_2-A feature in the earth's atmosphere at 760 nm, and the solar Fraunhofer lines at 393, 485, 588, 655, 855, and 867 nm. The second uses Spectralon diffusers doped with rare earth elements (erbium and holmium) as spectral references at 380, 448, 454, 522, and 1010 nm.

For the O_2-A spectral line, clear-sky Earth scenes over bright targets, such as high deserts, are analyzed to define the apparent wavelength of the spectral line. For the solar spectral lines, observations of the standard Spectralon SD are analyzed to define the apparent wavelengths of the spectral lines. For doped Spectralon diffusers, solar calibrations are performed using the doped diffusers in place of the standard diffusers. These modified solar observations are analyzed to define the apparent wavelengths of the spectral lines. For each of these spectral calibration methods, instrument-specific spectral models provide the apparent wavelengths of the spectral lines. The primary source of uncertainty in these calibrations is the accuracy with which the spectral line shapes can be determined in the instrument data.

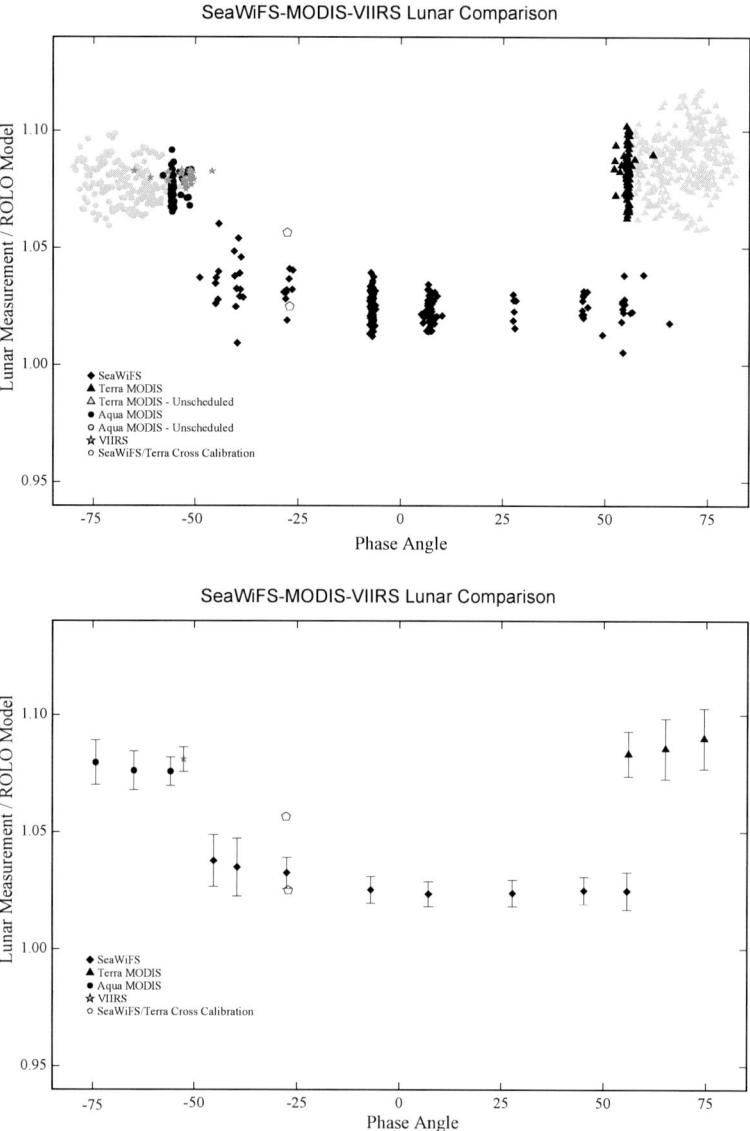

FIGURE 5 Lunar calibration residuals vs phase angle. For each instrument, the different colors of each data set are shown in the rows along the bottom of the plots. The scheduled observations are the nominal lunar calibrations. The unscheduled observations are instances when the moon was visible in the space view. The cross-calibration observation was a simultaneous observation of the moon by SeaWiFS and Terra MODIS.

5. VICARIOUS CALIBRATION

Early on in the era of satellite-based ocean color measurements [18,19] it was understood that the uncertainties in the measured TOA reflectance required to retrieve accurate water-leaving reflectances were not achievable with a prelaunch instrument calibration. At best, the current instrument calibration uncertainty for modern ocean color sensors is on the order of 2–3%. As the atmospheric contribution is on the order 80–99% of the TOA signal [18], an uncertainty of 2% at TOA translates into approximately 10–20% uncertainty on the water-leaving reflectance. Given a maximum of a 5% uncertainty on water-leaving reflectance in the blue spectral region of oligo- and mesotrophic waters in order to derive chlorophyll to an uncertainty of less than 30%, it was clear that a means of reducing the instrument uncertainties was necessary. Several methods have been employed to address this conundrum, including the use of on-board calibrated light sources and SDs, and while these have their place in the overall calibration to monitor the stability of the space sensor on orbit, the most effective method has been what is referred to as vicarious calibration.

For our purposes, the word vicarious is defined as "acting or done for another". The process of deriving calibration adjustment factors postlaunch is referred to as a "vicarious" calibration because the on-orbit calibration is not performed by the direct use of a National Metrological Institute (NMI, e.g., the U.S. National Institute of Standards and Technology) traceable source, as it was prelaunch. Often the source is an in situ radiometer, ideally itself with traceability to an NMI standard, but this is not always the case (e.g., the vicarious calibration in the near infrared or NIR). In addition to the use of a secondary (or even tertiary) source, vicarious calibration incorporates the atmospheric correction algorithm and is therefore a "system-level" calibration [18,20]. This is a key component to the vicarious calibration process, as the accurate retrieval of surface reflectances requires a means of removing the atmospheric path radiance from the TOA signal. The complexity of the atmospheric correction algorithm can by itself introduce biases and increase the uncertainties beyond acceptable limits. By including the atmospheric correction algorithm in the calibration process, it uncertainties are mitigated in the overall calibration process.

One of the earliest uses of a vicarious calibration method for a satellite-based radiometer was for the VIS/NIR channel of the European geostationary satellite Meteosat-1 [21]. The Koepke approach is predicated on the fact that given knowledge of the viewing geometry, extraterrestrial spectral irradiance from the Sun and the optical properties of the atmosphere and Earth's surface, the TOA radiance measured by the satellite based sensor can be accurately computed by solving the radiative transfer equation. In the absence of sun glint, and ignoring multiple scattering effects, the basic TOA reflectance equation can be approximated as:

$$\rho_{\text{TOA}}(\lambda) = \rho_{\text{r}}(\lambda) + \rho_{\text{a}}(\lambda) + t\rho_{\text{f}}(\lambda) + t\rho_{\text{w}}(\lambda) \qquad (20)$$

The terms $\rho_r(\lambda)$, $\rho_a(\lambda)$, and $\rho_f(\lambda)$ represent the Rayleigh scattering, aerosol (including molecular–aerosol interactions) and surface foam contributions respectively (with t accounting for the diffuse transmission from the surface to TOA). The remaining term, $\rho_w(\lambda)$, represents the water-leaving reflectance. The vicarious calibration coefficient for each wavelength is simply the ratio of the predicted $\rho_{TOA}(\lambda)$ to the at-sensor measured value:

$$g(\lambda) = \frac{\rho'_{TOA}(\lambda)}{\rho_{TOA}(\lambda)} \qquad (21)$$

Several of the earliest proposed methods [18,19,22–24] for a vicarious calibration of ocean color sensors implemented the basic approach outlined by Koepke, in that the components necessary to solve the radiative transfer equation were independently derived. The advantage to this approach is that, if the radiative transfer equation is complete, a direct relationship between the sensor measured radiance and the actual incoming radiance can be established. However, in the practical application of TOA reflectance to ocean color observations an atmospheric correction algorithm is a necessity. This atmospheric correction algorithm essentially reverses the radiative transfer equation, in that rather than derive the TOA radiance from the surface and atmospheric path radiance, it removes the path radiance to retrieve the surface radiance. For this to be done efficiently, the atmospheric correction algorithm uses a number of look-up tables that were themselves created using a radiative transfer model. Several of the atmospheric components necessary for the atmospheric correction can be estimated with high fidelity (e.g., Rayleigh reflectance, nonabsorbing aerosols and gaseous absorption). However, a number of the necessary components (e.g., meteorological information, concentration of absorbing gases) are obtained from relatively coarse resolution modeling or data assimilation products and still others require significant modeling efforts (e.g., aerosol reflectance). Errors in these components will reduce the absolute accuracy in the derived TOA reflectance for the sensor calibration, however their effect on the retrieved surface reflectance values is in a large part canceled by the use of the same atmospheric correction algorithm in the retrieval process [24].

An additional advantage of a vicarious calibration approach over the use of on-board calibration sources (e.g., internal lamps, SDs) is that the vicarious calibration includes the entire optical and electrical path of the instrument [21]. Onboard calibration sources can be critical to understanding and accounting for temporal drift in the absolute instrument calibration, but they are not sufficient to address all biases in the instrument measurement system. A vicarious approach using the same algorithm used for the retrieval of surface reflectances can also account for the errors introduced by the algorithm itself and is thus a much more robust method.

Most current atmospheric correction algorithms follow the method outlined by Gordon and Wang (GW94) [25]. This method employs a primary assumption that the surface reflectance over clear, open ocean waters in the near- and short-wave infrared (NIR/SWIR) wavelengths is a negligible component to the TOA reflectance in those bands. The algorithm uses two bands in the NIR/SWIR to estimate the aerosol reflectance via the use of look-up tables. Calibrating through the inverse application of such atmospheric correction algorithms requires a two-step approach. The first step is to calibrate the relative spectral response of the NIR/SWIR bands to ensure an accurate retrieval of the aerosol components. The second step uses these calibrated NIR/SWIR bands to retrieve the aerosol reflectance in the visible bands.

5.1 NIR/SWIR Band Calibration

Water is strongly absorbing in the near- and short-wave infrared wavelengths [26,27]. In the absence of suspended particulate matter, the water reflectance in these bands is negligible and does not make a significant contribution to the TOA reflectance. With this assumption and given that the Raleigh reflectance can be accurately calculated [28—30], the remaining signal can be entirely attributed to aerosol reflectance. The GW94 algorithm uses two bands in the NIR (or alternatively the SWIR [31,32]) in the determination of the aerosol component to the TOA signal. The spectral relationship between the aerosol reflectance in the two NIR/SWIR bands is used to select a candidate aerosol type from as set of precalculated aerosol models. The total signal in the longer of the two wavelengths is used to derive the baseline aerosol reflectance, which is then used in combination with the retrieved aerosol model to extrapolate this aerosol signal into the visible wavelengths. This method has proven very effective for non- or weakly absorbing aerosols.

When vicariously calibrating the NIR/SWIR bands, two additional assumptions are required. First, it is assumed that the calibration of the longest wavelength is sufficiently accurate and thus the vicarious calibration coefficient for this band is set to unity. Wang and Gordon [33] determined that if the absolute calibration of this band is known to within 5%, the GW94 algorithm could retrieve surface reflectance values of sufficient accuracy for ocean color.

Additionally, it is assumed that the aerosol type over the calibration target is known. With this assumption, the corresponding aerosol model can be used in combination with the retrieved aerosol reflectance in the longest wavelength to predict aerosol reflectance in the shorter NIR/SWIR wavelength [4]. This predicted reflectance allows for the prediction of the TOA radiance, which is compared to the measured radiance to derive a gain factor.

To accommodate the two required assumptions, the vicarious calibration of the NIR bands is performed in regions of the ocean where the dominant aerosols contribution is from maritime sources and the particulate concentration in the surface ocean is sufficiently small for the surface reflectance to be negligible (i.e., oligotrophic waters). Thus remote, clear open ocean waters such as those found in the South Pacific gyre are typically chosen.

The limitation of an assumed aerosol model can overcome if the satellite instrument has more than two NIR bands (e.g., MERIS) by using a single scattering approximation for the NIR reflectance. Such a method has been used for the calibration of MERIS [34]. Since this method employs more than one reference wavelength, it gives the additional benefit of providing a calibration for the all of the NIR bands.

5.2 Visible Band Calibration

With the calibration of the NIR/SWIR bands determined, the vicarious calibration of the visible wavelength bands can be performed. Operating the GW94 algorithm for the aerosol determination, the expected TOA radiance can be calculated. However, unlike the NIR/SWIR calibration, the visible calibration cannot make use of the assumption that the surface reflectance is negligible. Therefore, the surface reflectance values must be input into the inverse process to retrieve the expected TOA reflectance. To minimize the uncertainties contributed by the surface reflectance measurements to the final calibration, the surface reflectance values ideally should be measured using well-calibrated, preferably NMI-traceable, hyperspectral instrumentation in oligotrophic waters with spatially homogenous optical conditions under clear atmospheric conditions [28,35]. However, Bailey et al. [20] demonstrated that strict adherence to the ideal is not as critical the earlier studies suggested and, should practical necessities warrant, vicarious calibration can be reasonably accurate under certain nonideal conditions. The reason this is true stems from the fact that most of the TOA signal is from atmospheric contributions, i.e., Rayleigh and aerosol reflectance. As the most important component to the vicarious calibration process is the prediction of the atmospheric contributions [36], these contributions will remain the same regardless of the source of the water-leaving reflectance measurements.

A primary consideration of vicarious calibration being the reduction of the measurement uncertainties, adherence to the ideal measurement conditions is preferred. Each of these conditions shall now be addressed.

The use of a hyperspectral instrument allows for the spectral response function of the remote sensor to be matched by the in situ reflectance measurements. This ensures that any measurement uncertainty resulting from spectral out-of-band can be captured in the vicarious calibration process. There is also the practical benefit of having a single instrument provide a vicarious calibration target data for any remote sensor.

The use of NMI-traceable instrumentation provides a desired measure of credibility to the resulting calibration. This is of particular importance when the calibrated data are used in climate change studies.

The footprint of the satellite-based remote sensor is significantly larger than the area measured by the in situ instrumentation. To ensure that the in situ measurement is representative of the signal measured by the remote sensor, the surrounding waters should be spatially homogenous. While the vicarious calibration process can be performed at the level of an individual pixel for the remote sensor, it is typically performed using the average of a number of pixels surrounding the in situ location. Without the requirement of spatial homogeneity, this would simply add unwanted noise to the process.

Given that the largest contribution to the TOA signal arises from the atmosphere, it is critical to the vicarious calibration process that this component can be estimated with a high degree of fidelity. This is most readily accomplished under clear atmospheric conditions with a low aerosol load. Selection of scenes that minimize the influence of clouds, sun glint and whitecap foam reflectance is desired.

Dedicated vicarious calibration facilities are preferred over an ad-hoc approach, as with a dedicated program, a rigorous calibration and characterization of the instruments can be achieved. Since the mid-1990s, the ocean color community has been fortunate to have several such facilities. The Marine Optical Buoy (MOBy) [37] operated by the U.S. National Oceanic and Atmospheric Administration off the coast of Lanai, Hawaii was designed and located so as to satisfy the measurement and environmental requirements for vicarious calibration. In near- continual operation since 1996, MOBy has provided surface reflectance values for number of international missions including Ocean Color Temperature Scanner (OCTS), SeaWiFS, MODIS, MERIS, and VIIRS. Beginning in 2002, the BOUee pour l'aquiSition d'une Serie Optique a Long termE (BOUSSOLE) mooring [38] in the Ligurian Sea has provided similar calibration quality data that has been used for the official vicarious calibration of the MERIS mission. Several fixed platform sites, originally intended to provide a source of validation data for ocean color missions (e.g., AErosol RObotic NETwork Ocean Color (AERONET OC)) have also been evaluated for their potential to serve as a source of vicarious calibration data [39]. While in general these sites do not fulfill the ideal requirements for vicarious calibration, they do offer the possibility of retrieving a sufficient quantity of target data in a short period of time. This can be quite useful for the initial calibration of a newly launch instrument.

The requirement that the source for the surface reflectances be measured field instrumentation can itself, if the situation necessitates, be relaxed. Werdell et al. [40] successfully demonstrated the use of an ocean reflectance model for the purpose of vicarious calibration. If used in oligotrophic waters, the reflectace model can faithfully reproduce water-leaving reflectances similar to what would be measured.

In practice, the gains are computed for a number of pixels for satellite scenes throughout the mission lifetime. The spatial and temporal averaged gain is computed using the mean of the semi-interquartile range (a simple average of the data within the 25th to 75th percentiles) to minimize the impact of statistical outliers [4]. Environmental variability will introduce random noise into the computation of the gains; therefore a sufficient number of gain

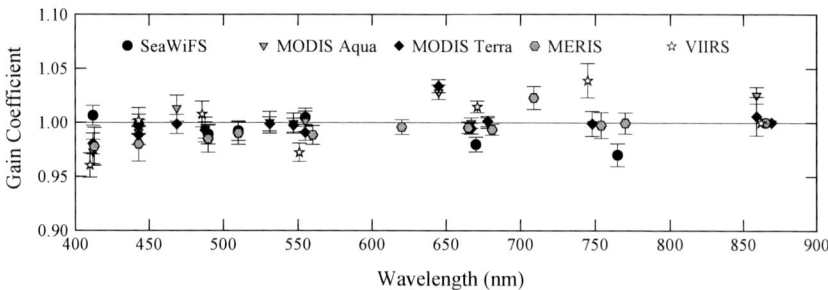

FIGURE 6 Vicarious calibration coefficients for SeaWiFS (●), Aqua MODIS (▽), MERIS (◉), and VIIRS (☆).

measurements are required to reduce the uncertainties resulting from this random error. Franz et al. [4] showed that between 30 and 40 measurements were required for SeaWiFS using MOBy data.

The approach presented here has been used to derive vicarious calibration coefficients used in the operational data processing for a number of ocean color missions, including but not limited to, Coastal Zone Color Scanner (CZCS), OCTS, POLarization and Directionality of the Earth's REflectances (POLDER), SeaWiFS, MODIS, MERIS, and VIIRS. Using NASA's multisensor processing code based on the GW94 algorithm, gains for several of these sensors computed. These are compared in Figure 6. The variability evident suggests that any systematic atmospheric correction bias is small compared to instrument artifacts. While the accuracy of vicarious calibration techniques is limited by the accuracy of the surface measurements and the atmospheric radiative transfer model as implemented in the atmospheric correction [41], these effects are sufficiently small in the overall uncertainty budget.

5.3 Alternative Approaches

Various alternative approaches to derive sensor gain coefficients have been put forth including a simple Rayleigh scattering method [42,43] and the use of Lambertian targets, i.e., cloud-top [44,45]; sun glint [46]; and deserts [47]. These methods have been used to assess the interband calibration (i.e., spectral consistency) and to cross-calibrate multiple sensor with some success. However, these methods suffer from deficiencies that make the less suitable for ocean color sensors than the approach presented here.

6. ON-ORBIT CALIBRATION UNCERTAINTIES

Having addressed uncertainties in the solar, lunar, and vicarious calibrations, we will now show how the various uncertainty estimates are merged to develop an overall uncertainty in calibrated TOA radiances. We will address these uncertainties in terms of accuracy (biases in the measurements), stability (changes in accuracy with time), and precision (scatter in the measurements). Since an extensive analysis has been performed for SeaWiFS, the SeaWiFS uncertainties will be used as a case study [5]. The chapter will conclude by comparing uncertainties across instruments.

6.1 Accuracy

The mission-averaged mean solar residuals or mean lunar residuals provide the best on-orbit estimates of the biases in the TOA radiances that are internal to a given instrument calibration. The measurement accuracy is defined as:

$$\text{Accuracy}(\lambda) = \left< \frac{\text{Instrument}(\lambda)}{\text{Reference}(\lambda)} - 1 \right>. \tag{22}$$

For lunar observations, the reference becomes the ROLO model and the accuracy is the mean bias between the instrument and the ROLO model. For solar observations, a stable reference is not as easily defined, since SDs change over time. For vicarious calibration the reference becomes the in situ radiometer, so the SeaWiFS vicarious calibration is relative to MOBy. As shown in Figure 4, the accuracy of the SeaWiFS TOA radiances is 2−3% relative to the ROLO model. The calibration bias between SeaWiFS and Terra MODIS or Aqua MODIS is 3−8%, while the bias between SeaWiFS and VIIRS is 1−10%. The bias in the vicarious calibration for SeaWiFS is 1−2%.

The key question that remains unanswered in the accuracy or bias determination is one of bias compared to which external reference? The specific scientific question being addressed determines the external reference for the calibration bias. The external reference can be a diffuser (e.g., MODIS, MERIS, and VIIRS, though diffuser reflectances degrade on orbit), the ROLO photometric model of the Moon (e.g., SeaWiFS, though the absolute uncertainty on the ROLO model is 10%), ground truth sites (though the atmospheric correction contributes uncertainties or another instrument on orbit).

6.2 Long-term Stability of the TOA Radiances

The long-term stability of the TOA radiances is a validation of the effectiveness of the radiometric correction of the instrument data. The uncertainty in the long-term stability arises from any residual time dependence in the TOA radiances or any residual periodic signals in the on-orbit calibration data. Accordingly, the limits on the residual time drift in a fully calibrated on-orbit calibration data set provide the best estimate of the long-term stability of the TOA radiances. The fully calibrated SeaWiFS lunar time series are shown in Figure 7. The standard deviation of the mean (or RMS error) of the mission-averaged lunar time series for each band defines the upper limit on the residuals time drift in the band, so the actual calibration stability could be better than the RMS error. The RMS error in the lunar data for SeaWiFS is

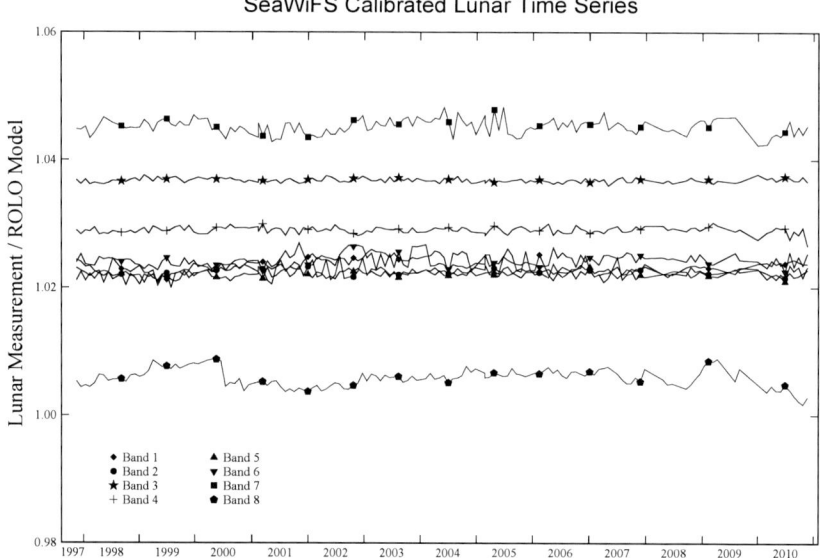

FIGURE 7 Fully calibrated SeaWiFS lunar time series.

0.033–0.13% per band, so a long-term stability of 0.13% is a reasonable estimate of the instrument performance across all bands [5].

The long-term stability of the vicariously calibrated SeaWiFS TOA radiances combines the uncertainty in the lunar-derived radiometric correction, the uncertainty in the vicarious gains (or atmospheric correction) and the uncertainty in the water-leaving radiances (or remote-sensing reflectances) measured by MOBy, then propagated to the top of the atmosphere, as in Table 3 [5].

6.3 Precision of the TOA Radiances

The scatter in the on-orbit measurements made by an instrument gives the precision of the TOA radiances. The SNRs determined from SD measurements, and the scatter in the lunar residuals, are two estimates of the precision of the on-orbit measurements. The scatter in the vicarious gains gives the precision of the vicarious calibration. The SeaWiFS on-orbit precision measurements are provided in Table 4 [5].

6.4 Combined Uncertainty Assessment

We have examined the uncertainties in the SeaWiFS TOA radiances in terms of accuracy, stability, and precision. The combined uncertainty estimates for each band are presented in Table 5 [5]. A similar uncertainty assessment is required for every ocean color instruments. As noted in the introduction to this section, these uncertainty estimates do not include the effects of vicarious calibration.

TABLE 3 SeaWiFS Vicarious Calibration Stability Estimates

Band	Radiometric Correction (%)	Vicarious Gain (%)	MOBy TOA Radiances (%)	Combined RMS Error (%)
1	0.124	0.07	0.24	0.28
2	0.0778	0.07	0.21	0.24
3	0.0334	0.07	0.24	0.26
4	0.0456	0.07	0.23	0.25
5	0.0578	0.07	0.24	0.26
6	0.0958	0.06	0.33	0.35
7	0.116	0.11		0.16
8	0.129			

TOP, top-of-the-atmosphere; MOBy, the marine optical buoy

TABLE 4 SeaWiFS On-Orbit Precision Estimates

Band	Solar SNR	Solar Precision (%)	Lunar Precision (%)	Vicarious Gain Precision (%)
1	646	0.155	0.124	0.07
2	794	0.126	0.078	0.07
3	976	0.102	0.0334	0.07
4	1013	0.0987	0.0456	0.07
5	953	0.105	0.0578	0.07
6	833	0.120	0.0958	0.06
7	857	0.117	0.116	0.11
8	767	0.130	0.129	

SNR, signal-to-noise ratio.

7. COMPARISON OF UNCERTAINTIES ACROSS INSTRUMENTS

As was discussed in the introduction, VIIRS views the SD and the moon at the same incidence angle on the mirror, allowing the VIIRS solar-derived and lunar-derived radiometric responses to be compared directly. The OBPG has taken advantage of this design feature by using the lunar-derived radiometric

TABLE 5 SeaWiFS TOA Uncertainty Assessment. The Overall Uncertainty is the Uncertainty for all Bands. The Units are Percent

Uncertainty	B1	B2	B3	B4	B5	B6	B7	B8	Overall
Accuracy (ROLO)	2.35	2.25	3.68	2.90	2.22	2.43	4.52	0.60	2–3
Accuracy (MOBy)	−0.656	0.170	1.09	0.766	−0.468	2.05	3.09		1–2
Stability (TOA)	0.124	0.0778	0.0334	0.0456	0.0578	0.0958	0.116	0.129	0.13
Stability (VC TOA)	0.28	0.24	0.26	0.25	0.26	0.35	0.16		0.30
Precision solar	0.155	0.126	0.102	0.0987	0.105	0.120	0.117	0.130	0.16
Precision lunar	0.124	0.078	0.0334	0.0456	0.0578	0.0958	0.116	0.129	0.13
Precision vicarious	0.070	0.070	0.070	0.070	0.070	0.060	0.11		0.10

TOA, top-of-the-atmosphere; ROLO, robotic lunar observatory.

trends to directly validate the solar-derived trends through a residual analysis of the two data sets. As an example of this analysis, Figures 8 and 9 shows the residuals of for fits to the solar and lunar calibration times series for bands M1 (412 nm)—M4 (555 nm) and for bands M5 (672 nm)—M7 (865 nm). Fits to the solar time series show mean residuals of 0.067—0.17% per band. Fits to

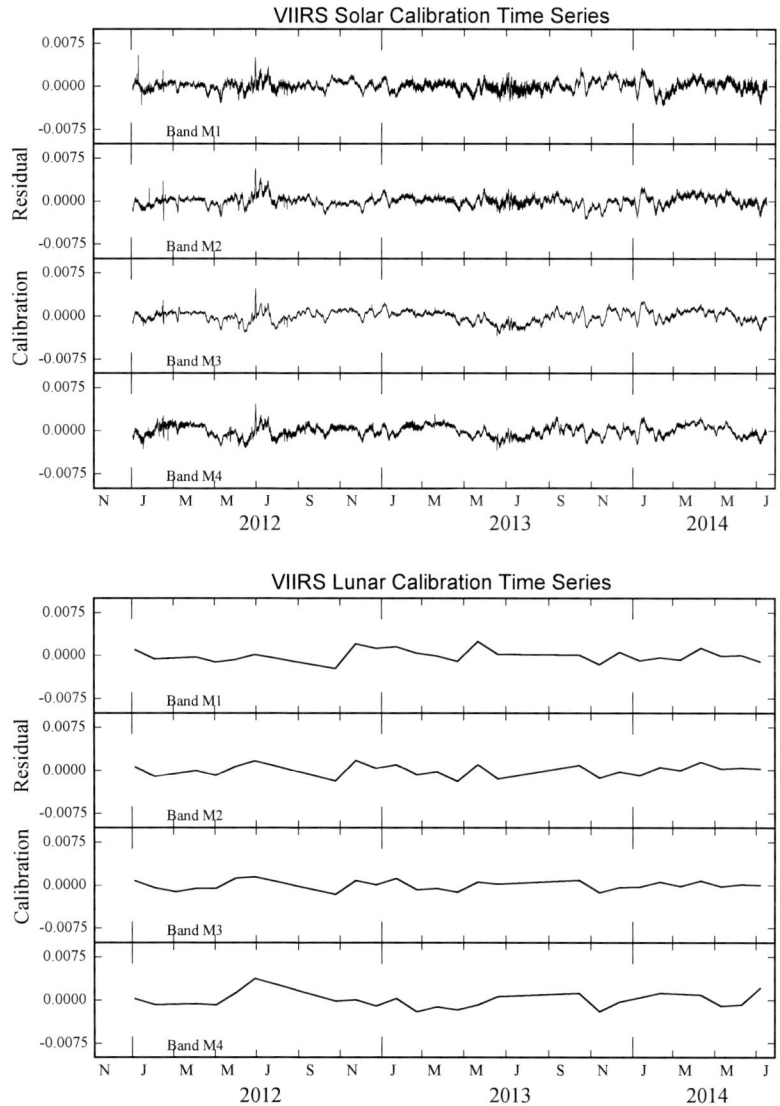

FIGURE 8 VIIRS solar and lunar Cal residuals.

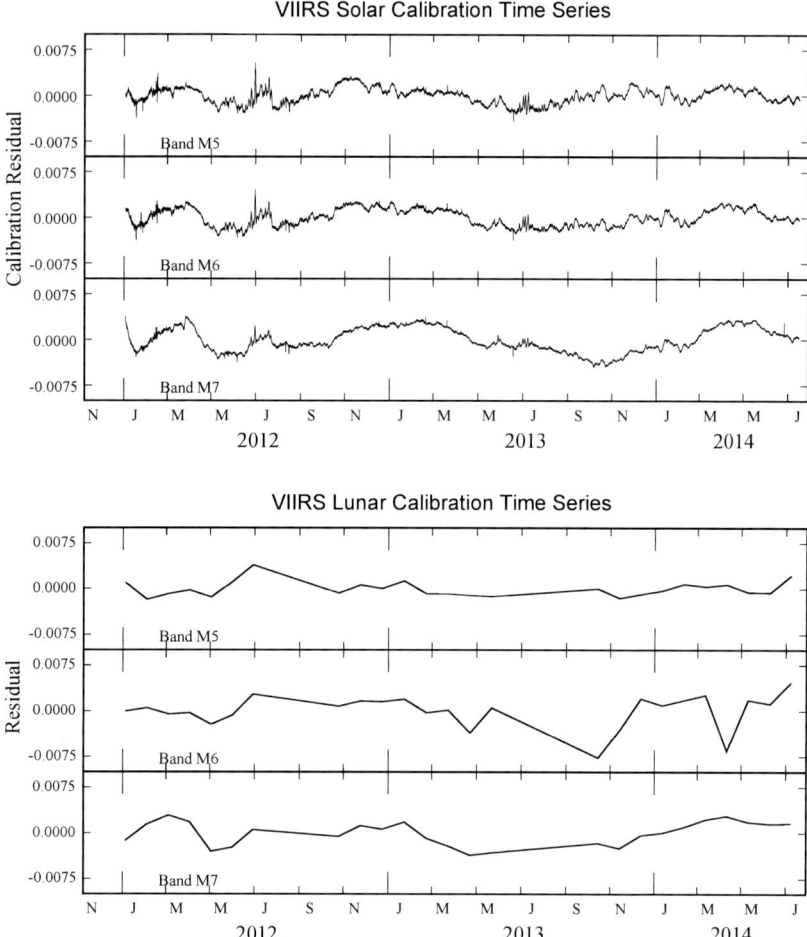

FIGURE 9 VIIRS solar and lunar Cal residuals.

the lunar time series show mean residuals of 0.069–0.20% per band. Figure 10 shows the differences in the solar and lunar time series for all seven visible and near-infrared bands. The mean differences in the solar and lunar time series are 0.097–0.22% per band. The uncertainties in the VIIRS solar and lunar calibrations are comparable to the mission-long uncertainties of 0.033–0.13% determined for SeaWiFS lunar calibrations, uncertainties of 0.38–0.94% determined for Terra MODIS lunar observations, and uncertainties of 0.30–0.58% determined for Aqua MODIS lunar observations. These comparisons are summarized in Table 6 [10].

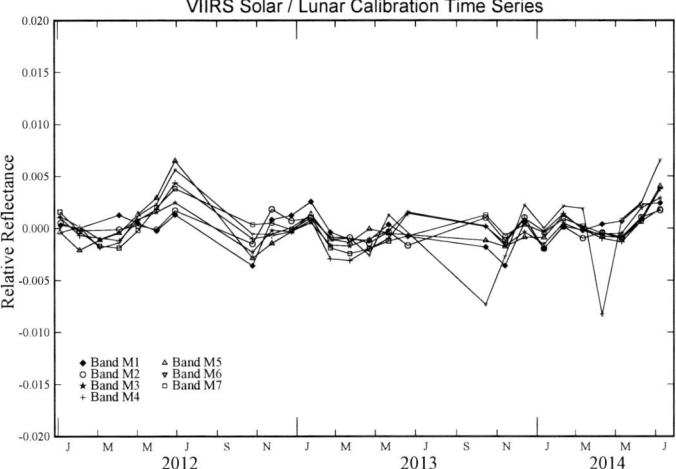

FIGURE 10 VIIRS solar/lunar Cal differences.

TABLE 6 On-Orbit Calibration Comparison

Instrument Calibration	Mean Residuals (per Band)
VIIRS solar	0.067–0.17%
VIIRS lunar	0.069–0.20%
SeaWiFS lunar	0.033–0.13%
Terra MODIS lunar	0.38–0.94%
Aqua MODIS lunar	0.30–0.58%

8. SUMMARY OF ON-ORBIT CALIBRATION

Experience with heritage instruments demonstrates that in order to meet the long-term radiometric stability necessary for produce CDRs from ocean color data requires the implementation of a mission-long on-orbit calibration/validation program for each instrument. Over time, the experience of the ocean color community in on-orbit instrument calibration develops as well. SeaWiFS was the first remote sensing satellite instrument to make full use of the Moon as a monitor for on-orbit radiometric performance. The on-orbit calibration strategies implemented for each new ocean color instrument build on the heritage experience. SeaWiFS, Terra MODIS, and Aqua MODIS have demonstrated the necessity of the ROLO photometric model of the Moon for on-orbit lunar calibration data analysis. Yet, the VIIRS experience shows the needs for the continued maintenance and development of the ROLO model. The methodologies of vicarious calibration have become more robust as more diverse in situ data sources become available.

Given the relatively short duration of satellite missions relative to the time scales of climate change, the production of ocean color CDRs requires a consistent time series of ocean color data across multiple missions. This is an important lesson as the production of ocean color data products transitions from research instruments and programs (e.g., SeaWiFS and MODIS) to operations instruments and programs (e.g., VIIRS) [48].

REFERENCES

[1] National Research Council, Climate Data Records from Environmental Satellites: Interim Report, The National Academies Press, Washington, D.C, 2004.

[2] C.R. McClain, W.E. Esaias, W. Barnes, B. Guenther, D. Endres, S.B. Hooker, G. Mitchell, R. Barnes, NASA Tech. Memo. 104566, in: S.B. Hooker, E.R. Firestone (Eds.), SeaWiFS Calibration and Validation Plan, vol. 3, NASA Goddard Space Flight Center, Greenbelt, Maryland, 1992.

[3] H.R. Gordon, Atmospheric correction of ocean color imagery in the earth observing system era, J. Geophys. Res. 102 (1997) 17081−17106.

[4] B.A. Franz, S.W. Bailey, P.J. Werdell, C.R. McClain, Sensor-independent approach to the vicarious calibration of satellite ocean color radiometry, Appl. Opt. 46 (2007) 5068−5082.

[5] R.E. Eplee Jr, G. Meister, F.S. Patt, R.A. Barnes, S.W. Bailey, B.A. Franz, C.R. McClain, On-orbit calibration of SeaWiFS, Appl. Opt. 51 (2012) 8702−8730.

[6] X. Xiong, J. Sun, W. Barnes, V. Salomonson, J. Esposito, H. Erives, B. Guenther, Multiyear on-orbit calibration and performance of Terra MODIS reflective solar bands, IEEE Trans. Geosci. Remote Sci. 45 (2007) 879−889.

[7] X. Xiong, J. Sun, X. Xie, W.L. Barnes, V.V. Salomonson, On-orbit calibration and performance of Aqua MODIS reflective solar bands, IEEE Trans. Geosci. Remote Sens. 48 (2010) 535−546.

[8] C. Cao, F.J. DeLuccia, X. Xiong, R. Wolfe, F. Weng, Early on-orbit performance of the visible infrared imaging radiometer suite onboard the Suomi national Polar-Orbiting Partnership (S-NPP) satellite, IEEE Trans. Geosci. Remote Sens. 52 (2014) 1142−1156.

[9] J. Sun, X. Xiong, W.L. Barnes, B. Guenther, MODIS reflective solar band on-orbit lunar calibration, IEEE Trans. Geosci. Remote Sens. 45 (2007) 2383−2393.

[10] R.E. Eplee Jr, K.R. Turpie, G. Meister, F.S. Patt, G. Fireman, B.A. Franz, C.R. McClain, A synthesis of VIIRS solar and lunar calibrations, in: J.J. Butler, X. Xiong, X. Gu (Eds.), Earth Observing Systems XVIII, Proc. SPIE 8866, vol. 88661L, 2013.

[11] H.H. Kieffer, T.C. Stone, The spectral irradiance of the Moon, Astron. J. 129 (2005) 2887−2901.

[12] [a] S. Delwart, R. Preusker, L. Bourg, R. Santer, D. Ramon, J. Fischer, MERIS in-flight spectral calibration, Int. J. Remote Sens. 28 (2007) 479−496;
[b] S. Delwart, L. Bourg, MERIS calibration: 10 years, in: J.J. Butler, X. Xiong, X. Gu (Eds.), Earth Observing Systems XVIII, Proc. SPIE 8866, vol. 88660Y, 2013.

[13] Labsphere, Spectralon Wavelength calibration standards (wavelength standards product sheet), Labsphere, Inc., North Sutton, N.H.
[14] X. Xiong, R.E. Eplee Jr, J. Sun, F.S. Patt, A. Angal, C.R. McClain, Characterization of MODIS and SeaWiFS solar diffuser on-orbit degradation, in: J.J. Butler, X. Xiong, X. Gu (Eds.), Earth Observing Systems XIV, Proc. SPIE 7452, vol. 74520Y, 2009.
[15] T.C. Stone, H.H. Kieffer, Use of the Moon to support on-orbit sensor calibration for climate change measurements, in: J.J. Butler, J. Xiong (Eds.), Earth Observing Systems XI, Proc. SPIE 6296, vol. 62960Y, 2006.
[16] T.C. Stone, Radiometric calibration stability and intercalibration of solar-based instruments in orbit using the Moon, in: J.J. Butler, J. Xiong (Eds.), Earth Observing Systems XIII, Proc. SPIE 7081, vol. 70810X, 2008.
[17] R.E. Eplee Jr, J.-Q. Sun, G. Meister, F.S. Patt, X. Xiong, C.R. McClain, Cross calibration of SeaWiFS and MODIS using on-orbit observations of the Moon, Appl. Opt. 50 (2011) 120−133.
[18] H.R. Gordon, Calibration requirements and methodology for remote sensors viewing the ocean in the visible, Remote Sens. Environ. 22 (1987) 103−126.
[19] H.R. Gordon, In-orbit calibration strategy for ocean color sensors, Remote Sens. Environ. 63 (1998) 265−278.
[20] S.W. Bailey, S.B. Hooker, D. Antoine, B.A. Franz, P.J. Werdell, Sources and assumptions for the vicarious calibration of ocean color satellite observations, Appl. Opt. 4 (2008) 2035−2045.
[21] P. Koepke, Vicarious satellite calibration in the solar spectral range by means of calculated radiances and its application to Meteosat, Appl. Opt. 21 (1982) 2845−2854.
[22] R.S. Fraser, Y.J. Kaufman, Calibration of satellite sensors after launch, Appl. Opt. 25 (1986) 1177−1185.
[23] K.L. Carder, P. Reinersman, R.F. Chen, F. Muller-Karger, AVIRIS calibration and application in coastal oceanic environments, Remote Sens. Environ. 44 (1993) 205−216.
[24] B. Fougnie, P.-Y. Deschanps, R. Frouin, Vicarious calibration of the POLDER ocean color spectral bands using in Situ measurements, IEEE Trans. Geo- Sci. Remote Sens. 3 (1999) 1567−1574.
[25] H.R. Gordon, M. Wang, Retrieval of water-leaving radiance and aerosol optical thickness over the oceans with SeaWiFS: a preliminary algorithm, Appl. Opt. 33 (1994) 443−452.
[26] R.M. Pope, E.S. Fry, Absorption spectrum (380−700 nm) of pure water. II. Integrating cavity measurements, Appl. Opt. 36 (1997) 8710−8723.
[27] L. Kou, D. Labrie, P. Chylek, Refractive indices of water and ice in the 0.65-2.5 m spectral range, Appl. Opt. 32 (1993) 3531−3540.
[28] H.R. Gordon, J.W. Brown, R.H. Evans, Exact rayleigh scattering calculations for use with the Nimbus-7 coastal zone color scanner, Appl. Opt. 27 (1988) 862−871.
[29] M. Wang, The Rayleigh lookup tables for the SeaWiFS data processing: accounting for the effects of ocean surface roughness, Int. J. Remote Sens. 23 (2002) 2693−2702.
[30] M. Wang, A refinement for the Rayleigh radiance computation with variation of the atmospheric pressure, Int. J. Remote Sens. 26 (2005) 5651−5653.
[31] M. Wang, W. Shi, The NIR-SWIR combined atmospheric correction approach for MODIS ocean color data processing, Opt. Express 15 (2007) 15722−15733.
[32] M. Wang, J. Tang, W. Shi, MODIS-derived ocean color product along the China east coastal region, Geophys. Res. Lett. 34 (2007). http://dx.doi.org/10.1029/2006GL028599.

[33] M. Wang, H.R. Gordon, Calibration of ocean color scanners: how much error is acceptable in the near infrared? Remote Sens. Environ. 82 (2002) 497−504.

[34] C. Lerebourg, C. Mazeran, J.P. Huot, D. Antoine, Vicarious Adjustment of the MERIS Ocean Colour Radiometry, ATBD 2.24, European Space Agency, 2011.

[35] D.K. Clark, H.R. Gordon, K.J. Voss, Y. Ge, W.W. Broenkow, C. Trees, Validation of atmospheric correction over the oceans, J. Geophys. Res. 102 (1997) 17209−17217.

[36] H.R. Gordon, T. Zhang, How well can radiance reflected from the ocean- atmosphere system be predicted from measurements at the sea surface, Appl. Opt. 35 (1996) 6527−6543.

[37] D.K. Clark, M.A. Yarbrough, M.E. Feinholz, S. Flora, W. Broenkow, Y.S. Kim, B.C. Johnson, S.W. Brown, M. Yuen, J.L. Mueller, MOBY, a radiometric buoy for performance monitoring and vicarious calibration of satellite ocean color sensors: measurement and data analysis protocols, in: J.L. Mueller, G.S. Fargion, C.R. McClain (Eds.), Ocean Optics Protocols for Satellite Ocean Color Sensor Validation, Revision 4, Volume VI: Special Topics in Ocean Optics Protocols and Appendices, NASA Goddard Space Flight Center, Greenbelt, MD, 2003, pp. 3−34. NASA/TM2003−211621/Rev4-Vol.VI.

[38] D. Antoine, M. Chami, H. Claustre, F. D'Ortenzio, A. Morel, G. Bécu, B. Gentili, F. Louis, J. Ras, E. Roussier, A.J. Scott, D. Tailliez, S.B. Hooker, P. Guevel, J.-F. Desté, C. Dempsey, D. Adams, BOUSSOLE : A Joint CNRS-insu, ESA, CNES and NASA Ocean Color Calibration and Validation Activity, NASA Technical memorandum N2006−214147, NASA/GSFC, Greenbelt, MD, 2006, p. 61.

[39] F. Melin, G. Zibordi, Vicarious calibration of satellite ocean color sensors at two coastal sites, Appl. Opt. 49 (2010) 798−810.

[40] P.J. Werdell, S.W. Bailey, B.A. Franz, A. Morel, C.R. McClain, On-orbit vicarious calibration of ocean color sensors using an ocean surface reflectance model, Appl. Opt. 46 (2007) 5649−5666.

[41] M. Dinguirard, P.N. Slater, Calibration of space-multispectral imaging sensors: a review, Remote Sens. Environ. 68 (1999) 194−205.

[42] E. Vermote, R. Santer, P.-Y. Deschamps, M. Herman, In-flight calibration of large field of view sensors at shorter wavelengths using Rayleigh scattering, Int. J. Remote Sens. 13 (1992) 3409−3429.

[43] O. Hagolle, P. Goloub, P.-Y. Deschamps, H. Cosnefroy, X. Briottet, T. Bailleul, J.M. Nicolas, F. Parol, B. Lafrance, M. Herman, Results of POLDER in-flight calibration, IEEE Trans. Geosci. Remote Sens. 37 (1999) 1550−1566.

[44] E. Vermote, Y.-J. Kaufman, Absolute calibration of AVHRR visible and near infrared channels using ocean and cloud views, Int. J. Remote Sens. 16 (1995) 2317−2340.

[45] B. Lafrance, O. Hagolle, B. Bonnel, Y. Fouquart, G. Brogniez, M. Herman, Interband calibration over clouds for POLDER space sensor, IEEE Trans. Geosci. Remote Sens. 40 (2002) 131−142.

[46] O. Hagolle, J.-M. Nicolas, B. Fougnie, F. Cabot, P. Henry, Absolute calibration of VEGETATION derived from an interband method based on the sun glint over ocean, IEEE Trans. Geosci. Remote Sens. 42 (2004) 1472−1481.

[47] F. Cabot, O. Hagolle, H. Cosnefroy, X. Briottet, Inter-calibration using desertic sites as a reference target, in: IGARSS'98, 6-10 July, Geosci. Remote Sensing Symposium Proceedings, vol. 5, 1998, pp. 2713−2715.

[48] National Research Council, Assessing Requirements for Sustained Ocean Color Research and Operations, The National Academies Press, Washington, D.C, 2011.

Chapter 2.3

Thermal Infrared Satellite Radiometers: Design and Prelaunch Characterization

David L. Smith
RAL Space, Science and Technologies Facilities Council, Harwell Oxford, Oxford, UK
E-mail: dave.smith@stfc.ac.uk

Chapter Outline

1. **Introduction** — 154
2. **Radiometer Design Principles** — 155
   - 2.1 Performance Model — 159
   - 2.2 Signal to Noise — 160
3. **Remote Sensing Systems** — 161
   - 3.1 Along Track Scanning Radiometers (ATSR) — 161
   - 3.2 Sea and Land Surface Temperature Radiometer (SLSTR) — 164
   - 3.3 Advanced Very High Resolution Radiometer (AVHRR) — 165
   - 3.4 MOderate Resolution Imaging Spectroradiometer (MODIS) — 166
   - 3.5 Visible Infrared Imaging Suite (VIIRS) — 167
   - 3.6 Spinning Enhanced Visible and Infrared Imager (SEVIRI) — 171
4. **Calibration Model** — 172
   - 4.1 Radiometric Noise — 174
   - 4.2 Nonlinearity — 174
   - 4.3 Offset Variations — 176
5. **On-Board Calibration** — 176
   - 5.1 Calibration Sources — 178
     - 5.1.1 Deep Space View — 178
     - 5.1.2 Black Plates or Structured Black Plate — 179
     - 5.1.3 Cavity Blackbodies — 180
6. **Pre-launch Characterization and Calibration** — 182
   - 6.1 Blackbody Calibration — 182
     - 6.1.1 Thermometry — 182
     - 6.1.2 Emissivity — 183
   - 6.2 Instrument Radiometric Calibration — 184
     - 6.2.1 Calibration Facility — 186
     - 6.2.2 Blackbody Sources — 186
     - 6.2.3 Calibration Test Procedures — 189
     - 6.2.4 Data Analysis — 191
     - 6.2.5 Test Results — 192
7. **Conclusions** — 197

References — 198

1. INTRODUCTION

Until the twentieth century, much of the understanding of the Earth environment was gained through localized observations from the ground. Our view was limited to the horizon and the equipment available to us. Even mariners had to perform all their navigation from the deck of a slow moving ship. Obtaining global measurements meant that we had to travel great distances by land or sea (latterly by air) to often inhospitable environments. With the advent of satellite technology and manned spaceflight, our view of the Earth was changed forever, as symbolized by the iconic "earthrise" photograph of the Earth taken from the Apollo 8 mission in 1968. For the first time in history, the entire globe could be seen in one go. Advances in imaging technology and computing have meant that the whole Earth environment can now be measured with unprecedented detail and accuracy in a very short time. For most people (even remote sensing scientists), images from space are a source of fascination and help us find directions to get from place to place. But, satellite observations are also providing important observations that are essential for our understanding of the Earth's climate system.

The views from space confirm that water covers over 70% of the Earth's surface. Because of the large heat capacity of water, the oceans act as a huge heat reservoir and have a significant impact on the Earth's climate. Also, the oceans contain dissolved gases that are essential for supporting aquatic life, which in turn is necessary for life on dry land. The sea surface temperature is therefore an important indicator of the state of the ocean and the Earth's climate. For climate monitoring and research it is necessary to measure global sea surface temperatures (SSTs) to an uncertainty <0.3 K with a stability better than 0.1 K per decade.

An obvious question to ask is how is it possible to measure SST from space? Historically, SST was typically measured by collecting seawater from a bucket and measuring the temperature with a thermometer. Although the methods and instrumentation developed over time, the approach was basically the same; to measure temperature a thermometer had to be in physical contact with the object.

The ability to measure the temperature of an object remotely is due to the property that all physical bodies emit electromagnetic radiation. Planck's radiation law describes the dependence of the intensity of emitted radiation on the temperature. For a blackbody in thermal equilibrium at an absolute temperature T, the spectral radiance $L(\lambda,T)$ at wavelength λ emitted, per unit area, per unit solid angle, per unit wavelength is given by

$$L(\lambda, T) = 2hc^2 \bigg/ \lambda^5 \left(\exp\left(\frac{hc}{\lambda k_b T}\right) - 1 \right) \qquad (1)$$

where

    h is Plank's constant $= 6.6260755 \times 10^{-34}$ J s;
    c is the velocity of light $= 299{,}792{,}458$ m s^{-1}; and
    k_b is Boltzmann's constant $= 1.380658 \times 10^{-23}$ J K^{-1}.

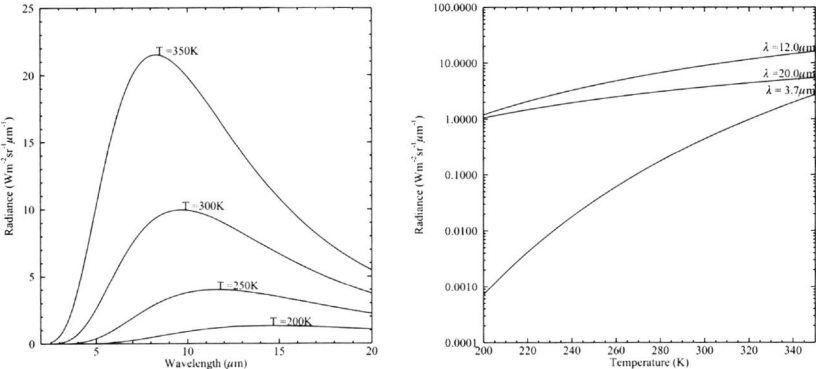

FIGURE 1 Planck function as a function of wavelength (left) at 200, 250, 300, and 350 K, and as a function of temperature at 3.7, 12, and 20 μm (right).

The Planck function means that it is possible to determine the temperature of an object by measuring the energy distribution of the emitted radiation. For example, the temperature of the Sun's surface is computed as ∼6000 K based on measurements of the solar spectrum which peaks at visible wavelengths. At the other extreme the cosmic microwave background has a spectrum corresponding to a near perfect black-body at 2.7 K. For temperatures that cover the typical range of Earth scenes from 200 to 350 K, Figure 1 we see that the peak of the distribution lies in the thermal infrared (IR) range between 8 and 15 μm. Also, we see that the radiances cover a wide dynamic range, particularly at wavelengths <5 μm which makes observation at these wavelengths particularly suited for accurate measurements of SST and LST.

In this section we will describe the design principles of a satellite radiometer operating in the thermal IR specifically designed for SST and LST observations, a survey of existing and future satellite sensors, and an overview of the prelaunch calibration activities. The examples provided will draw specifically from the experience of the advanced Along Track Scanning Radiometer ((A)ATSR) series that were specifically designed and built for climate monitoring, but applicable to all IR instruments.

2. RADIOMETER DESIGN PRINCIPLES

A fundamental requirement for any sensor to be used for climate quality measurements of geophysical parameters is to design a system that can be calibrated, such that any biases can be measured to a known accuracy either before launch or in-orbit. The flow down of requirements from SST to the sensor performance is illustrated in Figure 2. Here an instrument performance model takes the uncertainty requirements and breaks these down into specifications for the individual components. The sensor calibration is derived from measurements performed at component level through to testing

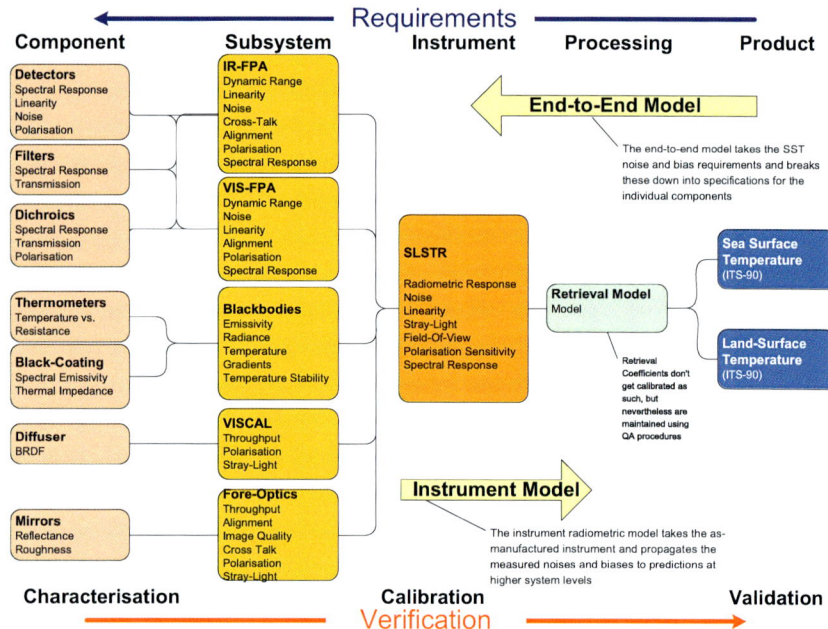

FIGURE 2 Diagram showing the concept of flow of the SLSTR performance requirements down to instrument, subsystem, and eventually component level accuracy requirements, and the subsequent calibration and characterization measurements to be performed to provide a fully calibrated system.

performed at instrument level to demonstrate the end-to-end calibration of the system. The calibration must then be maintained after launch through sustained postlaunch calibration and validation activities.

The basic components for a radiometer needed to accurately quantify Earth-scene radiances are:

- Detector + amplifier;
- Filters to select the required wavelength of interest;
- Optics to collect the signal and focus onto detector;
- Stray light control to minimize unwanted signals; and
- Calibration sources to provide a traceable reference standard.

A simple radiometer comprises a detector with filter, a light tight enclosure, and exit aperture as illustrated in Figure 3. The radiant flux, ϕ, incident at the detector element with peak response at wavelength λ, from a scene radiance L_{scene} is given by

$$\phi_\lambda = A\left(\Omega L_{\lambda,\text{scene}} + \pi - \Omega L_{\lambda,\text{radiometer}}\right) \qquad (2)$$

where A is the active area of the detector and Ω is the solid angle of the radiometer. In this basic concept, the etendue of the radiometer $A\Omega$ is defined

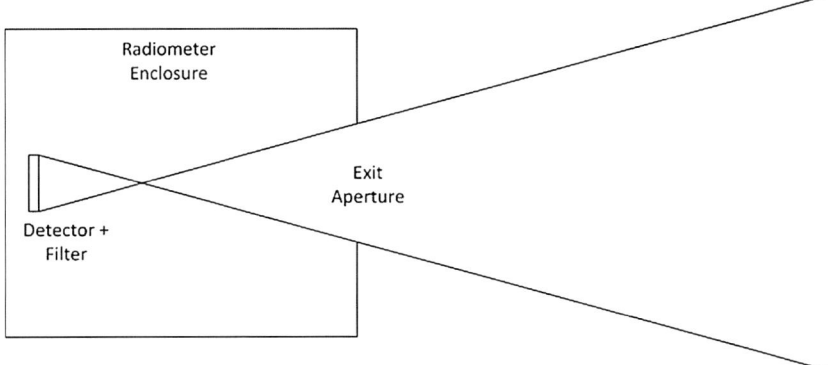

FIGURE 3 A very basic radiometer.

by the active area of the detector, the distance between the detector and aperture and the aperture area. It is important to note that $A\Omega$ is conserved throughout, even if it passes through optical systems through reflection and refraction.

For an instrument with a spectral response as a function of wavelength, $R_\lambda(\lambda)$, the band average radiance is given by

$$L_\lambda(T) = \int R_\lambda(\lambda) L(\lambda, T) d\lambda \bigg/ \int R_\lambda(\lambda) d\lambda \qquad (3)$$

The choice of detector depends on the wavelength range of interest. For the sensors described in this book, indium antimonide (InSb) and mercury cadmium telluride (HgCdTe) detectors are used because their peak sensitivity is close to the wavelengths of interest [1]. These are quantum detectors where electron-hole pairs are generated by the incident radiation and are either (1) photovoltaic where a voltage is generated or (2) photoconductive where the electrical conductivity changes. This is not to be confused with thermal detectors (e.g., bolometer, Golay cell) where the incident radiation is absorbed by the detector element resulting in a change of temperature. To achieve the optimum noise performance the detectors need to be operated at close to 77 K in an enclosure that is also cooled to minimize the background photon flux.

For IR instruments it is important to highlight the second term in the equation. For wavelengths <2.5 μm we can ignore this term because for a dark box, the radiated signal should be negligible compared with the scene radiance. At thermal IR wavelengths, the very fact that the enclosure has temperature means that it is a radiance source. If the enclosure is at the same temperature as the scene then it is no longer possible to distinguish between the two. There are two solutions to this problem; the first is to cool the enclosure to low temperatures so that the radiance can be neglected or to use a chopper to allow alternate views between the scene and the enclosure. In the latter case the difference between the thermal background and scene radiance is recorded. In practice the two are often used together since it is not always possible to cool the complete radiometer enclosure.

Although the basic design is used for many laboratory applications for wavelengths from visible to thermal IR, it is of very limited use for Earth observation. Here we first need to introduce a telescope to collect sufficient light and focus onto the detector. At IR wavelengths, reflecting telescopes are generally used to minimize the losses at each surface and eliminate chromatic aberrations. The most common approaches include Cassegrain, Newtonian and three-mirror anastigmat. For the (A)ATSR sensors, a single off-axis paraboloid mirror is used (Herschelian[1] telescope) to produce an intermediate image of the Earth scene at a field stop Figure 4. To achieve the multiple spectral bands, the signal is split into different wavelength ranges by a series of dichroic beam-splitters on the focal plane assembly. The radiation is then reimaged onto the detector via the passband filter by a powered mirror.

A scanning mirror is used to direct the signal from the Earth scene toward the telescope aperture. This also allows views of the on-board calibration targets. The optical enclosure should have adequate stray light baffling to ensure that there are no unwanted signals from the satellite and more importantly from reflected sunlight that can affect the thermal stability of the instrument.

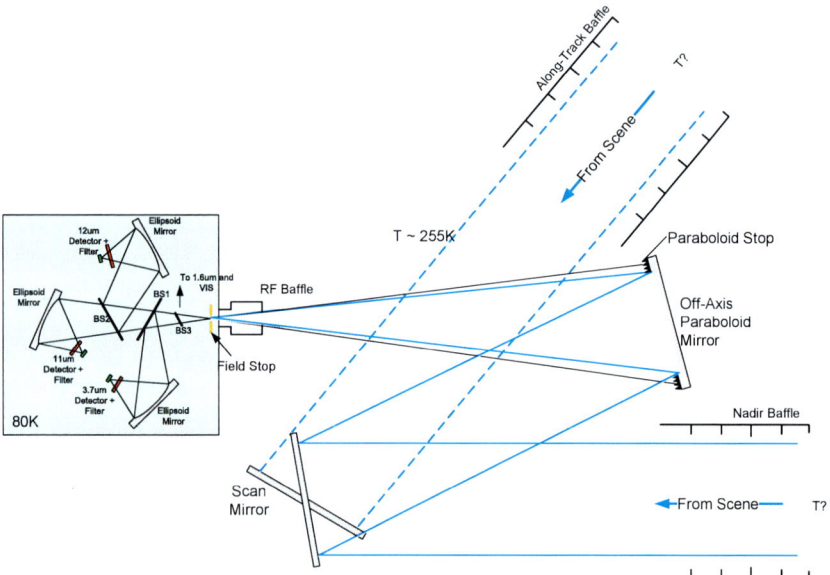

FIGURE 4 Schematic of the ATSR TIR channel optical chain.

---

1. After the astronomer William Herschel who discovered infrared radiation in 1800.

2.1 Performance Model

To understand the instrument performance and the impact on the calibration it is important to model the end to end response. For this we need to modify the radiometric model to account for all the optical components in the chain, remembering that for each surface the etendue is conserved, and more importantly each surface in the chain emits IR radiation.

$$\phi_\lambda = A_\lambda \left(\Omega_\lambda \left(\begin{array}{l} (\pi - \Omega_\lambda)L_{\text{FPA},\lambda} + \\ \tau_{\text{FPA},\lambda}\left(\begin{array}{l} (1 - \tau_{\text{FPA},\lambda})L_{\text{FPA}\lambda} + \\ \xi_\lambda \left(\begin{array}{l} (1 - \xi_\lambda)L_{\text{surr},\lambda} + \\ r_{\text{para},\lambda}\left(\begin{array}{l} (1 - r_{\text{para},\lambda})L_{\text{para},\lambda} + \\ r_{\text{scan},\lambda}\left(\begin{array}{l} (1 - r_{\text{scan},\lambda})L_{\text{scan},\lambda} + \\ r_{\text{scan},\lambda}L_{\text{scene},\lambda} \end{array} \right) \end{array} \right) \end{array} \right) \end{array} \right) \end{array} \right) \right) \quad (4)$$

where

$L_{\text{scene},\lambda}$ is the in-band integrated spectral scene radiance;

L_{scan}, L_{para}, L_{FPA}, and L_{surr} are respectively, the radiances from the scan mirror, telescope mirror, focal plane assembly (FPA) and telescope surround (assumed to be the same as the temperature of the FPA baffle and to have an emissivity of 1.0);

τ_{FPA} is the transmission of the FPA at the wavelength of peak response for each spectral channel;

$A\Omega$ is the total throughput of the detectors defined by the field-of-view of the detectors;

ξ is the fraction of the measured signal viewed via the primary aperture; and

r_{scan}, r_{para}, are the reflectances of the mirror coatings of the instrument fore optics.

Since the detectors and enclosure are cooled to cryogenic temperatures (~ 80 K) we can assume that the contribution from the cold FPA and optics to the total signal is negligible. Also, if the instrument is designed so that it can be assumed that the telescope and scan mirrors have roughly the same high reflectance and are at the same temperature such that $L_{\text{scan}} = L_{\text{para}} = L_{\text{surr}} = L_{\text{inst}}$, where L_{inst} is the blackbody radiance of the instrument fore-optics, the equation reduces to.

$$\begin{aligned} \phi_\lambda &= A_{i,\lambda}\Omega_{i,\lambda}\tau_{\text{FPA},\lambda}\left(\xi_\lambda r_\lambda^2 L_{\text{scene},\lambda} + \left(1 - \xi_\lambda r_\lambda^2\right)L_{\text{inst},\lambda}\right) \\ &= \phi_{\text{scene},i,\lambda} + \phi_{\text{inst},i,\lambda} \end{aligned} \quad (5)$$

It is vital to note that in order for the above calibration scheme to work, it is critical that there is no observable contribution to the measured radiance from other sources, notably mechanisms.

The signal measured by each channel is converted to a voltage and then digitized to counts, DN.

$$\mathrm{DN}_{\mathrm{scene}} = F_{\mathrm{ADC}}\left(V\{\phi_{\mathrm{scene},i,\lambda} + \phi_{\mathrm{inst},i,\lambda}\} + V_{\mathrm{off}}\right) \tag{6}$$

where

F_{ADC} = conversion factor for ADC;
τ_{opt} = transmission/reflectance of the FPA optics;
τ_{ome} = transmission/reflectance of the OME optics = ξr^2;
$A\Omega$ = throughput of optical chain;
V = voltage output as a function of photons at detector—this will be a function of detector + amplification and could be nonlinear with photon flux. The response could also be sensitive to instrument polarization; and
V_{off} — offset voltage.

This reduces to:

$$\mathrm{DN}_{\mathrm{scene}} = g(L_{\mathrm{scene}}) + \mathrm{offset} \tag{7}$$

2.2 Signal to Noise

The key figure of merit for a radiometer is the signal-to-noise ratio, which at IR wavelengths is usually expressed as the noise equivalent temperature difference, given by,

$$\mathrm{NE}\Delta T = \frac{\Delta V_{\mathrm{scene}}}{V_{\mathrm{scene}}} \left(\left.\frac{\partial V}{\mathrm{d}T}\right|_T\right)^{-1} \tag{8}$$

The total signal noise will be a combination of the shot, detector, quantization (digitization) and amplifier noise sources such that.

$$\Delta V_{\mathrm{tot}}^2 = \Delta V_{\mathrm{shot}}^2 + \Delta V_{\mathrm{det}}^2 + \Delta V_{\mathrm{amp}}^2 + \Delta V_{\mathrm{quant}}^2 \tag{9}$$

Ideally, the system noise should be shot noise limited, which is dependent on the number of electron–hole pairs produced by the photons arriving at the detector; hence it is sometimes referred to as photon noise. Since we are concerned with large numbers of photons the signal to noise ratio (SNR) is,

$$\mathrm{SNR} = \frac{V_{\mathrm{scene}}}{\Delta V_{\mathrm{shot}}} = \frac{N}{\sqrt{N}} = \sqrt{N} \tag{10}$$

For a given photon flux, ϕ_{det}, the number of electrons generated by the detector is dependent on the quantum efficiency of the device, η and the frequency bandwidth in Hz, f, such that.

$$N_{el} = \frac{\eta \phi_{det}(\lambda/hc)}{f} \qquad (11)$$

It is important to note that the shot noise is dependent on the total number of photons arriving at the detector. As discussed earlier this includes thermal emission from the radiometer. Hence, an essential factor in the design of any radiometer is to minimize the thermal background by using highly reflective optical surfaces and to control the optics temperature.

The intrinsic detector noise performance is usually given by the D-star ($D*$), or detectivity which is formally defined as the SNR from a detector when 1 W of radiant power is incident on a 1 cm^2 detector at a noise equivalent bandwidth of 1 Hz.

$$D* = \frac{\sqrt{A_{det}f}}{NEP} \text{ cm } \sqrt{Hz} \text{ W}^{-1} \qquad (12)$$

From this we can derive the SNR.

$$SNR = \frac{V_{scene}}{\Delta V} = \frac{\phi_{det}}{\sqrt{A_{det}f}} D* \qquad (13)$$

As a note of caution, when modeling the signal-to-noise performance, the $D*$ values should not include the shot noise but should refer to the noise for a dark (i.e., cold) signal.

3. REMOTE SENSING SYSTEMS

3.1 Along Track Scanning Radiometers (ATSR)

The Along Track Scanning Radiometers (ATSRs) formed a series of space-borne instruments specifically optimized to provide accurate remotely sensed measurements of SST. Since mid-1991 the three ATSR sensors provided global observations from the European Space Agency's Earth Observation satellites; namely ATSR-1 on ERS-1 (European Remote-sensing Satellite), then ATSR-2 on ERS-2 and then AATSR (advanced ATSR) on ENVISAT. The missions were operated with good overlaps between successive sensors. All three spacecraft were in Sun-synchronous orbits with a near 3 day repeat cycle and approximately 14 orbits per day. Each ATSR, with its combined day and night coverage and 500 km swath width, provided virtually complete global coverage every three days (Figure 5).

Each ATSR instrument had spectral channels corresponding to the AVHRR (Advanced Very High Resolution Radiometer) bands at 1.6, 3.7, 10.8 and 12 μm that enables "split" window sea surface temperature retrieval. ATSR-2 and AATSR were also equipped with visible (VIS) and near infrared (NIR) channels at 0.555, 0.660 and 0.870 μm for daytime cloud, aerosol, and vegetation monitoring. The nominal instantaneous field of view, IFOV, was 1.3 mrad (as for AVHRR) to give 1 km at nadir (Table 1).

FIGURE 5 The advanced along track scanning radiometer.

TABLE 1 Spectral Bands of ATSR

Band Center (μm)	Bandwidth (μm)	Function	ATSR-1	ATSR-2	AATSR	Detector Type
0.555	0.020	Chlorophyll	N	Y	Y	Si
0.659	0.020	Vegetation index	N	Y	Y	Si
0.870	0.020	Vegetation index	N	Y	Y	Si
1.600	0.060	Clouds, fire	Y	Y	Y	InSb
3.7		Night-time SST, fire	Y[a]	Y	Y	InSb
10.8		SST	Y	Y	Y	HgCdTe
12		SST	Y	Y	Y	HgCdTe

[a] The ATSR-1 3.7 μm channel failed after 100 days in orbit.

The main, unique design feature of the ATSR instruments was a conical scanning geometry that provided two views of the Earth: a nadir view and a corresponding inclined "along-track" view at 55° view zenith angle as illustrated in Figure 6. This geometry provides two views of the sea surface through two atmospheric paths to provide additional information for

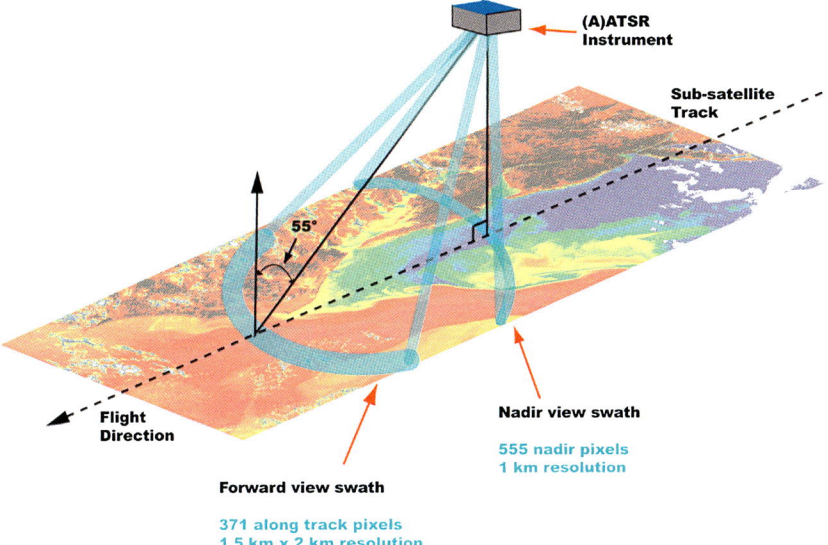

FIGURE 6 Scanning geometry of the ATSR.

atmospheric corrections. The dual view also provides capability for improved cloud and aerosol retrievals.

The (A)ATSR optical configuration is illustrated in Figure 4. The small IFOV of (A)ATSR is defined by a single 110 mm diameter, f5 off-axis parabolic mirror feeding to a single on-axis field-stop at the primary focus. An inclined scan mirror covers the telescope's slightly diverging beam, and it is rotated at 400 rpm with a constant angular velocity vector parallel to this beam's primary ray to generate a scanned cone. The angle of incidence at the scan mirror is $\sim 11.7°$ and four times this, the full scan cone angle, is $\sim 47°$. With its minimized optical surfaces this configuration is well suited to the IR, and the scan plus paraboloid mirrors only need to be cooled to approximately $-10\,°C$ for their emitted photon noise to fit within allocation in the noise budget. Low angles of incidence and high mirror reflectivity mean that these Earth imaging fore-optics are essentially nonpolarizing.

The IR channels centered at wavelengths of 1.6, 3.7, 10.8 and 12 μm are mounted on the IR-FPA which is cooled to 80 K by the sterling cycle cooler. All channels are optically co-aligned behind the common field stop which is also cooled to 80 K.

The conical scanning geometry allows views of the two on-board blackbody calibration targets which will be described in Section 4.2.1.

The detector signals are sampled at ~ 13 kHz and digitized at 12bit resolution. Since (A)ATSR is optimized for climate quality measurements of SSTs, the dynamic range or the IR channels is typically from 210 to 315 K.

NEΔT values vary slightly between instruments but are typically ~ 0.05 K at 3.7 µm and ~ 0.03 K for the 10.8 and 12 µm for a scene brightness temperature of 270 K [1].

3.2 Sea and Land Surface Temperature Radiometer (SLSTR)

The sea and land surface temperature radiometer (SLSTR) to be flown on ESA's Sentinel-3 mission is a multichannel scanning radiometer that will continue the 21 year data sets of the ATSR series. As its name implies, measurements from SLSTR will be used to retrieve global SSTs to an uncertainty of <0.3 K traced to international standards (Figure 7).

SLSTR shares many features of the ATSR sensors including: thermal IR spectral bands that are cooled using a Stirling cycle cooler, a dual view allowing the same terrestrial scene to be viewed through two atmospheric paths, a nadir view and an along-track view at 55° zenith angle, two blackbody sources to provide continuous calibration of the IR channels, and a diffuser based VISCAL source for calibrating the solar reflectance bands. The optical design of the instrument is a development of the ATSR scanning concept to provide a 1400 km nadir view and 750 km inclined view. The spectral bands for SLSTR listed in Table 2 are based on those from the ATSR series but include additional SWIR bands (S4, S6) for improved daytime cloud detection and thermal channels with dynamic range up to 500 K for fire

FIGURE 7 The structural thermal model of the sea and land surface temperature radiometer under preparation for thermal testing.

Thermal Infrared Satellite Radiometers Chapter | 2.3

TABLE 2 SLSTR Spectral Bands

Band Number	Central Wavelength (μm)	Bandwidth (μm)	Spatial Resolution at Nadir (km)	Function
S1	0.555	0.020	0.5	Chlorophyll
S2	0.659	0.020	0.5	Vegetation index
S3	0.870	0.020	0.5	Vegetation index
S4	1.375	0.015	0.5	Thin cirrus cloud detection
S5	1.610	0.060	0.5	Clouds, fire
S6	2.225	0.050	0.5	Clouds, fire
S7	3.700	0.380	1.0	Night-time SST, fire
S8	10.850	0.900	1.0	SST/LST
S9	12.000	1.000	1.0	SST/LST
S7F	3.700	0.380	1.0	Fire
S8F	12.000	0.900	1.0	Fire

detection (S7F, S8F). A more detailed description of SLSTR and its expected performance is described the paper by Coppo et al. [2].

3.3 Advanced Very High Resolution Radiometer (AVHRR)

The Advanced Very High Resolution Radiometer, AVHRR, is a polar orbiting scanning radiometer that has been operational since the 1970s and is still the workhorse instrument providing operational thermal IR and solar reflected (Figure 8). AVHRR was first launched on the Television Infrared Observation

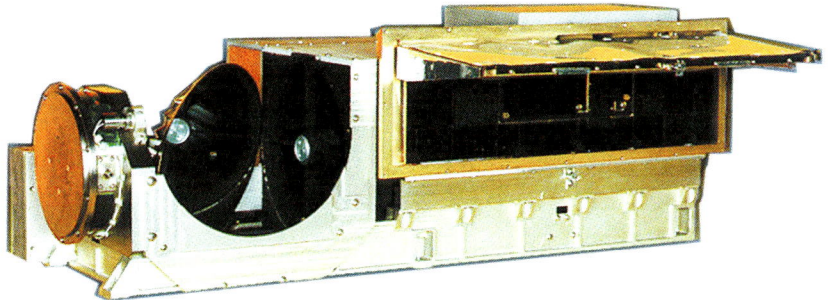

FIGURE 8 The advanced very high resolution radiometer. *Photo Courtesy NOAA.*

TABLE 3 AVHRR Spectral Bands

Band Number	AVHRR/1 TIROS-N NOAA-6, 8, 10	AVHRR/2 NOAA-7,9,11,12,14	AVHRR/3 NOAA-15,16,17,18,19 METOP-A,B,C	Detector Type
1	0.58–0.68 μm	0.58–0.68 μm	0.58–0.68 μm	Silicon
2	0.725–1.00 μm	0.725–1.00 μm	0.725–1.00 μm	Silicon
3A			1.58–1.68 μm	InGaAs
3B	3.55–3.93 μm	3.55–3.93 μm	3.55–3.93 μm	InSb
4	10.40–11.30 μm	10.40–11.30 μm	10.40–11.30 μm	HgCdTe
5	Band 4 —repeat	11.50–12.50 μm	11.50–12.50 μm	HgCdTe

Satellite-N (TIROS-N) in 1978 with spectral channels at 0.60, 0.86, 3.7, and 10.85 μm. There have been several developments since (see Table 3) with the present version, AVHRR/3 comprising six spectral channels, at 0.60, 0.86, 1.6, 3.7, 10.85, and 12 μm. In 2013 there were four AVHRRs in operation, two on NOAA and two on METOP to meet the operational requirements for daily global coverage. METOP is on a Sun synchronous orbit with a 21:31 local time at the Ascending Node Crossing (ANX). The NOAA satellites are on Sun synchronous orbits with ANX times of 19:15 and 13:30, so-called morning and afternoon crossing times.

AVHRR comprises a 20.3 cm main Cassegrain type telescope common to both the VIS detector assembly and the IR detector assembly. The IR detectors are radiatively cooled to 105 K by a two stage cooler. A rotating elliptical scan mirror mounted ahead of the 20 cm diameter telescope provides alternate views of the Earth, cold space and the Internal Calibration Target (ICT). The scan mechanism continuously rotates at 360 rpm to provide 2048 Earth view pixels with view zenith ranging ± 55°. The detector IFOV is nominally 1.3 mrad square to give a spatial resolution of ∼1.1 km at nadir. The detector signals are sampled at 40 kHz and transmitted at 10bit digitization. The dynamic range of the IR channels is from 180 to 335 K with NEdT ∼0.12 K.

The scanner also provides views of an ICT and a view to cold space to provide a two point on-board calibration.

3.4 MOderate Resolution Imaging Spectroradiometer (MODIS)

The MOderate Resolution Imaging Spectroradiometer (MODIS) is a wide swath (±55°) cross-track scanning radiometer with 20 reflective solar bands

FIGURE 9 MODIS *Photo courtesy NASA.*

with wavelengths from 0.41 to 2.2 µm and 16 thermal IR bands from 3.7 to 14.4 µm (Figure 9). The spectral bands have been chosen to meet specific observation requirements as listed in Table 4. Two MODIS instruments were in operation at the time of writing, MODIS Terra (EOS AM) with a 10:30 descending node crossing time, and MODIS Aqua (EOS PM) with a 13:30 ascending node crossing time. As for the AVHRR sensors, the two satellites provide global coverage in 1 to 2 days.

The detectors are mounted on four separate focal plane assemblies (FPAs) at VIS, NIR, short and mid-wave IR (SWIR/MWIR), and thermal IR (TIR). MODIS has an along track view of 10 km at nadir, which means that there are 10 detector elements for each of the 1 km bands, 20 elements for the 0.5 km bands, and 40 elements for the 0.25 km bands. Bands 13 and 14 are split into two arrays of 10 elements each. The IR channels are passively cooled to 83 K.

Light from the Earth is directed into a two mirror a focal telescope via a double sided scan mirror rotating at 20.3 rpm. The scanning mirror also allows views of the on-board calibration sources and a space view.

3.5 Visible Infrared Imaging Suite (VIIRS)

The Visible Infrared Imaging Suite (VIIRS) is intended as the replacement for AVHRR and MODIS (Figure 10). The first VIIRS instrument was launched in October 2011 on the Suomi National Polar-orbiting Partnership satellite.

TABLE 4 MODIS Spectral Bands

Band	Center Wavelength	Bandwidth	Function	IFOV at Nadir (km)	Required SNR/NEdT
1	0.645	0.050	Land/cloud/ aerosols boundaries	0.25	128
2	0.858	0.035		0.25	201
3	0.469	0.020	Land/cloud/ aerosols properties	0.50	243
4	0.555	0.020		0.50	228
5	1.240	0.020		0.50	74
6	1.640	0.024		0.50	275
7	2.130	0.050		0.50	110
8	0.412	0.015	Ocean color/ phytoplankton/ biogeochemistry	1.00	880
9	0.443	0.010		1.00	838
10	0.488	0.010		1.00	802
11	0.531	0.010		1.00	754
12	0.551	0.010		1.00	750
13	0.667	0.010		1.00	910
14	0.678	0.010		1.00	1087
15	0.748	0.010		1.00	586
16	0.869	0.015		1.00	516
17	0.905	0.030	Atmospheric water vapor	1.00	167
18	0.936	0.010		1.00	57
19	0.940	0.050		1.00	250
20	3.750	0.180	Surface/cloud temperature	1.00	0.05 K
21	3.960	0.060		1.00	2.00 K
22	3.960	0.060		1.00	0.07 K
23	4.050	0.060		1.00	0.07 K
24	4.466	0.065	Atmospheric temperature	1.00	0.25 K
25	4.516	0.067		1.00	0.25 K
26	1.375	0.030	Cirrus clouds water vapor	1.00	150(SNR)
27	6.715	0.360		1.00	0.25 K
28	7.325	0.300		1.00	0.25 K

TABLE 4 MODIS Spectral Bands—cont'd

Band	Center Wavelength	Bandwidth	Function	IFOV at Nadir (km)	Required SNR/NEdT
29	8.550	0.300	Cloud properties	1.00	0.05 K
30	9.730	0.300	Ozone	1.00	0.25 K
31	11.03	0.500	Surface/cloud temperature	1.00	0.05 K
32	12.02	0.500		1.00	0.05 K
33	13.34	0.300	Cloud top altitude	1.00	0.25 K
34	13.64	0.300		1.00	0.25 K
35	13.94	0.300		1.00	0.25 K
36	14.24	0.300		1.00	0.35 K

MODIS Web site.

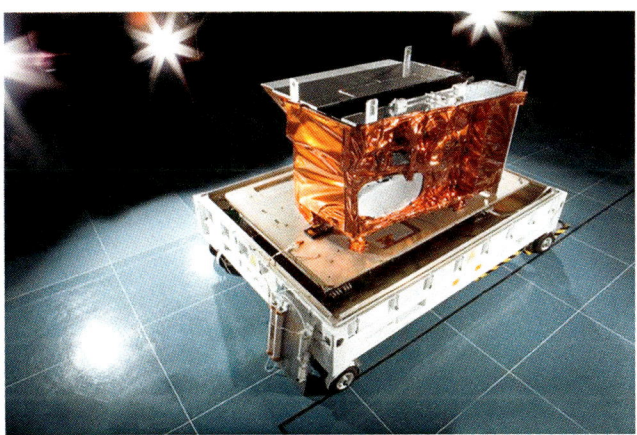

FIGURE 10 VIIRS. *Photo courtesy NASA.*

Suomi is acting as a gap-filler between NOAA's Polar Operational Environmental Satellites and the Joint Polar Satellite System that will supersede them.

Although VIIRS retains some of the spectral bands of MODIS and AVHRR needed for SST retrievals (Table 5), there are significant differences to the instrument architecture that are worth noting.

First, a rotating three mirror anastigmatic telescope is used instead of a fixed telescope and scanning mirror. A half angle mirror rotating at half the telescope speed is used to compensate for the image rotation. A rotating telescope has some advantage over the single scanner used for MODIS and

TABLE 5 VIIRS Spectral Bands

Band Number/ Gain	Center Wavelength	Bandwidth	Function	IFOV at Nadir along Track × across Track (km)
M1 dual	0.412	0.020	Ocean color, aerosols	0.742 × 0.259
M2 dual	0.445	0.020	Ocean color, aerosols	0.742 × 0.259
M3 dual	0.488	0.020	Ocean color, aerosols	0.742 × 0.259
M4 dual	0.555	0.020	Ocean color, aerosols	0.742 × 0.259
I1	0.640	0.080	Imagery, vegetation	0.371 × 0.387
M5 dual	0.672	0.020	Ocean color, aerosols	0.742 × 0.259
M6 single	0.746	0.015	Atmospheric correction	0.742 × 0.776
I2, single	0.865	0.039	Vegetation	0.371 × 0.387
M7 dual	0.865	0.039	Ocean color, aerosols	0.742 × 0.259
DNB, multiple	0.7	0.400	Imagery	0.742 × 0.742
M8, Single	1.24	0.020	Cloud particle size	0.742 × 0.776
M9, single	1.38	0.015	Cirrus cloud cover	0.742 × 0.776
M10, single	1.61	0.060	Snow fraction	0.742 × 0.776
I3, single	1.61	0.060	Binary snow map	0.371 × 0.387
M11, single	2.25	0.050	Clouds	0.742 × 0.776
M12, single	3.70	0.180	SST	0.742 × 0.776
I4, single	3.74	0.380	Imagery, clouds	0.371 × 0.387
M13, dual	4.05	0.155	SST, fires	0.742 × 0.259

TABLE 5 VIIRS Spectral Bands—cont'd

Band Number/ Gain	Center Wavelength	Bandwidth	Function	IFOV at Nadir along Track × across Track (km)
M14, single	8.55	0.300	Cloud top properties	0.742 × 0.776
M15, single	10.76	1.000	SST	0.742 × 0.776
I5, single	11.45	1.900	Cloud imagery	0.371 × 0.387
M16, single	12.01	0.950	SST	0.742 × 0.776

AVHRR since the full telescope can view the Earth scene and calibration sources without the need for additional optics.

VIIRS aggregates the signals from several detector subpixels to provide an approximately constant and square (approximately) 1 × 1 km pixel resolution across the full ±55° swath. This is intended to overcome the increasing size and distortion of the pixel footprint at wide view angles that is common to all scanning radiometers. Since the aggregation varies with view angle the effective $A\Omega$ of the instrument is no longer constant with view angle and has to be accounted for in the calibration algorithms.

3.6 Spinning Enhanced Visible and Infrared Imager (SEVIRI)

An example of a geostationary instrument used for SST measurements is the Spinning Enhanced Visible and Infrared Imager (SEVIRI) is the main optical payload on the Meteosat Second Generation satellites (Figure 11). SEVIRI comprises four visible/near IR bands from 0.4 to 1.6 μm and eight IR channels from 3.9 to 13.4 μm Table 6. The resolution at subsatellite point from 36,000 km is 1 km for the VIS-SWIR channels and 3 km for the IR channels.

A 50 cm diameter scan mirror directs the scene radiance to the main telescope which is focused onto the detectors. The Earth image is generated by using the satellite spin to collect the east–west image rows, and moving the scan mirror in the north south direction to produce 1527 scan lines [4]. A complete Earth disc image is obtained every 12 min. A "flip-flop" mechanism is activated at the end of each image acquisition to position the IR calibration source into the instrument field of view. The full cycle takes 15 min to complete.

The IR detectors are radiatively cooled and because the cooling is passive, there is a seasonal variation of the detector temperatures from 85 to 95 K. The signals are provided at 10bit resolution from 0 to 1023.

FIGURE 11 SEVIRI. *From Ref. [3].*

4. CALIBRATION MODEL

Level-1 radiometric data for thermal IR sensors are usually expressed as a top-of-atmosphere brightness temperature in Kelvin which equates the measured scene radiance to a blackbody at temperature T. Although this is a convenient unit to use in that it is easy to understand and gives a first order approximation of the scene temperature, at MWIR-TIR wavelengths radiance is not linear with temperature and calibration must be performed as a function of the scene radiance.

As discussed previously, the signal measured by each channel is converted to a voltage and then digitized.

We invert equation 7 to get the scene radiance as a function of DN such that

$$L_{scene} = g^{-1}\left(DN_{scene} - DN_{offset}\right) \qquad (14)$$

For most of the sensors described in this chapter, a two-point calibration scheme is usually adopted with sources of known radiance L_1 and L_2

TABLE 6 SEVIRI Spectral Bands

Band Number	Central Wavelength (μm)	Bandwidth (μm)	Spatial Resolution at Nadir (km)	Function
HRV	0.75	0.30	1	Imaging
VIS 0.6	0.635	0.15	1	Clouds, vegetation, imaging
VIS 0.8	0.81	0.14	1	Clouds, vegetation, imaging
NIR 1.6	1.64	0.28	1	Snow and ice cloud discrimination
IR6.2	6.25	0.88	3	Water vapor
IR3.9	3.92	1.80	3	SST, LST, clouds, fog
IR7.3	7.35	1.00	3	Water vapor
IR8.7	8.70	0.80	3	SST, LST, clouds
IR9.7	9.66	0.56	3	Ozone
IR10.8	10.80	2.00	3	SST, LST, clouds
IR12.0	12.00	2.00	3	SST, LST, clouds
IR13.4	13.40	2.00	3	Cirrus detection

producing average signals $\langle DN_1 \rangle$ and $\langle DN_2 \rangle$, so assuming that the response is linear with radiance (or at least adjusted for nonlinearity), we obtain

$$L_{\text{scene}} = XL_1 + (1 - X)L_2 + \Delta L_{\text{offset}} \tag{15}$$

where

$$X = \frac{DN_{\text{scene}} - \langle DN_2 \rangle}{\langle DN_1 \rangle - \langle DN_2 \rangle} = \frac{L_{\text{scene}} - L_2}{L_1 - L_2} \tag{16}$$

The combined uncertainty in the radiometric calibration can be derived analytically (assuming that the sources of uncertainty are uncorrelated at a first order) using

$$(\text{ur})^2 = \sum_{i=1}^{n} \left(\frac{\partial r}{\partial a_i} \text{ua}_i \right)^2 \tag{17}$$

The radiometric errors in terms of a brightness temperature error can then be derived using

TABLE 7 Components of IR Calibration Uncertainty Budget

Parameter	Term	Partial Derivative
Upper calibration source radiance	uL_1	$\frac{\partial L}{\partial L_1} = X$
Lower calibration source radiance	uL_2	$\frac{\partial L}{\partial L_2} = 1 - X$
Digital counts noise (NEΔT)	uDN	$\frac{\partial L}{\partial DN} = \frac{L_1 - L_2}{\langle DN_1 \rangle - \langle DN_2 \rangle}$
Upper calibration source noise	$u\langle DN_1\rangle$	$\frac{\partial L}{\partial \langle DN_1 \rangle} = X \frac{L_1 - L_2}{\langle DN_1 \rangle - \langle DN_2 \rangle}$
Lower calibration source noise	$u\langle DN_2\rangle$	$\frac{\partial L}{\partial \langle DN_2 \rangle} = (X - 1) \frac{L_1 - L_2}{\langle DN_1 \rangle - \langle DN_2 \rangle}$
Nonlinearity	uNL	$\frac{\partial L}{\partial NL} = \frac{L_1 - L_2}{\langle DN_1 \rangle - \langle DN_2 \rangle}$
Offset variations	ΔL_{offset}	$\frac{\partial L}{\partial \Delta L_{\text{offset}}} = 1$

$$uT = uL \left(\frac{\partial L}{\partial T} \bigg|_T \right)^{-1} \tag{18}$$

Applying this to the calibration equation we get the breakdown of uncertainties given in Table 7.

Although the main source of uncertainty in the measurement should be with the calibration sources which will be addressed in the following sections, we must carefully consider other sources since these can be significant if neglected.

4.1 Radiometric Noise

Radiometric noise is usually considered as a purely random white noise signal that will reduce with multiple samples. As such the noise signals from the calibration sources can be considered as negligible. However, this is not always true and some apparent "noise" sources have detectable signatures that can give rise to "features" in images. Therefore, when characterizing the noise performance of an instrument it is important to consider the time series of measurements to quantify any effect and determine the impact on the data quality.

An example is for ATSR-1 where an additional significant noise contribution from magnetic cooler drive pick-up was observed by careful inspection of ATSR images over uniform scenes and is clear in Fourier analysis of the blackbody signals (see Figure 12). Due to cooler operating frequency the signal is almost an odd number of half cycles per scan and drifts slowly in relation to the scan cycle.

4.2 Nonlinearity

So far we have assumed that the instrument response is linear with scene radiance. However, for photoconductive HgCdTe detectors used in the thermal IR

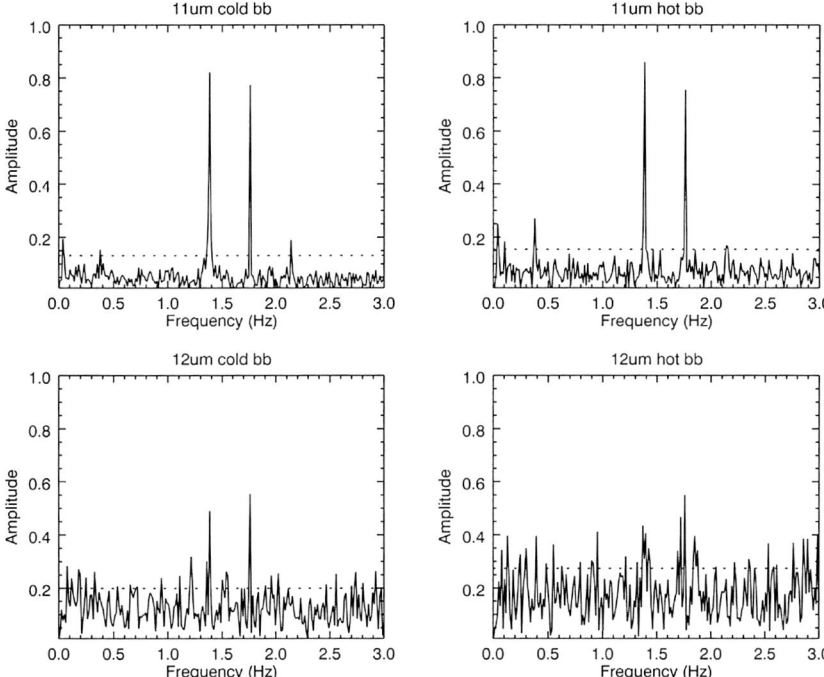

FIGURE 12 Fourier transform spectrum of ATSR blackbody noise signals. This analysis was performed over 512 scans and after the removal of background drifts due to the temperature variations of the blackbodies. It should be noted that because the cooler frequency is 43 Hz and the scan period is 6.7 Hz, the noise signal is aliased strongly.

region it is known that the detector response "falls off" with increasing photon flux (Refs [5,6]). Essentially the electron–hole recombination rate increases as the number of carriers (electrons and holes). A nonlinear response may also be due to the design and components used in the detector's amplifier circuit.

We can investigate the impact of nonlinearity on the calibration by simulating the nonlinear response vs. scene temperature and passing this through the calibration equations. In the example shown in Figure 13 we have simulated the effect on a two point calibration with a cold space view and blackbody at 285 K (as for AVHRR, MODIS) and with two on-board blackbodies at 255 and 300 K (as for AATSR). We have assumed a nonlinearity of ~5%. At the calibration target temperatures the brightness temperature errors are close to zero. Outside this range nonlinearity can give rise to significant errors. This was a problem for the earlier AVHRR sensors where nonlinearity was not accounted for and brightness temperature errors of up to 2 K were measured [7] and postprocessing corrections were provided by NOAA.

The approach used by (A)ATSR to minimize the effects of nonlinearity was to have the two calibration sources at either end of the temperature range

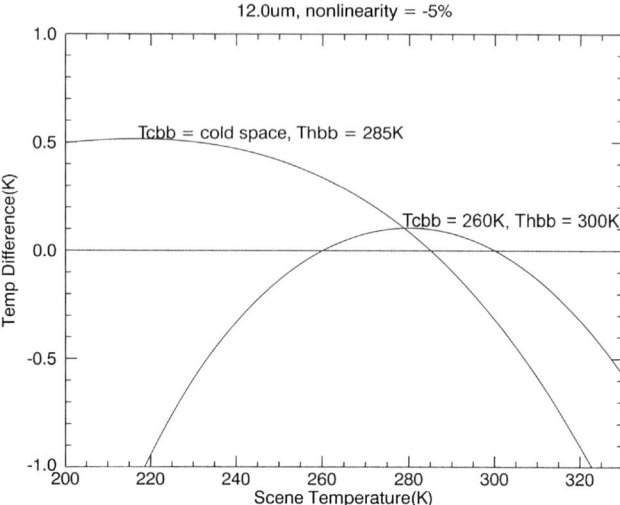

FIGURE 13 Effect of nonlinearity errors on calibration.

needed for SST measurements. Even with 5% nonlinearity the brightness temperature errors in this range are below 0.1 K. However, this does not completely solve the problem and the errors increase rapidly above and below this range.

Hence a key activity of the ground calibration activity is to characterize the response nonlinearity to provide a correction as described in Section 6.2.5.

4.3 Offset Variations

The stability of the sensor during the calibration and between calibration observations must be considered in the instrument design. In the performance model, a significant fraction of the measured signal comes from the thermal background of the instrument. Depending on the sensor design, the background signal can be between 10% and 20% of the total signal. Thus the stability of the thermal background is an integral part of the calibration scheme. To minimize the impact of these variations, the instrument thermal design should ensure that the main optical enclosure is stable around an orbit. Due to satellite design constraints however, it is not possible to completely eliminate temperature variations, so it is preferable to view the calibration sources as frequently as possible.

5. ON-BOARD CALIBRATION

As described earlier, different calibration schemes have been adopted in the various instruments. Although these are obvious technical differences they are assumed to be black-body sources. So before discussing the merits or otherwise of each approach, we will look at the basic physics of a blackbody source to understand the main sources of uncertainty.

For a perfect blackbody with emissivity $\varepsilon = 1.0$, the radiance can be derived from its temperature using the Planck function. In the real world, blackbodies are not completely black, i.e., $\varepsilon_\lambda < 1.0$ so the actual radiance will have a small reflected component so that

$$L_{\lambda,\text{BB}} = \varepsilon_\lambda L_\lambda(T_{\text{BB}}) + (1 - \varepsilon_\lambda) L_{\lambda,\text{back}} \tag{19}$$

L_{back}, is the background radiance from the instrument optics and structure corresponding to an instrument temperature T_{back}. Assuming that the instrument cavity is sufficiently black, $\varepsilon > 0.9$, it is possible to approximate the background radiance term using the Planck function for a temperature T_{inst}. Therefore the blackbody radiance becomes

$$L_{\lambda,\text{BB}} = \varepsilon_\lambda L_\lambda(T_{\text{BB}}) + (1 - \varepsilon_\lambda) L_\lambda(T_{\text{back}}) \tag{20}$$

For this to be valid, the target aperture must completely fill the instrument beam.

Using the same approach as for the calibration equation we obtain the following sources of uncertainty in the blackbody radiances.

From Table 8 we note that the uncertainty in blackbody radiance is dependent on several factors. For nonunity emissivity, uncertainties will be dependent on the thermal background in the view of the blackbody (L_{back}) and how well this is known (uL_{back}). If the enclosure supporting the blackbody is isothermal and is well measured, then it may be possible to neglect emissivity errors. However, if the instrument is not at the same temperature as the blackbody then errors due to nonblackness can become very significant,

TABLE 8 Components of Blackbody Source Uncertainty Budget

Parameter	Term	Partial Derivative
Emissivity errors	$u\varepsilon$	$\frac{\partial L}{\partial \varepsilon} = L_{\text{bb}} - L_{\text{back}}$
Errors in the background radiance due to blackbody emissivity < 1.0	uL_{back}	$\frac{\partial L}{\partial T_{\text{back}}} = (1 - \varepsilon) \frac{\partial L}{\partial T}\big|_{T\text{back}}$
Blackbody thermometry errors	uT_{cal}	$\frac{\partial L}{\partial T}\big|_{T\text{bb}}$
Blackbody temperature gradients	uT_{grad}	$\frac{\partial L}{\partial T}\big|_{T\text{bb}}$
Blackbody temperature stability	uT_{stab}	$\frac{\partial L}{\partial T}\big|_{T\text{bb}}$
Knowledge of instrument spectral response	$u\lambda$	$\frac{\partial L}{\partial \lambda}\big|_{T\text{bb}}$

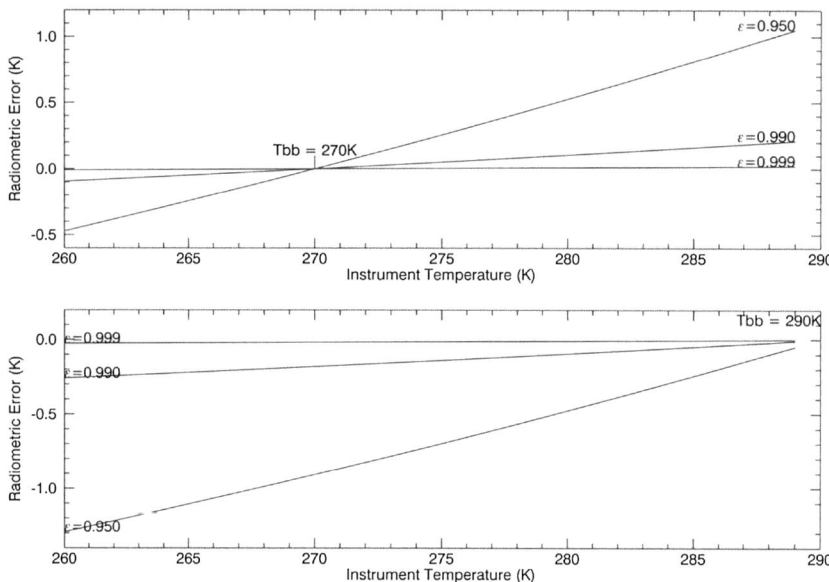

FIGURE 14 Radiometric errors at 11 μm due to nonblackness of target emissivity. The example shown here assumes a blackbody at 270 K and at 290 K.

particularly when overall radiometric errors <0.1 K are needed, see Figure 14. In this situation it is vital that the instrument temperatures in the view of the plate are at least measured to provide a better estimation of the blackbody radiances. But this alone is insufficient since we also need to know the emissivity of the target.

As well as the absolute temperature of the blackbody thermometers (uT_{cal}), the gradients across the baseplate (uT_{grad}) and stability during the calibration period (uT_{stab}) are also critical components. Simply quoting traceability to ITS-90 (or international standards) is insufficient, careful consideration has to be given to the thermal design of the instrument to ensure that other error sources are minimized. Particularly, since the emissivity is high and therefore the radiative coupling is strong, the thermal design of the instrument enclosure and blackbody source has to be considered together rather than independent entities.

5.1 Calibration Sources

5.1.1 Deep Space View

Adopted for AVHRR, MODIS, and VIIRS, a view to deep space provides a value for the radiometric offset, since it can be assumed that the scene radiance is zero (or close to zero). In many ways this is an ideal source since no electronics are needed for power or monitoring and has zero mass. Provided that the full optical chain and aperture are included in the view the uncertainty in this source should be negligible. The main drawback with this approach is that zero radiance is usually far lower than most terrestrial scene radiances and any residual nonlinearity errors are amplified.

5.1.2 Black Plates or Structured Black Plate

The main advantage of "black plate" type calibration source is that it can easily be accommodated within an instrument's structure without adding significant mass or volume. However, because the thermal environment of the blackbody is not uniform it is not possible to achieve high emissivities. Without any structure, the surface emissivity of a black painted plate is ~0.95. With some surface structure, emissivity ~0.99 can be achieved but this is conditional on the blackbody being at the same temperature as the instrument environment.

AVHRR's ICT is a flat plate with a black-painted carbon honeycomb surface with a stated emissivity of 0.992 [7]. The temperature is monitored by four platinum resistance thermometers, PRTs mounted in the plate. The PRTs are calibrated against NIST (National Institute for Standards and Technology) traceable PRTs over the range 278–298 K to cover the operational range. The overall uncertainty estimate of the blackbody source is 0.4 K and is mainly attributed to temperature gradients and measurement errors.

The MODIS and VIIRS blackbody sources comprise a v-grooved plate as illustrated in Figures 15 and 16. The BB is viewed every scan to provide a near continuous calibration. The combination black-coating and the 45° grooves

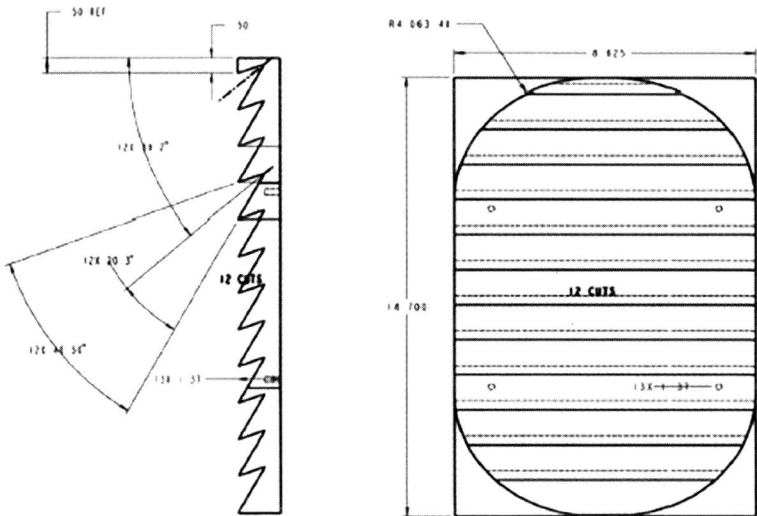

FIGURE 15 Schematic of MODIS on-board calibration blackbody. *From Ref. [8].*

FIGURE 16 MODIS on-board blackbody source. *From Ref. [9].*

provide an effective emissivity of ~ 0.992. The BB is monitored by 12 precision thermistors traceable to NIST standards to determine the temperature of the plate to an uncertainty of ± 0.1 K. The BB operates typically at 290 K, but is capable of operating from 274 to 320 K to permit additional calibration measurements, in particular to characterize nonlinearity.

5.1.3 Cavity Blackbodies

As illustrated by Figure 14, it is desirable to use a blackbody source with very high emissivity $\varepsilon \sim 1$ and where the emissivity can be characterized with very low uncertainty. The conventional and most reliable approach is to use an isothermal cavity with re-entrant cone geometry as illustrated in Figure 17.

With careful design, the blackbody emissivity becomes less sensitive to the properties of the surface coating and in principle more stable over time. As a general design rule, the cavity length needs to be much longer than the diameter and $D > 2H$ [10]. For ground-based blackbodies it is fairly straightforward to design a cavity with a good length to radius ratio with $\varepsilon \sim 1$, although consideration has to be given to the temperature gradients and stability. For flight blackbodies where mass and volume are at a premium, it is

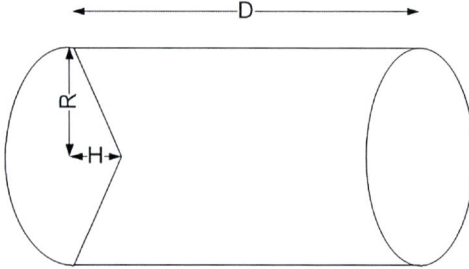

FIGURE 17 Basic re-entrant cone blackbody cavity with length D, radius R, and cone height H. *From Ref. [10].*

not always possible to achieve the optimum geometry and careful design analysis is needed to obtain the best trade-off. For the (A)ATSR blackbodies a length-to-radius ratio of 2.74 was used to provide emissivity of ~ 0.999.

The (A)ATSR instruments used two cavity blackbody sources viewed in turn every scan by all of the spectral channels, to provide the cold and hot reference radiances, Figures 18 and 19. One blackbody temperature "floats" cold at the optimized temperature of the instrument fore-optics enclosure, while the other is heated at a constant power to provide a hot reference. The blackbodies and optical system are designed such that the instrument's full optical beam has an unobstructed view of the baseplate; is not clipped by any aperture or the blackbody cavity walls. Each baseplate has been designed to maintain a uniform temperature. The cavities provide a very high effective emissivity ($\varepsilon > 0.999$) by a combination of Martin Marietta black coating and a re-entrant cone base geometry. These are described in detail in Mason et al. [11].

The temperatures of the blackbody bases are measured with high accuracy precision platinum resistance thermometers (PRTs). These are calibrated with their flight electronics against a transfer standard PRT traceable to ITS-90.

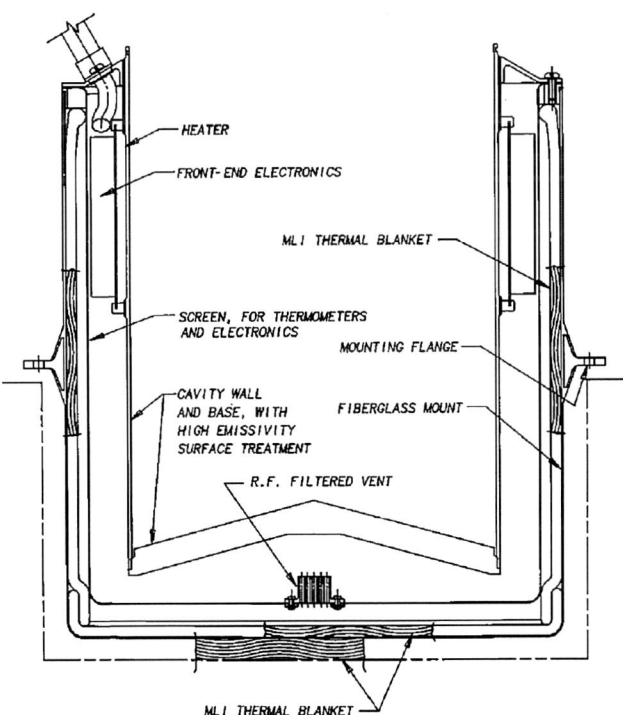

FIGURE 18 Schematic diagram of the ATSR on-board blackbody cavity. *From Mason et al. [11].*

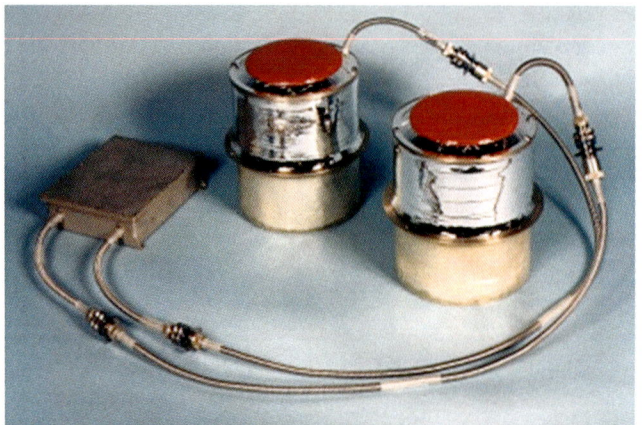

FIGURE 19 AATSR blackbody calibration sources. *Photo courtesy ABSL.*

6. PRE-LAUNCH CHARACTERIZATION AND CALIBRATION

6.1 Blackbody Calibration

If we are to provide climate quality radiometric measurements the radiances from the on-board sources need be traced to international standards to the required uncertainty levels. In the case of AATSR and more recently SLSTR, radiometric uncertainties below 0.1 K are needed. This is extremely challenging given that the thermal environment of the BB cavity is a key component of the measured radiance. Laboratory measurements of blackbody sources under controlled conditions have been performed using instruments such as the NPL amber radiometer [12] and the NIST TXR radiometer [13] with uncertainty estimates around 50 mK over a limited temperature range. These measurements are useful for validating the blackbody radiance model (Section 5) but cannot be considered as a calibration because the thermal environment is not truly representative of the flight conditions. For practical purposes the in-flight calibration utilizes the radiance model for which we need knowledge of the cavity temperatures, thermal environment, and the emissivity.

6.1.1 Thermometry

For instruments described in this chapter, the calibration source temperatures are typically measured by platinum resistance thermometers PRTs mounted in the baseplate and are usually quoted as being traceable to international standards. In principle, traceability of the calibration can appear to be straightforward, but the practice is quite complicated and needs a detailed understanding of the sources of uncertainty.

We first need to understand what we are trying to measure. For the radiance we need to know the thermodynamic temperature of the emitting surfaces.

Direct measurement of the thermodynamic temperature is not possible, and instead we refer to the International-Temperature-Scale of 1990 (ITS-90). The temperature scale is defined at various defined points that are based on thermodynamic states of pure chemical elements for example the triple point of water at 273.16 K. The scale is disseminated through standard platinum resistance thermometers (SPRTs).

Ideally one would use SPRTs directly with flight blackbodies, but their limited availability and their fragility mean that they are not at all suited for flight use. We, therefore, use SPRTs as a reference to cross-calibrate the flight PRTs that are generally more robust.

One approach is to calibrate the flight PRTs as stand-alone items by placing thermometers in a cryostat and cross-calibrate against SPRTs. Since PRTs are not mounted in their actual flight configuration we need to consider uncertainties due to their mounting in the flight units. This could be due to poor thermal contact, self-heating, mechanical stress, heat leaks down the wires, sensor drift resulting from handling.

An alternative is to calibrate the PRTs mounted in blackbody in isothermal enclosure, which is the approach used for the AATSR and SLSTR blackbodies [11]. Here SPRTs are temporarily mounted onto the cavity to provide the traceability to ITS-90. This approach has the advantage that any systematic errors such as self-heating, thermal contact, heat leaks will be calibrated out in the measurement.

In both approaches the calibration should be performed using the flight electronics and harnessed to minimize any systematic errors introduced by the electronics response functions.

6.1.2 Emissivity

The emissivity of the blackbody cavities is usually derived from the measurements of samples the black coating material and modeling to account for the cavity geometry. This is generally accepted to be more reliable and cost-effective than a direct measurement of the cavity emissivity. The measurement of the black coating reflectance is a standard measurement service performed by standards laboratories such as NPL and NIST. The measured reflectance is then incorporated into a geometric model blackbody cavity and a Monte Carlo Model simulation is employed to determine the effective cavity emissivity [14].

For the ATSR blackbodies, validation of the calculated emissivity was performed at MSSL by comparisons against a reference blackbody with an emissivity >0.999 [11]. The tests were conducted in a purpose built vacuum chamber to provide a controlled thermal environment and to eliminate any effects due to water vapor. The reference blackbody was maintained at a constant temperature $\sim 20\,°C$ and the thermal shroud at a much lower temperature to reduce the reflected component of the BB radiation. The signals from the BBs were measured by an IR radiometer that was chopped between the two BBs. When the signals from the two BBs were identical (i.e., a null signal), the measured temperature differences could be used to infer the average emissivity over the wavelength band, by analysis.

The results for the AATSR FM BBs in Table 9 below, reproduced from the test reports, show that the "measured" temperature differences are within 2σ of the calculated values thus giving confidence in the calculated values. It should be noted that direct calibration of the blackbody emissivity is a very difficult measurement to make and requires a full understanding of the background sources. The measurement errors are large (60 mK at 3.7 μm) compared to the required accuracy of the emissivity (35 mK). In addition, the measurements, which covered a wider spectral band (3–6.5 μm) than the 3.7 μm channel, were intended as a confirmation of the calculated value rather than an absolute measurement. It is, therefore, most likely that the calculated values are a better representation of the true target emissivity.

6.2 Instrument Radiometric Calibration

The radiometric calibration testing is a vital component of any satellite instrument program since it is often the only opportunity before launch to operate the complete instrument as a radiometer under flight representative conditions against reference standards. Although the instruments described in this chapter can be considered as "self" calibrating because they utilize on-board sources, it is essential to demonstrate the end-to-end calibration performance over a range of test conditions. The results from these tests provide critical information that is needed for the on-ground processing and to provide references against which to evaluate the in-orbit behavior. Even for so-called repeat builds of the same instrument type, design changes intended to improve the performance (for example, electronics, mechanisms, and optical coatings) can affect the radiometric data quality.

It is recognized that performing a rigorous calibration is a time-consuming and expensive activity. Since the calibration campaign is usually at the end of an instrument development phase, the pressure to drop some or all of the calibration testing can be an attractive proposition for a project manager needing to meet the launch schedule and tight budgets. This is a common problem. When devising a test plan, it is therefore important to consider what information can be only obtained through prelaunch testing at instrument level, and if a test at instrument level is not possible how the information might be obtained.

With these considerations in mind, an instrument level test campaign should aim to cover as a minimum the following:

- Verify the "on-board" radiometric calibration for a range of scene temperatures covering the expected dynamic range of the sensor.
- Characterize the response nonlinearity and derive appropriate corrections.
- Verify that the different channels produce self-consistent results. That is, the measured brightness temperatures in each channel agree.

TABLE 9 Calculated and Measured Emissivities for the AATSR Flight Model BBs

BBC No	Wavelength Band (μm)	Calculated ΔT (mK)	Calc. Weighted Emissivity	Measured ΔT (mK)	Calc. Weighted Emissivity	Estimated 3σ Uncertainty (mK)
FM01	12	50	0.99892	29	0.99937	30
	11	46	0.99898	26	0.99960	30
	3–6.5	71	0.99745	84	0.99697	60
FM02	12	50	0.99892	54	0.99883	30
	11	46	0.99898	41	0.99909	30
	3–6.5	71	0.99745	107	0.99614	60

- Measure the radiometric noise as a function of scene temperature (NEΔT).
- Determine and measure any view angle dependent variation in the radiometric performance.
- Verify the calibration with the on-board blackbodies at different temperatures.
- Investigate the radiometric performance under different thermal conditions, including simulated orbital transient thermal conditions.
- Characterize the radiometric performance at different detector temperatures.

This list is neither intended to be fully prescriptive nor is it exhaustive, and ultimately depends on the requirements for the instrument.

6.2.1 Calibration Facility

Calibration at thermal IR wavelengths is not simply a matter of placing a blackbody calibration source in front of an instrument. Because everything, including the instrument itself, is a source of IR radiation, the thermal environment of the test setup needs to be controlled and monitored. Also, the tests need to be performed under vacuum to achieve the operating temperature range for the instrument and the calibration sources, and to remove errors due to gaseous absorption (particularly CO_2 and H_2O).

For (A)ATSR, the tests were performed in a vacuum chamber with the instrument under test surrounded by temperature controlled panels to allow it to operate at temperatures close to those expected from the thermal model at flight conditions. To accomplish this goal, four main thermal zones in the test facility were controlled. These were an "Earth-shine" plate (ESP) which was used to simulate radiation from the Earth, a "Payload Electronics Module simulator" which mimicked the interfacing panel of the spacecraft, and a "cold box" to provide a uniform space temperature environment around the instrument and a "drum baffle" to shield the instrument to strays from the chamber walls. The ESP was used to support the external calibration targets which could be rotated about the instrument's scan cone, nominally set to be the tank's axis.

ATSR and ATSR-2 were calibrated in a purpose built facility at the Department of Atmospheric Oceanic and Planetary Physics at Oxford University [5]. Although of the same focal length and aperture, AATSR was a larger instrument than ATSR and ATSR-2, and could not be fitted within the Oxford test facility. The AATSR calibration activities were therefore performed in the larger space test chamber at the Rutherford Appleton Laboratory retaining the original design concepts and philosophy developed for the Oxford testing, and a subset of the test equipment, Figures 20–23, [15].

6.2.2 Blackbody Sources

The calibration sources that were used for the (A)ATSR prelaunch calibration and to be used for SLSTR were designed and built by the U.K. Meteorological

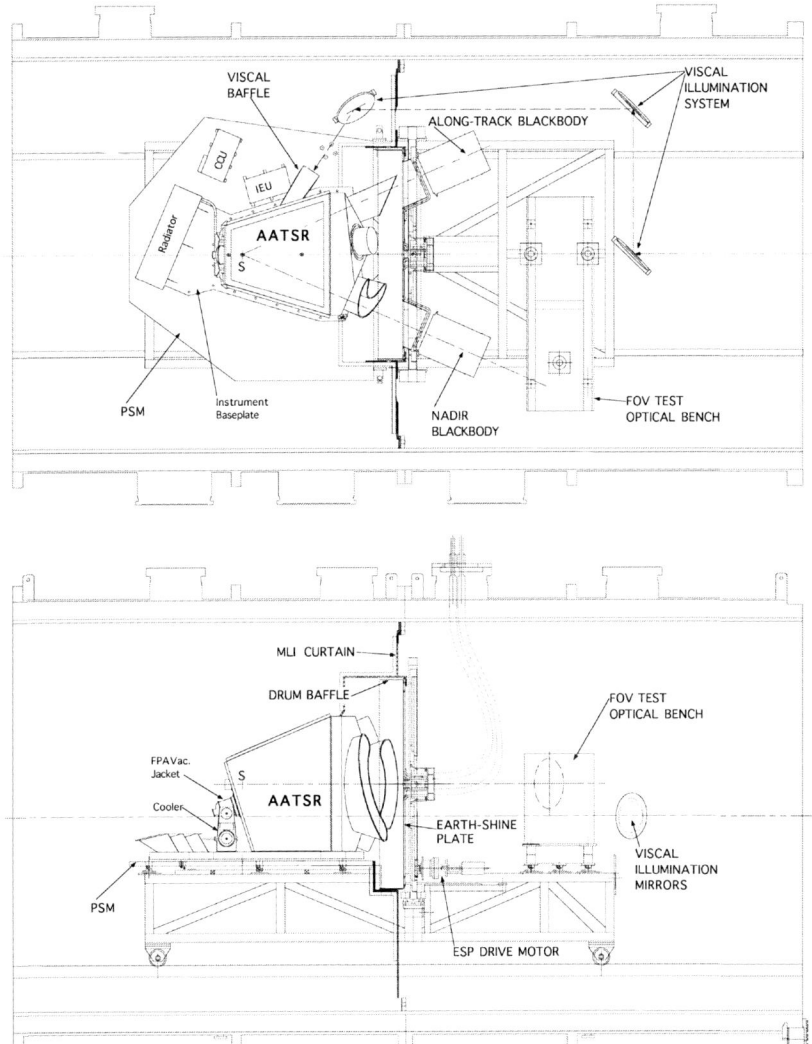

FIGURE 20 Layout of the AATSR calibration equipment in the RAL space test chamber.

Office. The targets consist of 350 mm high copper cylinders with a 250 mm diameter, structured aluminum alloy base and an elliptical entrance baffle plate (236 mm major axis and 160 mm minor axis). The baffles were slightly tapered having a diameter of 252 mm at the base and 240 mm at the entrance plate. The inside faces of the base plates were machined with circularly symmetric groves (10 mm wide, 15° half angle). All of the inside surfaces of the targets were coated with Nextel 101-C10 black paint. The target temperatures were controlled by circulating a refrigerant through the structured base

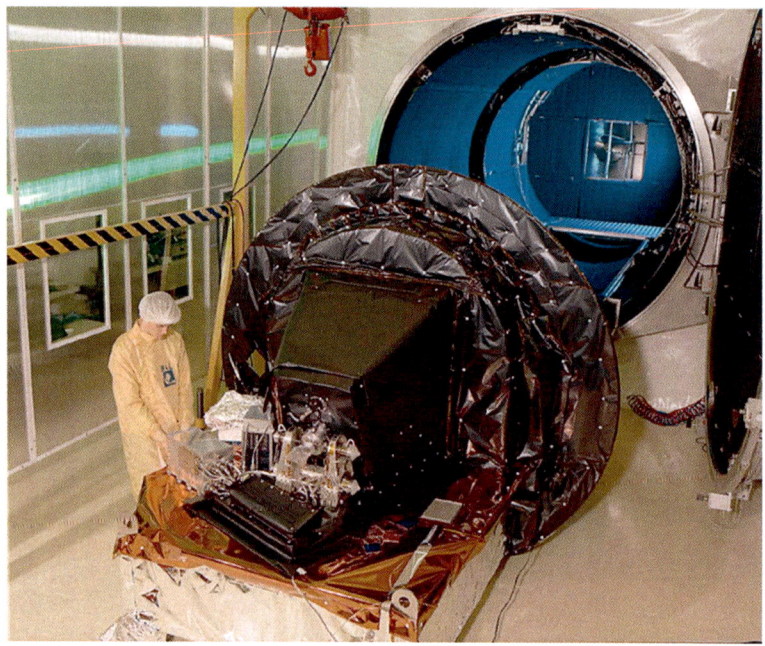

FIGURE 21 AATSR being prepared in the RAL space test facility.

plate and around the baffles. The along-track target had a second fluid loop around the baffle to allow it to be cooled to lower temperatures using liquid nitrogen. The targets are insulated with MLI and supported by aluminum cans.

The temperatures of each target are measured by six precision resistance thermometers, four mounted in the baseplate and two on the baffles. The baseplate sensors are mounted in probes to ensure good thermal contact with the blackbody. Originally PRTs were used, but these showed significant self-heating during ATSR-1 calibration so these were replaced by rhodium-iron resistance thermometers supplied by Oxford Instruments. When calibrated these sensors are accurate to ± 0.01 K traceable to ITS-90 with self-heating of less than 0.001 K when measured by an AC resistance bridge.

Two Huber closed cycle bath refrigerators were used to control the refrigerant temperature. The units used for AATSR were a CC-90 unit having 600 W cooling power at 200 K and ± 0.1 K stability and an HS-80 unit with 250 W cooling at 200 K and ± 0.05 K stability was used for the forward view target.

As for the flight blackbodies the measured radiance is derived in a similar way to those of the flight blackbodies from the Planck function. For the ATSR series, the target emissivity was calculated using a geometric model of the target and the spectral emissivity of the paint. These values have been validated by comparison against the ATSR-2 and AATSR on-board blackbodies as shown in Table 10.

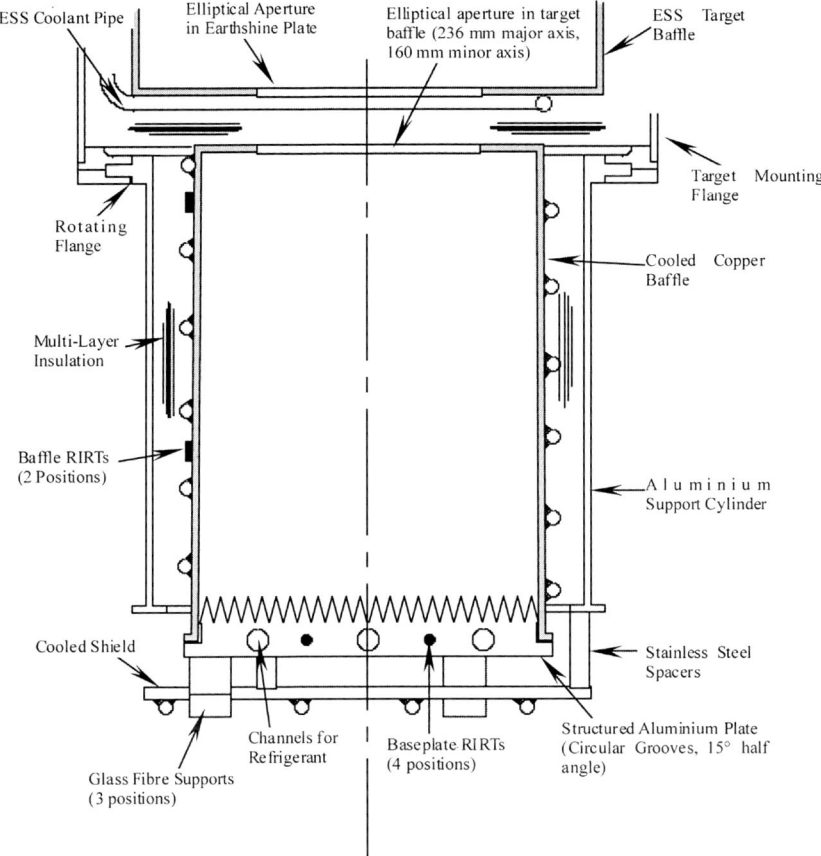

FIGURE 22 Sketch of blackbody source to be used for the ground calibration of the thermal IR channels of (A)ATSR and to be used for SLSTR.

6.2.3 Calibration Test Procedures

Here we describe the basic procedure for performing an IR calibration measurement. The starting point of the measurement is that the instrument and thermal environment temperatures are at their required set-points and stable, and the instrument in its main operational mode. For (A)ATSR this means that the IR detectors are controlled at their operating temperatures of 80 K and switched on, the flight blackbodies are operating at their nominal operating conditions, the scan mirror is scanning and there is continuous acquisition of science data (all spectral bands).

It should be noted that the calibration activities have priority over other instrument activities. In other words the instrument configuration (e.g., changes to software databases) should not be changed during a calibration run,

FIGURE 23 Blackbody calibration source mounted in the AATSR calibration facility at RAL.

as this can affect the calibration analysis. If configuration is changed during a measurement sequence then the test may need to be stopped and restarted. Ideally, the instrument should be in its final flight configuration before starting the calibration, although this is not always possible. Any subsequent changes to the instrument configuration need to be tracked and the impact on the calibration results must be assessed.

Once the required instrument configuration and environment conditions have been met the test can proceed.

The first step in the procedure is to set the external blackbody temperatures. For the AATSR blackbodies, the temperatures are controlled by circulating a refrigerant through the structured baseplate and around the baffles. The eventual temperature reached will be dependent on the difference between the refrigerant temperature and the environment temperatures. For most cases it is sufficient to allow the target to settle at whatever temperature is reached.

TABLE 10 Calculated and Measured Emissivity of External Blackbodies

	Emissivity		
	Calculated	Measured	Difference
3.7 μm	0.99899 ± 0.00035	0.99911 ± 0.00055	0.00012
11 μm	0.99847 ± 0.00036	0.99870 ± 0.00040	0.00023
12 μm	0.99871 ± 0.00037	0.99871 ± 0.00032	0.00000

However, if a specific target temperature is required, e.g., to match the flight blackbody temperatures, some fine tuning of the set point may be required. The transition time between set-points depends on the temperature interval, the required stability and the size of the blackbodies. Typically this would take about 1 h for the ATSR blackbodies.

The measurement sequence is performed by cycling the blackbody over the required temperature range at set intervals (typically 5—10 K). In the case of (A) ATSR with two Earth views, one of the blackbodies would remain as a fixed reference at 280 K. Alternatively the fixed reference could be a cold (liquid nitrogen) target to simulate a space view. The calibration sequence should also include measurements with the external target set at the same temperatures as the on-board blackbodies. This allows a direct comparison of the on-board blackbodies against the reference and removes any uncertainties due to nonlinearity.

Before recording the radiometric calibration measurements it is important to allow blackbody temperatures to stabilize. To achieve the radiometric accuracy requirement for (A)ATSR it was necessary the blackbody temperatures shall drift at no more than 10 mK over a 5 min period. Determining when the steady-state condition is best performed using a simple software test. Typically a status flag is set to indicate the status of each thermometer when the following conditions have been met.

- Red—temperature drift >50 mK over 5 min
- Amber—temperature drift >10 mK and <50 mK over 5 min
- Green—temperature stability has been reached.

Ideally measurements should only be performed when the drift criteria for all baseplate sensors has been achieved. If the blackbodies are to be scanned across the instrument view (as for AATSR), the source's temperature stability should be recovered after it has been moved to each set position.

When the temperature stability criteria have been met, data can be acquired and analyzed to confirm measurements are valid. Ideally the measurement should be performed several times (at least three) to confirm results before proceeding to the next temperature step. This may take several minutes at each set point. Results should be manually checked to verify that the results are repeatable!

The process is repeated for each set point temperature and/or blackbody position.

In all steps of the measurement, a careful log of all activities must be recorded.

6.2.4 Data Analysis

An important element is the calibration analysis tools that are needed to process the measured data through the instrument calibration algorithms. During the test campaign, the instrument and test facility will generate

192 Optical Radiometry for Ocean Climate Measurements

significant amounts of digital data (many GBytes per day). These data need to be reduced to obtain the necessary information to allow interpretation of the results. The calibration tools should be run as the measurement is progressing to allow a "quick look" of the test results.

The basic data processing is as follows.

- Read instrument and test facility source packets for corresponding time window. Note that we assume that instrument and test facility data are time synchronized.
- Unpack data and process to engineering units.
- Check for data errors—checksum errors, scan jitter, data out of limits etc...
  - If data errors are detected then do not use data and repeat measurement;
  - Check stability flags;
  - Only process data where instrument, environment, and source was stable;
  - Apply flight IR calibration algorithm to instrument data;
  - That is, Convert Digital Counts → Radiance → Brightness Temperatures;
  - Use same number of samples as for flight processing; and
  - Compute mean + NEdT for external blackbody targets.

All records, Raw source packets and processed data must be saved in a format that can easily be recovered! This may sound obvious, but in the 1990s, data storage and computing power were at a premium and it was not possible to record everything electronically. Also, many storage devices that seemed to guarantee long-term solutions (e.g., DAT (Digital Audio Tape)) have been superseded and no longer supported. Even with significantly increased computing power and data storage, the principal still applies.

6.2.5 Test Results

In Figure 24 we see a typical result from the AATSR calibration campaign that was performed in 1998. In the top plot we see a comparison of the measured brightness temperature in each spectral band against the actual brightness temperature of the calibration target derived from the baseplate thermometer readings. Remember that the calibration is a function of radiance rather than the physical temperature of the target. The lower plot shows the difference between the measured and the actual brightness temperature, $\Delta T = BT_{meas} - BT_{actual}$ for each spectral band and shows the effect of the nonlinear response of the signal channels.

Where the external blackbody temperature is close to that of an on-board blackbody, errors due to nonlinearity become insignificant and only residual errors remain. So for each of the calibration runs, measurements were taken with the external blackbody temperature matched to the on-board target temperatures to investigate these residuals, without the need to allow for detector nonlinearity. The differences between the brightness temperatures

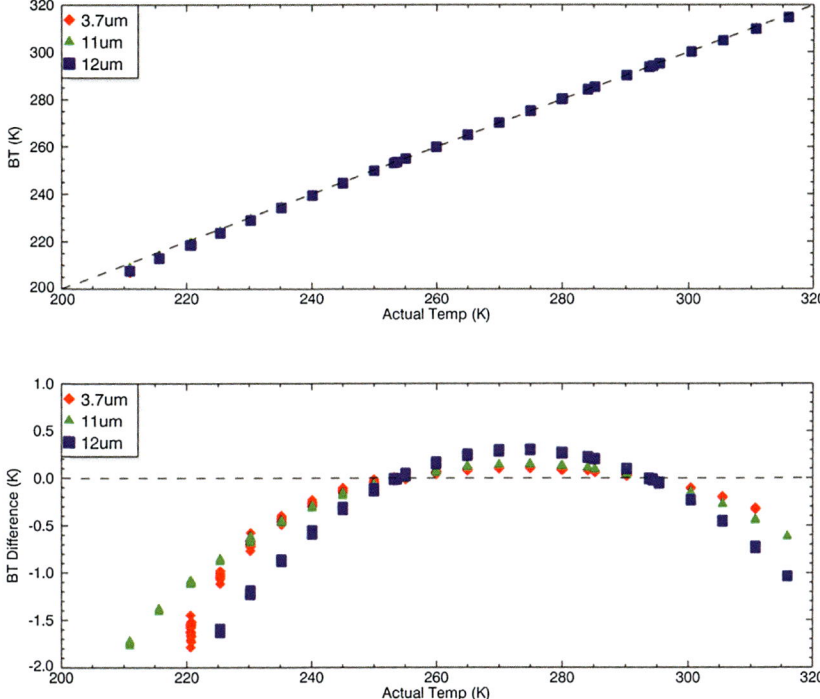

FIGURE 24 Results from radiometric calibration at the center of nadir-view for BOL thermal balance conditions. The upper plot shows the brightness temperature of the external calibration target as measured by AATSR against the actual target temperature measured by the rhodium iron resistance thermometers. The lower plot shows the brightness temperature minus actual target brightness temperature. No correction for nonlinearity has been made at this stage.

and actual target temperatures, for all tests at BOL thermal environment, are shown in Figure 25

The approach for determining and correcting for the effects of nonlinearity was established for ATSR [5] and has been used for the other two sensors. First the actual and measured radiances are normalized to the radiance at 320 K.

$$X = \frac{L_\lambda(T)}{L_\lambda(320 \text{ K})} \text{ and } Y = \frac{L_{\lambda,\text{meas}}(T)}{L_\lambda(320 \text{ K})} \tag{21}$$

First a polynomial fit is performed on the data to obtain Y as a function of X so that

$$Y = \sum_{i=0}^{n} a_{i,\lambda} X^i \tag{22}$$

From this, we derive the linear response function $g(X) = a_1 X$ [16]. The nonlinearity is then computed

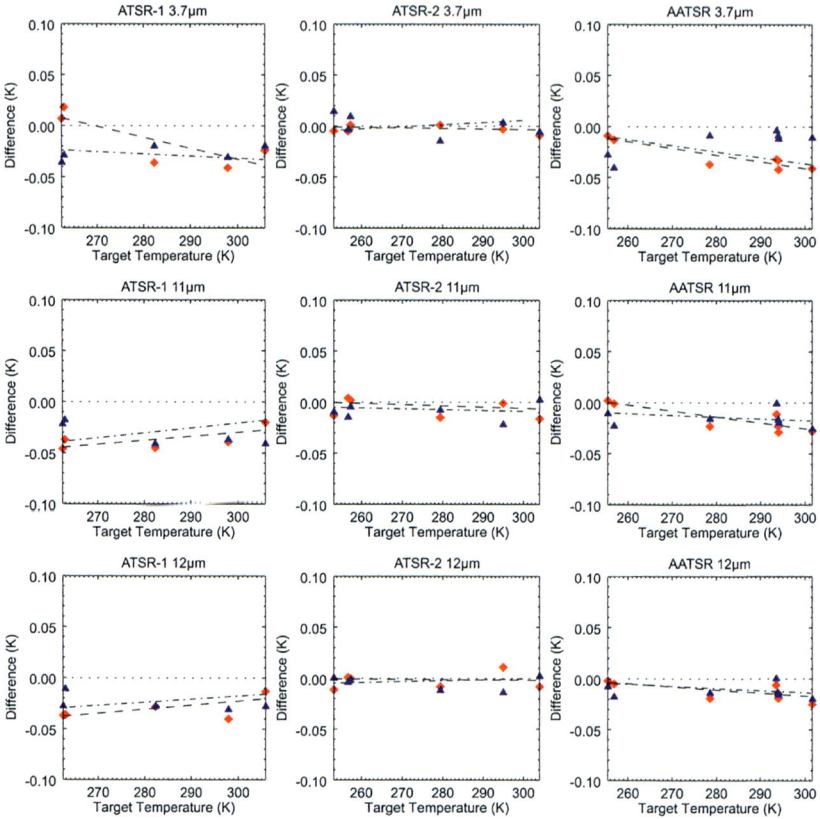

FIGURE 25 Brightness temperature differences with the external blackbodies set at the same temperatures as the on-board blackbodies for ATSR (left), ATSR-2 (center), and AATSR (right). Data for the +XBB are shown as red diamonds and the −XBB are shown as blue triangles.

$$NL_\lambda(T) = (Y - a_0)/g(X) - 1.0$$

$$= \sum_{i=2}^{n} a_{i,\lambda} X^{i-1} \bigg/ a_{1,\lambda} \qquad (23)$$

$$= \sum_{j=1}^{n} b_{j,\lambda} X^j$$

This fall-off shape is applied within the calibration algorithm by considering a corrected radiance, $L'_\lambda(T)$ where:

$$L'_\lambda(T) = (1 + NL_\lambda(T))L_\lambda(T) \qquad (24)$$

Strictly speaking the correction should be applied as a function of the output voltage of the detector since nonlinearity is a function of the photon flux at the detector rather than scene radiance. However, the approach used was considered to be sufficient for (A)ATSR over the range of SST measurements (Figures 26 and 27).

As well as establishing the radiometric bias errors, the radiometric noise was measured as a function of scene temperature. The NEΔTs are derived from the standard deviation of the detector counts over the blackbody source (assuming that the source is stable and isothermal).

$$\text{NE}\Delta T = \frac{\langle \text{DN}_1 \rangle - \langle \text{DN}_2 \rangle}{L_1 - L_2} \sigma \text{DN} \left(\frac{\partial L}{\partial T} \bigg|_T \right)^{-1} \quad (25)$$

These are shown for each ATSR instrument as a function of target temperature for the BOL calibrations at the center-of-nadir view Figure 28. It should be noted that the apparent rise in NEΔT as the target temperature decreases is purely related to $\partial L/\partial T$ as a function of scene temperature. For

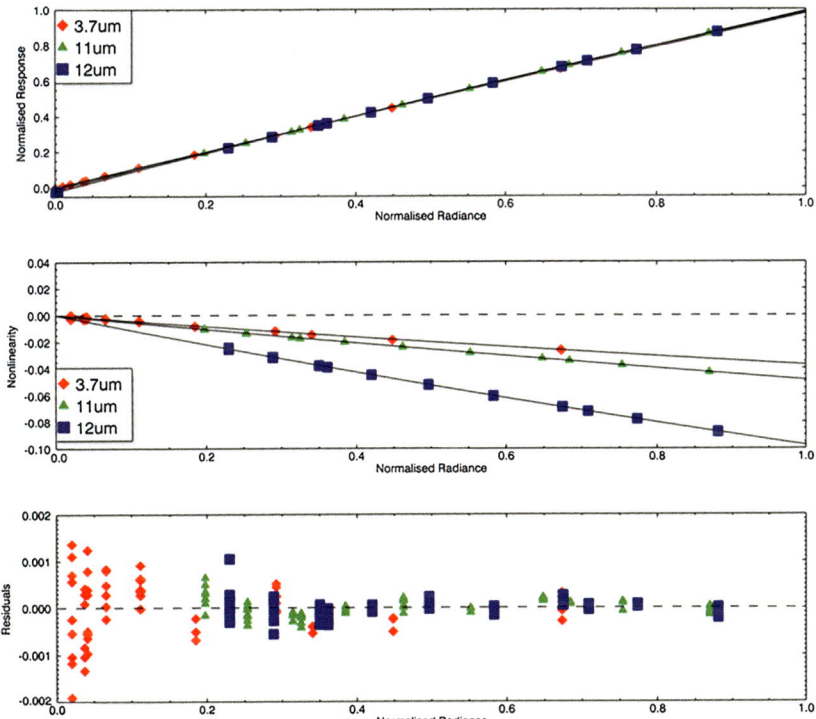

FIGURE 26 (Top) Measured AATSR scene radiance (normalized) as a function of the expected radiance at 320 K. The solid line shows a polynomial fit through the data. (Center) Fractional nonlinearity in the detector response plotted against the expected radiance. (Bottom) Residuals from least squares fitting process.

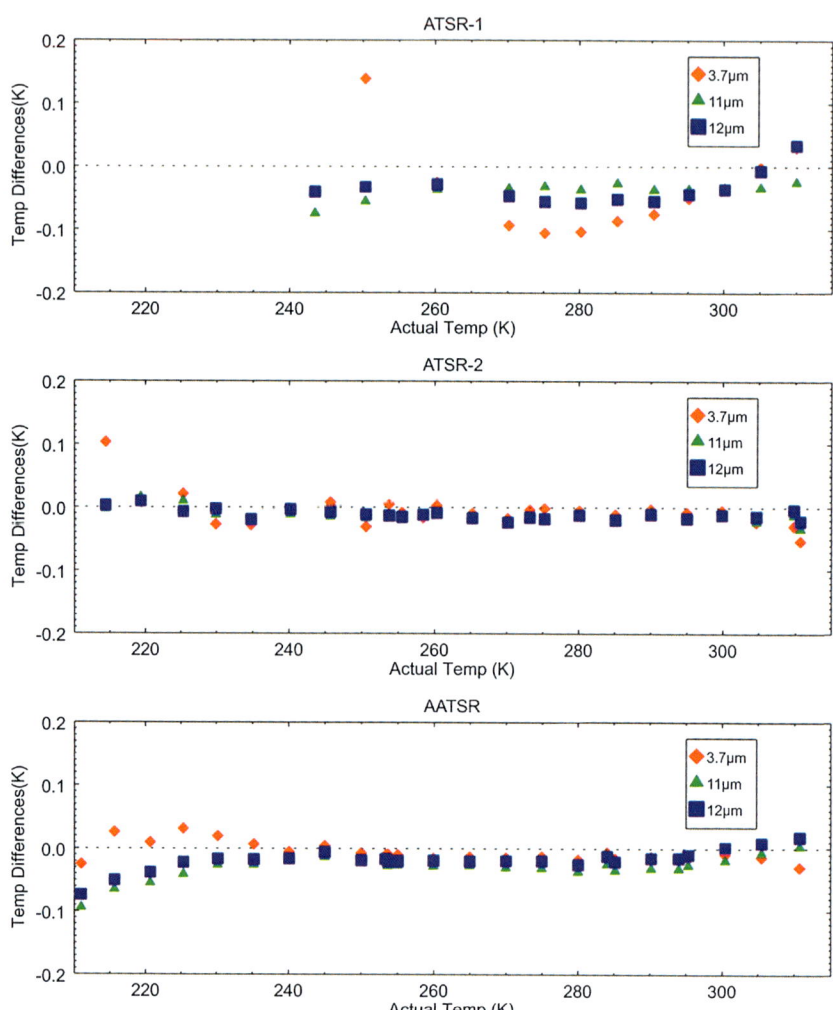

FIGURE 27 Results from the radiometric calibration at the center of nadir-view for BOL thermal balance conditions for ATSR (top), ATSR-2 (middle), and AATSR (bottom). Each plot shows, as a function of target temperature, the differences between the brightness temperature as measured by the instrument and the actual target brightness temperature as measured by the resistance thermometers.

the 11 and 12 μm channels the dominant noise source is detector noise, which remains almost constant for all scene temperatures. At 3.7 μm, preamplifier noise dominates for low photon fluxes ($T_{\text{scene}} < 250$ K) while at higher photon fluxes the noise becomes dominated by statistical photon signal noise.

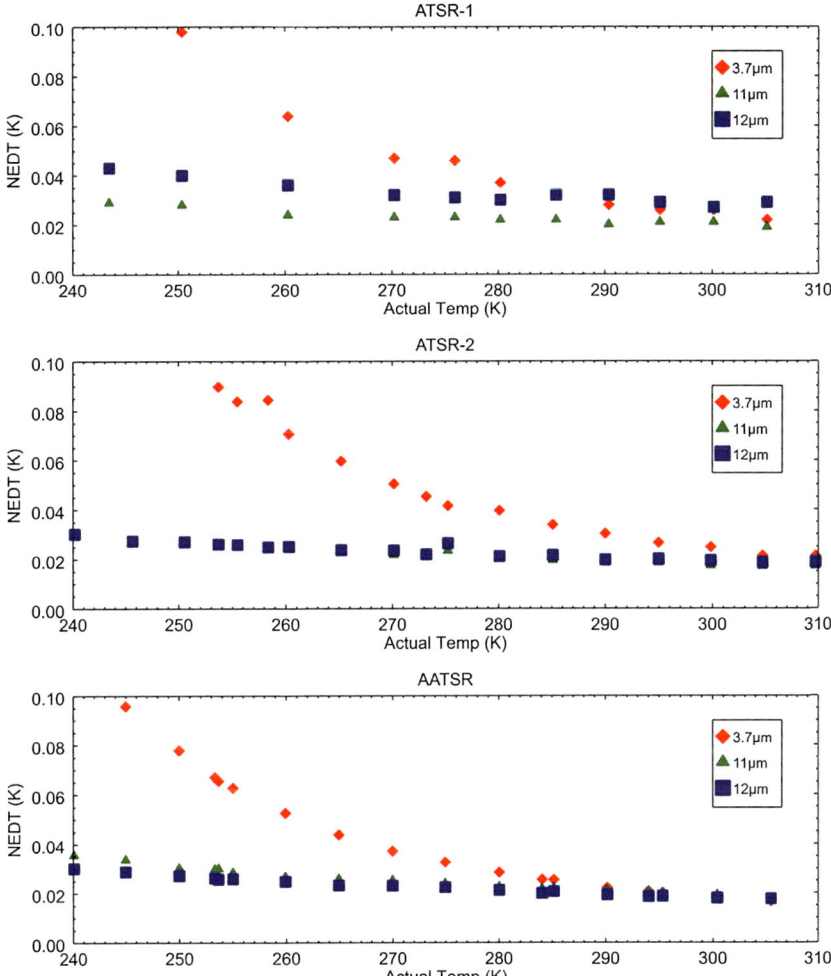

FIGURE 28 NEΔT as a function of target temperature for ATSR (top), ATSR-2 (middle), and AATSR (bottom) at BOL thermal conditions.

7. CONCLUSIONS

In this chapter we have examined the design and calibration principles necessary for an IR radiometer for measurement of sea and land surface temperatures from space. Achieving the measurement accuracy and stabilities needed for climate monitoring can only be achieved only by paying careful attention to the many sources of uncertainty that will affect the measurements. These include:

- Cooled detectors to ensure good signal to noise;
- High accuracy and stable two point calibration system covering the temperature range for SST and LST observations;
- Frequent on-board calibration measurements;
- Stable on-orbit thermal environment of optical enclosure to ensure stable calibration;
- Thorough preflight calibration and characterization of the instrument end-to-end performance; and
- On-orbit monitoring of the instrument performance.

There are several IR sensors that have been or are in use that are used to provide SST measurements to support the general observational needs of the meteorological services for short term weather forecasting (i.e., AVHRR, MODIS, VIIRS, SEVIRI). While these have on-board calibration sources, the instruments were primarily designed with a wide range of applications in mind and inevitably some compromises were made in radiometric performance.

In contrast, the (A)ATSR instruments were specifically designed and built for climate quality measurements of SST and LST. The key distinguishing feature of the radiometers is the dual view of the Earth scene to enable a robust atmospheric correction. As a consequence the (A)ATSR series has provided a 21 years stable and accurate data set as explored in the following chapters. Although some compromises were made that made (A)ATSR less suited for operational meteorology, the accuracy of the SST data has been used to underpin the measurements made by the other sensors described in this chapter.

Many of these design features have been incorporated into the SLSTR instruments due to fly on the Copernicus Sentinel-3 missions where it is expected that the accurate SST record from AATSR will be continued for a further 20 years.

REFERENCES

[1] D. Smith, C. Mutlow, J. Delderfield, B. Watkins, G. Mason, ATSR infrared radiometric calibration and in-orbit performance, Remote Sens. Environ. 116 (2012) 4−16.
[2] P. Coppo, B. Ricciarelli, F. Brandani, J. Delderfield, M. Ferlet, C. Mutlow, et al., SLSTR: a high accuracy dual scan temperature radiometer for sea and land surface monitoring from space, J. Mod. Opt. 57 (2010) 1815−1830.
[3] D.M.A. Aminou, MSG's SEVIRI Instrument, ESA Bull. 111 (2002) 15−17.
[4] J. Schmid, The SEVIRI instrument, in: Assembly, 2000, pp. 1−10.
[5] G. Mason, ATSR Test and Calibration Report, Oxford, 1991.
[6] E. Theocharous, J. Ishii, N.P. Fox, Absolute linearity measurements on HgCdTe detectors in the infrared region, Appl. Opt. 43 (2004) 4182−4188.

[7] M. Weinreb, G. Hamilton, Nonlinearity corrections in calibration of Advanced Very High Resolution Radiometer infrared channels, J. Geophys. 95 (1990) 7381.
[8] X. Xiong, N. Chen, S. Xiong, K. Chiang, W. Barnes, Performance of the terra MODIS on-board blackbody, in: J.J. Butler (Ed.), Opt. Photonics 2005, International Society for Optics and Photonics, 2005, pp. 58820U–58820U–10.
[9] X. Xiong, B.N. Wenny, A. Wu, W.L. Barnes, MODIS onboard blackbody function and performance, IEEE Trans. Geosci. Remote Sens. 47 (2009) 4210–4222.
[10] K.H. Berry, Emissivity of a cylindrical black-body cavity with a re-entrant cone and face, J. Phys. E 14 (1981) 629–632.
[11] I.M. Mason, P.H. Sheather, J. a Bowles, G. Davies, Blackbody calibration sources of high accuracy for a spaceborne infrared instrument: the Along Track Scanning Radiometer, Appl. Opt. 35 (1996) 629–639.
[12] T. Theocharous, N. Fox, CEOS Comparison of IR Brightness Temperature Measurements in Support of Satellite Validation. Part II: Laboratory Comparison of the Brightness Temperature of Blackbodies, 2010.
[13] J.P. Rice, B.C. Johnson, The NIST EOS thermal-infrared transfer radiometer, Metrologia 35 (1998) 505–509.
[14] A. Prokhorov, N.I. Prokhorova, Application of the three-component bidirectional reflectance distribution function model to Monte Carlo calculation of spectral effective emissivities of nonisothermal blackbody cavities, Appl. Opt. 51 (2012) 8003–8012.
[15] D.L. Smith, J. Delderfield, D. Drummond, T. Edwards, C.T. Mutlow, P.D. Read, et al., Calibration of the AATSR instrument, Adv. Sp. Res. 28 (2001) 31–39.
[16] S. Yang, I. Vayshenker, X. Li, T. Scott, Accurate measurement of optical detector nonlinearity, in: Natl. Conf. Stand. Lab. Work, 1994.

Chapter 2.4

Postlaunch Calibration and Stability: Thermal Infrared Satellite Radiometers

Peter J. Minnett,[1,]* David L. Smith[2]

[1] *Meteorology & Physical Oceanography, Rosenstiel School of Marine and Atmospheric Science, University of Miami, Miami, FL, USA;* [2] *RAL Space, Science and Technologies Facilities Council, Harwell Oxford, Oxford, UK*
Corresponding author: E-mail: pminnett@rsmas.miami.edu

Chapter Outline

1. Introduction — 201
2. On-Board Calibration — 203
   2.1 (A)ATSR Radiometric Calibration — 204
   2.2 AVHRR Calibration — 209
   2.3 MODIS and VIIRS Radiometric Calibration — 213
   2.4 MODIS Spectroradiometric Calibration Assembly for On-Orbit Stability — 214
   2.5 MODIS Mirror Response versus Scan Angle — 216
3. Comparisons with Reference Satellite Sensors — 218
   3.1 Spatial Comparisons — 219
   3.2 Temporal Comparisons — 220
   3.3 Simultaneous Nadir Overpasses — 222
   3.4 Instruments on the Same Satellite — 223
4. Validating Geophysical Retrievals — 225
   4.1 Cloud Screening — 229
   4.2 Atmospheric Correction Algorithm — 230
   4.3 Geophysical Validation — 232
   4.4 Ship-Board Radiometers — 236
5. Discussion — 237
6. Conclusions — 239
References — 239

1. INTRODUCTION

The quantitative applications of satellite-derived data are guided by estimates of the uncertainties in the measurements; this, of course, is true for any dataset. Noisy or inaccurate measurements cannot be applied to problems where high accuracy, precision, and stability are required. Satellite radiometers are subjected to extensive and rigorous prelaunch calibration and characterization (Chapter 2.3), and this is an extremely important aspect of preparing the instruments for launch, but once operating in space, there are no opportunities to directly reassess the accuracy of the radiometric calibration or the instruments' characteristics. Monitoring the performance of the instruments throughout their missions is therefore through a combination of direct, on-orbit measurements, such as thermometers measuring the temperatures of critical components of the instrument, and indirectly by monitoring the appearance of the imagery to detect, for example, degradation of the bearings of the scan mirror, and by assessing the long-term uncertainty characteristics of the derived geophysical variables.

Here we focus on the measurement of sea-surface temperature (SST) using infrared radiometry as the SST requirements are the most stringent of many geophysical variables derived from satellite infrared radiometers in terms of absolute accuracy and stability, and the most stringent of these are requirements for climate research. Although the required temporal and spatial scales are not discussed, an absolute accuracy of ±0.1 K and a decadal stability of 0.04 K [1] or 0.03 K [2] are required and these are predicated by the anticipated magnitudes of climate-change signals. As such, we can presume that these requirements are attributable to spatial scales of order 10^4-10^6 km^2 and temporal scales of weeks to months. These ambitious accuracy targets, especially for stability, impose a need for continual assessment of the postlaunch performance of the satellite radiometers throughout their lifetimes, and also require a mechanism for generating long-time series of measurements from satellite over multiple missions.

The determination of surface temperatures from infrared radiometer measurements from spacecraft is accomplished by taking measurements in spectral intervals, where the atmosphere is relatively transparent to the propagation of the radiation. Such spectral intervals are often called "atmospheric transmission windows" or more simply "atmospheric windows" and SST is derived from measurements in two windows, one in the wavelength range of 3.5–4.1 μm and the other in the wavelength range of 10–13 μm. The window at shorter wavelengths is frequently referred to as the "mid-infrared" window and the longer wavelength interval as the "thermal infrared" window as it is close to the peak of the electromagnetic emission spectrum, given by Planck's Function, at terrestrial- and ocean-surface temperatures. For our purposes, we refer to both wavelength intervals as "thermal" as the signal of interest is the thermal emission which is closely related to the surface temperature.

In the next section of this chapter, we describe the on-board, on-orbit calibration procedures that result in the generation of calibrated radiances that are subsequently converted to brightness temperatures. Comparison between radiance and brightness temperatures measured simultaneously by radiometers on two satellites is a method of monitoring their relative accuracies and stabilities, and this is discussed in Section 3. The brightness temperatures are the inputs to algorithms to derive surface temperatures, and these are briefly described in Section 4, along with approaches to determine the accuracies by comparisons with independent temperature measurements.

2. ON-BOARD CALIBRATION

The calibration approach and derivation of the calibration equations for the infrared radiometers are given in detail in Chapter 2.3, along with the descriptions of the satellite radiometers used for the retrieval of SST. All use a two-point calibration approach in which the scan mirror, sometimes called the *scene mirror*, directs the field of view of the radiometer at targets of known temperature and emissivity. Outputs from the detectors during these calibration measurements are used with the known temperatures to generate a linear calibration function that is used to convert the detector outputs in to calibrated measurements of radiance during the earth view. The calibration measurements should be taken very frequently, and typically are included in every rotation of the scan mirror. The slight nonlinearities in the response of HgCdTe detectors are determined before launch and monitored on-orbit.

Ideally, the two in-flight calibration points should be provided by stable and well-characterized blackbody targets, whose temperatures are accurately measured by embedded thermometers. Also, ideally, the temperatures of the calibration targets should straddle the range of the SSTs, $-2\,°C$ to $35\,°C$, being measured in the earth view. This is the calibration configuration for the Along-Track Scanning Radiometer (ATSR) series, including the Advanced ATSR (AATSR), which accommodates the two blackbody cavities between the nadir and slant-view ports. For current and past wide swath imagers, however, there is insufficient physical space to mount two blackbodies in the $360°$ rotation of the scan mirror, of which a segment of typically $110°$ is used for the earth-view measurements. The approach is to use measurements of cold space as the cold calibration target, assuming a target temperature of $2.7\,K$—essentially zero radiance. The cold target measurement is taken through a space-view port on the side of the instrument that faces away from the sun. The space-view port serves also in the calibration of the solar reflectance (visible) bands by permitting measurements of the reflectivity of the moon when the moon is visible through the port (Chapter 2.2; [3]). Sometimes a spacecraft roll maneuver is required to bring the moon into the space-view port; during the lunar measurements, the space-view data are not suitable for calibrating the infrared bands.

There are several sources of uncertainties in the calibration processes of the infrared bands (Chapter 2.3) and these have to be assessed and monitored to determine the long-term stability of the infrared measurements.

2.1 (A)ATSR Radiometric Calibration

As described in Chapter 2.3, the ATSRs employ two cavity blackbody sources that are viewed on every rotation of the scan mirror. Because they are viewed at the same angle of incidence on the scan mirror as the nadir and forward earth views, no corrections for mirror polarization or additional optical components are necessary. The calibration of the blackbody radiances are traced through the cavity emissivity and the precision platinum resistance thermometers (PRT's) mounted on the baseplate and cavity wall. Each blackbody has multiple temperature sensors, each of which has its own precision amplifier before their signals are multiplexed. The thermometers are calibrated mounted in situ within the cavity and connected to the flight electronics referenced to Standard PRTs that are traceable to ITS-90 (Figure 1).

The preflight calibration activities ensure traceability of the radiometric calibration and performance up to the point of launch [4]. Soon after launch, the stability and uniformity of the temperatures of the blackbody cavities and the instrument sensitivity must be closely monitored to establish the Beginning of Life baseline, against which any degradation in performance through the lifetime can be referenced. For (A)ATSR, the on-board calibration relies on the stability of the thermometry and the black coating. Confidence in the thermometer calibration is first achieved by monitoring the stability and consistency of the blackbody readings. Table 1 shows a set of on-orbit readings compared against measurements taken during the prelaunch calibration. Although the in-flight readings are warmer by about 10 K, the differences between the individual sensor readings and the baseplate averages are well maintained. Typical orbital variations of the blackbody baseplates are ~ 0.2 K peak-to-peak demonstrating the stability of the thermal design. The good stability and uniformity ensures that uncertainties due to temperature variations and gradients are minimized. The trends for the full mission suggests that both the temperature differences across the blackbodies and the relative calibrations of the PRTs did not change significantly, linking well to pre-launch-temperature baselines (Figure 2).

A useful technique to monitor the radiometric stability is a "blackbody cross-over test." This test is performed by switching the heated blackbody from the +XBB to the −XBB (and vice versa) and allowing the temperatures to cross-over and stabilize. The basic idea is to compare the radiometric signals in the thermal channels when the two blackbodies are at identical temperatures. Any significant difference would imply a drift in the blackbody thermometer calibration or change in target emissivity caused by a deterioration of the black surface finish. The blackbody temperatures and radiometric signals during a typical cross-over test are shown in Figure 3.

The test was performed during commissioning and roughly annually thereafter. The results for AATSR indicate that, relative to each other, the brightness temperature errors from the blackbodies were typically <10 mK at

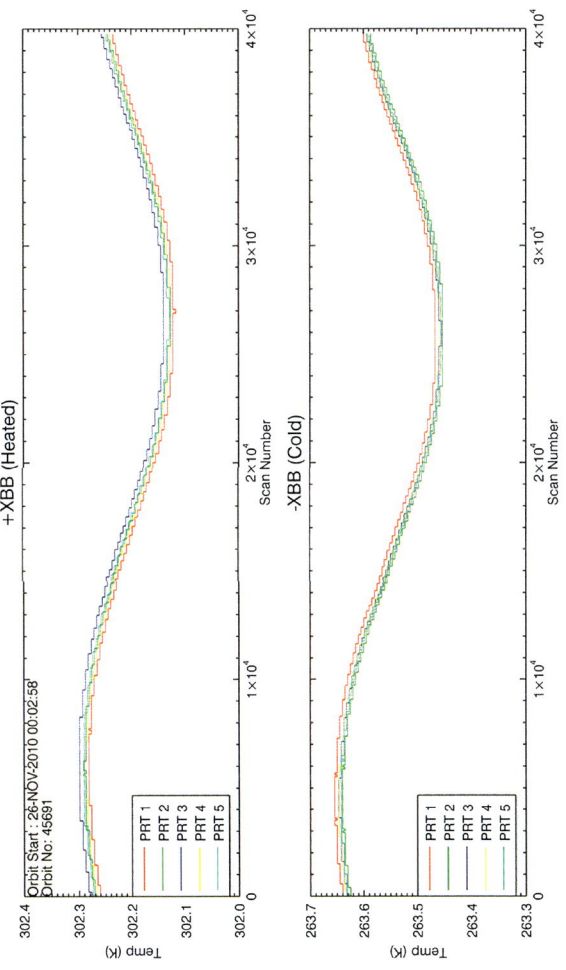

FIGURE 1 Typical orbital variation of the AATSR blackbody baseplate thermometers.

TABLE 1 Typical Blackbody Thermometer Readings for AATSR Taken on June 3, 2002 and the Final Readings on April 8, 2012

	+XBB Temperatures (K)				
	2002		2012 Last Reading		
	Reading	Difference	Reading	Difference	Prelaunch
Baseplate average	301.522	–	301.650	–	293.527
PRT1	301.513	−0.009	301.640	−0.010	−0.009
PRT2	301.518	−0.004	301.651	0.001	−0.002
PRT3	301.526	0.004	301.658	0.008	0.002
PRT4	301.525	0.003	301.643	−0.007	0.001
PRT5	301.530	0.008	301.660	0.009	0.006
PRT6	301.905	0.383	302.049	0.399	0.391
	−XBB Temperatures (K)				
	2002		2012 Last Reading		
	Reading	Difference	Reading	Difference	Prelaunch
Baseplate average	262.897	–	262.509	–	252.773
PRT1	262.898	0.001	262.518	0.009	0.001
PRT2	262.899	0.002	262.502	−0.007	0.000
PRT3	262.897	0.000	262.505	−0.004	0.000
PRT4	262.892	−0.005	262.510	0.001	−0.001
PRT5	262.897	0.000	262.509	0.000	−0.002
PRT6	262.882	−0.015	262.517	0.008	−0.017

The Top Row Shows the Average of the 5 Baseplate Sensors (PRT1-PRT5). PRT6 is the Baffle Temperature and is Not Used in the Average. The Difference Column Shows the Differences between the Individual Sensor Readings and the Average Temperature. The Final Column is a Typical Reading from the Prelaunch Calibration in December 1998.

11 and 12 µm, and below 20 mK at 3.7 µm. Comparing with earlier measurements, Figure 4, it can be seen that the 11 and 12 µm channels are stable over time, while there appears to be a very slow increase in the 3.7 µm channel of approximately 6 mK over the mission. Even for the trend in this channel, the one with the lowest blackbody emissivity, the apparent brightness temperature difference is still much smaller than the radiometric noise.

Postlaunch Calibration and Stability **Chapter | 2.4** 207

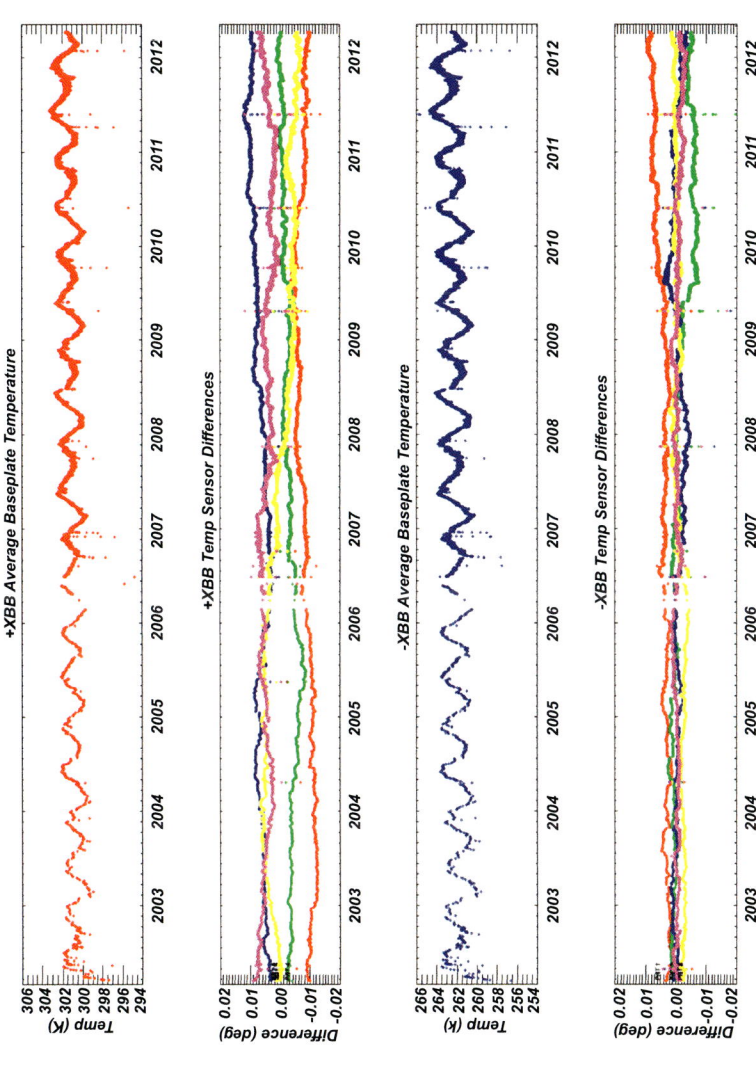

FIGURE 2 Trends for the AATSR mission of daily averages of the +XBB and −XBB baseplate mean temperatures, and the differences from the mean of the individual sensor readings.

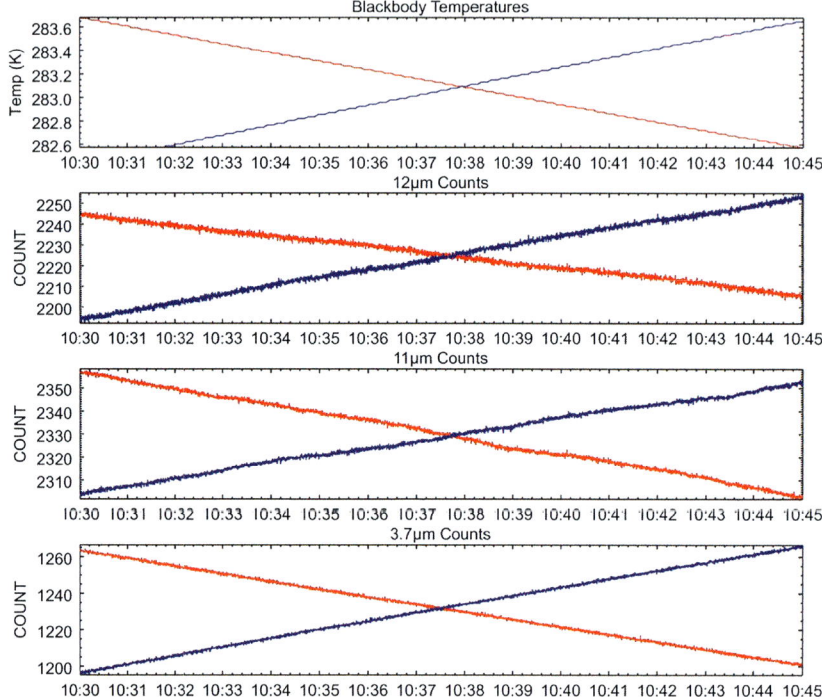

FIGURE 3 Blackbody temperatures and IR channel blackbody signals for AATSR blackbody cross-over test performed on April 21, 2009.

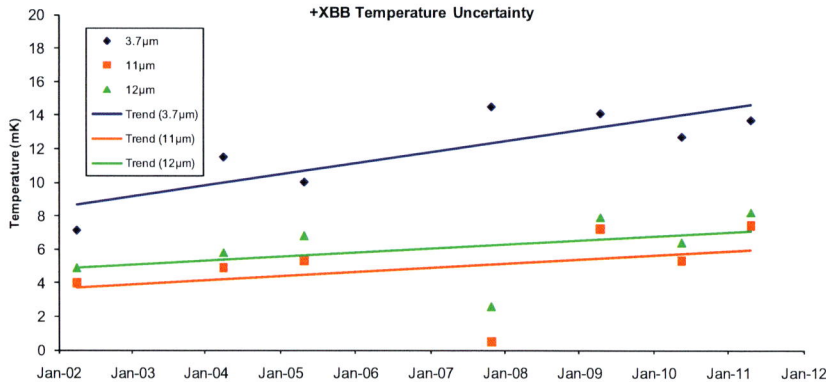

FIGURE 4 Temperature uncertainties for the +XBB from the cross-over tests.

It should be noted that the test is a comparison of a blackbody against the other with the assumption that the reference is stable, and therefore, does not provide an absolute calibration of the blackbodies. The radiometric calibration of the TIR channels is of course closely linked to the prelaunch tests and characterization of the instrument, which has to be done in a rigorous fashion, as after launch, there is no direct method to verify the absolute radiometric calibration of the on-board blackbody thermometry and emissivity.

For the TIR channels, the signal-to-noise ratio (SNR) is usually expressed as the Noise Equivalent Differential Temperature (NEΔT) and is formally defined as

$$\text{NE}\Delta\text{T} = \frac{L}{\text{SNR}} \left(\frac{dL}{dT}\bigg|_T \right)^{-1} \qquad (1)$$

where L is the radiance corresponding to a scene brightness temperature of T Kelvin.

Using the signals (detector counts)—C_{hbb} and C_{cbb}—and radiances—$L(T_{hbb})$ and $L(T_{cbb})$—of the two on-board blackbodies, we can derive the NEΔT from the noise measurements. For AATSR, we use the standard deviation of the signal channel counts from the on-board calibration sources to obtain values of NEΔT at the hot and cold blackbody temperatures. In Figure 5, we see that the NEΔTs for the thermal infrared channels have remained stable for the duration of the mission. The mean values over this time given in Table 2 are within the requirements and are comparable with the prelaunch calibration measurements.

There have been a few occasions where the noise has increased above the baseline level, possibly due to water–ice contamination or warmer thermal environment, although the noise has remained within the specified limits.

2.2 AVHRR Calibration

The first Advanced Very High-Resolution Radiometer (AVHRR; described in Chapter 2.3 and by Cracknell, 1997 [5]) flew on the TIROS-N satellite in 1978, and with minor modifications continues on the NOAA-n and MetOp series of polar orbiting satellites [6] to provide infrared images from which SST is derived. The design of the AVHRR (Figure 4 in Chapter 2.3) is such that the space view can be easily accommodated, but the "blackbody" calibration target being part of the baseplate of the instrument is poorly shielded against radiation from the sun or the earth, or other parts of the instrument, that can introduce unwanted temperature variations. The temperature of the AVHRR calibration target is monitored by four PRTs but cannot be controlled by heaters or by cooling, and so floats at an ambient temperature close to that of the instrument baseplate. An analysis of these temperature measurements revealed gradients across the target and significant changes with position around the orbit [7] as shown in Figure 6. The size of these spatial gradients and temporal changes, neither of which is desirable in a well-calibrated radiometer, can lead to uncertainties in the on-orbit calibration process that are much greater than the required accuracies in the derived SSTs. The drop of the temperatures up to about scan line 2500 in Figure 6 results from cooling of

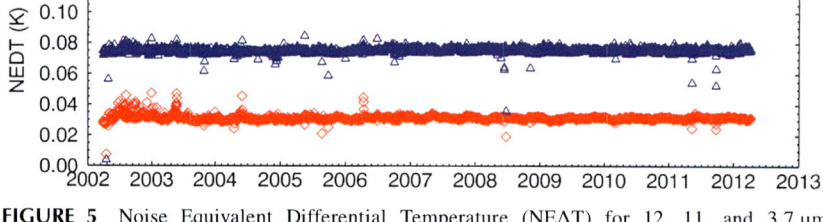

FIGURE 5 Noise Equivalent Differential Temperature (NEΔT) for 12, 11, and 3.7 μm channels for the AATSR on ENVISAT for the entire mission for the hot BB (red) and cold BB (blue).

the instrument and the satellite during the eclipse part of the orbit (in the shadow of the earth), and the subsequent sharp increase in temperature is caused by the satellite emerging into direct sunlight. The following drop in temperature is caused by the instrument being shielded from direct solar heating by the satellite itself. As angles to the sun change during an orbit, the shadowing effect of the spacecraft is removed and the sunlight falls on the instrument again and the thermometers. After the satellite has entered into eclipse, the temperature drops again. The heating of the target can be the result of the sunlight falling directly on its surface, or by sunlight warming other parts of the instrument, or spacecraft, and heat being conducted to the

TABLE 2 In-Flight Noise Equivalent Differential Temperature (NEΔT) Values Compared with Requirement and Prelaunch Measurements

		3.7 μm (K)	11 μm (K)	12 μm (K)
Requirement	T = 270 K	0.080	0.050	0.050
On-orbit average	T = CBB (251 K)	0.075	0.034	0.035
	T = HBB (301 K)	0.031	0.031	0.033
Prelaunch	T = 270 K	0.037	0.025	0.025
	T = CBB (253 K)	0.065	0.030	0.030
	T = HBB (295 K)	0.020	0.020	0.020

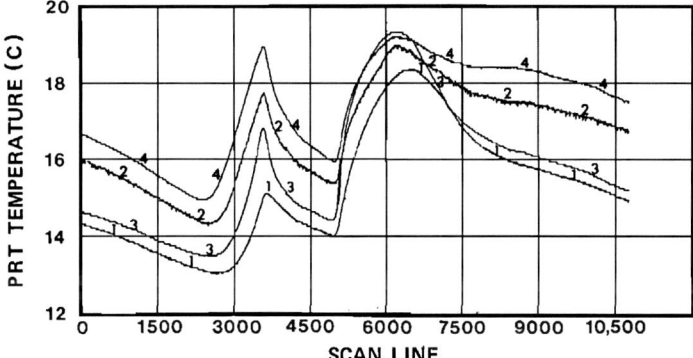

FIGURE 6 Changes in the temperature of the calibration target of the Advanced Very High-Resolution Radiometer (AVHRR) on NOAA-7 around a single orbit. The thermometers are identified by the numbers 1—4. *After Ref. [7].*

calibration target. Another source of measurement error is radiation falling onto detectors with an origin outside of the nominal field of view, called "strays," and which is emitted from a surface at an unknown, and not necessarily constant, temperature.

The magnitudes of the temperature variations in Figure 6 do not translate directly into calibration errors of the same size, as, in principle, the calibration process will compensate for variations in the temperature of the target. This requires the average, perhaps weighted, of the thermometer measurements to give an accurate estimate of the temperature of the emitting surface of the target at the time of the field of view of the detectors scans the target. However, it is not known whether the spatial gradients on the calibration target are such that an average of the thermometer outputs is a valid estimate of the mean temperature of that part of the target surface; given the shape of the spatial response of the detectors within the field of view is generally not well known, the effects of the temperature gradients cannot be determined with confidence.

The best way of limiting the calibration uncertainties due to spatial and temporal temperature changes is to have a spatially uniform and temporally constant calibration target temperature. This, of course, is a requirement of well-designed and -characterized radiometers (Chapter 2.3).

Trishchenko et al. [8] extended the analysis of the AVHRR in-flight calibration to later models on the NOAA polar-orbiting satellites, up to NOAA-16. They found similar orbital signals in the blackbody target thermometers for the

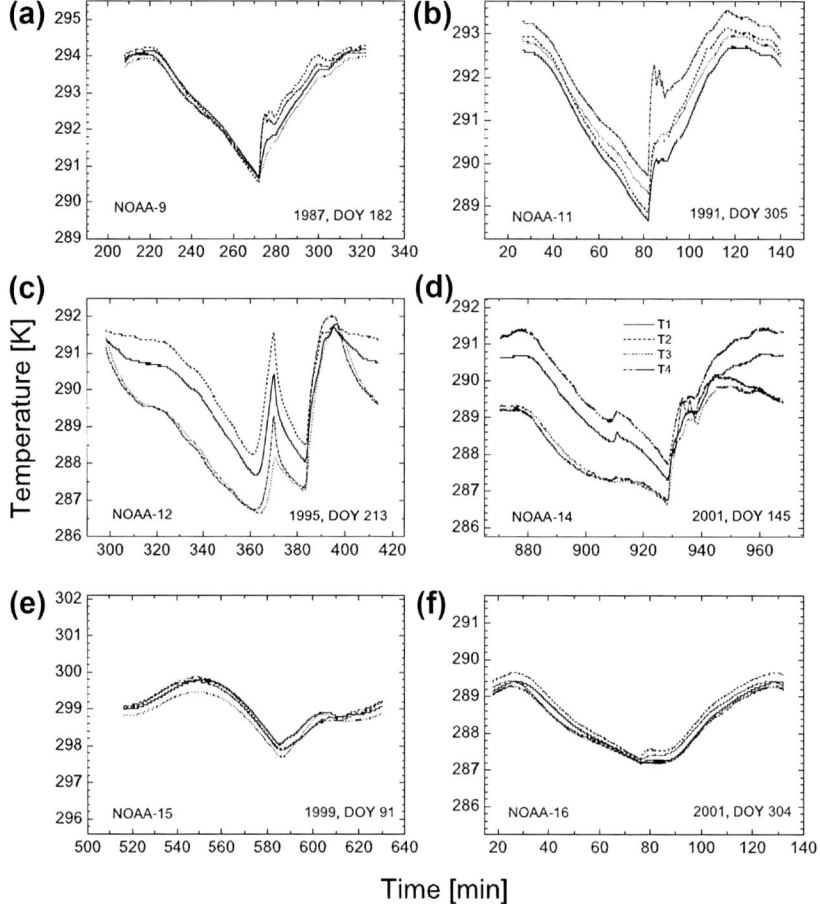

FIGURE 7 Variation in Advanced Very High-Resolution Radiometer (AVHRR) blackbody thermometer readings for single orbits for (a) NOAA-9, (b) NOAA-11, (c) NOAA-12, (d) NOAA-14, all AVHRR/2s, (e) NOAA-15, and (f) NOAA-16, which are AVHRR/3s. Solar contamination results in rapid changes in temperature. DOY, day of year. *From Ref. [8].*

AVHRRs on satellites up to NOAA-14 (Figure 7). There are differences between each model of the AVHRR on each satellite, some caused by variations in the properties of the individual instruments and some by differences in the orbits of the satellites, and hence differences in the solar illumination. The span of temperatures measured by thermometers in the calibration target was worst, >4 K, for the AVHRR on NOAA-12. This is likely to be caused by spatial temperature gradients across the blackbody. The variations in the thermal state of the AVHRR can introduce systematic errors in the calibrated brightness temperatures of >0.5 K. Trishchenko et al. [8] show a marked improvement in the stability of the AVHRR blackbody temperatures around the orbit and also better agreement between the thermometers for the AVHRRs on NOAA-15 and latter. NOAA-15 carried the first version of a third type of AVHRR, referred to as AVHRR/3, that benefited from a number of improvements over earlier models, including better shielding of the calibration target from solar radiation, and in thermal insulation of the target to variations in the satellite temperatures. The amplitudes, time rates of change, and the spatial gradients of the temperatures measured by the calibration target thermometers have all been reduced in the AVHRR/3.

In a thorough reassessment of the AVHRR prelaunch calibration and characterization [9] and on-orbit performance [10], Mittaz et al. identified weaknesses in the calibration equation used for the AVHRR infrared measurements. By developing a calibration algorithm that more realistically accounts for the conditions under which the prelaunch measurements were taken, and for the on-orbit conditions, Mittaz et al. were able to demonstrate residual mean errors in calibrated brightness temperatures of near-zero and scatter of ∼0.05 K, which is a marked improvement over the standard calibration procedures.

2.3 MODIS and VIIRS Radiometric Calibration

Although the fore-optics of MODIS (MODerate resolution Imaging Spectroradiometer) and VIIRS (Visible-Infrared Imaging Radiometer Suite) are very different (Chapter 2.3), the on-orbit calibration of the infrared bands is the same, relying on a space view and a grooved plate blackbody target (Figure 8; [11]). In both MODIS and VIIRS, the blackbody is enclosed in the instrument

FIGURE 8 Grooved plate blackbody targets for MODIS (left) and VIIRS (right). The numbers on the VIIRS target refer to the positions of the six embedded thermometers. The MODIS target has 12 thermometers, also in two rows. *From Ref. [11].*

and therefore well protected against orbital temperature changes driven by external factors, primarily solar illumination. The temperature is controlled to a predetermined value, which is that at which the prelaunch calibration and characterization was done; for Terra MODIS, Aqua MODIS, and S-NPP VIIRS, these are 290, 285, and 292.5 K. The emissivity of these blackbodies was designed to be >0.9995, and the embedded thermistors were calibrated prelaunch to the SI temperature scale to within 0.05 K [11].

The blackbody of the *Terra* MODIS shows larger temperature variations, especially for the thermistors at the extremes of the grooved plate than for the *Aqua* MODIS, for which the variations are usually <0.02 K (Figure 9). For S-NPP VIIRS, the blackbody temperatures are even more uniform during the night-time arcs of the orbit, but during the solar-illuminated daytime arcs, variations of ∼0.05 K are present in thermistors 3 and 6 which are located at one end of the plate. Since there are multiple detectors in each band (10 for MODIS, 16 for VIIRS), spatial gradients on the surface of the blackbody can lead to a degradation of the calibration as each detector samples different areas on the blackbody during the calibration procedure. Even so, the consequence for the gradients on the VIIRS black body on the calibration of the thermal infrared bands are expected to be <0.1% [11].

To assess any changes in the nonlinearity of the HgCdTe detectors in MODIS and VIIRS, the blackbody temperature can be changed through a programmed sequence of values. The "warm-up and cool-down" (WUCD) sequence takes the temperature from instrument ambient, operating value to a lower temperature and then heated to 315 K (Figure 10). This procedure is conducted about every 3 months for both MODIS's and VIIRS. The range of temperatures of the blackbodies achievable on-orbit is much smaller than that can be covered in the laboratory prior to launch, but these operations serve a very useful purpose in monitoring the stability of the nonlinear response of the detectors.

The WUCD procedures also allow the NEΔTs of the detectors as a function of the target temperatures, and all the bands used in SST retrievals are currently well below the specified values (Figure 11); note the VIIRS specifications are less stringent than those of MODIS [11]. An exception in the infrared atmospheric transmission window bands is MODIS Band 20 ($\lambda = 3.75$ μm) for both Terra and Aqua, which is marginally above the specification for target temperatures below ∼282 K [11].

2.4 MODIS Spectroradiometric Calibration Assembly for On-Orbit Stability

One of the MODIS internal, on-orbit calibration facilities is the spectroradiometric calibration assembly (SRCA) [12]. This is a built-in calibration instrument which has three main functions. The first is to measure the band-to-band registration of all MODIS bands, and to track any changes with time. The

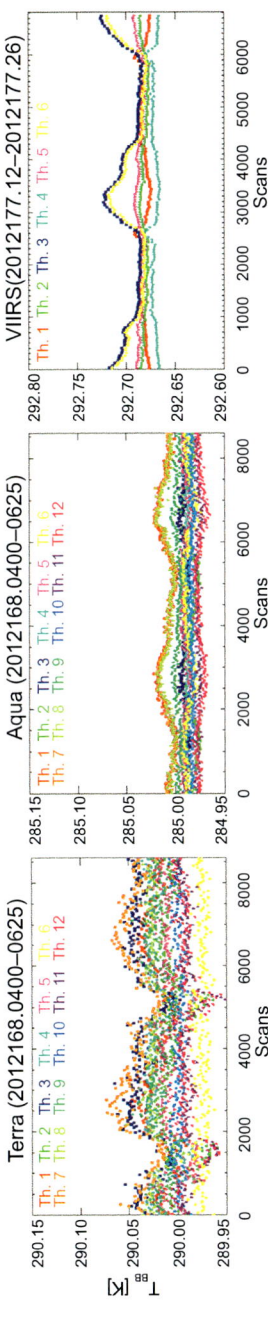

FIGURE 9 Temperatures measured by thermometers in the blackbody calibration targets of Terra MODIS (left), Aqua MODIS (center), and S-NPP VIIRS (right). Colors indicate the measurements of individual thermometers. Temperature variations around two orbits are shown. *After Ref. [11].*

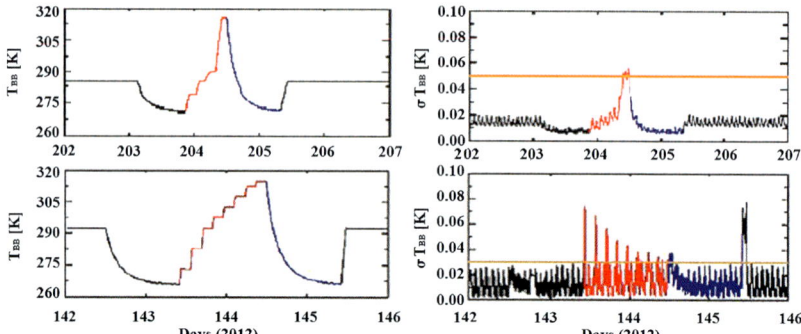

FIGURE 10 Temperatures of the internal blackbodies during a warm-up and cool-down procedure for MODIS on Aqua (top-left) and S-NPP VIIRS (bottom-left). Note the operating temperatures of the blackbody calibration targets are different for the two instruments: 285 K for Aqua MODIS and 292.5 K for S-NPP VIIRS. At right, the standard deviations of the multiple thermometer measurements are shown. *From Ref. [11].*

second and third are directed to the reflected solar bands and these are to measure the central wavelength shifts and to track changes in the radiometric gain. The SRCA on Terra showed the relative band-to-band registration shifts of the bands on the cold focal plane arrays with respect to Band 1 (which is not on a cold focal plane) agreed with the prelaunch calibration values to better than 50 m at the Earth's surface [13]. Long-term monitoring using the SRCA has shown that the band-to-band registration has remained very stable postlaunch for the Terra MODIS [14], and also for the MODIS on Aqua with shifts of <20 m between channels 31 and 32 [12].

2.5 MODIS Mirror Response versus Scan Angle

Unlike the AVHRR and the (A)ATSR in which the angle of the incident radiation on the scan mirror is constant and therefore any effects of imperfect reflectivity in the infrared is compensated for by the on-orbit calibration procedure, the large paddle-wheel scan mirror of MODIS (Chapter 2.3) means the angle of incidence on the mirror surface changes across the swath. The multilayer coating on the scan mirror surface introduces a wavelength dependence of the reflectivity on the incidence angle which becomes pronounced for $\lambda > \sim 7$ μm. This effect is referred to as "reflectivity versus scan angle," or rvs. The blackbody calibration measurements occur at a single angle of incidence on the scan mirror, and so the rvs effects are not removed by the on-orbit calibration procedure. Since the sum of reflectivity and emissivity is unity, changes in the rvs cause changes in the mirror surface emissivity, and so the correction for the rvs requires knowledge of the scan mirror temperature; this is estimated from a noncontact thermistor mounted in a cavity in the mirror along the axis of rotation. Prelaunch measurements

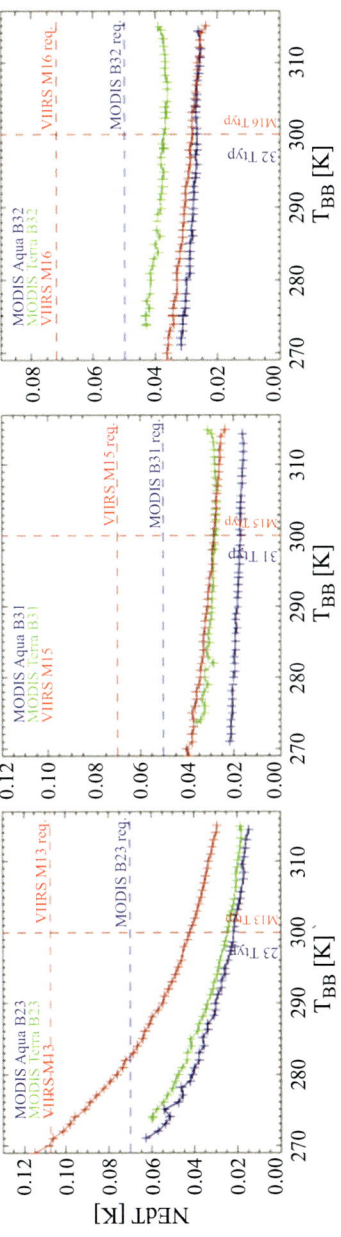

FIGURE 11 Measured NEΔTs of three MODIS and VIIRS spectrally matched bands used in SST retrievals as a function of the blackbody target temperatures during "warm-up cool-down" exercises. The vertical dashed lines indicate typical surface temperatures, and the horizontal dashed lines are the specified values. MODIS bands B23, B31, and B32 have central wavelengths at 4.05, 11.03, and 12.02 μm; VIIRS bands M13, M15, and M16 have central wavelengths of 3.70, 0.76, and 12.01 μm. *From Ref. [11].*

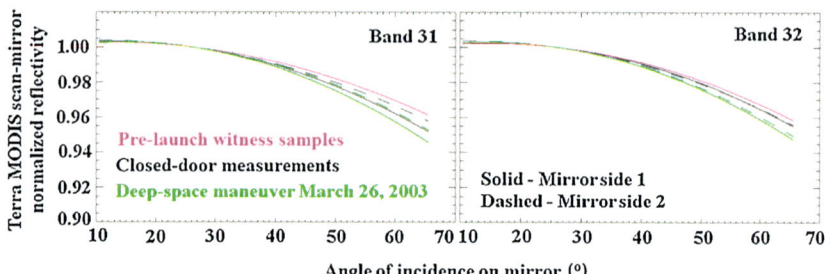

FIGURE 12 Normalized mirror emissivity for the Terra MODIS Bands 31 and 32 ($\lambda = 11.04$ and 12.02 µm) as a function of angle of incidence on the scan mirror surfaces. The red curves are derived from the prelaunch witness samples, black from the Terra MODIS closed door measurements, and green from the Terra deep-space maneuver in March 2003.

on witness samples of the mirror surfaces were used to determine the mirror reflectivity and corrections were developed to extend the calibration to the wide range of incidence angles experienced across the earth swath. However, postlaunch data revealed consistent steps in the derived SSTs where adjacent swaths overlapped, which were the result of imperfect rvs corrections. A fortuitous situation provided the measurements needed to improve the correction, in which the MODIS entered a safe-hold mode with the earth-view door closed, but the mirror continued to rotate and measurements continued to be taken. By assuming the temperature of the inside surface of the door as isothermal, the gradients in the measured temperatures were caused by the changing incidence angle on the mirror surface and hence provided a mechanism for improving the rvs correction (Figure 12). However, there were still residual rvs effects and the data needed to further improve the correction were derived by a "deep-space maneuver" on March 26, 2003. The deep-space maneuver involves increasing the pitch rotation rate of the Terra satellite during the eclipse part of an orbit, so that all the sensors that are normally directed toward the earth, including MODIS, are directed to cold space. The measured radiance in the infrared bands were then simply of emission from the mirror, and this led to a new rvs correction (Figure 12), which led to much improved SST retrievals [15].

3. COMPARISONS WITH REFERENCE SATELLITE SENSORS

Among the several approaches that permit appraisal of the performance of the on-orbit calibration procedures and the instrument stability are comparisons with other satellite instruments measuring the same or related variables. Additional comparisons are made with measurements from other types of sensors determining surface temperatures and that is the focus of the next section, but first we consider intersatellite instrument comparisons.

There is usually a delay of several weeks following injection into orbit before the infrared channels of a radiometer are chilled to operating temperatures. This delay is to allow gases, especially water vapor, to diffuse into space as otherwise they would condense on the chilled optical surfaces and detectors, reducing the sensitivity of the radiometer.

3.1 Spatial Comparisons

Comparison of the global SSTs from the new sensor with those from an established satellite instrument is an effective and quick way to establish relative confidence in the new sensor. Global, daily fields can be used to reveal spatial differences in the derived SSTs. Figure 13 shows the difference between VIIRS SSTs derived from the night-time orbits of a given day and SSTs derived from the microwave measurements of WindSat [16]. The geometry of the WindSat swaths requires the compiling of 5 days of measurements to generate nearly complete global fields [17]. The sources of uncertainties in the infrared and microwave SSTs are mostly uncorrelated: in the infrared, the main sources of errors are undetected clouds, aerosols, and anomalous vertical distributions of water vapor, whereas in the microwave errors are primarily caused by raining clouds and radiofrequency interference. Thus, the differences shown in Figure 13 reveal problems in either or both SST fields. For example, there are cold biases in regions off West Africa and the Arabian Sea, where aerosol contamination of the VIIRS SST is expected. The WindSat

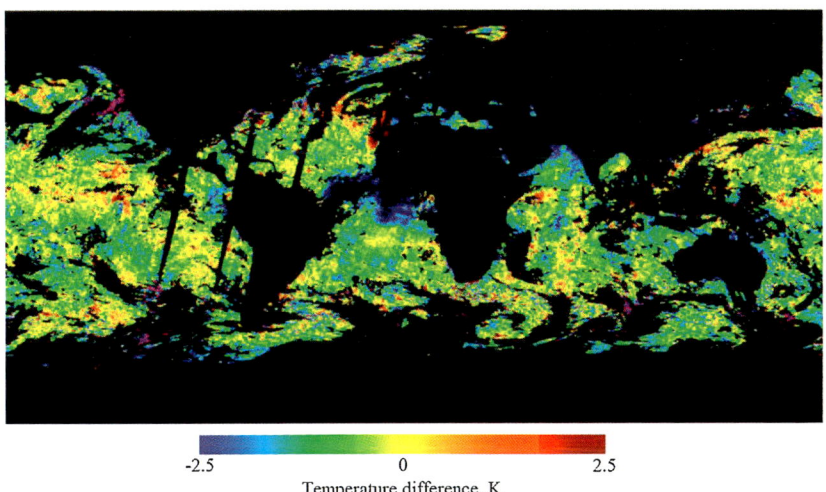

FIGURE 13 Difference between SSTs derived from the infrared measurements of VIIRS and the microwave measurements of WindSat. Black indicates land and clouds. The VIIRS SSTs are night time and the WindSat SSTs are a 5 day composite. Negative values indicate regions where VIIRS SSTs are cooler than those of WindSat. *From Ref. [16].*

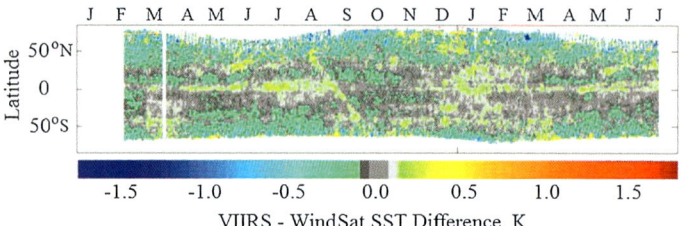

FIGURE 14 Hovmöller plot of the zonal average difference between night-time VIIRS skin SSTs and 5 day composites of WindSat SSTs. The horizontal axis shows time from January 2012 to the end of June 2013. The mid-gray color indicates the differences of <0.05 K, and the other gray tones show the differences between 0.05 and 0.1 K. *From Ref. [16].*

SSTs have been reduced by 0.17 K, i.e., a mean thermal skin effect, so that they can be compared to skin SST retrievals from VIIRS.

3.2 Temporal Comparisons

Temporal dependences of differences between the two SST fields are revealed by using Hovmöller diagrams. Figure 14 shows the daily, zonal averages of the differences between the zonal average differences of VIIRS night-time SSTs and 5 day composites of WindSat SSTs for the first 17 months of the VIIRS infrared mission. While there is a good correspondence between the two SST time series at the level 0.1 K and less, Figure 14 reveals the seasonal discrepancies along the Equator, where VIIRS SSTs are warmer than those of WindSat, and at high latitudes where they are cooler. Whether these are the results of errors in the infrared or microwave SSTs, or both, is difficult to determine from this type of analysis alone. There is also the possibility that some of the differences might result from the mismatch in temporal sampling, and the difference between SSTs derived from infrared and microwave measurements.

A way of comparing temporal relative accuracy between different infrared radiometers has been developed at NOAA-NESDIS-STAR using the Advanced Clear-Sky Processor for Oceans system [18–20]. Clear-sky ocean pixels are identified in the swath brightness temperatures derived operationally at NOAA, and the Community Radiative Transfer Model [21,22] is used to simulate top-of-atmosphere brightness temperatures using the National Centers for Environmental Prediction Global Forecast System atmospheric profiles of temperature and humidity with "Reynolds" daily first-guess SST fields as inputs. The time series of the global medians of the differences between the measured and modeled brightness temperatures is a useful diagnostic of potential problems with the satellite instrument. The upper panel of Figure 15 shows an example of multiyear differences between brightness temperatures in the spectral channels at $\lambda = 3.7$ μm measured by AVHRRs on

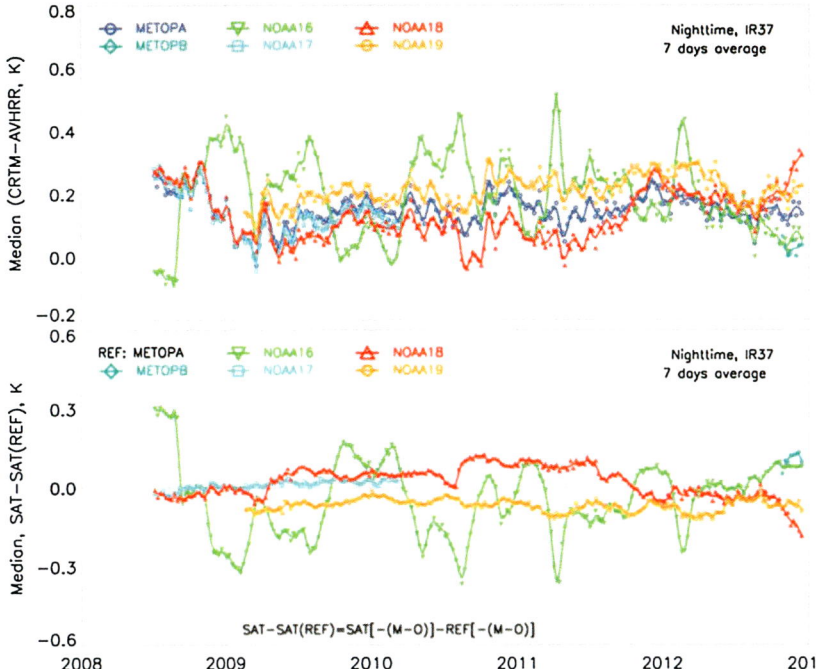

FIGURE 15 Time series of global night-time modeled-observed biases AVHRR measurements in the $\lambda = 3.7$ μm channel on NOAA-16, -17, -18, -19, and Metop-A (top), and double differences with the AVHRR/3 on MetOp-A used as a reference (bottom). *After Ref. [18].*

several satellites, each referenced to simulated values; the time series are based on global, night-time cloud-free measurements for 7 day periods. Apart from revealing that the AVHRR on NOAA-16 is "out of family," the figure shows that many of the fluctuations in the differences between the modeled and observed brightness temperatures are coherent between many satellite radiometers, indicating that the source of the discrepancies is common to many differences, and is very suggestive that the source is the radiative transfer modeling. This contribution to the time series can be removed by selecting one satellite instrument as a relative reference and calculating the modeled−observed differences with respect to the modeled−observed difference of the reference sensor. This approach is called "double differences" and an example is shown in the lower panel of Figure 15, where the time series of the upper panel are shown referenced to the modeled−observed values of the MetOp-A AVHRR/3. The double differences reveal a more consistent comparison of the relative stability of multiple satellite radiometers, and such information is critical to ensure stability and continuity of SSTs over multiple satellite missions [20], which is a prerequisite of a Climate Data Record. Some of the discrepancies in double differences might be due to different overpass times of the satellites, although the effects of diurnal heating and cooling should be small in the night-time examples given in Figure 15. Offsets in the discrepancies might be due to differences in the relative spectral response functions (see below). Other differences might be caused by artifacts in the on-board calibration procedures, and the anomalous behavior of the AVHRR on NOAA-16 is suggested by Liang and Ignatov [20]. This artifact could be the result of drift of the orbit over the period shown here from 4 to 7 a.m. local equator crossing time, accentuating the effects of contamination of the calibration target by solar radiation in a dusk-dawn orbit. NOAA-16 was launched in September 2000 into an orbit with a 2 p.m. local equator crossing time, with a 2 year planned mission, and so it had been in space many years before the period of the comparisons shown in Figure 15, indicating that the relatively consistent behavior of the calibration target temperatures shown in Figure 7 just over a year after launch cannot be presumed to be applicable over the entire length of an extended mission, especially if the orbit is not maintained.

3.3 Simultaneous Nadir Overpasses

To reduce the effects of differences in measurement times and geometries, it is desirable to compare data from two infrared radiometers when the measurements of each are collocated in space and contemporaneous in time, and made through the same atmospheric path length. Such opportunities occur when the nadir measurements of the two instruments on two satellites pass over the same place at the same time, or within acceptable limits of temporal and spatial coincidence and scan angle from nadir; acceptable meaning that the additional uncertainties do not contribute significantly to the differences in the satellite measurements. Such coincidences are referred to as Simultaneous Nadir Overpasses (SNOs). Such comparisons are valuable for establishing correspondence with SSTs measured from a new sensor and one that has been on orbit for sufficient time for its uncertainties and errors to have been characterized. For radiometers on polar-orbiting satellites, such SNOs tend to occur at high latitudes. Figure 16 shows the SNOs between one orbit of

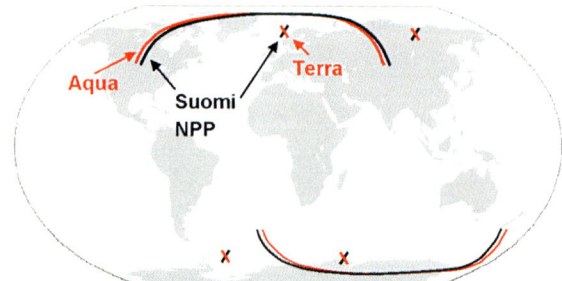

FIGURE 16 Positions of simultaneous nadir overpasses between VIIRS on S-NPP and MODIS on Terra and MODIS on Aqua. *From Ref. [23].*

measurements from the VIIRS on Suomi-NPP and the MODIS on Terra and the MODIS on Aqua [23]. The equator crossing time of Terra is 10:30 and for Suomi-NPP 13:30, and so only a couple of SNOs is to be expected in each hemisphere per orbit. The equator crossing time for Aqua is also 13:30 and even though Aqua and Suomi-NPP are in different orbits, the opportunities for SNOs are much greater. Such comparisons between brightness temperatures at the top of atmosphere have to take into account differences in the relative spectral response functions (RSRs) of the two radiometers (Figure 17 [24]), as two perfectly calibrated radiometers measuring the same scene would measure different brightness temperatures if the RSRs are different. This source of error can be reduced by comparing the measurements of a radiometer with RSRs defined by filters with spectra measured by spectrometers as the RSRs can be convolved with the measured spectra to simulate the measurement of the filter radiometer.

3.4 Instruments on the Same Satellite

The comparisons can be extended to the entire globe and not be limited to high latitudes if the filter radiometer and the spectrometer are both on the same satellite, such as Atmospheric Infrared Sounder (AIRS) and MODIS on Aqua, and Infrared Atmospheric Sounding Interferometer (IASI) and AVHRR/3 on the MetOp satellites.

The AIRS uses a set of grating spectrometers to measure parts of the infrared spectrum in the wavelength range of 3.7–15.4 μm, having 2378 infrared channels. It also has four visible and near-infrared channels, primarily for scene identification. The scan mirror directs the field of view across a

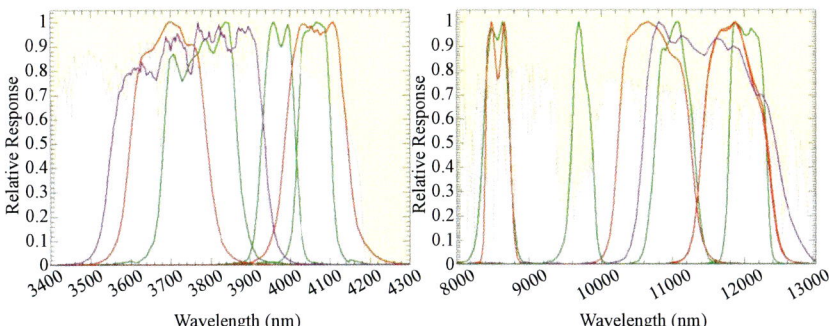

FIGURE 17 Relative spectral response (RSR) functions, normalized to have a peak value of unity, for the infrared atmospheric transmissions window bands for MODIS (green) and VIIRS (red). The purple lines are the broad-band RSR functions of the VIIRS imaging bands, which have a nadir ground resolution of 0.375 km. The background white is the atmospheric transmission spectrum for vertical propagation through a cloud-free standard atmosphere. The x-axis is having the wavelength in nanometers. *After Ref. [24].*

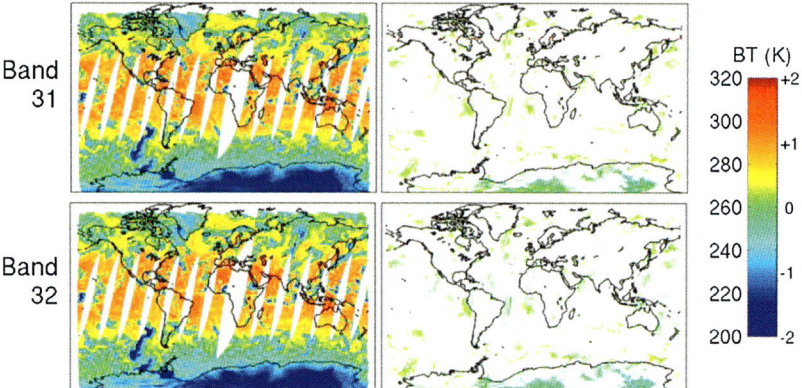

FIGURE 18 Night-time MODIS brightness temperatures (left) and AIRS – MODIS brightness temperature differences (right) on September 6, 2002 for MODIS bands 31 and 32. The left-hand tick marks of the color scale apply to the brightness temperatures, and those on the right hand are for the brightness temperature differences. *After Ref. [27].*

swath width of 1650 km, with a spatial resolution of 13.5 km at nadir [25]. AIRS spectra are well calibrated, both radiometrically and spectrally [26]. Despite the large difference in the footprint sizes, Tobin et al. [27] compared measurements of MODIS and AIRS by averaging the MODIS brightness temperatures in the pixels in each AIRS field of view, and convoluting the AIRS spectra with the MODIS RSRs. An example of the brightness temperatures measured in the MODIS spectral bands 31 and 32 that are used for SST retrieval both day and night is shown in Figure 18. The mean differences between AIRS and MODIS are <0.1 K for most of the MODIS bands, with the smallest differences being for the window channels. The AIRS–MODIS brightness temperature differences were found to have dependences on scene temperature and scan angle, but again these were found to be smallest for the atmospheric transmission window bands.

A similar analysis has been conducted comparing the spectral measurements of IASI with the data from the infrared channels of AVHRR on Metop-A [10,28]. The IASI is part of the scientific payload of the MetOp series of European meteorological polar-orbit satellites. IASI is a Fourier-Transform Infrared spectroradiometer operating as a Michelson interferometer producing spectra in the wavelength range of 3.7–15.5 µm, and it also has an imaging infrared radiometer with a broad spectral passband in the 10.3–12.5 µm spectral range [29]. The interferometer has a nadir spatial resolution is 25 km, and the 64 × 64 array of the imager each having a 0.8 km nadir resolution is centered on the field of view of the spectrometer to provide information on the scene content unresolved by the spectrometer. The swath width is about 2400 km. The radiometric accuracy of IASI has been assessed in field campaigns by comparison of coincident and collocated measurements during underflights with well-calibrated, SI-traceable Fourier Transform Spectrometers, such as NAST-I (National Polar-orbiting Operational Environmental Satellite System (NPOESS) Airborne Sounder Testbed-Interferometer), on high-altitude aircraft [30]. In the long-wave atmospheric transmission window region ($\lambda = 10.2-11.0$ μm), differences of 0.02 ± 0.18 K were found between the aircraft measurements and those of IASI, and of 0.09 ± 0.48 K in the mid-infrared window ($\lambda = 4.0-4.2$ μm) [30].

Building on the improved knowledge of the radiometric properties of the AVHRRs that resulted in a more accurate calibration procedure [9], Mittaz and Harries compared the on-orbit performance of the AVHRR on MetOp-A with the well-calibrated spectra measured by IASI on the same satellite [10]. The study demonstrated that the standard AVHRR calibration can introduce biases errors in the calibrated brightness temperatures of ~ 0.5 K, which is comparable to the result of an earlier analysis which showed differences (IASI − AVHRR) <0.4 K, with a standard deviation of ~ 0.3 K for AVHRR channels 4 and 5 [31], but using the new, physically based calibration scheme reduced the discrepancies with the IASI measurements to <0.05 K, at nadir. Differences between AVHRR and IASI up to 1.5 K remained at large zenith angles and for cold-scene temperatures (~ 210 K). However, in the range of temperatures characteristic of the sea surface, the changes caused by increasing satellite zenith angle are very small, ~ 0.02 K.

A great deal can be learned about the behavior of new satellite radiometers by comparisons with other satellite instruments that have been on orbit long enough for their characteristics and uncertainties to have been established. Continued comparisons over the lifetimes of the radiometers can provide useful information about the relative stabilities and possible degradation of performance, but all such comparisons are relative and other approaches are required to ascertain some absolute measure of accuracy.

4. VALIDATING GEOPHYSICAL RETRIEVALS

Having determined the performance of the radiometers on-orbit using diagnostics derived from internal data and through comparisons with other satellite instruments, additional approaches are needed to assess the absolute uncertainties and stabilities of satellite-derived SSTs, as is required for the generation of a Climate Data Record. This is achieved through comparisons with independent SST measurements. In some approaches, this has the added advantage of a rigorous measurement of the uncertainties and stability by using validating instruments that are repeatedly calibrated against SI temperature standards.

An approach that is intermediate between satellite-to-satellite comparisons and those that compare temperatures derived from satellite measurements with those taken at or close to the surface uses atmospheric radiative transfer modeling to simulate the satellite measurements. Because of uncertainties in the simulated to-of-atmosphere radiances caused by imperfections in the radiative transfer modeling and in the description of the atmosphere at the time of the satellite overpass, especially resulting from errors in the specification of the water vapor distribution, this approach is most successful when applied to conditions of a very dry, cloud-free atmosphere. Such conditions are met at Lake Tahoe in the western USA. Lake Tahoe is a large body of freshwater, about 32×16 km in size, so it is well-resolved by most infrared imaging radiometers, and being at a height of over 2 km means the water vapor content in the intervening atmosphere is very low. Therefore, the radiative transfer modeling is not seriously compromised by errors in the humidity profiles. Lake Tahoe frequently has cloud-free skies. A continuous series of measurements have been made at Lake Tahoe since 1999 using automated instruments on the surface floats of four moored buoys for the validation of the measurements of the Advanced Spaceborne Thermal Emission and Reflection Radiometer (ASTER) and MODIS [32], and now also for VIIRS. Mounted on each buoy is a self-calibrating radiometer developed at the Jet Propulsion Laboratory (Chapter 3.2) that measures the skin temperature of the lake surface. The measured skin temperature is used with an atmospheric profile derived from a weather prediction model as an input to the radiative transfer model to simulate the satellite measurements. The range of the lake-surface temperatures is about $5-25$ °C, and to extend this to higher temperatures, a similar monitoring station was set up in the Salton Sea. Figure 19 shows the

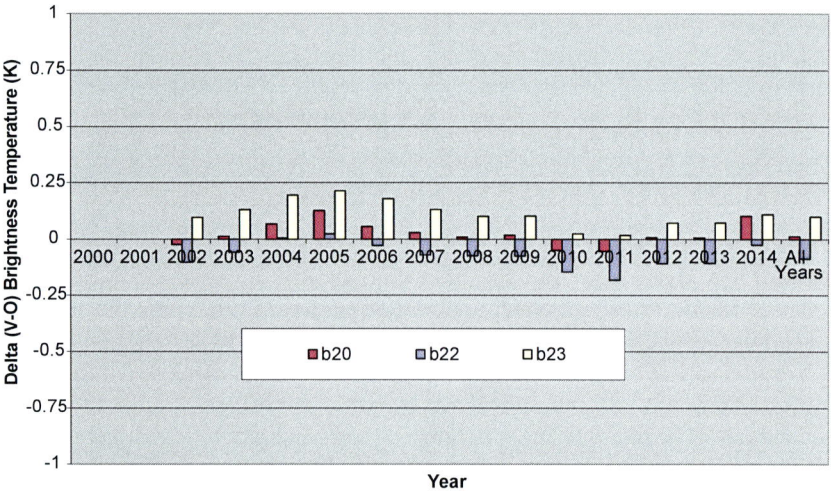

FIGURE 19 Comparisons between simulated (V) and measured (O) brightness temperatures for bands 20, 22, and 23 of the MODIS on Aqua based on measurements at Lake Tahoe and the Salton Sea for the Aqua mission up to April, 2014. Annual means are shown, with the last set of bars indicating all years. The results are for V5 of the MODIS Processing. *Figure courtesy of Dr. S. Hook, JPL.*

time series of comparisons between the simulated and measured Aqua MODIS measurements in bands 20, 22, and 23 for the mission so far. Annual means are shown. The comparisons are good, indicating the absence of major MODIS calibration issues, but an interesting interannual oscillation is revealed. When plotted as a function of the angle of incidence on the MODIS scan mirror, a detectable residual RVS signal is found, endorsing the validity of the approach (S. Hook, pers. comm. 2014).

For any infrared imager being used to derive skin SSTs from calibrated top-of-atmosphere brightness temperatures in the appropriate spectral bands, there are two distinct processing steps that need to be taken: first, identifying those pixels that are free of radiance from sources other than the gaseous components of the atmosphere, and then correction for the effects of the atmospheric gases. The first step is conventionally called "cloud screening" as the primary objective is to identify pixels that include radiance from clouds, and the second step is referred to as the "atmospheric correction." These algorithms are based on the combination of information from different spectral bands and this inevitably leads to growth in the uncertainties in the variables in going from brightness temperatures to derived SST, as illustrated in Figure 20 [33]. The right-hand column of Figure 20 shows the progression from Level 0 (L0) data, the raw measurements, through to remapped, gap-filled SST fields (L4) with the processing steps between. The boxes on the left enclose sources of uncertainties, many of which are time-dependent and instrument-specific, and therefore ought to be evaluated for each sensor so that consistent and accurate, long-term records, spanning multiple missions and sensors can be generated.

As many of the processing steps involve combining or comparing values in pixels measured at the same time and place by different detectors, the degree of band-to-band registration is important, especially if there is a possibility of it changing with time. The target coregistration is a part of the sensor specifications, and for MODIS, for example, this was 10% for the band-to-band registration in both along- and cross-track directions [34]. For a complex instrument, such as MODIS, with 36 spectral bands and multiple detectors per band, the coregistration of the fields of view of detectors on the same focal plane array is likely to be better than for bands with detectors on different focal plane arrays [34,35]. This was indeed found to be the case for the Aqua MODIS in prelaunch tests, where poorer coregistration was found between bands on the visible focal plane array, operating at ambient temperature, and those for the thermal infrared operating at cryogenic temperatures (Figure 21) [12]. Such misalignment can give rise to elevated uncertainties when a geophysical variable is derived by combining pixel values in different spectral bands, or in the classification of pixels (e.g., [36]). The consequences of misalignment will be scene dependent, in that they will be greater where there are large horizontal gradients in the relevant variable, and less important where the gradients are small.

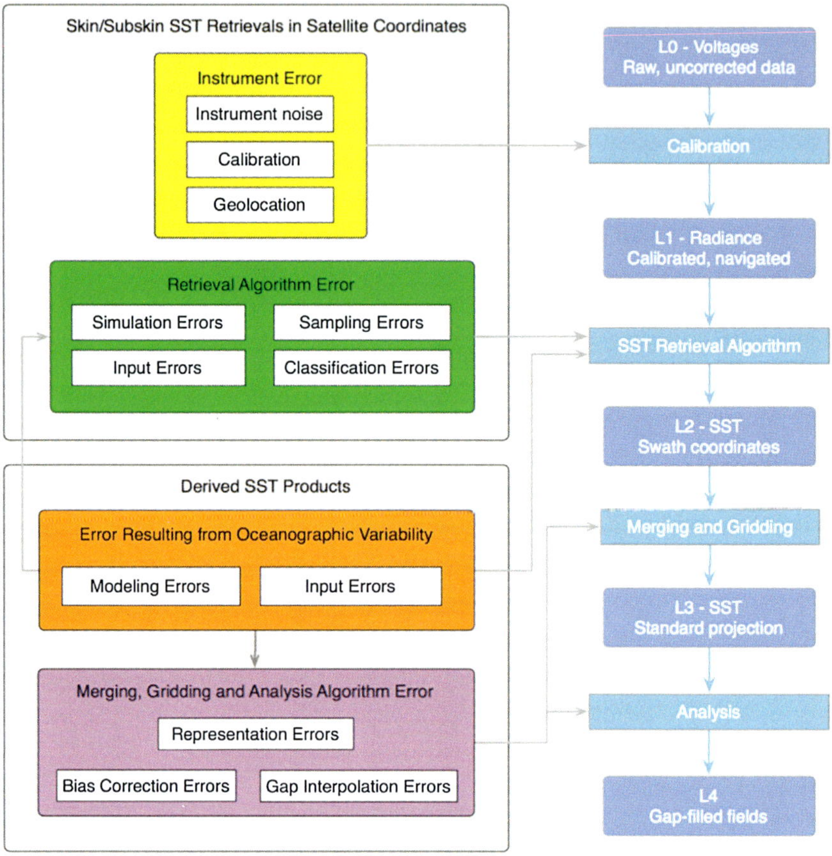

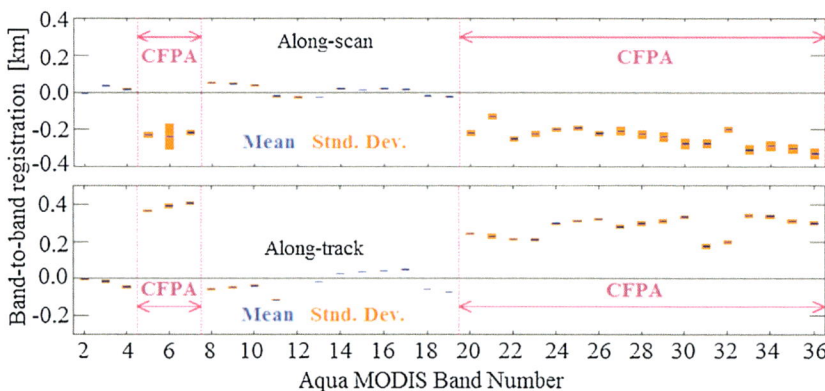

FIGURE 20 Sources of uncertainties, in the boxes at left, in deriving skin SST fields from satellite measurements of radiance through to mapped, gap-filled SST fields, going from top to bottom at right. *From Ref. [33].*

FIGURE 21 Band-to-band registration of all 36 bands of MODIS on Aqua, referenced to Band 1. Means and standard deviations are shown. The top panel shows along-scan misregistration, and the bottom panel shows along-track. CFPA indicates those bands on the cold focal planes. *After Ref. [35].*

4.1 Cloud Screening

A very important step in deriving skin SST from infrared satellite measurements is to identify pixels that are contaminated by the presence of clouds prior to correcting for the intervening atmosphere. One tried-and-tested approach to cloud screening is to grow a "decision-tree" cloud-identification algorithm. Based on the approach developed for the AVHRR Pathfinder program [37], this has subsequently been applied to MODIS [15] and VIIRS [16]. The scheme applies a series of tests to individual pixels, or small groups of pixels, to determine where their values lie with respect to thresholds that are chosen to distinguish between cloud-free and cloud-contaminated pixels. Because of the wide range of cloud properties, a single test does not suffice in all cases. For night-time data, tests are applied to the infrared measurements, but for daytime data, tests are also applied to measurements in the solar reflective bands. An example of a cloud-screening decision tree is shown in Figure 22.

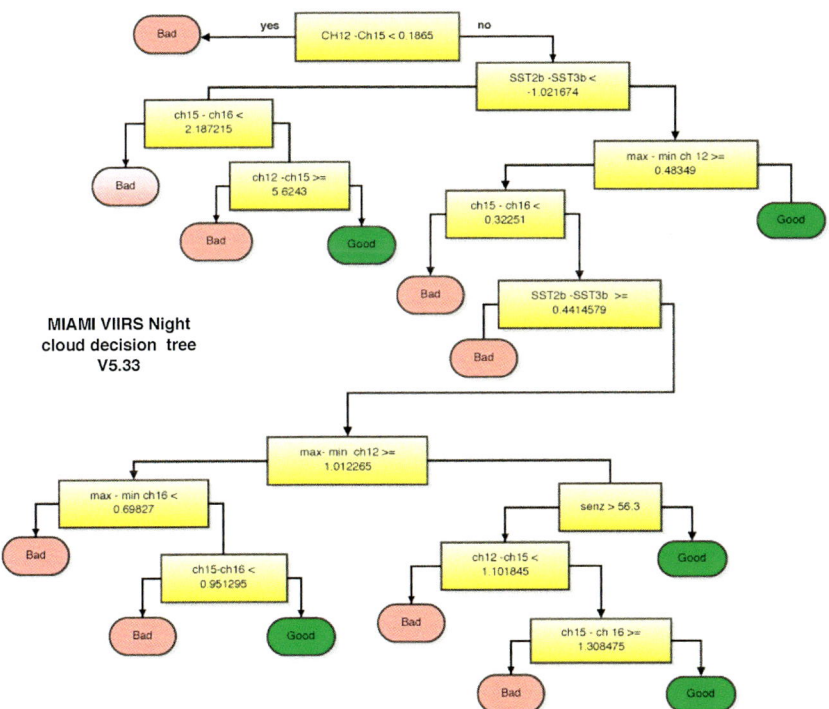

FIGURE 22 An example of a decision-tree for the identification of cloud-free pixels. Each test is derived from physical expectations based on the radiometric measurements in different VIIRS bands, and on the measurement geometry. This is for night-time measurements; a separate decision-tree (not shown) that includes information on the pixel reflectivity in the visible bands is used for day-time measurements. Only pixels that are designated "Good" are used in the derivation of skin SST. *From Ref. [16].*

An alternative approach uses a Bayesian classifier [38]. This probabilistic approach is based on radiative transfer simulations of clear-sky measurements to give joint probability density functions of brightness temperatures in two infrared channels, and empirically determined cloudy-sky joint probability density functions [39]. Analysis fields from numerical weather prediction models are used as prior information. A significant benefit of this approach is the determination of the degree of confidence in the classification of a pixel as cloudy or cloud-free.

All cloud-screening approaches have difficulties in certain conditions, such as in regions of large horizontal SST gradients, and optically thin clouds; no approach is perfect and undetected cloud effects are a source of error, at some level.

4.2 Atmospheric Correction Algorithm

The influence of reflected and scattered solar radiation means measurements of satellite radiometers taken in the mid-infrared atmospheric transmission window cannot be used in the day-time part of the orbit. The standard atmospheric correction algorithm used in both day and night conditions is therefore based on the measurements in the 10–13 μm atmospheric transmission window, often referred to as a "split window." The form of the algorithm for the nonlinear SST (NLSST [40]) is

$$\text{SST} = a_0 + a_1 T_{11} + a_2 (T_{11} - T_{12}) T_{\text{sfc}} + a_3 (T_{11} - T_{12})(\sec(\theta) - 1) \quad (2)$$

where a_0, a_1, a_2, and a_3 are coefficients; T_{11} is the brightness temperature measured in the band centered near $\lambda = 11$ μm; and T_{12} is the brightness temperature measured in the band centered near $\lambda = 12$ μm. T_{sfc} is a first guess or climatological SST that scales the coefficient multiplying the $T_{11} - T_{12}$ brightness temperature difference to account for the differing distribution of atmospheric water vapor that is correlated with the SST. θ is the sensor zenith angle and this fourth term compensates for the increasing path length when the scan is away from nadir. The values of the coefficients are dependent on the spectral characteristics of each infrared band and therefore are different for each satellite radiometer.

At night, the measurements from the mid-infrared window can be used since the scattering and surface reflection of solar radiation is, of course, absent. Similar multichannel combinations of the spectral measurements are used for AVHRR, (A)ATSR, and VIIRS including the data at ∼3.7 μm. MODIS has three spectral bands in the mid-infrared window and a combination of two (at wavelengths centered at 3.95 and 4.05 μm) provides the most accurate SSTs [15].

The method of deriving coefficients in the atmospheric correction algorithms is dependent on the time elapsed since launch. At-launch coefficients are derived by atmospheric radiative transfer modeling to simulate the spectral measurements through a large number of atmospheric profiles, measured by radiosondes [41] or from weather forecast models that represent a realistic distribution of atmospheric variability. Matchups with in situ validating data, such as from drifting or moored buoys or ships (see below), can begin very early in the mission, primarily to make an initial assessment of the effectiveness of the cloud screening and atmospheric correction algorithms, and as the elapsed time since launch increases, the number of matchup grows until it is feasible to generate the coefficients from the matchups themselves. The advantage of this is that any unknown or poorly characterized instrument artifacts that introduce error into the calibrated brightness temperatures would be indistinguishable from atmospheric effects, and can be compensated, at least in part, by the coefficients. So, if there is some time dependence to the instrumental artifacts, then by generating time-dependent coefficients, say on a monthly basis, some of the changes in the behavior of the radiometer can be offset by the coefficients, and the SST retrievals are rendered less sensitive to small instrumental changes. This approach of using time-dependent coefficients in the atmospheric correction algorithms places the onus of demonstrating whether the stability requirements have been met by the satellite retrievals on the in situ measurements. If it can be demonstrated that the in situ measurement data sets are stable over years and decades, to some level of accuracy, and the statistics of the differences between the satellite and in situ surface temperatures are also stable, then it can be inferred that the satellite data are stable within a demonstrable limit.

It is clear that the effects of the entire atmospheric variability cannot be compensated for by a statistically based combination of two or three measurements of brightness temperatures. In particular, the presence of anomalous vertical distribution of atmospheric water vapor, such as dry layers, can lead to larger errors in the derived SST than would be otherwise expected [42,43]. Knowledge of the vertical distribution of water vapor, derived from additional data, does lead to improved SST retrieval accuracies [44]. Making better use of external information about the atmospheric state is the foundation of an alternative approach to atmospheric correction using an Optimal Estimation approach; this is gaining favor. In this approach, simulated top-of-atmosphere brightness temperatures are derived by radiative transfer modeling through an estimation of the atmospheric state given by Numerical Weather Forecast models. Such an approach has been shown to bring benefit to retrievals from AVHRR on MetOp-A [45] and the Spinning-Enhanced Visible and Infra-Red Imager (SEVIRI; described in Chapter 2.3) on Meteosat 9 [46], indicating that the uncertainties in the derived SSTs resulting from those in the specification of the atmospheric state and in the radiative transfer simulations are now comparable to those that are inherent in the conventional statistical approach.

The diverse approaches to cloud screening and atmospheric corrections each have their own consequences on the uncertainties in the derived SSTs, and the determination of these uncertainties using independent measurements is a crucial step in assessing the suitability of satellite-derived SSTs for many applications such as the generation of Climate Data Records.

4.3 Geophysical Validation

A method of tracking the on-orbit stability of satellite radiometers over extended periods is to monitor the uncertainties of the retrieved surface temperatures by comparison with independent measurements. Temporal or regional changes in the uncertainties could indicate a degradation of the performance of the satellite radiometers.

A widely used source of independent temperature measurements are those from drifting buoys, which are deployed in all ocean areas to provide data for weather forecasting. The spatial and temporal distributions are not uniform, but the large numbers (1092 at the time of writing) render these a valuable resource. The thermometers are mounted below the waterline of the drifting buoy and measure a temperature at a depth of about 10–20 cm in calm conditions, probably deeper on average in rough conditions, which under conditions of low wind speed, can be decoupled from the surface skin temperature by vertical temperature gradients that result from solar heating (e.g., [47–49]). Additionally, the thermal skin effect introduces a further difference between a temperature measured in the body of the water, and that of the surface (e.g., [50–52]). The buoys are expendable and so are not recovered at the end of their deployments for the calibration of the thermometer to be confirmed. As a result, stringent quality control procedures have to be applied to ensure that changes in the calibration or other sources of error in the measurements from individual buoys do not go undetected to cause erroneous estimates of the satellite SST retrieval errors [53]. Nevertheless, they are a useful tool to assess the stability of satellite radiometers (Figure 23), but not at the level required for the generation of Climate Data Records (CDRs).

A shortcoming of the comparisons between satellite retrievals with measurements from drifting or moored buoys, or between pairs of satellite-derived SSTs, is that it is difficult to attribute the sources of uncertainties to one or the other measurement set. However, by combining comparisons between pairs of three sets of measurements, and making assumptions that include the errors and uncertainties in each being uncorrelated with those of the others, estimates of the uncertainties in each data set can be made [54]. By comparing pairs of measurements from the AATSR and Advanced Microwave Scanning Radiometer for Earth Observing System (EOS; AMSR-E) and those from buoys, all taken in 2003, O'Carroll et al. [54] found that the spatially averaged night-time AATSR dual-view three-channel SST retrievals, adjusted for the thermal skin effect, had a standard deviation of error of 0.16 K, whereas the values for the buoy SSTs were 0.23 K and for AMSR-E SSTs were 0.42 K. The standard deviation of the buoy measurements, which has subsequently

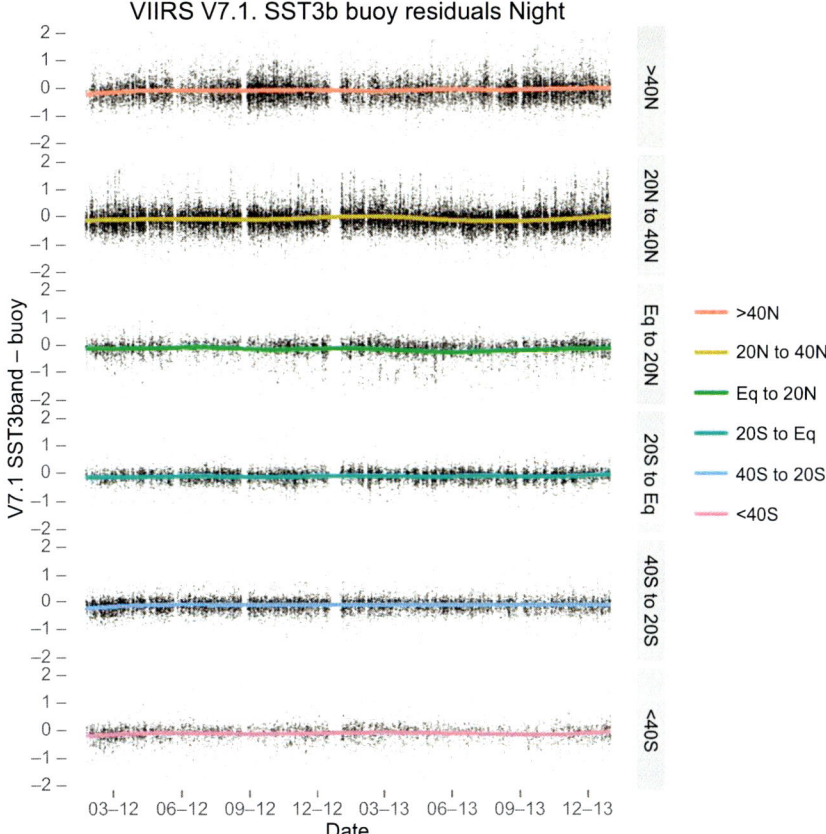

FIGURE 23 Time series of differences, divided into latitude bands, between VIIRS sea-surface temperatures (SSTs), derived with a night-time 3-band atmospheric correction algorithm, and collocated buoy measurements. The colors indicate the daily median errors, and the dots indicate the individual comparisons. The increase in the scatter at higher latitudes is apparent. *From Ref. [16].*

been confirmed by others (e.g., [55]), was much larger than the previously assumed uncertainty of ~ 0.1 K and this led to the realization that the buoys were making a large contribution to the discrepancies between the satellite and in situ matchups that had previously been attributed to the satellite SST retrievals. The utility of the drifting buoys in determining the uncertainties in satellite SST retrievals would be greatly improved if the accuracy of the buoy thermometers were to be improved to be much better than 0.1 K.

Another contribution to the estimate of the satellite SST accuracies comes from the nonstationary nature of the ocean temperature, as the ocean warms during the day by the absorption of solar radiation and cools at night, and accepting collocated matchups that are not contemporaneous introduces

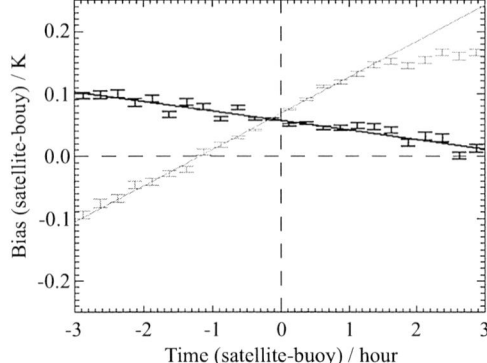

FIGURE 24 AATSR retrieval bias errors as a function of satellite-buoy time difference for daytime (gray) and night-time (black) matches. Solid lines show linear best fit to data, using time differences <1.5 h for day-time matches. The AATSR on Envisat has a mean ascending node time of 10:00 h. *From Ref. [56].*

inaccuracies, as shown in Figure 24, which shows the mean rates of change of the AATSR dual-view SST retrievals compared to in situ measurements from drifting buoys, as a function of the time interval between the satellite and buoy data [56]. The night-time rate of cooling is between -0.01 and -0.02 K h^{-1}, whereas the day-time warming is up to 0.1 K h^{-1}. The contributions of the changing temperatures to the uncertainty budget of the satellite SSTs will be strongly dependent on the time of day as well as the acceptance window of the satellite-buoy measurement times. When a three-way comparison involves measurements from instruments on different satellites, there are contributions from the time differences between all pairs of measurements. These can be reduced if the two satellite radiometers are on the same spacecraft, as is the case for MODIS, AMSR-E, and AIRS on Aqua, and AVHRR and IASI on MetOp.

A three-way comparison between surface temperatures derived from MetOp-A AVHRR and IASI data, and those measured from drifting buoys indicated that the standard deviations of error are 0.29 K for IASI, 0.13 K for AVHRR, and 0.23 K for drifting buoys [28]. All the AVHRR pixels in a 21 × 21 array in an IASI footprint that were identified as cloud-free were used in the analysis. The spatial distributions are shown in Figure 25, in which the top row shows the spatial distribution of the average differences between each pair of the three data sets, and the bottom row shows the standard deviations of the errors attributed to each. As expected, there are regional changes in the errors in each data source and indicate that IASI SSTs are more accurate than those of AVHRR in some regions, but not in others. For both satellite instruments, the larger standard deviations and uncertainties are in regions with higher SST spatial variability implying that sampling errors are contributing to estimated mean SSTs derived from both IASI and AVHRR.

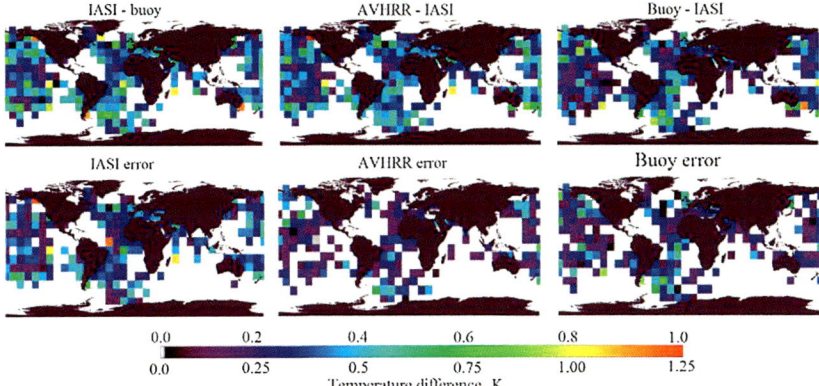

FIGURE 25 Standard deviation of global 5° gridded IASI-buoy, AVHRR-IASI, buoy-AVHRR temperature differences for October 2010 to July 2011 (top), and standard deviations of errors of global 10° gridded IASI, AVHRR, and buoy sea-surface temperature (SST) observations for the same period (bottom). *After Ref. [28].*

An alternative source of near-surface ocean temperatures measured from drifting buoys are those from thermometers on moored buoys, such as those of the Global Tropical Moored Buoy Array (GTMBA) that span the Equatorial and tropical zones of the world's oceans [57,58]. At a depth of ~1 m, the shallowest thermometers are deeper than those of the drifting buoys, but the GTMBA thermometers are better calibrated, and, very importantly, are periodically recovered after deployment for recalibration and refurbishment [59,60]. The accuracies of the thermometers on the GTMBA moorings in the Pacific Ocean (often referred to as the Tropical Ocean Atmosphere, or TAO, array) were found to be ~0.03 K [57] and the calibration drifts during deployments were typically 0.003 K or less over a deployment, typically lasting about a year [60]. In low wind-speed conditions conducive to the growth and decay of diurnal temperature gradients, the 1 m temperature, while being an accurate measurement, may not represent the skin SST needed to assess the accuracy of the satellite-derived SSTs. There is evidence that the thermometers on the earlier deployments of the TAO array were sensitive to direct solar heating [61]; but because of the shading of the mooring float, this was not pronounced in the 1 m temperatures.

As a result of their long duration, high accuracy, and stability, the subsurface temperature measurements have been used successfully to demonstrate the stability of the skin SSTs derived from the (A)ATSR series of satellite radiometers in the tropics, even though models are needed to bring the subsurface temperature measurement to an estimate of the skin SST [56]. Because the GTMBA moorings include a suite of meteorological sensors, the measurements for driving such models are available close to the time of the satellite overpass.

The second group of moored buoys is the ones deployed in coastal regions, mainly around the USA. In principle, these could be used to extend the results from the GTMBA to mid-latitudes, but it has been demonstrated [62] that the near-shore temperatures show elevated levels of variability, and that comparisons with satellite-derived skin SSTs are noisier than the comparisons with the GTMBA. This is likely to be the result of small-scale temperature variability that is not resolved by the satellite measurement.

Thermometers on the autonomous profilers of the Argo program provide a mechanism for extending high-accuracy in situ measurements to higher latitudes than those covered by the GTMBA. The Argo profilers support thermometers as part of a CTD instrument mounted on the top end-cap of the cylindrical float that take temperature profiles over a depth of about 2000 m to about 5–10 m [63]. The floats remain parked at a depth of about 1000 m and advect with the currents at this depth; every 10 days, the floats descend to 2000 m and then begin an ascent to the surface, taking measurement of undisturbed water on the way. The standard profiles are terminated at a depth between 5 and 10 m to preserve the calibration of the sensors that might occur if they were in use as the float breaks surface. The data are transmitted to land by satellite before the float descends again to 1000 m to park. A second generation of floats has been recently developed, primarily for validation of surface salinity measurements from the microwave radiometer Aquarius [64], that have a second set of sensors that continue to operate through the uppermost several meters of the water column. The first Argo profilers were deployed in the year 2000, and the target of 3000 floats in operation was reached in 2007. There are currently about 3500 floats deployed, meaning ~350 temperature profiles are taken each day. As the time series of Argo profilers lengthens, it will become an additional source of measurements potentially capable of providing information to assess the stability of satellite-derived SSTs over decades [63], even though collocated meteorological measurements needed to drive models to extend the subsurface measurements to the skin SST are not taken by the floats themselves.

4.4 Ship-Board Radiometers

Dealing with the uncertainties of extending subsurface temperatures to the skin SST can be avoided by using infrared radiometers mounted on ships to measure the skin SST. Such instruments have the added advantages of using internal blackbody calibration targets and the calibration procedure can be checked against laboratory calibration targets before and after each deployment [65]. Through a series of international workshops, the laboratory calibration facilities have been characterized using reference transfer radiometers to provide traceability to SI standards [66]. The SI traceability of the ship-board radiometers provides a mechanism for generating climate data records from satellite radiometers [67]. The ship-board radiometers include instruments with wavelength passbands determined by filters, such as the Infrared SST Autonomous Radiometer [68] and the Scanning Infrared Sea Surface Temperature Radiometer (described by Barton et al., [69]), or spectroradiometers, such as the Marine-Atmospheric Emitted Radiance Interferometers, which are Fourier Transform InfraRed interferometers that take measurements in the 3–18 μm wavelength range [70]. These instruments and their deployments are described and discussed in Chapter 3.2.

5. DISCUSSION

Satellite radiometers are not recovered for recalibration after their on-orbit mission and so it is not possible to check directly whether the prelaunch calibration has been maintained over the duration of their operations. It is not trivial to ascertain the degree to which on-board calibration procedures sustain the accuracy of radiometric measurements of satellite radiometers over the years, sometimes over a decade, they function on orbit. It requires careful and thorough calibration and characterization of the instrument prior to launch to allow the construction of an instrument model which can be used to assess the performance of the instrument postlaunch. Adequate monitoring of the thermal characteristics of the satellite instrument requires many thermometers (themselves well calibrated) to be built into the instruments and their measurements to be transmitted to ground along with the calibration and earth view data.

Despite the difficulties in establishing the accuracy and stability of the satellite-derived SSTs, several approaches have been developed and applied in recent years that indicate the demanding requirements for generating CDRs are within reach, and already may have been attained by some sensors. The required length of SST CDRs demands that time series from multiple consecutive missions be used, and overlap between new and old instruments is needed to provide comparisons on regional and global scales. Such comparisons can be used to demonstrate the useful continuity (or otherwise) of the SST time series, but the outcome is primarily a determination of consistency rather than accuracy.

The assessment of the accuracy of satellite radiometers requires comparisons with independent measurements with traceable calibration. Until high-spectral resolution spectroradiometers with SI-standard calibration are on obit (e.g., [71–73]), the determination of accuracies and stabilities must rely on comparisons with measurements of SST. And this is indeed fundamental to generating CDRs of SST. However, additional sources of uncertainties are introduced, including those from imperfect cloud screening and atmospheric corrections. Disentangling the various sources of uncertainties in the comparisons is not straightforward and can require years of measurements to separate trends in SSTs from the effects of instrumental artifacts.

For many years, the discrepancies in the temperature measurements from satellite and those from in situ thermometers, usually mounted on drifting buoys, were interpreted as errors in the satellite SST retrievals. But, as improvements in the performance of successive satellite instruments, better monitoring their on-orbit performance and enhancements in the atmospheric correction algorithms led to smaller discrepancies, it became apparent that the inaccuracies in the validating measurements and in the method of the comparison were making significant contributions to the uncertainties, meaning that the accuracies in the satellite SSTs were better than could be demonstrated in a simple way. This became a serious concern once it was shown that the accuracies of the thermometers on drifting buoys were much poorer than previously thought. The large number of buoys renders them an important resource for validating satellite SSTs, especially since it is expected that they will be deployed well into the future to provide data for weather forecasting, and therefore should not be discarded. At the time of writing, efforts are underway to improve the accuracies of the buoy temperature measurements to enhance their utility on the assessment of satellite-derived SSTs.

Alternative sources of validating measurements are being exploited because of their accuracy, stability, and long-time series, such as some of the GTMBA moorings, and others of more recent development, such as the Argo profilers. And while the sensors on these are of demonstrably higher accuracy than the thermometers on the drifting buoys, their use in validating satellite-derived SSTs necessitates a careful evaluation of all of the contributions to the discrepancies so that the accuracies of the satellite retrievals can be correctly assessed. This, of course, should also be the case when drifting buoy data are used for satellite validation.

All the subsurface temperature measurements are decoupled to some degree from the skin SST by the vertical temperature gradients, primarily due to diurnal heating and cooling and the thermal skin effect, and these contribute to the total discrepancies when compared to the satellite measurements. These uncertainties can be eliminated by using ship-board radiometers that measure the skin SST, and therefore can be used to make a "like with like" comparison. A further advantage of the use of ship-board radiometers is that they have internal calibration procedures comparable to those used by the satellite radiometers and the internal calibration can be repeatedly checked in the laboratory, before and after deployment on ships, against calibration facilities with SI-traceability. Such traceability provides a basis for satisfying the requirements for the generation of a temperature CDR.

As the on-orbit performance of the satellite radiometers becomes better understood, it is important that this knowledge be used, perhaps with improved interpretations of the prelaunch calibration and characterization, to lead to better processing algorithms to provide more accurate SST retrievals. Such improvements can only be realized if measurements from entire missions are reprocessed to generate new satellite-derived time series. Fortunately, with the rapid progress in computing power and online data storage, this is not as formidable a task as was the case only a few years ago.

6. CONCLUSIONS

The fact that some satellite radiometers have operated for over a decade (e.g., MODIS's on Terra, launched in December 1999, and Aqua, launched in May 2002) and others nearly a decade (AATSR) without significant degradation that would severely restrict the accuracy and stability of the measurements, is a great credit to design and construction of the instruments and to the on-orbit monitoring of the performance to correct and compensate for inevitable instrumental degradation.

Decadal time series of consistent SST retrievals from multiple satellite radiometers of the same type is feasible, as shown by the AVHRR Pathfinder SST [37]. Merging data from different types of infrared radiometers flying on different satellites at the same times offers a way of providing highly desired overlaps between the time series from individual sensors, but to effectively combine data taken at different overpass times requires compensating for effects of diurnal heating. For this purpose, the concept of a foundation temperature, i.e., the temperature below the influence of diurnal heating on any given day, was introduced [74]. However, deriving the foundation temperature from an individual satellite measurement requires modeling the diurnal heating over the course of that day, up to the time the satellite measurement was taken, and this introduces challenges and uncertainties.

The use of accurate, subsurface temperatures from thermometers on the GTMBA array that are recovered for postdeployment calibration, and of skin SSTs measured by ship-borne radiometers, provides a mechanism for underpinning the stability of the satellite SSTs as well as contributing to the assessment of their accuracies, in a way that can lead to the generation of CDRs of SST from satellite measurements. But no single approach is adequate in insolation and a suite of instruments and techniques are required.

Monitoring the performance of the satellite instruments throughout their missions, and assessing the long-term characteristics of the SST retrieval uncertainties, requires continuous commitments on the part of funding agencies and of both operational practitioners and research groups to ensure that the time series currently under development are extended and improved in accuracy to contribute to our understanding of the earth's climate system as well as to provide useful information for weather and ocean forecasting.

REFERENCES

[1] G. Ohring, B. Wielicki, R. Spencer, et al., Satellite instrument calibration for measuring global climate change: report of a workshop, Bull. Am. Meteorol. Soc. 86 (9) (2005) 1303–1313.

[2] GCOS, Systematic Observation Requirements for Satellite-based Data Products for Climate. Supplemental Details to the Satellite-based Component of the Implementation Plan for the Global Observing System for Climate in Support of the UNFCCC–2011 Update, World Meteorological Organization (WMO), Geneva, Switzerland, 2011.

[3] X. Xiong, A. Wu, B. Guenther, et al., Applications and results of MODIS lunar observations, in: Sensors, Systems, and Next-generation Satellites XI, SPIE, Florence, Italy, 2007. http://dx.doi.org/10.1117/12.736787, 67441H-10.
[4] D. Smith, C. Mutlow, J. Delderfield, et al., ATSR infrared radiometric calibration and in-orbit performance, Remote Sens. Environ. 116 (0) (2012) 4−16.
[5] A.P. Cracknell, The Advanced Very High Resolution Radiometer, Taylor and Francis, London, UK, 1997, 534 pp.
[6] D.K. Klaes, M. Cohen, Y. Buhler, et al., An introduction to the EUMETSAT polar system, Bull. Am. Meteorol. Soc. 88 (7) (2007) 1085−1096.
[7] O.B. Brown, J.W. Brown, R.H. Evans, Calibration of advanced very high resolution radiometer infrared observations, J. Geophys. Res.: Oceans 90 (C6) (1985) 11667−11677.
[8] A.P. Trishchenko, G. Fedosejevs, Z. Li, et al., Trends and uncertainties in thermal calibration of AVHRR radiometers onboard NOAA-9 to NOAA-16, J. Geophys. Res.: Atmos. 107 (D24) (2002) 4778.
[9] J.P.D. Mittaz, A.R. Harris, J.T. Sullivan, A physical method for the calibration of the AVHRR/3 thermal IR channels 1: the prelaunch calibration data, J. Atmos. Oceanic Technol. 26 (5) (2009) 996−1019.
[10] J. Mittaz, A. Harris, A physical method for the calibration of the AVHRR/3 thermal IR channels. Part II: an in-orbit comparison of the AVHRR longwave thermal IR channels on board MetOp-a with IASI, J. Atmos. Oceanic Technol. 28 (9) (2011) 1072−1087.
[11] X. Xiong, J. Butler, A. Wu, et al., Comparison of MODIS and VIIRS onboard blackbody performance, in: R. Meynart, S.P. Neeck, H. Shimoda (Eds.), Proc. SPIE 8533, Sensors, Systems, and Next-generation Satellites XVI, 853318, November 19, 2012. http://dx.doi.org/10.1117/12.977560.
[12] X. Xiong, W. Barnes, X. Xie, et al., On-orbit Performance of Aqua MODIS Onboard Calibrators. In Sensors, Systems, and Next-generation Satellites IX, SPIE, Brugge, Belgium, 2005, pp. 59780U−59789U. http://dx.doi.org/10.1117/12.627619.
[13] H. Montgomery, N. Che, J. Bowser, Determination of MODIS band-to-band co-registration on-orbit using the SRCA. in geoscience and remote sensing symposium, 2000, in: Proceedings. IGARSS 2000. IEEE 2000 International, vol. 5, 2000, pp. 2203−2205. http://dx.doi.org/10.1109/IGARSS.2000.858356.
[14] X. Xiong, C. Nianzeng, W. Barnes, Terra MODIS on-orbit spatial characterization and performance, Geosci. Remote Sens. IEEE Trans. 43 (2) (2005) 355−365.
[15] K.A. Kilpatrick, G. Podestá, S. Walsh, et al., A decade of sea surface temperature from MODIS: current status and future directions, Remote Sens. Environ. (2014) in review.
[16] P.J. Minnett, R.H. Evans, G.P. Podestá, et al., Suomi-NPP VIIRS sea surface temperature retrievals; algorithm evolution and an assessment of uncertainties, Remote Sens. Environ. (2015) in preparation.
[17] P.W. Gaiser, K.M. St Germain, E.M. Twarog, et al., The WindSat spaceborne polarimetric microwave radiometer: sensor description and early orbit performance, Geosci. Remote Sens. IEEE Trans. 42 (11) (2004) 2347−2361.
[18] X.-M. Liang, A. Ignatov, Y. Kihai, Implementation of the community radiative transfer model in advanced Clear-Sky processor for oceans and validation against nighttime AVHRR radiances, J. Geophys. Res. 114 (2009).
[19] B. Petrenko, A. Ignatov, Y. Kihai, et al., Clear-Sky Mask for the advanced Clear-Sky processor for oceans, J. Atmos. Oceanic Technol. 27 (10) (2010) 1609−1623.
[20] X. Liang, A. Ignatov, AVHRR, MODIS, and VIIRS radiometric stability and consistency in SST bands, J. Geophys. Res.: Oceans 118 (6) (2013) 3161−3171.

[21] Y. Han, P.V. Delst, Q. Liu, et al., JCSDA Community Radiative Transfer Model (CRTM)—version 1, NOAA Tech. Rep. NESDIS 122, NOAA, Camp Springs, MD, 2006, p. 40.
[22] Y. Chen, Y. Han, F. Weng, Comparison of two transmittance algorithms in the community radiative transfer model: application to AVHRR, J. Geophys. Res.: Atmos. 117 (D6) (2012) D06206.
[23] A. Wu, X. Xiong, NPP VIIRS and Aqua MODIS RSB Comparison Using Observations from Simultaneous Nadir Overpasses (SNO), 2012, pp. 85100P-85100P-85112P.
[24] C. Moeller, J. McIntire, T. Schwarting, et al., VIIRS F1 "best" relative spectral response characterization by the government team, in: J.J. Butler, X. Xiong, X. Gu (Eds.), SPIE 8153, Earth Observing Systems XVI, 81530K, San Diego, California, USA. September 13, 2011. http://dx.doi.org/10.1117/12.894552.
[25] H.H. Aumann, M.T. Chahine, C. Gautier, et al., AIRS/AMSU/HSB on the Aqua Mission: design, science objectives, data products, and processing systems, IEEE Trans. Geosci. Remote Sens. 41 (2) (2003) 253−264.
[26] D.C. Tobin, H.E. Revercomb, R.O. Knuteson, et al., Radiometric and spectral validation of atmospheric infrared sounder observations with the aircraft-based scanning high-resolution interferometer sounder, J. Geophys. Res.: Atmos. 111 (D9) (2006) D09S02.
[27] D.C. Tobin, H.E. Revercomb, C.C. Moeller, et al., Use of atmospheric infrared sounder high−spectral resolution spectra to assess the calibration of moderate resolution imaging spectroradiometer on EOS Aqua, J. Geophys. Res.: Atmos. 111 (D9) (2006) D09S05.
[28] A.G. O'Carroll, T. August, P. Le Borgne, et al., The accuracy of SST retrievals from MetOp-A IASI and AVHRR using the EUMETSAT OSI-SAF matchup dataset, Remote Sens. Environ. 126 (0) (2012) 184−194.
[29] D. Blumstein, G. Chalon, T. Carlier, et al., IASI Instrument: Technical Overview and Measured Performances, 2004, pp. 196−207, http://dx.doi.org/10.1117/12.560907.
[30] A.M. Larar, W.L. Smith, D.K. Zhou, et al., IASI spectral radiance validation inter-comparisons: case study assessment from the JAIVEx field campaign, Atmos. Chem. Phys. 10 (2) (2010) 411−430.
[31] L. Wang, C. Cao, On-orbit calibration assessment of AVHRR longwave channels on MetOp-a using IASI, Geosci. Remote Sens. IEEE Trans. 46 (12) (2008) 4005−4013.
[32] S.J. Hook, R.G. Vaughan, H. Tonooka, et al., Absolute radiometric in-flight validation of mid infrared and Thermal infrared data from ASTER and MODIS on the terra spacecraft using the Lake tahoe, CA/NV, USA, automated validation site, Geosci. Remote Sens. IEEE Trans. 45 (6) (2007) 1798−1807.
[33] ISSTST, Interim, Sea Surface Temperature Science Team. Sea Surface Temperature Error Budget: White Paper, NASA, 2010. http://www.sstscienceteam.org/white_paper.html.
[34] K. Yang, A.J. Fleig, R.E. Wolfe, et al., MODIS band-to-band registration, in: T.I. Stein (Ed.), Proceedings of the International Geoscience and Remote Sensing Symposium, Honolulu, HI, 2000, pp. 887−889. http://dx.doi.org/10.1109/IGARSS.2000.861735.
[35] X. Xiong, B. Wenny, J. Sun, et al., Overview of Aqua MODIS 10-year on-orbit calibration and performance, in: S.P.N. Roland Meynart, Haruhisa Shimoda (Eds.), Proc. SPIE 8533, Sensors, Systems, and Next-generation Satellites XVI, 853316, November 19, 2012, pp. 853316−853316-9.
[36] L. Wang, X. Xiong, J.J. Qu, et al., Impact assessment of Aqua MODIS band-to-band misregistration on snow index, J. Appl. Remote Sens. 1 (1) (2007) 013531−013531-11.
[37] K.A. Kilpatrick, G.P. Podestá, R.H. Evans, Overview of the NOAA/NASA pathfinder algorithm for sea surface temperature and associated matchup database, J. Geophys. Res. 106 (2001) 9179−9198.

[38] M.J. Uddstrom, W.R. Gray, R. Murphy, et al., A bayesian cloud mask for sea surface temperature retrieval, J. Atmos. Oceanic Technol. 16 (1) (1999) 117−132.
[39] C.J. Merchant, A.R. Harris, E. Maturi, et al., Probabilistic physically based cloud screening of satellite infrared imagery for operational sea surface temperature retrieval, Quart. J. R. Meteorol. Soc. 131 (2005) 2735−2755.
[40] C.C. Walton, W.G. Pichel, J.F. Sapper, et al., The development and operational application of nonlinear algorithms for the measurement of sea surface temperatures with the NOAA polar-orbiting environmental satellites, J. Geophys. Res. 103 (1998) 27999−28012.
[41] A.M. Závody, C.T. Mutlow, D.T. Llewellyn-Jones, A radiative transfer model for sea surface temperature retrieval for the along-track scanning radiometer, J. Geophys. Res.: Oceans 100 (C1) (1995) 937−952.
[42] P.J. Minnett, A numerical study of the effects of anomalous North Atlantic atmospheric conditions on the infrared measurement of sea-surface temperature from space, J. Geophys. Res. 91 (1986) 8509−8521.
[43] M. Szczodrak, P.J. Minnett, R.H. Evans, The effects of anomalous atmospheres on the accuracy of infrared sea-surface temperature retrievals: dry air layer intrusions over the tropical ocean, Remote Sens. Environ. 140 (0) (2014) 450−465.
[44] I.J. Barton, Improving satellite-derived sea surface temperature accuracies using water vapor profile data, J. Atmos. Oceanic Technol. 28 (1) (2010) 85−93.
[45] C.J. Merchant, P. Le Borgne, A. Marsouin, et al., Optimal estimation of sea surface temperature from split-window observations, Remote Sens. Environ. 112 (5) (2008) 2469−2484.
[46] C.J. Merchant, P. Le Borgne, H. Roquet, et al., Sea surface temperature from a geostationary satellite by optimal estimation, Remote Sens. Environ. 113 (2) (2009) 445−457.
[47] A. Soloviev, R. Lukas, Observation of large diurnal warming events in the near-surface layer of the western equatorial Pacific warm pool, Deep Sea Res. Part I 44 (6) (1997) 1055−1076.
[48] B. Ward, Near-surface ocean temperature, J. Geophys. Res. 111 (2006) C02005.
[49] C.L. Gentemann, P.J. Minnett, P. Le Borgne, et al., Multi-satellite measurements of large diurnal warming events, Geophys. Res. Lett. 35 (2008) L22602.
[50] C.J. Donlon, P.J. Minnett, C. Gentemann, et al., Toward improved validation of satellite sea surface skin temperature measurements for climate research, J. Clim. 15 (2002) 353−369.
[51] P.J. Minnett, Radiometric measurements of the sea-surface skin temperature − the competing roles of the diurnal thermocline and the cool skin, Int. J. Remote Sens. 24 (24) (2003) 5033−5047.
[52] P.J. Minnett, M. Smith, B. Ward, Measurements of the oceanic thermal skin effect, Deep Sea Res. Part II 58 (6) (2011) 861−868.
[53] C.P. Atkinson, N.A. Rayner, J. Roberts-Jones, et al., Assessing the quality of sea surface temperature observations from drifting buoys and ships on a platform-by-platform basis, J. Geophys. Res.: Oceans 118 (7) (2013) 3507−3529.
[54] A.G. O'Carroll, J.R. Eyre, R.W. Saunders, Three-way error analysis between AATSR, AMSR-e, and in situ sea surface temperature observations, J. Atmos. Oceanic Technol. 25 (7) (2008) 1197−1207.
[55] J.J. Kennedy, R.O. Smith, N.A. Rayner, Using AATSR data to assess the quality of in situ sea-surface temperature observations for climate studies, Remote Sens. Environ. 116 (0) (2012) 79−92.
[56] O. Embury, C.J. Merchant, G.K. Corlett, A reprocessing for climate of sea surface temperature from the along-track scanning radiometers: initial validation, accounting for skin and diurnal variability effects, Remote Sens. Environ. 116 (0) (2012) 62−78.

[57] M.J. McPhaden, A.J. Busalacchi, R. Cheney, et al., The tropical ocean-global atmosphere observing system: a decade of progress, J. Geophys. Res.: Oceans 103 (C7) (1998) 14169−14240.
[58] J.J. Kennedy, A review of uncertainty in in situ measurements and data sets of sea surface temperature, Rev. Geophys. Vol 51 (2014), p. 2013RG000434.
[59] H.P. Freitag, Y. Feng, L. Mangum, et al., Calibration procedures and instrumental accuracy estimates of Tao temperature, relative humidity and radiation measurements, in: NOAA Technical Memorandum, NOAA, 1994, p. 32.
[60] H.P. Freitag, M.E. McCarty, C. Nosse, et al., COARE Seacat Data: Calibrations and Quality Control Procedures, NOAA Technical Memorandum, 1999, p. 89.
[61] P.N. A'Hearn, H.P. Freitag, M.J. McPhaden, ATLAS Module Temperature Bias Due to Solar Heating, NOAA Technical Memorandum, 2002, p. 24.
[62] S.L. Castro, G.A. Wick, W.J. Emery, Evaluation of the relative performance of sea surface temperature measurements from different types of drifting and moored buoys using satellite-derived reference products, J. Geophys. Res. 117 (C2) (2012) C02029.
[63] D. Roemmich, G. Johnson, S. Riser, et al., The Argo program: observing the global ocean with profiling floats, Oceanography 22 (2009) 34−43.
[64] G. Lagerloef, F.R. Colomb, D.L. Vine, et al., The aquarius/SAC-D mission: designed to meet the salinity remote-sensing challenge, Oceanography 21 (2008) 68−81.
[65] P.J. Minnett, The validation of sea surface temperature retrievals from spaceborne infrared radiometers, in: V. Barale, J.F.R. Gower, L. Alberotanza (Eds.), Oceanography from Space, Revisited, Springer Science+Business Media B.V, 2010, pp. 273−295.
[66] J.P. Rice, J.J. Butler, B.C. Johnson, et al., The Miami2001 infrared radiometer calibration and intercomparison: 1. Laboratory characterization of blackbody targets, J. Atmos. Oceanic Technol. 21 (2004) 258−267.
[67] P.J. Minnett, G.K. Corlett, A pathway to generating climate data records of sea-surface temperature from satellite measurements, Deep Sea Res. Part II 77−80 (0) (2012) 44−51.
[68] C. Donlon, I.S. Robinson, M. Reynolds, et al., An infrared sea surface temperature autonomous radiometer (ISAR) for deployment aboard volunteer observing ships (VOS), J. Atmos. Oceanic Technol. 25 (1) (2008) 93−113.
[69] I.J. Barton, P.J. Minnett, C.J. Donlon, et al., The Miami2001 infrared radiometer calibration and inter-comparison: 2. Ship comparisons, J. Atmos. Oceanic Technol. 21 (2004) 268−283.
[70] P.J. Minnett, R.O. Knuteson, F.A. Best, et al., The marine-atmospheric emitted radiance interferometer (M-AERI), a high-accuracy, sea-going infrared spectroradiometer, J. Atmos. Oceanic Technol. 18 (6) (2001) 994−1013.
[71] J.A. Dykema, J.G. Anderson, A methodology for obtaining on-orbit SI-traceable spectral radiance measurements in the thermal infrared, Metrologia 43 (3) (2006) 287.
[72] F.A. Best, D.P. Adler, S.D. Ellington, et al., On-orbit absolute calibration of temperature with application to the CLARREO mission, 2008. http://dx.doi.org/10.1117/12.795457, pp. 70810O-70810O-10.
[73] P.J. Gero, J.A. Dykema, J.G. Anderson, A blackbody design for SI-traceable radiometry for earth observation, J. Atmos. Oceanic Technol. 25 (11) (2008) 2046−2054.
[74] C.J. Donlon, K.S. Casey, I.S. Robinson, et al., The GODAE high-resolution sea surface temperature pilot project, Oceanography 22 (3) (2009) 34−45.

Chapter 3

In Situ Optical Radiometry

Craig J. Donlon,[1,*] Giuseppe Zibordi[2]
[1] *European Space Agency/ESTEC, Noordwijk, The Netherlands;* [2] *Institute for Environment and Sustainability, Joint Research Centre, Ispra, Varese, Italy*
*Corresponding author: E-mail: craig.donlon@esa.int

Long-term optical satellite data used to create climate data records are composed of measurements derived from multiple space instruments having quite different design and performances, operated in different orbits, with different spectral and ground sampling characteristics, and different revisit and coverage. These data records may include some intervals when satellite measurement capability was either degraded significantly or altogether absent. Furthermore, the retrieval algorithms used to derive ocean geophysical properties from top-of-the-atmosphere radiance evolve in time, complicating the merging of satellite data products into a single climate data record. Thus, when considering the fundamental importance of ground-based (i.e., in situ high-quality) reference measurements for the development of geophysical retrieval algorithms applicable to the generation of satellite data products, and their successive assessment, it is essential that both satellite- and ground-based measurements are fully anchored to the International System of Units (SI).

The most accurate ground-based reference measurements traceable to SI and with fully quantified uncertainties making them suitable to supporting satellite applications are from field-deployed radiometers. In this context, satellite ocean color and thermal infrared satellite missions supporting climate change investigations require specific *ground-based reference measurements* (*Fiducial Reference Measurements* (FRM)) derived from optical radiometry data of exceptional traceability, accuracy, long-term stability, and cross-site consistency. The following two chapters focus on field optical radiometer design, calibration, measurement uncertainties, and applications.

The first chapter reviews field visible and near-infrared ocean color radiometers and summarizes the key aspects for the characterization, calibration, and their deployment. Additionally, basic elements on data reduction are presented and discussed in view of outlining the state-of-art of marine field optical radiometry and areas requiring further community effort. Emphasis is placed on uncertainties, which need to be considered as an integral part of measurements. Finally, a few examples are presented to illustrate typical applications of FRM ocean color data in the characterization of the in-water light field, biooptical modeling for the determination of pigments concentration from radiometric data products, validation of satellite-derived quantities, and vicarious calibration of the space system.

The chapter on thermal infrared radiometry (TIR) is concerned with the practical realization of ship-borne radiometers used to produce FRM for the validation of satellite-derived Sea Surface Temperature. A review of the key design and deployment choices for in situ TIR radiometers that are used to validate optical satellite measurements contributing to the climate data record is presented showing the evolution of field radiometer design. Optical materials and calibration concepts that have resulted in modern ship-borne capability for FRM meeting requirements for climate applications are discussed. In particular, science requirements are first reviewed from a Climate Data Record perspective and the fundamental measurement equation is applied to define the basic ship-borne radiometer instrument design. These elements are further elaborated using examples from a variety of systems successfully deployed in the field. Finally, the chapter concludes with a consideration of innovations expected in the coming years.

Chapter 3.1

In situ Optical Radiometry in the Visible and Near Infrared

Giuseppe Zibordi,[1,*] Kenneth J. Voss[2]
[1] European Commission, Joint Research Centre, Ispra, Italy; [2] Physics Department, University of Miami, Coral Gables, FL, USA
*Corresponding author: E-mail: giuseppe.zibordi@jrc.ec.europa.eu

Chapter Outline

1. Introduction and History 248
2. Field Radiometer Systems 249
   - 2.1 General Classification: Multispectral and Hyperspectral 249
   - 2.2 Irradiance Sensors 250
   - 2.3 Basic Radiance Sensors 252
      - 2.3.1 Gershun Tube 252
      - 2.3.2 Systems for Radiance Distribution Measurements 253
      - 2.3.3 Systems for Radiance Polarization Measurements 253
3. System Calibration 254
   - 3.1 Linearity Response 255
   - 3.2 Temperature Response 255
   - 3.3 Polarization Sensitivity 256
   - 3.4 Stray Light Perturbations 257
   - 3.5 Spectral Response 257
   - 3.6 Angular Response of Irradiance Sensors 258
   - 3.7 Rolloff of Imaging Systems 260
   - 3.8 Immersion Effects 260
      - 3.8.1 Irradiance Sensors 260
      - 3.8.2 Radiance Sensors 262
   - 3.9 Absolute Response 263
4. Measurement Methods 264
   - 4.1 In Water Systems 265
      - 4.1.1 Fixed-depths 265
      - 4.1.2 Profiling 266
      - 4.1.3 Radiance Distribution Systems 266
   - 4.2 Above Water Systems 267
   - 4.3 Radiometric Data Products 268
      - 4.3.1 Products from In-Water Measurements 268
      - 4.3.2 Products from Above-Water Measurements 269
      - 4.3.3 Normalized Water-Leaving Radiance 270
5. Errors and Uncertainty Estimates 273
   - 5.1 Calibration Specific Sources of Uncertainties 274

5.1.1 Irradiance Standards 274	5.4 Examples of Uncertainty Budget for Radiometric Products 282
5.1.2 Radiance Standards 275	6. **Applications** **285**
5.2 Instrument Specific Sources of Uncertainties 276	6.1 Sky and Sea Radiance Distribution 285
5.2.1 Effects of Cosine Error 276	6.2 In-water Light Field Polarization 287
5.2.2 Immersion Factor 277	6.3 Bio-Optical Models 289
5.3 Methods and Field Specific Sources of Uncertainties 277	6.4 Validation of Satellite Radiometric Products 291
5.3.1 In Water Methods 277	6.5 In situ Data and System Vicarious Calibration 293
5.3.2 Above Water Methods 281	7. **Summary and Outlook** **294**
	References 295

1. INTRODUCTION AND HISTORY

Measurements of marine light were attempted as early as the beginning of the twentieth century [1—4]. However, due to constraints in available technology, these measurements were often qualitative. Quantitative marine optical radiometry began in the 1950s with advances in light detector technology and availability of standards for absolute spectral calibrations [5—8]. Nevertheless, the period between these two phases was crucial for the development of methods for marine optics and was additionally marked by theoretical advances in the modeling of the in-water light field [9,10].

The first ocean color satellite sensor, Coastal Zone Color Scanner (CZCS) [11], provided a large impetus to advance marine optical radiometry. In particular, in situ radiometry became a key element for the creation of algorithms relating primary satellite data products (i.e., the water-leaving radiance) to the optically significant constituents suspended or dissolved in the water [12—14]. During the satellite ocean color missions that followed, in situ optical radiometry acquired further relevance in view of the high accuracy requirements for field data applicable to bio-optical modeling, validation of space derived radiometric products, and vicarious calibration of satellite sensors (e.g., the Sea-viewing Wide Field-of-view Sensor (SeaWiFS), the Moderate Resolution Imaging Spectroradiometer (MODIS), the Global Imager (GLI), the Medium Resolution Imaging Spectrometer (MERIS), and Visible Infrared Imaging Radiometer (VIIRS)). The need to comply with accuracy requirements for those applications led to advances in the characterization of field radiometers [15], the development and assessment of measurement methods [16], and the quantification of uncertainties for in situ radiometric data products [17].

This chapter provides an overview of the state-of-the-art of in situ marine optical radiometry with particular reference to its relevance and applications.

2. FIELD RADIOMETER SYSTEMS

Optical radiometers provide the capability of measuring the radiant energy in the spectral regions ranging from the ultraviolet to the infrared. Radiometers are basically composed of three parts: (1) the optical system which collects the radiant energy through an aperture, spectrally filters or disperses it, and finally focus it on a field stop; (2) the detector that converts the radiant energy into an analog electrical signal; and (3) electronics which translates the analog output of the detector into digital counts.

The spectral range of ocean color radiometers is generally restricted to the visible and near infrared where the optically significant water constituents (i.e., phytoplankton pigments, detrital particles and colored dissolved organic matter) may significantly affect the spectral signature of light.

2.1 General Classification: Multispectral and Hyperspectral

Radiance and irradiance sensors are commonly classified according to their capability of spectrally resolving the light field. Specifically, optical radiometers can be separated into hyperspectral, multispectral, or broad-band in the order of decreasing spectral resolution.

Hyperspectral radiometers are characterized by a large number (generally more than 20) of narrow spectral bands typically less than 10 nm wide distributed continuously through the spectrum (e.g., see Clark et al. [15]). Modern hyperspectral instruments use a dispersive optical element (i.e., diffraction grating or prism) with a one- or two-dimensional detector array that collects the light in discrete portions of the spectrum and converts this into an electrical signal. When characterizing these instruments, it is important to distinguish between the spectral resolution of each band, and the sampling interval at which the spectrum is measured. Typically the sampling interval has much higher resolution than the spectral resolution of each spectral band. With these radiometers it is essential to characterize the stray light due to scattering or reflections in the optical system, which cause light from one region of the spectrum to interfere with light from another region [18]. Also, since the dispersive elements are often sensitive to polarization and natural light fields are polarized, the radiometer sensitivity to polarization must be determined [19].

As opposed to hyperspectral radiometers, multispectral radiometers measure the light field at some number (generally less than 20) of discrete spectral bands (e.g., Abbot et al. [20]). Typically these bands are 10 nm wide, which is generally appropriate for ocean color applications because many of the atmospheric and marine spectral signatures do not have sharp features. In these radiometers the spectral selection is often performed with interference filters, but occasionally absorption filters are used for wider bands. Similar to the stray light problems of dispersive systems, the spectral responsivity of multispectral radiometers must be carefully characterized to identify possible spectral regions of response away from the central band (out-of-band response). In fact, if the filter is not sufficiently blocked to reduce the out-of-band response [21], it may introduce errors in radiometric measurements with varied spectral signatures such as between a red-rich calibration source and the blue-rich clear ocean.

The last category of radiometers is represented by broadband instruments. Typically these have single spectral bands larger than 50 nm dedicated to the measurement of a specific physical-biological property. An example is provided by photosynthetic available radiation sensors [22], which should ideally return a signal with a constant quantum efficiency from 400 to 700 nm [23]. Another example is offered by broadband sensors for ultraviolet (UV) measurements of long- (UVA) and medium-wavelength (UVB) light components [24]. These radiometers are difficult to calibrate accurately as most often the spectral signature of the calibration source is much different than that of the light field being measured.

2.2 Irradiance Sensors

Irradiance is a measure of the light flux per unit surface area. Typical quantities measured in the field are the downward or upward irradiance, which are the light energy per unit time going through a flat horizontal surface with a given area either in the downward or upward directions. An example of a perfect irradiance sensor would be one which collects all the light entering a 1 cm^2 hole, irrespective of the incidence angle (e.g., an integrating sphere with a 1 cm^2 entrance aperture) or a 1 cm^2 flat detector with collection efficiency independent of the angle of incidence. In reality, it is not practical to operate an integrating sphere in the field without an optical window in front of the entrance aperture. But, once surfaces are introduced in the foreoptics through either an optical window or a flat detector, processes such as Fresnel reflection at the surface produce an angular-dependent efficiency. So specific designs must be developed to realize an irradiance collector with 180° field-of-view (equivalent to 2π sr solid angle) that compensates for the increased loss at greater angles of incidence due to its reflectance.

For almost all irradiance sensors that attempt to achieve an ideal cosine response, the practical solution is to use a diffuser with a *top hat* design (see Figure 1) [25]. In this case, when moving away from the normal to the top surface, the projected area of the top of the diffuser in the measuring direction decreases proportional to the cosine of the incidence angle. The lateral surface of the diffuser provides extra projected area that increases with incidence angle

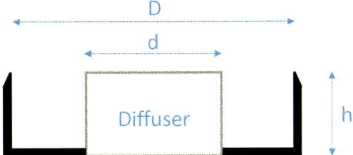

FIGURE 1 Diagram of an irradiance collector. The central portion is made of diffusing material. Dimensions D, d, and h are varied depending on diffusing material and external medium (water or air) to optimize the angular response of the irradiance collector.

and compensates for the reduced collection efficiency resulting from the greater Fresnel reflectance of the diffuser surface. A lateral shield (side block) located at some distance from the diffuser limits the extra projected area and ensures that at 90° incidence angle there is no light collection. The diameter, d, and height, h, of the top hat diffuser and the diameter of the side block, D, need to be experimentally determined to optimize the angular response of the collector. These dimensions are dependent on the diffuser material and its relative index of refraction with respect to that of the external medium (i.e., air or water) and additionally on the detector design behind the collector. Because of this, changing diffuser material generally requires a redesign of the collector. It should also be pointed out that, since the optical properties of the diffuser vary with wavelength, the accuracy of the angular response of the irradiance collector also varies spectrally.

Diffusers are commonly made of plastic material which absorbs UV light. Because of this, diffusers for UV applications are constructed with a thin layer of polytetrafluoroethylene above a lens [26] or in front of an integrating sphere [27].

For any irradiance collector the spectral angular response must be measured to determine the error in the cosine response. The effects of this error in irradiance measurements depend both on the spectral angular response of the collector and the radiance distribution of the light field being measured (see Section 3.6).

One important characteristic, which will be detailed later, is the immersion factor that quantifies the change in responsivity of a sensor as a function of the refractive index of the external medium [7,28,29]. Since most instruments have their absolute calibration performed in air, the immersion factor accounts for the difference in their responsivity while in water. For irradiance sensors with the design detailed above, this change originates from the difference in transmission between the air-diffuser with respect to water-diffuser interfaces. Because the gradient in the index of refraction between water and diffuser is less than that between air and diffuser, the Fresnel transmission coefficient is greater in-water than in-air. While this means more light can enter the diffuser from the water, perhaps increasing the signal measured by the detector, this effect is overwhelmed by light more easily leaving the diffuser and being remitted into the water rather than reaching the detector. So the immersion factor for these types of irradiance collectors has to account for the decreased collection efficiency while in water with respect to in air (see Section 3.8).

2.3 Basic Radiance Sensors

Radiance is the flux per unit area within a specified solid angle centered in a given direction. It is generally measured by limiting the field-of-view of the radiometer and assuming radiance is spatially invariant, or at the least slightly changing, over the projected solid angle. The measurement of radiance over many, or ideally all, directions provides the capability of determining the radiance distribution. When spectral resolution and polarization is added to the radiance distribution, the resulting spectral polarized radiance distribution comprehensively describes the light field at a given point. The most common radiance sensors are briefly described in the following subsections.

2.3.1 Gershun Tube

The simplest radiance sensor, commonly referred to as a Gershun tube radiometer [9] is formed by combining an irradiance collector and a tube restricting its field-of-view as illustrated in Figure 2. It is noted that the nominal viewing angle in degrees is defined by $\omega = 2\tan^{-1}[(D)/2h]$ where D is the front aperture of the optics and h the distance between aperture and detector. The related solid angle field-of-view in sr is given by $\Omega = 2\pi(1-\cos(\omega/2))$.

One advantage of the radiometer illustrated in Figure 2 that only requires the detector to have uniform response over ω is its constant solid angle field-of-view regardless of the medium in which it is operated. Hence the calibration difference between in-air and in-water has to take into account only the difference in collection efficiency of the irradiance collector when in water versus in air, and not a change in Ω. However, in general, an optical window is placed in front of the Gershun tube and thus the measured radiance is affected by

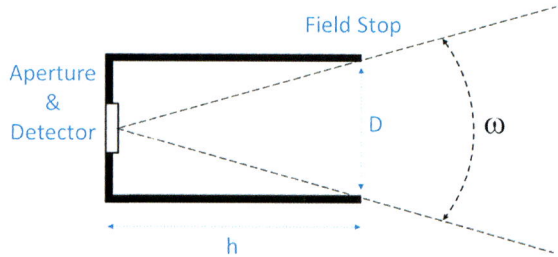

FIGURE 2 Diagram of the Gershun tube radiometer.

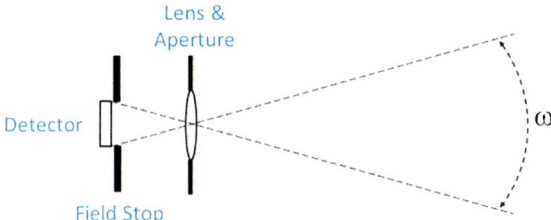

FIGURE 3 Diagram of an imaging radiometer.

immersion effects which depend on both a change in the reflectance between air-window and water-window interfaces, and a decrease in the acceptance viewing angle of the instrument due to refraction (see Section 3.8).

An alternative way to realize a radiance sensor for a single direction is using a simple design relying on a lens with an aperture at its focal plane. In this case the area of the aperture combined with the lens focal length, defines the viewing angle (Figure 3). These systems are always protected in a waterproof housing, thus they require the determination of an immersion factor to take into account both the reflection and refraction changes which happen at the window interface when the instrument is used in water.

2.3.2 Systems for Radiance Distribution Measurements

As mentioned earlier, the radiance distribution is determined through measurements, ideally simultaneously performed, in a number of directions. This can be approximated by using a radiance collector and quickly scanning it over different directions [30–32]. Another approach, pioneered by Smith et al. [33], is the use of a fisheye lens and a camera, which collect radiance data from a hemisphere. In recent systems the fisheye lens projects radiance from a hemisphere onto a two-dimensional array of a solid state camera [34,35]. To allow a wide angle lens such as the fisheye to operate properly in the water, requires a hemispherical dome window. The related immersion effects, which are complicated by the lens aperture and window curvature, can only be characterized through laboratory measurements [36].

2.3.3 Systems for Radiance Polarization Measurements

The in-water and sky radiances are polarized in almost all viewing directions, thus polarization adds to the quantities needed to comprehensively describe the light field. The polarization of an incoherent light beam can be completely described by the four elements of the Stokes vector [37]. The first element indicates the intensity of light (i.e., radiance), the next two elements describe the linear polarization of the light field, while the last describes its circular component. The Stokes vector of light can be determined by a minimum of four measurements with the correct combination of linear and circular polarization sensors. In practice, both the sky and in-water light fields have very low levels of circular polarization, so the number of measurements can be reduced to three by assuming the circular polarization is negligible. When considering that a clear sky is relatively stable over periods of minutes, measurements of polarized skylight can be performed by sequentially moving the polarizers over the detector and collecting data at each position [38]. However, measurements of polarized light in the water must be done simultaneously because the light field varies quickly due to wave focusing. This requires four (or three, if circular polarization is ignored) matched sensors, each detecting a different orthogonal polarization state of the light field [39].

The same constraints apply to measurements of the polarized radiance distribution. Sequential sky radiance measurements can be made through scanning or fisheye based systems. Polarizers can be inserted into the optical path of the system and sequential data combined to obtain the complete polarization state of the sky radiance distribution. In the water, simultaneous measurements of different polarization states are necessarily required. These can be performed either with multiple radiometers or fisheye-based systems. In the case of multiple radiometers, the radiance must be simultaneously measured for all polarization states in each direction [32,40]. When relying on a fisheye design, measurement of different polarization states can be performed by (1) combining data from multiple fisheye systems acquiring data simultaneously [41], (2) combining in a single image the data from multiple fisheye lenses [42], or (3) using a camera array with polarization filters on the array itself [43]. This last method has not been used with fisheye optics to date because of limitations experienced in the polarization of the microarray at large incidence angles.

3. SYSTEM CALIBRATION

Calibration relates the instrument output to the input radiometric quantity through sensor responsivity. Calibration should account for: absolute in-air responsivity to the radiometric source (either radiance or irradiance); in-water responsivity changes due to differences between the refractive index of air and that of water; and, additionally, correction factors to account for nonideal performances of the radiometer such as nonlinearity, temperature dependence, sensitivity decay with time, and deviation from ideal angular response (mostly for irradiance sensors or radiance sensors combining array detectors and large field-of-view optics). The calibration concept is implemented through the application of the measurement equation [44] yielding the radiometer output for a given source configuration. Specifically, the conversion from relative to physical units of the radiometric quantity $\Im(\lambda)$ (either $E(\lambda)$ or $L(\lambda)$) at wavelength λ is performed through.

$$\Im(\lambda) = C_\Im(\lambda)\, I_f(\lambda)\, \aleph(\lambda)\, \mathrm{DN}(\Im(\lambda)) \qquad (1)$$

where $C_\Im(\lambda)$ is the in-air absolute calibration coefficient (i.e., the absolute responsivity), $I_f(\lambda)$ is the immersion factor accounting for the change in responsivity of the sensor when immersed in water with respect to air, and $\aleph(\lambda)$ (for simplicity only expressed as a function of λ) corrects for any deviation from the ideal performance of the measuring system. Specifically, in the case of an ideal radiometer $\aleph(\lambda) = 1$, but in general

$$\aleph(\lambda) = \aleph_i(i(\lambda)) \, \aleph_j(j(\lambda))...\aleph_k(k(\lambda)) \qquad (2)$$

where $\aleph_i(i(\lambda))$, $\aleph_j(j(\lambda))$, ..., and $\aleph_k(k(\lambda))$ are correction terms for different factors indexed by i, j, ..., k affecting the nonideal performance of the considered radiometer.

Finally, the term $DN(\Im(\lambda))$ indicates the digital output corrected for the dark value (i.e., the actual digital output $DN(\Im(\lambda))^*$ from which the dark value $D0(\lambda)$, measured by obstructing the entrance optics, has been subtracted).

In the following sections the term $DN(\Im(\lambda))$ will also be used for laboratory measurements performed to characterize and calibrate radiometric instruments which may be affected by ambient light due to inappropriate shielding and baffling of light sources. In these cases $DN(\Im(\lambda)) = [DN(\Im(\lambda))^* - D0(\Im(\lambda))] - [DA(\Im(\lambda)) - D0(\Im(\lambda))] = DN(\Im(\lambda))^* - DA(\Im(\lambda))$, where $DA(\Im(\lambda))$ is the ambient value measured by occulting the direct source [66]. For ideal measurement conditions the values of $DA(\Im(\lambda))$ equals that of $D0(\lambda)$. Additionally, the removal of ambient light is also applied during the calibration of lamps used as irradiance working standards (see Section 3.9). Thus, for consistency the quantification of ambient light is a necessary step when performing absolute calibrations.

3.1 Linearity Response

Electronics and detector response may affect the linearity of radiometer responsivity. This is quantifiable by varying the flux at the entrance optics over the operating range of the instrument. A simple nonlinear response can be minimized by correcting the radiometer output through

$$\aleph_L(\Im(\lambda)) = l_1[DN(\Im(\lambda))]^{l_2} \qquad (3)$$

where l_1 and l_2 are coefficients determined from the least square fit of the radiometer output as a function of the input flux during its characterization. In the case of instruments with complicated and highly nonlinear responses [45], a full lookup-table mapping of the radiometer output to input flux may be required.

3.2 Temperature Response

The optics, electronics, and mechanics of a radiometer may be temperature dependent. Two examples of optical components sensitive to temperature changes and affecting the radiometer responsivity, are silicon detectors and interference filters. Silicon detectors exhibit appreciable changes in response with temperature in the near-infrared while interference filters may exhibit a decrease of the center-wavelength combined with an increase of the bandwidth with temperature.

In the absence of any thermal stabilization of the sensor, radiometers may exhibit changes in the dark signal and responsivity with temperature. While the dark signal is an additive term that can be quantified and removed by performing regular measurements with the entrance optics obstructed, changes in responsivity with temperature require laboratory characterizations.

Tests of temperature dependence can be performed by operating the radiometer in a temperature-controlled chamber while looking at a stable source.

Similar to nonlinear response, any appreciable temperature dependence can be minimized by correcting the radiometer output with

$$\aleph_T(\Im(\lambda), \Delta T) = c_1 \left[\mathrm{DN}(\Im(\lambda), \Delta T) \right]^{C_2} \tag{4}$$

where c_1 and c_2 are coefficients determined through the least square fit of the radiometer outputs as a function of temperature differences ΔT between the internal radiometer temperature during absolute calibration and during its thermal characterization.

3.3 Polarization Sensitivity

Natural light has a degree of polarization varying with the concentration of optically significant constituents in the water and sky conditions (as defined by cloudiness and aerosols). Even when polarization is not the aim of the measurement, surfaces of the optical components constituting the radiometer (e.g., optical windows, lenses, dispersive elements) may be sensitive to the polarization state of the light field. Thus, at a minimum, the polarization sensitivity of optical radiometers must be determined. Considering that the radiance distribution in the atmosphere and natural waters has almost no circular polarization, the simplest test to determine the linear polarization sensitivity is to incrementally rotate a linear polarizer positioned between a nonpolarized source $\Im(\lambda)$ and the entrance optics of the radiometer. The polarization sensitivity, in percent, can be expressed as

$$P(\lambda) = 100 \left[\mathrm{DN}(\Im_M(\lambda)) - \mathrm{DN}(\Im_m(\lambda)) \right] / \left[\mathrm{DN}(\Im_M(\lambda)) + \mathrm{DN}(\Im_m(\lambda)) \right] \tag{5}$$

where $\mathrm{DN}(\Im_m(\lambda))$ and $\mathrm{DN}(\Im_M(\lambda))$ indicate the minimum and maximum values recorded while rotating the polarizer.

The matrix which describes how the Stokes vector is transformed in the atmosphere, ocean, or optical instrument, is called the Mueller matrix [46]. If the Mueller matrix of the radiometer has been determined through laboratory characterization, a correction can be made for polarization sensitivity [47]. However, this correction is often unpractical because it requires knowledge of the exact polarization state of the light field. Therefore, best practice suggests reducing the polarization sensitivity to the maximum extent possible.

3.4 Stray Light Perturbations

Stray light is light which has been scattered or reflected inside the radiometer and interferes with the measurement. For radiometers which measure in a single direction, stray light occurs when light from outside the desired collection solid angle still contributes to the detected radiance with errors increasing with the inhomogeneity of the light field (e.g., such as in the case of measurements of the sun aureole). Single angle radiometers do not provide information on the shape of the radiance distribution around the measurement direction which would aid in making corrections. It is thus extremely important to minimize stray light through instrument design, for instance by inserting baffles between the external aperture and the detector of the radiance sensor.

For systems which image the radiance distribution, spatial stray light is often referred to as the point spread function (PSF) of the instrument [48]. These systems, which provide information on the shape of the radiance distribution, offer the capability to correct for stray light perturbations by deconvolving the radiance measurements with the PSF of the instrument [49].

In hyperspectral instruments spectral stray light can be significant and cause the mixing of light from other regions of the spectrum [50]. Its characterization requires the illumination of the optics with a sequence of sources with well-resolved spectral characteristics, most often narrowband lasers [51]. Once stray light is spectrally quantified, a correction can be made to the measurements using the characterization data [52,53].

3.5 Spectral Response

The spectral response $S_r(\lambda)$ of radiometers depends on the performance of each optical component of the system. In the case of a narrowband multispectral radiometer, $S_r(\lambda)$ is mainly determined by the spectral response of the interference filters of each spectral band, and to a lesser extent by the spectral response of the detector and foreoptics. The design of other radiometer components such as the field-of-view, may also have a role in the spectral responsivity of filter-based radiometers. For example, increasing the angle of incidence of light on an interference filter lowers its transmission wavelength. Thus, with a large field-of-view, there are contributions from light which strike the interference filter significantly off axis leading to a broadening in the spectral bandpass and a shift to lower wavelengths of the nominal filter center-wavelength. Because of this, $S_r(\lambda)$ should be characterized for the whole system by using a monochromatic source that extends over the field-of-view of the instrument and covers its spectral range. In the case of multispectral radiometers, relevant quantities to determine are: (1) the full-width at half-maximum indicating the bandwidth at 50% of the peak transmittance T_p of each considered band; and (2) the center-wavelength λ_b indicating the wavelength at the center of the bandwidth at

50% of T_p. Additional quantities, are the exact center-wavelength λ_c which takes into account the total $S_r(\lambda)$, through

$$\lambda_c = \frac{\int_0^\infty \lambda\, S_r(\lambda)\,d\lambda}{\int_0^\infty S_r(\lambda)\,d\lambda} \qquad (6)$$

and the effective center-wavelength λ_e which accounts for both $S_r(\lambda)$ and the spectral response of the source $\mathfrak{J}(\lambda)$ through

$$\lambda_e = \frac{\int_0^\infty \lambda\, \mathfrak{J}(\lambda) S_r(\lambda)\,d\lambda}{\int_0^\infty \mathfrak{J}(\lambda) S_r(\lambda)\,d\lambda}. \qquad (7)$$

Values of λ_b and λ_c exhibit larger differences with an increase of the spectral features (e.g., asymmetries) of $S_r(\lambda)$, while differences between λ_c and λ_e depend on the spectral features of both $\mathfrak{J}(\lambda)$ and $S_r(\lambda)$.

As previously mentioned, for filter radiometers it is important to extend the spectral characterization significantly beyond the nominal spectral band of the instrument to capture possible out-of-band responses. Because sky and in-water radiances vary strongly with wavelength and generally exhibit spectral distributions markedly different than those of the calibration sources, large errors can occur when the out-of-band response is not blocked on the order of 10^{-5} to 10^{-4} below the response determined at the center-wavelength.

The characterization of the spectral responsivity of hyperspectral radiometers requires determining the spectral resolution and the center-wavelength of each detector element, by accounting for stray light effects.

It is finally recalled that the spectral responsivity of the sensor may change with temperature. This would suggest repeating spectral characterizations of the radiometer at different temperatures.

3.6 Angular Response of Irradiance Sensors

Irradiance sensors always exhibit angular response deviating from the ideal cosine function (i.e., the response to a parallel radiant flux is not proportional to the cosine of the angle between the normal to the collector faceplate and the direction of the flux). Thus, as discussed earlier, irradiance measurements are affected by the so-called cosine error, which is a measure of this deviation as a function of the angle of incidence.

This error can be quantified by illuminating the irradiance collector with a collimated source $E(\theta,\phi,\lambda)$ at varying incidence angles, θ, and azimuths, ϕ, and comparing the response of the instrument $DN(E(\theta,\phi,\lambda))$ with the value at $\theta=0$, $\mathrm{DN}(E(0,0,\lambda))$. From these, the cosine error $f_c(\theta,\phi,\lambda)$ is defined by

$$f_c(\theta,\phi,\lambda) = 100\left[\frac{\mathrm{DN}(E(\theta,\phi,\lambda))}{\mathrm{DN}(E(0,0,\lambda))\cos\theta} - 1\right] \quad (8)$$

where $\mathrm{DN}(E(0,0,\lambda))\cos\theta$ is the expected value at incidence angle θ for an ideal collector.

Radiometric corrections for the cosine error are obtained with

$$\aleph_c(\varepsilon_c(\theta_0,\lambda)) = 1 - \varepsilon_c(\theta_0,\lambda)/100 \quad (9)$$

where $\varepsilon_c(\theta_0,\lambda)$ is the percent correction determined from $f_c(\theta,\phi,\lambda)$.

By neglecting the sky light (i.e., with direct light only) and assuming the cosine error is azimuthally independent (i.e., $f_c(\theta,\phi,\lambda) \approx f_c(\theta,\lambda)$), then $\varepsilon_c(\theta_0,\lambda) = f_c(\theta_0,\lambda)$. This approximation is, however, challenged by the anisotrophy of skylight which varies significantly with λ and θ_0. A practical scheme for the determination of $\varepsilon_c(\theta_0,\lambda)$ accounting for realistic illumination conditions is proposed in Section 5.2.

As illustrated in Figure 4, it should be noted that an irradiance collector designed to work in air, if used in water (or vice versa) may lead to a large error in the cosine response, as the external medium is an important element of the collector design.

Finally, the cosine response may spectrally change over time as a result of aging of the diffuser material, and may also vary from instrument to

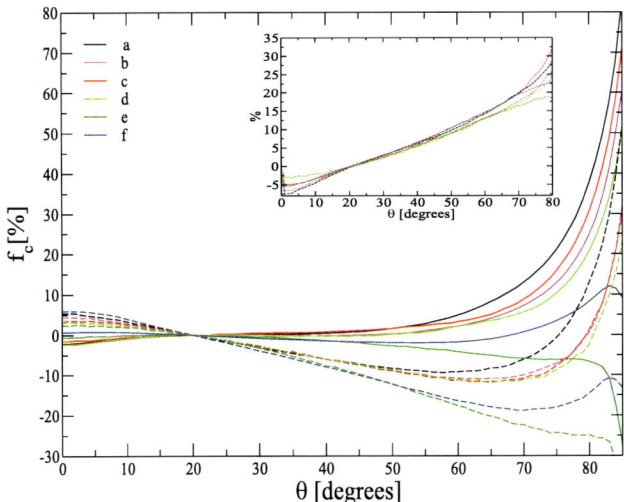

FIGURE 4 Average in-air (continuous lines) versus in-water (dashed lines) $f_c(\theta,\lambda)$ values (data are normalized at $\theta = 20°$) for a number of hyperspectral radiometers a–f from the same series. The inner panel shows an almost linear change with θ of the difference between in-air and in-water cosine errors for the various radiometers (redrawn from Mekaoui and Zibordi [54]).

instrument as a result of mechanical differences between collectors or changes in the optical properties of the diffuser material utilized for their construction. This imposes the need for individual characterizations of each collector and likely successive verifications over time.

3.7 Rolloff of Imaging Systems

Lenses may progressively attenuate radiance with increasing viewing angle. This effect, commonly called rolloff, is particularly relevant in wide field-of-view optics such as fish-eye lenses [36,55]. Conventional lenses exhibit a rolloff attenuation factor following a power of the cosine of the viewing angle [8]. However, wide-field-of-view optics, which generally comprise multiple lenses, require the experimental characterization of rolloff attenuation. This can be performed by sequentially imaging a stable and uniform radiance source while rotating the measuring system to view the source at different angles [55].

Rolloff effects can then be corrected through

$$\aleph_R(\theta, \lambda) = 1 + r_1(\lambda)\theta + r_2(\lambda)\theta^2 + \ldots + r_n(\lambda)\theta^n \tag{10}$$

where $r_1(\lambda)$, $r_2(\lambda)$, ... and $r_n(\lambda)$, are the spectral coefficients of the polynomial function fitting the angular rolloff measurements normalized to the value at viewing angle $\theta = 0$. In the case of achromatic lenses, the coefficients of the polynomial function are spectrally independent.

3.8 Immersion Effects

Radiometers are typically calibrated in air. As mentioned earlier, when operated in another medium (i.e., water), the instrument responsivity changes due to the refractive index of the external medium being different from that of air. In the case of radiometers used in water, the decrease in responsivity is roughly 40% for irradiance sensors and 70% for radiance sensors. The immersion factor, $I_f(\lambda)$, is the correction factor for this effect. Unfortunately the values of $I_f(\lambda)$ may vary among the various series of instruments and often exhibit appreciable differences from instrument to instrument in the same series.

3.8.1 Irradiance Sensors

Beginning in the 1930s, the immersion effects for irradiance sensors were the subject of investigations which led to a rigorous description of the refraction-reflection processes characterizing optics immersed in the water [28,56–58]. As already discussed, the responsivity of a wet irradiance sensor decreases as a result of the increase in light which escapes back into the water because the index of refraction of the diffuser is closer to that of water than air.

Since it is difficult to obtain a highly accurate optical characterization of irradiance collectors, which generally have complex geometries and sometimes are composed of layers of different diffusing materials, the theoretical computation of an accurate $I_f(\lambda)$ for irradiance sensors is unfeasible.

Comprehensive methods for the experimental determination of the immersion factor of in-water irradiance sensors were detailed by Tyler and Smith [7] and by Aas [29]. In both cases the radiometers are placed vertically in a water tank looking at a stable light source located in air at fixed distance above the collector and aligned with respect to its optical axis. While the method proposed by Tyler and Smith [7] relies on a collimated light source, the method by Aas [29] makes use of a point source. This latter method has been successively applied by Petzold and Austin [59], Mueller [60], and Zibordi et al. [61] and has later been revised mostly aiming at increasing the reproducibility of measurements [62].

Following Aas [29], the immersion factor, $I_f(\lambda)$, of an irradiance sensor is determined from

$$I_f(\lambda) = \frac{DN(E(0^+, \lambda))}{DN(E(0^-, \lambda))} t_{wa}(\lambda) \qquad (11)$$

where $DN(E(0^+, \lambda))$ is the digital value corresponding to the irradiance measured with the instrument in-air, $t_{wa}(\lambda)$ is the transmittance of the air-water interface determined from the Fresnel reflectance for a vertically incident light beam, and $DN(E(0^-, \lambda))$ the digital value corresponding to the irradiance measured with the instrument in-water. This latter term, which cannot be directly measured, is determined from the least square fit—as a function of the water level z_i above the sensor—of log-transformed in-water irradiance measurements corrected for the geometric perturbation induced by the finite distance between source and collector, $\ln[DN(E(z_i, \lambda))/G(z_i, \lambda)]$. The correction term $G(z_i, \lambda)$, required for a point source, minimizes the effects of radiant flux change at the collector as a function of the water level z_i above it, the source-collector distance d and refractive index of water $n_w(\lambda)$, and is given by Aas [29]

$$G(z_i, \lambda) = \left[1 - \frac{z_i}{d}\left(1 - \frac{1}{n_w(\lambda)}\right)\right]^{-2}. \qquad (12)$$

Independent determinations from different laboratories of $I_f(\lambda)$ for the same series of radiometers indicated average reproducibility values of 0.6%. However, sensor-to-sensor differences of several percent [60,61] indicated the need for individual sensor characterizations as opposed to the application of class-based factors.

3.8.2 Radiance Sensors

The responsivity of radiance sensors operated in water with respect to air, is mostly effected by the decrease in solid angle field-of-view and additionally by an increase in the transmittance of the optical window when in water versus air. For radiance sensors, the value of $I_f(\lambda)$ is commonly determined theoretically applying the general equation proposed by Austin [63] based on Fresnel transmission, which requires the knowledge of $n_w(\lambda)$ and of the refractive index of the optical window $n_g(\lambda)$

$$I_f(\lambda) = \frac{n_w(\lambda)\left[n_w(\lambda) + n_g(\lambda)\right]^2}{\left[1 + n_g(\lambda)\right]^2}. \tag{13}$$

This equation is strictly based on the design of Gershun tube radiometers, with an optical window between the tube and the external medium. Efforts have been made to experimentally characterize $I_f(\lambda)$ for more complex optical designs and the different series of optical radiometers [64,65], and also for radiance distribution cameras [36] where the interaction of light with the various components of the optical system may significantly affect the responsivity.

Following the scheme proposed by Zibordi [64], the experimental characterization of $I_f(\lambda)$ for radiance sensors is made through in-air and in-water radiance measurements successively performed with constant sensor-source distance and the sensor looking vertically at a stable, homogeneous and Lambertian source virtually immersed in the water. Specifically, $I_f(\lambda)$ is determined from

$$I_f(\lambda) = \frac{DN(L(0^+,\lambda))}{DN(L(0^-,\lambda))} \frac{\Omega_a}{\Omega_w(\lambda)} \frac{1}{t_{wa}(\Omega_w,\lambda)} \tag{14}$$

where $DN(L(0^+,\lambda))$ is the digital value related to the above-water radiance. This term is computed as the intercept of the least squares fit—as a function of the distance of the optical window from the water surface—of in-air measurements made with different water levels z_i and corrected for the different air-water optical paths. The term $DN(L(0^-,\lambda))$ is the digital value related to the in-water radiance measured with the instrument immersed. The terms Ω_a and $\Omega_w(\lambda)$ are the in-air and in-water solid angle field-of-views, respectively (their exact values are not required because their ratio is known: $\Omega_a/\Omega_w(\lambda) = n_w^2(\lambda)$), while $t_{wa}(\Omega_w,\lambda)$ indicates the water-air transmittance averaged over the solid angle $\Omega_w(\lambda)$.

As opposed to irradiance sensors, the experimental characterization of $I_f(\lambda)$ for sample radiometers from the same series, did not show appreciable sensor-to-sensor dispersion [64]. However, theoretical and experimental determinations exhibited appreciable differences for complex optics design [65]. These findings suggest that (1) class values of $I_f(\lambda)$ can be applied to radiance sensors, still, (2) the experimental characterization of $I_f(\lambda)$ for sample radiance sensors of each class is required to quantify possible differences between actual immersion factors and their theoretical determinations.

The reproducibility of $I_f(\lambda)$ values for both radiance and irradiance sensors determined experimentally can be increased by using pure water. But the application of derived values in field measurements requires correction factors to account for differences in the refractive indices between pure and natural waters as a function of salinity and temperature [61,64]. In the case of irradiance sensors, this is only possible through the determinations of $I_f(\lambda)$ for different solutions of pure water and salt to obtain different salinities.

3.9 Absolute Response

In-air absolute calibration of irradiance sensors is generally performed using a spectral irradiance source $E_0(\lambda)$ traceable to a reference standard. Common working standards are calibrated 1000 W quartz-halogen tungsten coiled filament (FEL) lamps [66,67]. These have a relative spectral distribution equivalent to that of a blackbody at approximately 3000 K and are commonly used as a source in the 250–2500 nm spectral range.

With the assumptions of (1) point-source, (2) point-detector, and (3) narrow bandwidth centered at the wavelength λ, the calibration coefficient $C_E(\lambda)$ for an irradiance sensor is determined from the output $DN(E(\lambda))$ related to the input irradiance $E(\lambda)$ by applying Eqn (1) with $I_f(\lambda) = 1$. For an irradiance sensor with the faceplate of the collector normal to the source

$$E(\lambda) = E_0(\lambda) \frac{d_0^2}{d^2} \qquad (15)$$

where d is the distance between source and sensor, and d_0 the distance at which the value $E_0(\lambda)$ was determined.

Note that Eqn (15) is only exact for a point irradiance source and a point detector. For an extended source such as the FEL lamp, there is an ambiguity about the measurement point for d [68]. In fact, while the distance d_0 applies from the plane tangent to the near surface of the lamp terminal posts during its absolute radiometric calibration [66], a few mm offset should be considered to account for the actual position of the center of the filament behind the posts during its use when $d \neq d_0$. Similarly, d is typically measured from the faceplate of the irradiance collector, but actual irradiance collectors have some physical thickness. Because of this, when neglecting the offset between the faceplate and the actual reference plane behind it identifying the receiving aperture, the relationship between detector signal and $1/d^2$ is nonlinear. This source of nonlinearity is solved by including in Eqn (15) the spectrally dependent offsets, which determined experimentally, account for the effective center of the filament and reference plane of the collector [69,70].

In the case of radiance sensors, the in-air absolute calibration coefficient $C_L(\lambda)$ is determined using a known radiance source $L(\lambda)$. This can be obtained through a calibrated integrating sphere or a system composed of a reflectance standard (i.e., a reflectance plaque with calibrated directional-directional reflectance) illuminated by an irradiance standard (e.g., a calibrated 1000 W FEL lamp). This solution offers the ability to reduce uncertainties in some data products determined from the combined use of irradiance and radiance data (see section 5.4) because of correlations between uncertainties of absolute calibration coefficients determined with the same irradiance standard.

Focusing on the lamp-plaque system, $C_L(\lambda)$ is determined with Eqn (1) from the output $DN(L(\lambda))$ for the related input radiance $L(\lambda)$. With the lamp positioned on axis and normal to the reflectance plaque, assuming a radiance sensor with (1) narrow bandwidth centered at wavelength λ and additionally (2) a narrow field-of-view, the radiance $L(\lambda)$ for the sensor viewing the center of the reflectance plaque at an angle θ with respect to the normal is given by

$$L(\lambda) = E(\lambda)\rho_d(\lambda, 0, \theta)\pi^{-1} \qquad (16)$$

where $\rho_d(\lambda, 0, \theta)$ is the directional-directional reflectance of the plaque for the specific viewing configuration (generally $\theta = 45°$) and $E(\lambda)$ is given by Eqn (15) with distance d between lamp and reflectance plaque. Equivalent to the case of irradiance sensors, if $d \neq d_0$, the distance d need to be corrected for the offset between the plane tangent to the near lamp terminal posts and the center of the filament.

It is critical that the radiance source fills the field-of-view of the radiance sensor uniformly. Because of this, in the case of appreciable deviation from the basic assumption of narrow field-of-view, the distance d needs to be chosen to satisfy both intensity and homogeneity requirements for the sensor under calibration (i.e., an increase of d tends to increase the homogeneity of the radiance across the plaque, but it also produces a decrease in the irradiance at the plaque).

Frequently, the directional-hemispherical reflectance $\rho_h(\lambda, \theta)$ is provided with the reflectance plaque instead of the directional-directional reflectance $\rho_d(\lambda, 0, \theta)$. This requires the application of a conversion factor to $\rho_h(\lambda, \theta)$ [71].

4. MEASUREMENT METHODS

In situ radiometric measurement methods supporting satellite ocean color are largely focused on the determination of those quantities essential for the vicarious calibration of space sensors, the validation of primary radiometric data products, and the development and assessment of bio-optical algorithms.

The required measurements are performed with above- and in-water optical radiometers. In-water systems offer advantages such as minimum wave perturbations in the upward radiometric quantities (e.g., upwelling radiance and upward irradiance), ability to determine a number of apparent optical properties in the water column, restricted temperature excursions, and direct radiance measurement in the nadir direction. Disadvantages include instrument self-shading and biofouling during long-term deployments. Above-water systems do not produce self-shading perturbations and are not affected by biofouling, but require accurate correction of the skylight reflected in the field-of-view by the air-sea interface, operate with nonnadir viewing geometry, and need deployment platforms whose generally large superstructures perturb the nearby light field. The basic elements characterizing in- and above-water methods are summarized separately in the following subsections.

4.1 In Water Systems

Following early developments [72–74], in-water radiometers can be roughly separated into fixed-depth and profiling systems. Both rely on radiometric measurements performed in the water column at different depths and as well as on measurement of the above-water downward irradiance [75,76]. While in-water data allows for the extrapolation to 0-depth (i.e., just below the water surface) of radiometric quantities that cannot be directly measured because of wave perturbations, the above-water downward irradiance data allows the effects of illumination changes on in-water data during the measurement period to be removed.

In addition to the radiometers directed toward the measurement of spectral irradiance and radiance, in-water imaging systems can be used to map the spectral radiance distribution of the upward and downward fields [34]. These measurements describe the variation with viewing angle of the in-water radiance field, i.e., its bidirectional structure [77–80]. Profiles of the radiance distribution can be used to derive inherent optical properties (IOP) such as absorption [81] and scattering [82].

4.1.1 Fixed-depths

Bio-optical buoys specifically designed to support satellite ocean color applications [83–87] rely on the basic concept of multiple radiometers, or fiber optic collectors, deployed at different fixed depths [15,72]. In combination with the above-water downward irradiance, $E_d(0^+, \lambda)$, fixed-depths systems provide the capability of measuring the upwelling nadir radiance, $L_u(z_i, \lambda)$, downwelling irradiance, $E_d(z_i, \lambda)$, and sometimes the upwelling irradiance, $E_u(z_i, \lambda)$, at two or more discrete depths z_i generally set between 1 and 10 m (see sample data in Figure 5).

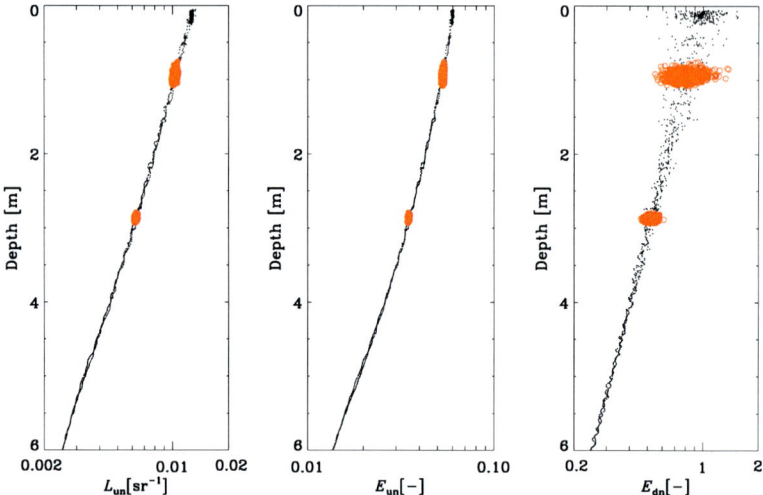

FIGURE 5 Sample $L_u(z, \lambda)$, $E_u(z, \lambda)$, and $E_d(z, \lambda)$ data from fixed-depths (empty circles) clustered at approximately 1 and 3 m, and from continuous profiles (dots), normalized with respect to $E_d(0^+, \lambda)$ at $\lambda = 555$ nm (i.e., L_{un}, E_{un}, and E_{dn} in the left, center and right panels, respectively). Profiles were collected on September 9, 2005 on vertically inhomogeneous seawater conditions, with 20 cm average wave height and diffuse attenuation coefficient $K_d = 0.31$ m^{-1} at 490 nm.

4.1.2 Profiling

In-water radiometric profiling is generally performed through winched and free-fall systems offering the capability of measuring $L_u(z, \lambda)$, $E_d(z, \lambda)$, and optionally $E_u(z, \lambda)$ as a function of depth z (see sample data in Figure 5), in addition to $E_d(0^+, \lambda)$. While winched systems were extensively used for many years [88] and are still occasionally applied on specific deployment platforms [89], since the late 1980s free-fall systems have become popular [90,91]. In fact they provide the capability of performing continuous profiles at extended distances from ships and consequently minimize perturbations due to ship-shading and ship-motion.

The determination of accurate in-water data products with profiling systems requires: (1) sampling near the surface, especially in coastal regions due to possible vertical inhomogeneity in the seawater optical properties; and (2) the capability of producing a suitably large number of measurements per unit depth to minimize the effects of wave focusing and defocussing [92,93].

4.1.3 Radiance Distribution Systems

Electro-optics cameras are commonly used to determine the angular variation of downwelling and upwelling radiance fields [34,36,94]. These systems, which are a technological development of an underwater camera equipped with a fisheye lens and photopic filter [33,95], are a major metrological advance with respect to early systems based on single field-of-view radiometers requiring successive measurements over various directions to map the in-water radiance distribution [30–32].

Recently, systems which allow rapid profiling of the in-water radiance distribution have also been developed [96,97].

4.2 Above Water Systems

Starting in the 1980s' quantitative above-water radiometry was established [98–101] through the definition and application of new measurement methods. These early developments were followed by theoretical investigations of the physical principles governing above-water radiometry [102,103] and led to the development of comprehensive measurement protocols. Successive experimental work [104–107] was then essential for the actual implementation, assessment, and refinement of protocols for operational measurements.

Basically, above-water radiometry relies on the use of radiometers not affected by sea spray, manually or autonomously operated from ships or fixed platforms at locations minimizing superstructure perturbations (e.g., shading or disturbing the sea surface). The target quantity is the water-leaving radiance, $L_w(\lambda)$, (i.e., the radiance leaving the sea and quantified just above the surface). This is determined from measurements of the total radiance from above the sea, $L_T(\theta, \phi, \lambda)$, (which includes contributions from $L_w(\lambda)$, sky-glitter and sun-glint), the diffuse radiance from the sky, $L_i(\theta', \phi, \lambda)$ (i.e., sky radiance), and $E_d(0^+, \lambda)$. The radiance measurements are performed with a specific observation geometry defined by the azimuth angle ϕ and the viewing angle θ specular to θ' [107–109]. An essential quantity required for the determination of $L_w(\lambda)$ is the sea surface reflectance ρ, theoretically quantified as a function of sea state [102]. The minimization of glint perturbations in $L_T(\theta, \phi, \lambda)$, due to sky- and sun-glint, is the main challenge for any above-water method. In fact, the time scale (tens of milliseconds to seconds) and spatial extent of L_T measurements (generally varying from a few up to several hundreds of cm^2, depending on the field-of-view and height above the water), reduce the effectiveness of any statistical modeling of the sea surface reflectance as a function of the sea state [110]. Additionally sky-glint contains sky radiance contributions from a variety of zenith and azimuth angles, and not only from the specular reflection of an ideal flat sea surface [102]. In certain cases sun-glint and foam contributions may add to sky-glint which become the source for significant spectral dependence of ρ [111]. Still, by neglecting the dependence of ρ on λ, the application of filtering schemes to sequences of $L_T(\theta, \phi, \lambda)$ measurements [105,106] has been shown effective in minimizing the sky- and sun-glint perturbations.

It is noted that most of current methods, excluding those specifically relying on the use of polarizers [103], neglect sky-radiance polarization, that itself depends on the aerosol/molecular ratio and measurement geometry. This may appreciably affect the determination of the actual sea-surface reflectance (see section 5.3).

4.3 Radiometric Data Products

Basic data products from above- and in-water radiometry are $L_w(\lambda)$ and derived radiometric products such as the normalized water-leaving radiance $L_{wn}(\lambda)$ or the remote sensing reflectance $R_{rs}(\lambda)$. In-water radiometry allows the capability of generating additional data products such as the irradiance reflectance, the diffuse attenuation coefficients, and the Q-factor (an index for the nonisotropy of radiance in the water). Consolidated data reduction schemes for both above- and in-water radiometry measurements are summarized in the following subsections.

4.3.1 Products from In-Water Measurements

Key quantities are the values just below the surface which cannot be directly measured due to wave perturbations. Using $\Im(z, \lambda, t)$ to represent both profiling and fixed-depth radiometric data at wavelength λ, depth z, and time t (i.e., $L_u(z, \lambda, t)$, $E_u(z, \lambda, t)$, and $E_d(z, \lambda, t)$), the first step in data reduction is to account for the effects of changes in the incident light field during data collection. This is performed using the above-water downward irradiance $E_d(0^+, \lambda, t)$ according to

$$\Im_0(z, \lambda, t_0) = \frac{\Im(z, \lambda, t)}{E_d(0^+, \lambda, t)} E_d(0^+, \lambda, t_0) \qquad (17)$$

where $\Im_0(z, \lambda, t_0)$ is the radiometric quantity normalized to the incident light field at t_0, $E_d(0^+, \lambda, t_0)$, with t_0 generally chosen to coincide with the beginning of the acquisition sequence.

Omitting the dependence with time and assuming that measurements satisfy the requirement of linear decay of $\ln \Im_0(z, \lambda)$ with depth in the extrapolation interval identified by $z_0 < z < z_1$, the subsurface values $\Im_0(0^-, \lambda)$ (i.e., $L_u(0^-, \lambda)$, $E_d(0^-, \lambda)$, and $E_u(0^-, \lambda)$) are determined as the exponential of the intercepts from the least squares linear regressions of $\ln \Im_0(z, \lambda)$ versus z. The negative values of the slopes of the regression fits are the so-called diffuse attenuation coefficients $K_\Im(\lambda)$ (i.e., $K_L(\lambda)$, $K_d(\lambda)$, and $K_u(\lambda)$) for the selected extrapolation interval.

It is mentioned that the value of $E_d(0^+, \lambda)$ could be determined from $E_d(0^-, \lambda)$, however, these derived values are generally largely affected by wave perturbations and should be used with caution.

Measurements near the surface can be heavily impacted by surface effects and, as such, many measurements are desirable in the very near surface for the regressions, or conversely in the case of fixed-depth measurements for averaging over a time scale longer than that of surface fluctuations. Best practice also suggests that regressions are performed using measurements in the top attenuation length (i.e., between the surface and $1/K_d(\lambda)$). Regression results may be challenged in shallow waters or in areas with vertical gradients in optically significant constituents. In any case, the accuracy of the sensor depth measurements is critical to obtaining an accurate subsurface value.

It should be noted that the linearity of the log-transformed radiometric profiles is an approximation because of changes in the radiance distribution in the near surface due to wave perturbations [112] and, additionally, because of inelastic scattering processes such as Raman scattering [113] and chlorophyll-a fluorescence [114]. These combined effects are more important in the red spectral region.

Derived quantities of interest for remote sensing applications are the dimensionless irradiance reflectance at depth 0^-, $R(0^-, \lambda)$, defined as $E_u(0^-, \lambda)/E_d(0^-, \lambda)$, and the Q-factor at nadir, $Q_n(0^-, \lambda)$ in units of sr, defined as $E_u(0^-, \lambda)/L_u(0^-, \lambda)$.

The water-leaving radiance $L_w(\lambda)$ in units of W m^{-2} nm^{-1} sr^{-1}, considered the fundamental quantity in ocean color studies obtainable from satellite observations, is given by

$$L_w(\lambda) = 0.543 L_u(0^-, \lambda). \qquad (18)$$

where the factor 0.543, computed assuming n_w is independent of wavelength [115], accounts for the reflection and refraction effects at the air−sea interface.

4.3.2 Products from Above-Water Measurements

The basic data product from above-water radiometry is $L_w(\lambda)$, determined from measurements of $L_T(\theta, \phi, \lambda)$ and $L_i(\theta', \phi, \lambda)$ with a viewing geometry defined by θ, θ' and ϕ (where θ and θ' are directions relative to the zenith and ϕ indicates the relative azimuth with respect to the sun). Given the sun azimuth ϕ_0, measurement geometries commonly applied to minimize the effects of wave perturbation are defined by $\theta = 40°$, $\theta' = 140°$, and $\phi = \phi_0 \pm 135°$ or alternatively $\phi = \phi_0 \pm 90°$. Specifically, $L_w(\lambda, \theta, \phi)$ for the considered viewing geometry is computed from

$$L_w(\theta, \phi, \lambda) = L_T(\theta, \phi, \lambda) - \rho(\theta, \phi, \theta_0, W) L_i(\theta', \phi, \lambda) \qquad (19)$$

where $\rho(\theta, \phi, \theta_0, W)$ is the sea surface reflectance expressed as a function of the measurement geometry identified by θ, ϕ, θ_0, and of the sea state conveniently expressed through the wind speed, W. Note that Eqn (19) neglects the geometric effects of the facets created by waves and thus the glint contribution is assumed to depend only on the sky-radiance in the direction identified by θ' and ϕ, and is expected to represent the mean radiance around this direction. $L_T(\theta, \phi, \lambda)$ and $L_i(\theta', \phi, \lambda)$ are commonly determined from the average of n-independent measurements satisfying filtering criteria aiming at removing those values affected by significant wave, foam and cloud perturbations [110]. Additionally, as with in-water data, the individual measurements contributing to the determination of $L_T(\theta, \phi, \lambda)$ and $L_i(\theta', \phi, \lambda)$, should be corrected for illumination changes using corresponding $E_d(0^+, \lambda)$ values. When $E_d(0^+, \lambda)$ measurements are not available [116], the method described is still effective with the assumption of stable illumination during each sequence of $L_T(\theta, \phi, \lambda)$ and $L_i(\theta', \phi, \lambda)$ measurements.

A number of alternative methods, which could be seen as a development of that formulated through Eqn (19), have been proposed to account for residual glint. These provide an additional correction to the sea surface reflectance term by assuming a negligible or quantifiable $L_w(\lambda)$ in the near-infrared [99,117,118].

Regardless of the above-water method applied, the derived $L_w(\theta, \phi, \lambda)$ requires correction for the viewing angle dependence. This is performed through

$$L_w(\lambda) = L_w(\theta, \phi, \lambda) \frac{\Re_0}{\Re(\theta, W)} \frac{Q(\theta, \phi, \theta_0, \lambda, \tau_a, \text{IOP})}{Q_n(\theta_0, \lambda, \tau_a, \text{IOP})} \qquad (20)$$

where the ratio of $\Re_0$ (i.e., $\Re(\theta, W)$ at $\theta = 0$) to $\Re(\theta, W)$, accounts for changes in surface reflectance and refraction, and the ratio of $Q(\theta, \phi, \theta_0, \lambda, \tau_a, \text{IOP})$ to $Q_n(\theta_0, \lambda, \tau_a, \text{IOP})$, i.e., the Q-factors at viewing angle θ and at nadir (i.e., $\theta = 0$), minimize the effects of the anisotropic radiance distribution of the in-water light field as a function of the water IOPs, the observation and illumination geometries defined by θ, ϕ, and θ_0, and the atmospheric optical properties defined through the aerosol optical thickness τ_a.

In chlorophyll-a dominated waters, commonly referred to as Case-1 waters [119], the IOPs can be solely expressed as a function of the chlorophyll-a concentration, Chla [79]. For this specific case, values of $\Re(\theta, W)$ and of the Q-factors are obtainable from tabulated data determined through simulations [79]. It is mentioned that, while the values of $\Re(\theta, W)$ are directly provided, the ratio $Q(\theta, \phi, \theta_0, \lambda, \tau_a, Chla)/Q_n(\theta_0, \lambda, \tau_a, Chla)$ needs to be determined from tabulated values of $f(\theta_0, \lambda, \tau_a, Chla)/Q(\theta, \phi, \theta_0, \lambda, \tau_a, Chla)$ and of $f(\theta_0, \lambda, \tau_a, Chla)/Q_n(\theta_0, \lambda, \tau_a, Chla)$. Since $f(\theta_0, \lambda, \tau_a, Chla)$ relates the irradiance reflectance to IOPs through the ratio of backscattering to absorption coefficients, $b_b(\lambda)/a(\lambda)$, it does not depend on θ. Because of this, the ratio of $f(\theta_0, \lambda, \tau_a, Chla)/Q_n(\theta_0, \lambda, \tau_a, Chla)$ to $f(\theta_0, \lambda, \tau_a, Chla)/Q(\theta, \phi, \theta_0, \lambda, \tau_a, Chla)$, applied as an alternative to $Q(\theta, \phi, \theta_0, \lambda, \tau_a, Chla)/Q_n(\theta_0, \lambda, \tau_a, Chla)$, does not affect Eqn (20). It is finally mentioned that the dependence on τ_a of both $Q(\theta, \phi, \theta_0, \lambda, \tau_a, Chla)$ and $f(\theta_0, \lambda, \tau_a, Chla)$, is small with respect to that of the other quantities. Because of this, the tabulated values of $Q(\theta, \phi, \theta_0, \lambda, \tau_a, Chla)$ and $f(\theta_0, \lambda, \tau_a, Chla)$ [79] are solely provided assuming a maritime aerosol with $\tau_a = 0.2$ at 550 nm.

4.3.3 Normalized Water-Leaving Radiance

Gordon and Clark [120] introduced the concept of normalized water-leaving radiance, $L_{wn}(\lambda)$, generally in units of $W\,m^{-2}\,nm^{-1}\,sr^{-1}$, to express the water-leaving radiance that would occur with no atmosphere, the sun at the zenith, and at the mean sun–earth distance

$$L_{\text{wn}}(\lambda) = \frac{L_{\text{w}}(\lambda)}{E_{\text{d}}(0^+, \lambda)} E_0(\lambda) \tag{21}$$

where $E_0(\lambda)$ is the mean extra-atmospheric solar irradiance in units of W m^{-2} nm^{-1} [121] and the ratio $L_{\text{w}}(\lambda)/E_{\text{d}}(0^+,\lambda)$ is commonly referred to as the remote sensing reflectance $R_{\text{rs}}(\lambda)$ in units of sr^{-1}.

In the case $E_{\text{d}}(0^+,\lambda)$ measurements are not available, or are affected by large uncertainties (e.g., large cosine errors or ship motion), the ratio $E_0(\lambda)/E_{\text{d}}(0^+,\lambda)$ can be replaced by $[D^2 t_{\text{d}}(\lambda)\cos\theta_0]^{-1}$ [116], where D^2 accounts for the variation in the sun-earth distance as a function of the day of the year, and $t_{\text{d}}(\lambda)$ is the atmospheric diffuse transmittance computed from measured or estimated values of $\tau_a(\lambda)$ [120].

As previously indicated, both $L_{\text{wn}}(\lambda)$ and $R_{\text{rs}}(\lambda)$ are quantities which take into account illumination effects such as sun–earth distance, atmospheric transmittance and to some extent the sun zenith angle [122]. However, this initial definition was based on the assumption of a constant Q-factor, and it did not allow a correction for the bidirectional effects of an anisotropic radiance distribution.

Corrections for bidirectional effects, determined through model parameters, were introduced by Morel and Gentili [78] with the concept of *exact normalized water-leaving radiance*, $L_{\text{WN}}(\lambda)$. By applying the proposed correction scheme to nadir-view $L_{\text{wn}}(\lambda)$ values, $L_{\text{WN}}(\lambda)$ is given by

$$L_{\text{WN}}(\lambda) = L_{\text{wn}}(\lambda) \frac{f(0, \lambda, \tau_a, \text{IOP})}{Q_n(0, \lambda, \tau_a, \text{IOP})} \left[\frac{f(\theta_0, \lambda, \tau_a, \text{IOP})}{Q_n(\theta_0, \lambda, \tau_a, \text{IOP})}\right]^{-1} \tag{22}$$

where the ratios $f(0, \lambda, \tau_a, \text{IOP})/Q_n(0, \lambda, \tau_a, \text{IOP})$ and $Q_n(\theta_0, \lambda, \tau_a, \text{IOP})/f(\theta_0, \lambda, \tau_a, \text{IOP})$ account for the effects of the anisotropic radiance distribution principally due to $\theta_0 \neq 0$. It is recalled that for chlorophyll-*a* dominated waters, tabulated values of $f(0, \lambda, \tau_a, Chla)/Q_n(0, \lambda, \tau_a, Chla)$ and $f(\theta_0, \lambda, \tau_a, Chla)/Q_n(\theta_0, \lambda, \tau_a, Chla)$ were produced by Morel et al. [79].

$L_{\text{WN}}(\lambda)$ or the equivalent $R_{\text{RS}}(\lambda)$, are the essential quantities that should be applied in all ocean color applications. Additionally, when the normalized water-leaving reflectance $\rho_{\text{WN}}(\lambda)$ would be considered a more convenient quantity, it can be determined from $L_{\text{WN}}(\lambda)$ through

$$\rho_{\text{WN}}(\lambda) = \frac{\pi L_{\text{WN}}(\lambda)}{E_0(\lambda)}. \tag{23}$$

$L_{\text{WN}}(\lambda)$ determined from in-water measurements performed in different European Seas are displayed in Figure 6 to illustrate spectral differences in various water types. These include the very blue waters of the Eastern Mediterranean Sea generally exhibiting maxima at 412–443 nm and negligible

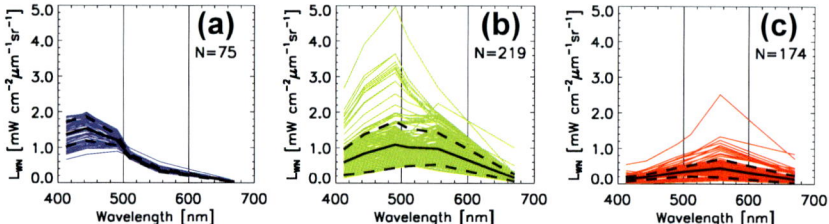

FIGURE 6 Spectra of L_{WN} determined from in-water radiometric profiles in different European Seas: (a) Eastern Mediterranean Sea; (b) Black Sea; and (c) Baltic Sea. Black continuous lines indicate averages while dashed lines indicate ±1 standard deviations. N is the number of spectra.

values at 665 nm (panel a). Conversely, the Baltic Sea waters dominated by colored dissolved organic matter, show maxima at 555 nm and minima at 412 nm (panel c). The Black Seas coastal waters dominated by suspended sediments and colored dissolved organic matter, may display maxima at either 490 or 555 nm (panel b).

A comprehensive intercomparison of radiometric data products from in- and above-water measurements performed during clear sky conditions in moderately sediment dominated coastal waters, is presented in Figure 7 to illustrate equivalence of measurement methods. Results exhibit spectrally averaged values of −1% and 8% for the bias and dispersion of $L_W(\lambda)$, respectively, in the 412−670 nm spectral region. These values are largely preserved for $L_{WN}(\lambda)$ (i.e., 0% and 8%) and fall within the combined values determined from the uncertainties independently determined for these above- and in-water data products [110]. The agreement between spectrally averaged values of measured and computed $E_d(0^+,\lambda)$ is also remarkable, exhibiting biases of −1% and dispersion of 4%.

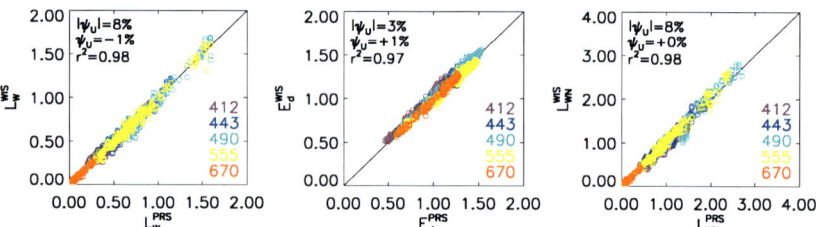

FIGURE 7 Scatter plots of in-water (WIS) versus above-water (PRS) data products for 1390 matching cases at the 412, 443, 490, 555 and 670 nm center-wavelengths. Specifically: $L_W(\lambda)$ in units of mW cm^{-2} μm^{-1} sr^{-1} (left panel); $E_d(0^+, \lambda)$ value measured (E_d^{WIS}) and computed (E_d^{PRS}) according to Zibordi et al. [116], in units of mW cm^{-2} μm^{-1} 10^{-2}; and $L_{WN}(\lambda)$ in units of mW cm^{-2} μm^{-1} sr^{-1} (right panel). The quantity ψ_u is the mean of percent differences between in- and above-water values, $|\psi_u|$ refers to unsigned differences, and r^2 is the determination coefficient.

5. ERRORS AND UNCERTAINTY ESTIMATES

Errors and uncertainties indicate distinct quantities. Differences between values of measured quantities and the true values of measurands are indicated as errors. These may comprise (1) systematic components indicating biases due to lack of accuracy, and (2) random components indicating dispersion due lack of precision. Bias components are generally minimized through corrections.

Uncertainties quantify the incomplete knowledge of the measurand through the available information. Thus, a *measurement of any kind is incomplete unless accompanied with an estimate of the uncertainty associated with that measurement* [123].

Uncertainties are generally classified into type A when determined through statistical methods and type B when determined by means other than statistical (e.g., models, published data, calibration certificates, or even experience). Type A and type B uncertainties can additionally be separated into additive (i.e., independent of the measured value such as the values related to the dark signal) or multiplicative (i.e., dependent on the measured value such as those related to the absolute responsivity of the radiometer). All uncertainties contribute to the overall measurement uncertainty through their combined values. When the various uncertainties are independent, the combined uncertainty can be determined as the quadrature sum (i.e., the root square sum) of the various contributions. The level of confidence of each uncertainty, defined by the coverage factor k, should be provided with the uncertainty estimate. Standard uncertainties refer to a confidence level of 68% determined by $k=1$, while expanded uncertainties defined by $k>1$ refer to confidence levels of approximately 95% ($k \approx 2$) or 99% ($k \approx 3$).

Uncertainties, when possible, should be provided in both relative (i.e., in percent) and physical units. The range of values for which the uncertainties are proposed should also be reported together with details on measurement conditions. In fact, uncertainties determined for a specific range of values may not necessarily be the same for other ranges or different measurement conditions.

The quantification of uncertainties of in situ measurements, should comprehensively address contributions from the calibration source and its transfer, the performance of the radiometer and of any model applied for data reduction, effects of environmental variability, and perturbations by the instrument housing and deployment platform.

The uncertainty threshold of 5% was originally defined for satellite derived $L_{WN}(\lambda)$ in the blue spectral region to restrict to within 35% the uncertainties in chlorophyll-*a* concentrations determined in oligotrophic waters with existing bio-optical algorithms [120,124]. This 5% uncertainty threshold was then set as the target for $L_{WN}(\lambda)$ determined from the SeaWiFS [125] and following ocean color missions, regardless of the wavelength and with an uncertainty requirement of 5% for the absolute top-of-atmosphere radiance [126]. The maximum uncertainty values given for $L_{WN}(\lambda)$ unavoidably prompted the need for uncertainties better than 5% for in situ optical radiometry data [127].

5.1 Calibration Specific Sources of Uncertainties

Uncertainties strictly connected with the calibration process of field radiometers are those related to the absolute radiometric calibration and the characterization of the instrument. The following analysis is limited to those uncertainties affecting the absolute radiometric calibration. These are the best case scenarios for laboratories carefully following the calibration protocols and working to obtain the highest performance in accuracy and precision. In fact, it is easy to introduce biases and dispersion in measurements if care is not taken through the entire instrument calibration and characterization process.

5.1.1 Irradiance Standards

Irradiance standards expanded uncertainties ($k = 2$) in 1000 W FEL primary standards determined with respect to the irradiance scale of the National Institute of Standards and Technology (NIST), vary from 1.1% at 350 nm to 0.5% at 900 nm [128]. These uncertainties generally increase by at least an additional 1% for FEL lamps calibrated with respect to working standards [129].

Assuming accurate mechanical positioning and baffling of laboratory stray light, additional sources of uncertainties in irradiance calibration are due to: (1) the offset between the center of the filament and the plane tangent to the near surface of the lamp posts; and (2) the interpolation of calibration values provided at discrete wavelengths.

As already discussed, the correction for the offset between the center of the filament and the plane tangent to the near surface of the lamp posts is necessary when the calibration distance d is different from the distance d_0 applied for the calibration of the lamp. This offset, which is lamp specific, is on the order of a few mm [70]. For instance, neglecting an offset of 0.3 cm with $d_0 = 50$ cm, would lead to an underestimate of approximately 1% of the irradiance value at the distance $d = 200$ cm.

Often irradiance values of calibrated lamps are provided with a low spectral resolution. Thus, their practical use requires the interpolation of calibration values [130]. For FEL values of irradiance, Walker et al. [66] suggested the following fitting equation

$$E(\lambda) = \left(a_0 + a_1\lambda + a_2\lambda^2 + a_2\lambda^3 + \ldots + a_n\lambda^n\right)\lambda^{-5}e^{(i_1 + i_2/\lambda)} \quad (24)$$

where the coefficients i_1 and i_2 are determined from the least square fit of

$$\ln(E(\lambda)\, \lambda^5) = i_1 + \frac{i_2}{\lambda} \qquad (25)$$

and the coefficients a_i are successively determined through the least square regression using $E^{-2}(\lambda)$ as weighting values.

This fitting procedure is expected to provide interpolated values with uncertainty increased by less than 0.5% with respect to that declared for $E(\lambda)$. Higher accuracy is obtained when the proposed procedure is separately applied to the UV and visible spectral ranges. Alternative solutions have also been proposed with improved performance in the UV spectral region [131].

A drawback of the FEL lamps, also common to integrating sphere sources, is a lower flux in the visible spectral region with respect to the near-infrared. This is generally responsible for higher uncertainties in the absolute radiometric calibration of radiometers in the blue spectral region.

5.1.2 Radiance Standards

Reflectance targets (e.g., plaques) with characterized bidirectional reflectance are often used as reflectance standards. Commonly used Spectralon (Labsphere, North Sutton) reflectance plaques with 99% directional hemispherical reflectance $\rho_h(\lambda, 8°)$, almost constant in the 400–1800 nm spectral interval, have a standard uncertainty of 0.005. By recalling that laboratory applications for radiance calibration require the directional-directional reflectance $\rho_d(\lambda, 0, 45°)$, the conversion factor $\rho_d(\lambda, 0, 45°)/\rho_h(\lambda, 6°) \approx 1.025$ (assumed equally applicable to $\rho_h(\lambda, 8°)$) was determined for Spectralon at 800 nm [71]. The combined expanded uncertainty ($k=2$) of $\rho_d(\lambda, 0, 45°)$ for Spectralon 99% reflectance plaques is approximately 1.1% when accounting for the contribution of the conversion reflectance factor.

The accuracy of laboratory absolute calibrations is additionally affected by the alignment of the radiometric source, positioning of the radiometer to be calibrated, blocking and reducing stray light, and the stability of the light source. All the former sources of uncertainty were investigated by focusing on a specific series of multispectral radiometers [132]. Results indicated that typical minimum and maximum uncertainties for irradiance calibrations are 1.1% and 3.4%, respectively. Corresponding values determined for radiance calibration are 1.5% and 6.3%, respectively.

Integrating spheres, which are a source of uniform radiance, are an alternative to the use of calibrated lamps and reflectance plaques. However, a difficulty with integrating spheres is their absolute calibration, which must be done as a system including the source, the reflectance properties of the internal coating, and areas of the internal surface and of the exit aperture. An uncertainty value for an accurately characterized integrating sphere is approximately 3% [133], comparable to that achievable with FEL calibrated lamps and reflectance plaques.

5.2 Instrument Specific Sources of Uncertainties

Instrument specific sources of uncertainties may include nonlinear response, responsivity changes with temperature, stray light and out-of-band perturbations. Uncertainties should also include contributions from drifts in the dark values due to temperature changes during field activities and changes in radiometric responsivity over the deployment period due to aging of optical components or contamination of foreoptics by particles or biofouling. The minimization of uncertainties strictly due to system performance can only be done through extensive laboratory characterizations and the application of adequate correction factors (see Section 3). The minimization of instrument specific uncertainties related to field operation may require extra in situ operations. For instance, best practice suggests regular determination of the dark signal during field measurements. Similarly, responsivity changes with time should be traced through regular checks of the radiometer during its deployment period. At the least, pre- and postdeployment calibrations should be performed and utilized during data reduction.

In the following subsections, the discussion on instrument related uncertainties is restricted to the effects of noncosine response of irradiance sensors and to the quantification of uncertainties affecting immersion factors, chosen as representative for in situ optical radiometry.

5.2.1 Effects of Cosine Error

Measurement error due to noncosine response of irradiance sensors should be determined to duly quantify the uncertainty budgets of measurements. The determination of the correction terms $\varepsilon_c(\theta_0,\lambda)$ and the related uncertainties are restricted here to $E_d(0^+,\lambda)$ because of its major relevance for the determination of $L_{WN}(\lambda)$.

Given the cosine error $f_c(\theta,\lambda)$, its effects can be quantified for clear sky conditions by assuming an isotropic distribution of the sky radiance and a known diffuse to direct irradiance ratio $I_r(\theta_0,\lambda)$, with

$$\varepsilon_c(\theta_0,\lambda) = \langle f_c(\lambda) \rangle \frac{I_r(\theta_0,\lambda)}{I_r(\theta_0,\lambda)+1} + f_c(\theta_0,\lambda)\frac{1}{I_r(\theta_0,\lambda)+1} \qquad (26)$$

where

$$\langle f_c(\lambda) \rangle = \int_0^{\pi/2} f_c(\theta,\lambda)\sin(2\theta)d\theta. \qquad (27)$$

This correction scheme was assessed through accurate radiative transfer simulations [134]. For the considered irradiance sensors and values of θ_0 lower than 70°, results indicated maximum differences of 1% between $\varepsilon_c(\theta_0,\lambda)$ values computed with Eqn (26) and those determined with simulations.

5.2.2 Immersion Factor

Experimental determination of $I_f(\lambda)$ for irradiance sensors has shown repeatability better than 0.5% [61]. This value was found much lower than the variability of immersion factors among sensors of the same class which may reach several percent.

In the case of radiance sensors, differences lower that 0.3% were observed between experimental and theoretical values of $I_f(\lambda)$ for radiometers with foreoptics based on a Gershun tube design [64]. However, differences up to several percent were observed for radiometers with more complex optics [65].

5.3 Methods and Field Specific Sources of Uncertainties

In addition to calibration and radiometer specific uncertainties, field operations and data reduction may also become the source of uncertainties. The most relevant are separately addressed in the following subsections for in-water and above-water methods.

5.3.1 In Water Methods

The accuracy of radiometric data products obtained from in-water measurements are affected by environmental perturbations (e.g., wave effects) and additionally by limits linked to specifics of the measuring system (e.g., acquisition rate and deployment speed for profiling radiometers). The possible sources of uncertainty in in-water radiometric data products are reviewed with the objective of providing an estimate of their effects and when possible suggestions for their minimization.

5.3.1.1 Wave Focusing and Defocussing

Surface waves are the cause of light field perturbations in the upper sea layer. Specifically, focusing and defocusing of sunlight refracted by surface waves produce large light fluctuations whose amplitude, frequency, and depth extension have been addressed both theoretically [92,135–139] and experimentally [140–143].

Given that light field perturbations affect the determination of subsurface values, increased precision of radiometric data products from fixed-depth in-water systems is achieved through averaging of data over time [89]. As already stated, in the case of profiling systems, increased precision is obtained by increasing the number of measurements per unit depth in the near surface, which allows more robust extrapolation of subsurface values [92]. This can be achieved by increasing the acquisition rate and decreasing the deployment speed [144,145] or by combining successive profiles collected close in time (i.e., multicasting [93,139]).

Finally, it is pointed out that the extrapolation of in-water log-transformed data should always be regarded with care. In fact, the averaging effects, which take place through the fit of log-transformed data, can introduce biases in subsurface values. An evaluation of the problem has shown that techniques relying on nonlinear fits on the untransformed data may produce improved subsurface values [146]. Specifically, a dedicated analysis based on a variety of profiles showed differences of the order of 1−2% between values of the key quantity $L_u(0^-, \lambda)$ determined from linear fits of log-transformed data and from nonlinear fits of the untransformed data [146]. The same analysis performed for $E_d(0^-, \lambda)$ exhibited differences easily exceeding 5% due to higher sensitivity to wave perturbations.

5.3.1.2 Self-Shading

Errors in the measured upwelling radiance and upward irradiance are brought by perturbations in the light field due to the nonnegligible size of the in-water radiometers, commonly referred to as self-shading. As first theoretically quantified by Gordon and Ding [147] and experimentally assessed by Zibordi and Ferrari [148], self-shading errors due to an ideal instrument composed of a disk of given radius with the detector located in its center, may range from a few up to several tens percent as a function of the size of the radiometer, the absorption coefficient of the medium, and the type of illumination (see data plotted in Figure 8 for clear sky conditions [148]). Specifically, for a given radiometer and illumination condition, the self-shading error increases markedly in the near-infrared due to the strong absorption of pure water, and can increase in the visible with increasing concentrations of absorbing particles and colored dissolved organic matter.

A number of studies have addressed the effects of asymmetry in the shape of radiometers [149–152] highlighting the complexity of the three-dimensional problem and suggesting the fundamental need for minimizing

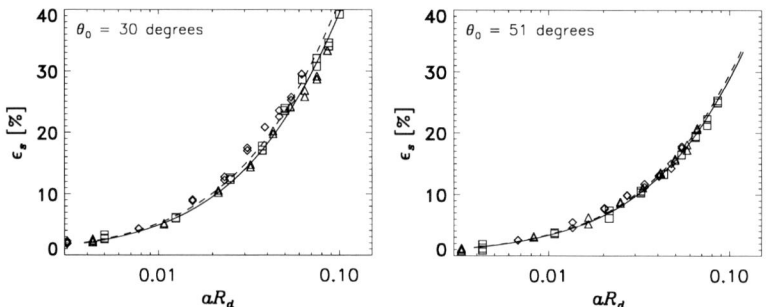

FIGURE 8 Self-shading error ε_s (in percent) in radiance data as a function of seawater absorption, a, times the radius of the radiometer case, R_d, at different sun zeniths, θ_0 (redrawn from Zibordi and Ferrari [148]): 30° (left panel) and 51° (right panel). Symbols indicate experimental data at 550 (diamond), 600 (triangle) and 640 nm (square). Continuous lines indicate the best fit of experimental data while dashed lines indicate the theoretical values computed according to Gordon and Ding [147].

self-shading errors through the development of smaller in-water radiometer systems [36,153]. Still, regardless of the size of the instrument, uncertainties introduced by self-shading should be always included in data analysis.

5.3.1.3 Tilt Effects

Instrument tilt can affect both radiance and irradiance measurements due to the nonuniformity of the sky and in-water radiance distribution. While tilt corrections may be attempted, measurements should be performed with minimum tilt perturbations.

Current free-fall systems can be operated with tilts generally lower than 2° except in conditions of high sea state. Ship motion may also affect measurements of $E_d(0^+,\lambda)$, generally collected from the deck of ships. This suggests that both in-water and deck units should be equipped with X-Y tilt sensors to reject data affected by elevated tilts.

5.3.1.4 Superstructure Perturbations

Ship-shading perturbations affect in-water radiance measurements when the radiometer is not deployed sufficiently far from the ship, and in an advantageous geometry relative to the sun [154–159]. Assuming the radiometer is always deployed from the sunny side of the ship (or other deployment platform), Mueller and Austin [16] suggested determining the minimum distance between ship and radiometer as a function of the water diffuse attenuation coefficient. Specifically they indicated that a distance larger than $3K_L(\lambda)$ is equally applicable to measurements of $L_u(z, \lambda)$, $E_d(z, \lambda)$, and $E_u(z, \lambda)$.

When making measurements from a fixed platform it is often impossible to avoid perturbations by the deployment structure. In such a case, corrections should be applied to the radiometric data products. Example of operational correction scheme implemented for a specific case is provided in Doyle and Zibordi [160] for the Acqua Alta Oceanographic Tower (AAOT). This scheme uses look-up tables of correction factors determined from Monte Carlo simulations of perturbed and nonperturbed radiometric quantities accounting for the measurement geometry, and the marine and atmospheric optical properties. Other similar schemes have been proposed for the Marine Optical Buoy (MOBY) site [161].

Correction estimated with the proposed methodology for time-series of $L_u(0^-, \lambda)$, $E_d(0^-, \lambda)$ and $E_u(0^-, \lambda)$ determined at the AAOT under clear-sky conditions, exhibited values lower than $+2\%$ at 665 nm, but ranging from $+2\%$ up to approximately $+6\%$ at 443 nm as a function of the measurement conditions.

5.3.1.5 Biofouling

The growth of biofilms on surfaces exposed to the marine environment (i.e., biofouling) occurs in several stages comprising the formation of organic molecules and the successive growth of bacteria followed by microalgae and finally by macro-organisms [162]. This growth can affect the long-term deployments of in-water radiometers. Corrections for these perturbing effects are challenging because of the irregular time scale of biofouling growth and additionally by its irregular distribution on exposed surfaces. This makes it difficult to minimize biofouling perturbations simply using interpolation of radiometer responsivity measurements performed during the deployment period.

A number of approaches, unfortunately with mixed results, have been investigated to avoid biofouling on the optics of immersed radiometers (see Manov et al. [163] and references therein). Currently, the most efficient solution appears the use of exposed copper in close vicinity to the optical surfaces and shutters made of copper offering the additional capability to keep the foreoptics in the dark.

5.3.1.6 Modeling

The determination of radiometric products such as $L_{WN}(\lambda)$ and $R_{RS}(\lambda)$ requires the application of models to correct for the bidirectional effects of sea surface and water. The derived corrections are thus affected by a number of model assumptions and approximations leading to uncertainties which are often difficult to quantify. For instance, Morel et al. [79] proposed a correction method which is strictly valid for Case-1 (i.e., *Chla* dominated) waters and whose application in different water types may become the source of unpredictable uncertainties in radiometric data products. Alternative correction methods have been proposed for optically complex waters [164–166], however they require a number of input quantities whose values are not always known and whose arbitrary definition may also lead to unpredictable uncertainties. Among these methods, that proposed by Lee et al. [165] was specifically developed to operationally support the reduction of ocean color radiometry data likely collected in any water type. This method, whose accuracy depends on assumptions on the scattering phase function, relies on the Quasi-Analytical Algorithm (QAA) approach for the determination of absorption and back-scattering coefficients from $L_w(\theta,\phi,\lambda)/E_d(\lambda)$ and from these the directional dependence required to determine $R_{RS}(\lambda)$. An experimental assessment of correction methods for bidirectional effects [80] indicated that, as expected, the Morel et al. [79] scheme is more accurate in *Chla* dominated waters, but the Lee et al. [165] scheme appears quite appropriate in optically complex waters.

5.3.1.7 Inelastic Scattering

Inelastic scattering in natural waters, due to Raman scattering, fluorescence of chlorophyll-a and fluorescence by colored dissolved organic matter, affect the accuracy of radiometric data products. Raman scattering is the largest source of inelastic light. Its relative importance with respect to elastic scattering is particularly pronounced in the red part of the spectrum (tentatively beyond 500 nm

[113]) and it can become a larger fraction of the natural light field at larger wavelengths and depths [167,168]. Fluorescence effects due to chlorophyll-a and colored dissolved organic matter vary with their concentration and absorption coefficient, respectively. Chlorophyll-a effects are restricted to a relatively narrow spectral region (approximately 50 nm wide) centered near 685 nm [114,169], while colored dissolved organic matter largely perturbs the blue spectral region with relative contributions decreasing with wavelength [169].

The previous general observations indicate that inelastic scattering may significantly challenge the assumption of linear decay of log-transformed radiometric data in the water column. Thus, inelastic scattering spectrally affects the determination of diffuse attenuation coefficients and subsurface values in the blue, red, and near infrared spectral regions. The problem is generally not particularly pronounced in the very first few meters below the surface for wavelengths below approximately 575 nm. Removal of perturbing effects in radiometric data products, however, requires modeling elastic and inelastic scattering in the water as a function of its IOPs [86].

5.3.2 Above Water Methods

Similar to in-water methods, above-water methods are challenged by a number of potential sources of perturbations where waves have a prominent position.

5.3.2.1 Wave Perturbations

The determination of above-water radiometric data products is perturbed by wave effects which challenge the removal of the glint component from $L_T(\lambda)$ by affecting the accurate determination of ρ. A practical solution for the minimization of wave effects is provided by the application of filtering techniques, which remove those measurements most likely affected by significant glint perturbation and additionally by whitecaps or foam [105,106]. Still, the capability of minimizing glint perturbations through data filtering decreases with an increase of the field-of-view and integration time. Additionally, this scheme is simply based on empirical thresholds, which do not have universal validity. Further, the use of the average of the lowest radiance values instead of the average of all values, may lead to an overcorrection of the sky-glint perturbation. In the case of data collected in moderately sediment dominated waters, the underestimate of $L_{WN}(\lambda)$ over a wide range of measurement conditions was quantified ranging from approximately 2% at 412 nm to 0.5% at 551 nm, and increasing to 2.5% at 667 nm [116].

5.3.2.2 Sky-Light Polarization

Recent investigations have shown that neglecting the polarization state of sky-light may lead to appreciable uncertainties in the determination of the sea surface reflectance.

Specific results from the analysis of the effects of sky-light polarization for a flat sea surface and the measurement geometries commonly applied for above-water radiometry, indicate extreme variations ranging from -20% to more than $+50\%$ of ρ at 555 nm as a function of the sun zenith angle and to a lesser extent the aerosol optical thickness [170]. For the specific measurement geometry defined by $\theta = 40°$ and $\phi = 90°$, Harmel et al. [170] presented an intercomparison of L_{WN} data corrected and noncorrected for the effects of sky radiance polarization. Results indicate positive increase in $L_{WN}(\lambda)$ from less than 1% in the green—red spectral regions up to 3% at 442 nm, and 6% at 413 nm where the effects of sky polarization are more pronounced.

5.3.2.3 Superstructure Perturbations

Equivalent to in-water radiometry, the minimization of perturbations by deployment platforms is another fundamental aspect of above-water radiometry [171]. A comprehensive quantification of the measurement uncertainties related to platform shading and reflections should entail 3-D radiative transfer simulations specific for each deployment structure, radiometer, and site. Because of this, in view of avoiding such a complex task, measurements are generally performed at locations minimizing platforms perturbations.

Considering that each deployment platform is an individual case and generalizations are difficult, the location of above-water radiometers onboard deployment structures are often guided by common sense and by a limited number of focused investigations. In general, deployment locations are chosen so that the above-water sensor views the surface in areas uncontaminated by waves created by the platform (in the case of ships this location is often the bow) and at a distance not shorter than the height of the superstructure [172,173]. The application of these simple rules, however, cannot assure perturbations are always negligible or spectrally independent.

5.3.2.4 Modeling

Equivalent to in-water methods, anisotropy of sea surface reflectance and in-water light distribution affect the accuracy of derived products. In fact a modeling problem specific to above-water radiometry is the need to remove the viewing angle dependence due to nonnadir view. The uncertainties in these corrections may be of the order of a few percent [109] and depend on the suitability of the model applied for the removal of bidirectional effects of surface and water, and on the accuracy of the input quantities utilized for computations.

5.4 Examples of Uncertainty Budget for Radiometric Products

Measurement uncertainties allow the quantitative evaluation of the results following from the application of measurements. Because of this, the

TABLE 1 Uncertainty Budget (in Percent) for L_{WN} Determined from In-Water Profile Data

Uncertainty Source	443	555	665
Absolute calibration of L_u	2.7	2.7	2.7
Immersion factor	0.4	0.4	0.4
Self-shading correction	0.5	0.3	1.3
Absolute calibration of above-water E_d	2.3	2.3	2.3
Cosine response correction	1.0	1.0	1.0
Environmental effects	2.1	2.2	3.2
Quadrature sum	4.3	4.3	5.1

TABLE 2 Uncertainty Budget (in Percent) for L_{WN} Determined from Above-Water Data

Uncertainty source	443	555	670
Absolute calibration	2.7	2.7	2.7
Sensitivity change	0.2	0.2	0.2
View. angle and bidir. corrections	2.0	2.9	1.9
t_d	1.5	1.5	1.5
ρ	1.3	0.7	2.5
Environmental effect	2.1	2.1	6.4
Quadrature sum	4.4	4.8	7.8

determination of uncertainty budgets should be an ongoing effort of any experimental activity.

As already anticipated, uncertainty budgets for radiometric data products should include estimates of uncertainties for calibration sources and their transfer, performance of the radiometer and of any model applied for data reduction, effects of environmental variability, and, perturbations by the instrument housing and the deployment platform. Accounting for examples provided in literature [116,174–176], generic uncertainty budgets are proposed in Tables 1 and 2 for $L_{WN}(\lambda)$ data determined from in-water radiometric profiles and above-water measurements performed in moderately sediment dominated coastal waters with the methods presented in § 4.3.

The contributions considered in Table 1 for $L_{WN}(\lambda)$ determined with a specific multispectral in-water optical profiler [93] are: (1) uncertainty of the absolute in-air radiance calibration of the L_u sensor [132]; (2) uncertainty in the theoretical determination of the immersion factor [64]; (3) uncertainty of the correction factors applied for removing self-shading perturbations computed as 25% of the applied corrections for a 5 cm diameter radiometer; (4) uncertainty of the in-air irradiance calibration of the above-water E_d sensor [132]; (5) uncertainties of the correction applied for the noncosine response of the related irradiance collectors [134]; (6) uncertainty in the extrapolation of subsurface values due to wave perturbations and uncertainties due to changes in illumination and seawater optical properties during profiling cumulatively quantified as the average of the variation coefficient of $L_{WN}(\lambda)$ from replicate measurements [89].

Estimated values, quantified assuming each uncertainty contribution is independent from the others, and neglecting sources expected not to be significant because of the application of accurate corrections, are in the range of 4–5% in the selected spectral region. It is noted that the proposed uncertainty analysis accounts for fully independent calibrations of E_d and L_u sensors (i.e., as obtained with different lamps and laboratory set-ups). The use of the same calibration lamp and set-up would lead to a reduction of approximately 1% of the quadrature sum of spectral uncertainties for $L_{WN}(\lambda)$, explained by correlations between absolute calibration uncertainties of $E_d(\lambda)$ and $L_u(\lambda)$.

The contributions included in Table 2 for $L_{WN}(\lambda)$ determined with a specific above-water radiometer [116] are: (1) uncertainty resulting from the in-air absolute calibration assumed equal for L_T and L_i [132]; (2) sensitivity change during long-deployments also assumed the same for L_T and L_i; (3) uncertainty in the correction factors applied to remove viewing angle dependence and bidirectional effects computed as 25% of the applied corrections; (4) uncertainty in the determination of the diffuse atmospheric transmittance $t_d(\lambda)$ used to compute $L_{WN}(\lambda)$ in the absence of measured values of $E_d(0^+, \lambda)$ (the applied spectrally independent value is a guessed minimum value, in fact the sole uncertainty of 0.02 in τ_a leads to an uncertainty of approximately 0.4% in $(D^2 t_d(\lambda)\cos\theta_0)^{-1}$ and consequently in $L_{WN}(\lambda)$); (5) uncertainty in the determination of the sea surface reflectance $\rho(\theta,\phi,\theta_0,W)$ as a result of uncertainties in the wind speed, modeled ρ and of the filtering applied to $L_T(\theta,\phi,\lambda)$ to minimize the wave effects [110,116]; (6) environmental variability resulting from the combination of wave induced perturbations with changes in seawater optical properties and illumination conditions during measurements cumulatively quantified as the average of the variation coefficient of $L_{WN}(\lambda)$ from replicate measurements [116] (this latter perturbation source is implicitly assumed to include uncertainties in the determination of $L_i(\theta',\phi,\lambda)$ while polarization effects of skylight are ignored).

Similar to in-water data, the combined uncertainties for $L_{WN}(\lambda)$ from above-water measurements are close to the 5% target except for the values at 670 nm which reaches 8% mostly because of the effects of wave perturbations.

It is important to point out that the estimates provided in Tables 1 and 2, have been determined for a given site and for specific measurement methods and systems, thus, they should not be considered equally applicable to any site and above- or in-water radiometric data product.

6. APPLICATIONS

In situ radiometric data are essential for a number of applications including (but not restricted to): (1) the study of light interaction with optically significant seawater constituents; (2) the development of bio-optical algorithms; (3) the assessment of satellite derived primary products; and (4) the vicarious calibration of satellite sensors. In the following subsections specific applications of in situ optical radiometric measurements are briefly discussed.

6.1 Sky and Sea Radiance Distribution

In-water bidirectional effects are key elements for an accurate analysis of satellite ocean color data collected with different illumination and observation geometries in different water types. In situ radiance distribution measurements are thus essential for investigating bidirectional effects and assessing prediction models [177]. Examples of sky and in-water upwelling radiance distributions are hereafter presented.

Sky radiance distributions recorded in the Santa Barbara Channel at 520 nm for two different sun zenith angles θ_0 are shown in Figure 9. The distributions are obtained from equidistant projections of fisheye sky images with the zenith direction and horizon corresponding to the center and edge of the circle, respectively. In agreement with theoretical determination [178], distributions show the enhanced radiance near the horizon caused by multiple

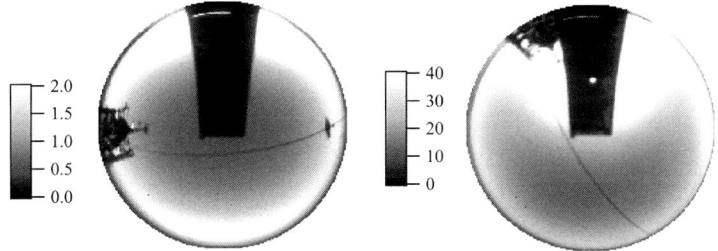

FIGURE 9 Angular distribution of sky radiance at 520 nm measured in the Santa Barbara Channel on September 22, 2008. Data, in units of $\mu W\,cm^{-2}\,sr^{-1}\,nm^{-1}$, were collected with $\theta_0 = 88°$ (left panel) and $\theta_0 = 34°$ (right panel). The aerosol optical depth at 500 nm was 0.16 during measurements.

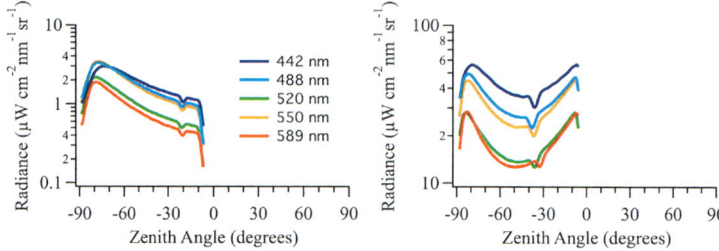

FIGURE 10 Spectral angular variations along the principal plane for the sky radiance data displayed in Figure 9. Because of the occulter, radiances are only available for the portion of the principal plane opposite of the sun.

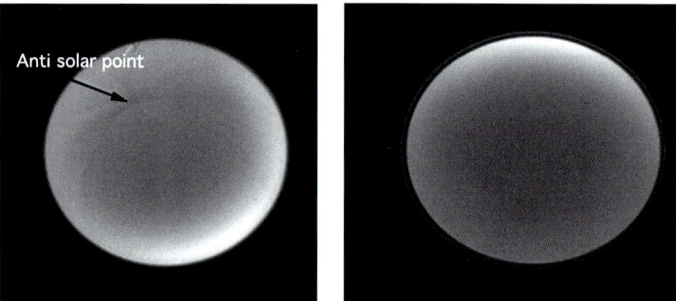

FIGURE 11 Upwelling radiance distribution at 436 nm from clear water off of Hawaii collected $\theta_0 = 59°$ (left panels) and from relatively turbid water in the Ligurian Sea collected with $\theta_0 = 49°$ (right panel).

scattering, more pronounced at low sun angles. Not surprisingly, the capability of producing a full radiance distribution is limited by the superstructure of the ship and additionally by the occulter blocking the light from the sun and its aureole. This figure points out the difficulties and importance of correctly positioning the above water irradiance sensor for $E_d(0^+,\lambda)$ measurements on deployment platforms.

Figure 10 shows the spectral variation of the sky radiance distribution along the principal plane for the two cases displayed in Figure 9. The radiance plots illustrate the effects of Rayleigh scattering, with the largest radiance values in the blue wavelengths, decreasing toward the red. Both cases exhibit only slight spectral variations in the shape of the radiance distribution.

Examples of upwelling in-water radiance distributions are presented in Figure 11. The two cases refer to clear waters from near Hawaii and relatively turbid waters in the Ligurian Sea.

The radiance image for the clear water example shows the light rays caused by wave refraction at the surface, which seem to originate from the antisolar point. This is a common effect in clear waters and clear sky conditions. The light rays also appear more distinct in low wind conditions, and tend to disappear in

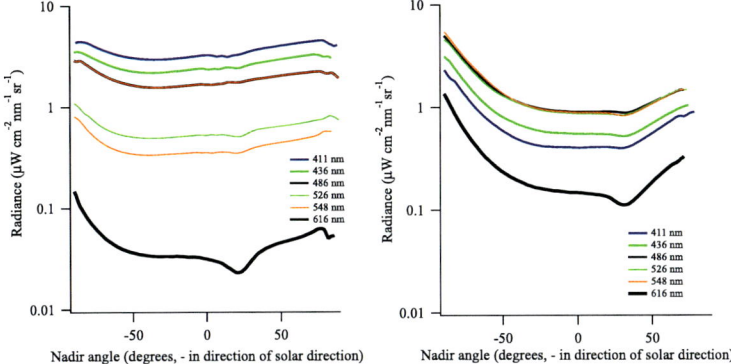

FIGURE 12 Spectral angular variations along the principal plane for the upwelling radiance data displayed in Figure 11.

higher sea states as the surface becomes more geometrically disordered. The brighter side toward the lower right is in the direction of the sun.

The spectral variation of the radiance distribution along the principal plane is shown in Figure 12. As expected, in these clear waters the upwelling radiance is larger in the blue and decreases toward the red wavelengths. The shape of the radiance distribution, displayed with a logarithmic scale, is quite consistent for wavelengths from 411 to 486 nm. As the water absorption increases, the shape of the radiance distribution changes and becomes more accentuated toward the edges (horizon). For the blue wavelengths (i.e., 411−486 nm), the ratio between the maximum and minimum radiance is less than 1.5. At larger wavelengths (e.g., 616 nm) the ratio becomes closer to a factor of three or more. The graphs also clearly show the effects of instrument self-shading. This is only slightly evident in the blue wavelengths, but becomes pronounced at 616 nm where it appears as a large dip in the antisolar region. Another feature shown by these radiance distributions is the minimum (in the absence of self-shading) occurring toward the sun on the side of nadir. This is the effect of the Rayleigh scattering phase function of water, which is minimum at the 90° scattering angle.

6.2 In-water Light Field Polarization

Natural light is polarized [179], and this polarization is a further component needed to comprehensively describe the sky and in-water light fields. Examples of the sky and in-water upwelling polarized radiance distributions are thus presented in this subsection.

As discussed previously, the polarized light field can be described in terms of a four-component Stokes vector. Its first term is the radiance, I, the next two components, Q and U, describe the linear polarization of the light field along

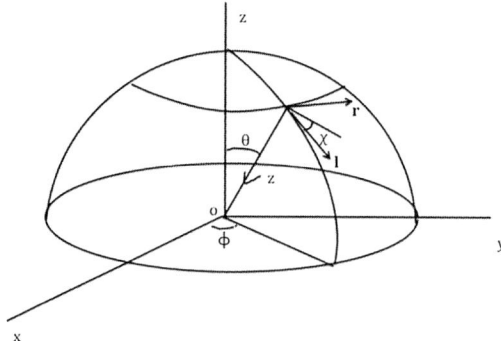

FIGURE 13 The illustration of the coordinate system in the sky frame. Light is propagating along the $z = r \times l$ direction and χ is the angle of the polarization plane measured from l to r such that r and l are perpendicular and parallel, respectively, to the meridian plane (containing the view and zenith direction).

two axis oriented 45° to each other, while the last component, V, describes the circular polarization which is generally negligible in natural light fields [180].

Making reference to Figure 13, a common reference frame describing the polarized radiance distribution in polar coordinates is provided by the meridian plane identified by the vector in the directions of zenith and light propagation. The polarization of the light ray can then be resolved into two components: in the directions parallel (identified by l) to the frame and perpendicular (identified by r) to it. By using this reference frame, Q/I is $+1$ when the light is fully polarized along the unit vector l and -1 when fully polarized along the unit vector r. U/I is then $+1$ when it is along an axis 45° with respect to l, in the direction of r, and -1 when the light is fully polarized along a direction $-45°$ from l.

Using this reference frame, Figure 14 displays the Stokes vector components Q/I and U/I for the sky radiance distributions presented in Figure 9. Much of the pattern seen in the sky polarization components are explained by the simple Rayleigh single scattering Mueller matrix [181]. The maximum of these components varies with aerosol loading, which tends to slightly depolarize the light field.

The degree of linear polarization ($DoLP$), suitable to summarize the polarization state of in-water and sky radiances, is defined as $DoLP = (Q^2 + U^2)^{1/2}/I$, with $I \geq (Q^2 + U^2 + V^2)^{1/2}$ where, as discussed, V is generally negligible in the sky and water. In Figure 15 the sky spectral variation in $DoLP$ is shown for the radiance data presented in Figure 10. The degree of polarization is fairly spectrally constant in the sky data for both cases. In general, the maximum $DoLP$ decreases toward the blue wavelengths because of multiple scattering, and decreases toward the red wavelengths as aerosols have a greater influence.

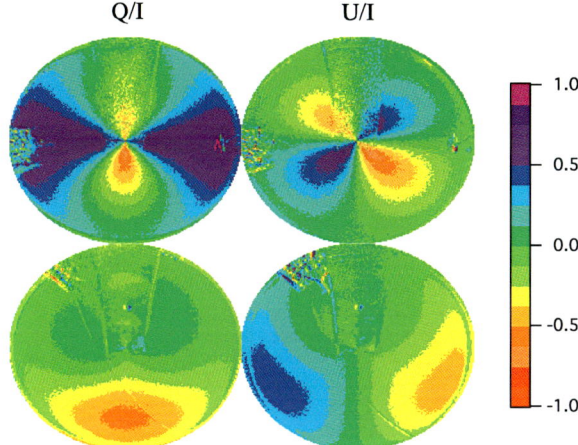

FIGURE 14 The angular distribution of *Q/I* and *U/I* for the sky data presented in Figure 9 with $\theta_0 = 88°$ (upper panels) and $\theta_0 = 34°$ (lower panels).

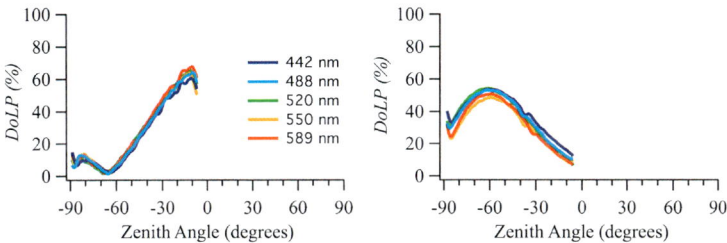

FIGURE 15 *DoLP* for the sky data presented in Figure 10.

A representative example for the polarization of the upwelling light field for the clear water off of Hawaii is presented in Figure 16. The pattern of *Q/I* and *U/I* can be explained by a Rayleigh scattering Mueller matrix, which follows the scattering of water and particulates [182]. As shown in the right panel of Figure 16, the light field can be highly polarized with the *DoLP* reaching 60–70% in clear water. The area of maximum *DoLP* is at a scattering angle of 90° from the refracted solar direction. In turbid waters the pattern is similar, but with a lower maximum *DoLP*.

6.3 Bio-Optical Models

Bio-optical models are generally separated into analytical and empirical. Analytical models are commonly used to predict the radiometric quantities from the modeled or measured IOPs of water constituents, but also to support the inversion of radiometric data to determine IOPs. In contrast, empirical

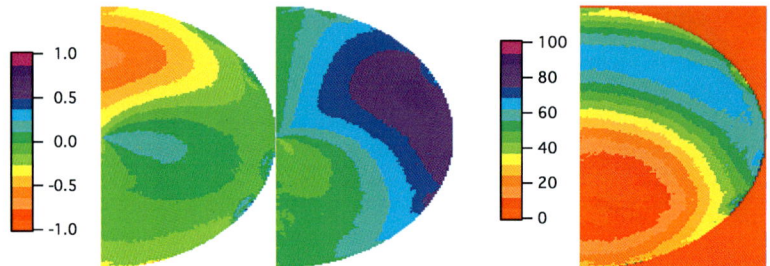

FIGURE 16 Polarization parameters determined for upwelling radiance in clear water off of Hawaii at 0.5 m depth with $\theta_0 = 53°$. From left to right the images display Q/I, U/I, and $DoLP$. The distributions result from the average of three images taken within a 10-min interval. Because of symmetry in the radiance distribution with respect to the sun plane, only half of the image is displayed. In these distributions, the antisolar point is toward the bottom (i.e., in the wide area of low $DoLP$).

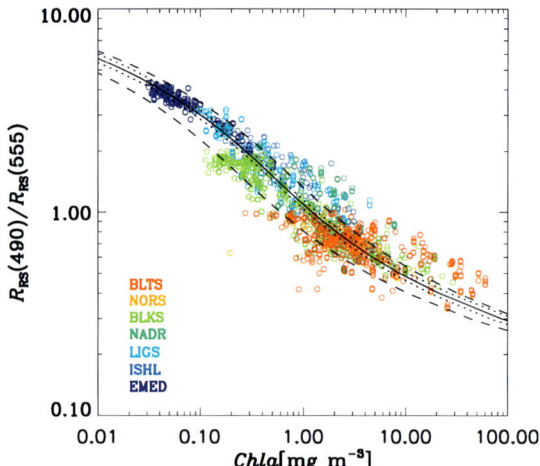

FIGURE 17 Empirical algorithm for *Chla* as a function the $R_{RS}(490)/R_{RS}(55)$ spectral ratio, represented by the solid line (third order polynomial fit). Circles indicate experimental data from different European Seas (i.e., Easter Mediterranean Sea (EMED), Iberian Shelf (ISHL), Ligurian Sea (LIGS), northern Adriatic Sea (NADR), Black Sea (BLKS), North Sea (NORS) and Blatic Sea (BLTS). The dotted lines indicate the polynomial fit as obtained by applying a ± 5% bias to $R_{RS}(490)/R_{RS}(555)$, while the dashed lines indicate ±51% in *Chla* values with respect to the reference band-ratio algorithm.

models are commonly derived from statistical analysis of field measurements and are used to describe the bio-optical state of water [183]. An example of these empirical models (usually called algorithms) is given by ratios of remote sensing reflectance at various wavelengths as a function of the concentration of the optically significant constituents (see Figure 17). These specific algorithms find their rationale in the spectral variations of $L_{WN}(\lambda)$ as a function of the concentration of water constituents: this can be evidenced by the different shape of sample spectra for waters likely related to different concentrations of the optically significant constituents (see Figure 6). Unlike analytical models, which are expected to have more universal applicability, empirical algorithms are mostly valid for the specific waters where measurements were performed.

Bio-optical modeling at large, makes use of radiometric quantities such as L_{WN}, R, R_{RS}, Q_n, K_d and spectral ratios of the same quantities. Thus their uncertainties affect the accuracy of models and consequently that of the higher level products derived by applying those models to satellite ocean color radiometry data. To investigate these effects, a sensitivity analysis has been performed for the algorithm presented in Figure 17 for *Chla* by applying a bias of 5% in R_{RS} ratios. Results from this analysis for *Chla* ranging from 0.1 to 20 mg m^3, indicate changes of $\pm 15\%$ and $\pm 20\%$, respectively, at each end of the considered range. These results are well below the standard uncertainty of 51% determined for the proposed algorithm in complex waters and indicate that accuracy requirements are ultimately defined by the target application. In particular, a 5% bias in R_{RS} ratios does not largely affect the accuracy of the derived *Chla* for the algorithm considered.

6.4 Validation of Satellite Radiometric Products

Comprehensive assessment of satellite derived $L_{WN}(\lambda)$ through in situ data is a fundamental requirement for any ocean color mission (see Chapter 6.1 for details). In view of satisfying such a requirement, most ocean color validation programs rely on data sets constructed combining measurements from many different and fully independent sources [184]. This solution provides the advantage of working with large in situ data likely representative of the various marine water types. On the other hand, it creates the condition for inconsistencies among measurements performed with various field radiometers, diverse sampling methods, a variety of calibration sources and protocols, and assorted data reduction schemes. This suggests that validation activities would benefit from a network of standardized instruments operating at sites representative of different water types, relying on a consistent and assessed measurement protocol, calibration source, and processing code. This would certainly offer optimal conditions for a thorough assessment of uncertainties and regular reprocessing of in situ data to account for advances in data analysis and better characterization of the deployed radiometers [116].

The assessment of satellite versus in situ data for the generic quantity $\mathfrak{J}$ can be presented through the average of percent differences, ψ, and the average of unsigned percent differences, $|\psi|$, of N matchups (i.e., pairs of satellite and in situ data for the same site collected close in time to minimize perturbations induced by the temporal variability of the sea and atmosphere).

Specifically, the value of ψ is computed with

$$\psi = \frac{1}{N} \sum_{i=1}^{N} \psi_i \qquad (28)$$

where i is the matchup index, and ψ_i is

$$\psi_i = 100 \frac{\Im^S(i) - \Im^R(i)}{\Im^R(i)} \qquad (29)$$

with the superscripts S and R indicating the satellite derived and the in situ reference data, respectively. The absolute values of ψ_i, $|\psi_i|$, are applied to determine the average of unsigned percent differences $|\psi|$ through.

$$|\psi| = \frac{1}{N} \sum_{i=1}^{N} |\psi_i|. \qquad (30)$$

The statistical index ψ determines the bias, while $|\psi|$ indicates the dispersion of data points between the two data sources. The root mean square of differences, rmsd, and the determination coefficient, r^2, are additional quantities, which commonly complement the previous statistical indices.

Figure 18 illustrates the application of in situ above-water radiometric data to the validation of satellite ocean color products. Specifically, it displays scatter plots of MODIS and in situ L_{WN} matchups for the AAOT in the northern Adriatic Sea (see Chapter 6.1 for details on match-up construction). In situ data are from the Ocean Color component of the Aerosol Robotic Network (AERONET-OC) [116]. It is mentioned that a band-shift correction has been applied to in situ $L_{WN}(\lambda)$ to minimize uncertainties due to differences in center-wavelengths between data products [116,185].

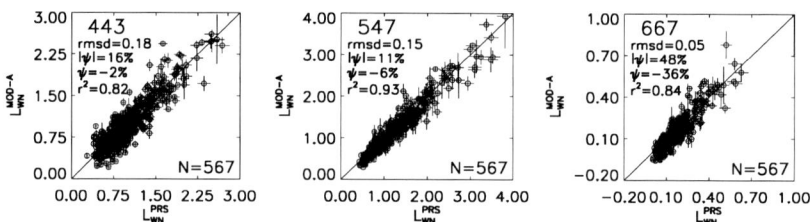

FIGURE 18 Scatter plots of MODIS (MOD) and in situ (PRS) $L_{WN}(\lambda)$ matchups (in units on mW cm^{-2} μm^{-1} sr^{-1}) for the AAOT in the Adriatic Sea at the 443, 547 and 667 nm center-wavelengths. N indicates the number of match-ups, $L_{WN}(\lambda)$ and rmsd are in units of mW cm^{-2} μm^{-1} sr^{-1}, $|\psi|$ is the mean of unsigned percent differences while ψ is the mean of percent differences, and r^2 is the determination coefficient. Horizontal bars indicate ± the estimated uncertainties in in situ L_{WN} while vertical bars indicate ±1 standard deviation of the 3 × 3 values centered on the in situ measurement site and utilized for computing the average MODIS L_{WN}.

Scatter plots in Figure 18 show spectrally changing biases between satellite and in situ data, with higher values at the red wavelength likely due to larger uncertainties introduced by the atmospheric correction process in the specific coastal area. As opposed to the case of band-ratio algorithm for coastal European waters, this application indicates the importance of highest accuracy for radiometric in situ $L_{WN}(\lambda)$ to confidently assess satellite data products.

A critical assumption for the proposed intercomparison is the equivalence of radiometric products determined at very different spatial resolutions. Such an assumption commonly applies to open sea regions, but it is often questioned in coastal regions likely affected by high spatial and temporal variability. In general, intercomparison exercises are considered still applicable to spatially inhomogeneous waters when the analyses are supported by a number of matchups capable of capturing the effects of random changes in the optical properties of water. Best practice would, however, suggest an effort to estimate the satellite subpixel spatial variability to determine how this affects matchup analysis [185].

6.5 In situ Data and System Vicarious Calibration

System vicarious calibration of satellite ocean color sensors is the determination of the radiometric calibration factors through the simulation of radiance at the space sensor using: (1) highly accurate in situ measurements traceable to the International System of Units (SI), selected to serve as in situ radiometric standards; and (2) the same models and algorithms applied for the atmospheric correction of satellite data. This calibration scheme (see Chapter 2.2. for details and examples) is referred to as *system vicarious calibration* because it minimizes the combined effects of: (1) the uncertainties due to the absolute preflight radiometric calibration of the space sensor corrected for sensitivity change with time; and (2) the inaccuracy of models applied for the atmospheric correction process. The fundamental need for system vicarious calibration is documented by the necessity to meet uncertainty requirements for satellite radiometric products (e.g., 5% for L_w). This can be easily proven through an uncertainty analysis relying on a simplified measurement equation (i.e., by neglecting gaseous absorption and, sun-glint and foam perturbations) relating total radiance L_T at the space sensor to water leaving radiance L_w

$$L_T = L_R + L_A + L_w t_d \tag{31}$$

where L_R and L_A indicate the Rayleigh and aerosol atmospheric radiance contributions, and t_d the diffuse atmospheric transmittance. Following the *Guide to the Expression of Uncertainty in Measurement* [186] (see also Chapter 1.2 for details) and assuming that for any given observation condition and wavelength the terms L_R and L_A are exactly computed, the absolute uncertainty in L_T, $u(L_T)$, is expressed as a function of the sole uncertainty in L_w, $u(L_w)$, according to

$$u^2(L_T) = \left(\frac{\partial L_T}{\partial L_w}\right)^2 u^2(L_w). \tag{32}$$

Accounting for Eqn (31) and solving partial derivatives

$$u^2(L_T) = t_d^2 \, u^2(L_w) \tag{33}$$

then, dividing the square root of terms in Eqn (33) by the product $L_T L_w$ and rearranging them, the relative uncertainty of L_w, $u(L_w)/L_w$, is determined by

$$\frac{u(L_w)}{L_w} = \frac{u(L_T)}{L_T} \frac{L_T}{L_w} \frac{1}{t_d}. \tag{34}$$

Equation (34) indicates that $u(L_w)/L_w$ does not only depend on $u(L_T)/L_T$, but also on the ratio L_T/L_w and on $1/t_d$. For instance, as a first approximation assuming $t_d = 1$, a relative uncertainty as low as 0.5% for $u(L_T)/L_T$ would lead to values of $u(L_w)/L_w$ of approximately 5%, 10% and 50% with L_w equal to 10%, 5%, and 1% of L_T, respectively (tentatively indicating spectral values from the blue toward the red for oligotrophic waters). These results show that, even assuming an uncertainty $u(L_T)/L_T$ which is not currently practical, the ability to determine L_w from L_T with an uncertainty less than 5% is most likely impossible in the red, and very challenging in the blue-green spectral regions for most of the world seas.

Equation (34) also indicates that assuming an uncertainty of 5% for L_w and $t_d = 1$, $u(L_T)/L_T$ would exhibit values of 0.5%, 0.25% and 0.05% with L_w equal to 10%, 5% and 1% of L_T, respectively. This indicates that (1) the preflight absolute radiometric calibration uncertainties, which are of the order of 2–3%, largely fail to meet science requirements and that (2) system vicarious calibration, even though not literally leading to the absolute calibration of the space sensor, is the only viable solution to determine L_w from L_T with the required uncertainty for observation conditions close to those considered for the vicarious calibration process. This also entails that the in situ L_w data utilized for system vicarious calibration must have at least the same requirements as defined for satellite derived L_w.

7. SUMMARY AND OUTLOOK

In situ optical radiometry is an essential component of any satellite ocean color mission. In fact it is the key for system vicarious calibration (i.e., the combined calibration of space sensor and data reduction models/algorithms), for the assessment of satellite derived L_{WN} from which higher level data products are derived, and finally for the development of bio-optical algorithms for the generation of high level products from satellite radiometric data.

Recent advances in ocean color field radiometry were mostly driven by the need to support satellite missions devoted to climate change investigations and meet the 5% target uncertainty commonly required for L_{WN} (or equivalent quantity) determined from space. Still, the actual achievement of such a target uncertainty requires a major effort in terms of characterization and calibration of field radiometers, correct deployment of instruments, and successive data reduction. The various sections of this chapter, have addressed each one of the previous elements with the objective of summarizing the state of the art in marine optical radiometry, but also to indicate areas for which further development is needed in view of ensuring progress. Example of these areas of improvement is the need to increase the accuracy of L_{WN} determined from both in-water and above-water radiometry through methods minimizing the perturbing effects of surface waves. In the case of in-water radiometry this entails the development of new extrapolation techniques for the determination of subsurface values from profile data. In the case of above-water radiometry, this requires further progress in the determination of the spectral value of the sea surface reflectance.

Uncertainties are integral part of measurements and their complete quantification and combination into uncertainty budgets, is only possible through deep knowledge of the performance of the measuring system, correct operations in the field, truthful understanding, and application of data reduction schemes.

Typical applications of in situ optical radiometry data have been included in the work by addressing examples on radiance distribution, light field polarization, bio-optical algorithms, validation of satellite radiometric data products, and requirements for system vicarious calibration. These basic examples further confirm the importance of supporting measurements with uncertainty budgets in the view of providing confidence into applications.

REFERENCES

[1] M. Kundsen, On measurements of the penetration of light into the sea, Pub. de Circ. 76 (1922) 1−15 (Cons Perm Internat Explor Mer).
[2] W. Shoulejkin, On the Color of the Sea, Phys. Rev. 23 (1924) 744−751.
[3] H. Pettersson, S. Landberg, Submarine daylight, Medd. Oceanogr. Inst. Göteborg 6 (1934) 1−13.
[4] N.G. Jerlov, G. Liljequist, On the angular distribution of submarine daylight and the total submarine illumination, Sven Hydrogr − Biol. Komm Skr, Ny Ser. Hydrogr 14 (1938) 1−15.
[5] N.G. Jerlov, Optical studies of ocean water, Rep. Swedish Deep-sea Expedition 3 (1951).
[6] N.E. Steeman, Conditions of light in the fjord, Medd Denmarks Fiskeri-og Havunders 5 (1951) 21−27.
[7] J.E. Tyler, R.C. Smith, Measurements of Spectral Irradiance Underwater, Gordon and Breach Science Publishers, 1970.
[8] P.N. Slater, Remote Sensing: Optics and Optical Systems, Addison-Wesley Publishing Company, 1980.

[9] A. Gershun, The light field, J. Math. Psychol. 18 (1939) 51–151.
[10] Y. Le Grand, La pénétration de la lumière dans la mer, Ann. Inst. Ocèanogr 19 (1939) 393–436.
[11] W.A. Hovis, D.K. Clark, F. Anderson, R.W. Austin, W.H. Wilson, E.T. Baker, D. Ball, H.R. Gordon, J.L. Mueller, S.Z. El-Sayed, B. Sturm, R.C. Wrigley, C.S. Yentsch, Nimbus-7 coastal zone color scanner: system description and initial imagery, Science 210 (1980) 60–63.
[12] J.T.O. Kirk, Light & Photosynthesis in Aquatic Ecosystems, Cambridge University Press, 1983.
[13] C.D. Mobley, Light and Water. Radiative Transfer in Natural Waters, Academic Press, 1994.
[14] R.W. Spinrad, K.L. Carder, M.J. Perry, Ocean Optics, Oxford University Press, 1994.
[15] D.K. Clark, M.E. Feinholz, M.A. Yarbrough, B.C. Johnson, S.W. Brown, Y.S. Kim, R.A. Barnes, Overview of the radiometric calibration of MOBY, in: Proc. Earth Observing Systems VI, vol. 4483, SPIE, 2002.
[16] J.L. Mueller, et al., Ocean Optics Protocols for Satellite Ocean Color Sensor Validation, Revision 4, NASA Tech. Memo 211621, Goddard Space Flight Center, 2003.
[17] G. Zibordi, K.J. Voss, Field radiometry and ocean color remote sensing, in: V. Barale, J. Gower, L. Alberotanza (Eds.), Oceanography from Space, Springer, New York, 2010.
[18] Y. Zong, S.W. Brown, B.C. Johnson, K.R. Lykke, Yoshi Ohno, Simple spectral stray light correction method for array spectroradiometers, Appl. Opt. 45 (2006) 1111–1119.
[19] H.J. Kostkowski, Reliable Spectroradiometry, Spectroradiometry Consulting, La Plata, MD, 1997.
[20] M.R. Abbott, K.H. Brink, C.R. Booth, D. Blasco, L.A. Codispoti, P.P. Niiler, S.R. Ramp, Observations of phytoplankton and nutrients from a Lagrangian drifter off northern California, J. Geophys. Res. 95 (C6) (1990) 9393–9409.
[21] W.R. McCluney, Introduction to Radiometry and Photometry, Artech House Publ, Boston, London, 1994.
[22] G.P. Harris, Photosynthesis, productivity and growth, the physiological ecology of phytoplankton, Arch. HydrobioL Beih. Ergebn. Limnol. 10 (1978) 1–171.
[23] P. Fielder, P. Comeau, Construction and testing of an inexpensive PAR sensor, Res. Br. Min. For. (2000) 32. Victoria, BC. Working Paper 53.
[24] D. Conde, L. Aubriot, R. Sommaruga, Changes in UV penetration associated with marine instrusions and freshwater discharge in a shallow coastal lagoon of the Southern Atlantic Ocean, Mar. Ecol. Prog. Ser. 207 (2000) 19–31.
[25] R.C. Smith, An underwater spectral irradiance collector, J. Mar. Res. 27 (1969) 341–351.
[26] R.C. Smith, R.L. Ensminger, R.W. Austin, J.D. Bailey, G.D. Edwards, Ultraviolet submersible spectroradiometer, in: Proc. Ocean Optics VI, SPIE 0208, 1980. http://dx.doi.org/10.1117/12.958268 JL.
[27] G. Bernard, C.R. Booth, J.C. Ehramjian, Comparison of UV irradiance measurements at Summit Greenland: Barrow, Alaska: and South Pole, Antarctica, Atmos. Chem. Phys. 8 (2008) 4799–4810.
[28] D.F. Westlake, Some problems in the measurement of radiation under water: a review, Photochem. Photobiol. 4 (1965) 849–868.
[29] E. Aas, On Submarine Irradiance Measurements, Technical Report 6, Institute of Physical Oceanography, University of Copenhagen, Copenhagen, Denmark, 1969.
[30] N.G. Jerlov, M. Fukuda, Radiance distribution in the upper layers of the sea, Tellus 12 (1960) 348–355.

[31] J.E. Tyler, Radiance distribution as a function of depth in an underwater environment, Bull. Scripps Inst. Oceanogr. 7 (1960) 363–412.
[32] E. Aas, N.K. Højerslev, Analysis of underwater radiance observations: apparent optical properties and analytic functions describing the angular radiance distribution, J. Geophys. Res. 104 (1999) 8015–8024.
[33] R.C. Smith, R.W. Austin, J.E. Tyler, An oceanographic radiance distribution camera system, Appl. Opt. 9 (1970) 2015–2022.
[34] K.J. Voss, Electro-optic camera system for measurement of the underwater radiance distribution, Opt. Eng. 28 (1989a) 241–247.
[35] J. Shields, M.E. Karr, R.W. Johnson, A.R. Burden, Day/night whole sky imagers for 24-h cloud and sky assessment: history and overview, Appl. Opt. 52 (2013) 1605–1616.
[36] K.J. Voss, A.L. Chapin, Upwelling radiance distribution camera system NURADS, Opt. Express 13 (2005) 4250–4262.
[37] G.C. Stokes, On the composition and resolution of streams of polarized light from different sources, Trans. Cambridge Philos. Soc. 9 (1852) 399–416.
[38] Z. Li, P. Goloub, O. Dubovik, L. Blarel, W. Zhang, T. Podvin, A. Sinyuk, M. Sorokin, H. Chen, B. Holben, D. Tanré, M. Canini, J-P. Buis, Improvements for ground-based remote sensing of atmospheric aerosol properties by additional polarimetric measurements, J. Quant. Spectrosc. Rad. Transfer 110 (2009) 1954–1961.
[39] B. Lundgren, On the Polarization of the Daylight in the Sea, Report N.17, Institute of Physical Oceanography, University of Copenhagen, Copenhagen, Denmark, 1971.
[40] A. Tonizzo, J. Zhou, A. Gilerson, M. Twardowski, D. Gray, R. Arnone, B. Gross, F. Moshary, S. Ahmed, Polarized light in coastal waters: hyperspectral and multiangular analysis, Opt. Express 17 (2009) 5666–5683.
[41] K.J. Voss, N. Souaidia, POLRADS: polarization radiance distribution measurement system, Opt. Express 18 (2010) 19672–19680.
[42] P. Bhandari, K.J. Voss, L. Logan, An instrument to measure the downwelling polarized radiance distribution in the ocean, Opt. Express 19 (2011) 17609–17620.
[43] V. Gruev, R. Perkins, T. York, CCD polarization imaging sensor with aluminum nanowire optical filters, Opt. Express 18 (2010) 19087–19094.
[44] C.L. Wyatt, Radiometric, Calibration: Theory and Methods, Academic Press, 1978.
[45] J. Wei, R. Van Dommelen, M.R. Lewis, K.J. Voss, A new instrument for measuring the high dynamic range radiance distribution in near-surface sea water, Opt. Express 20 (2012) 27024–27038.
[46] C. Brosseau, Fundamentals of Polarized Light: A Statistical Optics Approach, Wiley, New York, 1998.
[47] G. Meister, E.J. Kwiatkowska, B.A. Franz, F.S. Patt, G.C. Feldman, C.R. McClain, Moderate-resolution imaging spectroradiometer ocean color polarization correction, Appl. Opt. 44 (2005) 5524–5535.
[48] L.K. Huang, R.P. Cebula, E. Hilsenrath, New procedure for interpolating NIST FEL lamp irradiances, Metrologia 35 (1998) 381–386.
[49] H. Du, K.J. Voss, Effects of point-spread function on calibration and radiometric accuracy of CCD camera, Appl. Opt. 43 (2004) 665–670.
[50] G. Zonios, Noise and stray light characterization of a compact CCD spectrophotometer used in biomedical applications, Appl. Opt. 49 (2010) 163–169.
[51] K.R. Lykke, P.-S. Shaw, L.M. Hanssen, G.P. Epppeldauer, Development of monochromatic, uniform source facility for calibration of radiance and irradiance detectors from 0.2 um to 12 um, Metrologia 35 (1998) 479–484.

[52] S.W. Brown, B.C. Johnson, M.E. Feinholz, M.A. Yarbrough, S.J. Flora, K.R. Lykke, D.K. Clark, Stray-light correction algorithm for spectrographs, Metrologia 40 (2003) S81.
[53] M.E. Feinholz, S.J. Flora, M.A. Yarbrough, K.R. Lykke, S.W. Brown, B.C. Johnson, Stray light correction of the Marine optical system, J. Atmos. Ocean. Technol. 26 (2009) 57–73.
[54] S. Mekaoui, G. Zibordi, Cosine error for a class of hyperspectral irradiance sensors, Metrologia 50 (2013) 187–199.
[55] K.J. Voss, G. Zibordi, Radiometric and geometric calibration of a spectral electro-optic "fisheye" camera radiance distribution system, J. Atmosph. Ocean. Tech. 6 (1989) 652–662.
[56] W.R.G. Atkins, H.H. Poole, The photo-electric measurement of the penetration of light of various wavelengths into the sea and the physiological bearing of results, T. Phil. Trans. Roy. Soc. London (B) 222 (1933) 129–164.
[57] F. Berger, Uber die ursache des "oberflächeneffekts" bei lichtmessungen unter wasser, Wetter U Leben 10 (1958) 164–170.
[58] F. Berger, Uber den "taucheffekt" bei der lichtmessung ber and unter wasser, Arch. Meteorol. Wien. (B) 11 (1961) 224–240.
[59] T.J. Petzold, R.W. Austin, "Chracterization of MER-1032, Tech. Memo. EN-001–88T, Visibility Laboratory of the Scripps Institution of Oceanography, University of California, San Diego, 1988.
[60] J.L. Mueller, Comparison of irradiance immersion coefficients for several Marine environmental radiometers (MERs), in: Case Studies for SeaWiFS Calibration and Validation, vol. 27, NASA Goddard Space Flight Center, Greenbelt, 1995, p. 46. TM-1995-104566, part 3.
[61] G. Zibordi, S.B. Hooker, J.L. Mueller, S. McLean, G. Lazin, Characterization of the immersion factor for a series of in water optical radiometers, J. Atmos. Oceanic Technol. 21 (2004b) 501–514.
[62] S.B. Hooker, G. Zibordi, Advanced methods for characterizing the immersion factor of irradiance sensors, J. Atmos. Oceanic Technol. 22 (2005a) 757–770.
[63] R.W. Austin, Air–Water Radiance Calibration Factor, Technical Memorandum ML - 76–004T, Scripps Institution of Oceanography, La Jolla, CA, 1976.
[64] G. Zibordi, Immersion factor of in–water radiance sensors: assessment for a class of radiometers, J. Atmos. Oceanic Technol. 23 (2006) 302–313.
[65] G. Zibordi, M. Darecki, Immersion factor for the RAMSES series of hyper-spectral underwater radiometers, J. Opt. a: Pure Appl. Opt. 8 (2006) 252–258.
[66] J.H. Walker, R.D. Saunders, J.K. Jackson, D.A. McSparron, NBS Measurement Services: Spectral Irradiance Calibrations, NBS SP 250-20, 1987, 102 pp.
[67] F. Grum, R.J. Becherer, Optical Radiation Measurements, Academic Press, 1979.
[68] H.W. Yoon, G.D. Graham, R.D. Saunders, Y. Zong, E.L. Shirley, The distance dependences and spatial uniformities of spectral irradiance standard lamps, in: Proc. Earth Observing Systems XVII, vol. 8510, SPIE, 2012, p. 13.
[69] P. Manninen, J. Hovila, L. Seppala, P. Kärhä, L. Ylianttila, E. Ikonen, Determination of distance offsets of diffusers for accurate radiometric measurements, Metrologia 43 (2006) S120–S124.
[70] P. Manninen, P. Kärhä, E. Ikonen, Determining the irradiance signal from an asymmetric source with directional detectors: application to calibrations of radiometers with diffusers, Appl. Opt. 47 (2008) 4714–4722.
[71] H.W. Yoon, D.W. Allen, G.P. Eppeldauer, B.K. Tsai, The Extension of the NIST BRDF scale from 1100 nm to 2500 nm, in: SPIE Optical Engineering+Applications, International Society for Optics and Photonics, 2009, p. 745204.

[72] J. Dera, W. Wensierski, J. Olszewski, A two-detector integrating system for optical measurements in the sea, Acta Gephysica Polonica 20 (1972) 3−159.
[73] N.G. Jerlov, Marine Optics, vol. 14 of Oceanography, Elsevier, 1976.
[74] J.E. Tyler, Light in the Sea, Dowden, Hutchinson and Ross, Inc., 1977.
[75] R.C. Smith, K.S. Baker, The analysis of ocean optical data, in: Proc. Ocean Optics VII, vol. 478, SPIE, 1984, pp. 119−126.
[76] R.C. Smith, K.S. Baker, Analysis of ocean optical data II, in: Proc. Ocean Optics VIII, vol. 637, SPIE, 1986, pp. 95−107.
[77] A. Morel, B. Gentili, Diffuse reflectance of oceanic waters. II Bidirectional Aspects, Appl. Opt. 32 (1993) 6864−6879.
[78] A. Morel, B. Gentili, Diffuse reflectance of ocean waters III: Implication of bidirectionality for the remote-sensing problem, Appl. Opt. 35 (1996) 4850−4862.
[79] A. Morel, D. Antoine, B. Gentili, Bidirectional reflectance of oceanic waters: accounting for raman emission and varying particle scattering phase function, Appl. Opt. 41 (2002) 6289−6306.
[80] A.C.R. Gleason, K.J. Voss, H.R. Gordon, M. Twardowski, J. Sullivan, C. Trees, A. Weidemann, J.-F. Berthon, D.K. Clark, Z.P. Lee, A detailed validation of the bidirectional effect in various case I and case II waters, Opt. Express 20 (2012) 7630−7645.
[81] K.J. Voss, Use of the radiance distribution to measure the optical absorption coefficient in the ocean, Limn. Oceanogr. 34 (1989b) 1614−1622.
[82] J.R.V. Zaneveld, New developments of the theory of radiative transfer in the oceans, in: N.G. Jerlov, E. Steemen Nielsen (Eds.), Optical Aspects of Oceanography, Academic Press, New York, 1974, pp. 121−134.
[83] D.K. Clark, H.R. Gordon, K.J. Voss, Y. Ge, W. Broenkow, C. Trees, Validation of atmospheric correction over the oceans, J. Geophys. Res. 102 (1997) 17209−17217.
[84] M. Kishino, J. Ishizaka, S. Saitoh, J. Senga, M. Utashima, Verification plan for ocean color and temperature scanner atmospheric correction and phytoplankton pigment by moored optical buoy system, J. Geophys. Res. 102 (1997) 17197−17207.
[85] M.H. Pinkerton, J. Aiken, Calibration and validation of remotely-sensed observations of ocean colour from a moored data buoy, J. Atmos. Oceanic Technol. 16 (1999) 915−923.
[86] D. Antoine, P. Guevel, J.F. Desté, G. Bécu, F. Louis, A.J. Scott, P. Bardey, The "BOUSSOLE" Buoy—A new transparent-to-swell taut mooring dedicated to marine optics: design, tests, and performance at sea, J. Atmos. Oceanic Technol. 25 (2008a) 968−989.
[87] V.S. Kuwahara, G. Chang, X. Zheng, T. Dickey, S. Jiang, Optical moorings of opportunity for validation of ocean color satellites, J. Oceanogr. 64 (2008) 691−703.
[88] R.C. Smith, C.R. Booth, J.L. Star, Oceanographic biooptical profiling system, Appl. Opt. 23 (1984) 2791−2797.
[89] G. Zibordi, J.-F. Berthon, D. D'Alimonte, An evaluation of radiometric products fixed−depth and continuous in−water profile data from a coastal site, J. Atmos. Oceanic Technol. 26 (2009) 91−186.
[90] M.R. Lewis, W.G. Harrison, N.S. Oakey, D. Herbert, T. Platt, Vertical nitrate fluxes in the oligotrophic ocean, Science 234 (1986) 870−873.
[91] K.J. Waters, R.C. Smith, M.R. Lewis, Avoiding ship−induced light-field perturbation in the determination of oceanic optical properties, Oceanography (November−1990) 18−21.
[92] J.R.V. Zaneveld, E. Boss, A. Barnard, Influence of surface waves on measured and modeled irradiance profiles, Appl. Opt. 40 (2001) 442−1449.
[93] G. Zibordi, D. D'Alimonte, J.-F. Berthon, An evaluation of depth resolution requirements for optical profiling in coastal waters, J. Atmos. Oceanic Technol. 21 (2004) 1059−1073.

[94] K.J. Voss, A. Morel, Bidirectional reflectance function for oceanic waters with varying chlorophyll concentrations: measurements versus predictions, Limn. Oceanogr. 50 (2005) 698−705.

[95] R.C. Smith, Structure of solar radiation in the upper layers of the sea, in: Optical Aspects of Oceanography, Academic Press, 1974.

[96] J. Wei, R. Van Dommelen, M.R. Lewis, S. McLean, K.J. Voss, A new instrument for measuring the high dynamic range radiance distribution in near-surface sea water, Opt. Express 20 (2012) 27024−27038.

[97] D. Antoine, A. Morel, E. Leymarie, A. Houyou, B. Gentili, S. Victori, J.-P. Buis, S. Meunier, M. Canini, D. Crozel, B. Fougnie, P. Henry, Underwater radiance distributions measured with miniaturized multispectral radiance cameras, J. Atmos. Oceanic Technol. 30 (2013) 74−95.

[98] A. Morel, In−water and remote measurements of ocean color, Bound-Layer Meteorol. 18 (1980) 177−201.

[99] K.L. Carder, R.G. Steward, A remote sensing reflectance model of a red tide dinoflagellate off West Florida, Limnol. Oceanogr. 30 (1985) 286−298.

[100] J. Rhea, W. Davis, A comparison of the SeaWiFS chlorophyll and CZCS pigment algorithms using optical data from the 1992 JGOFS equatorial pacific time series, Deep Sea Res. Part Topical Stud. Oceanogr. 44 (1997) 1907−1925.

[101] M. Sydor, R.A. Arnone, R.W. Gould, G.E. Terrie, S.D. Ladner, C.G. Wood, Remote-sensing technique for determination of the volume absorption coefficient of turbid water, Appl. Opt. 37 (1998) 4944−4950.

[102] C.D. Mobley, Estimation of the remote sensing reflectance from above−water methods, Appl. Opt. 38 (1999) 7442−7455.

[103] B. Fougnie, R. Frouin, P. Lecomte, P.Y. Deschamps, Reduction of skylight reflection effects in the above−water measurement of diffuse marine reflectance, Appl. Opt. 38 (1999) 3844−3856.

[104] D.A. Toole, D.A. Siegel, D.W. Menzies, M.J. Neumann, R.C. Smith, Remote-sensing reflectance determinations in the coastal ocean environment: impact of instrumental characteristics and environmental variability, Appl. Opt. 39 (2000) 456−469.

[105] S.B. Hooker, G. Lazin, G. Zibordi, S. McClean, An evaluation of above and in−water methods for determining water leaving radiances, J. Atmos. Oceanic Technol. 19 (2002) 486−515.

[106] G. Zibordi, S.B. Hooker, J.-F. Berthon, D. D'Alimonte, Autonomous above−water radiance measurement from an offshore platform: a field assessment experiment, J. Atmos. Oceanic Technol. 19 (2002) 808−819.

[107] P.Y. Deschamps, B. Fougnie, R. Frouin, P. Lecoomte, C. Verwaerde, SIMBAD: a field radiometer for satellite ocean-color validation, Appl. Opt. 43 (2004) 4055−4069.

[108] S.B. Hooker, G. Zibordi, J.F. Berthon, J.W. Brown, Above−water radiometry in shallow coastal waters, Appl. Opt. 21 (2004) 4254−4268.

[109] G. Zibordi, F. Mélin, S.B. Hooker, D. D'Alimonte, B. Holben, An autonomous above−water system for the validation of ocean color radiance data, IEEE Trans. Geosci. Remote Sensing 42 (2004) 401−415.

[110] G. Zibordi, Comment on Long Island sound coastal Observatory: assessment of above-water radiometric measurement uncertainties using collocated multi and hyperspectral systems, Appl. Opt. 51 (2012) 3888−3892.

[111] Z.P. Lee, Y.-H. Ahn, C. Mobley, R. Arnone, Removal of surface-reflected light for the measurement of remote-sensing reflectance from an above-surface platform, Opt. Express 18 (2010) 26313−26324.

[112] M.R. Lewis, J. Wei, R. Van Dommelen, K.J. Voss, A quantitative estimation of the underwater radiance distribution, J. Geophys. Res. 116 (2011) C00H06. http://dx.doi.org/10.1029/2011JC00727.

[113] R.H. Stavn, A.D. Weidemann, Optical modeling of clear ocean light fields: raman scattering effects, Appl. Opt. 27 (1988) 4002–4011.

[114] H.R. Gordon, Diffuse reflectance of the ocean: the theory of its augmentation by chlorophyll a fluorescence at 685 nm, Appl. Opt. 18 (1979) 1161–1166.

[115] R.W. Austin, The remote sensing of spectral radiance from below the ocean surface, in: Optical Aspects of Oceanography, Academic Press, 1974.

[116] G. Zibordi, B. Holben, I. Slutsker, D. Giles, D. D'Alimonte, F. Mélin, J.-F. Berthon, D. Vandemark, H. Feng, G. Schuster, B. Fabbri, S. Kaitala, J. Seppälä, AERONET-OC: a network for the validation of ocean color primary radiometric products, J. Atmos. Oceanic Technol. 26 (2009) 1634–1651.

[117] R.W. Gould, R.A. Arnone, M. Sydor, Absorption, scattering and remote-sensing reflectance relationships in coastal waters: testing a new inversion algorithm, J. Coastal Res. 17 (2001) 328–341.

[118] K.G. Ruddick, V. De Cauwer, Y.J. Park, Seaborne measurements of near infrared water-leaving reflectance: the similarity spectrum for turbid waters, Limnol. Oceanogr. 51 (2006) 1167–1179.

[119] A. Morel, L. Prieur, Analysis of variations in ocean color, Limnol. Oceanogr. 22 (1977) 709–722.

[120] H.R. Gordon, D.K. Clark, Clear water radiances for atmospheric correction of coastal zone color scanner imagery, Appl. Opt. 20 (1981) 4175–4180.

[121] G. Thuillier, M. Herse, D. Labs, T. Foujols, W. Peetermans, D. Gillotay, P.C. Simon, H. Mandel, The solar spectral irradiance from 200 to 2400 nm as measured by the SOLSPEC spectrometer from the ATLAS and EURECA missions, Solar Phys. 214 (2003) 1–22.

[122] J.R.V. Zaneveld, A theoretical derivation of the dependence of the remotely sensed reflectance of the ocean on the inherent optical properties, J. Geophys. Res. 100 (1995) 13135–13142.

[123] J.M. Palmer, B.G. Grant, The Art of Radiometry, SPIE Pres, Bellingham, 2010.

[124] H.R. Gordon, D.K. Clark, J.W. Brown, O.B. Brown, R.H. Evans, W.W. Broenkow, Phytoplankton pigment concentrations in the Middle Atlantic Bight: comparison of ship determinations and CZCS estimates, Appl. Opt. 22 (1983) 20–36.

[125] Joint EOSAT/NASA SeaWiFS Working Group, System concept for Wide-Field-of-View observations of ocean phenomena from space, Rep. Jt. EOSAT/NASA SeaWiFS Working Group, National Aeronautics and Space Administration, 1987, 92 pp.

[126] S.B. Hooker, W.E. Esaias, G.C. Feldman, W.W. Gregg, C.R. McClain, An overview of SeaWiFS and ocean color, in: S.B. Hooker, E.R. Firestone (Eds.), NASA Tech. Memo, vol. 1, NASA Goddard Space Flight Center, Greenbelt, MD, 1992, p. 104566.

[127] J. L. Mueller and R. W. Austin, Ocean optics protocols for SeaWiFS validation, TM-1995-104566, vol. 5 Of SeaWiFS Technical Report series, NASA Goddard Space Flight Center, Greenbelt, 1995, 46 pp.

[128] H.W. Yoon, C.E. Gibson, P.Y. Barnes, Realization of the National Institute of Standards and Technology detector based spectral irradiance scale, Appl. Opt. 41 (2002) 5879–5890.

[129] B.C. Johnson, G.D. Graham, R.D. Saunders, H.W. Yoon, E.L. Shirley, Validation of the dissemination of spectral irradiance values using FEL lamps, in: Proc. SPIE 8510, Earth Observing Systems XVII, 2012. International Society for Optical Engineering, http://dx.doi.org/10.1117/12.930801.

[130] H.W. Yoon, C.E. Gibson, NIST measurement services, spectral irradiance calibrations, NIST SP 250-89, 2011, 132 pp.
[131] L.K. Huang, R.P. Cebula, E. Hilsenrath, New procedure for interpolating NIST FEL lamp irradiances, Metrologia 35 (1998) 381–386.
[132] S.B. Hooker, S. McLean, J. Sherman, M. Small, G. Lazin, G. Zibordi, J.W. Brown, The Seventh SeaWiFS Intercalibration Round-robin Experiment (SIRREX-7), TM-2003-206892, vol. 17, NASA Goddard Space Flight Center, Greenbelt, 2002, p. 69.
[133] J.J. Butler, S.W. Brown, R.D. Saunders, B.C. Johnson, S.F. Biggar, E.F. Zalewski, B.L. Markham, P.N. Gracey, J.B. Young, R.A. Barnes, Radiometric measurement comparison on the integrating sphere source used to calibrate the moderate resolution imaging spectroradiometer (MODIS) and the Landsat 7 enhanced thematic m plus (ETM+), J. Res.-N. I. S. T. 108 (2003) 199–228.
[134] G. Zibordi, B. Bulgarelli, Effects of cosine error in irradiance measurements from field ocean color radiometers, Appl. Opt. 46 (2007) 5529–5538.
[135] H. Schenck, On the focusing of sunlight by ocean waves, J. Opt. Soc. Am. 47 (1957) 653–657.
[136] R.L. Snyder, J. Dera, Wave-induced light-field fluctuations in the sea, J. Opt. Soc. Am. 60 (1970) 1072–1079.
[137] D. Stramski, J. Dera, On the mechanism for producing flashing light under a wind disturbed water surface, Oceanologia 25 (1988) 5–21.
[138] R.E. Walker, Marine Light Field Statistics, John Wiley & Sons, Inc., 1994.
[139] D. D'Alimonte, G. Zibordi, T. Kajiyama, J.C. Cunha, Monte Carlo code for high spatial resolution ocean color simulations, Appl. Opt. 49 (2010) 4936–4950.
[140] J. Dera, D. Stramski, Maximum effects of sunlight focusing under a wind-disturbed sea surface, Oceanologia 23 (1986) 15–42.
[141] A. Weidemann, R. Hollman, M. Wilcox, B. Linzell, Calculation of near surface attenuation coefficients: the influence of wave focusing, in: Proc. Ocean Optics X, vol. 1302, SPIE, 1990, pp. 492–504.
[142] J. Dera, S. Sagan, D. Stramski, Focusing of sunlight by the sea surface waves: new results from the Black sea, Oceanologia 34 (1993) 13–25.
[143] M. Darecki, D. Stramski, M. Sokólski, Measurements of high-frequency light fluctuations induced by sea surface waves with an Underwater Porcupine Radiometer System, J. Geophys. Res. 116 (2011) C00H09. http://dx.doi.org/10.1029/2011JC007338.
[144] J.H. Morrow, C.R. Booth, R.N. Lind, S.B. Hooker, The compact-optical profiling system (C-OPS), NASA Tech. Memo. 2010–215856, in: J.H. Morrow, S.B. Hooker, C.R. Booth, G. Bernhard, R.N. Lind, J.W. Brown (Eds.), Advances in Measuring the Apparent Optical Properties (AOPs) of Optically Complex Waters, NASA Goddard Space Flight Center, Greenbelt, Maryland, 2010, pp. 42–50.
[145] S.B. Hooker, J.H. Morrow, A. Matsuoka, Apparent optical properties of the Canadian Beaufort Sea – Part 2: the 1% and 1 cm perspective in deriving and validating AOP data products, Biogeosciences 10 (2013) 4511–4527.
[146] D. D'Alimonte, E.B. Shybanov, G. Zibordi, T. Kajayama, Regression of in-water radiometric profile data, Opt. Express 21 (2013) 27707–27733.
[147] H.R. Gordon, K. Ding, Self-shading of in–water optical instruments, Limnol. Oceanogr. 37 (1992) 491–500.
[148] G. Zibordi, G. Ferrari, Instrument self-shading in underwater optical measurements: experimental data, Appl. Opt. 34 (1995) 2750–2754.

[149] J.P. Doyle, K.J. Voss, 3D instrument self-shading effects on in−water multi−directional radiance measurements, in: Proc. Ocean Optics XV, Monte Carlo, Arlington, VA, 2000 available from the Office of Naval Research.

[150] J. Piskozub, A.R. Weeks, J.N. Schwarz, I.S. Robinson, Self-shading of upwelling irradiance for an instrument with sensors on a sidearm, Appl. Opt. 39 (2000) 1872−1878.

[151] R.A. Leathers, T.V. Downes, C.D. Mobley, Self−shading correction for upwelling sea−surface radiance measurements made with buoyed instruments, Opt. Express 8 (2001) 561−570.

[152] R.A. Leathers, T.V. Downes, C.D. Mobley, Self-shading correction for oceanographic upwelling radiometers, Opt. Express 12 (2004) 4709−4718.

[153] C.R. McClain, G.C. Feldman, S.B. Hooker, An overview of the SeaWiFS project and strategies for producing a climate research quality global ocean biooptical time-series, Deep-sea Res. 51 (2004) 5−42.

[154] H.R. Gordon, Ship perturbation of irradiance measurements at sea. Part 1: Monte Carlo simulations, Appl. Opt. 24 (1985) 4172−4182.

[155] K.J. Voss, J.W. Nolten, G.D. Edwards, Ship shadow effects on apparent optical properties, in: Proc.. Ocean Optics VIII, vol. 637, SPIE, 1986, pp. 186−190.

[156] W.S. Helliwell, G.N. Sullivan, B. Macdonald, K.J. Voss, Ship shadowing: model and data comparison, in: Proc. Ocean Optics X, vol. 1302, SPIE, 1990, pp. 55−71.

[157] Y. Saruya, T. Oishi, K.K.M. Kishino, Y. Jodai, A. Tanaka, Influence of ship shadow on underwater irradiance fields, in: Proc Ocean Optics XIII, vol. 2963, SPIE, 1996.

[158] C.T. Weir, D.A. Siegel, A.F. Michaels, D.W. Menzies, In situ evaluation of a ships shadow, in: Proc.. Ocean Optics XII, vol. 2258, SPIE, 1994, pp. 815−821.

[159] J. Piskozub, Effect of ship shadow on in−water irradiance measurements, Oceanologia 46 (2004) 103−112.

[160] J.P. Doyle, G. Zibordi, Optical propagation within a 3-dimensional shadowed atmosphere-ocean field: application to large deployment structures, Appl. Opt. 41 (2002) 4283−4306.

[161] J. L. Mueller, Self-shading corrections for MOBY upwelling radiance measurements. Final Rep. NOAA Grant NA04NESS4400007, 2007, 33 pp.

[162] A. Kerr, M.J. Cowling, C.M. Beveridge, M.J. Smith, A.C.S. Parr, R.M. Head, J. Davenport, T. Hodgkiess, The early stages of marine biofouling and its effect on two types of optical sensors, Environ. Int. 24 (1998) 331−343.

[163] D.V. Manov, G.C. Chang, T.D. Dickey, Methods for reducing biofouling of moored optical sensors, J. Atmos. Oceanic Technol. 21 (2004) 958−968.

[164] Y.J. Park, K. Ruddick, Model of remote-sensing reflectance including bidirectional effects for case 1 and case 2 waters, Appl. Opt. 44 (2005) 1236−1249.

[165] Z. Lee, K. Du, K.J. Voss, G. Zibordi, B. Lubac, R. Arnone, A. Weidemann, An IOP-centered approach to correct the angular effects in water-leaving radiance, Appl. Opt. 50 (2011) 3155−3167.

[166] S. Hlaing, A. Gilerson, T. Harmel, A. Tonizzo, A. Weidemann, R. Arnone, S. Ahmed, Assessment of a bidirectional reflectance distribution correction of above-water and satellite water-leaving radiance in coastal waters, Appl. Opt. 51 (2012) 220−237.

[167] V. Haltrin, G.W. Kattawar, A.D. Weidemann, Modeling of elastic and inelastic scattering effects in oceanic optics, in: Proc. Ocean Optics XIII, International Society for Optics and Photonics, 1997, pp. 597−602.

[168] M. Schroeder, H. Barth, R. Reuter, Effect of inelastic scattering on underwater daylight in the ocean: model evaluation, validation, and first results, Appl. Opt. 42 (2003) 4244−4260.

[169] J.F.R. Gower, G.A. Borstad, The information content of different optical spectral ranges for remote chlorophyll fluorescence measurements, in: J.F.R. Gower (Ed.), Oceanography from Space, Plenum, New York, 1981, pp. 329–338.

[170] T. Harmel, A. Gilerson, A. Tonizzo, J. Chowdhary, A. Weidemann, R. Arnone, S. Ahmed, Polarization impacts on the water-leaving radiance retrieval from above-water radiometric measurements, Appl. Opt. 51 (2012) 8324–8340.

[171] G. Zibordi, G.P. Doyle, S.B. Hooker, Offshore tower shading effects on in–water optical measurements, J. Atmos. Oceanic Technol. 16 (1999) 1767–1779.

[172] S.B. Hooker, A. Morel, Platform and environmental effects on above–water determinations of water–leaving radiances, J. Atmos. Oceanic Technol. 20 (2003) 187–205.

[173] S.B. Hooker, G. Zibordi, Platform perturbation in above–water radiometry, Appl. Opt. 44 (2005) 553–567.

[174] S.W. Brown, S.J. Flora, M.E. Feinholz, M.A. Yarbrough, T. Houlihan, D. Peters, Y.S. Kim, J. Mueller, B.C. Johnson, D.K. Clark, The Marine Optical BuoY (MOBY) radiometric calibration and uncertainty budget for ocean color satellite sensor vicarious calibration, in: Proc. SPIE Optics and Photonics: Sensors, Systems and Next Generation Satellites XI 6744, 2007. International Society for Optical Engineering, http://dx.doi.org/10.1117/12.737400.

[175] K.J. Voss, S. McLean, M. Lewis, C. Johnson, S. Flora, M. Feinholz, M. Yarbrough, C. Trees, M. Twardowski, D. Clark, An example crossover experiment for testing new vicarious calibration techniques for satellite ocean color radiometry, J. Atmosph. Ocean. Tech. 27 (2010) 1747–1759.

[176] G. Zibordi, K. Ruddick, I. Ansko, G. Moore, S. Kratzer, J. Icely, A. Reinart, In situ determination of the remote sensing reflectance: an inter-comparison, Ocean Sci. 8 (2012) 567–586.

[177] K.J. Voss, A. Morel, D. Antoine, Detailed validation of the bidirectional effect in various Case 1 waters for application to ocean color imagery, Biogeosciences 4 (2007) 781–789.

[178] G. Zibordi, K.J. Voss, Geometric and spectral distribution of sky radiance: comparison between simulations and field measurements, Rem. Sens. Environ. 27 (1989) 343–358.

[179] T.H. Waterman, Polarization patterns in submarine illumination, Science 3127 (1954) 927–932.

[180] A. Ivanoff, T.H. Waterman, Elliptical polarization of submarine illumination, J. Mar. Res. 16 (1958) 255–282.

[181] K.L. Coulson, Polarization and Intensity of Light in the Atmosphere, A. Deepak Publishing, Hampton, VA, 1988.

[182] K.J. Voss, E.S. Fry, Measurement of the Mueller matrix for ocean water, Appl. Opt. 23 (1984) 4427–4439.

[183] J.E. O'Reilly, S. Maritorena, B.G. Mitchell, D.A. Siegel, K.L. Carder, S.A. Garver, M. Kahru, C.R. McClain, Ocean color chlorophyll algorithms for SeaWiFS, J. Geophys. Res. 103 (1998) 24937–24953.

[184] P.J. Werdell, S. Bailey, G. Fargion, C. Pietras, K. Knobelspiesse, G. Feldman, C.R. McClain, Unique data repository facilitates ocean color satellite validation, Eos Tran 84 (377) (2003) 387.

[185] G. Zibordi, F. Mélin, J.-F. Berthon, Comparison of SeaWiFS, MODIS, and MERIS radiometric products at a coastal site, Geophys. Res. Lett. (2006) L06617. http://dx.doi.org/10.1029/2006GL025778.

[186] Joint Committee for Guides in Metrology (JCGM), Evaluation of measurement Data—Guide to the expression of uncertainty in measurement, JCGM 100 (2008) 2008.

Chapter 3.2

Ship-Borne Thermal Infrared Radiometer Systems

Craig J. Donlon,[1,*] Peter J. Minnett,[2] Andrew Jessup,[3] Ian Barton,[4] William Emery,[5] Simon Hook,[6] Werenfrid Wimmer,[7] Timothy J. Nightingale,[8] Christopher Zappa[9]

[1] *European Space Agency/ESTEC, Noordwijk, The Netherlands;* [2] *Meteorology & Physical Oceanography, Rosenstiel School of Marine and Atmospheric Science, University of Miami, Miami, FL, USA;* [3] *Applied Physics Laboratory, University of Washington, Seattle, WA, USA;* [4] *CSIRO Marine and Atmospheric Research, Hobart, Tasmania, Australia;* [5] *Aerospace Engineering Sciences Department, University of Colorado, Boulder, CO, USA;* [6] *NASA Jet Propulsion Laboratory, California Institute of Technology, Pasadena, CA, USA;* [7] *Ocean and Earth Science, University of Southampton, European Way, Southampton, UK;* [8] *RAL Space STFC Rutherford Appleton Laboratory, Harwell, Oxford, Didcot, UK;* [9] *Ocean and Climate Physics Division, Lamont-Doherty Earth Observatory of Columbia University, Palisades, NY, USA*
*Corresponding author: E-mail: craig.donlon@esa.int

Chapter Outline
1. Introduction and Background 306
2. TIR Measurement Theory 311
   2.1 General Considerations 311
   2.2 SST_{skin} Ship-Borne Radiometer Measurement Challenges 317
   2.3 Practical Measurement of SST_{skin} from a Ship-Borne Radiometer 320
3. TIR Field Radiometer Design 321
   3.1 TIR Detectors 328
      3.1.1 Quantum Detectors 329
      3.1.2 Thermopile Detectors 332
      3.1.3 Pyroelectric Detectors 333
      3.1.4 Microbolometer 334
      3.1.5 Commercial Radiometer "Head" Detectors 335
   3.2 TIR Radiometer Spectral Definition 336
   3.3 Beam Shaping and Steering 341
      3.3.1 Beam Shaping 341
      3.3.2 Beam Positioning 345
   3.4 Thermal Control System 350
   3.5 An Environmental System to Protect and Thermally Stabilize the Radiometer 351
   3.6 Instrument Control and Data Acquisition 353
   3.7 A Calibration System 354
      3.7.1 External Stirred Water Bath Calibration 355

3.7.2 Self Calibrating Radiometers Using On-board Reference Radiance Sources 358
3.7.3 NNR Calibration 361
3.8 Summary 361
3.9 Additional Comments 363
4. Examples of FRM Ship-Borne TIR Radiometer Design and Deployments 363
4.1 The DAR-011 Filter Radiometer 363
4.2 The SISTeR Filter Radiometer 364
4.2.1 SISTeR Operation 366
4.2.2 Skin SST Measurements 366
4.2.3 SISTeR Mounting and Support 366
4.3 NASA JPL NNR 368
4.3.1 JPL NNR Operation 368
4.3.2 JPL NNR Calibration 369
4.4 The Calibrated Infrared In situ Measurement System 371
4.5 ISAR—Quasi Operational Ocean Field Radiometers 375
4.6 Use of Unmanned Airborne Vehicles BESST Radiometer 380
4.7 Spectroradiometers 382
4.7.1 SST_{skin} Determination Using a Spectro-Radiometer 382
4.8 Derivation of Air Temperature Using a Spectroradiometer 387
4.9 TIR Cameras 389
5. Future Directions 393
6. Conclusions 395
Acknowledgments 395
References 395

1. INTRODUCTION AND BACKGROUND

Since 1800 when Sir William Herschel discovered infrared radiation (IR), technology has gradually but systematically improved to fully utilise this part of the electromagnetic (EM) spectrum. Most notable are thermal infrared (TIR) radiometers carried by earth observation satellites operating in near-polar and geostationary orbits [1–8] that use the ~3.5–4.1 μm and ~10–12.5 μm wavebands to measure thermal emission at the top of the earth's atmosphere (TOA). Multiple missions have operated with global 1–2 day repeat coverage for more than three decades. TIR satellite data and derived sea surface temperature (SST) maps are fundamental for climate monitoring [9], numerical ocean prediction (NOP) [10], and numerical weather prediction (NWP) [11] applications. Because of this, the investment committed by many nations developing and operating TIR satellite instruments is considerable.

Thermal emission from the sea surface is related to sea surface temperature [8]. To retrieve an estimate of SST from satellite TOA TIR measurements, an algorithm [12] is required to compensate for the impact of the intervening atmosphere between the sea surface and satellite radiometer [4,13,14], the nonunity value of surface emissivity and its variability [15], and satellite instrument effects [16]. It is a complex process requiring specific algorithms that are tuned to the characteristics of each satellite instrument. SST retrieval algorithms typically use a combination of mid-infrared (MIR) (MIR: ~3.0–5.0 μm) and TIR (~8.0–12.5 μm) multispectral measurements to compensate for atmospheric emission and the atmospheric attenuation of the water-leaving signal that would otherwise introduce errors (as water vapor is opaque to TIR radiation). Cloud-free conditions are required and a combination of visible and near-infrared (NIR) wavelengths is typically used to determine and flag cloud contamination of satellite TIR measurements using a variety of statistical and probabilistic techniques [17–19]. SST maps from satellites have been produced operationally since ~1980 by several international centers for use by NOP and NWP services.

The most stringent uncertainty requirements for SST climate data records (CDRs) stem from the need for evidence-based climate monitoring [20,21]. Global-average surface warming trends (combined land-surface air temperatures and SST) are estimated to be ~0.165 K per decade when computed from data between 1979 and 2005 although significant hemispheric differences are apparent: N. hemisphere ~0.235 K per decade, S. hemisphere ~0.09 K per decade [22]. Trends computed for the period 1901–2005 yield estimates <0.1 K per decade, with little difference between hemispheres. Assuming a global surface temperature change signal of 0.1 K per decade, a global average temperature time series should be stable to much better than 0.1 K per decade in order to distinguish a warming signal from the instability of the time series. To detect such small trends, it is prudent to aim for an SST time-series stability of at least 0.03 K per decade and, if funds and technology allow, ideally 0.01 K per decade. It is understood that this level of stability is not achievable by current measurement systems, but it remains the target [21]. SST measurement stability should be sought at local spatial scales of ~1000 km or even better in addition to global averages [21]. The Global Climate Observing system (GCOS) [21] requires SST with absolute uncertainties of 0.1 K and a stability of 0.03 K per decade both over space scales of ~100–1000 km. These are challenging targets for any measurement system. Nevertheless, satellite TIR measurements of SST have proven to be one of the most reliable CDRs derived from space [23–28].

However, do satellite estimates of SST meet these challenging targets as set out by GCOS? In 1995 at the 20th Conférence Générale des Poids et Mesures [29], a recommendation was made that:

> *those responsible for studies of Earth resources, the environment, human well-being and related issues ensure that measurements made within their programs are in terms of well- characterized SI units so that they are reliable in the long term, are comparable world-wide and are linked to other areas of science and technology through the world's measurement system established and maintained under the Convention du Metre.*

This recommendation is the basis of the feasibility of producing SST CDR from any measurement of SST, because by referencing to traceable to SI standards, SST measurements from different sources taken over a period of time can be combined in a meaningful manner [30]. A critical element of this process is to ensure that prelaunch characterization and calibration of satellite TIR radiometers is comprehensive and fully traceable to SI standards (see Smith Chapter 2.3). After launch, on-board calibration systems are the only way to maintain the calibration of each satellite instrument throughout the mission (see Minnett and Smith, Chapter 2.4).

Once on-orbit, the uncertainty characteristics of the end-to-end satellite retrieval process can only be established via independent validation activities. The Committee for Earth Observation Satellites defines validation as a process of independently assessing the uncertainty of the data derived from the system outputs. Without validation, the geophysical retrieval methods, algorithms, and SST measurements derived from satellite measurements cannot be used with confidence because meaningful uncertainty estimates cannot be provided with derived measurements. It is somewhat obvious that validation is a core component of a satellite mission that cites climate as an objective (and thus should be planned for accordingly as part of the satellite mission) and validation must start at the moment satellite instrument data begin to flow until the end of the mission.

Comparing satellite measurements of SST to independent, colocated and, contemporaneous ground-based fiducial reference measurements (FRM) of SST is the practical realization of the validation process. FRM are the suite of essential and independent ground measurements that provide maximum return on investment for a satellite mission by delivering, to users, the required confidence in data products, in the form of accurate and independent validation results and estimates of uncertainty, for the duration of the mission. Ground measurements of SST colocated and contemporaneous with satellite derived measurements are the obvious FRM to maintain confidence in satellite SST measurements—especially for CDRs where several satellite missions, each with specific calibration characteristics, must be used together. *Chapter 5.2* provides a more detailed discussion of FRM and deployment strategies for ship-borne radiometers.

Yet SST is a difficult parameter to define exactly because the upper ocean (~ 10 m) has a complex and variable vertical temperature structure that is related to ocean turbulence and the air-sea fluxes of heat, moisture and momentum [31]. Standard definitions for "SST" have been agreed by the International Community [8] and are shown schematically in Figure 1. The hypothetical idealized vertical profiles of temperature in low wind speed conditions during the night and day shown in the Figure encapsulate the effects of the dominant heat transport processes and time scales of variability associated with distinct vertical and volume regimes (horizontal and temporal variability is implicitly assumed). The interface temperature (SST_{int}) is

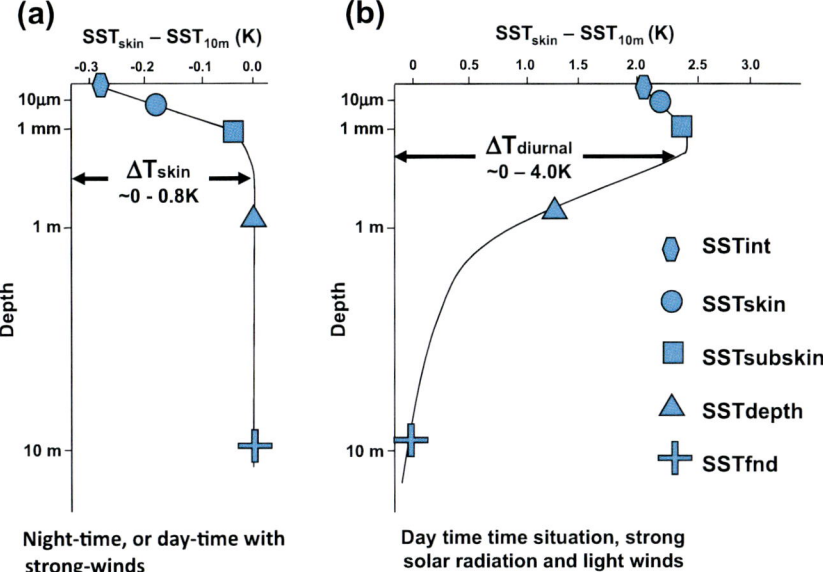

FIGURE 1 Definitions of sea surface temperature in the upper 10 m of the ocean. (a) An idealized vertical temperature profile during night—time/early morning conditions and (b) an idealized vertical profile during the early afternoon following intense solar inputs in low wind speed conditions.

a theoretical temperature at the precise air—sea interface. It represents the hypothetical temperature of the topmost layer of the ocean water and could be thought of as an even mix of water and air molecules. SST_{int} is of no practical use because it cannot be measured using current technology. However, it is important to note that it is the SST_{int} that interacts with the atmosphere.

The sea surface skin temperature (SST_{skin}) is the temperature measured by an infrared radiometer typically operating at wavelengths in the range 3.7—12 μm. It represents the temperature within the conductive diffusion-dominated sublayer [32] at a depth of ∼10—20 μm below the air—sea interface (depending on the spectral wavelength used to measure the SST_{skin}). SST_{skin} is subject to a large potential diurnal temperature cycle including cool skin layer effects (especially at night under clear skies and low wind speed conditions [31]) and warm layer effects in the daytime [33]. This definition has been chosen for consistency with the majority of infrared satellite and ship-borne radiometer measurements.

The sea surface subskin temperature ($SST_{subskin}$) is the temperature at the base of the conductive laminar sublayer of the ocean surface, that is, at a depth of approximately 1—1.5 mm below the air—sea interface. For practical purposes, this quantity can be well approximated to the measurement of surface temperature by a microwave radiometer operating in the 6−11 GHz frequency range [33], but the relationship is neither direct nor invariant to changing physical conditions or to the specific geometry of the microwave measurements. Measurements of $SST_{subskin}$ are also subject to a large potential diurnal cycle due to thermal stratification of the upper ocean layer in low wind speed high solar irradiance conditions.

All measurements of water temperature beneath the $SST_{subskin}$ are referred to as depth temperatures (SST_{depth}) that is measured using a wide variety of platforms and sensors such as drifting buoys, vertical profiling floats, or deep thermistor chains. These temperature measurements are distinct from those obtained using TIR or passive microwave radiometers (SST_{skin} and $SST_{subskin}$ respectively) and must be qualified by a measurement depth, z, in meters (e.g., SST_z e.g., SST_{5m}). Finally, the foundation SST, SST_{fnd}, is defined as the temperature of the water column free of diurnal temperature variability (daytime warming or nocturnal cooling). SST_{fnd} provides a connection with the historical concept of a "bulk" SST considered representative of the oceanic mixed layer temperature and represented by any SST_{depth} measurement within the upper ocean over a depth range of 1−20 m.

SST_{depth} measurements from the drifting buoy array [34] have been used to validate satellite SST retrievals in an operational context for many years [35]. The much larger number of drifter SST matchups compared to other in situ sources allows the inherent resolution and accuracy limitations (0.1 and 0.2 K, respectively) of drifter SST to statistically improve accuracy (assuming all drifters are measuring a "statistically stationary" ocean). However, drifting buoy SST_{depth} was never designed for satellite SST validation activities: it is not traceable to SI standards and cannot currently meet climate requirements [20,21,35]. Furthermore, near-surface temperature gradients in the upper ocean [33,36] (Figure 1) complicate the interpretation of subsurface drifter SST_{depth} (typically measured at a depth of ∼0.2 m) when compared to satellite SST_{skin} [31].

From this discussion the most appropriate FRM measurements for validating satellite SST_{skin} measurements are TIR radiometers that measure, at source, the same SST_{skin} that is measured from space, after once compensated for atmospheric effects [30]. Given the rapidly varying thermal character of the air−sea interface (time scales <10 s, [37]), SST validation requires ship-borne (or other platform) SST_{skin} measurements that are matched with satellite observations within narrow spatial and temporal limits [38]. Ideally, ship-borne measurements should be obtained at regular short time (10 min or less) intervals as block averages in order to properly sample the spatial and temporal variations of the SST field in a similar way to bulk aerodynamic flux estimates [39]. They should also properly sample the various atmospheric conditions (including both horizontal and vertical structures) for which SST retrieval algorithms are expected to function.

For satellite SST validation, GCOS requests [21] that "ship-mounted [TIR] radiometers, accurately calibrated before and after each deployment to traceable national standards, must be maintained as a truly independent reference data set for inter-calibration of follow-on satellite missions; this is particularly important where gaps in data exist between follow-on missions; a modest global array of ∼10 repeat lines in different atmospheric regimes is required; in situ radiometer sampling strategies must consider the variable nature of SST_{skin} dynamics."

Over the last 40 years a steady development of ship-borne radiometer designs has taken place leading to the present generation of instruments that are capable of target measurements with an accuracy of better than 0.1 K. To qualify as an acceptable FRM for the SST CDR, TIR ship-borne radiometers must be self-calibrating (i.e., include on-board autonomous calibration subsystems) to maintain the quality of data and be traceable to SI reference standards.

This chapter is concerned with the practical realization of ship-borne TIR radiometers that are used as FRM for satellite derived SST_{skin} measurements that contribute to the SST CDR. Separate chapters review strategies to deploy TIR field radiometers (Chapter 5.2) and the process of satellite validation (Chapter 6.2). We begin here with a summary introduction of TIR measurement theory tailored to ship-borne TIR radiometry. This is followed by a detailed discussion of best practice and engineering choices used by different authors to develop TIR field radiometer systems for use at sea. Emerging future directions are then discussed and our conclusions presented. Our aim is to provide the reader with a detailed and practical review of the key "lessons learned" over the past two decades of ship-borne TIR radiometry and assist teams contemplating ship-borne TIR FRM radiometer deployments, and instrument evolution.

2. TIR MEASUREMENT THEORY

2.1 General Considerations

Infrared radiation is emitted from all surfaces that have a temperature above 0 K ($-273.15\ °C$) and the strength of emitted radiation depends on the surface temperature (higher temperatures have greater radiant energy). IR radiation has the same optical properties as visible light, being capable of reflection, refraction, and forming interference patterns. Four general IR spectral regions can be distinguished as follows:

- NIR spectral region ranges from 0.7 to 1.4 μm.
- SWIR extends from 1.4 to 3 μm.
- MIR extends from 3 to 8 μm. The atmosphere includes many absorption lines (e.g., of carbon dioxide and water vapor) in parts of this spectral region and exhibits strong absorption.
- Long-wavelength infrared (LWIR) ranges from 8 to 15 μm, followed by the far infrared, which ranges up to ∼100 μm.

Total quantities, symbols and units provide the theoretical foundation for the measurement of infrared radiation [40] as follows (and shown schematically in Figure 2):

- Radiant energy, Q, is the total energy radiated from a point source in all directions in units of Joules (J).
- Radiant flux, $\phi = dQ/dt$, is the flux of all energy radiated in all directions from a point source in units of Watts (W).
- Emittance, $M = d\phi/dA$, is the radiant flux density from a surface area, A, in units of W m^{-2}. This is an integrated flux (i.e., independent of direction) and will therefore vary with orientation relative to a nonuniform source.
- Radiant intensity, $I = d\phi/d\omega$, is the radiant flux of a point source per solid angle ω (steradian, sr) and is a directional flux in units of W sr^{-1}.
- Radiance, $L = dI/d(A \cos \theta)$, is the radiant intensity of an extended source per unit solid angle in a given direction, θ, per unit area of the source projected in the same θ. It has units of W sr^{-1} m^{-2}.

Spectral quantities are obtained by restricting each to a specific spectral waveband.

The emittance, M, of a perfectly emitting surface (called a blackbody) at a temperature, T, in Kelvin is described by Planck's law. It is the radiant flux (ϕ)

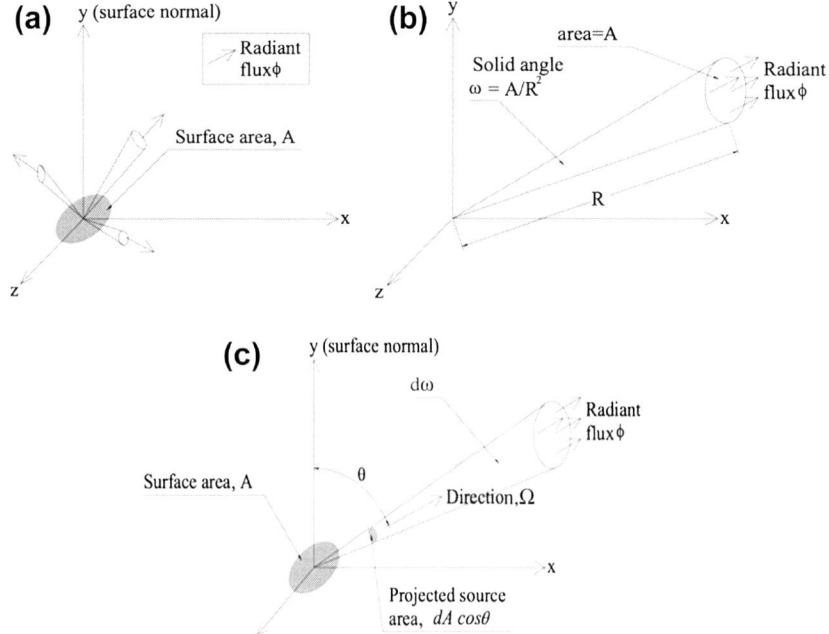

FIGURE 2 Schematic definitions of total quantities for (a) emittance, M, (b) radiant intensity, I, and (c) radiance, L. *After Ref. [41]*.

per unit bandwidth centered at wavelength, λ, leaving a unit area of surface in any direction in units of W m^{-2} m^{-1}:

$$M(\lambda, T) = \frac{2\pi hc^2}{\lambda^5 \left(e^{hc/\lambda kT} - 1\right)} \quad (1)$$

where $h = 6.626 \times 10^{-34}$ J s^{-1} is Planck's constant, $c = 2.998 \times 10^8$ m s^{-1} is the speed of light and $k = 1.381 \times 10^{-23}$ J K^{-1} is Boltzmann's constant. Figure 3 shows $M(\lambda,T)$ computed for the range of global ocean SST ($\sim$271.3 at the ice edge to $\sim$305 K the highest likely SST in an enclosed sea with extremely strong local thermal stratification e.g., Ref. [42]) as a function of wavelength. Also included are temperatures below 273 K representing cloud free (clear sky) atmospheric emission. Figure 3 is the foundation of TIR radiometer design requirements in terms of detector dynamic range, spectral range and sensitivity requirements. The spectral dependence of blackbody emission shows that the maximum emission for SST lies in the region $\sim$10.0–12.5 µm (Wein's displacement law: $\lambda_{max}T$ = 2897 µm K) and varies significantly with temperature. Emission at $\sim$3.0–5.0 µm has a larger sensitivity (change of signal per unit increase in T) when compared to $\sim$10–12 µm but with smaller spectral radiance. This is attractive for measuring SST$_{skin}$ but, it is important to recognize that in the $\sim$3.0–5.0 µm spectral region, solar radiation can be reflected at the sea surface which, under certain radiometer viewing conditions, significantly complicates the use of this spectral bandwidth. In any case, care must be taken to avoid direct solar reflection into a radiometer field of view (FoV).

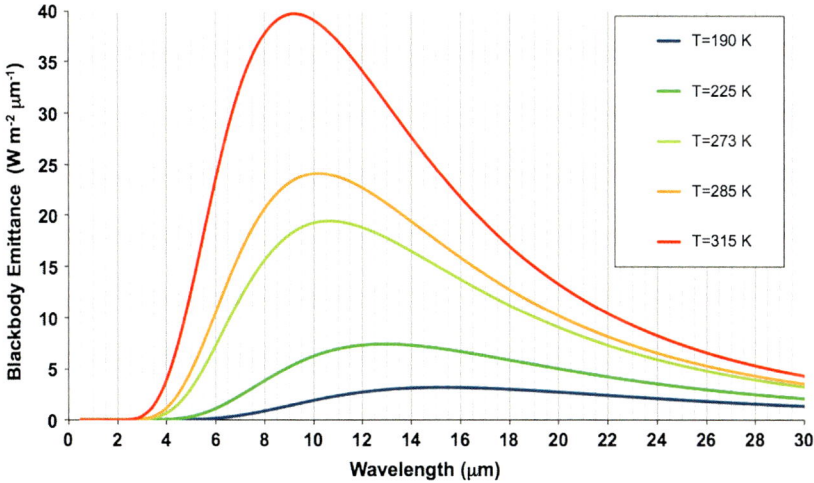

FIGURE 3 Infrared emission spectra for a blackbody at temperatures between 190 K, typical for a radiometer viewing a clear sky with a spectral band pass of 9–11 µm, and 315 K representing an extreme SST expected in an enclosed sea during extremely strong thermal stratification.

The high absorbance of water at TIR wavelengths [43,44] leads to a very small effective *optical depth* (from which most TIR emission originates) of ~10 μm at a wavelength of 10–12 μm. The optical depth increases to ~65 μm at a wavelength of ~3.5–4.1 μm [32]. The spectrally dependent TIR effective optical depth is feature that has been be used to "sound" the vertical temperature structure of the water "skin" in the laboratory [43,44]. However, in the field, this approach is difficult due to the long integration times required to make such measurements with sufficiently high signal-to-noise characteristics, and the almost ubiquitous presence of wind and waves that continuously disturb the water surface.

In the case of satellite TIR radiometers designed to measure SST from space, atmospheric attenuation due to water vapor, ozone, CO_2 and other absorbing lines [45] is significant. Figure 4 shows modeled [46] normalized atmospheric spectral transmittance for the 0–28 μm spectral region and the significant components of normalized spectral atmospheric transmittance for the 0–20 μm. Transmittance in the 3.5–4.1 μm and 10.0–12.5 μm spectral interval is high (leading to the term atmospheric "windows" [45]) which is why satellite instruments use these spectral intervals for SST measurements.

For a Lambertian source emitting uniform radiance in all directions the spectral radiance $L(\lambda)$ can be related to $M(\lambda)$ by

$$L(\lambda) = \frac{M(\lambda)}{\pi} \qquad (2)$$

Equation (1) is defined for ideal thermodynamic principles representing a perfect blackbody emitter at an actual temperature, T, and wavelength, λ. If the sea surface behaved as a perfect radiator then the absolute temperature could be determined simply by measuring $L(\lambda)$ over a finite spectral bandwidth, and inverting the Planck equation. But in practice the sea surface does not behave as a blackbody (it is slightly reflective in the infrared) and therefore its spectral and geometric properties need to be considered. By measuring $L(\lambda)$ using Eqn (2) and inverting Eqn (1) the *spectral brightness temperature*, $B(T,\lambda)$, is determined rather than the actual T of the target.

The emittance of a perfect emitter at the actual temperature T, wavelength λ, and view angle, θ is given by:

$$M_{(T,\lambda,\theta)} = \frac{\pi L_{(T,\lambda,\theta)}}{\varepsilon_{(\lambda,\theta)}} \qquad (3)$$

where the spectral emissivity, $\varepsilon(\lambda,\theta)$, can be calculated using:

$$\varepsilon(\lambda,\theta) = \frac{M_{(T,\lambda,\theta)} \text{ measured}}{M_{(T,\lambda,\theta)} \text{ blackbody}} \qquad (4)$$

$\varepsilon(\lambda,\theta)$ has a strong dependence on wavelength and viewing geometry [47]. An effective emissivity, ε, can be defined that integrates $\varepsilon(\lambda,\theta)$ over

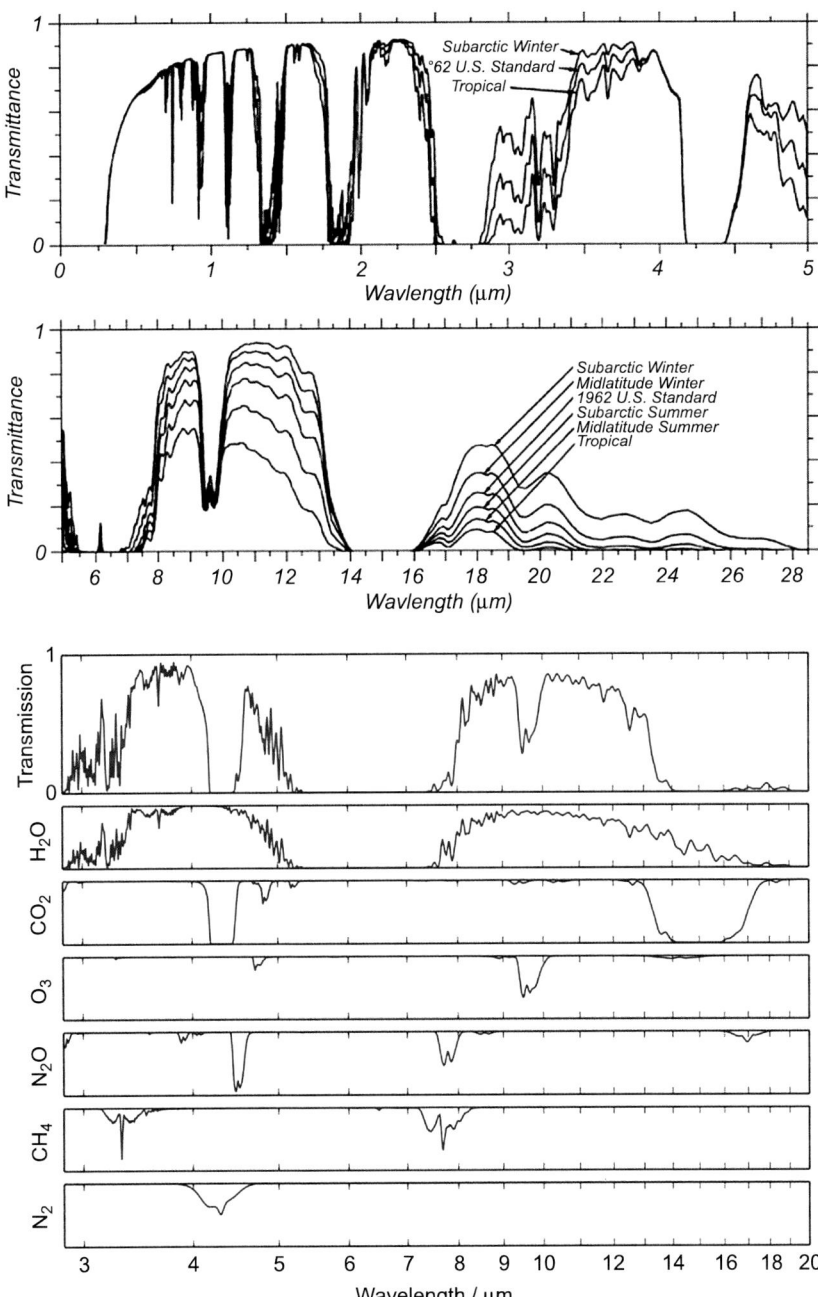

FIGURE 4 (Top) Atmospheric transmittance for a vertical path to space from sea level for six model atmospheres with very clear, 23 km, visibility, including the influence of molecular and aerosol scattering [46]. (Bottom) Components of atmospheric transmission derived from the Reference Forward Model (RFM) *(http://www.atm.ox.ac.uk/RFM/ (with permission, O. Embury))*. Note the different scales on the *x*-axis.

all wavelengths of interest for a specific view angle θ. Following Kirchoff's law:

$$1 = \int_{\lambda_i}^{\lambda_j} \rho(\theta) + \tau(\theta) + \varepsilon(\theta) \tag{5}$$

where ρ is the spectral reflectivity, τ is the spectral transmissivity and λ_i, λ_j set limits on the spectral interval of interest. For radiation incident on an opaque surface (i.e., $\tau = 0$, which to first order is true for the sea surface) ε can be calculated using

$$\varepsilon(\theta) = 1 - \rho(\theta) \tag{6}$$

Figure 5 shows the emissivity calculated for pure water using Eqn (6) as a function of both radiometer viewing angle from nadir, θ, and wavelength, λ. Pure water emissivity is slightly less than unity with a dependence on θ and λ. It has a minimum value at a wavelength of ~ 11 μm ($\rho_\lambda \approx 0.0015$) [48]. As the emissivity of the sea surface is close to unity across the wavelengths of interest, the temperature of the thin skin layer largely determines the intensity of TIR radiation leaving the sea surface that is measured by a TIR radiometer.

The difference between the reflective properties of pure water and artificial salt solutions representing seawater has been investigated by Ref. [49] at near-normal incidence angles. Results suggest that the 8–12 μm window is most affected by typical seawater solute [50] concentrations. In fact [50] go as far as to recommend that the 8–12-μm waveband should *not* used to measure SST_{skin} due to the poor characterization and sensitivity of the refractive index of sea water to temperature at larger θ. However, the effects of salinity on sea

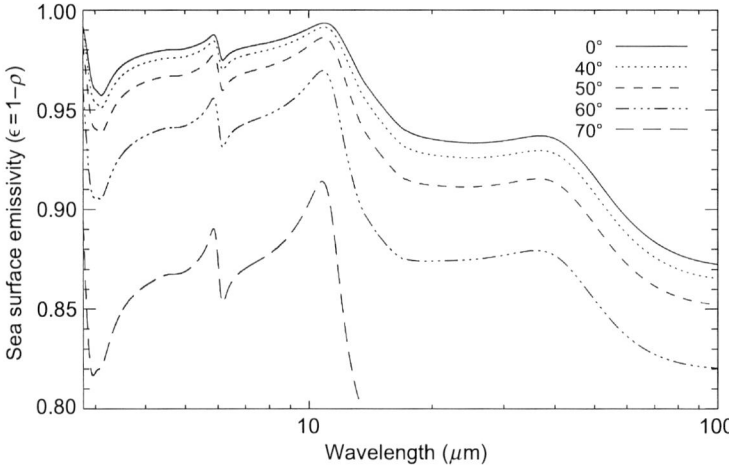

FIGURE 5 The emissivity of pure water as a function of viewing incidence angle (in the absence of surface roughness). The strong impact of incidence angles <40° and peak emissivity at ~ 11 μm is clearly visible.

surface emissivity are found to be well modeled [51] using standard refractive index corrections proposed in Ref. [52] with a significant temperature dependence is evident in the 11.5–13 μm region (although interestingly, not at ~10.5 μm as suggested by Ref. [50]).

While the spectral emission angle properties are known for *still* water surfaces at $\theta < 40°$ [48], they are poorly quantified for a *roughened* sea surface. When the sea surface is rough, radiance from many parts of the sky can be specularly reflected from appropriately oriented facets of surface waves into the radiometer FoV [53]. Numerical models [47,54–57] have been developed to consider the uncertainty associated with the variation of $\varepsilon(\lambda,\theta)$ through a sea state and wind speed dependence. For $\theta > 40°$, emissivity decreases significantly [47,54–57] although this is contested by Ref. [50] who suggest that emissivity remains constant at $\theta = 40°$ for wind speeds of 3–13 m s^{-1}.

We note that a 1% change in $\varepsilon(\lambda,\theta)$ corresponds to a change in retrieved SST$_{skin}$ of 0.66 K (at $\lambda = 10$ μm), 0.73 K (at $\lambda = 12$ μm), or 0.24 K (at $\lambda = 3.5$ μm) [50]. To approach the SST$_{skin}$ measurement accuracy required for climate research, $\varepsilon(\lambda,\theta)$ *for each measurement* must be known to better than 0.05% uncertainty in the 8–12 μm wavelength region, and better than 1% in the 3.5–4.5 μm window [50]. This is a challenge given our current knowledge of how to practically determine $\varepsilon(\lambda,\theta)$ while at sea. Thus, it is clear that more work is required to develop better knowledge of sea surface emissivity. The recent developments to address this issue include the generation of simplified schemes to calculate seawater emissivity [58], further study of polarization and cross-polarization impacts at large (>50) angles can be taken into account [59,60].

From ship-borne TIR radiometer design perspective, the above discussion suggests an optimal SST$_{skin}$ FRM measurement will be obtained when viewing a calm sea surface at θ of 15–40° (to minimize emissivity variations) in the 3.5–4.1 μm and/or the 10.5–12.5 μm spectral waveband.

2.2 SST$_{skin}$ Ship-Borne Radiometer Measurement Challenges

Because seawater emissivity is slightly less than unity, a small proportion of radiation originating from the atmosphere is reflected at the sea surface into the FoV of the radiometer complicating a simple measurement approach. If no allowance were made for reflected sky radiation the resulting SST$_{skin}$ retrieval would be erroneous: if a nonunity value for seawater emissivity is used but no correction for reflected atmospheric radiance under clear sky conditions was made, the SST$_{skin}$ would be too warm (except under low-overcast cloud conditions where the emission at the temperature SST$_{skin}$ could be close to the sky radiance). If no account of emissivity at all were made, then the SST$_{skin}$ would be too cool (because a proportion of very "cold" sky radiance reflected at the sea surface would be included in the measurement).

The previous discussion reveals that to measure SST_{skin} accurately from a ship, contemporaneous radiometric measurements of both the sea surface radiance and the downwelling atmospheric radiance must be obtained at the appropriate view-angles and, the value of seawater emissivity must be known accurately.

Consider a TIR radiometer mounted on a ship or a platform (Figure 6) at height, h, above the sea surface viewing a sea surface at temperature T_s and view angle θ. The spectral radiance components that must be considered when measuring the SST_{skin} include:

- $L_{sea}(\lambda,\theta)$: the radiance originating from the sea surface (the desired signal);
- $L_{refl}(\lambda,\theta)$: a proportion of $L_{sky}(\lambda,\theta)$ (the downwelling radiance emitted from the atmosphere) *directly* reflected at the sea surface into the radiometer FoV;
- $L_{scat}(\lambda,\theta)$: the component of reflected sky radiance from the atmospheric layer beneath the radiometer that is reflected into the sea-viewing measurement, but not included in the sky-view measurement;
- $L_{atm}(\lambda,\theta)$: radiance originating from atmospheric emission between the sea surface and radiometer at height h above the sea surface.

To measure SST_{skin} by a radiometer measuring the total upwelling radiance, $L_{up}(\lambda,\theta)$ at a view angle θ, the contribution of each spectral radiance component must be accounted for.

Assuming the atmospheric path is homogenous over the atmospheric depth h and the transmittance, τ_h, of that path is close to unity, the downwelling radiance, $L_{down}(\lambda,\theta)$, incident on the sea surface is given by:

$$L_{down}(\lambda, \theta) \approx \tau_h L_{sky}(\lambda, \theta) + (1 - \tau_{path})\overline{B(T_{air}, \lambda)} \qquad (7)$$

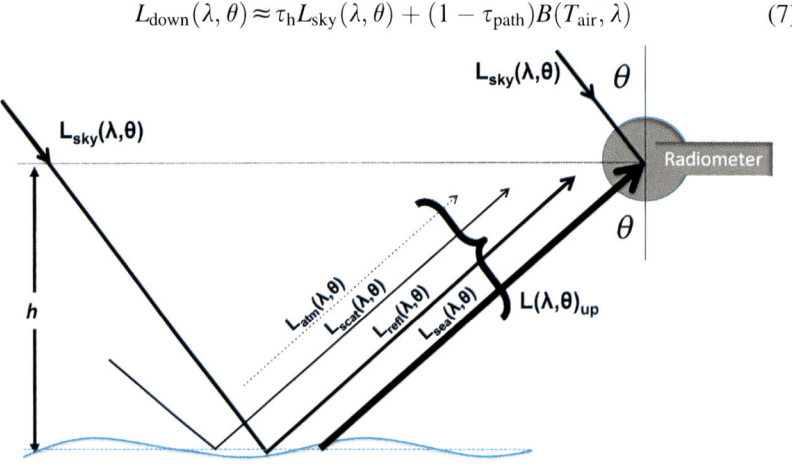

FIGURE 6 Geometrical arrangement of radiative components that must be considered when measuring the radiative temperature of the ocean surface. Symbols are explained in the text.

where $B(T,\lambda)$ is the Planck function providing the spectral radiance emitted by a blackbody and $\overline{B(T_{air},\lambda)}$ is the spectral radiance emitted at air temperature T_{air}, averaged over the atmospheric path. Typically, in order to proceed practically, $L_{scat}(\lambda,\theta)$ is neglected and $L_{refl}(\lambda,\theta)$ is assumed to represent the average of all reflected sky radiance (i.e., that from the direction that reflects in a calm sea surface). This approach has proven satisfactory for many authors [31,38,39,61−74].

The upwelling radiance, $L_{up}(\lambda,\theta)$ from the sea surface is given by

$$L_{up}(\lambda,\theta) = \varepsilon(\lambda,\theta)B(SST_{skin},\lambda) + (1-\varepsilon(\lambda,\theta))L_{down}(\lambda,\theta) \tag{8}$$

The spectral radiance arriving at the radiometer aperture in the direction of the sea surface is then:

$$L_{scene}(\lambda,\theta) \approx \tau_h L_{up}(\lambda,\theta) + (1-\tau_{path})\overline{B(T_{air},\lambda)} \tag{9}$$

$$L_{scene}(\lambda,\theta) = \tau_{path}\varepsilon(\lambda,\theta)B(SST_{skin},\lambda) + \tau_{path}^2(1-\varepsilon(\lambda,\theta))L_{sky}(\lambda,\theta) \\ + (1-\tau_h)[(1-\varepsilon(\lambda,\theta))\tau_h + 1]\overline{B(T_{air},\lambda)} \tag{10}$$

As τ_h approaches unity, $L_{sea}(\lambda,\theta)$ is given by:

$$L_{scene}(\lambda,\theta) = \varepsilon(\lambda,\theta)B(SST_{skin},\lambda) + (1-\varepsilon(\lambda,\theta))L_{sky}(\lambda,\theta) \tag{11}$$

If $h < \sim 40$ m, considered an upper bound for ship and platform deployments [75], and the relative humidity is below 95%, τ_h is very close to unity for IR measurements in the region 8−12 μm. A multiband radiometer or spectroradiometer could be used to account for the effect of $L_{atm}(\lambda,\theta)$ explicitly using a multispectral difference algorithm (e.g., $a_0 + a_1(11.0-12.0$ μm) $(\sec(\theta)-1))$ if the system were sufficiently sensitive [76,77]. The assumption $\tau_h = 1$ in Eqn (12) when the radiometer is close (<15 m) to the sea surface is considered valid [77] but, when radiometer deployment heights are much above this, the assumption introduces errors of into SST_{skin} retrievals. While this error is small, it is not insignificant when the goal is an uncertainty of 0.1 K.

Equation (11) requires that the emissivity for a given spectral interval and viewing geometry is known accurately. The approach described above assumes that the ocean surface is flat and that $L_{refl}(\lambda,\theta)$ originates from angle θ [50,53]. Discuss the emissivity and SST_{skin} errors associated with poor knowledge of radiometer viewing geometry related to wind speed, cloud cover and sea state effects. For radiometer θ of 55° or greater (particularly in the 8−12 μm wavelength range) an angular offset of ±3−4° can result in SST errors of up to 0.6 K [50] highlighting the sensitivity of emissivity values at large θ. If θ is $\sim$15−40° the impact of a wind roughened surface and ship movement will be considerably reduced [53] except in heavy seas. Under clear sky or overcast sky conditions errors associated with poor knowledge of emissivity are limited by the assumed homogenous emission from the atmosphere. However, even at small θ, if scattered clouds are present, significant errors of up to 0.3 K may still occur unless truly contemporaneous measurements of the sea surface and sky are made, ship/platform movements are small and, the time difference between such measurements is very small. Even then, the distribution of reflected sky radiance signals on the sea surface (which will originate from an area) depends on the roughness of the sea surface that dictates the atmospheric source. Further work is required to systematically reduce uncertainties in the values used for $\varepsilon(\lambda,\theta)$, particularly when the sea surface is rough (almost always). This remains the largest source of uncertainty in the determination of SST_{skin} from ship-borne radiometers (assuming a well calibrated radiometer).

For these reasons, ship-borne TIR radiometers typically (but not exclusively) observe the sea surface at $\theta = \sim 15-55°$. At lower θ, direct reflection of the ship superstructure at the sea surface and into the radiometer FoV can be significant [37]. Furthermore, it is difficult to view an undisturbed area of the sea surface that is free of the ships bow wave and wake. When $\theta > 55°$, sea surface emissivity is dramatically reduced (Figure 5) and interpretation of TIR measurements critically depends on accurate knowledge of both θ and emissivity.

2.3 Practical Measurement of SST_{skin} from a Ship-Borne Radiometer

Consider the signal measured by a single channel TIR field radiometer detector designed to measure SST_{skin}. The radiometer spectral response function, $\zeta(\lambda)$, is defined by the combined detector, pass-band filter and all optical components (e.g., mirrors, protective windows, detector band-pass etc.). The signal output from the detector, in output-units per unit radiance, S_{sea}, when viewing the sea surface is then:

$$S_{sea} = \int_{\lambda_1}^{\lambda_2} \zeta(\lambda)\left[\varepsilon(\lambda,\theta)B(SST_{skin},\lambda) + (1-\varepsilon(\lambda,\theta))L_{sky}(\lambda)\right]d\lambda \quad (12)$$

where the limits of integration are chosen to span the spectral bandwidth $\zeta(\lambda)$. The detector output when viewing the sky, S_{sky} is:

$$S_{sky} = \int_{\lambda_1}^{\lambda_2} \zeta(\lambda)L_{sky}(\lambda)d\lambda \quad (13)$$

Assuming a narrow waveband radiometer (e.g., 10.5–12.5 μm) viewing the sea surface at an angle <40°, $\varepsilon(\lambda,\theta)$ and $B(\lambda,T)$ vary only slowly with wavelength. Equation (12) can then be separated, to a good approximation, into a combination of the band-averaged values $\varepsilon_B(\theta)$, L_{sky} and $B_B(T)$ giving:

$$S_{sea} = \zeta_B \left[\varepsilon_B(\theta) B_B(SST_{skin}) + (1 - \varepsilon_B(\theta)) L_{B,sky} \right] \quad (14)$$

where

$$\zeta_B = \int_{\lambda_1}^{\lambda_2} \zeta(\lambda) d\lambda \quad (15a)$$

$$\zeta_B \varepsilon_B(\theta) = \int_{\lambda_1}^{\lambda_2} \zeta(\lambda) \varepsilon(\lambda, \theta) d\lambda \quad (15b)$$

$$\zeta_B L_{B,sky} = \int_{\lambda_1}^{\lambda_2} \zeta(\lambda) L_{sky}(\lambda) d\lambda \quad (15c)$$

$$\zeta_B B_B(T) = \int_{\lambda_1}^{\lambda_2} \zeta(\lambda) B(\lambda, T) d\lambda \quad (15d)$$

Equation (13) can be written as.

$$S_{sky} = \zeta_B L_{B,sky} \quad (16)$$

so that, finally

$$\zeta_B B_B(SST_{skin}) = \frac{S_{sea} - (1 - \varepsilon_B(\theta)) S_{sky}}{\varepsilon_B(\theta)} \quad (17)$$

In Eqn (17) two fundamental measurements, S_{sea} and S_{sky} in one or more spectral bands, are required to determine the SST_{skin}. Both must be obtained near-contemporaneously by viewing the sea surface at the incidence angle θ and the atmosphere at the zenith angle θ. As discussed by Ref. [53], the time difference between S_{sea} and S_{sky} measurements must be small to limit errors associated with rapidly changing atmospheric radiance conditions (i.e., varying clouds of different species and height have different radiative temperatures that are reflected at the sea surface into the radiometer FoV).

3. TIR FIELD RADIOMETER DESIGN

This section is devoted to a critical appraisal of successful "best-practice" design employed by TIR field radiometer engineers using examples of SST radiometers throughout. Our aim is to demonstrate to the reader the strategies employed to deliver an accurate, well-understood instrument that can be relied upon to provide an FRM for satellite validation activities of a standard suitable for use in the generation of CDRs.

Ship-borne TIR radiometers can be categorized as filter radiometers measuring radiance at one or more spectral waveband "channels," spectroradiometers that measure the spectral power distribution of radiance at fine spectral resolution over a large waveband, and thermal camera/imaging systems that generate a two-dimensional thermal image "map" of the sea surface for a specified spectral interval. There is a trade-off between the measurement capability of an instrument (e.g., a spectroradiometer provides considerably more information than a filter radiometer and an imager provides a 2D field of SST_{skin} that can provide an animation of SST_{skin} dynamics (e.g., Ref. [78])), its deployment autonomy (an autonomous instrument will not require an engineer to accompany it during a deployment), its physical size (large instruments cannot easily be deployed from some ships and cost more to transport) and its cost (to procure, deploy, and maintain).

Thus, the marine infrared instrument engineer has various options available to implement a field radiometer design trading solutions based on cost, capability, performance, durability in the field and ease of use. For example, the development of an accurate spectroradiometer leads to a high performance, high capability instrument design that is typically complex (with related maintenance-at-sea issues) and large in size (and thus can be challenging to deploy). Although modern spectroradiometer designs are durable, they are costly. The design remit for such an instrument would be for research measurements in support of air—sea interaction studies and satellite SST validation. If the focus is to develop an instrument specifically for satellite SST validation activities that is capable of long (3—6 months) autonomous deployments at sea, performance must be maintained, but the design has to be robust against the harsh marine environment. The instrument capability can be reduced (e.g., by the use of just one or two spectral channels) and the size can typically be much smaller. Costs are then low to moderate for such an instrument. The researcher requiring specific local measurements may even choose to use an "off the shelf" solution such as hand-held TIR radiometers [62] or thermal cameras [37] that can be deployed manually when conditions are favorable.

Figure 7 provides a view of several ship-borne TIR radiometer designs that have been developed and used successfully over the past 30 years to collect ship-borne measurements of SST_{skin}.

The most notable ship-borne TIR radiometer designs include the DAR011 radiometer [85], the Rutherford Appleton Laboratory (RAL)/Satellites International Radiometer (SIL [65,86]), the Scanning Infrared Sea surface Temperature Radiometer (SISTeR [87]), the Calibrated Infrared in situ Measurement System (CIRIMS [63]), the NASA/JPL Near Nulling Radiometer (NNR) [80—82] the Infrared SST Autonomous Radiometer (ISAR

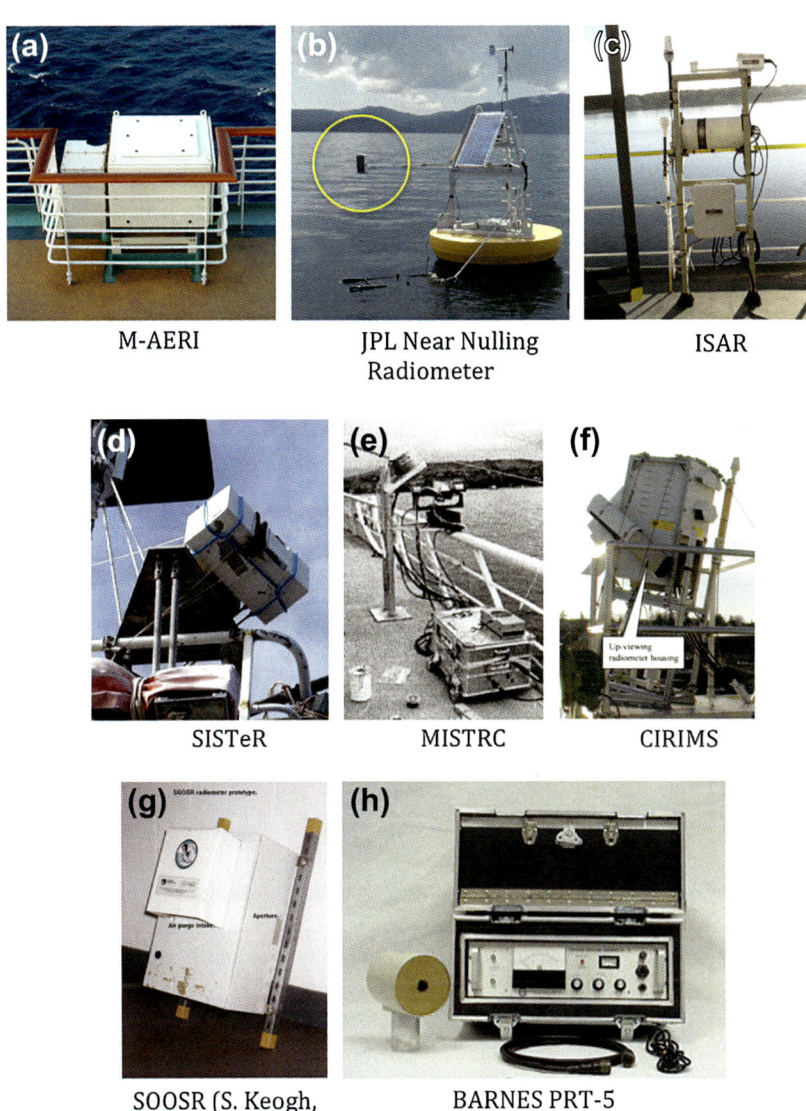

FIGURE 7 A selection of different design thermal infrared radiometers used for SST validation activities. (a) The marine atmospheric emitted radiance interferometer (M-AERI [79]), (b) the JPL nulling radiometer [80–82] mounted on a moored buoy (c) the infrared SST autonomous radiometer (ISAR [62]), (d) The scanning infrared sea surface temperature radiometer (SISTeR) radiometer installed on the foremast of the RRS James Clark Ross, (e) The multichannel infrared sea truth radiometric calibrator (MISTRC) [74] experimental setup aboard the R/V Vickers showing the radiometer head supported by deck mounting, calibration bucket subsystem located on the ship handrail, and main electronics unit located on the deck, (f) the calibrated infrared radiometer in situ measurement system (CIRIMS [63]), (g) the SOOSTR radiometer design [62], (h) Barnes model PRT-5 radiometer [83,84] showing radiometer head and electronics unit. *Image from Pyrometer Instrument Company, http://www.pyrometer.com, with permission.*

[38,62]), the Multi-channel Infrared Sea Truth Radiometric Calibrator (MISTRC [74]), the Marine Atmospheric Emitted Radiance Interferometer (M-AERI [79]), and the Ship of Opportunity SST Radiometer (SOOSR [62,88]). A notable design is the Ball Experimental SST radiometer (BESST [89]) which is a compact lightweight design for use on Unmanned Autonomous Vehicles (UAV) drones, which is discussed in Section 4. The main characteristics of each design are summarized in Table 1.

Seven fundamental subsystems are common to all infrared radiometer instruments used to measure SST_{skin} in the marine environment:

1. A detector to measure spectral radiance,
2. A means to define the spectral wavelength(s) of measurement (filter, spectrometer),
3. A fore-optics system to shape the beam, direct the radiometer FoV to a target and to focus the radiance measured by the radiometer onto the detector,
4. An electronics system to control the radiometer and log measurements.
5. A thermal control system,
6. Mounting and environmental protection from the marine environment,
7. A calibration system to quantify the detector output.

Figure 8 shows a schematic overview of the generic relationship between each of these subsystems that are "hierarchically embedded" within each other. The radiometer needs a housing and protection from the marine environment. The instrument thermal control system (often including the housing design) must be capable of managing the internal thermal environment of the instrument at a temperature that does not rapidly fluctuate to ensure good instrument calibration characteristics. The calibration system, which for climate applications must be traceable to SI standards, has to calibrate the instrument in an end-to-end manner so that all optical, spectral, detector (full aperture) and electronic subsystems impinge on the calibration subsystem design.

Each of these elements will be considered in separate sections below referring to different design choices used by ship-borne TIR radiometers. As a basis for this discussion, Figure 9 shows a schematic overview of a common TIR filter radiometer design. A parabolic focusing mirror is used to focus a collimated beam from the target radiance onto a chopped pyroelectric detector system protected by a window and band-pass filter. In this case, the parabolic mirror acts as the field stop defined by the physical limits of the mirror itself. The detector has an integrated aperture stop. A rotating plane mirror, angled at 45° to the radiometer beam, mounted on a rotating shaft is used to select a target view of sea surface, sky or one of two calibration blackbody reference cavities. The instrument uses a series of baffles to limit stray light originating outside of the main beam.

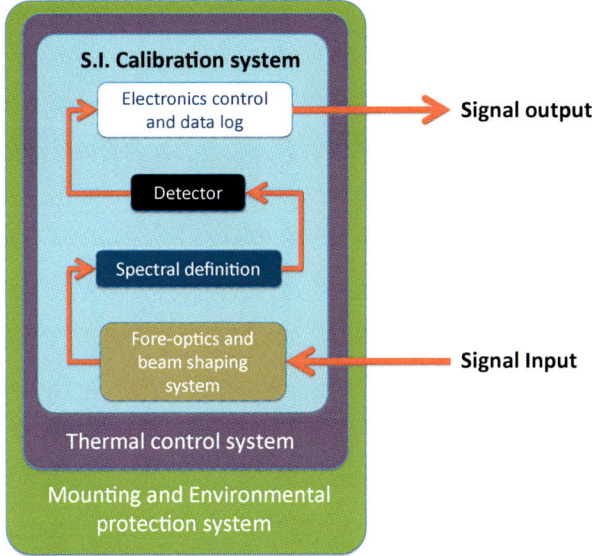

FIGURE 8 Schematic overview of the "embedded" relationship between ship-borne radiometer subsystems.

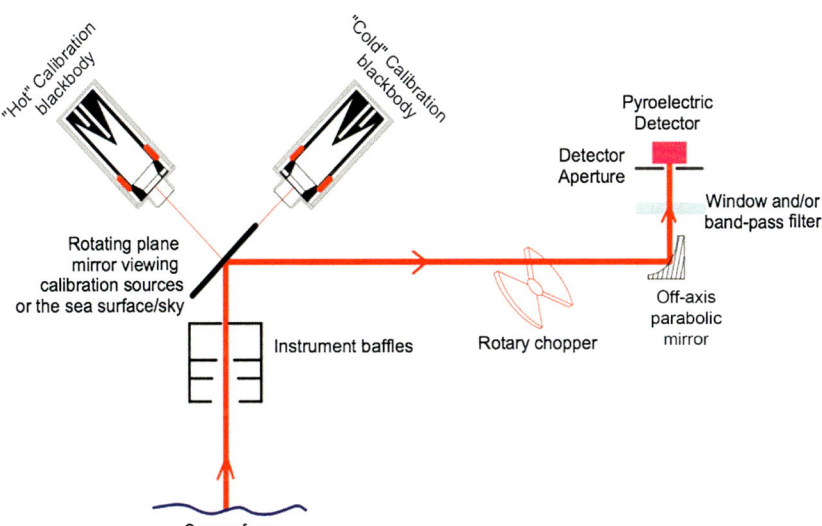

FIGURE 9 Simplified schematic overview of the signal path and components of a thermal infrared filter radiometer design.

TABLE 1 Basic Design Characteristics of Example Ship-Borne Thermal Infrared Radiometers Used for Satellite SST Validation

Instrument Name and Date	Spectral Definition	Passband (µm)	Detector Type	Beam Forming	Beam Steering	Calibration System	Typical View Nadir Angle(s)	Comments
PRT-5 [83,84,90] (1970)	Filter	Single channel 9.5–11.5	Bolometer	Objective lens	Non (manual pointing)	Stirred water bath	53°	Requires external calibration (often a water bath).
RAL/SIL [65,86,91] (1988)	Filter	Two channels 10–12	Pyroelectric	Gershun tube and off-axis parabolic mirror	Non (manual pointing)	Three internal black body cavities	Mechanically set (typical >15° and <40°)	No sky view unless radiometer manually pointed
DAR-011 [85] (1992)	Filter	SinIge channel 10.5–11.5	Pyroelectric	Gershun tube and off-axis parabolic mirror	Non (manual pointing)	Three internal black body cavities	Mechanically set (typical 45° this was the angle almost exclusively used on deployments to view outside the ship's wake.)	Sky view on opposite side from the sea view i.e., not at the most optimal θ
SISTeR [87] (1995)	Filter	Up to five channels 10.3–11.3	Pyroelectric	Off-axis ellipsoid mirror	Programmable plane rotating mirror	Two internal black body cavities	Programmable (typical >15° and <40°)	Research and autonomous deployment
MISTRC [74] (1995)	Filter: polarized filters available for the 3.7 and 4.02 channels	Four channels 3.728, 4.025, 10.75, 11.78	HgCdTe	Objective lens	Non (manual pointing)	External stirred seawater bath	Mechanically set (typical >15° and <40°)	Very large experimental demonstration system. Water bath calibration system challenging to deploy.

Instrument	Spectral selection	Channels (μm)	Detector	Optics	Scanning	Calibration	View angles	Notes
SOOSR [62,88] (1998)	Filter	Single channel 8.0–12.0	TASCO THI-500L thermopile	Ge AR lens	Non (manual pointing)	Two internal black body cavities	Mechanically set (typical >15° and <40°)	Low cost experimental demonstration system
M-AERI [79] (1996)	Fourier-transform interferometer	3.0–18.0	Cooled HgCdTe, InSb	Steerable plane mirror	Programmable plane rotating mirror	Two internal black body cavities	Programmable (typical 55°)	Large advanced research spectroradiometer
JPL-NNR [80] (2002)	Filter	One channel 7.8–13.6	Thermopile	Ge lens	Non (manual pointing)	One internal black body cavity	Mechanically set at 0–45°	No sky view, deployed on moorings
ISAR [62] (2004)	Filter	One channel 9.6–11.5	Heitronics KT15.85D	Optical lens focused to diameter 6 mm at focal length of 98 mm	Programmable plane rotating mirror	Two internal black body cavities	Programmable (typical >15° and <40°)	Autonomous operational deployment for up to 3 months
CIRIMS [63] (2005)	Filter	One channel 9.6–11.5	Heitronics KT11.85	Optical lens, two separate optical chains	Non (manual pointing)	One internal black body cavity	Mechanically set (typical >50° and <50°)	Large operational system with autonomous capability
BESST [89] (2013)	Filter	Three channels 10.8, 12.0, 8.0–12.0	320 × 256 pixel thermal imaging microbolometer	Ge objective lens	Programmable plane rotating mirror	Two internal black body cavities	Mechanically set nadir and azimuth	Lightweight design suitable for flight on drones

3.1 TIR Detectors

A detector system generates an output proportional to the target radiance incident on the detector. IR detectors have specific spectral responsivity, spectral response, linearity, long-term stability, environmental temperature dependencies (that may necessitate active cooling) [92], active detection areas and shapes, the number of elements (i.e., single element or multidimensional arrays) that must be considered by a TIR radiometer instrument engineer. There are two main types of detector in use: wavelength dependent *quantum* detectors that respond to a wavelength specific photon flux with excellent sensitivity and response times (but typically require cryogenic cooling) and thermal *energy* detectors that respond to direct heating. As thermal detectors respond to changes in temperature they have a relatively slow response and low sensitivity (influenced by the heat capacity of detector structure) compared to quantum detectors. In general, thermal detectors have a response that is weakly dependent on wavelength and can be operated at ambient temperatures.

The noise equivalent power, NEP (W/Hz$^{1/2}$) of a detector describes the quantity of incident radiation equal to the intrinsic noise level of a detector:

$$\text{NEP} = \frac{PA}{\left(\frac{S}{N}\sqrt{\Delta f}\right)} \tag{18}$$

where P is the incident energy (W cm^{-2}), A is the detector active area (cm^2), S is the signal, N is the noise output (in Volts), and Δf is the noise bandwidth (Hz).

The detectivity, D^* (cm Hz$^{1/2}$W^{-1}), defined as the photo sensitivity per unit active area of the detector, is often used to compare different detectors and can be calculated using.

$$D^* = \frac{\frac{S}{N}\sqrt{\Delta f}}{P\sqrt{A}} = \frac{\sqrt{A}}{\text{NEP}} \tag{19}$$

D^* is often expressed at a temperature of a radiant source or a wavelength, the chopping frequency and bandwidth. Larger D^* values indicate a better detector. Figure 10 shows the spectral characteristics for a variety of IR detectors as a function of D^*.

The $NE\Delta T$ of a detector represents the temperature change, for incident radiation, that gives an output signal equal to the rms. noise level. Detector $NE\Delta T$ and system $NE\Delta T$ are the same except for system losses. $NE\Delta T$ is defined as:

$$NE\Delta T = v_n \frac{\Delta T}{\Delta V_S} \tag{20}$$

where v_n is the rms noise and ΔV_S is the signal measured for the temperature difference ΔT.

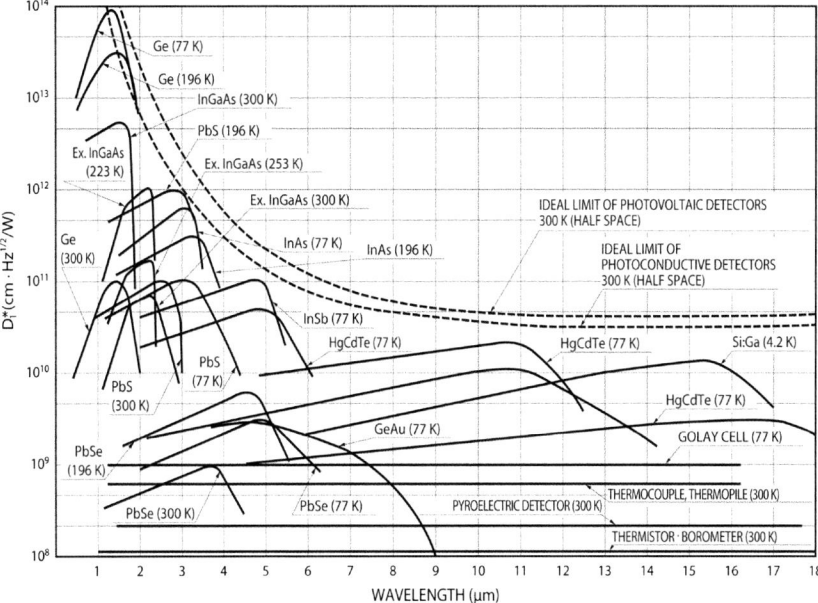

FIGURE 10 The spectral response characteristics of various infrared detectors. *From Ref. [93] with permission.*

It is essential that any nonlinearity of the detector output be fully characterized across the broad range of target temperatures. In the case of shipborne SST radiometers operating in the ~10 μm spectral waveband, clear sky temperatures <150 K are possible with maximum SST of ~315 K. Due to the inherent limitations of on-board calibration systems available today, linearity knowledge of the lower range of temperatures is particularly important as these cannot be calibrated easily. Furthermore, the detector's spectral characteristics must be fully defined in terms of transmission and spectral responsivity to power.

3.1.1 Quantum Detectors

Intrinsic quantum detectors are a based on semiconductor materials that generate electron—hole pairs (and thus a current) across a semiconductor band gap. Photoconductive detectors photogenerate charge carriers causing a change in resistance proportional to the incident infrared radiation whereas photovoltaic devices use an internal potential barrier with an inherent electric field in order to separate photogenerated electron—hole pairs. Table 2 shows the characteristics of several useful quantum detectors for marine TIR radiometers. Indium antimonide (InSb) photovoltaic detectors for the 2—6 μm waveband provide fast responsivity, low noise with excellent uniformity, linearity and stability but require cryogenic cooling to an operating

TABLE 2 Characteristics of Different IR Detectors Useful in Marine Infrared Radiometers

Type		Detector	Spectral Response (μm)	Operating Temperature (K)	D^* (cm Hz$^{1/2}$ W^{-1})
Intrinsic quantum	Photo conductive	PbS	1.0–3.6	300	D^* (500,600,1) ~1×10^9
		PbSe	1.5–5.8	300	D^* (500,600,1) ~1×10^8
		InSb	2.0–6.0	213	D^* (500,1200,1) ~2×10^9
		HgCdTe	2.0–16.0	77	D^* (500,1200,1) ~2×10^{10}
	Photo voltaic	InSb	1.0–5.5	77	D^* (500,1200,1) ~2×10^{10}
		HgCdTe	2.0–16.0	77	D^* (500,1200,1) ~1×10^{10}
Thermal	Thermopile	e.g., SbBi junctions	Defined by window	300	D^* (λ,10,1) ~1×10^8
	Pyrolectric	PZT	Defined by window	300	D^* (λ,10,1) ~1×10^8
		LiTaO$_3$			D^* (λ,10,1) ~2×10^8
	Microbolometer	e.g., VOx, a-Si	Defined by window or by design	300	D^* (λ,30,1) ~1×10^9

D^* values are expressed at a temperature of a radiant source or a wavelength, the chopping frequency and bandwidth.

temperature of ~77 K (typically using a liquid nitrogen Dewar). InSb detectors suffer from an instability called InSb "flashing" in which occurs when shortwave radiation falls onto the detector. Trapped charges build up in the detector that eventually discharge with a short but large signal output. Care must be taken to ensure that the detector is properly protected—ideally with internal masking on the detector itself in addition to effective band pass filters to prevent shortwave radiation falling onto the detector in the first place. mercury cadmium telluride (HgCdTe) photoconductive detectors are suitable for the 2–16 μm waveband and have fast responsivity, low noise, excellent uniformity, linearity and stability. HgCdTe suffers from Random Telegraph Signal (RTS, sometimes called Popcorn noise) that is manifest as random and sudden jumps in output signal due to random trapping and release of charge carriers. Electronic screening is used to minimize the effect. The wavelength of peak response of these detectors depends on the specific alloy composition used and the detectors must be cooled to operating temperatures of ~77 K to significantly improve noise performance. When using cryogenically cooled detectors, care is required to ensure that the position of the detector after cool-down is properly controlled in the radiometer design: a typical change of ~0.5 mm in detector position is expected after cool-down [94].

Liquid nitrogen could be used to reach the required cryogenic temperatures but there are obvious dangers associated with handling this fluid while at sea—particularly as regular top-ups are required as the liquid evaporates. This approach is to be avoided. Alternative solutions are recommended including thermoelectric systems such as Peltier cooling devices [95] or mechanical Stirling Cycle coolers [96].

Peltier devices have no moving parts and do not require the use of chlorofluorocarbons, are small, easy to install, have good reliability and are virtually maintenance free. Their ability to heat and cool by a simple reversal of current flow is useful for applications where both heating and cooling is necessary or where precise temperature control is required. However their main drawback is the limited range of cooling that can be achieved. This is ~50 K relative to the ambient temperature when using specially designed fan-cooled heat sinks that can manage temperatures cooler than ~263 K. The MISTRC [74] radiometer uses two HgCdTe detectors covering the 3.0–5.0 μm and 10.0–12.0 μm wavebands. The detectors are cooled to reduce noise using a four-stage thermo-electric cooler that can provide up to 95 K cooling below the ambient temperature. A thermoelectric cooler solution was chosen above other approaches on the grounds of reliability [97].

A stirling cycle cooler is a mechanical heat engine that moves a working gas (e.g., He) around a closed-cycle circuit in such a way as to compress the gas in the cold part of the engine (allowing heat to flow out) and expand it in the hot part of the engine to remove heat. Heat is supplied and removed through the walls of the engine. A Stirling cycle cooler provides a very efficient approach with a large cooling power enabling very low (77 K) cryogenic temperatures to be achieved, long continuous operating periods and, high reliability. Thermal camera systems and some ship-borne radiometer designs make use of Stirling Cycle coolers (see Section 4).

In the humid marine environment care is needed to mange a cryogenic cooling system. The detector environment must ensure a dry atmosphere to prevent the accumulation of ice on the cooled detector assembly. Furthermore, the cryogenic design must ensure that any optical surface or surface in close proximity to optical elements of the radiometer remain free of condensate (that could drip onto an optical surface): if the optical surface is contaminated with a water condensate the radiometer will have poor or no throughput.

Despite the challenging task of cryogenically cooling quantum detectors, their high performance, wide spectral capability, and rapid responsivity mean that they are preferred for spectroradiometer instruments. The ship-borne MAERI [79] is a FTIR spectroradiometer that operates in the range of $\sim 3.3-18$ μm waveband and measures spectra with a resolution of ~ 0.5 cm^{-1}. It uses two infrared detectors to achieve this wide spectral range: an InSb detector covering the 3.3–5.5 μm waveband and a HgCdTe detector covering the 5.5–18 μm waveband cooled to ~ 78 K by a Stirling cycle cooler to reduce the $NE\Delta T$ to levels well below 0.1 K.

3.1.2 Thermopile Detectors

Thermopile detectors are formed as an array of miniature thermocouples. This type of detector does not require cooling and can operate at ambient temperature. Thermocouples join two dissimilar conductors that when heated create a voltage proportional to the temperature at the junction (the Seebeck effect [98]). Thermocouple materials include antimony (Sb) and bismuth (Bi). When TIR radiation is focused onto a thermocouple detector, its temperature increases and an output voltage is generated proportional to its temperature. For optimal sensitivity, an array of thermocouples is used as a thermopile. Thermopile detectors typically have an invariant spectral response (defined by the detector window material and, to a lesser extent, the coatings applied to the active area of the detector), have inherent low noise characteristics, and detectivity values similar to pyroelectric detectors. An advantage over the latter is that thermopile detectors do not require an optical chopper and have very low $1/f$ noise. However, thermopile detectors have a relatively low sensitivity and a long measurement integration time is required. Furthermore, amplification of the small signal must be designed with care—to mitigate the introduction of significant noise—by using appropriate amplifying electronics. The components used have to be of the highest quality and protected from environmental effects. To improve thermopile sensitivity, thermocouple junctions must be thermally isolated. Hot junctions define the active area of the detector and are often suspended on a thin membrane, thermally isolating them from the rest of the detector package. They are typically coated with an energy absorbing material in increase their performance. The cold junctions are thermally connected to the detector package. Thermopile sensors are typically nonlinear [80] and the output of the sensor is a function of the difference between radiative fluxes into and out of the sensor surface that is dependent on the temperature of the surroundings as well as the flux from the target. Depending on the detector application, spectral sensitivity is defined by a choice of optical band-pass filters often incorporated into the detector package.

The NASA/JPL NNR [80–82] uses a thermopile detector that is calibrated using the near nulling calibration approach (see Section 4.3). The NNR uses a local linear approximation to the true nonlinear response of the thermopile sensor thereby achieving accuracies similar to those achievable only by using much more demanding and expensive linear sensors. The simplicity, low cost, and robustness of the thermopile sensor is a major advantage to developing inexpensive and uncooled field instrument capable of delivering accurate measurements.

3.1.3 Pyroelectric Detectors

Modern pyroelectric detectors are widely used for a variety of applications as they are simple, robust and provide good performance. Pyroelectric detectors (e.g., lithium tantalate ($LiTaO_3$), lead zirconate titanate (PZT)) are formed as thin plate capacitor dipole elements which, when exposed to heat, alter the charge density at the electrodes. Care must be taken to minimize the microphonic effect (vibrations) [40] that induce a piezioelectric response from the detector. In addition $1/f$ or "popcorn" noise maybe generated in a pyroelectric detector caused by changes in detector working temperature leading to rapid changes in instantaneous output signal. Pyroelectric detectors respond to *variations* in temperature that are typically achieved by modulating the incident TIR radiation using an optical chopper. Chopping is the process of interrupting the optical beam rapidly and regularly with a device such as a tuning fork chopper or a rotating chopper disc. The electronic processing chain following the detector amplifies only the AC component of the signal (AC electrical signals are significantly easier to process), whose amplitude is determined by the difference between the radiative signals from the instrument view and the surface of the chopper. The DC component generated by self-emission from components between the chopper and detector is rejected [99]. The simplest practical approach is a rotary blade chopper that has been widely used within SST radiometer designs. If there is sufficient thermal stability in the chopper blades (i.e., the temperature of the blades is stable over short periods) then the chopper itself can be used as a reference radiance source removing the need for an additional reference radiance calibration blackbody (see Section 3.7.2). A well-designed chopper system will optimize the chopping frequency to the detector characteristics to maximize the quality of the detector output. The signal to noise performance is often improved—even though about half of the available target photon flux is obscured by the chopper [99]. Detector noise away from the chopping frequency, including low frequency $1/f$ noise, can be rejected with phase sensitive detection.

Across the TIR spectral waveband, pyroelectric detectors have an invariant spectral response and spectral discrimination is derived from the detector window material that is used (the window also serves to seal the detector from the external environment). The detector package and window define the detector FoV (the radiometer system field and aperture stops define the radiometer FoV), which should be optimally chosen to maximize the target radiance yet minimize background stray radiance that depends on the optical design of the instrument. High angles of incidence should be avoided as the physical cut-on and cut-off narrow-band window spectral characteristics can be shifted toward shorter wavelengths [100]. Furthermore, care should be taken to ensure compensation for temperature-dependent spectral shifts. Pyroelectric detectors have a much slower responsivity and lower sensitivity compared photon detectors but still with sufficient performance for use in filter radiometers. The major advantages of pyroelectric detectors are their relatively low cost versus performance and the fact that they can operate at room temperature. For these reasons, pyroelectric detectors are the most popular choice for marine SST radiometers and are used in many designs including DAR-011, SISTeR, ISAR, and SIL designs. As an example, the SISTeR radiometer uses a pyroelectric detector and preamplifier, mounted onto an assembly containing a concentric 6-position filter wheel and a black rotating chopper. By locating a low-noise preamplifier within the detector package the signal is transmitted at higher levels from a low impedance source, minimizing EM interference. The beam is chopped at 100 Hz, a compromise between the optimum noise performance of the detector and a fast filter response in the signal processing chain.

Modern pyroelectric detector packages offer user-defined multispectral capabilities integrated into a single package [101]. A recent development is a miniature Fabry−Perot interferometer integrated together with a pyroelectric detector package [102,103]. Such devices are spectrally tuned using a control voltage across a spectral range of 3.0−5.0 μm or 8.0−10.5 μm with a spectral bandwidth of 150 nm. Such a narrow spectral bandwidth has obvious implications for SNR and should ideally be much wider for practical use in shipborne TIR radiometers. However, such technologies if properly tuned, offer an innovative solution for ship-borne multispectral TIR radiometers although to date such devices have not been used for this purpose.

3.1.4 Microbolometer

A microbolometer measures the power of incident radiation that heats a material with a temperature-dependent resistance [104]. An absorptive element, such as a thin layer of metal is connected to a thermal reservoir (at constant temperature) through a thermal link: radiation falling on the absorbing element raises its temperature compared to that of the thermal reservoir proportional to the radiative power falling on the detector active area. The temperature change can either be measured directly using a thermometer or the resistance of the absorptive element itself can be designed as a thermometer. The thermal time constant of the detector is the ratio of the absorptive element heat capacity to the thermal conductance between the absorptive element and the thermal reservoir.

There has been a rapid development of microbolometers in recent years driven largely by military imaging. Miniature two-dimensional imaging arrays are now common [105] with dimensions of 1024 × 768 pixels (with a pitch of ∼17 μm) having thermal time constants of <10 ms for the 8−14 μm wavelength. The most common type microbolometer today is based on microelectromechanical (MEMs) technology. The two most common microbolometer detector materials are amorphous silicon (A-Si) and vanadium oxide (VOx) [105]. A-Si microbolometers exhibit higher effective thermal insulation and shorter thermal time constants (<10 ms at 30 Hz) compared to VOx (although the latter have better noise characteristics). Amorphous silicon detector arrays are available in a variety of packages and are ideally suited for many infrared imaging applications due to their high performance, high reliability and low cost [105]. Typical $NE\Delta T$ of 50−80 mK with time constant of 10 ms at 300 K and 30 Hz sampling in the 8−14 μm wavelength are common. A microbolometer does not require a mechanical chopper and can deliver a fast sample frame rate.

Microbolometer technology offers a promising detector technology for ship-borne TIR radiometry. The BESST radiometer [89] uses a two-dimensional uncooled VOx microbolometer detector array (324 × 256) with a pixel size of 38 μm. The choice of this detector technology was driven by the requirement to fly over the sea surface using a UAV drone to obtain a large area sample (as an image) of SST for satellite validation [89]. Thus the BESST instrument demanded a lightweight but accurate radiometer solution. The microbolometer chosen is an FLIR Photon Thermal Imaging Camera Core (www.flir.com) which is very small and light (∼6 cm^3 and 125 g).

3.1.5 Commercial Radiometer "Head" Detectors

Several ship-borne radiometer designs have chosen to use a commercial "off the shelf" radiometer "head" as a "detector" package. Many devices are available offering a variety of features and complexity that make them attractive propositions for a robust low-cost solution. Often, spectral definition, optical beam definition, detector thermal and chopping management, data conditioning, configuration, control and data logging are defined by the "head," some of which can often be configured via dedicated communications interfaces. When using this type of approach it is important to ensure that the "detector" output is configured to be proportional to radiance so that effective end-to-end calibration strategies can be developed to calibrate the radiometer. In practical terms, the radiometer is often designed around the constraints/advantages of the chosen head device. Of note is the use of TASCO THI-500, Heitronics KT15 and KT11 instrument heads used in the SOSSR, ISAR and CIRIMS SST radiometer designs respectively. A Bomen Fourier transform interferometer (FTIR) is used by the MAERI instrument and an FLIR Photon microbolometer core is used by the BESST radiometer. The JPL NNR uses an Apogee Infrared Radiometer head with JPL custom calibration system.

The reader is referred to [40,106] for an extensive and detailed discussion of TIR detectors.

3.2 TIR Radiometer Spectral Definition

The spectral definition of an SST radiometer depends on the fundamental design and application requirements for the instrument in question. For a filter radiometer spectral definition is achieved by accounting for the spectral characteristics of all components in the optical chain including focusing lenses, mirrors, protective windows, band-pass filters (if used) and the detector window. For a spectroradiometer spectral definition is largely achieved by measurement of small spectral intervals using an interferometer. However, the characteristics of mirrors, protective windows and the detector package itself still need to be accounted for.

Figure 11 shows the normalized spectral transmittance of materials commonly used for lenses, windows and band-pass filters in TIR radiometers. Several materials have separate curves to highlight the impact of wafer thickness. Germanium (Ge) has moderate transmission properties in the $2-16$ μm spectral waveband, a high refractive index, good dispersion properties and is nonhydroscopic. It can be easily cut and polished to form lenses and windows, but is brittle.

Because of the high refractive index a proportion of radiation is lost by reflection at each optical surface and Ge optical components require an antireflection (AR) coating [107]. Double-imaging and a deterioration of contrast caused by multiple reflections if several optical components (e.g., detector window and filter windows) may also be problematic. AR coatings are hard refractory coatings applied to the surface of optical components that minimize surface reflections within specified wavelength ranges. There are many different antireflection coatings although four common types are typically used:

1. Single layer, which offer a low reflection over a medium to wide bandwidth.
2. Narrow band, which offer a very low reflection over a narrow bandwidth.
3. Broadband, which offer a very low reflection over a wide bandwidth.
4. Extended wideband, which offer a low reflection over a very broad bandwidth.

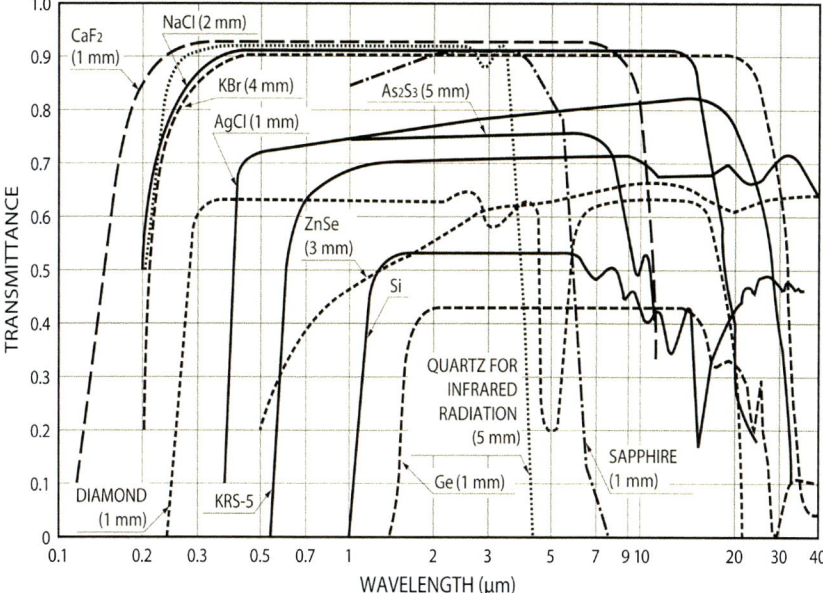

FIGURE 11 The spectrally dependent transmission characteristics of commonly used optical materials. Examples for different wafer thickness are marked onto the plot. *From Ref. [93] with permission.*

AR coatings can be designed for operation at angles of incidence other than normal, or optimized for a particular wavelength region. Broadband antireflective (BBAR) coatings are often used in SST radiometer components and consist of multiple layers, alternating between a high refractive index material and a low refractive index material. The layers are typically deposited on the substrate via electron-beam deposition. The thickness of the layers is optimized to produce destructive interference between reflected waves and constructive interference between transmitted waves. This results in an optic that has enhanced performance within a specified wavelength band as well as minimal internal reflections (ghosting). Many BBAR coatings provide good performance for a range of angles of incidence (up to ~30°). Other coatings provide windows that polarize the incident radiance signal such as optically thin interference coatings and wire grid diffraction polarizers [74].

The transmission properties of an AR coated Ge wafer 5 mm thick are >85% in 3.5–12 μm waveband but are highly temperature sensitive: the absorption becomes so large that Ge is nearly opaque at 303 K. This example serves as a reminder to the instrument engineer that any spectral temperature dependence of filter transmission properties must always be verified. Alternatives include silicon (Si), which is a low-cost lightweight material that is inert to marine environments with a high thermal conductivity. It is nonhydroscopic, harder than Ge and less brittle and has good transmission in the 1.0–6.5 μm spectral region. AR coatings can increase transmission to >99% in the 3.5–4.5 μm spectral region. However, transmission is poor at longer wavelengths. Si has a small spectral and transmission temperature dependence in the 250–350 K range. Zinc selenide (ZnSe) has better transmission properties compared to Ge or Si above 0.5 μm with little or no distortion of the transmitted signal. ZnSe is nonhydroscopic and has a high resistance to thermal shock but is easily scratched and must be handled with care (it is a hazardous material). There is also a small spectral temperature dependence that must be considered when using ZnSe. AR coatings in the 8–12 μm typically result in very high transmission ∼99% making ZnSe an excellent choice for SST radiometers. Zinc sulfide (ZnS) is harder and more chemically resistant than ZnSe with similar transmission properties.

Band pass filters are used in TIR filter radiometers to isolate specific regions of the TIR spectrum by simultaneously providing high transmission of desired spectral radiance and rejection of unwanted spectral radiance. Band pass filters are defined by four features:

- The center wavelength (CWL) which is the wavelength at the center of the pass band,
- The full width at half maximum (FWHM) which is the bandwidth at 50% of the maximum transmission,
- The peak transmission (T) that is the wavelength of maximum transmission,
- The blocking range which is the spectral region in which the filter does not transmit.

Narrow spectral band pass filters obviously reduce the signal reaching the detector decreasing the radiometer SNR. Note that it is important to recognise that the end-to-end spectral response function, $\zeta(\lambda)$, must also include the detector window spectral characteristics (Figure 12). AR coatings that reduce direct reflection of detector windows can be beneficial: the detector window will itself emit radiance that will increase the background noise of the detector. In the case of cooled detectors, narrow band pass filters should be placed inside the detector package so that the window itself is cooled down to the temperature of the detector.

The spectral characteristics of optical components vary from batch to batch and care must be taken to properly characterize the spectral definition of *each* individual radiometer. Figure 13(a) shows the variations between KT15.85D radiometer heads, used in the ISAR ship-borne radiometer, showing significant spectral differences and CWL between heads (although CWL, FWHM, and T all remain within the manufactures stated tolerances). Another problem is that of spectral "leaks" in parts of the waveband thought to be nontransmitting (blocking range). Figure 12 gives an example of an

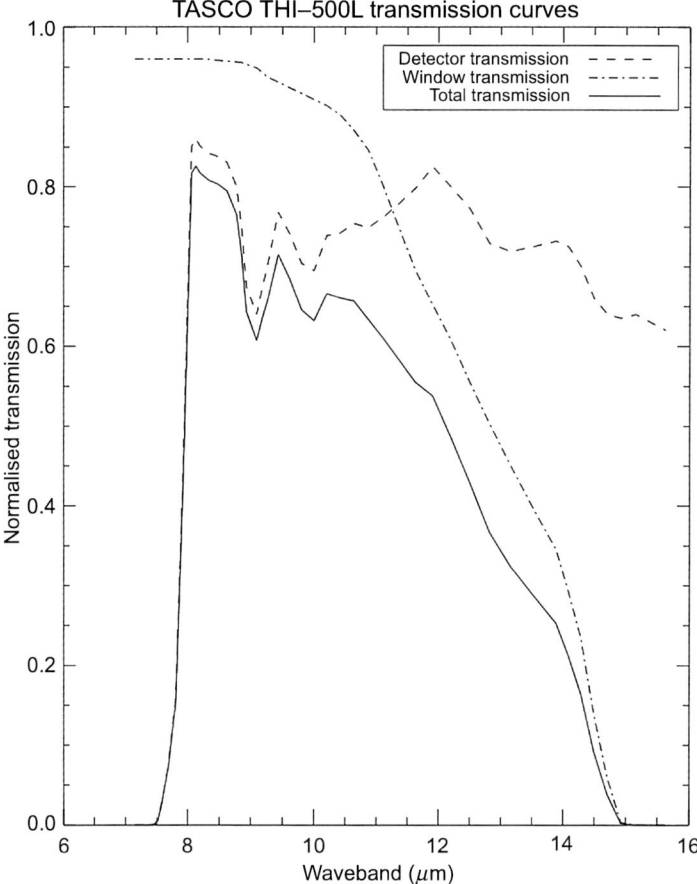

FIGURE 12 Spectral transmission for a TASCO THI500 radiometer [62] showing the spectral transmission of the detector window, focusing window and total spectral transmission.

out of band spectral leak at 19–26 µm which, in unaccounted for, will result in significant errors when measuring SST.

Spectral filters, windows, and lenses possess spectral properties that, together with the detector characteristics, define the overall spectral characteristics of a radiometer. It is essential that the transmission profile of instrument band pass (and therefore all relevant optical components), and the temperature dependence of that profile, be traceably measured in an optical beam of comparable speed to the beam in the instrument itself. In addition, the spectral calibration of any spectrometer, and the temperature dependence of that calibration, must be traceably measured [99]. Furthermore, care should be taken to properly characterize the transmission of a filter in spectral regions considered to be nontransmitting to determine if any spectral leaks exist.

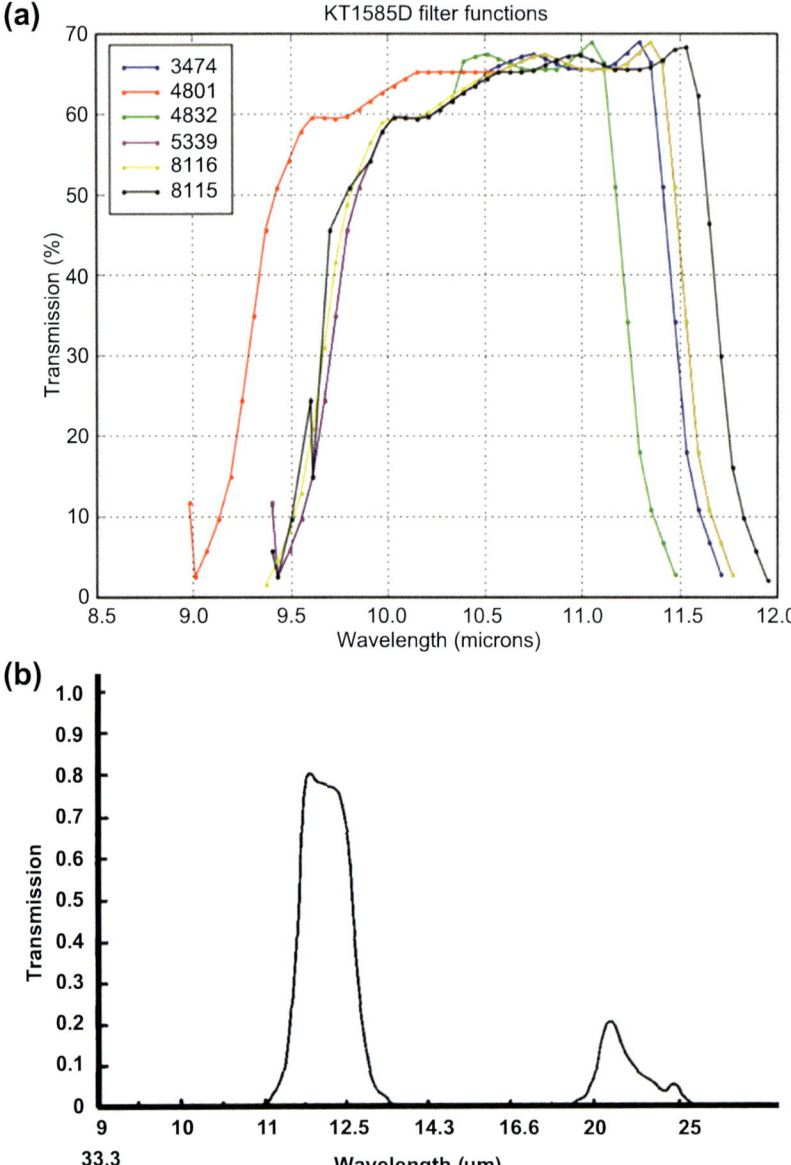

FIGURE 13 (a) Spectral transmission of six separate Heitronics KT15.85 radiometers used (serial numbers shown on inset) in the ISAR radiometer design showing small but important spectral differences between each individual unit. (b) SIL radiometer band pass showing a distinct spectral leak between 19–26 μm.

3.3 Beam Shaping and Steering

An optical system is required to *steer* the radiometer FoV to a target radiance source and to *shape* the beam in a manner that focuses the target radiance onto the detector. Figure 9 presents a simplified overview of a filter radiometer that summarizes the key elements of the optical system required to achieve these aims that are discussed below.

3.3.1 Beam Shaping

Optical beam shaping for a radiometer is required to accurately define the measurement solid angle from a target area within an extended source. This requires that both the optical path and geometry are well defined so that radiance from the source surrounding the instrument FoV viewed is properly attenuated. The simplest approach to produce a well-defined measurement angle is to use a Gershun tube [108]. This defines two apertures at each end of a tube, a known fixed distance apart, which if overfilled by an extended radiance source such as the sea surface, act as field stops. The two apertures then determine the radiance measurement solid angle. Baffles can be placed between the apertures to minimize stray radiance originating outside the FoV and any single-reflection radiation impinging onto the detector. The best approach is the place the inner field stop onto the detector itself, ideally beneath the detector window to minimize any gap between the detector and the aperture to minimize reflections between the window and detector arising from noncollimated beams and nonnormal beams [92]. Aperture edges should be thin and sharp to minimize radiation scattering with a sharp bevel edge included on the front face to minimize strays. Figure 14 shows a simplified overview of such a design for the case of a Gershun tube divergent beam and a parallel-collimated beam defined by an off-axis parabolic mirror.

Assuming that the baffles are slightly oversize relative to the beam width defined by the detector aperture, d, and the instrument aperture, D, the unvignetted FoV, α, of the design is given by Ref. [108]

$$\alpha = 2\tan^{-1}\frac{D-d}{2s} \tag{21}$$

and the full radiance measurement angle, β, is then [108]

$$\beta = 2\tan^{-1}\frac{D+d}{2s} \tag{22}$$

With such a design the unvignetted target-spot diameter (the plateau of the output signal) will be $\sim 40\%$ of the full spot diameter measured by the radiometer [108]. The radiometer designs shown in Figure 14 provide enough

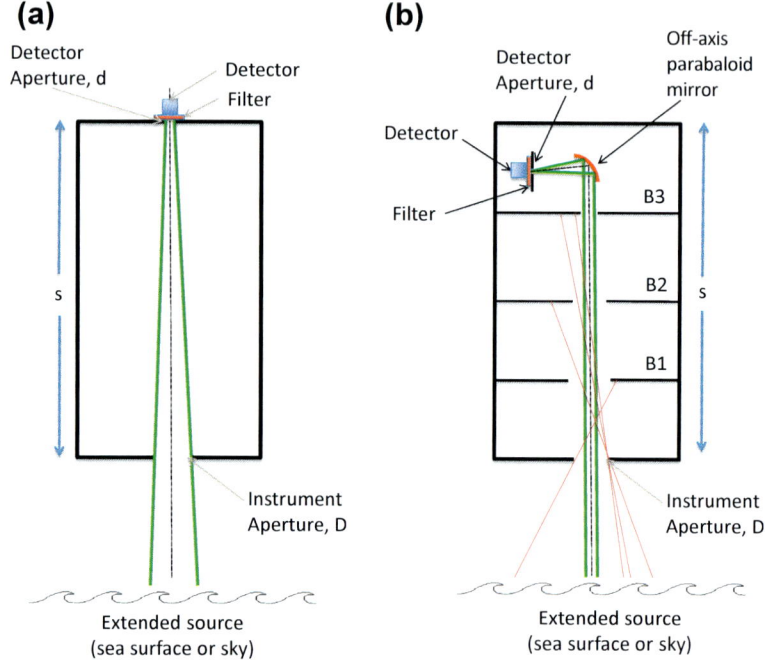

FIGURE 14 (a) Gershun tube radiometer with divergent beam viewing the sea surface extended source. D defines the size of the instrument aperture and d defines the size of the detector aperture separated by a distance. The filter is placed directly onto the detector. (b) Baffled luminance tube using an off-axis parabaloid focusing and fold mirror defining a collimated beam. B1, B2 and B3 are baffles. Green lines show the unvignetted FoV (α) and red lines are rays highlight the blocking role of baffles. Most of the signal in both cases is coming from the radiometer housing and the use of a chopper system or cryogenic cooling of the housing (to minimise stray contributions) is normally employed to account for this (see chapter 2.3). Choppers, coolers and windows have been omitted for clarity.

space to install calibration blackbody radiance sources, detector chopper systems and electronics. In the baffled Gershun tube case, a direct path between the instrument aperture and detector filter is evident—presenting a direct path for seawater that could damage the detector. The target source spot radius of the divergent beam is governed by the physical installation aboard a given ship. For ship-borne radiometers, typical full beam-widths of $\sim 6°$ are common. For a radiometer viewing the sea surface at an angle of $20°$ (in the plane of the ships length) and a height of 15 m the target spot size on the sea surface will be and ellipse of $\sim 1.5 \times 1.7$ m. For a look angle of $40°$ the target ellipse would be $\sim 1.5 \times 3.1$ m scaling linearly with height above the sea surface.

Most detector packages have active areas of $2.0-6.0^2$ mm and it is necessary to focus the target radiance onto this small area for maximum performance. Lenses manufactured from Ge, Si, ZnSe, or ZnS can be used for this purpose and many commercial "radiometer heads" (e.g., the Heitronics KT15 series [39,109], Heitronics KT11 series [63], TASCO TH500L [62,88]) follow this approach. Good AR coatings are required to increase transmittance and limit stray radiation. For extended target sources, such as the sea surface, an objective lens can be used to image the target and a second field lens used to focus the radiance onto a small detector active area with uniform radiation over the detector [40]. The BESST radiometer design uses a Ge lens focused onto a microbolometer (e.g., Ref. [110]) array for example. However, optical materials suitable for use in the TIR waveband are easy to damage (see Section 3.2) and can be costly.

A popular [65,85,87] alternative for marine SST radiometers is to use a mirror to shape the radiometer beam. Metallic reflectors are excellent broadband reflectors for TIR wavelengths and are also insensitive to polarization and angle of incidence. Vacuum deposited thin metal films are applied to a highly polished glass substrate such as Pyrex® (a low expansion borosilicate glass, resistant to thermal shock), Zerodor® (a unique glass-ceramic material with excellent thermal expansion for ultra-high stability), or Al. A metallic coating is relatively soft making it susceptible to damage and a protective hard layer over the metallic coating significantly improves durability. The surface quality of a reflector is described by its surface figure and irregularity. A surface flatness of $\lambda/4$ and low surface quality scratch-dig specifications (60−40) to minimize scattering is, in most cases, adequate for ship-borne TIR radiometers.

Good metallic reflectors for the TIR include aluminum (Al), rhodium (Rh), silver (Ag), and gold (Au). Polished Al is an excellent reflector in the 1−20 μm waveband but is prone to oxidation in the marine environment and should not be used without adequate protection (which is difficult to achieve). Rh also has excellent refection in the TIR and is extremely hard and durable (although it can be difficult maintain Rh coatings on Al substrates due to Al oxidation effects). Ag has excellent reflection in the TIR region but will tarnish if not protected from the marine atmosphere. Protected Ag (e.g., using a MgF_2 or SiO dielectric overcoat) provides an alternative solution with excellent reflectivity (99%) in the infrared regions but such coatings are not well suited in the humid marine environment. Gold (Au) reflectors with a hard protective coating (e.g., SiO) have the highest reflectance ($\sim$98%) in the 1.5−30 μm region and are very resistant to corrosion making these reflectors the preferred choice for ship-borne radiometers.

A parabolic mirror will tightly focus incoming collimated beams parallel to the mirror axis at the mirror focal point, F1. An off-axis parabolic mirror is cut as a circular section from one side of a full parabolic mirror: the focal point is now off the mechanical axis of the mirror allowing full use of the reflector focus area as shown in Figure 15(a). A detector with a small active area can then be positioned at F1. An off-axis parabolic mirror does introduce

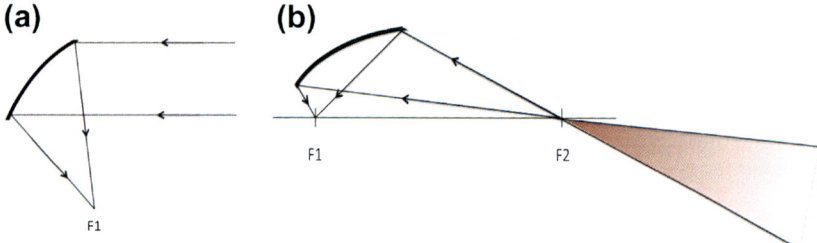

FIGURE 15 (a) A parabolic mirror reflects light from the focus into a collimated beam, or refocuses a collimated beam at the focus. The focus of an off-axis parabolic reflector is off the mechanical axis. (b) An ellipsoid mirror reflects light from one focus to a second focus that can be usually external. The focus of an off-axis ellipsoid reflector is off the mechanical axis.

significant aberrations for extended sources that results in a distorted spot rather than a sharp image focus at F1. For imaging cameras this presents a problem but for single element detectors this is not a major issue. This arrangement allows the detector and other delicate components to be positioned in a region of the radiometer housing that is better protected from the marine environment compared to the Gershun tube approach as shown in Figure 14(b).

A disadvantage of using an off-axis parabolic mirror is that, because the input beam is collimated (nondivergent), only very small target spot sizes (of the order of a few cm) commensurate with the size of the mirror itself can be achieved. Unlike a divergent beam, the target spot size will be invariant to the mounting height of the radiometer installed on a ship.

An alternative approach is to use an off axis ellipsoid mirror (or an appropriate optical lens configuration to achieve the same effect) as shown in Figure 15(b). Ellipsoid reflectors have two conjugate focci: a ray passing through one of the ellipsoid mirror foci (F2) will be reflected and emerge at the other focal point (F1). A detector placed at F1 will see the image at F2. For a pure point-source exactly at one focus of the ellipse, almost all of the energy is transferred to the other focus. However, if the target source is not exactly at the focus of the ellipse it will be magnified (by the ratio of each reflected beam length) and defocused at the image. Because of this effect, ellipsoids are most useful when coupled with a small source and a system that requires a strong radiant source without concern for particularly good imaging (which is not a major issue for single-element detectors used in many ship-borne SST radiometers). Key advantages of using an off-axis ellipsoid is the fact that a divergent beam is available to sample a large target area and second focus can be used to minimize the instrument aperture allowing significant protection from the marine environment (the latter can, of course, be achieved using lenses, as in the case of the ISAR radiometer). The SISTeR radiometer uses an off-axis ellipsoid mirror in its design which allows a very small instrument aperture of 7 mm (with a detector image magnification to 4 mm at the exit aperture) to be used providing significant protection from the marine environment.

The *étendue* or "light grasp," $A\Omega$, (where A is the active area of the detector and Ω is the solid angle of the radiometer) of an optical system is conserved throughout the optical system can be approximated by the field stop area, multiplied by the solid angle to the aperture stop (see Chapter 2.3). $A\Omega$ can be used in the trade-off between either a divergent beam with a small exit aperture (ellipsoid) or a collimated beam with a large exit aperture (parabolic). For mirror sections of the same area at the same distance from the detector (presuming they define the respective aperture stops), the noise performance of both systems will be identical. Note that even a parabolic beam will diverge a little so that $A\Omega$ is conserved—although for the distances considered by ship-borne radiometers this can be effectively ignored: for a satellite instrument (e.g., the Sentinel-3 SLSTR [111]) a 10 cm collimated beam at the instrument scan mirror becomes a 1 km nadir FoV at the earth's surface ~815 km below.

Care should be taken in the radiometer design to facilitate the optical alignment of all optical components. The use of lasers and positive alignment dowels, jigs, and other adjustable mounting arrangements will greatly assist the optical alignment of the instrument and ensure that the alignment remains true during use in the field. A certain amount of robustness needs to be built-in to the design: the working environment aboard a ship is considerably different to that of the laboratory. All optical components must be positively mounted and secured in a manner that can cope with constant vibrations from ships engines and mechanical shock that can occur in heavy seas. Prealigned "plug and play" components and subassemblies can be useful in this respect, especially if repairs have to be made while at sea.

3.3.2 Beam Positioning

Beam positioning refers to the pointing of a radiometer beam that has been shaped according to the optical arrangements implemented in a TIR radiometer. The simplest form of beam positioning is a hand held radiometer that is manually pointed at the sea surface to make single measurements. For a more sustained measurement campaign, and to ensure consistent radiometer pointing, the radiometer should be attached to the ship superstructure using dedicated mounts—some examples are shown in Figure 9 and 16.

The RAL/SIL radiometer [86] was a ship-borne radiometer that was designed using the approach shown in Figure 14(b). The instrument was installed on a ship (typically on the fore-mast [31,65,112]) with $15° < \theta < 40°$. The installation was performed when the ship was at berth and care taken to ensure that the radiometer optical beam cleared the entire ships superstructure. This was done at night using a laser attached to the

FIGURE 16 (a) Installation of the SIL radiometer on the foremast of the RRS James Clark Ross. The radiometer (white tube to the left) used a yoke-cradle arrangement. (b) Close up of the SISTeR radiometer (white box installed on the hand-rail second from left) on the fore-mast of the RRS Charles Darwin.

radiometer. A significant challenge when using the SIL radiometer was the inability to measure the downwelling sky radiance without unbolting the radiometer and pointing it towards the sky manually. This was clearly impractical and, instead, estimates of the downwelling sky temperature were made using a separate low-cost THI-500L radiometer [112] pointing at the appropriate zenith angle θ. The THI-500L was then spectrally matched to the spectral response function of the SIL radiometer during postprocessing. The CIRIMS radiometer also relied on a physical instrument mounting to set the radiometer sea surface pointing angle (see Figure 9) and a separate sky-viewing radiometer that was mounted in an external housing attached to the side of the main radiometer pointing at the appropriate zenith angle. The SOOSR radiometer used two TASCO THI-500 radiometer heads, pointing at the appropriate zenith and nadir angles, enclosed in a large protective box with external baffles. The JPL near-nulling radiometer used an innovative approach in which the thermopile detector itself is moved between different target views. However, all of these designs are limited in their ability to cater for different ship installations, and, in several cases, are not able to provide a calibrated measurement of sky radiance.

Use of a rotating plane mirror at an angle of 45° to the main radiometer target beam allows the radiometer to steer the beam electronically (e.g., using a stepper motor or using a position encoder of some description) to any number of target views around the arc of the rotating mirror. This "scanning mirror" approach (Figure 17) has been used by several SST radiometer designs (including SISTeR, ISAR, MAERI, DAR-011, and BESST) to great advantage. Such an approach allows the detector to view a variety of external targets (depending on the design of the radiometer) and internal calibration targets (Section 3.7.2) via the same optical chain. As discussed previously, metallic reflectors provide the best mirrors when used with appropriate AR and hardening coatings. Importantly the ability to "park" the mirror in a safe position and seal the radiometer form the environment during bad weather is also possible. If the radiometer beam shape is fashioned in such a way as to focus the beam to a small spot size at the exit aperture i.e., a short distance on from the scan mirror (e.g., using an off-axis ellipsoid mirror), the plane mirror can be housed in a protective scan drum with a very small aperture greatly minimizing contamination. For instruments with collimated beams a larger mirror is required (e.g., as in the case of MAERI) which can be difficult to protect from the marine environment and they are not easily incorporated into a scan drum arrangement.

Figure 17(a) shows the beam steering arrangement used by the SISTeR and ISAR radiometers. In position (a) the scan drum is rotated to view the sky while in position (b) it is rotated to view the sea surface. The DAR-011 and BESST radiometers use a simplified scan mirror arrangement that *only* allow a nadir and zenith view via dedicated ports cut into the radiometer casing. This approach, while somewhat simpler to manufacture, does not allow the radiometer to easily view the sky at the most appropriate angle required for SST_{skin} measurements. The mounting of the instruments and their view ports could be modified to improve on the design in this respect. This introduces errors due to mispointing sky views—especially in the presence of a mixed cloudy sky [53]. For this reason, this arrangement should be avoided.

It should be noted that even if a small scan drum aperture is used the corrosive marine atmosphere may still have an impact. While a certain amount of dry mirror contamination will be managed by the radiometer calibration system (NaCl has a good TIR transmission properties) the SNR will decrease. Care is also required in developing preventative maintenance of ship-borne radiometers. For example, some ISAR scan mirrors suffered serious degradation during initial deployments: the optical coating and gold surface blistered and was extensively corroded. In one case the mirror was so badly damaged that the ISAR data were unusable. The shattered gold mirror shown in Figure 17(d) was due to the long-term effect of thermal stresses. The fault was traced back to a change of glass cleaning solution used in the manufacturing process. The gold mirror coating did not bond properly with the glass substrate: changes in solar illumination during field deployments and consequent thermal expansion and contraction eventually shattered the mirror gold. In 2006 the commercial provider and manufacturing process of the ISAR

(a)

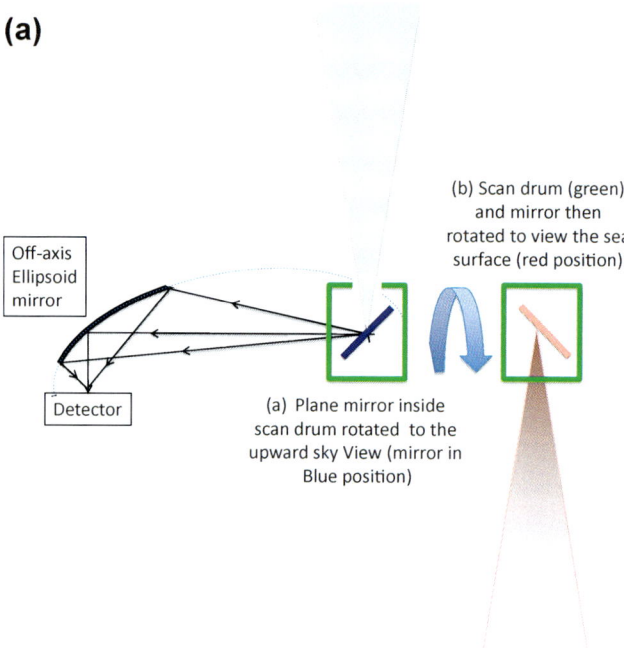

(b)

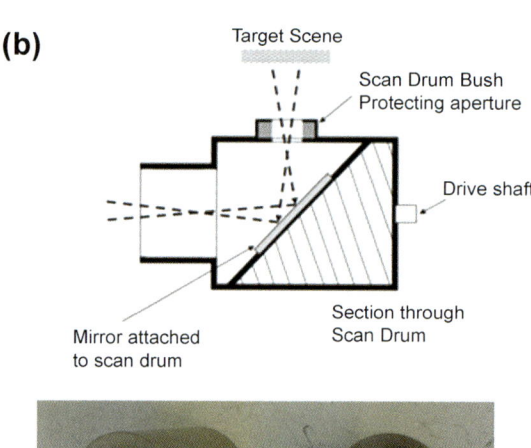

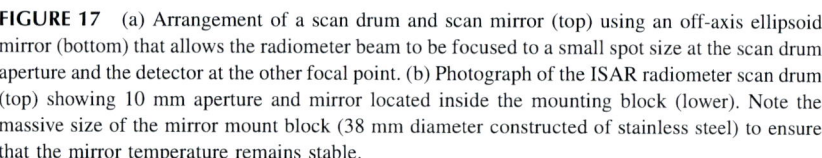

FIGURE 17 (a) Arrangement of a scan drum and scan mirror (top) using an off-axis ellipsoid mirror (bottom) that allows the radiometer beam to be focused to a small spot size at the scan drum aperture and the detector at the other focal point. (b) Photograph of the ISAR radiometer scan drum (top) showing 10 mm aperture and mirror located inside the mounting block (lower). Note the massive size of the mirror mount block (38 mm diameter constructed of stainless steel) to ensure that the mirror temperature remains stable.

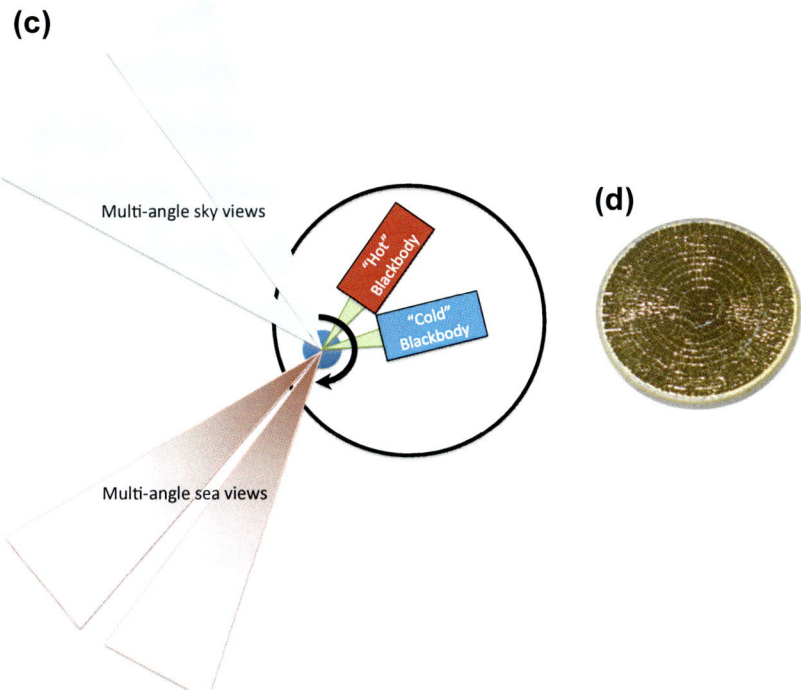

FIGURE 17 (c) Scan drum/mirror arrangement showing how multiple beams can be used to make measurements at different sea and sky locations and internally to view black body calibration targets. (d) Cracked and corroded gold primary mirror shown in the ISAR radiometer [39] scan drum housing after 3 months in the marine environment. The degradation of the mirror was due to thermal stresses during field deployments. The mirror design and gold deposition process was subsequently modified to resolve the problem.

protected-gold on glass-substrate scan-mirror was changed. The lesson learned is that extensive tests should be made using the specific mirror intended for use in the system to ascertain its degradation characteristics prior to any scientific deployments.

To conclude this section, as noted by the SST radiometers for CDRs International Space Studies Institute study group [99], for SST FRM radiometers it is essential [99] that:

- The axis of rotation of the scan mirror be aligned with the optical axis of the optical chain, so that the same part of the scan mirror surface is used in all views, and at the same angle of incidence on the mirror surface,
- The scan mirror has a high reflectance in the TIR part of the spectrum (typically, this is achieved by using an uncoated gold surface),

- The mirror's optical surface and substrate be both robust to a humid, saline environment,
- The scan mirror can select views to any black body and to any required view of the external scene,
- The angular position of the scan mirror shall be reported.

It is also recommended [99] that:

- The temperature of the scan mirror, or a close proxy, be recorded,
- The scan mirror has an optical and environmental baffle that rotates with it. This can also be used as a shutter to isolate the black bodies when they are not being observed,
- The optical system has well-defined field- and aperture-stops so that the envelope of the optical beam is well defined (to facilitate optical alignment) and well understood,
- As much as possible of the optical system, including the detector and spectral selection mechanism, be contained within a sealed enclosure that is fitted with a suitable high-transmission infrared window,
- The surface coating and substrate of the infrared window and any reflectors must be robust to a humid saline environment,
- Extensive tests should be carried out prior to active measurements to determine the degradation characteristics of optical components,
- In filter radiometers, the window's transmission be characterized and included in the instrument's composite spectral response and any out-of-band transmission,
- The optical system should be well baffled where appropriate to control stray light.

3.4 Thermal Control System

Thermal stability is a critical element of any TIR design. Without thermal stability it will be extremely difficult (if not impossible) to maintain the calibration of the detector signal at an appropriate quality. As internal calibration is generally performed periodically against on-board reference blackbody cavities, the goal is to ensure that all instrument temperature change is small and takes place over a much longer time interval than the instrument calibration cycle.

In general, a ship-borne radiometer is maintained at ambient air temperature. The main challenge is to then to manage thermal shock from direct solar illumination. At night, assuming a homogenous air mass, a ship-borne radiometer will in general follow the slowly varying temperature of the atmosphere. During an overcast cloud conditions during the day, solar illumination is via diffuse radiance and will tend to vary slowly (assuming an homogenous cloud thickness) although significant temperature excursions can be expected. However, in mixed cloud conditions, solar thermal shock via direct and immediate solar illumination as clouds pass by must be managed. The use of reflective paint and instrument shading is a first approach. Multiwalled housings with good insulating properties can also be used. Another approach is to mount sensitive components (e.g., detectors, calibration black bodies) within a massive housing having a high heat capacity to absorb thermal shock (as in the case of the ISAR instrument design). Active cooling increases power demands and introduces a risk of condensation build up and contamination of optical surfaces with moisture resulting in zero throughput of the radiometer. For these reasons it has been avoided by most ship-borne radiometers and elevated temperature control is preferred if thermal management is required [92]. Active temperature management is in some cases unavoidable (e.g., when InSb or MCT cooled detectors are used) and appropriate design is required to ensure a practical and functional solution. What is important is that thermal management is considered in the TIR design at the outset which is often best accomplished when developing the radiometer housing.

3.5 An Environmental System to Protect and Thermally Stabilize the Radiometer

For any optical instrument intended for use in the harsh marine environment, adequate environmental protection is critical. Rain, seawater spray, and high humidity can ruin calibration systems and rapidly destroy all poorly protected components and fore-optics [39]. The marine environment aboard a ship—particularly towards the bow area—is characterized by high humidity and large volumes of marine aerosol (composed almost exclusively of water and salt). During bad weather seawater thrown up by the ships' bow (Figure 18) and in some regions strong solar illumination during the day.

Such an environment demands that all electronic circuitry be enclosed in a waterproof housing. The ubiquitous deposition of salt while at sea further requires that instrument optical components and calibration radiance sources be all effectively insulated from the external environment. This has the further advantage that during periods of inclement weather, the instrument is fully protected and a minimal amount of attention by an operator is required. The latter consideration should not be underestimated as weather changes extremely quickly at sea and typical radiometer installations are in relatively challenging locations (such as a ship's fore mast) and may be difficult and dangerous to reach. In certain ocean regions there may be extremes of environmental temperature in which the instrument is required to operate. Tropical regions will have a high amount of insolation during clear sky conditions that can appreciably heat the instrument and in some cases result in a null calibration. In higher latitude regions freezing and subzero temperatures may require the use of active heating to prevent ice accumulation.

352 Optical Radiometry for Ocean Climate Measurements

FIGURE 18 The RRS James Clark Ross making way highlighting the amount of spray and water from the ship that must be considered when developing a ship-borne FRM radiometer. The SISTeR radiometer is installed on the foremast to the left.

It is critical that any optical surface within the radiometer does not become completely wet; otherwise the optical system will have no throughput. However, it is unrealistic to expect that an instrument deployed on a ship with a clear view of the sea surface will not, at some point, get wet. If an operator is available with a sharp eye for bad weather, the radiometer can be manually protected although this solution is not without risk (both to the radiometer and to the operator). Alternative strategies that employ rain sensors and automatic systems to seal the radiometer in bad weather are preferable. This is the approach taken by the ISAR and SISTeR radiometers. In addition, the optical design must allow for a limited amount of rain water or sea spray during the time taken to completely seal the instrument. Contamination of optical surface in the marine atmosphere by the deposition of dust or salt on their surfaces presents an unavoidable problem. Aerosol dust (e.g., Saharan dust in particular) or severe salt contamination of optical surfaces remains a challenge. However, using a calibration system that includes the entire instrument optical path and internal reference black bodies ensures that moderate dry contamination of the optical surfaces can be tolerated since this will only decrease the signal relative to the noise of the system rather than introducing significant calibration bias.

In summary, when mounted on a ship, an SST radiometer is exposed to all weathers, and, sometimes, to challenging mechanical environments. Further, it is subjected to frequent shipping, installation and deinstallation. As ship-borne SST radiometers are required to view the sea surface free of the influence of the ship, a typical radiometer installation will view the sea surface with the aperture pointing towards the ships' bow. In heavy weather significant amounts of seawater are likely to be thrown up and enter the radiometer wetting the filter surface and reducing the radiometer throughput to zero. It is essential [99] that:

- The instrument has a stiff, robust structure to maintain its integrity, optical alignment and calibration during shipping and handling,
- The complete instrument be enclosed in a robust casing that is watertight, except for a viewing aperture,
- The viewing aperture can be sealed when rain or spray is present.

It is recommended [99] that:

- The instrument has a significant thermal mass, to damp environmental temperature changes,
- The instrument has a white overall external coating to reduce solar heating.

3.6 Instrument Control and Data Acquisition

A dedicated instrument control and data acquisition system is required to operate a ship-borne TIR radiometer. The instrument control computer may be on-board the instrument or via a remote computer or a combination of both approaches. RS-232 and RS-485 serial communications protocols are common. The advantage of using an on-board system is that installation on-board a ship is minimized limiting potential problems associated with long cable runs. Typical installations at sea use custom regulated uninterruptable power supplies (UPS) having surge protection devices. However, such devices cannot compensate for low current−voltage brownouts that often occur on a ship, which in many cases will have an unpredictable effect. In addition, the EM noise levels found on typical ship installations are often high. Sources include high power radio transmitters, radar equipment and mechanical noise generated by the ships engines and various attachments often found on ships such as cranes, derricks, pumps, etc. To minimize these effects heavy duty shielded cables should be used at all times especially over the long cable runs characteristically encountered on ship installations. Digital connection to the instrument is strongly advised for the same reasons. Cables can cause problems if not earthed correctly or are so long that power loss occurs. The challenge of effectively grounding a ship to earth and the nonstationary nature of ships' power supplies must all be considered with care for every deployment.

Data storage should be designed for the typical duration of a deployment with sufficient margins in case the deployment is extended. An electronic watchdog-timer can reboot an instrument if no activity has been recorded for some time. This feature provides an "auto-self protection" system if the first operation following reboot is to place the instrument in a safe-mode. Electronics should be of the highest quality in terms of analog to digital (ADC) conversion with robust accurate bridge circuits to ensure that all temperature sensors are sampled properly. Adequate power and signal conditioning is required to control on-board stepper motors, shaft encoders, thermistors etc. Redundant systems may also be considered for operational autonomous instruments if appropriate. Care must be taken to ensure that all time-stamps are referenced to the appropriate time system in use by other supporting instruments. It is essential [99] that:

- The excitation and readout electronics for the black body thermometers have a traceable performance that does not significantly limit the accuracy of the thermometric measurements,
- The detector preamplifier and signal processing electronics do not add significant nonlinearity or noise to the detector signal,
- The complete instrument state, including all available housekeeping data, be recorded regularly,
- All data must be time stamped.

It is recommended [99] that:

- Digitized data from the readout electronics not be quantization limited,
- All electronic systems be integrated within the instrument,
- Power consumption be carefully managed to limit instrument self-heating,
- For safety, the instrument should be powered from a low voltage DC source.

Simple "machine independent" formatting of data is preferred to facilitate data processing. Several systems use a National Marine Electronics Association (NEMA) 0183-like ASCII text string approach for data formatting—if manageable depending on data rates. This provides easy access to data and ensures that data are easy to interpret without additional processing. NEMA style data can be easily compressed using any number of routines to minimize data volumes (if a problem). Spinning hard-disk drives should be tested aboard ships before use to determine if the continuous movement of the ship has any impact on the drive reliability and performance. Solid-state drives may provide a better solution. It is useful to include all configuration parameters used by the radiometer in the header of every data file written to ensure that the correct parameters are used when postprocessing radiometer data.

3.7 A Calibration System

The detector of an infrared radiometer will respond not only to radiation emitted from a target source but also stray radiation reflected into the radiometer FoV, reflected from the detector and fore-optics and, changes of the detector temperature itself (i.e., stray radiation will originate both from within the instrument itself and by reflection at the sea surface). If the temperature of the detector and band-pass filters are not regulated, as is the case on many in situ instruments, their response function may vary as the instrument temperature varies. This results in a gradual drift of the instrument calibration.

The role of a calibration system is to quantify the radiometer detector output in terms of the measured target radiance incident on the detector. In order to correct for instrument calibration drift, calibrations against a known reference radiance source need to be made frequently. For an FRM TIR radiometer, calibration must be traceable to SI standards if it is to be used to validate satellite measurements that from part of the SST_{skin} CDR [28]. Proper calibration which, ideally includes calibration of the full dynamic range of signal (i.e., sky radiances and sea radiances), accounts for the following primary sources of error:

- Detector nonlinearity,
- Compensation of fore-optics (i.e., mirrors, lenses, windows) and their temperature dependencies,
- Unavoidable drifts in detector gain and bias
- Temperature dependence of the end-to-end system (strays)
- Long-term degradation of optomechanical and electronic components.

In general, the calibration of employed by different FRM SST ship-borne radiometers falls into two distinct categories:

1. *Externally calibrated* radiometers in which an external black body source is periodically viewed to maintain calibration.
2. Those using internal reference black body cavities that are periodically viewed and are thus *self-calibrating*.

3.7.1 External Stirred Water Bath Calibration

An external calibration system for ship-borne SST radiometers was developed in 1975 [113,114] that was designed to compensate for the primary sources of calibration error. In this calibration method, known as the 'stirred tank method', a well-stirred tank of seawater is periodically viewed ($\sim 2-5$ min intervals) by the radiometer alternately with a sea surface target measurement. Seawater has a very high emissivity and, in effect, the stirred tank acts as an external black body source. The temperature of the water in the tank is accurately determined and, assuming a total breakdown of the thermal skin due to turbulent motion driven by vigorous stirring of the bucket water, the radiometer can, in theory, be absolutely calibrated by relating the measured water bath temperature to the radiometer signal. An accuracy of ± 0.05 K is cited when using this method [84,90,113−115]. The approach claims [90] to compensate for (1) the deviation of the sea surface emissivity from unity including variations in surface reflectivity for off-nadir pointing, (2) radiance contributions reflected at the sea surface from the clear sky and cloud measured by the radiometer, (3) contamination of the radiometer fore-optics by sea spray, (4) temperature drift of the radiometer and any reference blackbody cavities used for internal calibration and, (5) the drift of the instrument electronics where (4) and (5) are largely driven by direct solar heating.

A major drawback is the implementation of this system is that the radiometer or the stirred tank needs to be physically moved on a regular basis. Most authors using this approach [84,90,113−115] choose an automated system in which the stirred tank, rather than the radiometer, is moved into the FoV of the radiometer. Figure 7(e) shows the calibration system used by the OPHIR Multi-Band Infrared Sea-truth Radiometric Calibrator (MISTRC) system aboard the Research Vessel (R/V) Vickers [74,115]. In this system two nested buckets are used in which seawater is pumped up to the base of the inner bucket through small holes producing strong vertical jets of water, which it is assumed, destroy the surface skin layer. Water then spills over the side of the inner bucket and is carried away via the outer bucket. The temperature in the inner bucket is continuously monitored with a PRT and a precision thermistor and the whole unit is mounted on a dual axis gimbal frame to maintain a constant zenith angle view by the MISTRC. The bucket assembly is automatically carried into the MISTRC FoV using a rail track assembly every minute.

The assumption that the skin of a water bath (and therefore the thermal skin temperature deviation from that measured in the water itself) is completely destroyed by vigorous stirring is the crux of this method. Laboratory experiments [112,116] collected detailed measurements of a stirred tank using infra-red cameras and concluded that extremely vigorous stirring of the bucket was required to completely destroy the skin temperature deviation. This was in part due to the role of wind on the bucket water surface (which typically fully exposed on a ship installation and as the ship is moving there is always a relative wind acting on the bucket surface) driving quasi-instantaneous [112] evaporative fluxes resulting in a variable the skin temperature deviation. A second criticism of this calibration method is the assumption that the downwelling sky radiance reflected both at the sea surface and bucket water surface the radiometer is fully accounted for. The stirred-bucket method, in theory, implicitly accounts for the sky-reflection by viewing the exposed water bath and sea surface at the same angles with the advantage that no direct measurement of sky radiance is required. However, this critically assumes a uniform cloud cover (overcast or clear sky which is rarely the case) for both sea and calibration bucket measurment cycles if errors are not to be introduced into the final measurements [53]. Another criticism of this calibration approach is the fact that the radiometer only has an effective single point calibration because the water bath temperature is often slightly warmer than that of the sea surface—warming having taken place as the water is pumped through the ships internal pipework. This means that the calibration gain relationship remains largely undetermined for local measurements and a time series of data is required to derive gain values via data post-processing. This approach then assumes linearity of the calibration over time. Finally, it must also be recognized that the stirred-bucket approach is a "fair weather" calibration approach. As wind speeds increase, spray is blown from the bath surface onto the radiometer head often wetting fore-optics with the consequence that the radiometer measures the temperature of the wetted optical surface and never actually views the sea surface (contrary to the claims of Ref. [90]).

An interesting measurement approach exploits Brewster's angle. On reflection at the sea surface, diffuse sky radiance is polarized [117] and, at Brewster's angle ($\sim 53°$ from nadir at a wavelength of 11 µm), the horizontally polarized (h-pol) radiance component is negligible for a given wavelength (Figure 19). Thus, at Breswter's angle only the vertical polarization (v-pol) component remains and, if a radiometer's spectral response is optimized for v-pol, negligible reflected sky radiance (h-pol) is measured by the radiometer. The MISTRC radiometer [74] implements this approach using a horizontal grid polarizing filters as part of the short-wave optical chain that are particularly effective in removing "sunglint" effects. In practice, because Brewster's angle is very sensitive to the geometry of a particular deployment (limits of approximately $\pm 2°$) this technique is best suited to deployments from fixed platforms and/or when the sea surface is relatively calm. Further, the use of a polarizing filter may reduce the signal falling on the detector increasing the signal-to-noise (SNR) ratio of the radiometer.

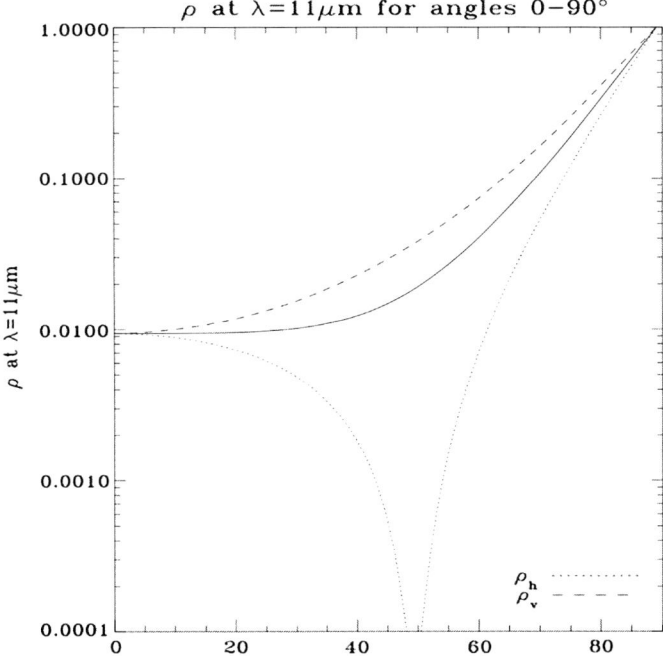

FIGURE 19 Polarization of reflected sky radiance at 11 µm as a function of viewing angle. The total polarization is shown as a solid line.

The challenges and debatable assumptions underpinning this calibration method [74] together with the logistical arrangements of installing and maintaining a stirred bucket system aboard a ship, means that this calibration method has now been largely superseded by an internal instrument calibration based on internal reference radiance cavities and a second measurement of downwelling sky radiance [74].

3.7.2 Self Calibrating Radiometers Using On-board Reference Radiance Sources

Self-calibration is the process of illuminating a radiometer's complete optical chain with two (or more) known sources of infrared radiation [99]. From two (or more) such points, a linear (or higher order) relation can be derived between detector signal and external radiances for the current instrument state. With this relation, the detector signal can then be converted to radiance for subsequent observations of an external scene. This process is repeated at a rate that is fast enough to capture and characterize any changes in the instrument's self-emission and responsivity. Typically, calibration and scene measurement cycles are between 1 and 10 min long.

In most ship-borne TIR radiometer designs two precision temperature controlled reference calibration blackbodies are used to provide a two-point (i.e., a "hot" and a "cold") calibration over a limited range of temperatures. This can also be achieved by dynamically changing the temperature of a single blackbody. A simple approach is to allow one blackbody to "float" at the ambient instrument temperature that is always close to the temperature of the sea surface and the other cavity is heated above the ambient temperature (i.e., warmer than the actual SST). With this arrangement the accuracy of calibration is optimized in the temperature range of the target SST_{skin} because linearity is only assumed over a small range of temperature and the effects of any nonlinearity in the calibration system are minimized. Clearly the instrument calibration is degraded at the much colder brightness temperatures (<180 K) recorded when viewing a clear dry atmosphere. As $<2\%$ sky radiance is typically reflected at the sea surface, the impact of calibration errors due to extrapolation on the overall accuracy of SST_{skin} determination is minimal even when considerable errors of >5 K are evident. A viable solution using a cryogenic cold reference radiance source for on-board calibration of ship-borne TIR radiometers has yet to be found.

Figure 20(a) shows a section view of a typical re-entrant-cone blackbody reference radiance cavity design similar to those used by the ISAR, SISTeR, BEST, SOOSR, and DAR-011 radiometers. Figure 20(b) shows an alternative large aperture cavity design used by the MAERI (with a similar cavity profile also used by the CIRIMS and JPL NNR designs).

A re-entrant cone and a partially closed aperture design which, combined with a high emissivity surface finish (e.g., Nextel velvet black) and critical internal geometry, ensure that the black body cavities have an emissivity of >0.999 in the TIR waveband [118]. Different radiometer designs call for

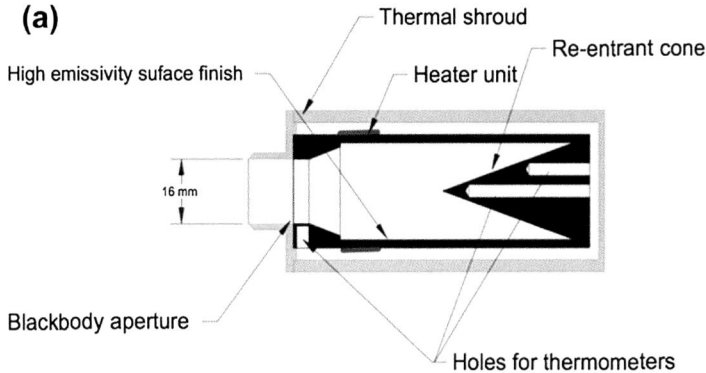

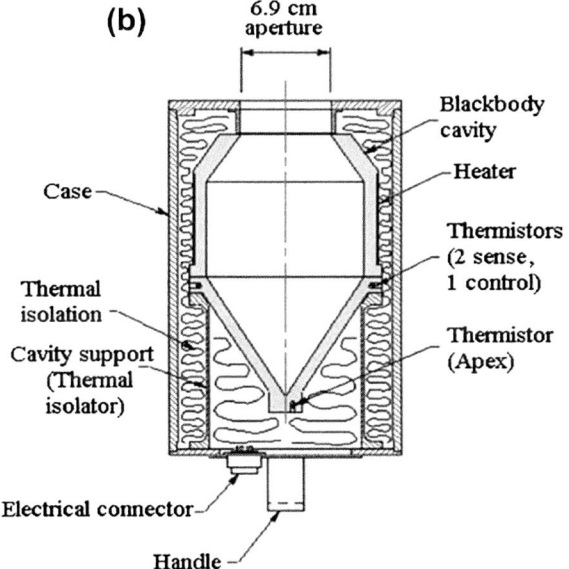

FIGURE 20 (a) Section through an ISAR radiometer [39] calibration blackbody calibration cavity showing the re-entrant cone design, thermal shroud and location of thermistors used to determine the radiative temperature of the blackbody. The inner surfaces of the blackbody are coated with NEXTEL® suede 3103 paint. The cavity emissivity of this design is 0.9993. (b) The M-AERI blackbody cavity [79]. Two such cavities are used in the routine calibration of the interferometer; one being heated to 330 K and the other floating close to ambient temperatures. To validate the calibration, a third cavity is attached to the instrument at the position of the zenith view, the temperature of which can be programmed to a number of set points spanning the SST range.

different blackbody aperture sizes: the MAERI has a large collimated beam of several cm diameter and requires relatively large blackbody apertures (670 mm) as shown in Figure 20(b). Care must be taken in the radiometer optical design to ensure that the radiometer beam *under-fills* the aperture of the blackbody to avoid calibration errors.

Thermistors, RhFe thermometers, or Platinum Resistance Thermometers (PRTs) having an SI traceable calibration are used to monitor the temperature of each black body. Thermistors located in the cavity base provide the primary measurement and additional thermistors are installed along the cavity axis to detect and monitor any thermal gradients when operated in a heated mode. The radiance cavity is often housed in an insulating shroud (e.g., leaving a small insulating air gap between the outer wall and the shroud or using insulation material) to inhibit heat loss and maintain temperature uniformity. During operation, one black body cavity remains at the ambient temperature while the heater of the other is allowed to equilibrate to a higher temperature. A constant-power resistive heating element wrapped around the outer diameter of the radiant cavity is used in the Figure 20(a) design to raise the temperature of the cavity. The design used in Figure 20(b) has an adjustable heater circuit allowing a larger temperature range to be utilised.

A calibration black body must be positioned at the end of the radiometer optical path so that the radiometric impact of any optical components is identical for all instrument views (i.e., all target views and all blackbody views). As the blackbody is generally exposed to the marine atmosphere (although the CIRIMS radiometer design is a notable exceptions to this arrangement see Section 4.4), active cooling of black body cavities is not considered as condensation may occur on inner surfaces leading to erroneous calibration data (there would be a skin temperature deviation across the condensate on the blackbody surface decoupling the blackbody temperature sensors from the radiating surface of the condensate). Furthermore, active cooling would significantly increase instrument power consumption and require careful thermal design to ensure that all heat is rapidly and effectively conducted away from the radiometer. In many ship-borne radiometer designs blackbodies remain fixed and a rotating mirror is used to select the appropriate target. In the case of the JPL NNL and the CIRIMS, the radiometer detector head is moved to view a reference blackbody or target. The RAL/SIL and SOOSR radiometers used an alternative approach in which black bodies are periodically moved into the radiometer FoV using a stepper motor and rotary arm holding each blackbody.

Considering the simple filter radiometer described in Figure 9, a typical SST_{skin} measurement cycle involves rotating the target selection mirror to measure four different radiances in a measurement "scan" sequence: (1) radiance from the sea surface $L_{scene}(\lambda,\theta)$, (2) radiance from the sky $L_{sky}(\lambda,\theta)$, (3) radiance from the ambient temperature black body, L_{bba}, and, (4) radiance from the heated black body, L_{bbh}. The detector output is related to the incoming radiance integrated over the radiometer's frequency pass band by a linear relationship. Radiometric calibration of the instrument consists of determining the linear relationship between detector output and the integrated incoming radiance.

The signal measured by the detector when viewing the ambient blackbody (BB_{amb}) is given by:

$$C_{\text{amb}} = G\zeta_B B_B(T_{\text{amb}}) + O_o \quad (23)$$

where T_{amb} is the thermometric temperature of BB_{amb} and the signal when viewing the hot blackbody (BB_{hot}) is given by

$$C_{\text{hot}} = G\zeta_B B_B(T_{\text{hot}}) + O_o \quad (24)$$

where T_{hot} is the thermometric temperature of BB_{hot}. The radiometric gain G, of the system can then be derived from Eqns (23) and (24) using

$$G = \frac{(C_{\text{hot}} - C_{\text{amb}})}{\zeta(B(T_{\text{hot}}) - B(T_{\text{amb}}))} \quad (25)$$

where C denotes measured detector counts, B is the Plank function, T is the radiometric temperature of the ambient or hot blackbody cavities. ζ is radiometer spectral response function. By substitution, the calibration offset can be found using

$$O_o = C_{\text{amb}} - G\zeta_B BB(T_{\text{amb}}) \quad (26)$$

This scheme assumes an emissivity of 1.0 for the black body cavities and that the output of the detector is proportional to the radiance. The radiance of each blackbody cavity is calculated using the temperatures measured by the embedded thermistors (ensuring full traceability to SI standards) and the radiometer specific radiance-to-temperature and temperature-to-radiance functions based on spectral integration of the Plank function across the radiometer spectral response function. These are specific to *each* radiometer.

The internal calibration of an FRM ship-borne TIR radiometer needs to be verified and traceable to SI standards on a regular basis as this is the foundation of the CDR. Several external precision reference blackbody radiance sources [87,119,120] that are traceable to SI standards [121−123] are used for this purpose as discussed in Chapter 5.2.

3.7.3 NNR Calibration

An alternative approach based on the nulling radiometer calibration [80] method is discussed in Section 4.3.

3.8 Summary

The complex deployment logistics and questionable assumptions of the stirred water-bath calibration technique lead the authors to conclude that it should not be used for FRM of SST that underpin the CDR. On-board reference blackbody radiance sources are preferred in a self-calibrating radiometer system.

When using blackbody reference cavities for FTM TIR SST$_{skin}$ radiometers it is essential [99] that:

- As much as possible of the radiometer optical system, including the detector and spectral selection mechanism, be contained within a sealed enclosure that is fitted with a suitable infrared window,
- The entire optical chain, including windows, lenses, mirrors etc. must be calibrated end-to-end against the instrument's reference radiance source.
- To determine the uncertainties with which the ship-borne radiometers are capable of measuring the SST$_{skin}$, the instruments have to be self-calibrating in the field. That is, they must have internal calibration targets against which the measurements of the sea can be compared. To assess the uncertainties in the internal calibration, and the stability of the internal calibration targets, external calibration of the entire radiometer system should be periodically undertaken using laboratory calibration facilities; typically before and after every field deployment. To fulfil the requirement of the BIPM that "measurements…are in terms of well-characterized SI units" the laboratory calibration facilities should have an unbroken chain of comparisons to SI standards.
- The calibration black body target emissivity must be high and well understood. The uncertainty in the emissivity should not dominate the instrument's error budget,
- The calibration black body target aperture should be large enough to completely contain the instrument view, including any diffraction fringes. Ideally, the aperture should not be significantly oversized, as this will reduce the emissivity of the cavity,
- The calibration black body target must be thermally isolated from its immediate surroundings,
- The calibration black body target must be isolated, as far as possible, from the external environment,
- The temperature of the calibration black body target emitting surface, particularly the part directly observed by the radiometer, be accurately measured with one or more thermometers,
- The calibration black body target thermometers must be recalibrated regularly and traceable to SI standards.

It is recommended [99] that:

- The calibration black body target should be mounted on external structures far removed from the directly observed surface. This will be near to the aperture,
- One calibration black body target should ideally be maintained at a temperature cooler (or near to) the SST brightness temperature to limit calibration gain sensitivity. In order to reduce the risk of condensation on the blackbody surface it can be allowed to float at ambient temperature and a second blackbody can be operated at an elevated temperature.
- The hot black body cavity be heated near the point where it is mounted, to reduce temperature gradients along the cavity walls.

3.9 Additional Comments

Additional sensors to support the primary SST_{skin} measurements have often included in the design of FRM ship-borne radiometers. A low-cost Global Positioning System (GPS) unit provides real-time position, course made good, speed made good, heading and UTC time for the radiometer. Oil-damped inclinometers provide measurements of ship roll and pitch: such measurements are essential for selecting the most appropriate value for seawater emissivity.

4. EXAMPLES OF FRM SHIP-BORNE TIR RADIOMETER DESIGN AND DEPLOYMENTS

4.1 The DAR-011 Filter Radiometer

The DAR-011 filter radiometer [85] is a single-channel, self-calibrating, infrared radiometer developed within the CSIRO Division of Atmospheric Research. The radiometer has a long heritage going back many years and is the culmination of developments leading to a reliable accurate instrument. Full details of the instrument are provided in Ref. [85].

A rotating 45° plane mirror sequentially views the sea, a hot black body calibration target, the sky, and finally an ambient temperature black body calibration target. The temperatures of the two calibration black bodies are accurately monitored providing good absolute radiometric accuracy. The 10-min operating cycle includes 7 min viewing the sea, and 1 min each viewing the hot black body calibration target, the sky, and the ambient temperature black body target. In all cases readings are taken once every 0.4 s and the sea-view data are averaged up to 1-min values. The sky radiances are interpolated with time to provide a value to be used for the sky correction of the sea-view measurements.

The incoming radiation is physically chopped against a second ambient temperature black body and the chopped radiation is focused with a 45° parabolic front surfaced mirror onto a pyroelectric detector. Before reaching the detector the radiation passes through an interference filter that passes radiation with wavelengths between 10.5 and 11.5 μm. The radiometer operates with a fixed spectral width of 1 μm centered on 11.0 μm. No other wavelength measurements are available. 150-mm-long circular tubes are attached to the two instrument apertures (sea- and sky-view) to help protect the internal optics and calibration black bodies from sea spray and other contaminants. Although the two calibration black bodies are open to the ambient environment they are located well inside the radiometer system.

The advantages of this design include:

- No windows are used between the sea- and sky-viewing apertures, the blackbody calibration targets, and the interference filter that is located in front of the pyroelectric detector. Thus no corrections are required for

window transmissivity and the reflectivity of the associated optical surfaces.
- Radiation from the two calibration black bodies and the sea and sky views all take the same optical path via a 45° reflection off the rotating plane mirror.
- The output from the detector is passed through an amplifier that is phase-locked to the physical chopper.
- The integration time used to smooth the chopped signal from the pyro-electric detector can be varied between 0.1 and 10 s, with most deployments using an integration time of 1 s.

The design is limited by the following choices:

- The sampling/calibration cycle provides 7 min of sea-view data followed by a calibration period of 3 min in which no SST data are available.
- The optics and black bodies are open to the environment and care is required to ensure that contamination is minimal. The radiometer thus requires careful calibration between deployments to ensure the integrity of the black body calibration system.
- Most contamination occurs on the 45° plane mirror. However, even with obviously visible contamination the black body calibration system and the phase-locked loop ensure that good measurements are still produced in these cases. Spare mirrors are available to replace those damaged by salt from sea spray.
- With the radiometer optics being open to the environment, and having no method for automatic covering, the system must be manually covered with a plastic casing and covers attached to the two open apertures whenever there is a possibility of rain or extreme sea conditions with excessive sea spray. The system thus requires constant attention and is not suitable for deployments on ships of opportunity unless a dedicated operator is available.
- With the physical structure of the radiometer the sky-viewing aperture is opposite to the sea-viewing aperture, causing the sky correction to be from the incorrect direction. Care must be taken with the radiometer installation to ensure that none of the ship superstructure obscures the (backward) sky view.

4.2 The SISTeR Filter Radiometer

The Scanning Infrared Sea Surface Temperature Radiometer (SISTeR) is a compact and robust chopped self-calibrating filter radiometer (Figure 21). The SISTeR has been designed to survive and maintain its calibration over extended periods in a maritime environment. It measures approximately $20 \times 20 \times 40$ cm and weighs about 20 kg. The instrument is divided into three compartments: one containing the fore-optics, a central compartment housing

FIGURE 21 The SISTeR radiometer and supporting equipment.

a scan mirror and reference black bodies, and third compartment containing a small-format PC with signal processing and control electronics. The fore-optics and electronics compartments are waterproof and the scan mirror and black bodies are carefully protected with interleaved baffles.

The fore optics compartment contains a deuterated L-alanine doped triglycine sulfate (DLATGS) pyroelectric detector and preamplifier, mounted onto assembly containing a concentric 6-position filter wheel and a black rotating chopper. The filter wheel presently contains three narrow-band filters centered at 3.7, 10.8, and 12.0 μm, matching those of the satellite Along Track Scanning Radiometer (ATSR) instruments. The beam is chopped at 100 Hz, a compromise between the optimum noise performance of the detector and a fast filter response in the signal processing chain. The main optical element is an ellipsoid mirror, by which the detector can view a 45° scan mirror through an antireflection coated ZnSe window.

The scan mirror can direct the detector's view to either of two internal black bodies or to the external scene at any point about an arc spanning 180° from nadir to zenith. Two concentric baffles surround the scan mirror. A field-plane is centered in an exit aperture in the outer baffle, so that the aperture can be made as small as possible to ensure the maximum internal protection. The full cone angle of the instrumental FoV is approximately 13°.

The self-calibrating design of SISTeR is intrinsically tolerant of contamination to its optics. The entire optical system is referred to two highly accurate reference black bodies. One floats near to ambient temperature and the other at approximately 15 K above ambient temperature. Embedded in each black body is a 27 Ω RhFe thermometer. The entire black body cavity can be installed in a specially-constructed calibration block maintained by Oxford University, and the thermometer calibrated to an absolute accuracy of better than 4 mK relative to ITS-90. As the ambient temperature on a boat or ship is generally very near to the SST, the cooler black body temperature always tracks that of the sea.

4.2.1 SISTeR Operation

All aspects of the SISTeR instrument, from the scan mirror position to the detector signal are accessible through variables defined in a C library. Control programs of arbitrary complexity can be written, but generally just a few lines of code are needed to define a scan sequence. When a control program is running, the complete instrument state is transmitted over a serial link to a laptop ground station after every measurement. All SISTeR measurement sequences contain repeated measurements of its two internal black bodies. In addition, to calculate the skin SST, the SISTeR is programmed to make measurements both of upwelling radiances from the sea surface and complementary downwelling sky radiances. In the SISTeR longwave channels, the measured noise temperature for a 1 s sample at typical SSTs is less than 30 mK. Measurements of an external CASOTS-I black body [87] before, during and after a typical one-month validation campaign showed that the SISTeR calibration remained repeatable to better than 20 mK, even though the scan mirror finish had deteriorated noticeably over the same period.

4.2.2 Skin SST Measurements

Typically, SISTeR radiances $L_{\text{scene}}(\lambda,\theta)$ and $L_{\text{sky}}(\lambda,\theta)$ are sampled every 0.8 s with the 10.8 µm filter. Skin SSTs are calculated from the upwelling ocean radiance samples following the approach described in Section 3.7.2.

4.2.3 SISTeR Mounting and Support

The SISTeR is generally mounted as far forward and as high as possible on the host ship, so that it is clear of "green water" and spray, and can view undisturbed water forward of the bow wave. Where possible, the viewing angle to

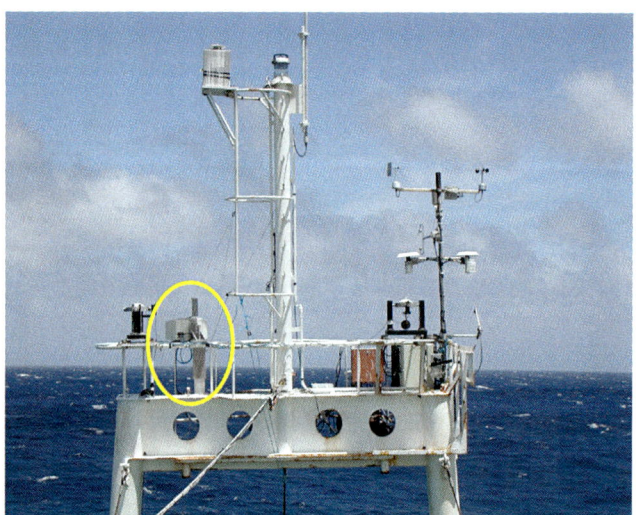

FIGURE 22 Installation of the SISTeR radiometer aboard the RRS Charles Darwin foremast.

the sea is kept within the range 15–40° from nadir [56]. The instrument also requires a clear view to the sky at the complementary angle from zenith.

The SISTeR is equipped with a quick release mount and is provided with a small turret, to which a mating bracket is attached. The turret can be mounted on a horizontal surface with a pattern of eight holes. A small horizontal platform, with the pattern predrilled, is also available and can be attached to handrails with U-bolts (see Figure 22). The instrument requires 24 V DC power and serial data connections. Instrument data is logged remotely on a laptop PC. Waterproof power supplies, serial modems and cable sets are available for runs of 100 m or more with terminations for a variety of mains outlets.

Figure 23 presents a typical example data set obtained using the SISTeR radiometer highlighting the multi-angle measurements of sky radiance that can

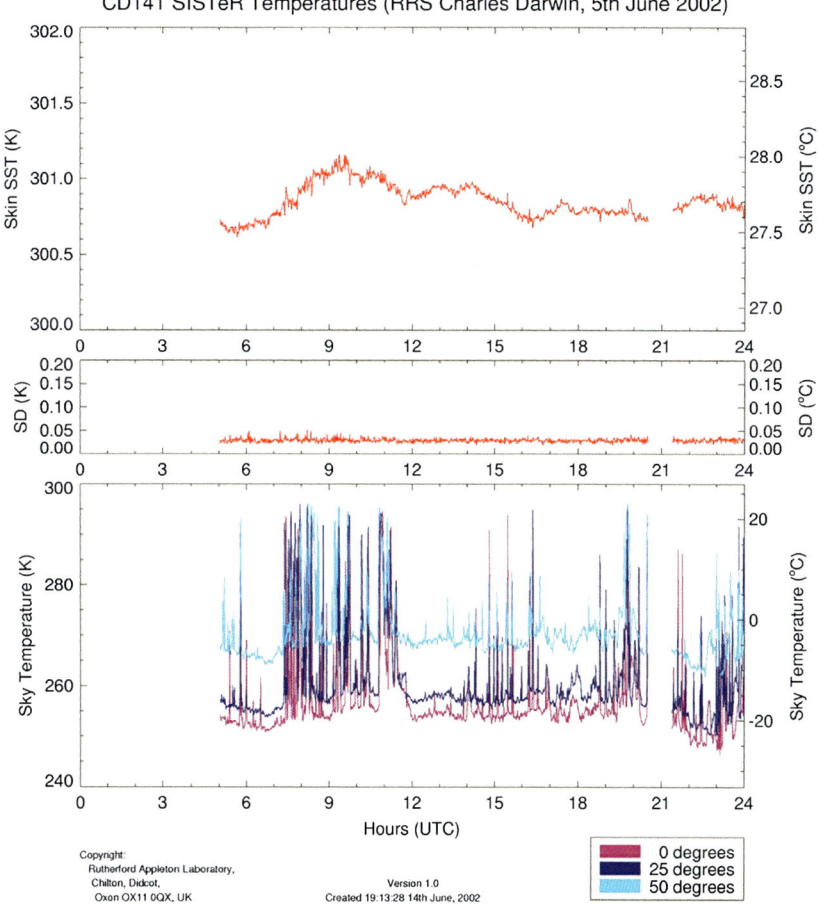

FIGURE 23 Example data set acquired from the SISTeR radiometer showing the SST_{skin} (top panel), the standard deviation of the SST_{skin} measurements (middle panel) and the sky radiant temperature measured at three different angles used to correct for sky radiance reflection at the sea surface.

be used to improve the corrections required for reflection at the sea surface. Note the highly variable nature of the sky radiance in the presence of clouds and the different mean offsets when looking at different angles—particularly the shallower 50° view that has a much longer atmospheric path length.

4.3 NASA JPL NNR

The NASA/JPL NNR is a low cost, compact, low power, highly accurate radiometer that uses a thermopile sensor [80–82]. The NASA/JPL NNR consists of two main parts. The first is a moveable thermopile (TIR sensor). The thermopile sensor is housed in a cylindrical enclosure that in turn is attached to the shaft of a motor. The motor can point the thermopile sensor in the direction of the internal blackbody or point it outside the unit (Figure 24). The second part is a blackbody with a platinum resistance thermometer and a pair of thermoelectric heater coolers.

A key design philosophy for the NASA/JPL NNR was to develop a radiometer that is robust, compact, low power and can be deployed autonomously on moored buoys. The NASA/JPL NNR's have been deployed on buoys/platforms at Lake Tahoe CA/NV and Salton Sea CA since 1999 and 2007 respectively. The key characteristics of the radiometers are shown in Table 3.

The deployed radiometers have measured the skin temperature of these water bodies, on a near continuous basis (every 2 min) and used to calibrate numerous satellite instruments including the Along Track Scanning Radiometer-2 (ATSR-2), Advanced Spaceborne Thermal Emission and Reflection Radiometer (ASTER), Moderate Resolution Imaging Spectroradiometer (MODIS) and Landsat [80–82].

4.3.1 JPL NNR Operation

The basic operation of the instrument is to first point a thermopile detector head at the desired target scene to obtain a radiance measurement. The

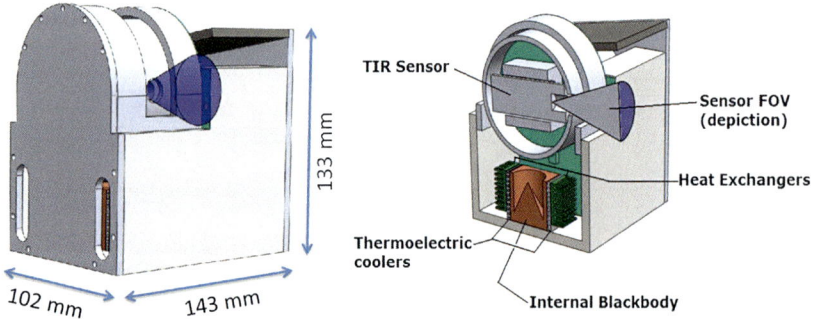

FIGURE 24 The NASA/JPL NNR. Left image: external view showing field of view, right image: cutaway showing internal blackbody.

TABLE 3 Key Characteristics of the JPL Near Nulling Radiometer (NNR)

NASA JPL NNR Specifications	
Length	5.625 inches (143 mm)
Width	4 inches (102 mm)
Height	5.25 inches (133 mm)
Weight	4 lb 3 oz (1.845 kg)
Voltage/current	12 V, typical 300 mA draw, maximum draw of 2.92 A in extreme conditions
Current draw	25 Ah per day
Voltage range	11.5 V stops working, 16 V short durations only
Communication	RS232 at 2400 baud N 8 1 with no flow control
Total field of view	200 series = 44°; 400, 500 series = 36°

thermopile is then rotated to point at an internal reference calibration blackbody. Since the thermopile is smaller and lighter than the blackbody assembly it is the logical choice for the moving component, but still a precision pointing system and flexible connections are required. From a reliability standpoint it would be desirable to avoid the flexible connections but any wireless scheme would add a large measure of complexity. In practice therefore the wired connection is not perfect, but close to optimal.

4.3.2 JPL NNR Calibration

The thermopile is calibrated using a near-nulling approach. A built-in platinum resistance thermometer accurately determines the temperature of the internal blackbody. Determining the brightness temperature (and radiance) of the scene requires calculating the temperature difference between the internal blackbody and the target scene based on the thermopile output difference. The smaller the thermopile output difference between the scene and the internal blackbody the more accurately this calculation reflects the radiance difference. A thermoelectric heating and cooling system is installed within the blackbody cavity to enable the instrument to minimize the thermopile output difference between the scene and the internal blackbody. If the scene and internal blackbody thermopile outputs are equal, the radiances and brightness temperatures from the scene and the internal blackbody are equal. Using the temperature of the internal blackbody as measured with the platinum resistance thermometer along with a laboratory measured instrument predeployment calibration relationship, the brightness temperature for the scene can be calculated. With a known scene emissivity, this brightness temperature can be used to accurately determine the kinetic temperature of the scene.

The near nulling approach, where the radiance measured by the sensor is matched to the radiance from a blackbody at a similar temperature to the scene, enables the use of an inexpensive thermopile sensor. Thermopile sensors are typically nonlinear and the output of the sensor is a function of the difference between radiative fluxes into and out of the sensor surface that is dependent on the temperature of the surroundings as well as the flux from the target. The near nulling approach solves this problem as only the thermopile is being used to match the radiance levels from the target and the blackbody, thereby transferring the measurement accuracy to the internal blackbody.

A true nulling instrument would require that the thermopile readings would be equal. In a field instrument the physical constraints placed on the design make perfect equalization of the two readings difficult. The reason is that the effect of dynamic environmental conditions on the internal blackbody would necessitate strong thermal driving of the blackbody mass, which in turn could cause unpredictable thermal gradients within the instrument. Such gradients would reduce the correlation between the blackbody PRT measurement and the radiance from the blackbody introducing errors. The NNR uses a local linear approximation to the true nonlinear response of the thermopile sensor thereby achieving accuracies similar to those achievable only by using much more demanding and expensive linear sensors. The main advantage of this approach is that a very simple, inexpensive and uncooled thermal IR sensor can be used to perform accurate measurements. For a field instrument, the simplicity and robustness of the thermopile sensor is a major advantage, but another important aspect is the sensor does not require cooling.

There are several challenges with this approach. First and foremost is the comparatively low sensitivity of thermopile sensors. Overcoming this requires amplification of the signal that could introduce significant noise without appropriate amplifying electronics. The components involved have to be of the highest quality and protected from environmental effects. For a field instrument, there are additional challenges due to the size limitations. The blackbody has to be compact and modest in its power demands and the temperature measurement of the blackbody has to be highly accurate and stable over time. Long-term drifts in all the other components including the thermopile sensor are compensated for with the near nulling design and have no impact. Using a modern embedded processor allows the various processes to be easily automated, e.g., heating/cooling the blackbody, rotating the drum.

Calibration of the NNR to National Institute of Standards (NIST) (NIST traceability) involves finding the relationship between the output from the blackbody temperature sensing system and the measured temperature of the internal blackbody. The reason the calibration procedure is necessary is that the emissivity of the small inverted cone blackbody units used in these field instruments has not been traced to NIST standards and is not as accurate as a much larger external NIST traceable blackbody. Thus the platinum resistance thermometer measures the kinetic temperature of the internal blackbody that approximates a true blackbody that is then referenced to a NIST traceable external blackbody.

4.4 The Calibrated Infrared In situ Measurement System

The CIRIMS radiometer [63] (Figures 7 and 25) was designed to operate autonomously on ships of opportunity for at least six months, to withstand harsh weather conditions, and obtain an accuracy goal of ±0.1 K. Unique design features of the CIRIMS included a constant temperature housing to

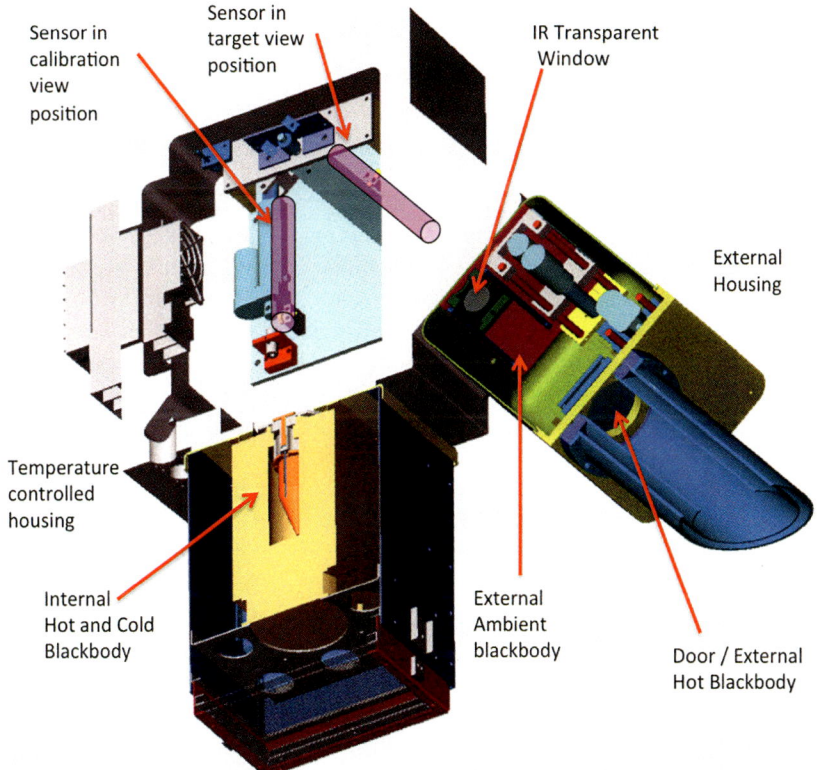

FIGURE 25 Schematic overview of the CIRIMS radiometers showing internal instrument configuration. Details are discussed in the text.

stabilize instrument drift, a two-point dynamic calibration procedure, separate up- and sea-viewing radiometers for simultaneous sea and sky measurements, and the ability to use an infrared transparent window for environmental protection. Extensive comparisons of ocean skin temperature measured by the CIRIMS and the M-AERI spectro-radiometer have been conducted [124].

The CIRIMS housing was insulated and maintained at a constant temperature within a range of approximately 0.1 K about the set temperature (standard deviation 0.1 K) by means of an integrated thermoelectric heater/cooler unit and circulation fan. This provided a stable, dry environment for the internal radiometer and the blackbody. The housing temperature was reset on a daily basis to 5 K above the highest air temperature of the previous day. This algorithm was based on minimizing errors in the correction for nonunity emissivity of the blackbody, the use of two different sky- and sea-viewing radiometers, and the window correction. The sky-viewing radiometer was contained in an unheated external housing attached to the side of the sensor housing. The sky radiometer housing was designed so that rain and spray would not reach the lens.

A common approach for providing a two-point calibration is to use a constant temperature hot target and an ambient (cold) temperature target. Because of the requirement of long deployments and associated uncertainty in the stability of the radiometer transfer function, the CIRIMS design used a different scheme referred to as a dynamic two-point calibration. In this approach, the hot and cold calibration temperatures follow the target sea surface temperature to be measured, bracketing it above and below by 2 K. This dynamic interval calibration technique makes linear interpolation possible and ensures consistent accuracy over a wide range of scene temperatures, albeit at additional complexity and cost. The technique was implemented by using a single blackbody calibration target immersed in a precision temperature-controlled water bath.

The rationale for using separate radiometers for the sea and sky measurements was based on the relative magnitude of the sea and sky contributions to radiance measured. In general, the sea contribution is nearly two orders of magnitude greater than the sky contribution because the emissivity is close to unity and the reflectivity is O(0.01) at moderate incidence. Since the sky contribution term is so much smaller than the sea contribution term, the accuracy requirement for the sky measurement is less stringent than that for the sea. The error from using two different radiometers was assessed by assuming a maximum calibration difference between the two radiometers and varying both the sea and sky radiance. Laboratory testing has shown that the maximum offset between two radiometers was roughly 2.5 K.

Protection of the radiometer and calibration blackbody is arguably the most challenging aspect of a practical design. A major design goal in the development of CIRIMS was to evaluate the use of an IR transparent window to provide complete protection of the optics and the blackbody during deployments when they are susceptible to spray. The motivation behind the use of a window is to ensure complete protection under all conditions because of the possibility of severe weather and sea conditions during a long deployment. This approach relies on the ability to correct for the variable effect of the contaminated window. The primary concerns regarding the use of a window are wetting, the effect of salt deposits on the transmission, and that the self-emission of the window is a function of ambient temperature.

The CIRIMS design included a high priority on the ability to continuously monitor the effect of the window in order to account for changes due to contamination or environmental conditions in an on-going fashion. The CIRIMS window mechanism design allowed the effect of the window to be determined by measuring the radiance of a simple flat-plate blackbody that is external to the window. First, the CIRIMS is put in a protected mode by closing a door between the optical path and the outside air. A two-point, heated, flat-plate blackbody is on the back of the door, which protects it from sea spray. The flat-plate blackbody cycles its temperature between 40 and 50 K. This design provides a method to correct for the effect of the window by making measurements of a two-point temperature target while the optics and primary calibration blackbody inside the main housing remain protected. This approach has been adapted in the design of a calibrated infrared sky-imaging camera [125].

The overall instrument measurement uncertainty can be divided into errors due to instrumentation and to environmental factors. Table 4 lists the primary source of both types of errors and an estimate of their magnitude. The primary sources of instrumentation error are the two different radiometers, the calibration uncertainty, and the IR transparent window correction if it is used. Sources of environmental error are changes in emissivity due to surface roughness and variation in local incidence angle produced by ship motion, and uncertainty in the sky correction due to variable sky conditions. Estimates of these individual errors are listed in Table 4(a) and their cumulative effect for different combinations of sky conditions and window use are summarized in Table 4(b). The calibration uncertainty is taken as the RMS error from the laboratory and incorporates uncertainty due to the sensor housing temperature, the stability of the blackbody temperature, the blackbody emissivity, and the radiometer stability. For uniform (clear or cloudy) sky conditions the overall errors are 0.064 and 0.110 K without and with the window, respectively. The overall errors increase for variable sky conditions to 0.081 and 0.121 K without and with the window, respectively. These results, combined with a three-way field comparison with the ISAR and M-AERI summarized in Table 5, indicate that the CIRIMS met the design goal of ±0.10 K accuracy. The knowledge gained during the development of the CIRIMS could be applied to a new, simpler design.

TABLE 4 Measurement Uncertainties for Calibrated Infrared Radiometer In situ Measurement System (CIRIMS). (a) Instrument and Environmental Factors and (b) Overall Errors as a Function of Sky Conditions and Use of Window

(a)

Source	Error (K)
Instrument	
Two radiometers	0.030
Calibration	0.018
Window	0.090
Environmental	
Incidence angle	0.053
Variable skies	0.030

(b)

Sky Conditions	Overall Error (K) No Window	Window
Clear/Cloudy	0.064	0.110
Variable	0.081	0.121

Specifically, the comparable performance of the ISAR and CIRIMS suggests that the unique features of the CIRIMS design may not be necessary to obtain the required accuracy.

Further results of CRIMS deployments are presented in Chapter 5.2.

TABLE 5 Comparison of Differences Between Skin Temperature Measured Concurrently by the CIRIMS (T_{CIRIMS}), the M-AERI (T_{MAERI}), and the ISAR (T_{ISAR}) During a Joint Cruise

Quantity	Mean (K)	Standard Deviation (K)	Minimum (K)	Maximum (K)
$T_{ISAR} - T_{CIRIMS}$	0.00	0.13	−0.64	0.52
$T_{MAERI} - T_{ISAR}$	0.08	0.15	−0.84	1.01
$T_{MAERI} - T_{CIRIMS}$	0.08	0.15	−1.15	1.10

CIRIMS, calibrated infrared radiometer in situ measurement system; MAERI, marine atmospheric emitted radiance interferometer; ISAR, infrared SST autonomous radiometer.

4.5 ISAR—Quasi Operational Ocean Field Radiometers

The ISAR is a self-calibrating instrument capable of measuring in situ SST_{skin} to an accuracy of 0.1 K [39]. ISAR is a fully autonomous infrared radiometer system that has been developed for satellite SST validation and other scientific programs and can be deployed continuously on ships of opportunity (Figure 26) without any service requirement or operator intervention for periods of up to 3 months. Ten ISAR instruments have been built and are in sustained use in the UK, Denmark, Australia, China and the USA.

The ISAR instrument is a compact (570×220 mm cylinder) system that employs two reference black body cavities to maintain the radiance calibration of a special Heitronics KT15.85D radiometer (hereafter simply called KT15) to an accuracy of ± 0.1 K. The ISAR instrument (Figure 27) consists of the following subsystems:

- A fore-optics system to route target radiance to the detector consisting of a plane mirror, a ZnSe window and a beam shaping lens,

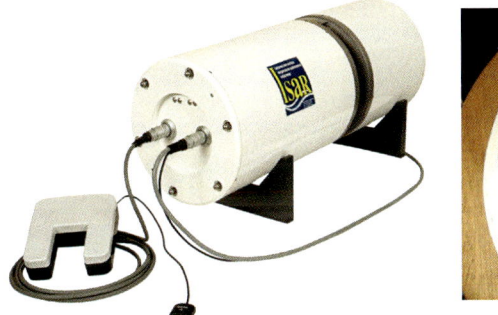

FIGURE 26 Location of the ISAR instrument installed on the P & O M/V Pride of Bilbao bridge wing. The ship operated regular routes between Bilbao (Spain) and Portsmouth (UK) between 2004 and 2010.

FIGURE 27 (Left) The ISAR Radiometer shown together with an OSi Optical rain gauge that is used to trigger the closure of a storm shutter (shown in the open position). Clearly seen is the silver colored scan-drum and central cutaway section providing access to a target position from nadir to zenith (i.e., 180°). (Right) The ISAR radiometer main bulkheads showing blackbodies, viewing ports and ZnSe window.

- A KT15.85 detector head,
- Two black body radiance sources maintained at different temperatures at the end of the instrument optical chain to maintain the instrument calibration,
- An internal control and data acquisition computer subsystem,
- An environmental protection subsystem incorporating an automatically closing/opening storm shutter triggered by an optical rain gauge,
- An external RS485 interface to which additional atmosphere and ocean sensors may be connected, powered and data collected.

The KT15 radiometer head incorporates a solid-state detector system and, in addition to a calibrated digital brightness temperature output, provides an analogue output of the detector signal proportional to the measured radiance. It employs an internal chopper and reference black body to maintain internal calibration stability. The KT15 has been modified from the normal factory configuration to allow brightness temperature (and corresponding radiance measurement) between 173 and 373 K. The unit uses focusing optics to reduce the target beam to a 5 mm diameter spot at a focal point 96 mm in front of the detector head (at 98.3% radiance). The KT15 has a single-channel spectral band-pass of 9.6–11.5 μm.

The detector views the target scene through a protective window via a plane mirror which is mounted at 45° on a steel block inside a protective scan drum. The scan mirror itself is a 3 mm thick hardened $1/4$ wave gold front surface mirror. This is mounted on a massive stainless steel mandrel in order to limit thermal gradients and rapid temperature changes that may otherwise occur across the mirror when looking at different targets. The scan drum and mirror rotate as a single unit driven by a small motor. An aperture port has been cut into the scan drum as a circular hole having a diameter of 10 mm that provides ample clearance for the KT15 beam. The aperture port is the only place that water may enter the ISAR instrument. The scan drum and mirror can be rotated 360° as a single unit and allows the FoV to be directed outside the instrument through a circumferential slot cut into the cylindrical ISAR casing. This design means that all target scenes (sea, sky and both black bodies) are viewed using exactly the same optical path. The scan drum-mirror assembly is connected to a 12-bit resolution absolute rotary position shaft encoder that can be programmed to view any angle in a vertical plane. The angular position of the scan mirror may be determined to an accuracy of 0.1° and the view angle can be changed quickly by the motor-encoder software. A mirror position change of 180° can be made in less than 3 s.

A 2 mm thick removable ZnSe plane window, which is set deep within the ISAR instrument, seals the instrument electronics housing from the external environment. A BBAR coating has been applied to both sides of the window, increasing transmission from approximately 70% to >90% while at the same time providing a protective "hard" coating. Two black bodies are housed in the main body of the ISAR instrument, which is a massive aluminum block, designed to protect the blackbodies from rapid thermal shock and minimize temperature gradients that could affect the instrument calibration. The absolute accuracy of the measurements produced by ISAR is determined by the effectiveness of the blackbody cavities as calibration targets. The Blackbody thermistor calibrations are S.I. (NIST) traceable and the end-to-end instrument calibration is verified using a CASOTS-II laboratory reference radiance source, also traceable to S.I. standards to an accuracy of ±0.1 K and a worst-case uncertainty of 75 mK [119].

Figure 28 plots a time series of blackbody temperatures for several days showing typical temperature excursions experienced by the blackbodies highlighting the need for regular calibration. In the case of ISAR calibration is performed ∼2 min.

The main challenge for an autonomous infrared radiometer deployed on ships is to protect the optical system from the effects of rain, seawater spray, and high humidity. It is critical that any optical surface (gold mirror, ZnSe window, black body cavities) within the ISAR does not become completely

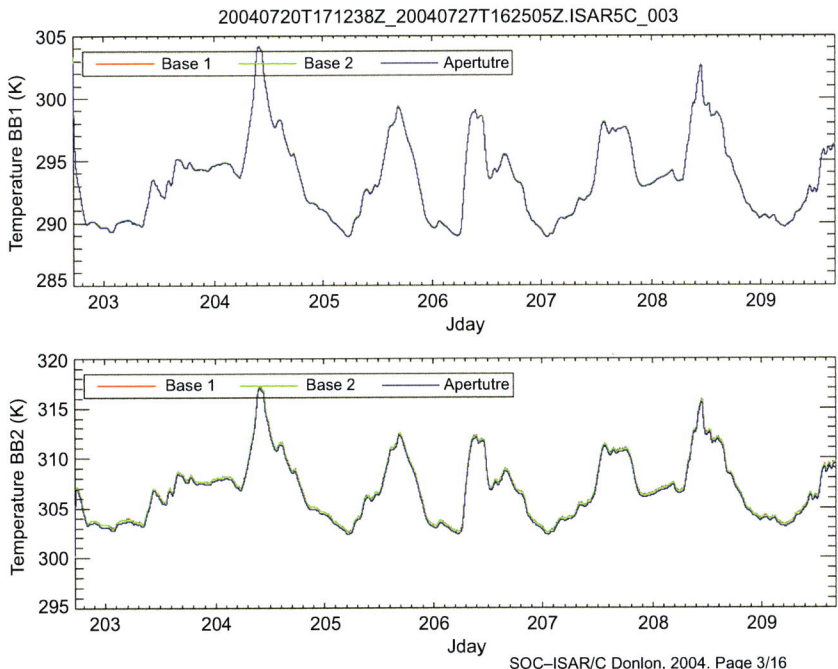

FIGURE 28 Example time series of ISAR blackbody thermistor temperatures from the cold (top) and hot (bottom) blackbody over a 7 day period. Three thermistor sensors are shown on each plot and highlight the diurnal temperature excursions experienced by the radiometer in mid latitudes.

wet, otherwise the optical system will have no throughput. The design of ISAR assumed that "ISAR will get wet aboard a ship." To address this design challenge, the ISAR system uses an optical rain detector and storm shutter arrangement that completely seals the instrument from the environment when the atmosphere contains dust or water droplets (from precipitation or ocean spray).

Figure 29 shows the ISAR shutter in the open and closed position. The shutter slides circumferentially around the main cylindrical body of the instrument, driven by a toothed belt drive located on the inner surface of the shutter. This record shows how the ISAR system maximizes the measurement time while safely protecting the instrument from wet marine environments. While no system is capable of providing 100% protection at sea (e.g., as a ship bow "digs in" and throws huge amounts of sea water into the air while the scan drum aperture is open), experience shows that the ISAR design provides a good working solution minimizing data loss while maximizing the protection of the instrument.

Figure 30 shows an example ship track and a time series of ISAR measurements collected over a 10-year period between the UK and north Spain. Breaks in the data are due to ship re-fit and availability. The development and international deployment of ISAR radiometers has provided a quasioperational SST_{skin} measurement program for more than eight years [11]. The low biases and standard deviation of matches between

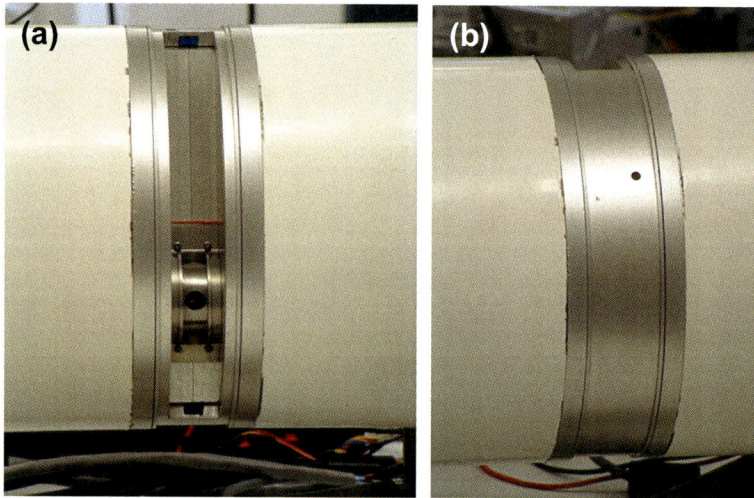

FIGURE 29 Photographs of the ISAR shutter in the open position (a) with the scan drum bush and view aperture exposed and, the shutter in its closed position (b). The small round black circle is one of two SmCo magnets used to control the angular position of the shutter assembly by actuating Hall effect switches.

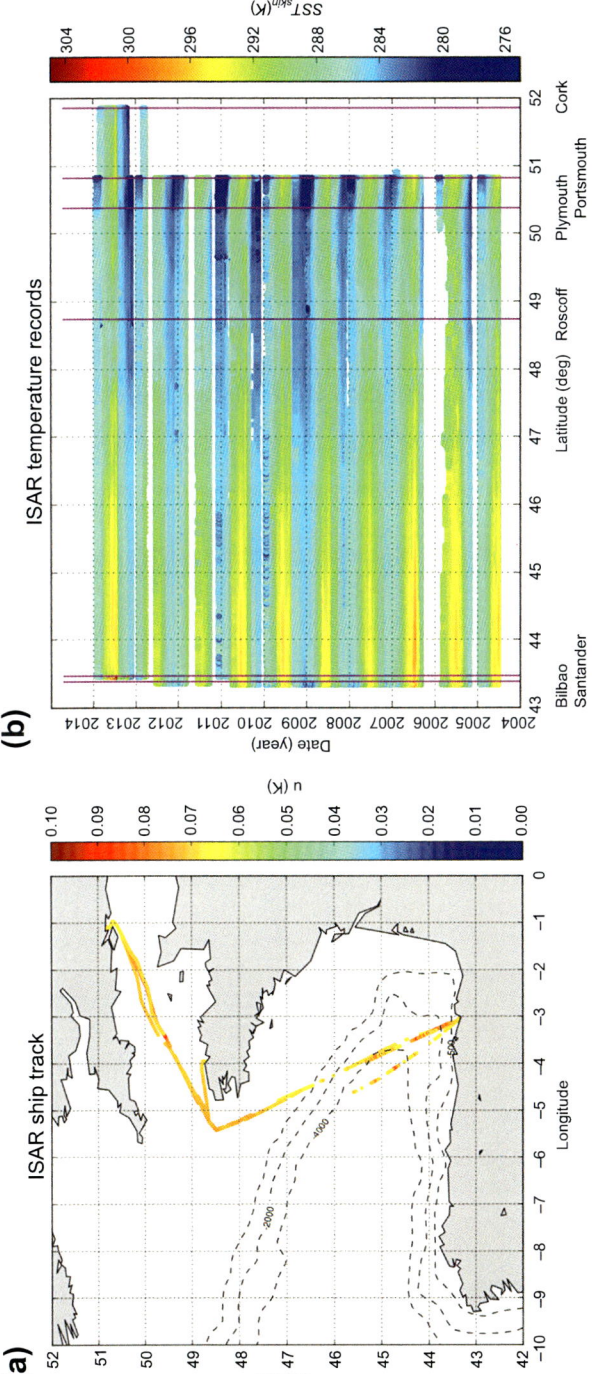

FIGURE 30 (a) Example ship track and (b) Hovmoller time series plot of ISAR measurements in the Bay of Biscay and English Channel.

satellite and ISAR SST_{skin} data have demonstrated the advantage of removing the uncertainty that arises in conventional validation programs in which satellite skin SST measurements are compared against in situ measurements of subsurface temperature (see Chapter 6.2). The capacity of the autonomous ISAR radiometer to sustain long periods at sea while delivering high quality data without the need for frequent operator intervention has been proven [11].

4.6 Use of Unmanned Airborne Vehicles BESST Radiometer

The development of the BEST [89] radiometer for use on a UAV is noteworthy as it presents a novel cost-effective approach. The BESST radiometer is a "push-broom" imaging radiometer that measures small scenes of infrared data as the UAV flies forward. The 320 × 256 pixel thermal imaging microbolometer can be equipped with filters in the 8−12 μm range. A 45° mirror is used to select a target view: either one of two blackbodies for self-calibration or one of two baffled view ports. Self-calibration is achieved by alternating the observed ocean scene and the two temperature-monitored blackbody references thus making it possible to construct a typical two-point calibration curve. The BESST FoV is 18° covering 200 pixels cross-track, while the swath width is about 1/3 of altitude. Pixels on the edge of the detector array (due to optical distortion) are not used to optimize performance and the image resolution depends on the altitude of the UAV that carries the BESST. For altitudes of about 600 m the surface resolution is about 1 m for a swath width of 200 m making BESST more than capable of resolving SST variability within the 1 km pixels typical of space borne infrared radiometers. Frequent calibration is made using two on-board blackbody cavities. BESST is constructed in a modular manner (Figure 31) allowing rapid reconfiguration. The whole instrument weight only 1.36 kg.

The instrument-mounting bracket is at the bottom and one of black baffles points down to the surface while the other looks up for the sky correction measurement. Foil (Mylar insulation) covers the black bodies. Through the target port in the downward looking baffle a single germanium lens collects incident radiation from the target scene focusing it onto a two-dimensional uncooled microbolometer. Three spectral bands are created by the use of individual interference filters positioned in front of the detector array. The channels selected are fairly narrow channels centered at 10.8 μm and another centered at 12 μm. The third channel is broad and covers 8.0−12.0 μm and provides a more sensitive SST channel by integrating a larger amount of thermal radiation. The filters are placed to view all scenes via the rotating mirror using the same optical path. The microbolometer uses an integration time of 0.3 s or less that is chosen depending on the platform speed over the ground. An anti-reflective coating on the germanium window

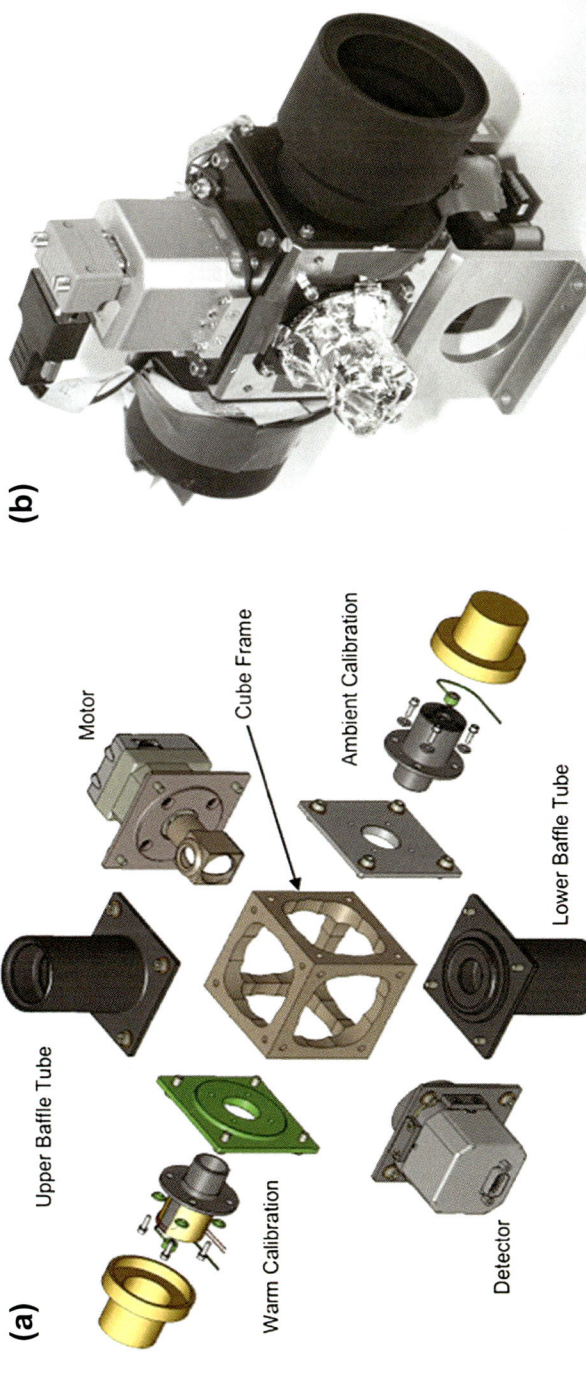

FIGURE 31 (a) Components and modular design of the BESST radiometer. (b) BESST-1 in its operational configuration. The mounting bracket is at the bottom and one of black baffles points down to the surface while the other looks up for the sky correction measurement. Foil (Mylar insulation) covers the black bodies.

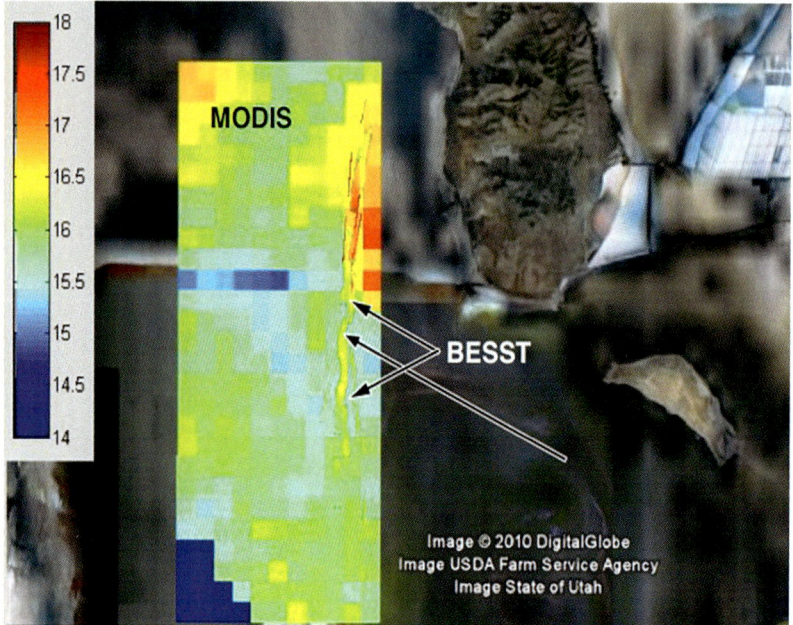

FIGURE 32 Google Earth image with overlays of a MODIS image (large pixels) and the nearly coincident BESST image strip over the Great Salt Lake, Utah.

into the focal plane array has a pass band of 8–12 μm and provides additional out-of-band blocking into the microbolometer. Figure 32 shows an example data take for the BESST radiometer flying over the Great Salt Lake Utah [89].

4.7 Spectroradiometers

4.7.1 SST_{skin} Determination Using a Spectro-Radiometer

The Marine-Atmospheric Emitted Radiance Interferometer (M-AERI; Minnett et al., 2001) is a Fourier-Transform Infrared (FTIR) Spectroradiometer that operates in the range of infrared wavelengths from ∼3 to 18 μm and measures spectra with a resolution of ∼0.5 cm^{-1}. It uses two infrared detectors to achieve this wide spectral range, and these are cooled to ∼78 K by a Stirling cycle cryocooler to reduce the noise equivalent temperature difference to levels well below 0.1 K. The M-AERI includes two internal blackbody cavities for accurate real-time calibration. A gold scan mirror, that can be programmed to step through a pre-selected range of angles, directs the FoV from the interferometer to either of the blackbody calibration targets or to the environment from nadir to zenith allowing a range of sea and sky view targets to be measured.

The interferometer integrates measurements over a preselected time interval, usually a few tens of seconds, to obtain a satisfactory signal to noise ratio, and a typical cycle of measurements including two view angles to the atmosphere, one to the ocean, and calibration measurements, takes about 5–12 min.

The absolute accuracy of the infrared spectra produced by the M-AERI is determined by the effectiveness of the blackbody cavities as calibration targets. The absolute accuracy of the spectral measurements of the M-AERI is better than 0.03 K [79]. The absolute uncertainties of the retrieved skin SST, determined by operating two M-AERI's side-by-side and by comparing M-AERI measurements with those from other well-calibrated radiometers, are less than 0.05 K [61,123], which are sufficiently small to give confidence in the use of such data in the validation of satellite SST retrievals.

The M-AERI uses a sandwich of two detectors to give the desired spectral range: an indium antimonide (InSb) detector is mounted in front of one of mercury cadmium telluride (HgCdTe). The InSb detector is for the so-called shortwave part of the spectrum from about 1800 to 3000 cm^{-1} ($\lambda = 3.3-5.5$ µm) and the HgCdTe detector for the long-wave part from about 550 to 1800 cm^{-1} ($\lambda = 5.5-18$ µm). The raw data stream from these detectors is captured by an interface card in the control computer that delivers four files comprising the raw spectra from each detector for the "forward" and "backward" movements of the mirror yoke of the Michelson interferometer (shown in Figure 33).

The normal model of deployment of the M-AERI's on ships comprises a series of mirror angles that direct the field-of-view of the interferometer towards the ocean surface, the sky at the same angle relative to zenith as the sea measurement is to nadir, and a zenith view. This sequence is preceded and followed by measurements of both internal blackbody targets, so the environmental measurement cycles are bracketed by calibration measurements.

A series of five processing steps are performed on the raw data:

1. A correction for detector nonlinearity is applied to the long-wave (HgCdTe) band,
2. The forward and backward Michelson scans for each of the long-wave and short-wave bands are calibrated individually,
3. The forward and backward interferograms are averaged for each band,
4. A finite FoV correction is applied to each calibrated spectrum, and
5. The spectra are resampled to a "standard" wavenumber scale common to all M-AERI's.

After the five processing steps a separate application reviews the processed data and compiles a file of summary information, including instrument performance characteristics and data quality flags. The final processing step is the computation of the SST from a combination of calibrated ocean and sky views.

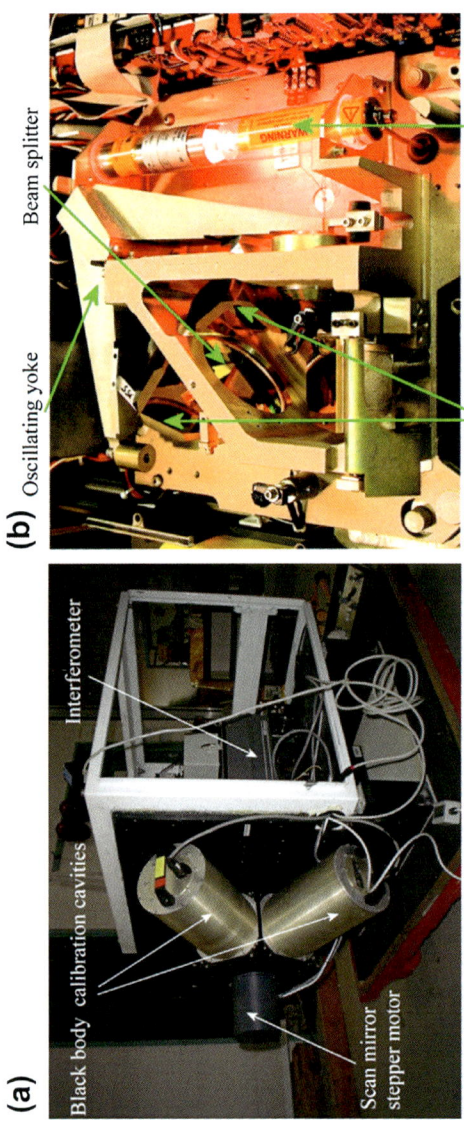

FIGURE 33 The M-AERI interferometer (a) View showing M-EARI blackbodies, scan mirror motor and interferometer (b) view showing the interferometer oscillating yoke, corner cube reflectors, HeNe laser (used for optical alignment and wavelength calibration) and the interferometer beam splitter.

The final data products made available for archive are the un-calibrated raw data; several intermediate product files, the final calibrated, corrected, and resampled radiance spectra; and the summary product file. These data files are incrementally extended throughout the course of a day as the new measurements are processed in real-time. At 00:00 UTC the control computer closes all files, checks there is enough disk space for the next day's data (if not the oldest data files are deleted), conducts a soft reboot and restarts the sampling process. The data files from the day just ended are copied to archive media by the operator.

Examples of the $L_{scene}(\lambda,\theta)$ and $L_{sky}(\lambda,\theta)$ spectra measured by the M-AERI are shown in Figure 34. These were taken in the high Arctic so the atmospheric emission in the "window" regions is very small as the atmosphere was very cold and dry. Figure 35 shows spectra obtained in the Tropical Pacific Ocean.

To derive SST_{skin}, the correction for the reflected sky radiance is achieved through measurements taken within a minute of the sea-view measurement, using values of the surface emissivity derived from the M-AERI measurements themselves [54,55,126]. The small correction for the emission from the air layer between the M-AERI and the sea surface is taken from a precomputed

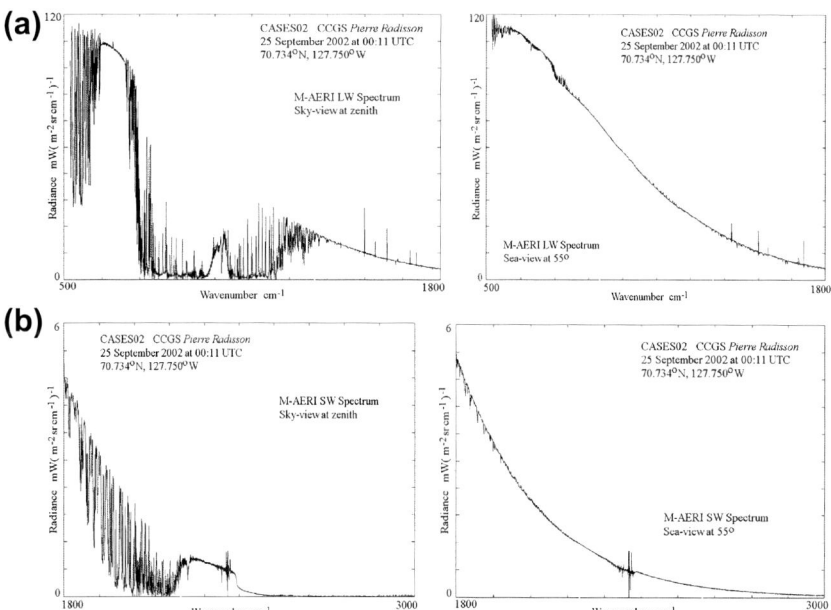

FIGURE 34 Examples of M-AERI radiance spectra derived from the long-wave detector (a) and short wave detector (b). Note the different vertical scales. The left-hand pair is from the atmosphere at a zenith, and the right-hand pair from the ocean view measurements at a nadir angle of 55°.

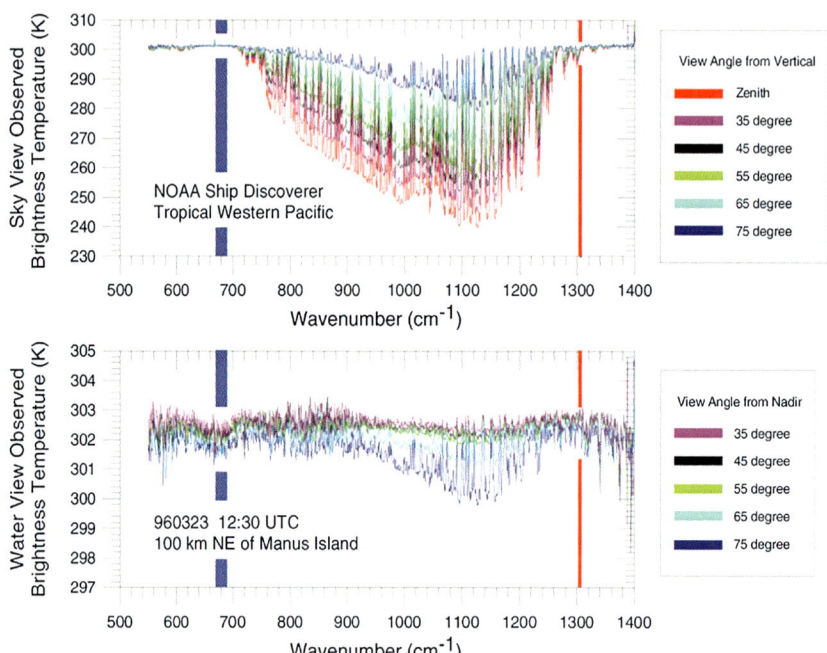

FIGURE 35 Examples of parts of spectra measured by the M-AERI, represented as temperature, and those intervals where the sky temperatures are smallest indicate where the atmosphere is most transparent. The spikes in atmospheric spectra are caused by emission lines. The blue bar shows which spectral region is used to measure air temperature, and the red bar skin sea-surface temperature. Note the change in temperature scales of the two panels. These data were taken in the Tropical Western Pacific during the Combined Sensor Program Cruise in 1996 [79].

look-up table with values sorted according to instrument height, air temperature and humidity [77]. The SST_{skin} is derived from the spectra in the vicinity of $\lambda = 7.7$ μm, where the atmospheric path length of the photons is significantly shorter than in the atmospheric window regions, rendering the retrieved SST_{skin} fairly insensitive to changing cloud conditions that could introduce an error in the reflected sky radiance correction [77] The spectral region for the measurement of air temperature, T_{air}, was chosen to be between $\lambda = 14.28-15.38$ μm where the atmospheric emission is dominated by that from CO_2, a sufficiently well-mixed gas in the lower atmosphere with only a small natural, largely seasonal, variability that contributes little to the uncertainties of the air temperature retrieval [126].

The M-AERI runs continuously under computer control, except for the brief period beginning at 0000 UTC each day. The most serious source of data loss results from the measures taken to avoid contamination of the scan mirror by heavy rain or sea spray. The mirror must remain clean and dry for the M-AERI to provide the required measurements. Contaminants on the mirror, wet or dry, act as irregularities that scatter stray radiation into the beam. The sources of these strays are not characterized and not necessarily the same for the sky and sea measurements. They are not likely to be the same for the calibration measurements of the blackbody cavities, so their influence is not necessarily removed by the real-time calibration procedure. To avoid such contamination, a rain sensor is mounted close to the M-AERI aperture, and when the output from this crosses a predetermined threshold, the mirror is moved into a "safe" position. In the safe mode, the mirror is directed to the ambient-temperature blackbody calibration cavity, and the back of the mirror cylinder is presented to the rain or spray. When the rain sensor output indicates good sampling conditions have returned, the mirror scan sequence is resumed. For prolonged periods of rain or rough conditions a tarpaulin is manually attached to cover the entire instrument for protection.

Episodically it is necessary to clean drying stains or small salt crystals from the surface of the mirror. This is done by rinsing the mirror with solvents and distilled water, and requires the operator to override the normal sequence of mirror rotation angles to position the mirror so the gold reflecting surface is exposed for non-tactile cleaning. During periods when the mirror is being cleaned, the instrument is in the mirror-safe mode, or covered by the tarpaulin, the detector outputs continue to be measured and the data stream processed and archived, but the data do not contain relevant information. Removing measurements taken during these periods is part of the quality assurance. Other instances of compromised data result from radio-frequency interference from the host ship, or, very occasionally, direct sunlight entering the instrument.

The M-AERIs have been deployed on many research ships, and on the cruise liner Explorer of the Seas of Royal Caribbean International [127] for the validation of MODIS SSTs. Figure 36 shows the tracks of the research vessels carrying MAERIs since the launch of Terra. There have been over 40 deployments of the three MAERIs on a large selection of research vessels encompassing a wide range of environmental conditions. There are now over ten-years (>3600 days at sea) of MAERI measurements.

4.8 Derivation of Air Temperature Using a Spectroradiometer

Direct measurements of air temperature, T_{air}, on board ships may contain significant errors if measurement thermometers are not carefully sited in well-exposed locations [128]. On sunny days, especially in the tropics, solar heating of the ship superstructure can elevate the measured T_{air}, even if the thermometer is screened from the solar radiation. Based on extensive u:measurements in the Pacific [77,126] radiometric measurements from M-AERI in the ~14.0—16.0 μm waveband are capable of retrieving a measurement of T_{air} aboard a ship that is not contaminated by ship warming effects

388 Optical Radiometry for Ocean Climate Measurements

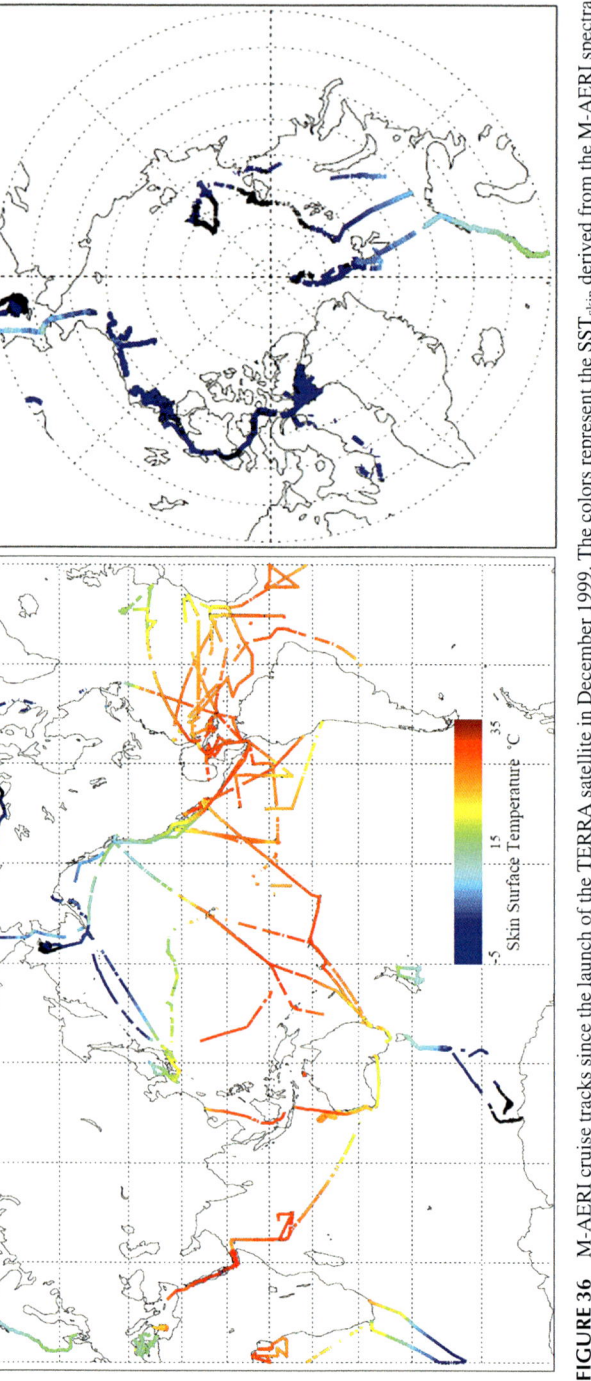

FIGURE 36 M-AERI cruise tracks since the launch of the TERRA satellite in December 1999. The colors represent the SST_{skin} derived from the M-AERI spectra.

over a diurnal cycle. Figure 37 shows significant disparities between conventionally measured and M-AERI measured air–sea temperature differences. The top panel shows Radiometric measurements of the T_{air}-SST_{skin} temperature difference. Data have been screened to remove situations where air warmed by the ships deck and superstructure could have been blown into the radiometer FoV. Differences are generally <2 K and T_{air} is nearly always cooler than the ocean, with some diurnal fluctuations, especially in clear sky conditions. The middle panel Figure 37 shows conventional measurements of the T_{air}-SST_{skin} temperature difference. There is a strong diurnal signal in conventional data, with a change in sign during the afternoon. This is mostly contamination of the measurements by direct radiative heating of the thermometer used to measure T_{air} and/or heat-island effects of the ship, and a smaller contribution from failure to sample oceanic diurnal thermocline. The lower panel Figure 37 shows difference between radiometric and conventional measurements of the air–sea temperature difference. In the R/V Mirai "Nauru99" data set, the spread of data is much greater for the conventional observations during the daytime; night-time measurements agree well. The radiometric measurements of T_{air} appear to be largely immune to the diurnal contamination that compromises conventional measurements, and which appears to be related to either direct solar heating of the thermometer or heat island effects of the ship [77,126].

The innovative use of channels centered on CO_2 absorption lines to measure T_{air} using a ship-borne radiometer system is an extremely important approach to develop as measurement biases evident in traditional measurements can be significantly reduced.

4.9 TIR Cameras

Measurements of SST_{skin} variability using infrared (IR) cameras from ships, towers, manned aircraft, and unmanned airborne systems (UAS) have been used since the 1960s [129,130] and are being made with more prevalence in the last decade. A time series of IR imagery measures the detailed microscale horizontal structure in SST_{skin} and can be used as a visualization tool for turbulence at water surfaces. Images can be generated at time steps of ~1 s. IR imaging techniques have quantified signatures of thermal variability that result from renewal processes such as large-scale wave breaking [131], microbreaking [78,132], near-surface shear and free-convective patchiness [133]. Similarly, rain will generate turbulence directly at the air–water interface and will contribute to, if not dominate, the disruption of the thermal ocean boundary layer [134,135]. Marmorino et al. [136] observed quasi-periodic, propagating variations in IR imagery and interpreted these as being associated with internal waves. Zappa and Jessup [137] also observed spatially periodic structure in IR imagery and used nearby mooring data to argue that the signal was due to (nonlinear) internal waves. Under conditions of weak winds and strong insolation (which favor formation of a diurnal warm layer), the data reveal a link

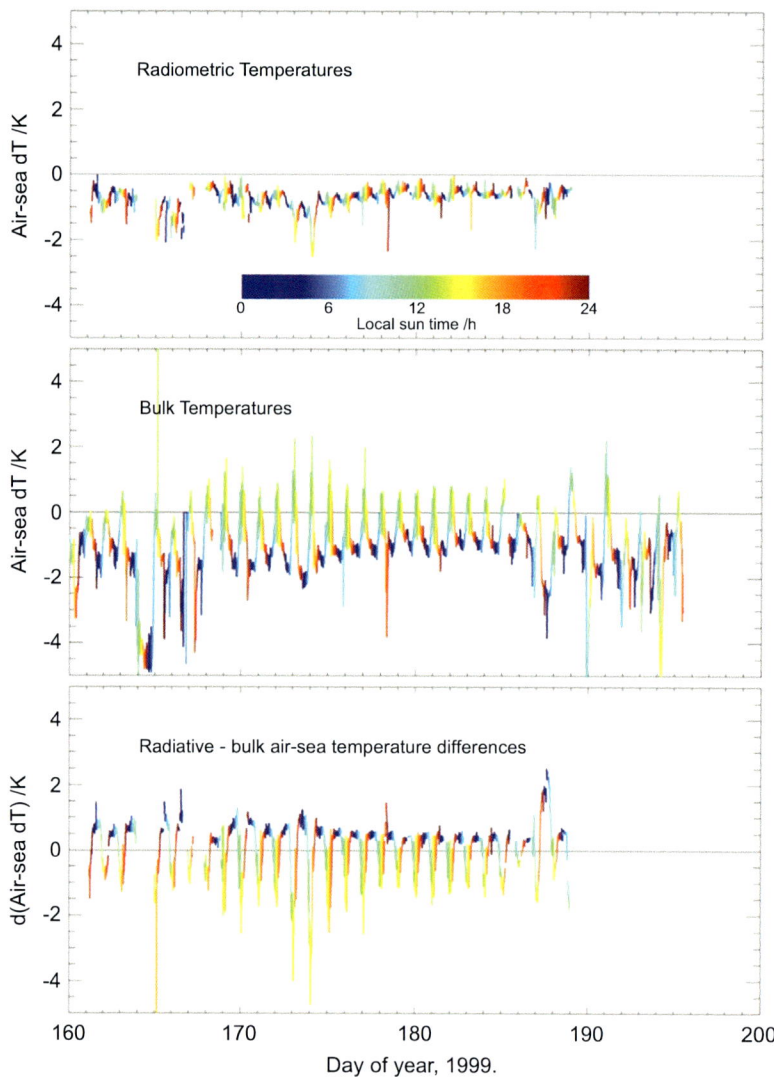

FIGURE 37 Time series of radiometric and conventional T_{air}-SST_{skin} temperature measurements from the R/V Mirai during the Nauru99 project. Most of the measurements are taken close to the Equator in the western Pacific Ocean. The T_{air} measurement is derived from the 14.28–15.38 μm band of the M-AERI Spectroradiometer. The SST_{skin} measurement is derived from the same instrument. The colors represent the local sun time according to the key. The conventional measurement of T_{air} is made using a WMO grade thermometer sensor mounted in a Stevenson's screen behind the wheelhouse that was poorly exposed.

between the spatially periodic SST fluctuations and subsurface temperature and velocity fluctuations associated with oceanic internal waves, suggesting that some mechanism involving the diurnal warm layer is responsible for the observed signal [138]. Infrared imagery has been shown a useful tool to investigate surface layer processes and especially near surface turbulence [139–143]. New techniques [143,144] based on passive thermography to estimate the heat flux and relate the scale of surface renewal events to wind speed and heat exchange, are now emerging. Thermographic techniques have also been used to investigate the features of near surface turbulence [145–151]. These example studies demonstrate the importance and uniqueness of IR thermal camera measurements to study the temporal and spatial characteristics of SST_{skin} variability and monitor a variety of upper-ocean processes.

TIR cameras typically use a cooled InSb ($\sim 3-5$ μm) or MgCdTe ($\sim 8-11$ μm) focal-plane array detector. The performance of such detectors' enables the determination of SST_{skin} variability of less than 20 mK when actively chilled by a Stirling cycle cooler. Many commercial units are available "off-the-shelf" with characteristic image capabilities of up to 1024 × 1024 pixels for MIR devices while LWIR cameras typically provide an image of 640 × 512 pixels for most commercially available systems. A downward-looking IR camera can be deployed on a ship or ocean tower mounted a height of 20–50 m above the sea surface. Depending on the desired FoV (which scales with height) a ground image resolution of $\sim 1-10$ cm and an image size of 10–50 m is generated. Aircraft deployments fly at a variety of heights but typically at an altitude of $\sim 300-850$ m provides an image resolution of $\sim 0.1-1$ m and an image size of ~ 100 m–1 km. Figure 38(a) shows a typical thermal camera setup and Figure 38(b) an example deployment aboard the R/V Floating Instrument Platform (FLIP) and (c) example thermal images of the sea surface at different wind speeds. In addition, thermal markers have been placed on the sea surface to investigate the characteristics of the air sea interface [151].

A thermal camera measurement system may include a second infrared camera that is directed skyward as well as a narrow FoV radiometer to complement the downward camera. The upward-looking instrumentation is used to discriminate real from apparent ocean surface temperature variability and correct for reflected sky radiance at the sea surface (see Section 2). Apparent temperature variability in the infrared imagery may arise on partly cloudy days when both the radiatively-cold sky and radiatively-warm clouds reflect into the imager FoV. It is important to install an inertial measurement unit (IMU) with a GPS system to provide accurate information of the motion performance of the platform such as its speed, course, heading, pitch, roll, and altitude are precisely measured to interpret and geolocate the imagery.

Many devices include a reference blackbody for the calibration and non-uniformity correction of the detector array. Cooled IR cameras employ non-uniformity correction (NUCs) algorithms for the focal-plane array detectors, which serve to homogenize the pixel response of the output.

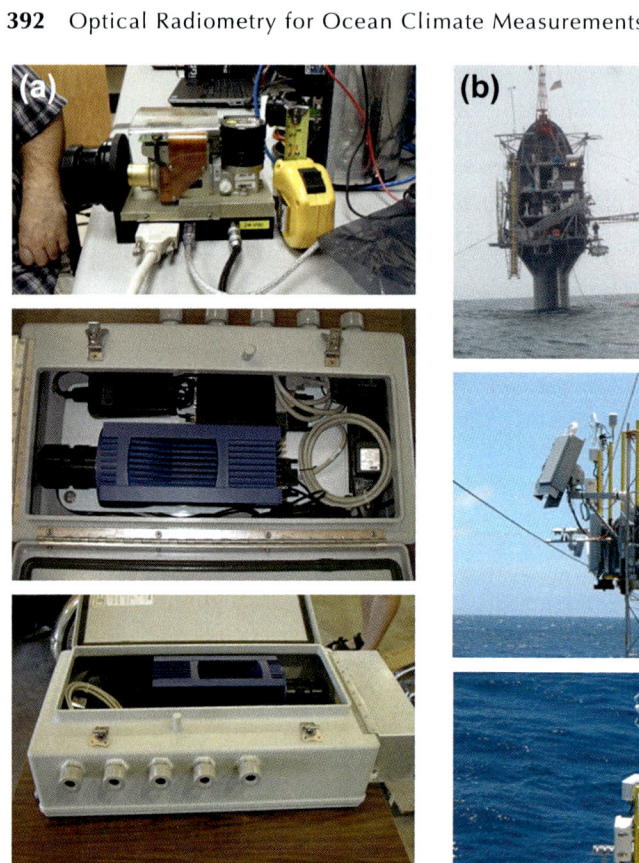
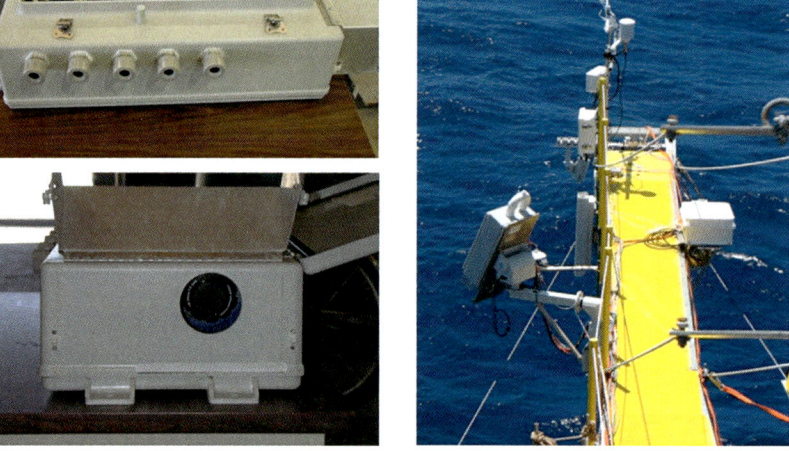

FIGURE 38 (a) (Top image) Internal components of TIR camera including a typical engine for a MgCdTe Stirling-cycle cooled detector with 20° FoV lens. (Lower images) Different views of an environmental enclosure for TIR imaging system including TIR camera, fiber optic extension, and power supply. Note shielded opening for clear viewing of the ocean surface. (b) Typical setup of a thermal imaging camera for measurements of SST_{skin} deployed on a boom from an oceanographic platform (see Ref. [152] for more details). R/P Flip with booms starboard boom deployed. TIR camera in environmental enclosure deployed from R/P Flip starboard boom. The boom was about 9 m above the mean water level and the incidence angles of the TIR camera were set up to view a patch of the water surface directly beneath the end of the boom. (c) Pairs of SST images taken 0.665 s apart. The images show Lagrangian heat markers (lighter spots) and are taken for wind speeds of (A), (B) 0.5 m s^{-1}; (C), (D) 3 m s^{-1} and (E), (F) 6.5 m s^{-1}. The arrow shows the wind direction. Note how the active Thermal Marking Velocity (TMV) pattern fades with time and is displaced, rotated, dilated, and sheared by the surface velocity field. Note also the development of linear temperature structures at the higher wind speeds (Ref. [151] with permission).

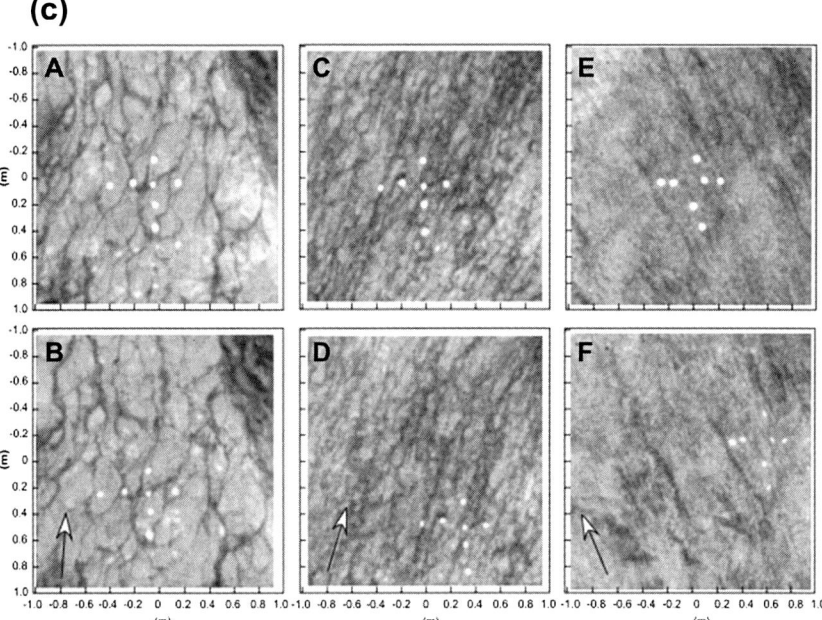

FIGURE 38 cont'd

The NUC algorithms are stored on-board the camera or in software, which also performs bad pixel replacement (BPR). NUC algorithms are dependent on housing/internal temperature so care must be taken to minimize rapid temperature changes (e.g., moving the camera from shadow to direct solar illumination). Cooled long-wave cameras are capable of thermography and precisely hold a calibration for a given internal temperature: a cooled long-wave camera can be laboratory calibrated for a suite of target temperatures using an SI traceable reference blackbody for different housing/internal temperatures. Figure 39 shows calibration curves for a range of housing/internal temperature.

The innovative use of thermal cameras aboard ships to study air−sea interaction processes [152] is important in the context of the SST CDR because these processes define the SST_{skin} temperature as measured by a satellite radiometer system. A detailed knowledge of the SST_{skin} dynamics and variability linked to processes will improve our understanding of the SST_{skin} measurements made from space and used to form the SST CDR.

5. FUTURE DIRECTIONS

The experience of the last 20 years together with the rapid development of new detector elements (microbolometer, pyroelectric), electronics and

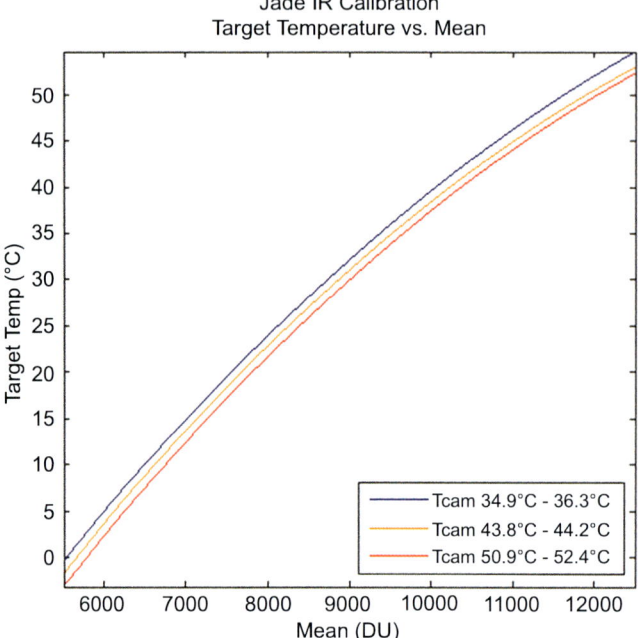

FIGURE 39 Laboratory calibration curves for a MgCdTe Stirling Cycle cooled detector with corresponding Low, Mid and High camera temperature ranges in the legend.

miniaturized computer technologies together with a sound understanding of requirements for FRM SST ship-borne SST radiometers means than progress towards smaller and more accurate radiometers can be expected. Importantly the cost of such instruments must be reduced in order to ensure a more widespread deployment of such instruments for satellite SST validation activities. The consolidation of design trends for SST filter radiometers is towards an autonomous system using two on-board blackbody cavities is clear. The ISAR and SISTeR systems are good examples. However, the miniaturized BESST radiometer system manages to miniaturize these basic design approaches into small package. Further work is required to protect the BESST radiometer from the marine environment aboard a ship as opposed to a UAV, but the experience of the ISAR brings viable solutions.

The wealth of information that can be derived from a spectro-radiometer is clear and the M-AERI system clearly demonstrates that the application of this technology aboard ships is quite mature. New detectors such as the InfraTec tuneable infrared detector with integrated micromachined Fabry—Perot spectrometer (FPS) [103] offer exciting possibilities in the 8.0—11.0 μm waveband. The detector integrates the miniature Fabry—Perot spectrometer above a pyroelectric detector: the FPS can then be tuned using a variable voltage.

The entire sensor package is 15 mm diameter and 6 mm high and could be integrated into a conventional ISAR system relatively easily offering exciting possibilities for SST measurements.

Of particular importance for FRM SST radiometers is the requirement to ensure SI traceability of all SST measurements. While instruments are able to deliver excellent accuracy and precision in the laboratory when viewing external reference radiance cavities, the challenge is to fully understand and compensate for the variability of the sea surface emissivity and the reflected sky component of the measurement. This is arguably the largest source of uncertainty in many measurements to date.

In addition, there is a clear need to solve the challenging calibration of detectors across the full dynamic range of target temperatures (i.e., ~ 150 K (sky) < Target < 303 K (sea surface)). Calibration of the low "sky temperature" remains a significant challenge for ship-mounted radiometers. Is it worth noting that the sky correction is only a few percent of the ocean signal, so larger uncertainties in the sky measurement can be tolerated.

6. CONCLUSIONS

Ship-borne TIR radiometers used to provide FRMs of sea surface skin temperatures are now an operational reality. Modern radiometers have taken considerable effort to ensure their designs are capable of delivering repeatable, accurate and reliable measurements with traceability to S.I temperature standards. The engineering design approaches taken over the last 20 years have evolved building on lessons learned and best practices that have ultimately established a number of instruments that are now routinely deployed aboard ships of opportunity. Future developments include miniaturization of technologies and better understanding of sea surface emissivity to improve the independent SST_{skin} measurements that provide an independent validation of satellite SST_{skin} retrievals that form the modern satellite SST CDR.

ACKNOWLEDGMENTS

The Authors would like to thank O. Embury for discussion production of Figure 4(b), the *Pyrometer Instrument Company* for permission to use Figure 7(h) and Hamamatsu for permission to use Figures 10 and 11. Thank you to R. Branch for reviewing the draft manuscript.

REFERENCES

[1] C. Donlon, B. Berruti, A. Buongiorno, M.-H. Ferreira, P. Femenias, J. Frerick, P. Goryl, U. Klein, H. Laur, C. Mavrocordatos, J. Nieke, H. Rebhan, B. Seitz, J. Stroede, R. Sciarra, The Global Monitoring for Environment and Security (GMES) Sentinel-3 mission, Remote Sens. Environ. 120 (2012) 27–57. http://dx.doi.org/10.1016/j.rse.2011.07.024.

[2] C. Welsch, H. Swenson, S.A. Cota, F. DeLuccia, J.M. Haas, C. Schueler, R.M. Durham, J.E. Clement, P.E. Ardanuy, VIIRS (Visible Infrared Imager Radiometer Suite): A Next-Generation Operational Environmental Sensor for NPOESS, in: Geoscience and Remote Sensing Symposium, 2001. IGARSS '01. IEEE 2001 International, vol. 3, 2001, pp. 1020−1022. http://dx.doi.org/10.1109/IGARSS.2001.976733.

[3] K.A. Kilpatrick, G.P. Podesta, R. Evans, Overview of the NOAA/NASA advanced very high resolution radiometer Pathfinder algorithm for sea surface temperature and associated matchup database, J. Geophys. Res. 106 (C5) (2001) 9179−9197. http://dx.doi.org/10.1029/1999JC000065.

[4] P.J. Minnett, O.B. Brown, R.H. Evans, E.L. Key, E.J. Kearns, K. Kilpatrick, A. Kumar, K.A. Maillet, G. Szczodrak, Sea-Surface Temperature Measurements from the Moderate-Resolution Imaging Spectroradiometer (MODIS) on Aqua and Terra, in: Geoscience and Remote Sensing Symposium, IGARSS '04. Proceedings 2004 IEEE International, vol. 7, 2004, pp. 4576−4579, 20−24. http://dx.doi.org/10.1109/IGARSS.2004.1370173.

[5] P. Le Borgne, H. Roquet, C.J. Merchant, Estimation of Sea Surface Temperature from the Spinning Enhanced Visible and Infrared Imager, improved using numerical weather prediction, Remote Sens. Environ. 115 (2011) 55−65.

[6] C.J. Merchant, L.A. Horrocks, J.R. Eyre, A.G. O'carroll, Retrievals of sea surface temperature from infrared imagery: origin and form of systematic errors, Q. J. R. Meteorol. Soc. 132 (2006) 1205−1223. http://dx.doi.org/10.1256/qj.05.143.

[7] X. Wu, W.P. Menzel, G.S. Wade, Estimation of sea surface temperatures using GOES-8/9radiance measurements, Bull. Am. Meteorol. Soc. 80 (1999) 1127−1138. http://dx.doi.org/10.1175/1520-0477(1999)080<1127:EOSSTU>2.0.CO;2.

[8] C. Donlon, N. Rayner, I. Robinson, D.J.S. Poulter, K.S. Casey, J. Vazquez-Cuervo, E. Armstrong, A. Bingham, O. Arino, C. Gentemann, D. May, P. LeBorgne, J. Piollé, I. Barton, H. Beggs, C.J. Merchant, S. Heinz, A. Harris, G. Wick, B. Emery, P. Minnett, R. Evans, D. Llewellyn-Jones, C. Mutlow, R.W. Reynolds, H. Kawamura, The global ocean data assimilation experiment high-resolution sea surface temperature pilot project, Bull. Am. Meteorol. Soc. 88 (2007) 1197−1213. http://dx.doi.org/10.1175/BAMS-88-8-1197.

[9] J.J. Kennedy, R.O. Smith, N.A. Rayner, Using AATSR data to assess the quality of in situ sea-surface temperature observations for climate studies, Remote Sens. Environ. 116 (2012) 79−92. http://dx.doi.org/10.1016/j.rse.2010.11.021.

[10] E. Lee, Y. Noh, N. Hirose, A new method to produce sea surface temperature using satellite data assimilation into an atmosphere−ocean mixed layer coupled model, J. Atmos. Oceanic Technol. 30 (2013) 2926−2943.

[11] C.J. Donlon, M. Martin, J.D. Stark, J. Roberts-Jones, E. Fiedler, W. Wimmer, The Operational Sea Surface Temperature and Sea Ice Analysis (OSTIA), Remote Sens. Environ. 116 (2011) 140−158. http://dx.doi.org/10.1016/j.rse.2010.10.017.

[12] I.J. Barton, A.M. Zavody, D.M. O'Brien, D.R. Cutten, R.W. Saunders, D.T. Llewellyn-Jones, Theoretical algorithms for satellite-derived sea surface temperatures, J. Geophys. Res. 94 (1989) 3365−3375. http://dx.doi.org/10.1029/JD094iD03p03365.

[13] C.J. Merchant, A.R. Harris, M.J. Murray, A.M. Závody, Toward the elimination of bias in satellite retrievals of sea surface temperature: 1. Theory, modeling and interalgorithm comparison, J. Geophys. Res. 104 (1999) 23565−23578. http://dx.doi.org/10.1029/1999JC900105.

[14] C.J. Merchant, P. Le Borgne, H. Roquet, G. Legendre, Extended optimal estimation techniques for sea surface temperature from the Spinning Enhanced Visible and Infra-Red Imager (SEVIRI), Remote Sens. Environ. 131 (2013) 287−297. http://dx.doi.org/10.1016/j.rse.2012.12.019.

[15] O. Embury, C.J. Merchant, M.J. Filipiak, A reprocessing for climate of sea surface temperature from the along-track scanning radiometers: basis in radiative transfer, Remote Sens. Environ. 116 (2012) 32–46. http://dx.doi.org/10.1016/j.rse.2010.10.016.

[16] E. Maturi, A. Harris, J. Mittaz, C. Merchant, B. Potash, W. Meng, J. Sapper, NOAA's sea surface temperature products from operational geostationary satellites, Bull. Am. Meteorol. Soc. 89 (2008) 1877–1888. http://dx.doi.org/10.1175/2008BAMS2528.1.

[17] B.B. Barnes, C. Hu, A Hybrid Cloud Detection Algorithm to Improve MODIS Sea Surface Temperature Data Quality and Coverage Over the Eastern Gulf of Mexico, in: Geoscience and Remote Sensing, IEEE Transactions on, vol. 51, 2013, pp. 3273–3285. http://dx.doi.org/10.1109/TGRS.2012.2223217.

[18] C.E. Bulgin, S. Eastwood, O. Embury, C.J. Merchant, C. Donlon, The sea surface temperature climate change initiative: alternative image classification algorithms for sea-ice affected oceans, Remote Sens. Environ. Available online January 27, 2014, ISSN: 0034-4257, http://dx.doi.org/10.1016/j.rse.2013.11.022.

[19] C.J. Merchant, A.R. Harris, E. Maturi, S. MacCallum, Probabilistic physically based cloud screening of satellite infrared imagery for operational sea surface temperature retrieval, Q. J. R. Meteorol. Soc. 131 (2005) 2735–2755.

[20] G. Ohring, B. Wielicki, R. Spencer, W. Emery, R. Datla, Satellite instrument calibration for measuring global climate change: report of a workshop, Bull. Am. Meteorol. Soc. 86 (2005) 1303–1313.

[21] WMO, Systematic Observation Requirements for Satellite-Based Products for Climate Supplemental Details to the Satellite-Based Component of the Implementation Plan for the Global Observing System for Climate in Support of the UNFCCC – 2011 Update, GCOS-154, 2011, p. 138. Available from: http://www.wmo.int/pages/prog/gcos/Publications/gcos-154.pdf.

[22] S. Solomon, D. Qin, M. Manning, Z. Chen, M. Marquis, K.B. Averyt, M. Tignor, H.L. Miller (Eds.), Contribution of Working Group I to the Fourth Assessment Report of the Intergovernmental Panel on Climate Change, Cambridge University Press, Cambridge, United Kingdom and New York, NY, USA, 2007.

[23] R. Hollmann, C.J. Merchant, R. Saunders, C. Downy, M. Buchwitz, A. Cazenave, E. Chuvieco, P. Defourny, G. de Leeuw, R. Forsberg, T. Holzer-Popp, F. Paul, S. Sandven, S. Sathyendranath, M. van Roozendael, W. Wagner, The ESA climate change initiative: satellite data records for essential climate variables, Bull. Am. Meteorol. Soc. 94 (2013) 1541–1552. http://dx.doi.org/10.1175/BAMS-D-11-00254.1.

[24] C.J. Merchant, D. Llewellyn-Jones, R.W. Saunders, N.A. Rayner, E.C. Kent, C.P. Old, D. Berry, A.R. Birks, T. Blackmore, G.K. Corlett, O. Embury, V.L. Jay, J. Kennedy, C.T. Mutlow, T.J. Nightingale, A.G. O'Carroll, M.J. Pritchard, J.J. Remedios, S. Tett, Deriving a sea surface temperature record suitable for climate change research from the along-track scanning radiometers, Adv. Space Res. 41 (2008) 1–11.

[25] J. Roberts-Jones, E.K. Fiedler, M.J. Martin, Daily, global, high-resolution SST and sea ice reanalysis for 1985–2007 using the OSTIA system, J. Clim. 25 (2012) 6215–6232. http://dx.doi.org/10.1175/JCLI-D-11-00648.1.

[26] A.G. O'Carroll, T. Blackmore, K. Fennig, R.W. Saunders, S. Millington, Towards a bias correction of the AVHRR Pathfinder SST data from 1985 to 1998 using ATSR, Remote Sens. Environ. ISSN: 0034-4257 116 (2012) 118–125. http://dx.doi.org/10.1016/j.rse.2011.05.023.

[27] C.J. Merchant, O. Embury, N.A. Rayner, D.I. Berry, G. Corlett, K. Lean, K.L. Veal, E.C. Kent, D. Llewellyn-Jones, J.J. Remedios, R. Saunders, A twenty-year independent

record of sea surface temperature for climate from Along Track Scanning Radiometers, J. Geophys. Res. 117 (2013) 12013. http://dx.doi.org/10.1029/2012JC008400.

[28] K.S. Casey, T. Brandon, P. Cornillon, R. Evans, The past, present, and future of the AVHRR Pathfinder SST program, in: V. Barale, J.F.R. Gower, L. Alberotanza (Eds.), Oceanography from Space, Springer, The Netherlands, 2010, pp. 273−287. http://dx.doi.org/10.1007/978-90-481-8681-5_16.

[29] BIPM, Comptes rendus de la 20e reunion de la Conference generale des poids et mesures, 1995. Available online at: http://www.bipm.org/en/CGPM/db/20/S.

[30] P.J. Minnett, G.K. Corlett, A pathway to generating Climate Data Records of sea-surface temperature from satellite measurements, Deep Sea Res. Part II Top. Stud. Oceanogr. 77−80 (2012) 44−51. http://dx.doi.org/10.1016/j.dsr2.2012.04.003.

[31] C.J. Donlon, I.S. Robinson, Observations of the oceanic thermal skin in the Atlantic Ocean, J. Geophys. Res. 102 (1997) 18585−18606. http://dx.doi.org/10.1029/97JC00468.

[32] Peter M. Saunders, The temperature at the ocean−air interface, J. Atmos. Sci. 24 (1967) 269−273. http://dx.doi.org/10.1175/1520-0469(1967)024<0269:TTATOA>2.0.CO;2.

[33] C.L. Gentemann, C.J. Donlon, A. Stuart-Menteth, F.J. Wentz, Diurnal signals in satellite sea surface temperature measurements, Geophys. Res. Lett. 30 (2003) 1140. http://dx.doi.org/10.1029/2002GL016291.

[34] D. Meldrum, E. Charpantier, M. Fedak, B. Lee, R. Lumpkin, P. Niller, H. Viola, Data buoy observations: the status quo and anticipated developments over the next decade, in: J. Hall, D.E. Harrison, D. Stammer (Eds.), Proceedings of OceanObs'09: Sustained Ocean Observations and Information for Society, vol. 2, Venice, Italy, September 21−25, 2009, ESA Publication WPP-306, 2010. http://dx.doi.org/10.5270/OceanObs09.cwp.62.

[35] W.J. Emery, D.J. Baldwin, P. Schlüssel, R.W. Reynolds, Accuracy of in situ sea surface temperatures used to calibrate infrared satellite measurements, J. Geophys. Res. 106 (2001) 2387−2405. http://dx.doi.org/10.1029/2000JC000246.

[36] S. Marullo, R. Santoleri, D. Ciani, P. Le Borgne, S. Péré, N. Pinardi, M. Tonani, G. Nardone, Combining model and geostationary satellite data to reconstruct hourly SST field over the Mediterranean Sea, Remote Sens, Environ 146 (25 April 2014) 11−23. http://dx.doi.org/10.1016/j.rse.2013.11.001.

[37] A.T. Jessup, C.J. Zappa, M.R. Loewen, V. Hesany, Infrared remote sensing of breaking waves, Nature 385 (1997) 52−55.

[38] W. Wimmer, I.S. Robinson, C.J. Donlon, Long-term validation of AATSR SST data products using shipborne radiometry in the Bay of Biscay and English Channel, Remote Sens. Environ. 116 (2012) 17−31.

[39] C. Donlon, I.S. Robinson, M. Reynolds, W. Wimmer, G. Fisher, R. Edwards, T.J. Nightingale, An infrared sea surface temperature autonomous radiometer (ISAR) for deployment aboard volunteer observing ships (VOS), J. Atmos. Oceanic Technol. 25 (2008) 93−113.

[40] W.L. Wolfe, G.J. Zissis (Eds.), The Infrared Handbook, Infrared Information Analyses Center, Ann Arbor, Michigan, USA, 1978, p. 1722.

[41] I.S. Robinson, Measuring the Oceans from Space, Springer/Praxis, 2004, ISBN 978-3-540-42647-9, 669 pp.

[42] M.Z. Moustafa, Z.D. Moustafa, M.S. Moustafa, Resilience of a high latitude Red Sea corals to extreme temperature, Open J. Ecol. 3 (2013) 242−253.

[43] W. McKeown, W. Asher, A radiometric method to measure the concentration boundary layer thickness at an air−water interface, J. Atmos. Oceanic Technol. 14 (1997) 1494−1501.

[44] W. McKeown, F. Bretherton, H.L. Huang, W.L. Smith, H.L. Revercomb, Sounding the skin of water: sensing air−water interface temperature gradients with interferometry, J. Atmos. Oceanic Technol. 12 (1995) 1313−1327.

[45] P.Y. Deschamps, T. Phulpin, Atmospheric correction of infrared measurements of sea surface temperature using channels at 3.7, 11 and 12 μm, Boundary-Layer Meteorol. 18 (1980) 131−143.

[46] J.E.A. Selby, R.A. McClatchey, Atmospheric Transmittance from 0.25 to 28.5 um: Computer Code LOWTRAN 3, Air Force Cambridge Research Laboratories, Air Force Systems Command, United States Air Force, 1975.

[47] K. Masuda, T. Takashima, Y. Takayama, Emissivity of pure and sea waters for the model sea surface in the infrared window region, Remote Sens. Environ. 24 (1988) 313−329.

[48] J.E. Bertie, Z. Lan, Infrared intensities of liquids: the intensity of the OH stretching band of liquid water revisited and the best current values of the optical constants of H_2O (l) at 25C between 15,000 and 1 cm^{-1}, Appl. Spectrosc. 50 (1996) 1047−1057.

[49] L.W. Pinkley, D. Williams, Optical properties of sea water in the infrared, J. Opt. Soc. Am. 66 (1976) 554−558.

[50] J. Hannafin, P. Minnett, Measurements of the infrared emissivity of a wind-roughened sea surface, Appl. Opt. 44 (2005) 398−411. http://dx.doi.org/10.1364/AO.44.000398.

[51] S.M. Newman, J.A. Smith, M.D. Glew, S.M. Rogers, J.P. Taylor, Temperature and salinity dependence of sea surface emissivity in the thermal infrared, Q. J. R. Meteorol. Soc. 131 (2005) 2539−2557. http://dx.doi.org/10.1256/qj.04.150.

[52] D. Friedman, Infrared characteristics of ocean water, Appl. Opt. 8 (1969) 2073−2078.

[53] C.J. Donlon, T.J. Nightingale, The effect of atmospheric radiance errors in radiometric sea surface temperature measurements, Appl. Opt. 39 (2000) 2392−2397.

[54] N.R. Nalli, P.J. Minnett, P. van Delst, Emissivity and reflection model for calculating unpolarized isotropic water surface-leaving radiance in the infra- red. I: theoretical development and calculations, Appl. Opt. 47 (2008a) 3701−3721.

[55] N.R. Nalli, P.J. Minnett, E. Maddy, W.W. McMillan, M.D. Goldberg, Emissivity and reflection model for calculating unpolarized isotropic water surface-leaving radiance in the infrared. 2: validation using Fourier transform spectrometers, Appl. Opt. 47 (2008b) 4649−4671.

[56] P. Watts, M. Allen, T. Nightingale, Sea surface emission and reflection for radiometric measurements made with the along-track scanning radiometer, J. Atmos. Oceanic Technol. 13 (1996) 126−141.

[57] X. Wu, W.L. Smith, Emissivity of rough sea surface for 8-13 μm: modeling and verification, Appl. Opt. 36 (1997) 2609−2619.

[58] R. Niclòs, V. Caselles, E. Valor, C. Coll, J.M. Sánchez, A simple equation for determining sea surface emissivity in the 3−15 μm region, Int. J. Remote Sens. 30 (2009) 1603−1619.

[59] H. Li, N. Pinel, C. Bourlier, Polarized infrared reflectivity of one-dimensional Gaussian sea surfaces with surface reflections, Appl. Opt. 52 (2013) 6100−6111.

[60] H. Li, N. Pinel, C. Bourlier, Polarized infrared reflectivity of 2D sea surfaces with two surface reflections, Remote Sens. Environ. ISSN: 0034-4257 147 (2014) 145−155. http://dx.doi.org/10.1016/j.rse.2014.02.018.

[61] I.J. Barton, P.J. Minnett, C.J. Donlon, S.J. Hook, A.T. Jessup, K.A. Maillet, T.J. Nightingale, The Miami 2001 infrared radiometer calibration and inter-comparison: 2. Ship comparisons, J. Atmos. Oceanic Technol. 21 (2004) 268−283.

[62] C.J. Donlon, S.J. Keogh, D.J. Baldwin, I.S. Robinson, T. Sheasby, I. Ridley, I.J. Barton, E.F. Bradley, T. Nightingale, W.J. Emery, Solid state radiometer measurements of sea surface skin temperature, J. Atmos. Oceanic Technol. 15 (1998) 774−776.

[63] A.T. Jessup, R. Branch, Integrated ocean skin and bulk temperature measurements using the calibrated infrared in situ measurement system (CIRIMS) and through-hull ports, J. Atmos. Oceanic Technol. 25 (2008) 579−597.

[64] E.T. Kent, T. Forrester, P.K. Taylor, A comparison of the oceanic skin effect parameterizations using ship-borne radiometer data, J. Geophys. Res. 101 (1996) 16649−16666.

[65] J.P. Thomas, R.J. Knight, H.K. Roscoe, J. Turner, C. Symon, An evaluation of a self calibrating infrared radiometer for measuring sea surface temperature, J. Atmos. Oceanic Technol. 12 (1995) 301−316.

[66] S.J. Keogh, I.S. Robinson, C.J. Donlon, T.J. Nightingale, The validation of AVHRR SST using shipborne radiometers, Int. J. Remote Sens. 20 (1999) 2871−2876.

[67] P. Schluessel, W.J. Emery, H. Grassl, T. Mammen, On the bulk-skin temperature difference and its impact on satellite remote sensing of sea surface temperature, J. Geophys. Res. 95 (1990) 13341−13356. http://dx.doi.org/10.1029/JC095iC08p13341.

[68] Gary A. Wick, W.J. Emery, L.H. Kantha, P. Schlüssel, The behavior of the bulk − skin sea surface temperature difference under varying wind speed and heat flux, J. Phys. Oceanogr. 26 (1996) 1969−1988.

[69] I.J. Barton, A.J. Prata, D.T. Llewellyn-Jones, The along track scanning radiometer − an analysis of coincident ship and satellite measurements, Adv. Space Res. ISSN: 0273-1177 13 (1993) 69−74. http://dx.doi.org/10.1016/0273-1177(93)90529-K.

[70] L. Guan, K. Zhang, W. Teng, Shipboard Measurements of Skin SST in the China Seas: Validation of Satellite SST Products, in: Geoscience and Remote Sensing Symposium (IGARSS), 2011 IEEE International, 2008. http://dx.doi.org/10.1109/IGARSS.2011.6049522.

[71] A.T. Jessup, R. Fogelberg, P. Minnett, Autonomous shipboard infrared radiometer system for in-situ validation of satellite SST, Proc. SPIE 4814 (2002) 222−229.

[72] G.K. Corlett, I.J. Barton, C.J. Donlon, M.C. Edwards, S.A. Good, L.A. Horrocks, D.T. Llewellyn-Jones, C.J. Merchant, P.J. Minnett, T.J. Nightingale, E.J. Noyes, A.G. O'Carroll, J.J. Remedios, I.S. Robinson, R.W. Saunders, J.G. Watts, The accuracy of SST retrievals from AATSR: an initial assessment through geophysical validation against in situ radiometers, buoys and other SST data sets, Adv. Space Res. 37 (2006) 764−769.

[73] I.J. Barton, Interpretation of satellite-derived sea surface temperatures, Adv. Space Res. 28 (2001) 165−170.

[74] M.J. Suarez, W.J. Emery, G.A. Wick, The multi-channel infrared sea truth radiometric calibrator (MISTRC), J. Atmos. Oceanic Technol. 14 (1997) 243−253.

[75] B.I. Moat, M.J. Yelland, R.W. Pascal, A.F. Molland, An overview of the airflow distortion at anemometer sites on ships, Int. J. Climatol. 25 (2005) 997−1006.

[76] E.P. McClain, W.G. Pichel, C.C. Walton, Z. Ahmad, J. Sutton, Multi-channel improvements to satellite-derived global sea surface temperatures, Adv. Space Res. ISSN: 0273-1177 2 (1982) 43−47. http://dx.doi.org/10.1016/0273-1177(82)90120-X.

[77] W.L. Smith, R.O. Knutsen, H.E. Rivercomb, F. Wentz, H.B. Howell, W.P. Menzel, N.R. Nali, O. Brown, J. Brown, P. Minnett, W. McKeown, Observations of the infrared radiative properties of the ocean—implications for the measurement of sea surface temperature via satellite remote sensing, Bull. Am. Meteorol. Soc. 77 (1996) 41−51.

[78] C.J. Zappa, W.E. Asher, A.T. Jessup, Microscale wave breaking and air−water gas transfer, J. Geophys. Res. 106 (5) (2001) 9385−9391.

[79] P.J. Minnett, R.O. Knuteson, F.A. Best, B.J. Osborne, J.A. Hanafin, O.B. Brown, The marine-atmosphere emitted radiance interferometer (M-AERI), a high-accuracy, sea-going infrared spectroradiometer, J. Atmos. Oceanic Technol. 18 (2000) 994−1013.

[80] S.J. Hook, A.J. Prata, R.E. Alley, A. Abtahi, R.C. Richards, S.G. Schladow, S.Ó. Pálmarsson, Retrieval of lake bulk-and skin-temperatures using along track scanning radiometer (ATSR) data: a case study using Lake Tahoe CA, J. Atmos. Oceanic Technol. 20 (2003) 534–548.

[81] S.J. Hook, R.G. Vaughan, H. Tonooka, S.G. Schladow, Absolute radiometric in-flight validation of mid infrared and thermal infrared data from ASTER and MODIS on the terra spacecraft using the Lake Tahoe, CA/NV, USA, automated validation site, IEEE Trans. Geosci. Remote Sens. 45 (2007) 1798–1807.

[82] J.R. Schott, S.J. Hook, J.A. Barsi, et al., Thermal infrared radiometric calibration of the entire Landsat 4, 5, and 7 archive (1982–2010), Remote Sens. Environ. 122 (2012) 41–49.

[83] M. Colacino, E. Rossi, F.M. Vivona, Sea-surface temperature measurements by infrared radiometer, Pure Appl. Geophys. 83 (1970) 98–110.

[84] P. Schluessel, H.-Y. Shin, W.J. Emery, H. Grassl, Comparison of satellite-derived sea surface temperatures with in situ skin measurements, J. Geophys. Res. 92 (1987) 2859–2874. http://dx.doi.org/10.1029/JC092iC03p02859.

[85] J.W. Bennett, CSIRO Single Channel Infrared Radiometer – Model DAR011, CSIRO Atmospheric Research Internal Paper, 1998, 19 pp.

[86] R.J. Knight, RAL Sea Surface Temperature Radiometer Operating Manual, Internal Report Rutherford Appleton Laboratory, Chilton, Didcot, Oxon, United Kingdom, 1988.

[87] C.J. Donlon, T.J. Nightingale, L. Fiedler, G. Fisher, D. Baldwin, I.S. Robinson, A low cost blackbody for the calibration of sea going infrared radiometer systems, J. Atmos. Oceanic Technol. 16 (1999) 1183–1197.

[88] S.J. Keogh, The use of infra red radiometers at sea and the development of a methodology for the correction of space borne SST measurements for the oceanic thermal skin effect (Ph.D. thesis), Department of Oceanography, University of Southampton, Southampton, United Kingdom, 1998.

[89] W.J. Emery, W. Good, W. Tandy, M. Izaguirre, P. Minnett, A microbolometer airborne calibrated infrared radiometer: the ball experimental sea surface temperature (BESST) radiometer, Geoscience and Remote Sensing, IEEE Transactions on 52 (12) (2014) 7775–7781. http://dx.doi.org/10.1109/TGRS.2014.2318683.

[90] P. Schluessel, W.J. Emery, H. Grass land, T. Mammen, On the bulk-skin temperature deviation and its' impact on remote sensing of the sea surface, J. Geophys. Res. 95 (1990) 13341–13356.

[91] C.L. Hepplewhite, Remote observation of the sea surface and atmosphere: the oceanic skin effect, Int. J. Remote Sens. 10 (1989) 801–810.

[92] G.P. Eppeldauer, S.W. Brown, K.R. Lykke, Transfer standard filter radiometers: applications to fundamental scales, in: A.C.R. Parr, U. Dalta, J.L. Gardner (Eds.), Optical Radiometry, Experimental Methods in Physical Sciences, vol. 41, Elsevier, 2005, pp. 155–211.

[93] Hamamatsu, Characteristics and Uses of Infrared Detectors, Technical Information SD-12, KIRD9001E04, Hamamatsu Photonics K. K, Solid State Division, 1126-1, Ichino-cho, Higashi-ku, Hamamatsu City, 435-8558, Japan, 2011, p. 43. http://www.hamamatsu.com/resources/pdf/ssd/infrared_techinfo_e.pdf.

[94] Sonalee Chopra, A.K. Tripathi, T.C. Goel, R.G. Mendiratta, Characterization of sol-gel synthesized lead calcium titanate (PCT) thin films for pyro-sensors, Mater. Sci. Eng. 100 (2003) 180–185.

[95] F.J. DiSalvo, Thermoelectric cooling and power generation, Science 285 (1999) 703–706.

[96] I. Urieli, D.M. Berchowitz, Stirling Cycle Engine Analysis, Hilger, Bristol, 1984.

[97] M.J. O'Brien, A Multi-Band Infrared Sea-Truth Radiometric Calibrator. Final Rep, Contract NAS5−32004, National Aeronautics and Space Administration, Goddard Space Flight Center, Greenbelt, MD 20770, 1993.
[98] A.W. Van Herwaarden, P.M. Sarro, Thermal sensors based on the Seebeck effect, Sens. Actuators 10 (1986) 321−346. http://dx.doi.org/10.1016/0250-6874(86)80053-1.
[99] Minnett, et al., Guidance for the Use of Radiometers in the Field for the Validation of Satellite-Derived Surface Temperatures, International Space Science Institute report XXX, Bern, Switzerland, 2014. pp XX.
[100] Infratec, InfraTec Catalog 2013 Pyroelectric and Multispectral Detectors, Infratec GmbH, Dresden, Germany, 2013.
[101] http://www.infratec-infrared.com/Data/LMM-244.pdf.
[102] M. Ebermann, N. Neumann, New MEMS Microspectrometer for Infrared Absorption Spectroscopy, Gasses and Instrumentation, September/Ocotober 2009, 18-21, 2009, available at http://www.gasesmag-digital.com/gasesmag/20090910?pg=18#pg18.
[103] N. Neumann, S. Kurth, K. Hiller, M. Ebermann, Tunable infrared detector with integrated micromachined Fabry-Perot filter, J. Micro/Nanolith. MEMS MOEMS 7 (2008). http://dx.doi.org/10.1117/1.2909206.
[104] R.A. Wood, Uncooled thermal imaging with monolithic silicon focal arrays, in: Infrared Technology XIX, Proc. SPIE, vol. 2020, 1993, pp. 322−329.
[105] Sofradir, White Paper: Uncooled Infrared Imaging: Higher Performance, Lower Costs, Edirion 10−12 rev. 02, Sofradir EC Resource Center, Sofradir, 373 US Hwy. 46W, Fairfield, NJ, USA, 2012, p. 11.
[106] A. Rogalski, Infrared Detectors, second ed., CRC press, Taylor Francis, 2011, ISBN 978-1-4200-7671-4, p. 850.
[107] M. Gilo, Low-Reflectance, Durable Coatings for Infrared Lenses, SPIE Newsroom, January 21, 2013. http://dx.doi.org/10.1117/2.1201212.004581.
[108] A. Gershun, The light field, J. Math. Psychol. 18 (1939) 51−151.
[109] W. Wimmer, Variability and uncertainty in measuring sea surface temperature (Ph.D. thesis), Department of Oceanography, University of Southampton, Southampton, United Kingdom, 2013.
[110] R.A. Wood, Uncooled thermal imaging with monolithic silicon focal planes, in: Proc. SPIE2020, Infrared Technology XIX, vol. 322, November 1, 1993. http://dx.doi.org/10.1117/12.160553.
[111] P. Coppo, B. Ricciarellia, F. Brandania, J. Delderfieldb, M. Ferletb, C. Mutlowb, G. Munrob, T. Nightingale, D. Smith, S. Bianchic, P. Nicolc, S. Kirschsteind, T. Hennigd, W. Engeld, J. Frerick, J. Nieke, SLSTR: a high accuracy dual scan temperature radiometer for sea and land surface monitoring from space, J. Mod. Opt. 57 (2010) 1815−1830.
[112] C.J. Donlon, An investigation of the oceanic skin temperature deviation (Ph.D. thesis), University of Southampton, Southampton, United Kingdom, 1994.
[113] H. Grassl, The dependence of the measured cool skin of the ocean on wind stress and total heat flux, Boundary-Layer Meteorol. 10 (1976) 465−474.
[114] H. Grassl, H. Hinzpeter, The Cool Skin of the Ocean, GATE Rep., 14, 1, WMO/ICSU, Geneva, 1975, pp. 229−236.
[115] G.A. Wick, W.J. Emery, L. Kantha, P. Schluessel, The behaviour of the bulk-skin temperature difference under varying wind speed and heat flux, J. Phys. Oceanogr. 26 (1996) 1969−1988.
[116] A.T. Jessup, Measurement of Small-Scale Variability of Infrared Sea Surface Temperature, Summary of Fall 1992 AGU Poster Session, 1992.
[117] J.A. Shaw, Degree of linear polarization in spectralradiances from water-viewing infrared radiometers, Appl. Opt. 38 (1999) 3157−3165.

[118] K.H. Berry, Emissivity of a cylindrical black-body cavity with a re-entrant cone end face, J. Phys. E Sci. Instrum. 14 (1981) 629–632.

[119] C.J. Donlon, W. Wimmer, I. Robinson, G. Fisher, M. Ferlet, T. Nightingale, B. Bras, A Second-Generation Blackbody System for the Calibration and Verification of Seagoing Infrared Radiometers, J. Atmos. Oceanic Technol. 31 (2014) 1104–1127. http://dx.doi.org/10.1175/JTECH-D-13-00151.1.

[120] J.B. Fowler, A third generation water bath based blackbody source, J. Res. Natl. Inst. Stand. Technol. 100 (1995) 591–599.

[121] [a] E. Theocharous, et al., Absolute measurements of black-body emitted radiance, Metrologia 35 (1998) 549.
[b] E. Theocharous, N.P. Fox, CEOS Comparison of IR Brightness Temperature Measurements in Support of Satellite Validation. Part II: Laboratory Comparison of the Brightness Temperature of Blackbodies, National Physical Laboratory, Teddington, Middlesex, United Kingdom, 2010, 43 pp.

[122] E. Theocharous, E. Usadi, N.P. Fox, CEOS Comparison of IR Brightness Temperature Measurements in Support of Satellite Validation. Part I: Laboratory and Ocean Surface Temperature Comparison of Radiation Thermometers, National Physical Laboratory, Teddington, Middlesex, United Kingdom, 2010, 130 pp.

[123] J.P. Rice, J.J. Butler, B.C. Johnson, P.J. Minnett, K.A. Maillet, T.J. Nightingale, S.J. Hook, A. Abtahi, C.J. Donlon, I.J. Barton, The Miami2001 infrared radiometer calibration and intercomparison: 1. Laboratory characterization of blackbody targets, J. Atmos. Oceanic Technol. 21 (2004) 258–267.

[124] R. Branch, A.T. Jessup, P.J. Minnett, E.L. Key, Comparisons of shipboard infrared sea surface skin temperature measurements from the CIRIMS and the M-AERI, J. Atmos. Oceanic Technol. 25 (2008) 1598–1606. http://dx.doi.org/10.1175/2007JTECO1480.1171.

[125] P.W. Nugent, J.A. Shaw, N.J. Pust, S. Piazzolla, Correcting calibrated infrared sky imagery for the effect of an infrared window, J. Atmos. Oceanic Technol. 26 (11) (2009) 2403–2412.

[126] J. Hanafin, P. Minnett, Measurements of the infrared emissivity of a wind-roughened sea surface, Appl. Opt. 44 (2005) 398–411.

[127] E. Williams, E. Prager, D. Wilson, Research combines with public outreach on a cruise ship, EOS Trans. AGU 83 (50) (2002) 590–596. http://dx.doi.org/10.1029/2002EO000404.

[128] D.I. Berry, E.C. Kent, The effect of instrument exposure on marine air temperatures: an assessment using VOSClim Data, Int. J. Climatol. 25 (2005) 1007–1022. http://dx.doi.org/10.1002/joc.1178.

[129] E.D. McAlister, Infrared-optical techniques applied to oceanography I. Measurement of total heat flow from the sea surface, Appl. Opt. 3 (5) (1964) 609–612.

[130] E.D. McAlister, W. McLeish, A radiometric system for airborne measurement of the total heat flow from the sea, Appl. Opt. 9 (12) (1970) 2697–2705.

[131] A.T. Jessup, C.J. Zappa, H. Yeh, Defining and quantifying microscale wave breaking with infrared imagery, J. Geophys. Res. 102 (C10) (1997) 23145–23154.

[132] C.J. Zappa, Microscale wave breaking and its effect on air-water gas transfer using infrared imagery (Ph.D. thesis), University of Washington, Seattle, 1999, 225 pp.

[133] C.J. Zappa, A.T. Jessup, H.H. Yeh, Skin-layer recovery of free-surface wakes: relationship to surface renewal and dependence on heat flux and background turbulence, J. Geophys. Res. 103 (C10) (1998) 21711–21722.

[134] D.T. Ho, C.J. Zappa, W.R. McGillis, L.F. Bliven, B. Ward, J.W.H. Dacey, P. Schlosser, M.B. Hendricks, Influence of rain on air-sea gas exchange: lessons from a model ocean, J. Geophys. Res. 109 (C08S18) (2004). http://dx.doi.org/10.1029/2003JC001806.

[135] C.J. Zappa, D.T. Ho, W.R. McGillis, M.L. Banner, J.W.H. Dacey, L.F. Bliven, B. Ma, J. Nystuen, Rain-induced turbulence and air-sea gas transfer, J. Geophys. Res. 114 (C07009) (2009). http://dx.doi.org/10.1029/2008JC005008.
[136] G.O. Marmorino, G.B. Smith, G.J. Lindemann, Infrared imagery of ocean internal waves, Geophys. Res. Lett. 31 (11) (2004) L11309. http://dx.doi.org/10.1029/2004GL020152.
[137] C.J. Zappa, A.T. Jessup, High resolution airborne infrared measurements of ocean skin temperature, Geosci. Remote Sens. Lett. 2 (2) (2005). http://dx.doi.org/10.1109/LGRS.2004.841629.
[138] J.T. Farrar, C.J. Zappa, R.A. Weller, A.T. Jessup, Sea surface temperature signatures of oceanic internal waves in low winds, J. Geophys. Res. Oceans 112 (C06014) (2007). http://dx.doi.org/10.1029/2006JC003947.
[139] R.A. Handler, G.B. Smith, R.I. Leighton, The thermal structure of an air-water interface at low wind speeds, Tellus 53A (2001) 233–244.
[140] H. Haußecker, S. Reinelt, B. Jähne, Heat as a proxy tracer for gas exchange measurements in the field: principles and technical realization, in: B. Jähne, E.C. Monahan (Eds.), Air-Water Gas Transfer, AEON Verlag & Studio, Hanau, 1995, pp. 405–413.
[141] B. Jähne, H. Haußecker, Air-water gas exchange, Annu. Rev. Fluid Mech. 14 (1998) 321–350.
[142] A.T. Jessup, W.E. Asher, M. Atmane, K. Phadnis, C.J. Zappa, M.R. Loewen, Evidence for complete and partial surface renewal at an air-water interface, Geophys. Res. Lett. 36 (L16601) (2009). http://dx.doi.org/10.1029/2009GL038986.
[143] U. Schimpf, C. Garbe, B. Jähne, Investigation of transport processes across the sea surface microlayer by infrared imagery, J. Geophys. Res. 109 (C08S13) (2004). http://dx.doi.org/10.1029/2003JC001803.
[144] C.S. Garbe, U. Schimpf, B. Jähne, A surface renewal model to analyze infrared image sequences of the ocean surface for the study of air-sea heat and gas exchange, J. Geophys. Res. 109 (C08S15) (2004). http://dx.doi.org/10.1029/2003JC001802.
[145] R.A. Handler, I. Savelyev, M. Lindsey, Infrared imagery of streak formation in a breaking wave, Phys. Fluids 24 (2012) 1070–6631.
[146] W.K. Melville, R. Shear, F. Veron, Laboratory measurements of the generation and evolution of Langmuir circulations, J. Fluid Mech. 364 (1998) 31–58.
[147] N.V. Scott, R.A. Handler, G.B. Smith, Wavelet analysis of the surface temperature field at an air-water interface subject to moderate wind stress, Int. J. Heat Fluid Flow 29 (2008) 1103–1112.
[148] F. Veron, W.K. Melville, Experiments on the stability and transition of wind-driven water surfaces, J. Fluid Mech. 446 (2001) 25–65.
[149] F. Veron, W.K. Melville, L. Lenain, Infrared techniques for measuring ocean surface processes, J. Atmos. Oceanogr. Technol. 25 (2) (2008) 307–326.
[150] F. Veron, W.K. Melville, L. Lenain, Measurements of ocean surface turbulence and wave–turbulence interactions, J. Phys. Oceanogr. 39 (2009) 2310–2323. http://dx.doi.org/10.1175/2009JPO4019.1.
[151] F. Veron, W.K. Melville, L. Lenain, Small scale surface turbulence and its effect on air-sea heat fluxes, J. Phys. Oceanogr. 41 (1) (2011) 205–220.
[152] C.J. Zappa, M.L. Banner, J.R. Gemmrich, H. Schultz, R.P. Morison, D.A. LeBel, T. Dickey, An overview of sea state conditions and air-sea fluxes during RaDyO, J. Geophys. Res. Oceans 117 (C00H19) (2012). http://dx.doi.org/10.1029/2011JC007336.

Chapter 4

Theoretical Investigations

Barbara Bulgarelli,[1,*] Menghua Wang,[2] Christopher J. Merchant[3]
[1] *European Commission, Joint Research Centre, Ispra, Italy;* [2] *NOAA Center for Satellite Applications and Research, College Park, Maryland, USA;* [3] *Department of Meteorology, University of Reading, Reading, UK*
*Corresponding author: E-mail: barbara.bulgarelli@jrc.ec.europa.eu

Simulation capabilities relying on radiative transfer models are central to contemporary remote sensing. In fact, theoretical simulations of satellite and field data are key means for investigating and interpreting remote sensing and in situ measurements and additionally assessing their quality. Specifically, theoretical simulations have been proven fundamental to investigate perturbations affecting field measurements, as well as to quantify the various contributions to the top-of-atmosphere (TOA) radiometric signal. This offers the capability to thoroughly address the perturbing effects in satellite observations allowing the development and implementation of correction schemes, as well as to refine protocols for in situ measurements.

The following three chapters address the essential contribution of modeling tools to in situ and satellite radiometry of the ocean in the visible, near-infrared (NIR), shortwave infrared (SWIR) and thermal infrared (TIR) spectral regions. Specifically, Chapter 4.1 provides an overview on radiative transfer simulations applied in the evaluation of the perturbing effects induced in field radiometric measurements by overstructures (such as deployment ships and buoys), instrument case itself, or sea-surface waves. The chapter summarizes results from recent studies and presents the methodologies based on theoretical evaluations to minimize measurement perturbations.

Chapter 4.2 presents and discusses theoretical estimations of the TOA radiance spectra over global open oceans and coastal and inland waters. Specifically, radiance contributions from atmosphere, i.e., contributions from air molecules (Rayleigh scattering) and aerosols and ocean surface, as well as from ocean waters (both Case-1 and Case-2 waters) are discussed in detail. It also provides simulated results for sensor-measured TOA radiances. Specifically, through histogram data analysis, typical TOA radiances over ocean from ultraviolet to visible, NIR, and SWIR wavelengths are derived. The simulated results are compared with those from the Sea-viewing Wide Field-of-view Sensor (SeaWiFS) and the Visible Infrared Imaging Radiometer Suite (VIIRS). In addition, the chapter describes modeling for the normalized water-leaving radiance by accounting for ocean surface and in-water bidirectional effects, the radiance contribution that interacted with the atmosphere and the water surface (the so-called path radiance), including modeling of the aerosol optical properties and of the atmospheric diffuse transmittance.

Finally, Chapter 4.3 discusses TIR simulations at the TOA for sea surface temperature retrievals, cloud detection (and classification), and related uncertainty estimates. This chapter addresses the capability to choose the most suitable numerical model for simulations of the radiative transfer processes at thermal wavelengths, to properly describe the thermal properties of the active atmospheric constituents and to select the most fitting model for surface emissivity and reflection. The application of the simulations to the tasks of image classification, surface temperature estimation, and uncertainty modeling are then discussed.

Chapter 4.1

Simulation of In Situ Visible Radiometric Measurements

Barbara Bulgarelli,[1,]* Davide D'Alimonte[2]
[1] European Commission, Joint Research Centre, Ispra, Italy; [2] Centre for Marine and Environmental Research, University of Algarve, Faro, Portugal
*Corresponding author: E-mail: barbara.bulgarelli@jrc.ec.europa.eu

Chapter Outline

1. Overview 407
2. The RTE and Its Solution Methods 408
     2.1 The Radiative Transfer Equation 408
     2.2 Deterministic Solutions of the RTE 410
     2.3 Monte Carlo Solutions of the RTE 410
3. Simulations of In Situ Radiometric Measurement Perturbations 413
     3.1 Overstructure Perturbations 414
       3.1.1 Ship-Shading Effect 415
       3.1.2 Tower-Shading Effects 420
       3.1.3 Self-Shading 423
     3.2 Perturbations Induced by Sea-Surface Waves 429
       3.2.1 Light Focused by Wind-Generated Waves 430
       3.2.2 Experimental Findings 431
       3.2.3 Statistical Modeling 433
       3.2.4 MC Case Studies 434
4. Summary and Remarks 441
References 442

1. OVERVIEW

This chapter concerns radiative transfer simulations applied to minimize perturbations of in-water data for the calibration of space optical sensors, the validation of space-borne derived products, and the development of bio-optical modeling for ocean color inversion schemes. The propagation of the radiance through a scattering, absorbing, and emitting medium is described by the radiative transfer equation (RTE). Analytical solutions of the RTE for marine optics exist under restrictive assumptions, whereas numerical schemes are required in many ocean color studies.

Numerical approaches can be divided into deterministic and stochastic. Deterministic solutions of the RTE are computationally efficient, not affected by statistical uncertainties, but they tend to be mathematically quite complex, physically nondirect and so generally not suited for complex three-dimensional (3D) geometries. Stochastic solutions of the RTE are based on Monte Carlo (MC) photon-transport algorithms. The MC scheme is mathematically simpler than the deterministic approach since physically direct, whereas main tradeoffs are its statistical uncertainties and its computing load. While deterministic solutions are more accurate for solving the RTE in cases of horizontal translational invariance, i.e., in plane-parallel systems, the power of the MC method is borne out when 3D geometries and/or time-dependent problems are involved.

The stochastic approach is thus the reference method to model the processes reviewed in this work, which require a multidimensional and time-dependent description of the propagating system. Cases of interest for this venue include perturbations due to deployment structures, instrument self-shading, and focusing effects of the sea-surface. Specific results detailed henceforth are those allowing for the implementation of correction schemes to remove biases from in situ radiometric data and support the refinement of field measurement protocols.

The additional parts of this chapter are organized as follows: Section 2 overviews the RTE and its solution methods; simulation techniques applied to analyze the perturbation effects in in-water radiometric measurements with correlated results are reviewed in Section 3; summary and remarks are finally reported in Section 4.

2. THE RTE AND ITS SOLUTION METHODS

The solution of the RTE for deriving regression parameters from in-water radiometric measurements at different depths is discussed hereafter. As well, fundamentals of radiative transfer modeling for marine optics are overviewed with specific reference to deterministic and MC techniques.

2.1 The Radiative Transfer Equation

The RTE describes the radiance propagation in an absorbing, emitting, and scattering 3D medium. Specifically, the radiance $L(r;\xi)$ at r propagating in direction ξ satisfies

$$\underbrace{(\xi \cdot \nabla) L(r;\xi)}_{(a)} = \underbrace{-c(r)L(r;\xi)}_{(b)} + \underbrace{\int_\Omega L(r;\xi')\beta(r;\xi' \to \xi) d\Omega'}_{(c)} + \underbrace{S(r;\xi)}_{(d)}, \quad (1)$$

where r is the position vector of Cartesian coordinates; $\xi = (\theta, \phi)$ is the direction unit vector, with θ and ϕ indicating the zenith and azimuth angles, respectively; ∇ is the gradient operator; $c(\mathbf{r})$ is the attenuation coefficient at $\mathbf{r}$; β is the Volume Scattering Function (VSF) defining the probability that radiance at $\mathbf{r}$ from any direction ξ' is scattered into direction ξ; and $S(\mathbf{r};\xi)$ is the source term. The wavelength dependence has been omitted for brevity. It is recalled that $c(\mathbf{r}) = a(\mathbf{r}) + b(\mathbf{r})$, where $a(\mathbf{r})$ is the absorption coefficient and $b(\mathbf{r}) = \int_{4\pi} \beta(\mathbf{r}; \xi' \to \xi) d\Omega'$ is the scattering coefficient.

In Eqn (1), term (a) denotes the variation of the radiance L per unit distance along ξ; term (b) describes the radiance loss due to attenuation; term (c) designates the radiance gain due to scattering of the radiance field at $\mathbf{r}$ into direction ξ; and term (d) represents the emission from the internal radiance source S. Equation (1) can then be written as

$$\frac{(\xi \cdot \nabla) L(\mathbf{r}; \xi)}{c(\mathbf{r})} = -L; (\mathbf{r}; \xi) + \omega(\mathbf{r}) \int_{4\pi} L(\mathbf{r}; \xi) \tilde{\beta}(\mathbf{r}; \xi' \to \xi) d\Omega' + \tilde{S}(\mathbf{r}; \xi), \quad (2)$$

where $\omega = b/c$ is the single-scattering albedo, $\tilde{\beta} = \beta/b$ is the scattering phase function describing the angular distribution of the scattered radiation, and $\tilde{S} = S/c$.

In a plane-parallel propagating medium and in the absence of internal sources (i.e., $S = 0$), Eqn (2) simplifies to

$$\frac{\cos\theta}{c(z)} \cdot \frac{dL(z; \xi)}{dz} = -L(z; \xi) + \omega(z) \int_{4\pi} L(z; \xi') \tilde{\beta}(z; \xi' \to \xi) d\Omega'. \quad (3)$$

An example of analytic solution of Eqn (3) applied in the processing of in situ radiometric measurements is the exponential decay law derived hereafter for the upwelling radiance.

In an infinitely deep and optically homogenous water volume (i.e., the asymptotical stationary case), the RTE for the nadir radiance L_u satisfies

$$\frac{dL_u(z)}{dz} = -c \cdot L_u(z) + \int_\Omega L(z; \xi') \beta(\xi' \to \mathbf{k}) d\Omega', \quad (4)$$

where $\mathbf{k}$ is an upward-oriented unitary vector. Using the probability density function $p(\xi)$ to factorize the radiance distribution, $L(z;\xi) = L_u(z) \, p(\xi)$ (i.e., assuming a depth-independent angular distribution of the radiance), allows for reformulating Eqn (4) as

$$\frac{dL_u(z)}{dz} = L_u(z) \left(c - \int_\Omega p(\xi') \beta(\xi' \to \mathbf{k}) d\Omega' \right). \quad (5)$$

In Eqn (5), the difference between the attenuation coefficient c and the convolution of the VSF with the probability density function is the attenuation coefficient K_L. Hence $K_L \leq c$, with equality holding in a purely absorbing medium. The differential expression $dL_u(z)/dz = -K_L \cdot L_u(z)$ is solved by

$L_u(z) = L_u(0^-)e^{-K_L \cdot z}$ (where $L_u(0^-)$ indicates the radiance value just below the water surface).

Analogous exponential decays are also valid for the upward, E_u, and downward, E_d, irradiances. By using $\Re$ to indicate E_d, E_u, or L_u, the radiometric value just below the sea—air interface and the attenuation coefficient can be denoted as $\Re_0$ and $K_\Re$, respectively. With this formalism,

$$\Re(z) = \Re_0 e^{-K_\Re \cdot z}, \tag{6}$$

and

$$\ln(\Re(z)) = \ln(\Re_0) - K_\Re \cdot z. \tag{7}$$

Equation (7) is used to determine $\Re_0$ and $K_\Re$ through the linear regression of log-transformed radiometric measurements of $\Re(z)$ performed at different depths [1,2]. Note also that the standard notation K_d, K_u, and K_L will still be adopted in the next sections when explicitly referring to E_d, E_u, or L_u, respectively.

The hypotheses underpinning the linear trend of log-transformed $\Re(z)$ values have limited validity and are not applicable for instance when it is necessary to account for optical stratification within the water column, in the absence of translational invariance, or if the probability density function of the radiance distribution is not stationary.

Core topics addressed in this chapter are numerical simulations to analyze perturbations affecting $\Re(z)$ values and consequent uncertainties of $\Re_0$ and $K_\Re$.

2.2 Deterministic Solutions of the RTE

There is a richness of mathematical methods for the deterministic numerical solution of the plane-parallel RTE (Eqn (2)) [3—5]. Examples include the adding and doubling technique [4,6], the discrete-ordinate method [4,7,8], the method of successive order of scattering [9—11], the invariant imbedding technique [12—15], and the finite-elements method [16,17].

Some of them were specifically applied to simulate the propagation of light into water [15,18—21] and to address uncertainty analysis for in situ optical radiometric measurements [22,23]. Namely, the invariant imbedding technique was utilized for the production of look-up tables (LUTs) to correct the dependence of radiometric quantities on illumination and measurement geometry [22], while the finite-element method was applied for the characterization of the nonideal cosine response of the irradiance sensors [23].

2.3 Monte Carlo Solutions of the RTE

MC methods define the state of a system based on the stochastic behavior of its constituents [24—26]. MC radiative transfer simulations of the in-water light field are performed by estimating what percentage of photons emitted by the sun and propagating through the atmosphere—ocean system can be detected by the radiometric sensor. Virtual photons are traced in their path from the source to the detectors, properly accounting for their response [27–32]. By knowing all relevant probabilities for each elementary event that the photon may undergo in its "life history," the probability for the entire sequence of events can be determined. In practice, one photon at the time is followed in its 3D path through the absorbing and scattering medium, and all possible interactions are defined by probability density functions and cumulative distribution functions.

It is briefly recalled that, given the probability density function $p(x)$ opportunely normalized to 1,

$$\int_{-\infty}^{\infty} p(x)dx = 1, \qquad (8)$$

the probability that a specific event occurs over the range x to $(x + dx)$ is given by $p(x)dx$; while the probability for the event to occur between the lowest possible value x_{min} and x is provided by the cumulative distribution function $P(x)$

$$P(x) = \int_{x_{min}}^{x} p(x)dx, \qquad (9)$$

with $0 \leq P(x) \leq 1$.

For each MC event, the x value is sampled by generating a random number n uniformly distributed between 0 and 1, and then solving $P(x) = n$ for x.

For the simulation of the in-water radiance field, probability functions are used to sample the photon optical distance traveled before interacting with the medium, $\tau = c \cdot r$ (where r is the geometrical distance to the next collision point); the nature of the interaction events; and the photon propagation directions.

Since the optical distance τ follows an exponential probability density function $p(\tau) = e^{-\tau}$, the free-flight optical distance of each photon at its starting point and after each scattering event is obtained by solving $\int_0^\tau e^{-\tau'} d\tau' = n$ for τ, leading to $\tau = -\ln n$.

The nature of each interaction with the medium is selected by comparing the sampled number n with the single-scattering albedo ω, which defines the probability for an interaction to be a scattering event: if $n \leq \omega$, a scattering event occurs, otherwise the photon is absorbed.

The direction of the photon after each scattering event is finally defined through the scattering phase function $\tilde{\beta}$, which is azimuthally invariant in natural waters and satisfies the normalization condition $2\pi \int_0^\pi \tilde{\beta}(\theta) \sin(\theta) d\theta = 1$. This allows for defining the probability density function of the scattering angle θ, $p(\theta) = 2\pi\tilde{\beta}(\theta)\sin(\theta)$, so that θ is obtained by solving $P(\theta) \equiv 2\pi \int_0^\theta \tilde{\beta}(\theta')\sin(\theta')\, d\theta' = n$. Conversely, the azimuth angle ϕ is sampled from a uniform distribution, $p(\phi)=1/2\pi$, leading to $\phi=2\pi n$.

The accuracy of the MC simulation procedure depends on the correct description of the propagating system, including its boundary conditions, and on the capability to constraint the random statistical noise. Indeed MC computations are inherently affected by a *statistical* uncertainty, which is a linear function of the square root of the number of initialized photons N_{pho}. This might impose long processing times, and depending on the case study, might require high-performance computing solutions (e.g., parallel execution on computer clusters [33]).

Most MC methods rely on variance-reducing techniques to inhibit photon loss, thus increasing computational efficiency and keeping the statistical variability relatively small. A widely applied variance-reducing technique is to force each interaction with the medium to be a scattering event, so that photons are not lost in absorption processes. This is performed by associating a statistical photon weight $w = 1$ at the source and by multiplying w by ω at each photon interaction with the medium. This is equivalent to consider a photon packet and to remove from it at each scattering event the photons that are absorbed. The process is heuristically ended when w becomes smaller than a given threshold, so that photons are traced only as long as they appreciably contribute to model radiometric quantities. The threshold might be predefined (e.g., in the order of 10^{-6} [34]) or a function of the seawater Inherent Optical Properties (IOPs) and of the MC simulation accuracy requirements [35]. Alternatively, the Russian Roulette technique can be applied [36,37]. A short mention is furthermore given to the extensively applied backward ray tracing technique. While in the so-far described forward MC method photons are started at the source and tracked to the detector, in the backward MC method, time reversal is applied and photons are tracked from the detector back to the source. At each scattering event, the contributions to the detected signal are deterministically computed, ensuring highly increased computational efficiency. Other variance-reducing techniques will be presented when discussing their application in the next sections.

The scattering phase functions commonly applied for the numerical modeling of the radiance field in marine optics mainly include the *KA* phase function derived by Gordon et al. [29] from the measurements performed by Kuellenberg [38] in the Sargasso Sea; an average phase function derived by Mobley [15] from the Petzold's data set [39] (hereafter shortly called the Petzold phase function); and several analytical parameterizations. Among them mention is given to those proposed by Henyey and Greenstein (HG) [40], including its two-terms formulation (TTHG) [41,42], by Kopelevich [43], and by Fournier and Forand [44–47]. Mentioned phase functions have been alternatively used to describe the scattering properties of the whole seawater or of its sole suspended matter (i.e., the hydrosols).

The phase function of the seawater can be modeled as a weighted sum of the phase functions of the pure water and of its suspended hydrosols. Molecular and hydrosol phase functions significantly differ in shape. The first—of well-known magnitude, angular pattern, and spectral dependence—is usually approximated by the Einstein—Smolouchowski formulation [15]. The second—showing a large natural variability—is typically characterized by high values at small angles, which is distinctive of scattering by particles with a diameter greater than the wavelength of the incident radiation.

The experimental measurement of the oceanic VSF is highly challenging [48], and until recently only sparse in situ measurements were available (e.g., Refs [38,39]). This set the premises for considering possible to apply a "typical" hydrosol function (classically described by the Petzold phase function), assuming that any variation in the angular distribution of the scattered light in natural waters only depended on the concentration of the suspended matter [15,20,49]. Successive theoretical studies [21,50—52] as well as experimental measurements of the VSF in natural waters (e.g., Refs [53,54]) pointed out the limitation of this assumption, suggesting the need to account for a wider natural variability of the hydrosol angular scattering properties, although highlighting a remarkably consistent shape of the VSF in the backward direction [53].

With its two-parametric expression, the Fournier—Forand analytical formulation is particularly suitable to model the hydrosol scattering phase function [21,52,54], even accounting for its asymptotic behavior at small scattering angles. The two free parameters, which can be tuned to fit experimental data, are moreover directly correlated to the physical properties of the scattering particles, representing their real refractive index and the Junge parameter of their size distribution. Comparisons with experimental measurements [54] indicated a good performance also for the analytical function proposed by Kopelevich, which expresses the global hydrosol phase function as a linear combination of the phase function for small and large particles, weighted on their respective contribution. Comparison results interestingly pointed out the potential to simply reconstruct the seawater VSF using the Kopelevich model in conjunction with a single in situ measurement of the backscattering coefficient $b_b = 2\pi \int_{\pi/2}^{\pi} \beta(\theta, \phi) \sin(\theta) d\theta$ at 555 nm [54].

3. SIMULATIONS OF IN SITU RADIOMETRIC MEASUREMENT PERTURBATIONS

The scope of this section is (1) to overview the radiative transfer simulation procedures applied in the evaluation of the perturbations induced in the radiometric measurements by overstructures and sea-surface waves; (2) to summarize the results achieved; and (3) to present the methodologies developed on the basis of these theoretical evaluations to actually minimize measurement perturbations. Addressed methodologies mainly include refined measurements protocols and data correction schemes.

It is given only short mention to a theoretical study, performed with MC simulations, evidencing land perturbations in measurements of sky radiance over the ocean when the radiometer is located on an island, and which are particularly relevant (up to 39%) at blue visible wavelengths [55]. Theoretical analyses of land-induced perturbations in ocean-color remote sensing have been so far limited to address uncertainties in satellite data [56–59]. Nevertheless, evidence of an environment impact on ground-based observations of the solar radiation, especially in the blue [60], might suggest a dedicated study enquiring adjacency effects in above-water downward irradiance measurements performed in the vicinity of the land.

3.1 Overstructure Perturbations

Both above- and in-water measurements might be perturbed by the presence in the proximity of the measurement point of large 3D structures, like deployment ships and towers, moored buoys, or the casing of the instrument itself. Photons interact with these objects through scattering and absorption processes, which alter their propagation from the sun to the sensor, thus modifying the radiance field detected by the instrument.

Overstructures have dimensions much larger than the incident wavelengths; consequently, the principles of geometric optics apply in the description of their interaction with light: the scattering and absorption processes are described in terms of reflection and refraction, respectively, and the Snell's law determines the propagation direction of reflected and refracted radiation. The opacity of the considered objects additionally implies refracted rays absorption, so that one of the most visible effects is the casting of a shadow. As so, perturbations induced by deployment ships and towers and by the instrument casing are traditionally termed *ship-shading, tower-shading* and *instrument self-shading* effects, respectively.

In optics, the term *shadow* generally refers to a region characterized by a total (*umbra*) or partial (*penumbra*) absence of direct illumination from a finite-dimension light source. In marine optics, the concept of shadowing is extended to the case of a diffuse illumination field, and *shadow* defines a region characterized by a general decrease of illumination as a consequence of the interception of the radiation by the light-obstructing object. In other words, the sky, and not only the sun, is considered as an illumination source.

Theoretical investigations have been carried out since the beginning of the last century to both quantify the overstructures perturbations and identify strategies for their minimization. The dimension, the location, and the optical properties of the different perturbing structures may vary considerably, hindering the capability to formulate a unique simulation approach, and implying the adoption of specific sun—atmosphere—ocean—structure—detector systems for the solution of the RTE. Nonetheless, some common modeling features are easily individuated.

- The atmosphere—ocean propagating medium is mainly modeled as a coupled plane-parallel system.
- The perturbing structure, geometrically represented with different degrees of sophistication, is mainly assumed as perfectly absorbing, implying no perturbations from radiance reflections at the structure.
- The finite size of the overstructure, breaking the translational invariance of the plane-parallel atmosphere—ocean system, requires the solution of the 3D formulation of RTE. This is most effectively achieved through the application of the MC technique, applied both in its forward and backward formulation, and always in conjunction with variance-reducing procedures.
- The measurement error caused by the presence of the perturbing structure is commonly expressed in terms of the percent relative difference $\varepsilon_\Re$ between radiometric quantities computed in the absence ($\Re$) and in the presence ($\widehat{\Re}$) of the structure:

$$\varepsilon_\Re = \frac{\Re - \widehat{\Re}}{\Re} * 100. \tag{10}$$

The dependence on depth z and wavelength λ is omitted for simplicity.
- The error estimation is generally performed for in-water profiles of E_d, E_u, and L_u. Very few evaluations exist of the shading effects in off-nadir radiance measurements, i.e., of $L_u(\theta)$, while no theoretical estimate has been done so far for overstructures perturbations in measurements above water.
- The geophysical dependencies of the shadowing perturbations are usually analyzed as a function of the illumination conditions, the IOPs of the propagating system, and the sensor—structure geometry.

The following sections detail the theoretical procedures used to estimate the perturbing effects induced by the different light-obstructing structures, the main results achieved, and the suggested minimization procedures. Following historical developments, the perturbation effects induced by the deployment structures are treated at first, followed by the description of the instrument self-shading.

3.1.1 Ship-Shading Effect

Owing to the pitching and rolling motions of the ship, measurement instruments were traditionally deployed only few meters outboard, where ship-induced perturbations could not be avoided.

Early observations of the ship-shading effect date back to the first half of the twentieth century. Poole and Atkins gave already a mention to it in 1926 in their analysis of light penetration in the seawater [61]. Few years later, they suggested the shading due to the ship to become "relatively more important on a dull day than on a sunny day when the greater proportion of the illumination is due to direct sunlight if due care is taken to keep the photometer out of the direct shadow." [62]. The first tentative theoretical estimate of the bias induced by ship-shading is provided by Poole in 1936. From the dimensions of the ship and the distance outboard of the point of suspension, the proportion of sea-surface area occupied by the ship is calculated at any given depth, and from it the fraction of the diffuse light that is cut off by the ship is found. The magnitude of the shading by the ship for diffuse light was estimated to slightly increase in the first 5 m depth up to about 10%, decreasing afterward. Shading effects on the direct light component were considered negligible [63].

Years later, Jerlov [64] analyzing the results from Aas [65] concluded that the reduction in downward irradiance measurements (made 5 m abeam of the ship rail, on a ship of length 52 m) due to the ship shadow was <10% for all depths under clear skies, but as much as 22% (at 20 m depth) under overcast skies.

It is only in 1985 that the first accurate theoretical quantification of the shading effect was carried out by Gordon [66], applying a backward 3D MC (hereafter indicated as G-MC). Owing to severe limitations of computing power, only $\sim 10^4$ photons were tracked, while several variance-reducing techniques were applied to improve efficiency and decrease statistical uncertainties. Namely, the forced-collision sampling technique to prevent photons leaving the medium, the phase function truncation technique to improve the handling of highly peaked phase functions [67,68], and the correlated sampling technique to allow simultaneously tracking the photon propagation in the presence and in the absence of the perturbing object [69].

Gordon assumed an idealized ship modeled as a flat and totally absorbing rectangle of fixed length l and width s (Table 1), floating on homogenous and infinitely deep clear oceanic waters, with no overlying atmosphere (Figure 1(a)).

Two extreme illumination conditions were considered: a collimated irradiance to represent direct sunlight in the absence of atmospheric scattering (collimated illumination) and an isotropic angular distribution of the radiance reaching the surface to represent a completely overcast sky, or to approximate the skylight component of the solar irradiance for clear skies (diffuse illumination). Perturbation estimates under realistic illumination conditions could be obtained by weighting results for a collimated and a diffuse illumination over the ratio of the direct to diffuse irradiance.

The assumption of negligible ship reflectance, impelled by the adoption of the correlation sampling technique, was justified by observing that the upwelling irradiance is typically <10% of the downwelling irradiance while

TABLE 1 Geophysical Parameters Selected by Gordon [66], by Doyle and Zibordi [70] and by Piskozub [71] for the Simulation of the Ship-Shading Effects

Geophysical Parameters	Selected Values by Gordon [66] and by Doyle and Zibordi [70]	Selected Values by Piskozub [71]
θ_0	Up to 70°	Up to 80°
ϕ_0	0°, 45°, 90°	0°
c (m^{-1})	0.1, 0.3, 0.5	Up to 1.0
Ω	0.5, 0.7, 0.9	Up to 0.9
$\tilde{\beta}^a$	δ-KA [29]	Petzold [39]
v (m s^{-1})	0	Up to 15
$\Delta\phi_w$	–	0° → 360°
l (m)	38.4	1 ÷ 100
s (m)	6.55	10
d (m)	~1.25 → ~21.75	0 → 30
z_m (m)	Ref. [66]: 5 → 50^b; 0^{-c}; Ref. [71]: 0$^-$ → 50	1 → 40

^{a}For the whole seawater.
^{b}For E_d simulations.
^{c}For E_u and L_u simulations.

most ship's hulls are dark (with a small albedo typically <10%), so that the downwelling irradiance from reflection by a typical hull would contribute by <1% to the computed irradiance (i.e., 10% of 10% of the maximum downwelling irradiance).

To further decrease the variance of the simulations, photons were prevented to interact both with the sea-surface and with the ship's hull. This was obtained by forcing collisions in upward-moving photons before reaching the seawater interface, and by modeling the ship as suspended just above the sea-surface (simulation method III in Ref. [66]). This choice was justified by demonstrating that alternatively allowing photons to interact with the sea-surface (simulation method II in Ref. [66]) or positioning the idealized ship beneath the water (simulation method I in Ref. [66]) tends to produce higher variance without any significant improvement of the simulated results, excluding cases characterized by shallow water depths.

Values of L_u and E_u just beneath the water surface and E_d in-water profiles were simulated as if measured by a point-like detector located at several distances d from the sunny side of the ship along a horizontal reference line

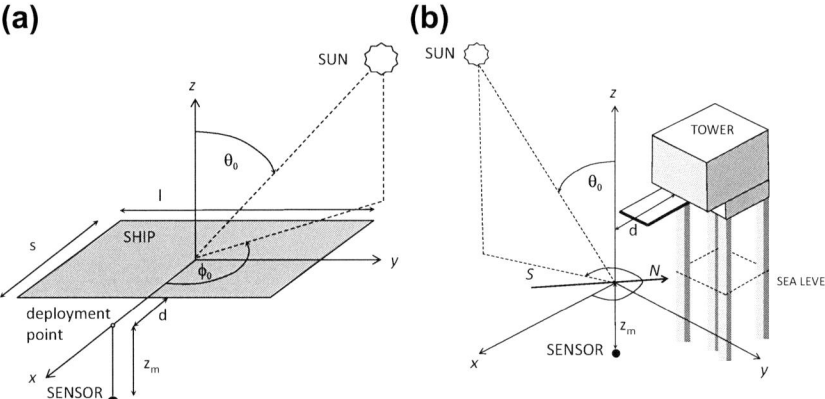

FIGURE 1 Shading effects from deployment structures. (a) Schematics of ship-shading effects as modeled in Ref. [66]. (b) Schematics of the measurement setting at the Aqua Alta Oceanographic Tower (AAOT) in the northern Adriatic Sea (AAOT, 45.31°N, 12.51°E) as modeled in Refs [70,82]. See the text for the definition of quantities expressing the measurement geometry.

(coincident with the positive x-axis, see Figure 1(a)) centered on and perpendicular to the longer side of the ship. An infinitesimally small nadir field of view (FOV) was assumed for L_u.

Simulations were performed assuming the δ-truncated form of the *KA* scattering phase function (δ-*KA* henceforth) for the water, and considering different combinations of (1) additional water IOPs (namely, c and ω); (2) solar illumination condition (determined by solar zenith and azimuth angles θ_0 and ϕ_0, and including overcast sky); (3) measurement depth z_m from the sea-surface ($z_m = 0^-$ indicates a measurement depth just below the water surface); and (4) distance d from the side of the superstructure (Table 1). It is noted that θ_0 is measured from the local vertical, while ϕ_0 is measured counterclockwise from the x-axis (Figure 1(a)).

Results confirmed a significant dependence of ε_{E_d} on the illumination conditions. For collimated illumination, ε_{E_d} increases with ω and c and shows minimal values (within 3%) for instrument deployment on the sunny side of the ship when $\Delta\phi = |\phi_v - \phi_0| < 45°$ and $\theta_0 \gtrsim 20°$. Perturbations increase when the shadowed zone approaches the measuring point, with ε_{E_d} likely exceeding 10% for $\Delta\phi \to 90°$ or $\theta_0 \to 0°$. On the contrary, for diffuse illumination, ship perturbations appear to be essentially geometrical, with negligible dependence on in-water IOPs, and ε_{E_d} can reach almost depth-invariant values up to 30%.

These findings are explained by observing that for collimated illumination, the in-water downward radiance field is highly peaked in the direction of the refracted solar beam, so that perturbations of radiance contributions from other directions (e.g., those from the shadowed area) are scarcely relevant when the instrument is deployed on the sunny side of the ship. A multiple scattering enhancement (as obtained by increasing ω or c) smoothes the angular distribution of the in-water radiance, and contributions from directions other than the direct beam acquire more importance. This increases the influence of the shadowed region. An isotropic angular distribution of the downward radiance impinging at the surface (as in the case of diffuse illumination) dampens the effect, and ε_{E_d} becomes proportional to the fraction of the sky obscured by the ship [63].

Simulations of L_u and E_u just beneath the water surface showed significant perturbations ($\sim 1-13\%$ and $\sim 5-20\%$, respectively) for any illumination condition. Estimated errors, generally larger for E_u, increase with θ_0 and for $\Delta\phi \rightarrow 90°$. They show a nonmonotonic dependence with c, with a negative second derivative, as well as negligible values when $1/a \ll d$.

As expected, ship perturbations decrease with increasing d, but (with the exception of ε_{E_d} in clear-sky conditions) they are still larger than $\sim 5\%$ at several meters off the ship.

Further MC estimations [70,71] performed with increased computational efficiency and a wider range of variation for the input parameters (Table 1) largely confirmed these previous findings, while additionally allowed to evidence (1) no influence of the sea-surface roughness on ship-shading perturbations [71]; and (2) the need to account for both interactions with the ship's hull and the sea-surface, as well as for a scattering atmosphere when simulating E_d in the first meters below the surface [70]. The first aspect was enquired [71] by assuming a statistical distribution of the sea-surface waves [72] for wind speed v up to 15 m s^{-1} and different wind directions $\Delta\phi_w = |\phi_0 - \phi_w|$, while neglecting the presence of whitecaps and bubbles, the effects induced by the ship on the surface roughness, and the seawater displacement over the still sea level. The second aspect was pointed out by extending E_d simulations up to $z_m = 0^-$, where a convergence of ε_{E_d} to an identical subsurface value, whatever the illumination conditions, was observed [70]. Notably, the interaction with the ship's hull allows accounting for second- and higher-order shadowing effects induced by absorption of random walk photons, whereas letting photons interact with the sea-surface allows accounting for a ship-induced decrease of internal reflection at the sea-surface, otherwise not taken into consideration. Accounting for these aspects is particularly important in the first meters below the surface for clear-sky conditions and small solar zenith angle.

Mention is additionally given to attempts to numerically reproduce in situ optical measurements affected by ship-shading [73,74]. Noteworthy, one of these studies [73] accounted for ship's hull reflectance and made use of a deterministic method for the solution of the RTE (i.e., the finite-difference method [73]) to describe the full angular radiance distribution, while imputed differences between theoretical and experimental data to the limit of the selected Petzold phase function in representing the actual experimental conditions.

Although a first-level correction scheme of ship-shading effects for the downwelling in-water irradiance could be possible, as suggested by previous investigations [63,66], no operational estimate for upwelling radiometric measurements was ever proposed since it would require expensive simulations. As so, performing the measurements at distance from the ship appeared a more feasible strategy to minimize the ship perturbations. In agreement with it, a stringent and conservative measurement protocol for avoiding ship-shadow perturbations in all radiometric measurements was formalized for the validation procedure of the SeaWiFS data [75]. The protocol suggested (1) to deploy the instrument from the stern, with the sun's relative bearing aft of the beam; and (2) to perform, under conditions of clear sunny skies, measurements for E_d, E_u, and L_u at a distance d away from the ship equal to $0.75/K_d(\lambda)$, $3/K_u(\lambda)$, and $1.5/K_L(\lambda)$, respectively. Increased distances were suggested for large vessels.

Accordingly, optical oceanographic instruments and deployment techniques were appositely designed to perform measurements at distance from the ship. An example is the optical free-fall instrument [76]. The use of remotely operated vehicle to deploy optical sensors was also suggested [75].

It is finally recalled that several experimental campaigns were carried out since the late 1980s to evaluate the ship-shading effects on in situ optical measurements, essentially confirming the herein described theoretical estimations [73,74,77–79].

3.1.2 Tower-Shading Effects

Fixed platforms can offer several advantages with respect to ships and buoys. Besides being often located in coastal regions, frequently equipped with ample power, manned, or frequently visited, and requiring lower maintenance costs (since they are maintained for other purposes) [80], their primary advantage is stability [81]. Optical instruments can be deployed on a tower with virtually no tilt (e.g., by using wire-stabilized systems [82]) so that the solar illumination geometry, required for a careful evaluation of shading effects, can be accurately determined. Nevertheless, tower-perturbing effects, which are unavoidable, are envisaged to be even higher than those induced by ships, and must be hence accurately estimated.

Due to the infrequent use in the past of offshore towers for in-water radiometric measurements, literature on their perturbing effects is scarce [83]. It is only at the turn of the twenty-first century that extensive and accurate theoretical investigations of platform-induced perturbing effects were carried out [70,82] with the specific aim to develop and implement an operational scheme to correct perturbations induced by the Aqua Alta Oceanographic Tower (AAOT) (45.31° N, 12.51° E) in in-water radiometric measurements performed with the Wire-Stabilized Profiling Environmental Radiometer system (WiSPER [82])

The need to implement a correction procedure to minimize tower-shading effects was first enquired with dedicated field measurements, which actually evidenced significant perturbations (up to ~20% for overcast sky conditions) at the WiSPER location. The same set of experimental data was further utilized to assess the feasibility to theoretically model actual tower perturbations with MC simulations [82]. The full 3D backward MC PHO-TRAN code [24] was utilized for this purpose. This code models the propagating system on a 3D grid delimiting the largest macroscopic volumes of uniform optical properties, and the solar source as a parallel monochromatic beam uniformly impinging on the top of the atmosphere. Overcast sky conditions are alternatively simulated as an isotropic radiance field impinging at the sea-surface. Within this reference frame, the point-like detectors are described by their location, FOV, and associated angular distribution function (ADF). Several variance-reducing techniques are applied, namely: semi-systematic sampling, restricting the angular direction of backward emitted or scattered photons to not yet sampled angular bins, allowing a more uniform selection of random directions [84]; forced absorptions and collisions, compelling photons to undergo scattering events before leaving the medium [85]; stacked reflections, allowing the tracking of both sea-surface refracted and sea-surface reflected photons; phase function truncation [66–68]; Russian roulette [36]; and correlated sampling [69].

Aiming at reproducing the actual experimental conditions, atmosphere and ocean (coupled by a flat foam-free interface) were divided into several plane-parallel layers to resolve the atmospheric aerosol, gas molecules, and ozone vertical distribution, as well as the experimentally measured in-water vertical profiles of c and a. The Gordon and Castaño maritime aerosol scattering phase function [86] and the *KA* phase function [29] were adopted for aerosols and hydrosols, respectively. The sea floor, at a depth of 17 m, was assumed Lambertian with albedo extrapolated from in situ measurements [87]. A detailed and realistic 3D description of the platform, assumed as totally absorbing, was implemented (Figure 1(b)). The radiance simulations were performed assuming an in-water 20° full-angle FOV and a unitary collection ADF, while the irradiance simulations assumed a 2π sr FOV with a cosine collection ADF.

The intercomparison between experimental and numerical results showed mean percent differences <3%, with highest values at 665 nm. A remarkable good agreement was observed for ε_{L_u}, with experimental data generally within the confidence limits of the simulated data. Conversely, results evidenced a slight systematic theoretical underestimate of ε_{E_d}, likely induced by the assumption of a totally absorbing tower (as required to apply the correlated sampling technique).

In order to select a set of representative parameters for building an operational LUT of shading correction factors, an extensive sensitivity analysis on $\varepsilon_\Re$ ($\Re = E_d, E_u, L_u$) dependences was further conducted [70]. A *standard* sun–atmosphere–ocean–AAOT–detector reference system was defined assuming *typical* geophysical values [82,88] (including a spectral TTHG aerosol phase function with parameters retrieved from experimental measurements at the AAOT [89]) and the actual WiSPER geometry of observation (i.e., $d = 7.5$ m in Figure 1(b)). The sensitivity analysis was carried out for representative variations from the standard values of sun position, wavelength, sensor location, aerosol optical thickness τ_a, and in-water IOPs (i.e., hydrosol absorption and scattering coefficients, a_{hyd} and b_{hyd}) [82,88]. Some considerations on $\varepsilon_{\Re}$ sensitivity to the hydrosol scattering phase function $\tilde{\beta}_{\text{hyd}}$ were also drawn accounting for the default Petzold phase function [39], its δ-truncated version, the δ-KA phase function [29], and a single-term HG phase function with $g = 0.95$ [40]. Simulations were performed for an instrument at subsurface level, except when appositely investigating the dependence on depth of the shadowing effects.

Results for $\varepsilon_{\Re}$ showed a slight increase with $|\phi_0|$ and a nonmonotonic dependence on θ_0, with a positive second derivative. The latter trend originates from the presence of a scattering atmosphere. In fact, although the shadowing perturbations induced by the direct light regime monotonically decrease with θ_0 (as already foreseen by Gordon [66]), the diffuse light regime (skylight) and its induced shadowing effects do increase with θ_0. For subsurface measurements, ε_{E_d} is mainly influenced by the atmospheric features, while ε_{E_u} and ε_{L_u} also depend on the optical properties of the water. As a consequence, the spectral dependence of ε_{E_d} exhibits an exponential decay in agreement with an increase of the direct light regime with the wavelength, while the spectral dependence of ε_{E_u} and ε_{L_u} appears more featured as a consequence of the more spectrally complex optical properties of the water. As well, while ε_{E_d}, ε_{E_u}, and ε_{L_u} all increase with the aerosol optical thickness, only shadowing effects in the upwelling light field show a significant dependence on in-water IOPs. Specifically, they decrease with a_{hyd} and slightly increase with b_{hyd}, while only ε_{L_u} shows sensitivity on $\tilde{\beta}_{\text{hyd}}$. Results additionally highlighted a negligible sensitivity on water and atmosphere stratification. As far as for the sensor location, (1) ε_{L_u} resulted almost depth invariant, while $\varepsilon_{E_d}(\varepsilon_{E_u})$ exhibited a slight decrease (increase) with depth; and (2) $\varepsilon_{\Re}$ showed a Gaussian-like decay with the logarithm of the x- and y-coordinates. The latter dependence evidenced distortions induced by the asymmetric variation of the sun–tower–sensor geometry with y.

On the basis of these results, a simplified and straightforward spectral remodeling of the propagating system was conducted to decrease its degrees of freedom, and an LUT of subsurface tower-shading correction factors $\eta_{\widehat{\Re}} = \Re/\widehat{\Re} = 1/(1 - \varepsilon_{\Re})$ was implemented [70].

The LUT approach was adopted to account for the multidimensional character of the problem. For each λ, factors $\eta_{\widehat{\Re}}$ were computed for all combinations of selected discretized values of the following parameters (see Table 2): solar zenith angle θ_0 and azimuth angle and ϕ_0 (including overcast sky conditions);

TABLE 2 Geophysical Parameters Selected by Doyle and Zibordi [70] to Compute Tower-Shading Correction Factors $\eta_{\tilde{\Re}}$

Geophysical Parameters	Selected Values
θ_0	25°, 30°, 40°, 50°, 60°, 70°
ϕ_0	−135°, −90°, −45°, 0°, 45°
λ (nm)	412, 443, 490, 510, 555, 665
τ_a	0.00, 0.05, 0.10, 0.50, 1.0, overcast conditions
z_m (m)	0
d (m)	7.5
a (m^{-1})	0.02, 0.05, 0.10, 0.30, 0.50
Ω	0.50, 0.70, 0.80, 0.85, 0.90, 0.95
$\tilde{\beta}_{hyd}$	δ-KA [29]
v (ms^{-1})	0

total seawater absorption coefficient a; total seawater single-scattering albedo ω; aerosol optical thickness τ_a, providing corresponding values of the diffuse over direct above-water irradiance ratio $r_E = E_{dif}/E_{dir}$. The latter quantity was introduced to describe the general illumination conditions in a practical way, hence accounting for inhomogeneities in the skylight distribution (such as in the presence of clouds). Simulations were performed assuming a single homogeneous atmospheric layer on top of a single homogeneous water layer. All other modeling features remained unchanged.

The appropriate correction factor $\eta_{\tilde{\Re}}$ for any actual subsurface radiometric product determined with the WISPER system is found by matching the actual values of the environmental parameters with those indexed in the LUT.

All correction factors showed a clear seasonal and spectral dependence, with the lowest values in summer and at 665 nm (generally <1.02), and the highest values in winter and at 443 nm (~1.02 ÷ 1.09). Remarkably, an experimental assessment of the proposed tower-shading correction scheme showed extremely good results, with absolute differences between measured and estimated values generally <2% [80].

3.1.3 Self-Shading

In addition to the perturbations induced by the deployment structures, optical measurements of upwelling radiance and irradiance can be affected by the shadow casted by the instrument casing. The higher is the absorption coefficient, the smaller and closer to the instrument is the accessible portion of the observed medium, and hence the more influential is the shadow casted by the casing. This implies that measurements performed in the water (where, particularly in the NIR, the absorption coefficient is very large) are likely to be significantly perturbed by self-shading effects. On the contrary, perturbations in above-water measurements can be considered negligible.

The first theoretical evaluation of the self-shading effect was performed in the early 1990s by Gordon and Ding [90] for a disk-shaped instrument floating just below the surface of deep oceanic waters. The G-MC code already used for the ship-shading analysis [66] was applied with simulation method III (i.e., photons were prevented to interact both with the instrument and with the sea-surface). The only modification was the shape of the obstacle, now assumed to be a flat disk of radius R_d with a point-like sensor of infinitesimal FOV positioned in its lower face. It is noted that the azimuthal symmetry of the light-obstructing object reduces the 3D radiative transfer problem to an azimuthally independent one. The self-shading error $\varepsilon_\Re$ (where $\Re = L_u$ or E_u) was analyzed as a function of the disk radius R_d, the in-water total IOPs, and the solar position, including overcast sky conditions (see Table 3). Perturbations proved to be mainly a function of θ_0 and of the product between a and R_d. Specifically, $\varepsilon_\Re$ decreases with θ_0 and with decreasing aR_d values, becoming <5% only for an instrument size considerably smaller than the water absorption length $1/a$ (e.g., $R_d \lesssim 1/60a$ for E_u and $R_d \lesssim 1/200a$ for L_u, for small θ_0). The dependence of $\varepsilon_\Re$ on b appeared instead negligible, particularly for small solar zenith angles.

Theoretical results were further used to formulate a correction method for reducing self-shading errors to $<\sim 5\%$ when measuring E_u and L_u with increased instrument size (i.e., R_d up to $1/12a$ and $1/60a$ for small θ_0, respectively). The method is based on a parametric relationship between $\varepsilon_\Re$, θ_0, and aR_d, deduced by observing that for collimated illumination conditions, $z_0 = R_d/\tan\theta_{0w}$ represents the depth at which the sensor FOV leaves the instrument shadow (Figure 2 (a), where θ_{0w} is the refracted solar zenith angle). Hence for $b \ll a$, it is $\widehat{L}_u(0) = L_u(z_0)e^{-az_0}$, which recalling Eqn (6) becomes

TABLE 3 Geophysical Parameters Selected by Gordon and Ding [90] for the Simulation of Self-Shading Effects for a Cylindrical Instrument

Geophysical Parameters	Selected Values
θ_0	10°, 30°, 70°(including overcast conditions)
R_d (m)	1
c (m^{-1})	0.01 → 30.0
ω	0.5, 0.7, 0.9, 0.95
$\widetilde{\beta}$	KA [29]

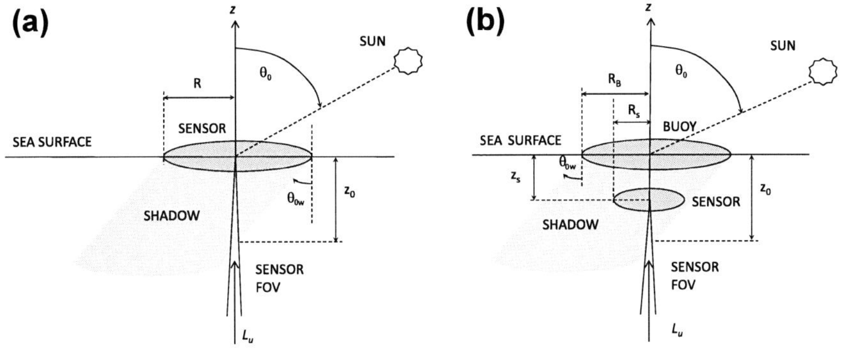

(a) **(b)**

FIGURE 2 Self-shading effects. (a) Schematics of instrument self-shading effects as modeled in Ref. [90]. (b) Schematics of self-shading effects for an instrument-buoy system as modeled in Ref. [30]. See the text for the definition of quantities expressing the measurement geometry.

$\widehat{L}_u(0) = L_u(0)e^{-(a+K_L)R_d/\tan\theta_{0w}}$ and finally $\widehat{L}_u(0) \sim L_u(0)e^{-kaR_d}$ for $K_L \approx \alpha \cdot a$ (with $\alpha \approx 1$).

By inserting this latter expression in Eqn (10), and by assuming that the same functional trend also applies to E_u and for diffuse illumination conditions, it is possible to express the self-shading errors $\varepsilon_{\Re_{dif}}$ and $\varepsilon_{\Re_{dir}}$, due to the diffuse and direct components of the solar beam respectively, as

$$\varepsilon_{\Re_{dif}} = 1 - \exp(-k_{\Re_{dif}} a R_d) \tag{11}$$

$$\varepsilon_{\Re_{dir}} = 1 - \exp(-k_{\Re_{dir}} a R_d). \tag{12}$$

The self-shading error $\varepsilon_\Re$ under actual illumination conditions simply becomes

$$\varepsilon_\Re = \left(\varepsilon_{\Re_{dir}} + \varepsilon_{\Re_{dif}} \tilde{r}_E\right) / (1 + \tilde{r}_E), \tag{13}$$

where $\tilde{r}_E$ can be approximately assumed equal to the ratio r_E between diffuse and direct solar irradiance. Coefficients $k_\Re$ were empirically inferred from a best fit of G-MC-simulated data and are provided in Ref. [90] as a function of θ_0 and of the sensor radius (i.e., point-like sensor versus finite sensor).

A useful parameterization for coefficients $k_{\Re_{dif}}$ and $k_{\Re_{dir}}$ was further derived [75,91] from tabulated G-MC data for $aR_d < 0.1$ and $30° < \theta_0 < 70°$. Coefficient values are given in Table 4 for the following parameterization:

$$k_{\Re_{dir}}(\lambda) = (1 - f_{R_d})k^p_{\Re_{dir}}(\lambda) + f_{R_d}k^e_{\Re_{dir}}(\lambda), \tag{14}$$

where f_{R_d} is the ratio of the sensor aperture-to-instrument case diameter, and $k^p_{\Re_{dir}}$ and $k^e_{\Re_{dir}}$ are for a point-like sensor and a finite sensor, respectively. It is noted that the slight spectral dependence of terms $k^p_{\Re_{dir}}$ and $k^e_{\Re_{dir}}$ derives from that of the refracted solar zenith angle $\theta_{0w} = \sin^{-1}(\sin\theta_0 / n_w(\lambda))$, where n_w is

TABLE 4 Functions for the Computations of Coefficients $k_{\Re_{dif}}$, $k^p_{\Re_{dir}}$, and $k^e_{\Re_{dir}}$

	L_u	E_u
$k_{\Re_{dif}}$	$4.61 - 0.87\ f_R$	$2.70 - 0.48\ f_R$
$k^p_{\Re_{dir}}$	$\dfrac{2.07 + 0.0056\theta_0}{\tan\theta_{0w}}$	$3.41 - 0.0155\ \theta_0$
$k^e_{\Re_{dir}}$	$\dfrac{1.59 + 0.0063\theta_0}{\tan\theta_{0w}}$	$2.76 + 0.0121\ \theta_0$

Refs [75,91].

the refractive index of the water. The operational correction of subsurface radiometric data is obtained by applying the corresponding multiplication factors $\eta_{\widehat{\Re}} = 1/(1 - \varepsilon_{\Re})$.

Further theoretical investigations, adopting the same instrument configuration (i.e., a disk of radius R_d with a point-like sensor at its center), pointed out the possibility to neglect the presence of sea-surface waves, but the need to account for the distance between instrument and sea bottom for measurements performed in optically shallow waters [92].

Experiments, appositely designed to assess the illustrated correction scheme, showed encouraging results [91,93]. In specific, Zibordi and Ferrari [91] found (for $25° \leq \theta_0 \leq 50°$, $0.001 < aR_d \leq 0.1$ and $\lambda = 500, 600$, and 640 nm) absolute differences between experimental and theoretical errors generally lower than 3% for the irradiances, and 5% for the radiances, when assuming an instrument FOV up to 18°. The latter findings suggested the feasibility to apply the correction scheme, developed for a narrow FOV, to relatively larger FOVs. Aas and Korsbø [93] found (for $40° \leq \theta_0 \leq 60°$, $aR_d \leq 0.4$ and $\lambda = 445, 514$, and 546 nm) differences between measured and estimated ε_{L_u} up to 7%, with observed errors always lower than corresponding G-MC simulations. This warned for potential theoretical misestimates of the self-shading error (particularly in the blue) likely due to neglecting the actual anisotropy of the sky radiance and the natural variability of the marine particles phase function [93].

On the basis of theoretical and experimental findings, recommendations were given to limit the size of the instrument [81], and the application of the described correction model was included in the ocean protocol for SeaWiFS validation, while recognizing the need for further extensions and verifications [75,94]. Indeed, besides the limited experimental confirmations, the correction model only applies to a "classical" system configuration consisting of a cylinder instrument with a sensor at its lower face and operating in optically deep waters. As far as for the instrument size, instrument diameter should not exceed 24 cm to assure postcorrection errors of 5% or less in measurements performed at $\lambda \leq 650$ nm with chlorophyll concentrations up to 10 mg m^{-3}

and $\theta_0 \geq 20°$ [75,94]. Instrument size constraints become even more restrictive for higher λ and lower θ_0 [75,94]. Additionally, it was acknowledged by the same authors of the model [90] (and demonstrated by successive simulation exercises [95,96]) that the assumption of a negligible instrument superstructure results in an underestimate of the self-shading effects.

As so, theoretical evaluations were further performed to estimate self-shading errors in alternative instrument configurations and measurement conditions, while accounting for the actual 3D casing dimension. Namely, instrument-induced perturbations were analyzed for the prototypal Southampton Underwater Multiparameter Optical Spectrograph System (SUMOSS) [97], aiming at minimizing self-shading effects by positioning the irradiance meter on a sidearm of the instrument [95]; for the submersible Radiance Distribution System (RADS) [98], able to capture the full angular distribution of the radiance field [96]; and for a commercially available buoyed instrument, operating in both optically deep and optically shallow waters [30,99].

A forward MC algorithm [95] was utilized to estimate maximum ε_{E_u} values in SUMOSS measurements as a function of the real optical parameters, the sea-surface roughness and the sun zenith angle. The instrument was represented by a black cylinder, with a protruding sidearm. A narrow atmospheric layer, composed by gas molecules alone, was included to allow for a partially submerged instrument casing, while the oceanic layer was assumed infinitely deep. Collimated and totally diffuse illuminations were considered, and the Petzold phase function [39] was adopted for the seawater. Simulations indicated that a sidearm configuration of the instrument generates lower self-shading effects with respect to the classical one. Nonetheless, because of the greater number of variables needed to describe the system, semi-empirical formulas for the estimate of the self-shading error would be more complicated, while the possibility to correct the self-shading effects utilizing the MC code in quasi-real time (as ventilated by the authors) would require coincident measurements of the sidearm azimuth angle.

The PHO-TRAN backward MC code [24] was instead applied to estimate self-shading effects in RADS measurements of $L_u(\theta)$ with the specific aim to interpret caveats in experimentally collected data [96]. An infinitely deep oceanic layer was assumed. The RADS instrument was modeled as a perfectly absorbing cylinder of actual dimensions and size, with point-like detectors located in its lower face, either centrally on the symmetry axis, or shifted from it along and across the sun plane. Results evidenced that anomalous dark zones detected in upwelling directional radiance measurements, especially around the antisolar direction, were due to self-shadowing effects. Results additionally highlighted that more refined self-shading instrument correction schemes should account for the detailed position of the sensor on the instrument casing, for their alignment with respect to the sun position and for the full 3D structure of the perturbing object, and suggested the use of quasi-real-time MC simulations to correct the data. To efficiently decrease self-shading errors, the instrument further developed in a smaller size version called the NURADS [100].

A backward MC code (L-MC) was finally employed to accurately estimate ε_{L_u} in measurements performed with a commercially available cylindrical instrument held by a flotation buoy [99]. The system was modeled as a plane-parallel oceanic layer bounded by a wind-roughed sea-surface (as described by Cox and Munk statistics [72]) and a flat seafloor of Lambertian reflectance. Collimated and totally diffuse illuminations were considered, and the Fournier-Forand [45] analytical phase function best reproducing the Petzold phase function [39] was assumed for the hydrosols. The instrument and the buoy were both modeled as black cylinders, while the portion of the buoy above the waterline was not included since expected of no significance. A point-like sensor was positioned in the instrument lower face.

As expected, simulated ε_{L_u} values laid in between those computed with Eqn (13) assuming either the radius of the buoy or that of the instrument casing. Specifically, they were larger for the instrument casing alone, due to the presence of the buoy, but were smaller than the error induced by the buoy alone, due to the optical distance between the buoy and the sensor. Additionally, because the optical distance between sensor and buoy increases with a, ε_{L_u} values got closer to that of the instrument casing alone for increasingly turbid waters. L-MC computations further evidenced a much more marked dependence of ε_{L_u} on θ_0 for a buoyed instrument since the buoy has a large effect only for small solar zenith angles (Figure 2(b)). Results confirmed a negligible influence of the sea-surface roughness, justified by an equivalent amount of photons refracted toward and away the shaded region, leading to a little overall effect. Results also highlighted a relevant influence of the seafloor when measuring in its proximity. An LUT of self-shading errors for discretized input values of θ_0 (up to 70°), a (ranging between 0.02 and 1.0 m^{-1}), and b (with $b/a = 1, 2, 3$) was further provided to accurately correct measurements performed in optically deep waters with the considered buoyed instrument.

By acknowledging the difficulty to develop ad hoc MC codes for each specific instrument configuration and measurement conditions, Leathers et al. [30,99] formulated an extension of the analytical expression provided for ε_{L_u} by Eqn (13) to include the cases of (1) buoyed instruments and (2) measurements performed in optically shallow waters [30], while strongly suggesting a better practice for instrument manufacturers to provide self-shading LUT or software together with their products.

With reference to Eqn (12), the extended analytical expression for the direct component of the shading error $\varepsilon_{\Re_{dir}}$ ($\Re = L_u$) was modeled as

$$\varepsilon_{L_u,\text{dir}} = \frac{L_{uw}}{L_u}\varepsilon_w + \frac{L_{uB}}{L_u}\varepsilon_B, \qquad (15)$$

where L_{uw} and L_{uB} represent the portion of L_u originating from light scattering within the water column and from radiance reflection at the seafloor, respectively, while ε_w and ε_B represent the self-shading error for a buoyed instrument in optically deep waters and affecting the sole radiance reflected by the sea bottom, respectively. The ratio L_{uw}/L_u may be numerically computed with a radiative transfer code or approximated as a function of the backscattering coefficient and the sea bottom albedo (as given in Ref. [30]).

The analytical expression of ε_w is formulated by modeling the system as a shading disk of radius R_B floating at the sea-surface and representing the buoy, above a shading disk of radius R_s and at depth z_s representing the instrument casing and hosting a small sensor with infinitesimal FOV in its lower face (Figure 2(b)):

$$\varepsilon_w = \begin{cases} 1 - \exp[-ka(R_B - z_s \tan\theta_{0w})], & \tan\theta_{0w} < (R_B - R_s)/z_s \\ 1 - \exp(-kaR_s), & \tan\theta_{0w} > (R_B - R_s)/z_s \end{cases} \quad (16)$$

where $k = 1/(\tan\theta_{0w}) + 1/(\sin\theta_{0w})$.

The analytical expression of ε_B is a slightly more complicated function of seafloor, buoy, and instrument depths; of instrument and buoy radius and their projected shadow on the sea bottom; and of the sensor FOV [30]. The diffuse component $\varepsilon_{Lu,dif}$ is simply assumed approximately equal to $\varepsilon_{Lu,dir}$ for $\theta_{0w} = 35°$. The theoretical assessment of the presented analytical expression of ε_{L_u} with numerical results from L-MC simulations gave excellent agreement, except for the combination of small solar zenith angles ($\theta_0 < 15°$) and highly scattering waters (experimental conditions where the shading error is extreme).

Findings reported in this section highlight the importance of theoretical simulations in both the analysis and the minimization of overstructure perturbations. Both aspects could not have been fully addressed by the only use of in situ data, which are constrained by measurement conditions and methods. It is the synergetic use of theoretical simulations—allowing extensive sensitivity analyses—and experimental measurements—allowing the validation of theoretical results—that permitted the development of effective operational strategies for the minimization of uncertainties induced by overstructures.

3.2 Perturbations Induced by Sea-Surface Waves

Wind-generated waves can perturb data collected by in-water radiometric sensors by focusing and defocusing the incident light (lensing effect, Figure 3). These perturbations can limit the precision of data products computed through the regression of in-water radiometric measurements (henceforth *data reduction* [101]), therefore limiting the applicability of field measurements for the calibration and validation of space sensors and for the development of ocean color inversion schemes. The scope of this section is to (1) provide an overview on fundamentals of wind-generated waves and the physical processes determining the light lensing in the marine environment [102–104]; (2) highlight experimental findings about focusing and defocusing effects on

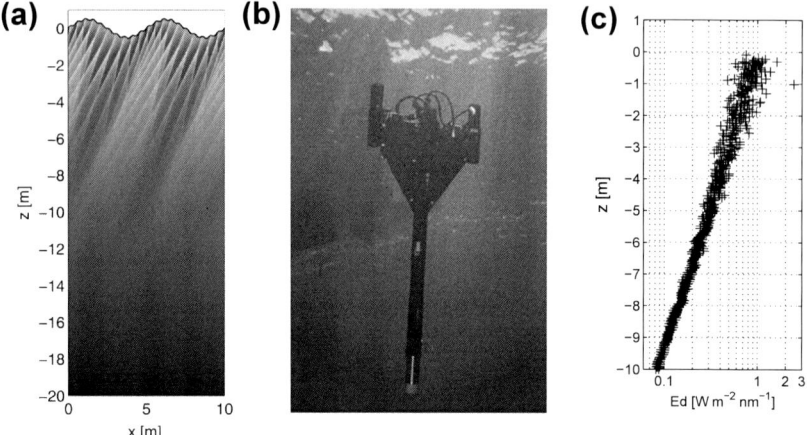

FIGURE 3 Irradiance distribution below a wavy sea-surface (Panel (a), MOX code simulations [34,101]). Example of focusing and defocusing patterns perturbing light-field measurements collected by an in-water free-fall radiometric system (Panel (b), courtesy of Scott McClean, Satlantic Inc., Halifax). Actual downward irradiance profile data recorded in the field.

in situ radiometric measurements [105–111] and discuss practices for enhancing data products accuracy [2,112,113]; (3) describe analytical and statistical models of focused light [103,114–117]; and finally (4) present radiative simulation results that allow for better understanding the perturbing effects and supporting the refinement of measurements protocols and data reduction schemes [34,101,118,119].

3.2.1 Light Focused by Wind-Generated Waves

Wind-generated sea state comprises both young and irregular short-crested waves (wind seas), and their evolution into long-crested regular waves (swells) that can travel for thousands of kilometers in open sea. Wind-generated waves are hence distinguished in (1) capillary waves due to the surface tension, with frequency above 4 Hz and wavelength up to 10 cm; and (2) gravity waves with frequency between 0.03 and 4 Hz, and wavelengths from 10 cm to 1500 m. Wind-generated sea-surface can be expressed in terms of Fourier series through the linear combination of harmonics with different amplitudes $H(\alpha,\nu)$ as a function of the frequency ν and propagation direction α. The energy of the sea-surface elevation with a single harmonic component is $W = \frac{1}{2}\rho g H(\alpha,\nu)^2$, where ρ is the seawater density and g is the acceleration due to gravity. The W distribution as a function of α and θ defines the wave spectrum [120–122].

Snell's law and Fresnell equations determine the directions of photons at the sea—surface interface and their transmission coefficients. Crests and troughs at the sea-surface focus and defocus the transmitted light, acting as planoconvex and planoconcave lenses that modulate the radiance distribution.

Analyses based on geometric optics show that (1) the depth and intensity of light foci depend on the wave shape; (2) short steep waves produce focal points closer to the surface than shallow waves; and (3) the time-normalized downwelling irradiance $\chi(z) = E_d(z)/<E_d(z)>$, where $<>$ indicates data averaging, can be up to 5—6 at the focal depth. Such strong variations of light intensity not only affect various biological processes of organism living in the photic oceanic zone, including primary productivity [110], but are also influential to the determination of in situ radiometric data products derived through the regression of in-water optical profiles.

Changes in the sea-surface height and slope of wind-driven waves induce variations in the photon propagation from the source to the sensor. Specifically, with respect to a still sea-surface, waves induce modifications of (1) the direction of the photons transmitted through the surface; (2) the photon path length to reach a given depth; (3) the transmittance of the radiant energy across the sea-surface and (4) the projection of a sea-surface element with respect to the direction of both incident and transmitted light [103]. However, it has been pointed out by Stramski and Dera that effects 2—4 induce only small variations on underwater irradiance values when compared to those due to the sea-surface curvature [104]. Underwater light fluctuations in the near-surface layer (e.g., up to 10 m depth) are mostly caused by wind-generated waves in the range between capillary waves of few centimeters and small gravity waves up to a few meters of length and 0.5 m height [105,109].

3.2.2 Experimental Findings

Light can be focused by a rough sea-surface in bright spots of few millimeters, appearing with occurrence of the order of 100 times per minute and lasting from milliseconds to a few tens of milliseconds [105,123,124]. Irradiance measurements performed in the water column at the spatial and temporal resolution of lensing effects show that the amplitude of irradiance fluctuations first increases up to a few meters below the sea-surface and then decreases deeper in the water column [124]. The frequency of fluctuations also changes as a function of depth. Specifically, analyses based on Fourier transform indicate that the frequency distribution of irradiance fluctuations becomes narrower a few meters below the sea-surface in correspondence with the formation of well-defined foci. This peak shifts toward lower frequency in deeper waters where the effect of waves with larger amplitude and period increases. Experimental measurements performed by Walker also indicated that the maximum of the power spectrum of irradiance fluctuations is at a higher frequency than the maximum of the wind-driven sea-surface wave spectrum [125]. Additional analysis reported that an increase of the energy in the short-period part of the wind-driven wave spectrum increases the amplitude of maximum irradiance fluctuations [126].

A pulse of underwater irradiance is denoted *flash* when χ exceeds a threshold η of at least 1.5 [124]. The analysis of irradiance measurements has displayed that (1) the distribution of flash duration depends on depth; (2) the duration of flashes reduces faster than the flashes frequency when η increases; and (3) the frequency of flashes decreases about exponentially as a function of η. Experimental results have also reported the presence of a right-skewed distribution of irradiance values with a mode about three times smaller than the average in the first couple of meters below the sea-surface. This peaked distribution becomes broader and eventually bimodal at increased depths [126]. Analyses of the spatial and temporal correlations of focusing effects also indicate that the dominant frequency of underwater irradiance fluctuations decreases about linearly as a function of the distance from the sea-surface [108].

When considering the dependence of focusing effects upon environmental conditions, Dera and Gordon [106] noted that maximum irradiance fluctuations appear at larger depths in waters with lower attenuation. Utmost focusing conditions occur for solar elevation above 40° and diffuse-to-direct sky irradiance below 40%. Both diffuse illumination and in-water attenuation of transmitted light diminish the focusing intensity. The amplitude of fluctuations reduces with depth because the light is scattered over a larger path length, and so the decrease of the flash frequency is faster in turbid than in clear waters [105,111,126]. The frequency and intensity of flashes are higher at winds of about $2-5$ m s^{-1} and reduces at greater speeds, whereas the duration of flashes tends to increase with the wind speed due to the formation of wider foci [107,124]. Irradiance fluctuations are larger in the red than in the blue portion of the visible spectrum, since the former is characterized by a stronger contribution of direct sun light [105]. Variations of water displacement due to sea-surface waves can partially affect the light distribution in the red and near-infrared due to the large attenuation coefficient at these wavelengths [109]. Different from the lensing effect, these fluctuations are directly related to the most energetic part of the sea-surface wave spectrum [127]. It is furthermore noted that variations of the amount of scattered light can produce fluctuations also in the upward irradiance and upwelling radiance, although these radiometric quantities are much less affected by lensing effects than the downward irradiance. Hence, E_u and L_u fluctuations occur at lower frequencies with respect to E_d since they are only indirectly influenced by the lensing effect produced in deeper water by larger gravity waves with longer period.

Uncertainties induced by the sea-surface lensing effects on the reduction results of the optical profiles (Eqn (7)) not only depend on operational specifications of the instrument (e.g., irradiance collector size, radiance sensor FOV, as well as integrating time and acquisition rate of the detector), but also are a function of the depth resolution of in-water measurements [2,113]. The number of samples per meter needed to minimize the effects of perturbations due to sea-surface light lensing is influenced by various environmental factors, such as illumination conditions, IOPs, and sea state. Besides, the extrapolation layer to derive subsurface values can be constrained to a few meters below the sea-surface in the presence of optical stratifications. Results derived by Zibordi et al. [113] with off-the-shelf instrumentations (e.g., collector size of the irradiance detector of 1 cm, FOV of the radiance sensor of 20°, integration time of tens of milliseconds, and acquisition rate of 6 Hz) in moderately complex conditions (e.g., wave height up to 0.5 m, downwelling irradiance attenuation up 0.14 m^{-1} at 490 nm, with extrapolation layer of 0.4–3.6 m; for more details, see Ref. [113]) indicate that the sampling density required to constrain perturbation of regression results below 2% is about 11, 40, 3, and 2 m^{-1} for L_u, E_u, E_d, and K_d, respectively (note that these figures represent minimal requirements). Multicast measurements forming a unique optical profile with an increased number of samples that group radiometric data measured at the same site and within a time interval in the order of few minutes have then been devised for enhancing the precision of regression data products [2,113,119,128]. When considering measurements collected through moored systems with in-water radiometers at fixed depths, perturbing effects due to sea-surface light focusing mostly affect E_d and K_d data products, whereas optical stratifications are especially relevant to the reduction of L_u and E_u data [112].

Findings based on analytical studies and field measurements analyses converge to identify several properties of lensing effects by a wavy sea-surface (e.g., manifold amplitude of flashes with respect to the mean irradiance, reduction of flash intensity because of diffuse illumination conditions or in-water photons scattering). Experimental results highlight additional features such as the exponential decay of the number of irradiance flashes when increasing the threshold intensity, and the linear decrease with depth of the dominant frequency of irradiance fluctuations to cite a few. Field experiments also give evidence of how uncertainties induced by the sea-surface lensing effects on data reduction products vary as a function of the depth resolution of radiometric profile measurements. A generalized analysis of these uncertainties only based on in situ data is however not practical because information derived from field experiments are constrained by measurement methods, sensor specifications, and environmental conditions at sea. On the other hand, complex boundary conditions at the sea–air interface limit the capability of analytical solutions to accurately describe heterogeneities of the underwater focused light. Statistical and numerical solutions discussed next are hence relevant to progress the understanding of light-focusing effects from a theoretical perspective, to investigate uncertainty budgets and give guidance for refining measurement protocols.

3.2.3 Statistical Modeling

Many factors can contribute to the complexity of underwater light fluctuation. For instance, power spectrum of sea-surface waves, illumination conditions, and seawater IOPs have different relevance depending on water depth. This challenges the possibility to represent all statistical properties of lensing effects with a unique model [114]. Irradiance flashes can significantly exceed the mean value only at very low rate inducing a heavy-tailed and right-skewed irradiance distribution in the first meters near the sea-surface. Probability density such as the log-logistic and the lognormal function can satisfactorily represent fluctuations measured in the field in this upper layer up to the 90th percentile of the irradiance distribution, but fail to describe the frequency of most intense flashes. The agreement with in situ data reduces when considering statistical models explicitly developed in the context of extreme value analysis (e.g., Gumbel [129] and Frecht [130] distributions). Studies of DeGroot and Schervish assessed that the skewness and excess kurtosis [131] reduce deeper in the water column, and tend to become constant at about two optical depths where multiple scattering of photons generates a Gaussian distribution of irradiance fluctuations. This indicates the presence of a critical depth, above which the irradiance distribution is asymmetric because of light focusing (sunny layer) and below which it is substantially symmetric (diffuse layer). Note that the critical depth varies spectrally due to the different fractions of scattered light at different wavelengths [114].

A statistical description of irradiance flashes has been obtained by Shen [115] directly modeling underwater-focused light instead of fitting fluctuations measured in the field with selected probability distributions. This requires computing the probability that multiple light rays refracted by the sea-surface can intercept the in-water radiometric sensor at the same sampling time. Assuming that the sea-surface is represented by small facets with independent and identically distributed Gaussian random slope, DeGroot and Schervish [131] concluded that underwater light fluctuations follow the nonhomogeneous Poisson distribution (i.e., large number of Bernoulli trials with a small probability of success). The resulting Gaussian—Poisson (GP) statistical model of irradiance fluctuations was further expanded to include scattering effects and to account for subfacet slope variations for overcoming the limit of flat facets. This scheme permits to identify the depth above which light diffusion is dominated by short waves at the sea-surface rather than by volume scattering. This model also succeeds in expressing the transition discussed above between the skewed and highly tailed irradiance regime in the layer close to the sea-surface to the normal distribution in deeper water. An additional theoretical finding is explaining how the probability of irradiance values manifolds larger than the variance diminishes asymptotically with depth faster than an exponential but slower than a Gaussian decay. The GP model however cannot explain the reduction of flash intensity very close to the sea-surface in virtue of approximations applied to obtain a closed-form solution. Nevertheless, it represents the state-of-the-arts statistical description of irradiance fluctuations due to sea-surface-focused light.

3.2.4 MC Case Studies

Results from field measurements and statistical evaluations are hereby complemented by the investigations of underwater-focused light based on MC radiative transfer simulations. Horizontal radiance gradient due to the waves lensing effect contravenes the plane-parallel assumption of the deterministic solutions of the RTE. MC is then the reference numerical approach to study the radiance distribution within a wavy sea. By providing a virtual environment where testing different hypotheses with fewer constraints than in situ measurements, MC radiative transfer modeling represents an effective approach to understand other specificities of data reduction results, for instance, due to differences in regression intervals, illumination conditions, and sea-surface geometries.

The Cox and Munk parameterization of the sea-surface slope distribution as a function of the wind speed [72] is a milestone component of MC studies to compute the mean transmittance and reflectance of a wavy air—sea interface [15,132]. However, the slope distribution of sea-surface waves alone (e.g., see Ref. [72]) does not permit to define the spatial autocorrelation of the seawater displacement above the still level required for studying focusing effects produced by wind-generated waves. Instead, MC simulations of underwater light fluctuations can be obtained by tracking photons over an extended simulation domain where the sea-surface is ideally defined by a limited number of harmonic components (Figure 3(a)) representing superimposed capillary and gravity wave [34,101,118,119]. The accuracy of simulation results can be enhanced modeling of the sea—air interface accounting for the harmonic components distribution defined by the wave spectrum [117,133—137].

The light distribution over the euphotic zone ($\sim 10^2$ m depth) can be due to relatively large gravity waves. The minimum focal depth of a sinusoid wave $\zeta = A \cos(\kappa \cdot x)$ is given by $z = [n_w/(n_w - 1)]/(A\kappa^2)$, where A is the wave amplitude, κ is the wave number, and n_w is the seawater refractive index [125]. Surface waves of about 100 m length and a few meters amplitude can still induce significant horizontal variability in the underwater light field. MC simulations executed in a 2D domain longitudinal to the wave propagation direction to represent the lensing effect of these large gravity waves needs then to be in the order of 100 m depth and hundreds of meter length [117,133,136]. Standard photon tracing schemes would take an exceeding time to draw the number of photons required to represent light focusing over such an extended domain when considering the resolution of a few millimeters needed to model finest irradiance fluctuations recorded in the field [124]. Different computational solutions can be applied to address this problem.

The first approach is to precompute a database of MC simulations of the light distribution generated by a single element of a flat sea-surface for different sun elevations. Individual contributions are then superimposed to model the overall light distribution below a wavy sea-surface. This scheme is at most effective when considering an idealized uniform sky radiance. In this case, in fact, the segments of the flat and wavy sea-surface are unpinned by the same diffuse radiance. It is noted that the anisotropy of the sky radiance in the yellow-red spectral interval tends to induce relatively minor uncertainties in modeled results because of the lower diffuse-to-direct irradiance ratio with respect to shorter wavelengths. Instead, the accuracy of this method slightly reduces in the blue spectral band where the diffuse-to-direct irradiance fraction is higher and the transmitted light creates focusing effects in deeper waters. Higher order light reflection–transmission processes at the sea–air interface not taken into account by precomputed MC simulations induce minor artifacts in modeled results.

The performance of the MC based on the superimposition of precomputed radiometric fields can be enhanced by modeling the light distribution generated by individual surface segments with a finer resolution close to the sea-surface and coarser resolution deeper in the water column. This is in agreement with the experimental findings by Dera and Stramski documenting that light flashes have longer duration and occur over larger areas at greater depths [124]. Additional performance improvements can be obtained by complementing results from superimposed radiometric fields with additional investigations based on direct ray tracing, enabling the evaluation of light distribution at high spatial resolution in the few meters close to the sea–air interface. Computing time can be further reduced by only considering the contribution of the direct light to wave-induced radiometric fluctuations in this upper layer.

The distribution of downward irradiance values as a function of depth derived with precomputed MC results shows statistical figures in agreement with field measurements. An example is the convergence of skewness and excess kurtosis to values typical of the normal distribution at increased distance from the sea-surface [114]. Another aspect is the shift toward lower fluctuation frequencies deeper in the water column [124]. The possibility to reuse the same database of MC simulations becomes very effective to compare light fluctuations generated by different wind-driven sea power spectra. In line with field observations [109] and of main interest to the reduction of in situ radiometric profiles [2,113] is the fact that light fluctuations in approximately the upper 10 m of water are mostly induced by sea-surface waves due to local wind [138]. Low wind speed between 3 and 5 m s^{-1} generates gravity-capillary waves with small curvature and length between 0.7 and 3 cm. MC simulation results show that corresponding focal points between 0.5 and 3 m depth produce E_d flashes up to 7 times more intense than average. Stronger local wind generates steeper gravity-capillary waves raising irradiance fluctuations closer to the sea-surface. Foci due to more developed gravity waves occur mostly below 10 m depth and can be responsible of irradiance flashes down to 30 m, or even deeper in presence of ideal favorable conditions [138].

Another approach to study the dynamics of underwater focused light with precomputed radiometric fields is the Hybrid Matrix Operational-Monte Carlo (HMOMC) scheme proposed by You [117]. This method divides the simulation domain in different layers. The radiative transfer process between layers is specified by the matrix operator method in terms of response to a given input [139,140]. The response is the radiance reflected or transmitted by a layer, which can then be the input to another layer. The radiance distribution for combined layers is hence formally defined through a sequence of transformations upon the initial input, and this permits to model the radiance and irradiance values at the selected points of the simulation domain corresponding to the locations of the in-water radiometer. Three layers are considered in the study case addressed here: the atmosphere, the sea—air interface, and the water volume. Their responses are static HMOMC components to be computed only once. Direct sunlight transmitted across the sea-surface needs instead to be computed based on the wave spectrum to model the dynamics of lensing effects [121]. The computational efficiency of this component is enhanced neglecting the water displacement above the still level, assuming that this induces minor effects on the underwater light distribution. The time-dependent irradiance field modeled with the HMOMC method can well capture overall statistical figures documented by in situ measurements [105—111] including (1) the asymmetry of the distribution of the normalized downwelling irradiance $\chi(z)$ with lower skewness and excess kurtosis in deeper waters; (2) the exponential decay of the frequency of flashes as a function of depth; (3) the flash duration; and (4) the power spectral density of irradiance values at different depths.

Theoretical analyses performed by Zaneveld and based on 2D forward MC have allowed for investigating additional radiometric features that are difficult to be quantified experimentally. An example is the study of the intensified backscattering where an increased amount of light is scattered upward due to coherent wave trains [118]. This effect can be described by first considering an ideal case where the incident light is a collimated beam perpendicular to a flat sea-surface and the radiance sensor is oriented to nadir. In a single-scattering approximation, only photons following the same return path of the incident light can be detected by the radiance sensor after leaving the sea-surface. However, the lensing effect of sea-surface waves produces focal points where underwater rays intersect, further increasing the number of return paths from the source to the detector as explained by the reciprocity principle [64]. Photon ray tracing allows for verifying how the presence of coherent capillary-gravity waves due to the surface tension at the crests of short gravity waves [141,142] foster the intensified backscattering by producing layers of foci (e.g., see Figure 1 in Ref. [118]). Swells are also characterized by coherent waves at the sea-surface that in principle could induce intensified backscattering. Focal points generated by gravity waves appear, however, much deeper in the water column with respect to those produced by small capillary-gravity coherent structures at the sea-surface. In the swell case, the fraction of photons that upon single scattering can reach the sea-surface at the right direction to be refracted toward the radiance sensor is largely depending on seawater IOPs lessening the intensified backscattering. This effect is further reduced when increasing the angle between the source and the detector and it hence affects at a larger extent the data collected by active sensors (e.g., lidar). Ocean color applications are based on passive remote sensing with large separation angles between the source and the detector to minimize sun-glint effects. Analyses based on forward MC simulations in a 2D domain allow for estimating that the intensified backscattering contributes much <5% to the water-leaving radiance in standard measurement geometries and environmental conditions [118].

Virtual optical profiles can be produced through forward MC simulations of radiometric fields in a 2D domain (e.g., Figure 3(a)). These may be designated to capture the variability of real measurements in the natural environment while accounting for the deployment speed, acquisition rate, and detector size of the profiling systems [34,101,119]. Virtual optical profiles can then be used to investigate uncertainty budgets induced by sea-surface-focusing effects on data products derived from the reduction of optical profile measurements in the field. Kajiyama [33,143] documented how simulating the radiative transfer process at the spatial scale and resolution of field measurements can require extensive computing power. The efficiency of forward MC simulations to derive virtual optical profiles can then be improved considering a seawater interface with periodic boundaries and intercepting photons that leave the domain from one side to reinjecting them at the opposite border [34,101,119]. Periodic boundary conditions also offer the advantage of allowing to consider virtual profiles that are diagonal under a static sea−air interface, instead of addressing the more realistic but computationally impractical modeling of vertical deployments below waves that travel at the sea-surface (e.g., Figure 3 in Ref. [34]). The underlying work hypothesis is the equivalence between spatial and temporal analysis (ergodicity) of underwater focusing effects, which is supported by the fact that statistical properties of light fluctuation do not significantly depend on different phase velocities of sea-surface wave components [144].

Of main interest to the validation of satellite data products is the fact that remote sensing observations average light fluctuations due to sea-surface lensing over the space-borne sensor footprint, which is of the order of 10^4 m^2 for most recent ocean color sensors. The footprint of in-water radiometric measurements in the field can only be up to a few square meters, about two orders of magnitude smaller than that of the sensor in the space [119]. The assumption underpinning the reliability of radiometric data products derived from in situ optical profiling systems for calibration and vicarious validation activities is that field measurements can still represent a valid ground truth to remote sensing observations despite of the difference between footprint sizes. Results based on the analysis of virtual radiometric profiles indicate that light focusing by sea-surface can generate in the first meters below the sea-surface a horizontal gradient in the downward irradiance typically larger than that induced vertically by absorption and scattering processes. A recommended practice to limit the effect of variability due to underwater focused light, and hence improve the quality of regression results derived from in situ measurements, is to adopt the multicasting method [34,101,113,119]. This is especially relevant when recalling that E_d is directly influenced by light focused beneath a wavy sea-surface. Regression methods alternative to the standard linear fit of log-transformed radiometric measurements as a function of depth can also be applied to further reduce uncertainties due to the horizontal variations of underwater light intensity. An example is the determination of K_d based on the upward integration of the irradiance values starting from a depth where the wave-focusing effect is almost negligible (see Ref. [119] for details).

The use of virtual profiles to estimate the coefficient of variation (CV; i.e., the ratio of the standard deviation and the mean of a set of samples) of data reduction products, such as $\Re_0$ and $K_\Re$, can be optimized by tracing a different number of photons N_{pho} according to the radiometric quantity under consideration [113]. The upward irradiance and the upwelling radiance are in fact much less affected by sea-surface light lensing than the downward irradiance. Knowing the number of photons needed to neglect the statistical variability of MC simulations is then critical to study the reduction of E_u and L_u profiles. An efficient approach to address this issue is performing complementary MC simulations with the same IOPs and illumination conditions of the wavy case, but with a flat sea-surface as boundary condition. The minimum number of photons needed to capture the effect of focused light on virtual optical profiles occurs when the CV of data reduction results in the flat sea-surface case is about one order of magnitude lower than that computed in wavy conditions (alternatively when the latter becomes negligible with respect to regression results accuracy requirements). The standard deviation of $\Re_0$ and $K_\Re$ results in the case of a flat sea-surface is a linear function of the square root of the photon population size. It is then possible to optimize the exploitation of computing resources by executing a minimum set of MC simulations to determine the linear CV trend in the presence of a flat sea-surface [33]. MC simulations indicate that the number of photons needed to quantify perturbations by the sea-surface-lensing effect in data products from in-water radiometric profiles is in the order of 10^6, 10^9, and 10^{10} for E_d, E_u, and L_u, respectively, when considering a swell with waves of 5 m width and 0.5 m height. These figures are presented only to highlight the variability of minimum N_{pho} values for different radiometric quantities since the exact photon population size has to be determined on a specific case-study basis. Virtual profiles derived from forward MC simulations document CV of $\Re_0$ due to light focusing in the range of 0.5–3.5% for E_{d0}, below 0.4% for E_{u0}, and up to 1.2% for L_{u0} when considering standard deployment speeds and sampling frequencies of commercial free-fall radiometer systems, as well as IOP values of moderately optically complex waters and typical illumination conditions (see Ref. [34] for details). A substantial agreement holds between these theoretical estimates and field measurements results [113].

MC simulations can further complement experimental findings with indications for refining measurement protocols. It is for instance recalled that analyses based on field measurements report that uncertainty budgets due to perturbations induced by underwater light fluctuations are reduced when increasing the depth resolution of optical profiles [112,113]. MC simulations farther highlight that for a given number of radiometric records per unit depth, the variability induced by wave focusing and defocusing effects over the extrapolation layer can be lessened more efficiently by reducing the deployment speed of the optical profiling system rather than increasing the data collection frequency. In practice, however, a deployment speed in the order of 0.2 m s^{-1} [2] is needed to limit additional sources of uncertainty, for instance, those due to the vertical instability of the profiler and illumination changes during the deployment. Both experimental and numerical results hence agree on recommending the use of the multicast measurement methodology to minimize the perturbing effect of underwater focused light [34,113,119].

Virtual optical profiles computed with 2D forward MC simulations can also be a valid mean for investigating reduction solutions to derive data products from in-water radiometric measurements. The standard approach is to compute $\mathfrak{R}_0$ and $K_\mathfrak{R}$ based on the linear regression of log-transformed profile data within the extrapolation layer (Eqn (7)), solving a set of linear equations to minimize the sum-of-squares error between samples and modeled values in a maximum likelihood framework. Data regression is affected by underwater flashes largely exceeding average radiometric values. Solutions presented in the literature for lessening perturbations induced by underwater fluctuation of $\mathfrak{R}(z)$ include removing most perturbed samples [128], binning data in regular depth intervals, and performing an incremental regression in successive layers [75], as already mentioned. Least-squares fits can also be applied to derive $K_\mathfrak{R}$ with Hermitian cubic polynomials [145]. These schemes have in common the retrieval of $\mathfrak{R}_0$ and $K_\mathfrak{R}$ from log ($\mathfrak{R}(z)$). The average of log-transformed values, however, tends to be smaller than the logarithm of their average due to the inequality between the arithmetic and the geometric means [146]. An alternative to avoid biasing caused by log-transformation [147] is applying nonlinear data reductions to compute $\mathfrak{R}_0$ and $K_\mathfrak{R}$ directly from $\mathfrak{R}(z)$ by solving a nonlinear optimization problem. Findings based on MC simulations indicate that the Trust Region algorithm [148,149] tends to be more efficient than other optimization methods such as the Levenberg–Marquardt scheme [150] in addressing the reduction of optical profile data below a wavy sea-surface. By performing a nonlinear optimization, these schemes cannot ensure global solution and it might hence be worth recomputing regression parameters using slightly different initializations to verify the validity of the results.

MC simulations confirm that log-transformed data tend to underestimate regression results with respect to those directly obtained from input radiometric values. The log-transformation is thus identified as a source of bias in $\Re_0$ and $K_\Re$. Differences between L_{u0} values computed with untransformed or log-transformed radiometric values are about 1−2%, which is well below the target uncertainty for data products from in situ measurements (i.e., 5%). Instead, difference can easily exceed 5% for E_{d0} because the bias due to log transformation depends on the amplitude and the vertical distribution of underwater light fluctuations. These findings are consistent with results derived from optical profiles measured in situ. Statistical figures derived from MC simulations to study differences between linear and nonlinear regression results also highlight the importance of the multicast method to minimize data products uncertainties by increasing the number of radiometric samples per unit depth. Additional case studies are theoretical investigations addressed to verify the effect of changes in the distribution of the photon-travelling direction due to underwater focused light which are difficult to be experimentally determined in the field (e.g., Figures 12 and 13 in Ref. [101]).

4. SUMMARY AND REMARKS

The collection of high-quality in situ measurements is prioritized by ocean color programs of international space agencies to support the validation and calibration of space-borne radiometric data and the development of inversion schemes for retrieving higher level product maps underpinning climate-change studies. In situ radiometric measurements to compute quantities such as the remote sensing reflectance are, however, subject to different perturbing factors. Scope of this chapter has then been discussing perturbations affecting radiometric profiles $\Re(z)$ and derived $\Re_0$ and $K_\Re$ values, as well as presenting correction schemes and/or measurement protocols for their minimization. Addressed case studies concern variations of the light-field distribution due to large deployment structures, such as ships and oceanographic towers, instrument self-shading, and sea-surface light focusing. The analysis of these perturbations on in situ measurements has been largely performed by means of MC radiative transfer simulations due to the lack of analytical RTE solutions and the limits of deterministic approaches in the absence of plane-parallel conditions.

In conclusion, this chapter has shown that numerical simulations are a key mean for addressing perturbations affecting radiometric field measurements. Most of the discussed analyses are based on the MC method. The stochastic nature of this scheme implies that the number of traced photons needs to be large enough to ensure the statistical noise affecting MC results to become negligible with respect to the analyzed perturbing effects. The capability to perform realistic simulations of the light distribution in the natural environment tends to require significant computing times (i.e., orders of magnitude larger than those typical of deterministic methods). Large-scale simulations, or investigations addressed to verify the effects of a high number of different environmental conditions, can benefit from different techniques to faster MC convergence (e.g., correlated sampling, backward ray-tracing, periodic boundary conditions), as well as from high-performance computing solutions such as parallel runs on computer clusters. At the light of the beneficial results characterizing radiative transfer simulations in the last decades, additional investigations based on the numerical modeling of data collected in the field should be continued in the future. Foreseen case studies include refined analyses of in-water radiometric systems by accounting for individual instrument specifications and measurement configurations, as well as the throughout analysis of uncertainties affecting above-water radiometric data.

REFERENCES

[1] S. Hooker, G. Zibordi, J.-F. Berthon, D. D'Alimonte, S. Maritorena, S. Mclean, J. Sildam, Results of the Second SeaWiFS Data Analysis Round Robin, (DARR-00), in: Ser. SeaWiFS Technical Report SERIES, vol. 15, NASA GSFC, Greenbelt, MD, USA, 2001, 206892, ch. 1, pp. 4–45.

[2] G. Zibordi, K. Voss, Field Radiometry and Ocean Colour Remote Sensing, Springer (2010) ch. 18, 307–334.

[3] A.A. Kokhanovsky (Ed.), Radiative Transfer Equation (RTE): Numerical Solution Methods – Introduction, Top. Part. Disp. Sci, 2008.

[4] K.N. Liou, An introduction to atmospheric radiation, Q. J. Roy. Meteorol. Soc. 129 (2003).

[5] H.C. Van De Hulst, Light Scattering by Small Particles (Structure of Matter Series.), Dover Ed, Dover Pubn Inc., 1981.

[6] F.X. Kneizys, G.P. Anderson, E.P. Shettle, L.W. Abreu, J.H. Chetwynd Jr, W.O. Selby, J.E.A. Gallery, S.A. Clough, LOWTRAN 7: status, review, and impact for short-to-long-wavelength infrared applications, in: AGARD, Atmospheric Propagation in the UV, Visible, IR, and MM-wave Region and Related Systems Aspects, March 1990, p. 11 (SEE N90–21907 15–32).

[7] B. Mayer, A. Kylling, Technical note: the libradtran software package for radiative transfer calculations - description and examples of use, Atmos. Chem. Phys. 5 (7) (2005) 1855–1877.

[8] K. Stamnes, S.-C. Tsay, W. Wiscombe, K. Jayaweera, Numerically stable algorithm for discrete-ordinate-method radiative transfer in multiple scattering and emitting layered media, Appl. Opt. 27 (12) (1988) 2502–2509.

[9] H.R. Gordon, M. Wang, Surface-roughness considerations for atmospheric correction of ocean color sensors i: the Rayleigh-scattering component, Appl. Opt. 31 (21) (1992) 4247–4260.

[10] S.Y. Kotchenova, E.F. Vermote, Validation of a vector version of the 6s radiative transfer code for atmospheric correction of satellite data. Part II. homogeneous Lambertian and anisotropic surfaces, Appl. Opt. 46 (20) (2007) 4455–4464.

[11] H.C. Van De Hulst, Multiple Light Scattering: Tables, Formulas, and Applications, Academic Press, New York, 1980, 2.

[12] R. Preisendorfer, P.M.E. Laboratory, Hydrologic Optics, in: Ser. Hydrologic Optics, vol. 3, U.S. Dept. of Commerce, National Oceanic and Atmospheric Administration, Environmental Research Laboratories, Pacific Marine Environmental Laboratory, 1976.

[13] R. Bellman, G. Wing, An Introduction to Invariant Imbedding, in: Ser. Classics in Applied Mathematics, Society for Industrial and Applied Mathematics, 1992.
[14] S. Chandrasekhar, Radiative Transfer, Dover Publications, Inc., 1960.
[15] C.D. Mobley, Light and Water. Radiative Transfer in Natural Waters, Academic Press, 1994.
[16] B. Bulgarelli, V.B. Kisselev, L. Roberti, Radiative transfer in the atmosphere-ocean system: the finite-element method, Appl. Opt. 38 (9) (1999) 1530−1542.
[17] V.B. Kisselev, L. Roberti, G. Perona, Finite-element algorithm for radiative transfer in vertically inhomogeneous media: numerical scheme and applications, Appl. Opt. 34 (36) (1995) 8460−8471.
[18] Z. Jin, K. Stamnes, Radiative transfer in nonuniformly refracting layered media: atmosphere-ocean system, Appl. Opt. 33 (3) (1994) 431−442.
[19] B. Bulgarelli, J.P. Doyle, Comparison between numerical models for radiative transfer simulation in the atmosphere−ocean system, J. Quantit. Spectrosc. Radiative Transfer 86 (2004) 419−435.
[20] C.D. Mobley, B. Gentili, H.R. Gordon, Z. Jin, G.W. Kattawar, A. Morel, P. Reinersman, K. Stamnes, R.H. Stavn, Comparison of numerical models for computing underwater light Fields, Appl. Opt. 32 (36) (1993) 7484−7504.
[21] B. Bulgarelli, G. Zibordi, J.-F. Berthon, Measured and modeled radiometric quantities in coastal waters: toward a closure, Appl. Opt. 42 (27) (2003) 5365−5381.
[22] A. Morel, D. Antoine, B. Gentili, Bidirectional reflectance of oceanic waters: accounting for Raman emission and varying particle scattering phase function, Appl. Opt. 41 (2002) 6289−6306.
[23] G. Zibordi, B. Bulgarelli, Effects of cosine error in irradiance measurements from field ocean color radiometers, Appl. Opt. 46 (22) (2007) 5529−5538.
[24] J.P. Doyle, H. Rief, Photon transport in three-dimensional structures treated by random walk techniques: Monte Carlo benchmark of ocean colour simulations, Math. Comput. Simulation 47 (2−5) (1998) 215−241.
[25] N. Metropolis, S. Ulam, The Monte Carlo method, J. Am. Stat. Assoc. 44 (1949) 335−341.
[26] A.V. Prokhorov, Monte Carlo method in optical radiometry, Metrologia 35 (1998) 465−471.
[27] D.J. Bogucki, J. Piskozub, M.-E. Carr, G.D. Spiers, Monte Carlo simulation of propagation of a short light beam through turbulent oceanic flow, Opt. Express 15 (21) (2007) 13988−13996.
[28] M. Gimond, Description and verification of an aquatic optics Monte Carlo model, Environ. Model. Software 19 (12) (2004) 1065−1076.
[29] H.R. Gordon, O.B. Brown, M.M. Jacobs, Computed relationships between the inherent and apparent optical properties of a flat homogeneous ocean, OSA 14 (2) (1975) 417−427.
[30] R.A. Leathers, T.V. Downes, C.O. Davis, C.D. Mobley, Monte Carlo Radiative Transfer Simulations for Ocean Optics: A Practical Guide, Naval Research Lab Washington DC Applied Optics Branch. Technical Report, September 2004.
[31] G.N. Plass, G.W. Kattawar, Polarization of the radiation reflected and transmitted by the Earth's atmosphere, Appl. Opt. 9 (5) (1970) 1122−1130.
[32] G.N. Plass, G.W. Kattawar, Monte Carlo calculations of radiative transfer in the earth's atmosphere-ocean system: I. flux in the atmosphere and ocean, J. Phys. Oceanogr. 2 (1972) 139−145.

[33] T. Kajiyama, D. D'Alimonte, J.C. Cunha, Statistical performance tuning of parallel Monte Carlo ocean color simulations, in: Parallel and Distributed Computing, Applications and Technologies 2012, Beijing, China, December 2012, pp. 761–766.

[34] D. D'Alimonte, G. Zibordi, T. Kajiyama, J.C. Cunha, Monte Carlo code for high spatial resolution ocean color simulations, Appl. Opt. 49 (26) (2010) 4936–4950.

[35] T. Kajiyama, D. D'Alimonte, J.C. Cunha, Performance prediction of ocean color Monte Carlo simulations using multi-layer perceptron neural networks, in: Procedia Computer Science, 4 (2011) 2186–2195. http://dx.doi.org/10.1016/j.procs.2011.04.239.

[36] H. Iwabuchi, Efficient monte carlo methods for radiative transfer modeling, J. Atmos. Sci. 63 (9) (2006) 2324–2339.

[37] S.A. Prahl, M. Keijzer, S.L. Jacques, A.J. Welch, A monte carlo model of light propagation in tissue, in: SPIE Proceedings of Dosimetry of Laser Radiation in Medicine and Biology, SPIE Press, 1989, pp. 102–111.

[38] G. Kullenberg, Scattering of light by sargasso sea water, Deep Sea Res. 15 (4) (1968) 423–432.

[39] T.J. Petzold, Volume Scattering Functions for Selected Ocean Waters, Scripps Institution of Oceanography. Technical Report, October 1972.

[40] L. Henyey, J. Greenstein, Diffuse radiation in the galaxy, Astrophys. J. 93 (1941) 70–83.

[41] V.I. Haltrin, One-parameter two-term Henyey-Greenstein phase function for light scattering in seawater, Appl. Opt. 41 (6) (2002) 1022–1028.

[42] G.W. Kattawar, A three-parameter analytic phase function for multiple scattering calculations, J. Quantit. Spectrosc. Radiative Transfer 15 (9) (1975) 839–849.

[43] O.V. Kopelevich, Small-parameter model of optical properties of sea water, in: A. Monin (Ed.), Ocean Optics, Physical Ocean Optic, vol. 1, Nauka Pub., Moscow, 1983.

[44] G.R. Fournier, Backscatter corrected Fournier-Forand phase function for remote sensing and underwater imaging performance evaluation, in: I.M. Levin, G.D. Gilbert, V.I. Haltrin, C.C. Trees (Eds.), Current Research on Remote Sensing, Laser Probing, and Imagery in Natural Waters, 6615, SPIE, 2007, pp. 6615–6622.

[45] G.R. Fournier, J.L. Forand, Analytic phase function for ocean water, in: Ocean Optics XII, 2558, SPIE, 1994, pp. 194–201.

[46] G.R. Fournier, M. Jonasz, Computer-based underwater imaging analysis, in: G.D. Gilbert (Ed.), Airborne and In-Water Underwater Imaging, 3761. 1, SPIE, 1999, pp. 62–70.

[47] W. Freda, J. Piskozub, Improved method of Fournier-Forand marine phase function parameterization, Opt. Express 15 (20) (2007) 12763–12768.

[48] M. Jonasz, Volume scattering function measurement error: effect of angular resolution of the nephelometer, Appl. Opt. 29 (1) (1990) 64–70.

[49] G.N. Plass, G.W. Kattawar, T.J. Humphreys, Influence of the oceanic scattering phase function on the radiance, J. Geophys. Res. Oc. 90 (C2) (1985) 3347–3351.

[50] Y.C. Agrawal, The optical volume scattering function: temporal and vertical variability in the water column off the New Jersey coast, Limnol. Oceanogr. 50 (2005) 1787–1794.

[51] M. Chami, E.B. Shybanov, T.Y. Churilova, G.A. Khomenko, M.E.-G. Lee, O.V. Martynov, G.A. Berseneva, G.K. Korotaev, Optical properties of the particles in the crimea coastal waters (black sea), J. Geophys. Res. Oc. 110 (2005). C11020.

[52] C.D. Mobley, L.K. Sundman, E. Boss, Phase function effects on oceanic light fields, Appl. Opt. 41 (6) (2002) 1035–1050.

[53] J.M. Sullivan, M.S. Twardowski, Angular shape of the oceanic particulate volume scattering function in the backward direction, Appl. Opt. 48 (35) (2009) 6811–6819.

[54] J.-F. Berthon, E. Shybanov, M.E.-G. Lee, G. Zibordi, Measurements and modeling of the volume scattering function in the coastal Northern Adriatic Sea, Appl. Opt. 46 (22) (2007) 5189−5203.

[55] H. Yang, H.R. Gordon, T. Zhang, Island perturbation to the sky radiance over the ocean: simulations, Appl. Opt. 34 (36) (1995) 8354−8362.

[56] B. Bulgarelli, V. Kiselev, G. Zibordi, Simulation and analysis of adjacency effects in coastal waters: a case study, Appl. Opt. 53 (8) (2014) 1523−1545.

[57] P.N. Reinersman, K.L. Carder, Monte carlo simulation of the atmospheric point-spread function with an application to correction for the adjacency effect, Appl. Opt. 34 (21) (1995) 4453−4471.

[58] R. Santer, C. Schmechtig, Adjacency effects on water surfaces: primary scattering approximation and sensitivity study, Appl. Opt. 39 (3) (2000) 361−375.

[59] A. Sei, Analysis of adjacency effects for two lambertian half-spaces, Int. J. Remote Sensing 28 (8) (2007) 1873−1890.

[60] B. Petkov, C. Tomasi, V. Vitale, A. di Sarra, P. Bonasoni, C. Lanconelli, E. Benedetti, D. Sferlazzo, H. Diemoz, G. Agnesod, R. Santaguida, Ground-based observations of solar radiation at three italian sites, during the eclipse of 29 march, 2006: Signs of the environment impact on incoming global irradiance, Atmos. Res. 96 (1) (2010) 131−140.

[61] H.H. Poole, W.R.G. Atkins, On the penetration of light into sea water, J. Marine Biol. Assoc. United Kingdom (New Series) 14 (3) (1926) 177−198.

[62] H.H. Poole, W.R.G. Atkins, Photo-electric measurements of submarine illumination throughout the year, J. Marine Biol. Assoc. United Kingdom (New Series) 16 (1929) 297−324.

[63] H.H. Poole, The photo-electric measurement of submarine illumination in off-shore waters, ICES Marine Sci. Symposia 101 (1936) 1−12.

[64] N.G. Jerlov, Marine Optics, in: Ser. Oceanography, 14, Elsevier, 1976.

[65] E. Aas, On Submarine Irradiance Measurements, in: Ser. Report (Københavns Universitet. Institut for Fysisk Oceanografi), 6, Københavns Universitet, Institut for Fysisk Oceanografi, 1969.

[66] H.R. Gordon, Ship perturbation of irradiance measurements at sea. 1: monte Carlo simulations, Appl. Opt. 24 (23) (1985) 4172−4182.

[67] J.E. Hansen, Exact and approximate solutions for multiple scattering by cloudy and hazy planetary atmospheres, J. Atmos. Sci. 26 (1969) 478−487.

[68] J.F. Potter, The Delta function approximation in radiative transfer theory, J. Atmos. Sci. 27 (1970) 943−949.

[69] J. Spanier, E.M. Gelbard, Monte Carlo Principles and Neutron Transport Problems, in: Ser. Addison-Wesley Series in Computer Science and Information Processing, Addison-Wesley, 1969.

[70] J.P. Doyle, G. Zibordi, Optical propagation within a three-dimensional shadowed atmosphere−ocean field: application to large deployment structures, Appl. Opt. 41 (21) (2002) 4283−4306.

[71] J. Piskozub, Effect of ship shadow on in-water irradiance measurements, Oceanologia 46 (1) (2004) 103−112.

[72] C. Cox, W. Munk, Measurement of the roughness of the sea surface from photographs of the sun's glitter, J. Opt. Soc. Am. 44 (11) (1954) 838−850.

[73] W.S. Helliwell, G.N. Sullivan, B. Macdonald, K.J. Voss, Ship shadowing: model and data comparison, in: Ocean Optics X, 1302, SPIE, 1990, pp. 55−71.

[74] Y. Saruya, T. Oishi, M. Kishino, Y. Jodai, K. Kadokura, A. Tanaka, Influence of ship shadow on underwater irradiance fields, in: S.G. Ackleson (Ed.), Ocean Optics XIII, 2963, SPIE, 1997, pp. 760−765.

[75] J.L. Mueller, R.W. Austin, Ocean Optics Protocols SeaWiFS for Validation, Revision 1, in: Ser. SeaWiFS Technical Report SERIES, 25, NASA GSFC, Greenbelt, MD, USA, 1995, 104566, ch. 6, pp. 48−59.

[76] K.J. Waters, R.C. Smith, M.R. Lewis, Avoiding ship-induced light-field perturbation in the determination of oceanic optical properties, Oceanography 3 (2) (1990) 18−21.

[77] R.W. Spinrad, E.A. Widder, Ship shadow measurements obtained from a manned submersible, in: Ocean Optics XI, 1750, SPIE, 1992, pp. 372−383.

[78] K.J. Voss, J.W. Nolten, G.D. Edwards, Ship shadow effects on apparent optical properties, in: Ocean Optics VIII, 0637, SPIE, 1986, pp. 186−190.

[79] C.T. Weir, D.A. Siegel, A.F. Michaels, D.W. Menzies, "In-situ evaluation of a ship's shadow, in: Ocean Optics XII, 2258, SPIE, 1994, pp. 815−821.

[80] J.P. Doyle, G. Zibordi, D. vanderLinde, Validation of an In-water, Tower-Shading Correction Scheme, in: Ser. SeaWiFS Technical Report SERIES, 25, Goddard Space Flight Center, Greenbelt, MD, 2003, p. 40. NASA Goddard Space Flight Center, TM-2003-206892.

[81] C.R. McClain, G.C. Feldman, S.B. Hooker, An overview of the SeaWiFS project and strategies for producing a climate research quality global ocean bio-optical time series, Deep Sea Res. Part Top. Stud. Oceanogr. 51 (1) (2004) 5−42.

[82] G. Zibordi, J.P. Doyle, S.B. Hooker, Offshore tower shading effects on in-water optical measurements, J. Atmos. Oceanic Tech. 16 (11) (1999) 1767−1779.

[83] E.G. Kearns, R. Riley, C. Woody, Bio-optical time series collected in coastal waters for SeaWiFS calibration and validation: large structure shadowing considerations, in: S.G. Ackleson, R.J. Frouin (Eds.), Ocean Optics XIII, 2963, SPIE, Halifax, NS, Canada, 1994, pp. 697−702.

[84] I. Lux, L. Koblinger, Monte Carlo Particle Transport Methods: Neutron and Photon Calculations, CRC Press, 1991.

[85] L. Roberti, Monte Carlo radiative transfer in the microwave and in the visible: biasing techniques, Appl. Opt. 36 (30) (1997) 7929−7938.

[86] H.R. Gordon, D.J. Castaño, Coastal zone color scanner atmospheric correction algorithm: multiple scattering effects, Appl. Opt. 26 (11) (1987) 2111−2122.

[87] G. Zibordi, J.-F. Berthon, J.P. Doyle, S. Grossi, D. van der Linde, C. Targa, L. Alberotanza, Coastal Atmosphere and Sea Time SERIES (CoASTS): A Long-term Measurement Program, in: Ser. SeaWiFS Postlaunch Technical Report SERIES, 19, NASA Goddard Space Flight Center, TM-2001-206892, Greenbelt, MD, 2002, pp. 1−29.

[88] J.-F. Berthon, G. Zibordi, J.P. Doyle, S. Grossi, D. van der Linde, C. Targa, Coastal Atmosphere and Sea Time Series (CoASTS): Data Analysis, in: Ser. SeaWiFS Postlaunch Technical Report, 20, NASA Goddard Space Flight Center, TM-2002-206892, Greenbelt, MD, 2002, pp. 1−25.

[89] B. Sturm, G. Zibordi, SeaWiFS atmospheric correction by an approximate model and vicarious calibration, J. Remote Sens. 23 (2002) 489−501.

[90] H.R. Gordon, K. Ding, Self-shading of in-water optical instruments, Limnol. Oceanogr. 37 (1992) 491−500.

[91] G. Zibordi, G. Ferrari, Instrumental self-shading in underwater optical measurements: experimental data, Appl. Opt. 34 (1995) 767−779.

[92] J. Piskozub, Effects of surface waves and sea bottom on self-shading of in-water optical instruments, in: Ocean Optics XII, vol. 2258, SPIE, 1994, pp. 300–308.
[93] E. Aas, B. Korsbø, Self-shading effect by radiance meters on upward radiance observed in coastal waters, Limnol. Oceanogr. 42 (5) (1997) 968–974.
[94] G.S. Fargion, J.L. Mueller, Ocean Optics Protocols for Satellite Ocean Color Validation, Revision 2, in: Ser. SeaWiFS Postlaunch Technical Report, vol. 20, NASA Goddard Space Flight Center, TM-2000-209966, Greenbelt, MD, USA, 2000, p. 194.
[95] J. Piskozub, A.R. Weeks, J.N. Schwarz, I.S. Robinson, Self-shading of upwelling irradiance for an instrument with sensors on a sidearm, Appl. Opt. 39 (12) (2000) 1872–1878.
[96] J.P. Doyle, K.J. Voss, Instrument self-shading effects on in-water multi-directional radiance measurements, in: Ocean Optics XV, SPIE, Monaco, 2000, pp. 16–20.
[97] A.R. Weeks, I.S. Robinson, J.N. Schwarz, K.T. Trundle, The southampton underwater multiparameter optical-fibre spectrometer system (sumoss), Meas. Sci. Technol. 10 (12) (1999) 1168.
[98] K.J. Voss, Electro-optic camera system for measurement of the underwater radiance distribution, Opt. Eng. 28 (3) (1989) 283241.
[99] R. Leathers, T.V. Downes, C. Mobley, Self-shading correction for upwelling sea-surface radiance measurements made with buoyed instruments, Opt. Express 8 (10) (2001) 561–570.
[100] K. Voss, A. Chapin, Upwelling radiance distribution camera system, nurads, Opt. Express 13 (11) (2005) 4250–4262.
[101] D. D'Alimonte, E.B. Shybanov, G. Zibordi, T. Kajiyama, Regression of in-water radiometric profile data, Opt. Express 21 (23) (2013) 27.
[102] J. Hilbert Schenck, On the focusing of sunlight by ocean waves, J. Opt. Soc. Am. 47 (7) (1957) 653–657.
[103] R.L. Snyder, J. Dera, Wave-induced light-field fluctuations in the sea, J. Opt. Soc. Am. 60 (8) (1970) 1072–1079.
[104] D. Stramski, J. Dera, On the mechanism for producing flashing light under a wind-disturbed water surface, Oceanologia 25 (1988) 5–21.
[105] M. Darecki, D. Stramski, M. Sokólski, Measurements of high-frequency light fluctuations induced by sea surface waves with an underwater porcupine radiometer system, J. Geophys. Res. Oc. 116 (C7) (2011) 16.
[106] J. Dera, H.R. Gordon, Light field fluctuations in the photic zone, Limnol. Oceanogr. 13 (4) (1968) 697–699.
[107] J. Dera, S. Sagan, D. Stramski, Focusing of Sunlight by Sea-surface Waves: New Measurement Results from the Black Sea, 1992, pp. 65–72.
[108] A. Fraser, R. Walker, F. Jurgens, Spatial and temporal correlation of underwater sunlight fluctuations in the sea, Oceanic Eng. IEEE J. 5 (3) (1980) 195–198.
[109] P. Gernez, D. Antoine, Field characterization of wave-induced underwater light field fluctuations, J. Geophys. Res. Oc. 114 (C6) (2009) 15.
[110] H.R. Gordon, J.M. Smith, O.B. Brown, Spectra of underwater light-field fluctuations in the photic zone, Bull. Marine Sci. 21 (2) (1971) 466–470.
[111] D.A. Siegel, T.D. Dickey, Characterization of downwelling spectral irradiance fluctuations, in: Proceedings of Ocean Optics IX, vol. 925, SPIE, Freemantle, Australia, October 1988, pp. 67–74.
[112] G. Zibordi, J.-F. Berthon, D. D'Alimonte, An evaluation of radiometric products from fixed-depth and continuous in-water profile data from moderately complex waters, J. Atmos. Oceanic Tech. 26 (2009) 91–106.

[113] G. Zibordi, D. D'Alimonte, J.-F. Berthon, An evaluation of depth resolution requirements for optical profiling in coastal waters, J. Atm. Ocean. Tech. 21 (7) (2004) 1059–1073.

[114] P. Gernez, D. Stramski, M. Darecki, Vertical changes in the probability distribution of downward irradiance within the near-surface ocean under sunny conditions, J. Geophys. Res. Oc. 116 (C7) (2011) 19.

[115] M. Shen, Z. Xu, D.K.P. Yue, A model for the probability density function of downwelling irradiance under ocean waves, Opt. Express 19 (18) (August 2011) 17528–17538.

[116] V. Weber, English Coefficient of variation of underwater irradiance fluctuations, Engl. Radiophys. Quantum Electron. 53 (1) (2010) 13–27.

[117] Y. You, D. Stramski, M. Darecki, G.W. Kattawar, Modeling of wave-induced irradiance fluctuations at near-surface depths in the ocean: a comparison with measurements, Appl. Opt. 49 (6) (2010) 1041–1053.

[118] J.R. Zaneveld, E. Boss, P. Hwang, The influence of coherent waves on the remotely sensed reflectance, Opt. Express 9 (6) (2001) 260–266.

[119] J.R.V. Zaneveld, E. Boss, A. Barnard, Influence of surface waves on measured and modeled irradiance profiles, Appl. Opt. 40 (9) (2001) 1442–1449.

[120] L.H. Holthuijsen, Waves in Oceanic and Coastal Waters, Cambridge University Press, 2007.

[121] W.J. Pierson, L. Moskowitz, A proposed spectral form for fully developed wind seas based on the similarity theory of s. a. kitaigorodskii, J. Geophys. Res. 69 (24) (1964) 5181–5190.

[122] I.R.I.R. Young, in: R. Bhattacharyya, M.E. McCormick (Eds.), Wind Generated Ocean Waves, Elsevier, 1999.

[123] J. Dera, S. Sagan, D. Stramski, Focusing of sunlight by sea-surface waves: new measurement results from the black sea, Oceanologia 34 (1993) 13–25.

[124] J. Dera, D. Stramski, Maximum effects of sunlight focusing under a wind-disturbed sea surface, Oceanologia 23 (1986) 15–42.

[125] R. Walker, Marine Light Field Statistics, in: Ser. A Wiley-interscience Publication, Wiley, 1994.

[126] J. Dera, J. Olszewski, Experimental study of short-period irradiance fluctuations under an undulated sea surface, Oceanologia 10 (1978) 27–49.

[127] M. Stramska, T.D. Dickey, Short term variability of optical properties in the oligotrophic ocean in response to surface waves and clouds, Deep Sea Res. 45 (1998) 1393–1410.

[128] D. D'Alimonte, G. Zibordi, The JRC Data Processing System, in: Ser. SeaWiFS Technical Report SERIES, vol. 15, NASA Goddard Space Flight Center, TM-2001-206892, Greenbelt,MD, May 2001, pp. 52–56.

[129] E.J. Gumbel, Statistics of Extremes, Columbia Univ. Press, New York, 1958.

[130] M. Fréchet, Sur la loi de probabilité de l'écart maximum, Ann. Soc. Polon. Math. 6 (1927) 93–116.

[131] M.H. DeGroot, M.J. Schervish, Probability and Statistics, fourth ed., Addison Wesley, Boston (MA), USA, 2012.

[132] R.W. Preisendorfer, C.D. Mobley, Albedos and glitter patterns of a wind-roughened sea surface, J. Phys. Oceanogr. 16 (7) (1986) 1293–1316.

[133] M. Hieronymi, A. Macke, O. Zielinski, Modeling of wave-induced irradiance variability in the upper ocean mixed layer, Ocean Sci. 8 (2) (2012) 103–120.

[134] R. Deckert, K.J. Michael, Lensing effect on underwater levels of UV radiation, J. Geophys. Res. Oc. 111 (2006) 8.

[135] M. Denis, W. Pierson, On the motion of ships in confused seas, Trans. Soc. Nav. Archit. 61 (1953) 280–357.

[136] M. Hieronymi, Monte carlo code for the study of the dynamic light field at the wavy atmosphere-ocean interface, JEOS: RP 8 (2013) 11.

[137] P.A. Hwang, O.H. Shemdin, The dependence of sea surface slope on atmospheric stability and swell conditions, J. Geophys. Res. Oc. 93 (C11) (1988) 13903–13912.

[138] M. Hieronymi, A. Macke, On the influence of wind and waves on underwater irradiance fluctuations, Ocean Sci. 8 (4) (2012) 455–471.

[139] G.W. Kattawar, G.N. Plass, F.E. Catchings, Matrix operator theory of radiative transfer 2: scattering from maritime haze, Appl. Opt. 12 (1973).

[140] G.N. Plass, G.W. Kattawar, J. John, A. Guinn, Radiative transfer in the Earth's atmosphere and ocean: influence of ocean waves, Appl. Opt. 14 (8) (1975) 1924–1936.

[141] M.S. Longuet-Higgins, A nonlinear mechanism for the generation of sea waves, Proc. Royal Soc. Edinburgh Sect. a Math. Phys. Sci. 311 (1969) 371–389.

[142] M.S. Longuet-Higgins, Capillary rollers and bores, J. Fluid Mech. Digital Arch. 240 (1992) 659–679.

[143] T. Kajiyama, D. D'Alimonte, J. Cunha, G. Zibordi, High-performance ocean color Monte Carlo simulation in the Geo-info project, in: R. Wyrzykowski, J. Dongarra, K. Karczewski, J. Wasniewski (Eds.), Parallel Processing and Applied Mathematics, Ser. Lecture Notes in Computer Science, vol. 6068, Springer, Wroclaw, Poland, 2010, pp. 370–379.

[144] J. N. Newman, Marine hydrodynamics, MIT Press, 1977.

[145] D.A. Siegel, Results of the SeaWiFS Data Analysis Round-robin, July 1994 (DARR-94), in: Ser. SeaWiFS Technical Report SERIES, vol. 26, NASA GSFC, Greenbelt, MD, USA, 1995, 104566, ch. 3, pp. 44–48.

[146] D. Schattschneider, Proof without words: the arithmetic mean-geometric mean inequality, Math. Mag. 59 (1) (1986) 11.

[147] J.J. Beauchamp, J.S. Olson, English Corrections for bias in regression estimates after logarithmic transformation, English Ecology 54 (6) (1973) 1403–1407.

[148] G.A.F. Seber, C.J. Wild, Nonlinear Regression, in: Ser. Wiley Series in Probability and Statistics, J. Wiley & Sons, 2003.

[149] Y. Yuan, A review of trust region algorithms for optimization, in: ICIAM 99, Oxford University, 2000, pp. 271–282.

[150] P.E. Gill, W. Murray, M.H. Wright, The Levenberg-Marquardt Method, Academic Press, 1981, ch. 4.7.3, pp. 136–137.

Chapter 4.2

Simulation of Satellite Visible, Near-Infrared, and Shortwave-Infrared Measurements

Menghua Wang
NOAA Center for Satellite Applications and Research, College Park, Maryland, USA
Email: Menghua.Wang@noaa.gov

Chapter Outline

1. Introduction — 452
2. Ocean–Atmospheric System — 455
3. Simulations — 457
   3.1 Ocean Radiance Contributions — 457
      3.1.1 Open Ocean Case-1 Waters — 457
      3.1.2 Coastal and Inland Typical Case-2 Waters — 458
      3.1.3 Ocean Bidirectional Reflectance Distribution Function — 463
   3.2 The TOA Atmospheric Path Radiance Contributions — 464
      3.2.1 Radiative Transfer Simulation for Ocean-Atmosphere System — 464
      3.2.2 Aerosol Models — 464
      3.2.3 Aerosol Spectral Reflectance — 466
      3.2.4 Atmospheric Path Reflectance $\rho_{path}(\lambda)$ — 468
   3.3 Atmospheric Diffuse Transmittance — 470
   3.4 Simulated and Satellite-Measured TOA Radiances — 471
      3.4.1 Simulated Typical TOA Radiances — 471
      3.4.2 Simulated Typical TOA Radiances Compared with Satellite Data — 476
4. Summary — 478
Disclaimer — 479
References — 479

1. INTRODUCTION

Satellite ocean-color remote-sensing products derived from the visible wavelengths have long been used to study and understand the global and regional ocean and atmospheric processes, as well as for monitoring the natural hazards. These include ocean's global-scale biological and biogeochemical variability [1–4], ocean's response to a short-term weather event [5–8], effects of volcano eruption on ocean environmental changes [9], mesoscale ocean processes [10,11], phytoplankton blooms in the open ocean [12], floating green-algae blooms [13,14], sea-surface temperature (SST) variability and ocean circulation [15–17], open ocean and coastal water primary productivity [4,18–20], ocean property variation in the Korean dump site of the Yellow Sea [21], environmental responses to a land reclamation project [22], coastal environmental changes and monitoring [14,21,23–27], harmful algae blooms [28–30], and monitoring inland freshwater environmental variations [31–33].

From a brief history, the initial systematic measurements of the ocean color from aircraft, carried out at about the end of the 1960s, were from those of George L. Clarke, Gifford C. Ewing, and Carl J. Lorenzen [34]. The remote spectroscopy of the light backscattered from the sea in coastal waters near Woods Hole and in offshore waters from the Sargasso Sea across Georges Bank demonstrated the possibility of detecting the chlorophyll-a concentration within the upper layers. Measurements from altitudes ranging from about 152 to 3050 m showed the effect of increasing "airlight" on the upward radiance received by the air-borne sensor, and thus the need for an "atmospheric correction" [35,36]. It was soon realized that the atmospheric path radiance (the atmospheric and ocean surface signal) was largely dominant compared to the marine signal in the visible part of the spectrum.

Simultaneously, field measurements of upward and downward spectral irradiance, $E_u(\lambda)$ and $E_d(\lambda)$, respectively, were performed by John E. Tyler and Raymond C. Smith in different water bodies [37]. Thereafter, systematic determinations of the same radiometric quantities were made during the SCOR WG-15 Discoverer 1970 Expedition (Data Report, 1973, J.E. Tyler Edit), and allowed the diffuse attenuation coefficients, $K_d(\lambda)$, and the irradiance reflectance, $R(\lambda) = E_u(\lambda)/E_d(\lambda)$, to be computed. The interpretation of this reflectance, a basic quantity in ocean-color science, was given in the frame of radiative transfer [38] and also in terms of optically significant substances present in the water [39,40].

Building upon these, satellite global ocean-color missions were started with the Coastal Zone Color Scanner (CZCS) on Nimbus-7 (launched in October 1978) [41–43] as a proof-of-concept mission, in which it demonstrated the feasibility of quantitative retrieval of the ocean near-surface optical and bio-optical properties. CZCS is a scanning radiometer viewing the ocean in five spectral bands, and was designed to measure the concentration of phytoplankton pigments in the open ocean [41–43]. The follows on satellite ocean color missions include NASA's sea-viewing wide field-of-view sensor (SeaWiFS) [44−46] and moderate-resolution imaging spectroradiometer (MODIS) on the Terra and Aqua satellites [47,48], and the European Space Agency's (ESA) medium-resolution imaging spectrometer (MERIS) on the Envisat [49]. Unfortunately, both SeaWiFS and MERIS had stopped collecting data, and MODIS sensors are long past their expected lifetime. There are also other satellite ocean-color missions [50−54], either lasted a short period of time or more in experimental missions, such as the German's modular optoelectronic scanner (MOS) [54], the Japanese ocean color and temperature scanner (OCTS) [51] and global imager (GLI), French polarization and directionality of the Earth's reflectances (POLDER) [50] and POLDER-2, etc. One can find a more complete list of the satellite ocean-color missions at the International Ocean-Colour Coordinating Group (IOCCG) website (http://www.ioccg.org).

The Suomi National Polar-Orbiting Partnership (SNPP) satellite was launched into an 824-km sun-synchronous polar orbit on October 28, 2011. SNPP carries the visible infrared-imaging radiometer suite (VIIRS) [55], a 22-band visible/infrared sensor that combines features of the NASA ocean-color sensors, SeaWiFS and MODIS, the NOAA advanced very high-resolution radiometer (AVHRR), and the Defense Meteorological Satellite Program (DMSP) Operational Linescan System (OLS). One of the primary goals for the VIIRS mission is to provide the data continuity for the science community with ocean-color products over global oceanic waters to enable assessment of climatic and environmental variability [45,46]. In fact, ocean-color product set is one of the key product suites derived from VIIRS [56].

For remote sensing of ocean properties from satellite sensors, the sensor spectral bands are usually located in atmospheric "windows," thereby maximizing sensitivity in sensor-measured radiance to the variation of the observing target and reducing atmospheric absorption effects that lead to uncertainty in the derived optical and geophysical products. Figure 1 provides an example of the transmittance of the atmosphere (looking toward the zenith) as a function of the wavelength (300−2200 nm). The atmospheric transmittance values are derived using LOWTRAN-7 for the 1976 U.S. standard atmosphere model [57,58]. Figure 1 shows absorption features from the main atmospheric constituent of O_2, O_3, H_2O, and CO_2 in the considered spectral interval. MODIS spectral bands are also plotted as vertical dashed lines in Figure 1, covering from visible (412−678 nm) to near-infrared (NIR) (748 and 869 nm) and to shortwave-infrared (SWIR) bands at around 1240, 1640, and 2130 nm. There is also a window band at the SWIR around 1000 nm. It is noted that MODIS SWIR bands at 1240, 1640, and 2130 nm are designed for the land and atmosphere applications with significantly poor sensor signal-to-noise ratio (SNR) performance at the SWIR bands, causing some problems for ocean-color applications [59,60]. VIIRS also has three SWIR bands at 1238, 1610, and 2250 nm similar to MODIS with also low sensor SNR values. For the ultraviolet (UV) spectral region, the useful radiance with the shortest

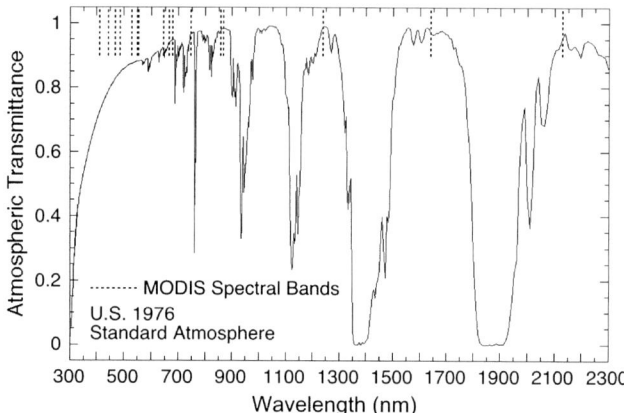

FIGURE 1 The atmosphere transmittance as a function of the wavelength (300–2300 nm) derived from LOWTRAN-7 using the 1976 U.S. standard atmosphere model. MODIS spectral bands are plotted as vertical dashed lines.

wavelength for ocean-color remote sensing is at around 340 nm due to extremely strong absorption by ozone at shorter wavelengths (i.e., <340 nm) (Figure 1). Thus, the UV wavelength region between 340 and 400 nm can be possibly added for the future ocean-color satellite sensors.

In deriving ocean optical, biological, and biogeochemical properties from satellite ocean-color sensors, accurately correcting/removing the signal (radiance) originating by the interaction of the solar radiation with the atmosphere and the ocean surface has been a great challenge [61]. In other words, it is important to accurately compute (or simulate) the sensor-measured top-of-atmosphere (TOA) radiance spectra because atmospheric correction is in fact removing unwanted radiance components from the TOA radiance spectra [61]. In early days, the problem of removing the atmospheric and ocean surface effects from satellite imagery of the ocean was examined [62–64], and the technique developed for that purpose [65] was then implemented into the NASA data processing system used with CZCS sensor. The problem that was encountered with this sensor was the absence of an appropriate channel in the NIR portion of the spectrum (Figure 1), where the open ocean can be reasonably considered as black (i.e., no ocean radiance contributions at the NIR wavelengths), so that the detected signal can be assumed of purely atmospheric and ocean surface origin. As a replacement, the CZCS band at 670 nm (red band) was used along with the assumption that the ocean radiance at this wavelength, $L_w(670)$, is negligible, or otherwise iteratively retrievable from an empirical relationship between $L_w(670)$ and the chlorophyll-a concentration [66,67], or from a *priori* knowledge when the chlorophyll content is sufficiently low (i.e., clear-water concept) [68].

During the CZCS era, the atmospheric contribution was predicted (and removed) based on the single-scattering approximation, i.e., single-scattering approximation was used for computing the TOA radiances. Within such an approach, the Rayleigh scattering by air molecules can be separately considered and computed in the absence of the aerosol, whereas the aerosol-scattering contribution can also be separately computed in the absence of molecules [43]. Later in the CZCS mission, this simplifying assumption was abandoned [69]. Particularly since SeaWiFS launch in 1997, the coupling of aerosol-molecules scattering was addressed [70], as well as the effect of polarization on the magnitude of the atmospheric contribution [71], including aerosol polarization effects [72]. The new schemes for atmospheric correction or for computing the sensor-measured TOA radiance spectra have been used since SeaWiFS mission [61,70,73−75]. With these new schemes, we have been routinely producing high-quality satellite ocean-color products [46].

In this chapter, we describe and discuss satellite-measured radiance spectra over global open oceans and coastal and inland waters. Specifically, radiance contributions from atmosphere, i.e., radiance contributions from air molecules (Rayleigh scattering), aerosols, and ocean surface, as well as from ocean waters (both Case-1 and Case-2 waters) will be discussed in detail. We will provide simulated results for sensor-measured TOA radiance spectra, and through histogram data analysis, typical TOA radiance spectra over ocean from UV to visible and NIR, and to SWIR wavelengths will be derived. Finally, the simulated TOA radiance values are compared with those measured by SeaWiFS and VIIRS.

2. OCEAN−ATMOSPHERIC SYSTEM

In this document, we define the reflectance $\rho(\lambda)$, at a given wavelength λ and for a specific solar-zenith angle of θ_0, to be related to the radiance $L(\lambda)$ through $\rho(\lambda) = \pi L(\lambda)/[F_0(\lambda) \cos\theta_0]$, where $F_0(\lambda)$ is the extraterrestrial solar irradiance [76]. Thus, the radiance and reflectance are interchangeable based on this definition. The purpose of the atmospheric correction for the remote retrieval of ocean properties is to remove the atmospheric and water-surface effects from the satellite-sensor-measured signals, thereby deriving the radiances (reflectances) coming from the ocean waters. For the ocean−atmosphere system, the TOA reflectance $\rho_t(\lambda)$ (or radiance $L_t(\lambda)$) can be linearly partitioned into various distinct physical contributions [61,70,77]:

$$\rho_t(\lambda) = \rho_r(\lambda) + \rho_a(\lambda) + \rho_{ra}(\lambda) + t(\lambda)\rho_{wc}(\lambda) + T(\lambda)\rho_g(\lambda)$$
$$+ t(\lambda) t_0(\lambda)\rho_{wN}(\lambda), \text{ or}$$
$$L_t(\lambda) = L_r(\lambda) + L_a(\lambda) + L_{ra}(\lambda) + t(\lambda)L_{wc}(\lambda) + T(\lambda)L_g(\lambda)$$
$$+ t(\lambda)t_0(\lambda)\cos\theta_0\, nL_w(\lambda),$$

(1)

where $\rho_r(\lambda)$ and $\rho_a(\lambda)$ (or $L_r(\lambda)$ and $L_a(\lambda)$) are, respectively, the reflectance (radiance) due to scattering by air molecules (Rayleigh scattering) in the absence of aerosols [71,78,79] and scattering by aerosols in the absence of air molecules [70,80], $\rho_{ra}(\lambda)$ (or $L_{ra}(\lambda)$) is the multiple interaction reflectance (radiance) term between air molecules and aerosols [81,82] over a Fresnel reflecting ocean surface, $\rho_{wc}(\lambda)$ and $\rho_g(\lambda)$ (or $L_{wc}(\lambda)$ and $L_g(\lambda)$) are the components of reflectance (radiance) due to whitecaps on the sea surface and the specular reflection of direct sunlight off the sea surface (sun glitter) [83−85], respectively, and $\rho_{wN}(\lambda)$ (or $nL_w(\lambda)$) [61] is the normalized water-leaving reflectance (radiance) due to photons that penetrate the sea surface and are backscattered out of the water. The quantities $t_0(\lambda)$ and $t(\lambda)$ are the atmospheric diffuse transmittances [86] from the sun to the water surface and from the water surface to the sensor, respectively. $T(\lambda)$ is the direct transmittance from the water surface to the sensor [84]. As it is shown, Eqn (1) applies to both reflectance and radiance (note an extra factor of $\cos\theta_0$ in the radiance equation in the last term). Based on the reflectance definition, the normalized water-leaving reflectance $\rho_{wN}(\lambda) = \pi\, nL_w(\lambda)/F_0(\lambda)$, where $nL_w(\lambda)$ is the normalized water-leaving radiance (more discussions later). It should be noted that Eqn (1) is valid when the target is large spatially, i.e., effects of the target environment can be neglected. The goal of the atmospheric correction is to retrieve the normalized water-leaving reflectance $\rho_{wN}(\lambda)$ accurately from the spectral measurements of the TOA reflectance (or radiance) $\rho_t(\lambda)$ (or $L_t(\lambda)$) at the satellite altitude. In effect, satellite sensor-measured TOA radiance (Eqn (1)) has to be computed accurately. We can further define the TOA atmospheric path reflectance/radiance (including contributions from both atmosphere scattering and surface reflection) as

$$\rho_{path}(\lambda) = \rho_r(\lambda) + \rho_a(\lambda) + \rho_{ra}(\lambda), \text{ or}$$
$$L_{path}(\lambda) = L_r(\lambda) + L_a(\lambda) + L_{ra}(\lambda) \quad (2)$$

and then Eqn (1) becomes

$$\rho_t(\lambda) = \rho_{path}(\lambda) + t(\lambda)\rho_{wc}(\lambda) + T(\lambda)\rho_g(\lambda) + t(\lambda)t_0(\lambda)\rho_{wN}(\lambda), \text{ or}$$
$$L_t(\lambda) = L_{path}(\lambda) + t(\lambda)L_{wc}(\lambda) + T(\lambda)L_g(\lambda) + t(\lambda)t_0(\lambda)\cos\theta_0\, nL_w(\lambda). \quad (3)$$

From now on, we will use either reflectance or radiance in discussions with understanding these two are interchangeable based on their definitions (Eqns (1)−(3)). The principal challenge in atmospheric correction is the estimation and removal of $\rho_{path}(\lambda)$ from $\rho_t(\lambda)$. In other words, we have to compute $\rho_{path}(\lambda)$ accurately. In Case-1 waters, $\rho_{path}(\lambda)$ contributes about 90% of the TOA reflectance in the blue and a higher fraction in the green and red. The estimation of the diffuse transmittances is next in the order of importance, and its main difficulty with their estimation relies on the dependence of the transmittance on the angular distribution of the radiance just beneath the sea surface [86]. Sun glint reflectance $\rho_g(\lambda)$ can be rendered as small as desired by avoiding the region surrounding the specular image of the sun using the sun glint model [84,85,87], and whitecap reflectance $\rho_{wc}(\lambda)$ can be estimated from the estimations of the surface wind speed [83,88—91].

3. SIMULATIONS

In this section, we describe the TOA radiance contributions from individual components according to Eqn (3) (or Eqn (1)) with specific focus on reflectance contributions from $\rho_{wN}(\lambda)$ and $\rho_{path}(\lambda)$ for various typical cases, e.g., water types, aerosols, atmospheric diffuse transmittance, ocean bidirectional reflectance distribution function (BRDF) effects, etc.

3.1 Ocean Radiance Contributions

The ocean radiance contributions for open ocean Case-1 waters and coastal typical Case-2 waters (i.e., sediment-dominated and yellow substance-dominated waters), which can be used for the TOA reflectance computations ($\rho_{wN}(\lambda)$ component in Eqn (3) or equivalently $nL_w(\lambda)$), are described below.

3.1.1 Open Ocean Case-1 Waters

Typical normalized water-leaving reflectance spectra $\rho_{wN}(\lambda)$ for the Case-1 water can be derived using a semianalytical model [92]. Revised open ocean $\rho_{wN}(\lambda)$ model for the Case-1 water has also been developed [93]. Here we briefly describe basic approach for modeling the Case-1 water using Gordon et al. (1988) [92] as presented and discussed in IOCCG (2010) [61]. The $\rho_{wN}(\lambda)$ in the Gordon et al. (1988) model is given by [92]

$$\rho_{wN}(\lambda) = \pi \left[\frac{(1 - \rho_f)(1 - \overline{\rho}_f)}{m^2} \right] \times \left[\frac{1}{1 - rR} \right] \times \frac{R}{Q}, \qquad (4)$$

where ρ_f is the Fresnel reflectance of the sea surface at normal incidence, $\overline{\rho}_f$ is the Fresnel reflectance for irradiance from the sun and sky, m is the refractive index of water, R is the irradiance reflectance just beneath the surface, r is the water-to-air surface reflectance for diffuse light ($r \approx 0.48$), and $Q = E_u/L_u$, with E_u is the upwelling irradiance just beneath the sea surface and L_u is the upwelling radiance in the same position propagating toward the zenith. Assuming a totally diffuse upwelling light field just beneath the surface, a nominal value of Q is π. The quantity R/Q can be related to the inherent optical properties (IOPs) of the water through [92]

$$\frac{R}{Q} \approx 0.11 \frac{b_b}{K}, \text{ and } K \approx 1.16(a + b_b), \qquad (5)$$

where the absorption coefficient of the medium is given by $a = a_w + a_c$ and the backscattering coefficient by $b_b = (b_b)_w + (b_b)_c$, where the subscripts "w" and "c" refer to water and its constituents, respectively. K is the attenuation coefficient of downwelling irradiance just beneath the water surface and can be related to $(a + b_b)$ [38].

In the Gordon et al. (1988) model [92], K is divided according to $K = K_w + K_{Chl} + K_{ys}$, where the subscripts "w," "Chl," and "ys," stand for water, chlorophyll, and yellow substance, respectively. Although this separation is not precisely correct because K is not rigorously sumable over constituents, it has been shown that for most purposes this is a valid approximation [94]. The spectral values of K_w and K_{Chl} are taken from Smith and Baker (1981) [95] and Smith and Baker (1978) [40], respectively. K_{ys} is taken to be zero here. This is equivalent to assuming that any background yellow substance is accounted for in K_{Chl}. The backscattering coefficient of the constituents (phytoplankton in Case-1 waters) is written as $(b_b)_c = A(\lambda) \cdot Chl^{B(\lambda)}$, where Chl is the chlorophyll concentration. The coefficients $A(\lambda)$ and $B(\lambda)$ were chosen in a manner that provided the best fit to the spectral radiance data collected [96]. Given these quantities, R/Q in Eqn (5) is determined as a function of Chl. Then assuming $Q = \pi$ (the value for a totally diffuse upwelling light field), R is determined for the computation of $(1-rR)$, providing $\rho_{wN}(\lambda)$ as a function of Chl through Eqn (4).

Figure 2(a) shows the normalized water-leaving reflectance $\rho_{wN}(\lambda)$ as a function of the wavelength for chlorophyll-a concentrations of 0.03, 0.1, 0.3, and 1.0 mg m^{-3}. These $\rho_{wN}(\lambda)$ spectra are representative of those in Case-1 waters and sufficiently accurate for the purposes of the TOA radiance simulations.

3.1.2 Coastal and Inland Typical Case-2 Waters

Some examples of coastal and inland $\rho_{wN}(\lambda)$ contributions from the Case-2 waters, which correspond to a typical sediment-dominated and a yellow substance (or colored dissolved organic matter (CDOM))-dominated waters, are discussed here for the TOA radiance computations. Two examples specifically for typical sediment-dominated and yellow substance-dominated waters are from IOCCG (2010). There are some other examples in coastal and inland waters from satellite and in situ measurements shown here.

The first example from IOCCG (2010) [61] corresponds to sediment-dominated Case-2 waters. It derives from actual irradiance measurements (upward and downward irradiances just below the surface, E_u and E_d), which were performed at a station off Cap Corveiro. The high particle load results from the resuspension of bottom sediments due to wave action generated by strong trade winds. The scattering coefficient, $b(\lambda)$, at 550 nm $b(550)$ was between 3.1 and 3.9 m^{-1}, and the chlorophyll-a concentration, Chl, ranged from 1.1 to 1.9 mg m^{-3}. Such turbid milky waters were encountered along this arid coast within a stripe delimited by the shoreline and offshore by the isobaths 40 or 50 m [97].

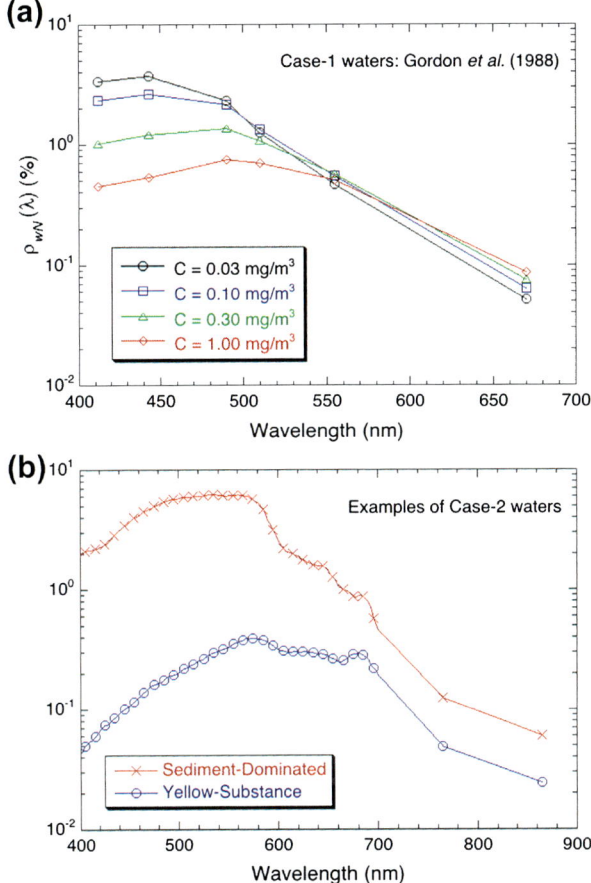

FIGURE 2 Examples of the ocean normalized water-leaving reflectance $\rho_{wN}(\lambda)$ (%) as a function of the wavelength for (a) the Case-1 waters for chlorophyll concentrations of 0.03, 0.1, 0.3, and 1.0 mg m^{-3} and (b) the Case-2 waters for a sediment-dominated water observed off the Mauritanian coast and a yellow substance-dominated water acquired from an inlet in Vancouver Island, respectively. *These are reproduced from IOCCG (2010) Ref. [61].*

The second example from IOCCG (2010) is for typical yellow substance-dominated Case-2 waters. The irradiance reflectance was determined in the Saanich inlet (Sidney, Vancouver Island, BC, Canada) on August 7, 1979. The water was dark brownish, almost black with $b(550)$ around 1.2 m^{-1} and Chl about 4.5 mg m^{-3}. The high yellow-substance content of terrigenous origin originates from the drainage of the forest soil. For both cases, normalized water-leaving reflectance $\rho_{wN}(\lambda)$ spectra were derived from the irradiance reflectance spectra.

Figure 2(b) provides $\rho_{wN}(\lambda)$ spectra as a function of the wavelength for a sediment-dominated water observed off the Mauritanian coast and a yellow substance-dominated water acquired from an inlet in Vancouver Island. As expected, $\rho_{wN}(\lambda)$ for a sediment-dominated water shows strong contributions at the green and red wavelengths, while for a yellow substance-dominated water, $\rho_{wN}(\lambda)$ has very low (almost negligible) contributions at the blue bands (Figure 2(b)). Both Case-2 waters, particularly for a sediment-dominated water, $\rho_{wN}(\lambda)$ has important contributions at the NIR bands.

Some other examples of sediment-dominated Case-2 waters from highly turbid China's east coastal region [98–100] are shown in Figures 3 and 4 (from both satellite and in situ measurements). Figure 4 shows the $\rho_{wN}(\lambda)$ spectra from satellite and in situ measurements corresponding to locations indicated in MODIS-Aqua true color image acquired on October 19, 2003 in Figure 3 [101]. During the spring and fall of 2003, there were extensive field campaigns along the East China Sea and the Yellow Sea regions for purposes of collecting various in situ physical, optical, and biological ocean data. In particular, in situ ocean water-leaving reflectance spectrum data from 350 to 1050 nm were measured using an Analytical Spectral Devices, Inc. FieldSpec Dual UV/VNIR spectrometer, for which the instrument spectral sampling interval is 1.4 nm for its entire spectral coverage (350–1050 nm). The in situ data collection and processing were carried out following the procedures outlined in the NASA ocean-optics protocols [102]. Specifically, in the in situ data processing, the radiance that is contributed by the ocean surface reflection

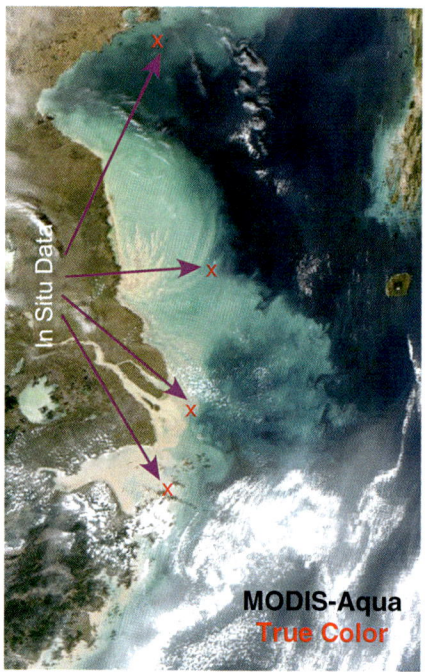

FIGURE 3 MODIS-Aqua true color image acquired on October 19, 2003 along the China's east coastal region around (33.0°N, 123.0°E). *Reproduced from Wang et al. (2007) Ref. [101].*

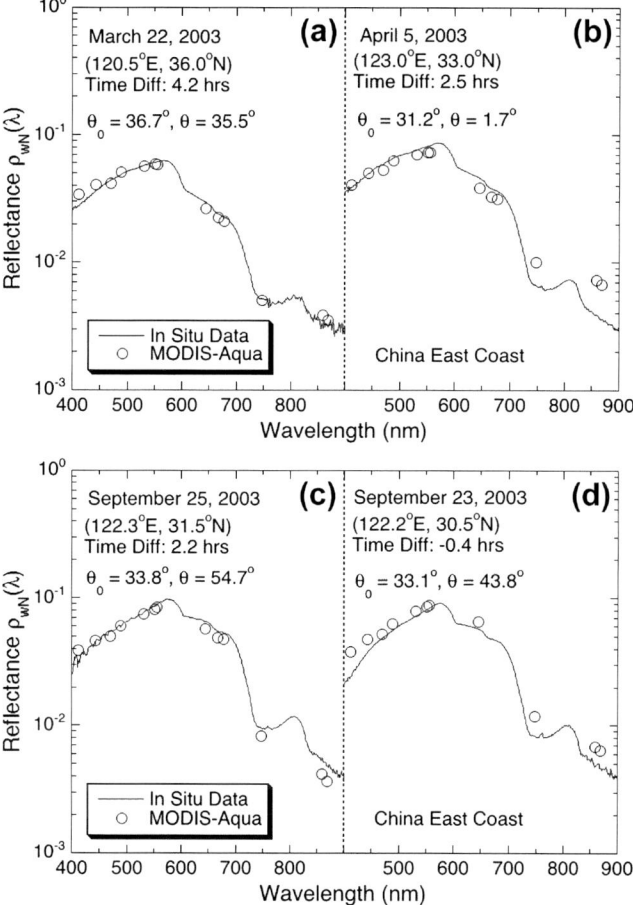

FIGURE 4 MODIS-Aqua-derived normalized water-leaving reflectance $\rho_{wN}(\lambda)$ spectra compared with the in situ data that were acquired on (a) March 22, 2003, (b) April 5, 2003, (c) September 25, 2003, and (d) September 23, 2003. The locations (marked in Figure 3) in latitude and longitude and time difference between MODIS and in situ measurements are indicated in each plot. *Reproduced from Wang et al. (2007) Ref. [101].*

from the sky radiance to the instrument detector has been calculated and removed. Here, we show the MODIS-Aqua-derived $\rho_{wN}(\lambda)$ spectra together with the corresponding in situ measurements. Figure 4 provides four examples of $\rho_{wN}(\lambda)$ for sediment-dominated China east coastal waters (Figure 3). Figure 4(a)–(d) shows the $\rho_{wN}(\lambda)$ data that were collected at the locations indicated in Figure 3 (from the North to the South), i.e., the corresponding latitudes for data obtained in Figure 4(a)–(d) are 36°N (March 22, 2003), 33°N (April 5, 2003), 31.5°N (September 25, 2003), and 30.5°N (September 23, 2003), respectively. These sediment-dominated $\rho_{wN}(\lambda)$ spectra from

China's east coastal region are quite similar to the sediment-dominated case in Figure 2(b), e.g., $\rho_{wN}(\lambda)$ spectra peak at the green-red spectral region.

Examples of MODIS-Aqua-derived $nL_w(\lambda)$ spectra maps from highly turbid inland Lake Taihu in China are shown in Figure 5 [31,103]. Specifically, using all MODIS-Aqua measurements from 2002 to 2010 in China's Lake Taihu, climatology maps of $nL_w(\lambda)$ spectra from visible to the NIR and SWIR bands were generated. Figure 5(a)−(f) provides the Lake Taihu's climatology maps of $nL_w(\lambda)$ at wavelengths of 443, 488, 555, 645, 859, and 1240 nm, respectively. Climatology $nL_w(\lambda)$ maps in Figure 5 have shown overall spatial distribution of water optical properties for Lake Taihu. Across the radiance spectrum from the blue to the NIR, high $nL_w(\lambda)$ contributions are found in most of the central lake regions (Figure 5(a)−(e)). In contrast, high SWIR $nL_w(1240)$ values are most present in the west coast of Lake Taihu (Figure 5(f)). Obviously, high $nL_w(\lambda)$ distributions for the NIR and SWIR (1240 nm) bands for the lake are different [103]. The NIR $nL_w(859)$ distribution and variation are mainly influenced by total suspended matter in the water, while the SWIR $nL_w(1240)$ seasonal and spatial patterns are primarily due to significant amounts of algae (surface-flowing algae) distributions [13,14,103]. Thus, in addition to already demonstrated usefulness of the NIR $nL_w(\lambda)$ products, the results show some important applications using satellite-measured SWIR $nL_w(1240)$ data [3,100,103] for characterizing and

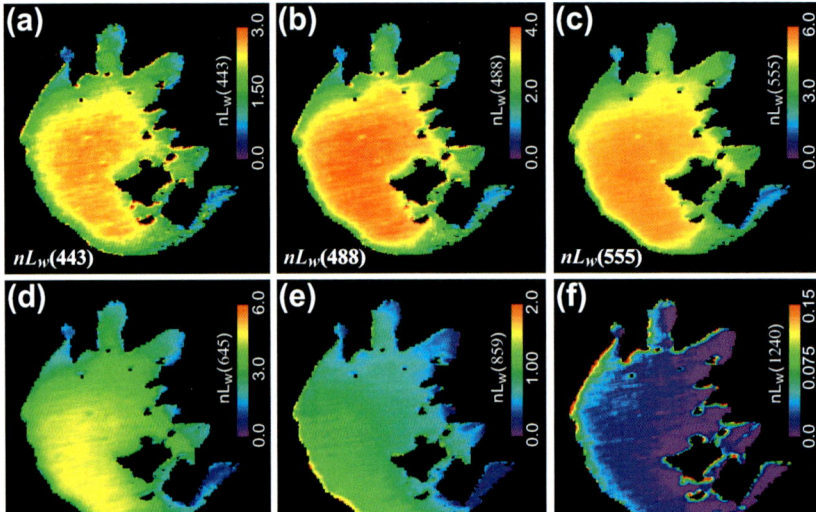

FIGURE 5 MODIS-Aqua-derived climatology $nL_w(\lambda)$ spectra maps for extremely turbid inland Lake Taihu in China for (a)−(f) as $nL_w(\lambda)$ at wavelengths of 443, 488, 555, 645, 859, and 1240 nm, respectively. *These are reproduced from Wang et al. (2013) Ref. [103].*

monitoring extremely turbid coastal and inland lake waters. Again, for highly turbid inland Lake Taihu, $nL_w(\lambda)$ spectra peak at the green-red bands, consistent with characteristics of sediment-dominated waters.

3.1.3 Ocean Bidirectional Reflectance Distribution Function

3.1.3.1 Ocean Surface BRDF Effects

Developed through a series of studies over many years, the ocean BRDF of ocean water-leaving radiances ($L_w(\lambda)$) has been studied by many scientists [68,104–108]. Key in this development is the theoretical concept of the *normalized* water-leaving radiance $nL_w(\lambda)$, defined as the water-leaving radiance $L_w(\lambda)$ that would be viewed at nadir when the Sun is at the zenith and mean Earth–Sun distance, with no intervening of atmosphere. The initial definition of the normalized water-leaving radiance by Gordon and Clark (1981) [68] accounts for variations between mean extraterrestrial solar irradiance and irradiance incident on the sea surface due to solar-zenith angle, atmospheric diffuse transmittance, and effects of variations in Earth–Sun distance during the year. Morel and Gentili (1991, 1993, 1996) [105–107] and Morel et al. (2002) [109] extended the definition to account for additional effects due to angular variations in reflection and refraction at the sea surface and for the in-water BRDF. Recently, it has been found that for cases with solar- and satellite sensor-zenith angles up to ~60°, the solar angle-related surface BRDF effect depends only on the properties of the ocean surface and not the atmosphere [108], while the sensor angle-related term can be accurately computed with a flat ocean surface (i.e., independent of the wind speed) [104]. Thus, the effects in angular variation of the surface reflection and refraction on the normalized water-leaving radiance can be easily calculated [104,108].

3.1.3.2 In-Water BRDF Effects

Within a water body, particularly close to the interface, the upward radiance field is not isotropic, so that a BRDF has to be taken into account [109]. The anisotropic in-water radiance distribution is basically related to the abundance of particles in the water and the corresponding particle optical properties. The first consideration deals with the average number of scattering events undergone by photons before they come back and reach the surface. The second one is related to the shape of the volume scattering function, particularly the shape of its backward lobe (which is particularly, but not exclusively, involved in the formation of the backward flux). For open ocean (Case-1) waters, the water BRDF effects are understood quite well and can be accounted for reasonably accurate [110–113]. On the other hand, the situation is by far less favorable in various coastal and inland Case-2 waters (sediment-dominated or yellow substance-dominated Case-2 waters), in spite of some preliminary attempts [114] that cannot be generalized. The lack of a reliable prediction and the absence of generic parameterizations for the IOPs in such complex waters remain presently a serious obstacle to go farther, except on a case-by-case basis.

3.2 The TOA Atmospheric Path Radiance Contributions

3.2.1 Radiative Transfer Simulation for Ocean-Atmosphere System

There are various techniques for solving the radiative transfer equation [115]. The integral−differential equation has been solved analytically only for isotropic and Rayleigh scattering [116]. For all other scattering phase functions, it is difficult to find analytical solutions, and thus various numerical techniques for finding solutions have been developed [115], e.g., the successive order of scattering (SOS) method, the matrix operator method (adding−doubling method), Monte-Carlo method, etc. The SOS code [115] was used to generate the TOA atmospheric path radiance $L_{path}(\lambda)$ (or $\rho_{path}(\lambda)$) as in Eqn (2). The SOS code was developed for the purpose of development of the advanced atmospheric correction algorithm for satellite ocean-color sensors, e.g., SeaWiFS, MODIS [70,78,82]. Typically, the SOS code can produce the TOA path radiance within uncertainty of $<\sim 0.1\%$, e.g., compared with results from the forward Monte-Carlo radiative transfer code [78,82,117]. They are generally agreed with each other in the reflectance within $\sim 5 \times 10^{-4}$, e.g., used for IOCCG (2010) [61] simulations. In addition, the accuracy of the SOS code has also been validated and compared with results from various other approaches [115,117−121]. The simulations were carried out for a two-layer plane-parallel atmosphere (PPA) model with 78% of molecules occupying the top layer (aerosols mixed with 22% of molecules confined in the bottom layer), overlying a flat Fresnel-reflecting ocean surface (black ocean). The PPA model is generally valid for the solar and sensor zenith angles $\sim <80°$ [122], and the effects of the aerosol vertical distribution on results of the TOA atmospheric path radiance computation are negligible for nonabsorbing and weakly absorbing aerosols [70,77].

3.2.2 Aerosol Models

In development of the advanced atmospheric correction and aerosol retrieval algorithms for SeaWiFS, MODIS, and the Multi-angle Imaging Spectroradiometer (MISR) [123−125], Gordon and Wang (1994) [70] adopted the external mixture hypothesis [126] for aerosol components, i.e., the aerosol is assumed to be composed of a collection of distinct, noninteracting species, each of which has a characteristic size distribution. In this case, the combined aerosol size-frequency distribution can be written as the summation of individual aerosol size-frequency distribution from various aerosol species, and the individual aerosol component is log-normally distributed [126,127].

The aerosol optical properties of each component can be determined from its corresponding size distribution and refractive index using Mie theory for spherical particles [128] or using scattering theory appropriate to the actual shape of the aerosol particles, e.g., nonspherical particles [129,130]. This allows computations for all the optical properties of the individual aerosol components. These aerosol optical properties can then be used to compute the TOA path radiance $L_{path}(\lambda)$ (or reflectance $\rho_{path}(\lambda)$) contributions.

The actual models used for the individual aerosol components are those from Shettle and Fenn (1979) [127], i.e., their oceanic and tropospheric models [127] for further generating other appropriate aerosol models for satellite ocean-color and aerosol applications [70,77,80,123,125]. It is noted that because the aerosol particles in the Shettle and Fenn (1979) [127] modes are hygroscopic, the aerosol size distribution (the aerosol mode parameter) depends on the relative humidity (RH). Specific aerosol models generated and used for satellite ocean-color data processing are the Oceanic model with RH of 99% (O99), the Maritime model with RH of 50, 70, 90, and 99% (M50, M70, M90, and M99), the Coastal model with RH of 50, 70, 90, and 99% (C50, C70, C90, and C99), and the Tropospheric model with RH of 50, 90, and 99% (T50, T90, and T99) [70,73,77]. In fact, these 12 aerosol models are used in generating aerosol lookup tables for producing ocean-color products from various satellite ocean-color sensors, e.g., SeaWiFS [80], the German's MOS [52], the Japanese OCTS, and the French POLDER [53], MODIS [31,103,131,132], and recently for VIIRS [56], as well as the Korean geostationary ocean-color imager (GOCI) [133−135].

However, for testing and evaluation algorithms, aerosol models that are different from those used to generate aerosol lookup tables are usually used [61,70,73]. Same as in the previous works, the Maritime aerosol model with RH of 80% (M80) and the Tropospheric model with RH of 80% (T80) are used here for generating the TOA radiance $L_t(\lambda)$ spectra. The M80 and T80 models represent nonabsorbing and weakly absorbing aerosols with the aerosol single-scattering albedo values of 0.99 and 0.95 at 865 nm, respectively. Their aerosol size distributions are also different with large particle size from the M80 model and fine aerosols from the T80 model. These two aerosol models are similar, but not identical, to the 12 aerosol models used in the aerosol lookup tables for the ocean-color data processing.

It is noted that NASA Ocean Biology Processing Group (OBPG) has now used other aerosol models for satellite ocean-color data processing [136]. Some evaluation results show that these two sets of aerosol models have both advantages and disadvantages compared with aerosol optical properties derived from the ground-based measurements [80,137−141]. However, it should be emphasized that for satellite ocean-color remote sensing, we are really chasing after aerosol reflectance $\rho_A(\lambda) = \rho_a(\lambda) + \rho_{ra}(\lambda)$, not aerosol optical properties, i.e., the focus is on the correction of aerosol reflectance, not on the aerosol properties. In fact, aerosol reflectance $\rho_A(\lambda)$ is proportional to the product of aerosol single-scattering albedo, aerosol effective phase function, and aerosol optical thickness (discussed in the next section).

3.2.3 Aerosol Spectral Reflectance

We define the aerosol single-scattering epsilon (SSE) parameter $\varepsilon(\lambda,\lambda_0)$ as the ratio of the aerosol single-scattering reflectance $\rho_{as}(\lambda)$ between two spectral bands [70,142,143], i.e.,

$$\varepsilon(\lambda,\lambda_0) = \frac{\rho_{as}(\lambda)}{\rho_{as}(\lambda_0)} = \frac{\omega_a(\lambda)c_{ext}(\lambda)p_a(\Theta,\lambda)}{\omega_a(\lambda_0)c_{ext}(\lambda_0)p_a(\Theta,\lambda_0)}, \tag{6}$$

where $\omega_a(\lambda)$, $c_{ext}(\lambda)$, and $p_a(\Theta,\lambda)$ are the aerosol single-scattering albedo, the aerosol extinction coefficient, and the aerosol effective scattering phase function (at the scattering angle Θ) [70,142], respectively. The aerosol effective scattering phase function $p_a(\Theta,\lambda)$ includes three scattering contributions [70,82,142,143]: one term for the direct backward scattering from solar to the sensor direction and two terms for the forward scattering through the ocean surface reflections. Specifically, the two forward single-scattering terms include one with the forward single-scattering in the atmosphere toward the ocean surface followed by reflection from the surface and another with first reflection of the direct solar beam from the ocean surface followed by the forward single scattering in the atmosphere [82]. Therefore, the SSE $\varepsilon(\lambda,\lambda_0)$ depends only on the aerosol IOPs (i.e., aerosol model) and not the aerosol optical thickness. It should be noted that aerosol reflectance $\rho_A(\lambda)$ is proportional to (or even approximated by) $\rho_{as}(\lambda) = \omega_a(\lambda) p_a(\Theta,\lambda) \tau_a(\lambda)$, where $\tau_a(\lambda)$ is the aerosol optical thickness.

The SSE parameter can thus be used to characterize the aerosol reflectance spectral variation for various aerosol models [70,142,143]. Figure 6 provides the examples of the SSE $\varepsilon(\lambda,\lambda_0)$ as a function of the wavelength (from the UV to NIR and to various SWIR bands) for the 12 aerosol models (i.e., O99, M50, M70, M90, M99, C50, C70, C90, C99, T50, T90, and T99). Figure 6(a)−(d) shows the SSE $\varepsilon(\lambda,\lambda_0)$ distributions (Eqn (6)) at the reference wavelengths λ_0 of 865, 1240, 1640, and 2130 nm, respectively. They are all for the case of a solar-zenith angle of 60°, sensor-zenith angle of 20°, and relative-azimuth angle of 90°. As expected, the $\varepsilon(\lambda,\lambda_0)$ value has significant variations corresponding to various reference wavelength λ_0 values, showing aerosol spectral reflectance distribution with various aerosol models. Between the O99 and the T50 models (corresponding to the aerosol models for the lowest and the highest $\varepsilon(\lambda,\lambda_0)$ values), Figure 6 shows that the SSE values at the UV band for $\varepsilon(340,865)$, $\varepsilon(340,1240)$, $\varepsilon(340,1640)$, and $\varepsilon(340,2130)$ are in the range of 0.8−2.6, 0.7−4.8, 0.7−9.2, and 0.7−22.6, respectively. In particular, the NIR and SWIR SSE values of $\varepsilon(765,865)$, $\varepsilon(1000,1240)$, $\varepsilon(1240,1640)$, $\varepsilon(1240,2130)$, and $\varepsilon(1640,2130)$, which are used for the selection of aerosol

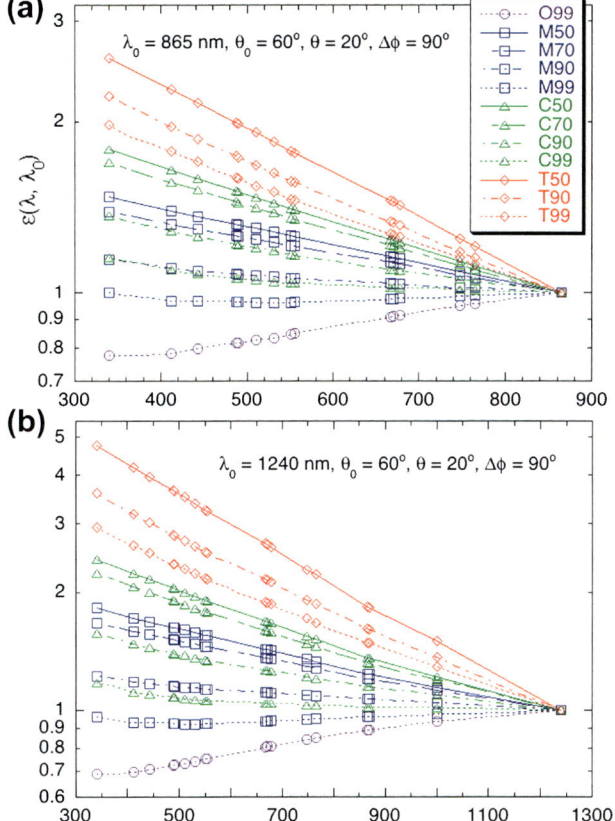

FIGURE 6 Single-scattering epsilon $\varepsilon(\lambda, \lambda_0)$ as a function of the wavelength for the 12 aerosol models and for the reference wavelength at (a) 865 nm, (b) 1240 nm, (c) 1640 nm, and (d) 2130 nm. *Reproduced from Wang (2007) Ref. [73].*

models in atmospheric correction [70,73], are in the range of 0.96–1.21, 0.93–1.50, 0.95–1.94, 0.98–4.76, and 1.04–2.46, respectively. Obviously, there are significantly higher measurement sensitivities in the SSE with the SWIR bands, where there is a substantially larger apart of wavelength distance between two bands, e.g., $\varepsilon(1240,2130)$.

It is quite clear from Figure 6 that spectrally aerosol reflectance has the highest contribution for the T50 model (or in general for the Tropospheric models with small aerosol particles) and the lowest aerosol reflectance contribution for the O99 model (or in general low $\rho_A(\lambda)$ values for the Oceanic and Maritime aerosols). Thus, aerosol reflectance for the T80 model has larger TOA reflectance contributions than those from the M80 aerosol model.

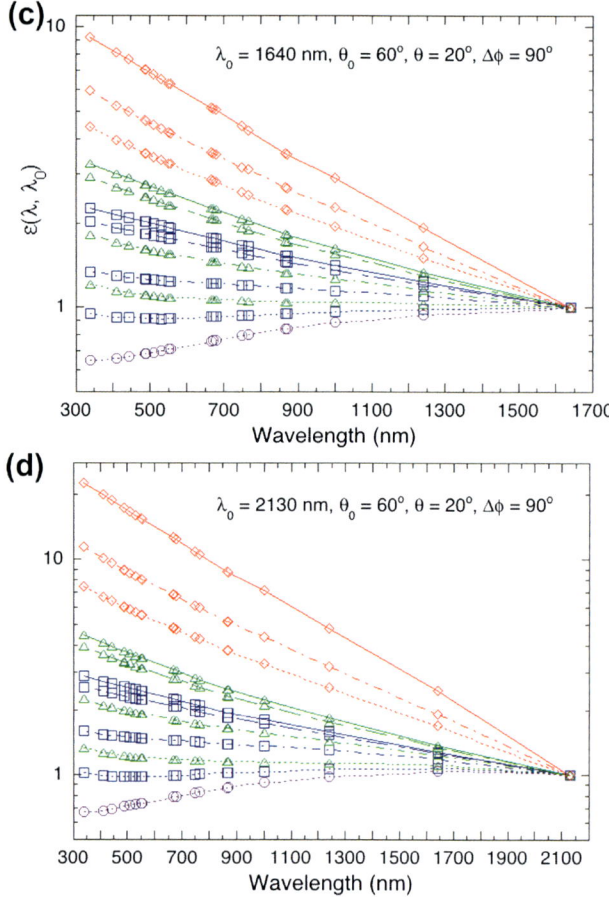

FIGURE 6 cont'd

3.2.4 Atmospheric Path Reflectance $\rho_{path}(\lambda)$

Figure 7(a) provides examples of the simulated TOA path reflectance spectra $\rho_{path}(\lambda)$ from 340 to 2130 nm for the case of the black ocean for aerosol models of the M80 and T80. Again, the M80 and T80 models represent nonabsorbing and weakly absorbing aerosols, respectively. In Figure 7(a), the reflectance spectra $\rho_{path}(\lambda)$ were simulated using the scalar radiative transfer computation (neglecting the polarization effects) for a two-layer atmosphere overlying a flat Fresnel-reflecting ocean surface for the M80 and T80 aerosol models with aerosol optical thicknesses at 865 nm $\tau_a(865)$ of 0.1 and 0.2. This is the case for a specific solar-sensor geometry with the solar-zenith angle of 60°, sensor-zenith angle of 45°, and a relative-azimuth angle of 90°. Figure 7(a) shows that the TOA path

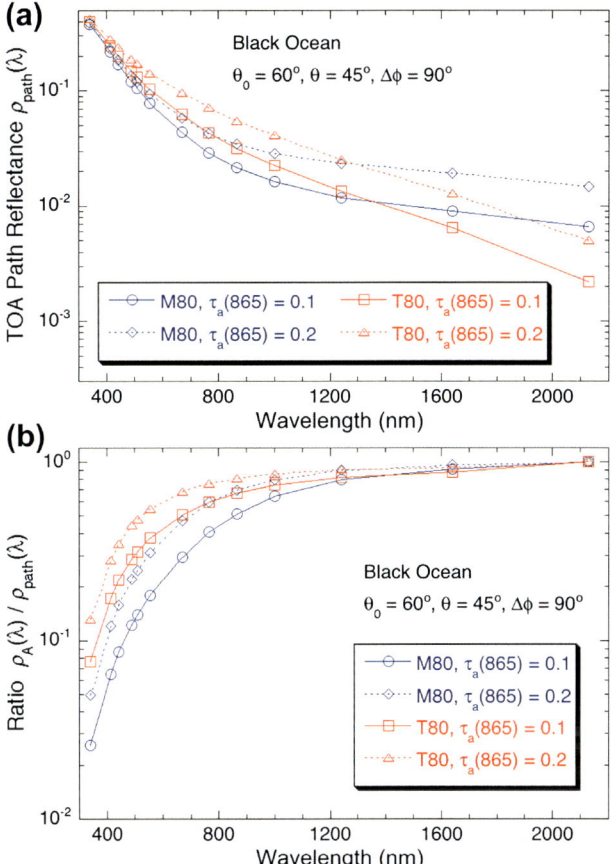

FIGURE 7 Simulated (a) TOA path reflectance $\rho_{path}(\lambda)$ and (b) ratio of the TOA aerosol reflectance to the TOA path reflectance $\rho_A(\lambda)/\rho_{path}(\lambda)$ as a function of the wavelength from 340 to 2130 nm for the case of the black ocean and for aerosol models of the M80 and the T80 with aerosol optical thicknesses at 865 nm of 0.1 and 0.2. *Reproduced from Wang (2007) Ref. [73].*

reflectance simulated with the T80 model has a significant spectral variation than that from the M80 model (as also shown in Figure 6). Indeed, for the T80 model with $\tau_a(865)$ of 0.1, ratio of the TOA reflectance $\rho_{path}(\lambda)$ between wavelengths at 340 and 2130 nm is ~ 200, while for the M80 model, the same reflectance ratio is ~ 60. It is noted that, however, at the wavelength 340 nm, the simulated TOA reflectance $\rho_{path}(\lambda)$ values are similar for various aerosol optical properties (model and aerosol optical thicknesses) due to significantly dominated Rayleigh reflectance contribution at the UV bands (Figure 7(a)).

Figure 7(b) provides the ratio of the TOA aerosol reflectance to the TOA path reflectance $\rho_A(\lambda)/\rho_{path}(\lambda)$ as a function of the wavelength from the UV to the SWIR wavelengths, corresponding to cases in Figure 7(a), where again $\rho_A(\lambda) = \rho_a(\lambda) + \rho_{ra}(\lambda)$. Figure 7(b) shows that at the UV and short visible bands, the sensor-measured signals are really dominated from contributions of air molecules (Rayleigh scattering). For the case of the M80 model with $\tau_a(865)$ of 0.1 and assuming a black ocean (for the TOA path reflectance), the aerosol reflectance $\rho_A(\lambda)$ contributes about 2.6, 6, 9, 12, 18, 29, and 51% in the TOA path reflectance $\rho_{\text{path}}(\lambda)$ for wavelengths at 340, 412, 443, 490, 555, 670, and 865 nm, respectively, while for the T80 model, the corresponding values are about 7.6, 17, 22, 29, 38, 51, and 67%, respectively. With the inclusion of the ocean reflectance contributions in $\rho_t(\lambda)$, the ratio of $\rho_A(\lambda)/\rho_t(\lambda)$ values in the UV and visible wavelengths are even smaller. This underscores the importance of accurately computing the Rayleigh reflectance $\rho_r(\lambda)$, in particular, for the UV and the visible wavelengths.

3.3 Atmospheric Diffuse Transmittance

In the remote sensing of ocean color, we are really after the water-leaving radiance $L_w(\lambda, \theta, \phi)$, the component of the radiance leaving the sea surface that was transmitted through the interface from below the ocean surface (to the above). The radiance just below the sea surface is $L_u(\lambda, \theta', \phi)$, where θ' and θ are related by the law of refraction, i.e., $\sin\theta = m\sin\theta'$. $L_w(\lambda, \theta, \phi)$ and $L_u(\lambda, \theta', \phi)$ are related by

$$L_w(\lambda, \theta, \phi) = \frac{T_f^{(w-a)}(\theta, \theta')}{m^2} L_u(\lambda, \theta', \phi) \tag{7}$$

where $T_f^{(w-a)}(\theta, \theta')$ is the Fresnel transmittance of the air–sea interface (from water to air, "w–a"). The diffuse transmittance $t(\lambda,\xi)$ for the water-leaving radiance $L_w(\lambda,\xi)$ exiting at the TOA direction ξ can be defined as

$$t(\lambda, \xi) = \frac{L_w^{(TOA)}(\lambda, \xi)}{L_w(\lambda, \xi)} \tag{8}$$

where $L_w^{(TOA)}(\lambda, \xi)$ and $L_w(\lambda,\xi)$ are water-leaving radiance at the TOA and ocean surface, respectively. Based on the reciprocity principle, Yang and Gordon (1997) [86] provided a rigorous framework to compute $t(\lambda,\xi)$. They showed that $t(\lambda,\xi)$ is not just a function of the atmospheric properties, but also depends on the angular distribution of $L_w(\lambda,\xi)$ itself [86]. Thus, if one knows $L_w(\lambda,\xi)$ distribution, one can easily compute $t(\lambda,\xi)$ from its definition of Eqn (8). However, $L_w(\lambda,\xi)$ distribution is generally unknown. In practice, simplified assumptions have been used to effect this computation. Assuming that the upwelling radiance just beneath the sea surface is uniform [70], Yang and Gordon (1997) [86] showed that in a reciprocal process for the case with solar-zenith angle θ_0, the atmospheric diffuse transmittance of $t(\lambda,\theta_0)$ can be computed as

$$t(\lambda, \theta_0) = \frac{E_d^-(\lambda, \theta_0)}{F_0(\lambda)\cos\theta_0 T_f^{(a-w)}(\theta_0)} \quad (9)$$

where $E_d^-(\lambda, \theta_0)$ is the downward irradiance just below the sea surface specified when the TOA is illuminated with the solar beam (irradiance) $F_0(\lambda)$ with solar-zenith angle θ_0, and $T_f^{(a-w)}(\theta_0)$ is the Fresnel transmittance of the air—sea interface (from air to water, "a—w"). It is found that the assumption of uniform beneath surface water-leaving radiance distribution in computing the atmospheric diffuse transmittance only introduced negligible error ($< \sim 1\%$) for the satellite-derived $nL_w(\lambda)$ as long as sensor-zenith angle $<60°$ [144]. There are also simple approximation formulas to compute atmospheric diffuse transmittance reasonably accurately [145,146]. For a pure Rayleigh atmosphere, the atmospheric diffuse transmittance $t_R(\lambda,\theta)$ can be approximated as

$$t_R(\lambda, \theta) = \exp[-C_R(\lambda, \theta)\tau_R(\lambda)/\cos\theta] \quad (10)$$

where $\tau_R(\lambda)$ is the Rayleigh optical thickness and $C_R(\lambda,\theta) \approx 0.5$ [145] and can be further refined to improve $t_R(\lambda,\theta)$ accuracy [146]. For an atmosphere composed by both air molecules and aerosols bounded by a Fresnel-reflecting ocean surface, the diffuse transmittance $t(\lambda,\theta)$ can be approximated by

$$\begin{aligned}t(\lambda, \theta) &= t_R(\lambda, \theta)t_a(\lambda, \theta) \\ t_a(\lambda, \theta) &= \exp[-a_0(\lambda)(1 + \omega_a(\lambda)C_a(\lambda, \theta))/\cos\theta]\end{aligned} \quad (11)$$

where $a_0(\lambda)$ is related to aerosol optical properties [146], $\omega_a(\lambda)$ is aerosol single-scattering albedo, and $C_a(\lambda,\theta)$ is a coefficient to best fit the model [146]. Comparison results show that approximations are quite accurate with errors within $\sim 0.5\%$ for most of cases [146].

3.4 Simulated and Satellite-Measured TOA Radiances

3.4.1 Simulated Typical TOA Radiances

Simulations have been carried out to estimate the typical TOA radiance spectra over global open oceans using Eqn (3) with assumptions that sun glint and whitecaps terms are ignored, i.e.,

$$\begin{aligned}L_t(\lambda) &= L_{path}(\lambda) + t(\lambda)t_0(\lambda)\cos\theta_0\, nL_w(\lambda), \text{ or} \\ \rho_t(\lambda) &= \rho_{path}(\lambda) + t(\lambda)\,t_0(\lambda)\rho_{wN}(\lambda),\end{aligned} \quad (12)$$

where $L_t(\lambda)$ and $\rho_t(\lambda)$ are the TOA radiance and the corresponding reflectance, respectively. Atmospheric diffuse transmittance terms $t_0(\lambda)$ and $t(\lambda)$ are calculated using the Yang and Gordon (1997) [86] formula with appropriate aerosol models (M80 and T80). We ignore the contributions of $L_{wc}(\lambda)$ and $L_g(\lambda)$ because for ocean-color remote sensing, the term $L_g(\lambda)$ is generally masked out, while $L_{wc}(\lambda)$ contribution was found to be less important than originally expected [83,91].

3.4.1.1 Examples of Simulated TOA Radiances

Figure 8 shows some examples of the simulated TOA reflectance $\rho_t(\lambda)$ as a function of the wavelength for various cases. Figure 8(a) provides examples of the simulated $\rho_t(\lambda)$ for three typical water types, i.e., the Case-1 water

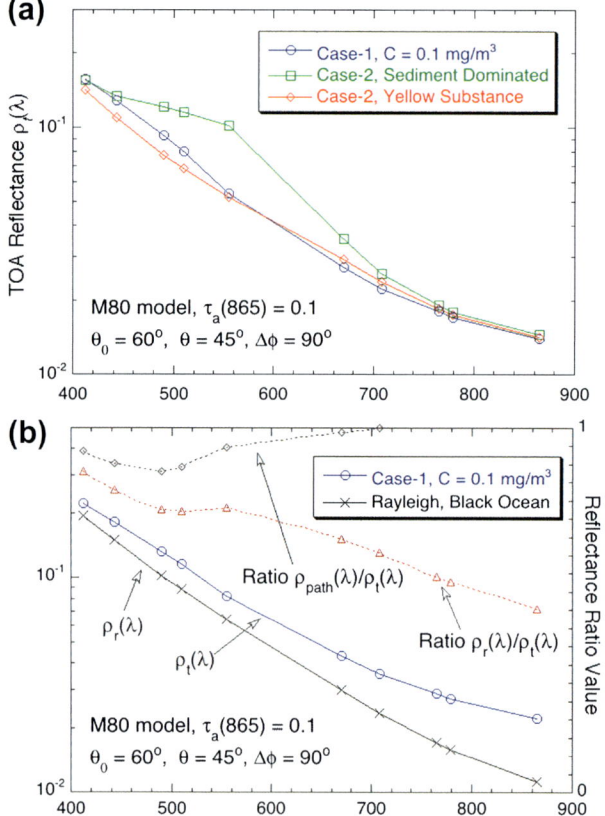

FIGURE 8 Examples of the simulated TOA reflectance $\rho_t(\lambda)$ as a function of the wavelength for the M80 aerosol model with aerosol optical thickness at 865 nm $\tau_a(865)$ of 0.1 and for (a) cases with both Case-1 and Case-2 waters (i.e., $\rho_{wN}(\lambda)$ data from Figure 2), (b) Case-1 water ($\rho_{wN}(\lambda)$ from Figure 2(a)) with chlorophyll concentration of 0.1 mg/m³, (c) Case-2 with sediment-dominated waters ($\rho_{wN}(\lambda)$ from Figure 2b), and (d) Case-2 with yellow substance-dominated waters ($\rho_{wN}(\lambda)$ from Figure 2b). Plots (b)–(d) also present results of ratio values for $\rho_{path}(\lambda)/\rho_t(\lambda)$ and $\rho_r(\lambda)/\rho_t(\lambda)$ (scaled in the right side). *These are reproduced from IOCCG (2010) Ref. [61].*

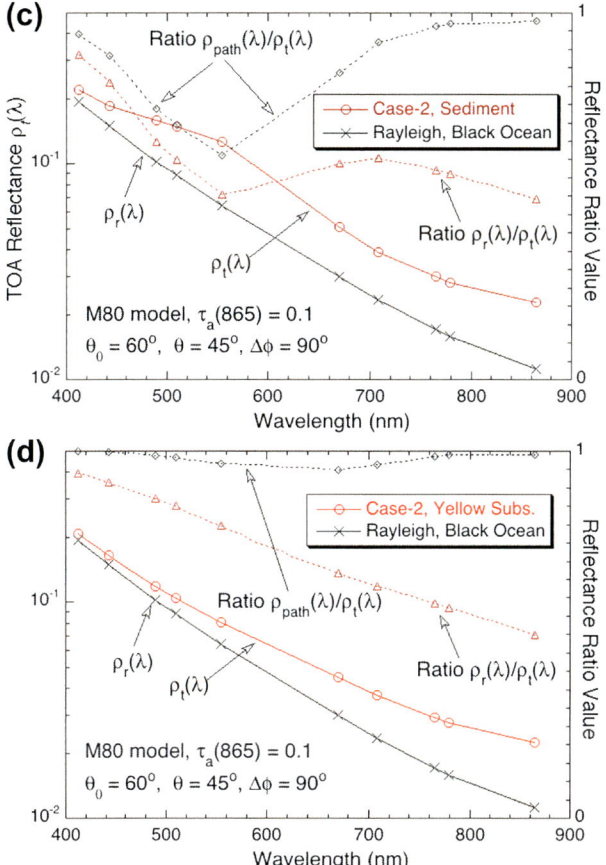

FIGURE 8 cont'd

with chlorophyll-a concentration of 0.1 mg m^{-3}, the sediment-dominated water (Case-2), and the yellow substance (or CDOM)-dominated water (Case-2), for the M80 aerosols with aerosol optical thickness at 865 nm $\tau_a(865)$ of 0.1, while Figure 8(b)–(d) provides the TOA total reflectance $\rho_t(\lambda)$ and the Rayleigh reflectance contribution $\rho_r(\lambda)$, as well as more detailed reflectance contribution partitions for each of these three cases. These are all for a case of the solar-zenith angle of 60°, sensor-zenith angle of 45°, and relative-azimuth angle of 90°. The right side of Figure 8(b)–(d) is the scale for the reflectance ratio value indicated in the plot. For example, results in Figure 8(b) show that for the case of the M80 model with $\tau_a(865)$ of 0.1 and with chlorophyll-a concentration of 0.1 mg m^{-3}, the TOA atmospheric path reflectance $\rho_{path}(\lambda)$ contributes about 93.6, 90.2, 87.9, 89.2, 94.5, and 98.7% in the TOA reflectance for wavelengths at 412, 443, 490, 510, 555, and

670 nm, respectively. In other words, the corresponding TOA water-leaving reflectance $t(\lambda)t_0(\lambda)\rho_{wN}(\lambda)$ contributes about 6.4, 9.8, 12.1, 10.8, 5.5, and 1.3% in the sensor-measured TOA reflectance at these wavelengths. On the other hand, for the sediment-dominated waters, Figure 8(c) shows that the TOA water-leaving reflectance $t(\lambda)t_0(\lambda)\rho_{wN}(\lambda)$ contributes about 5.8, 11.8, 26.3, 30.6, 39.0, 16.4, 8.2, 3.7, 3.0, and 2.4% in the TOA reflectance for wavelengths at 412, 443, 490, 510, 555, 670, 708, 765, 779, and 865 nm, respectively, while for the yellow substance-dominated waters, Figure 8(d) shows that these values are reduced dramatically to 0.2, 0.4, 1.2, 1.7, 3.6, 5.3, 3.8, 1.5, 1.2, and 1.0%, respectively.

Results in Figure 8 demonstrate that the ocean reflectance contributions only comprise a small portion of the sensor-measured signals in the visible wavelengths. In particular, for the yellow substance (or CDOM)-dominated waters, the ocean reflectance contributions to the TOA reflectance are very small. Therefore, for such waters, sufficiently accurate atmospheric correction in the blue and green bands is generally extremely difficult if not impossible. Results in Figure 8 also indicate the importance of the vicarious calibration [147–150] for the satellite ocean-color remote sensing. It requires the calibration accuracy of ~0.1%. Examples of the Case-2 waters in Figure 8(c) and (d) also show that the NIR black ocean assumption is not valid for these ocean waters [151–156].

3.4.1.2 Simulated Typical TOA Radiances over Open Oceans

To derive realistic typical TOA radiances, simulations were carried out for aerosol models of the M80 and T80 with aerosol optical thickness of 0.1 at 865 nm, chlorophyll-a concentration of 0.1 mg m^{-3} (for $nL_w(\lambda)$ computation), solar-zenith angles of 0–70° at a step of 5°, sensor-zenith angles of 0–60° at a step of 5°, and relative-azimuth angles of 0–180° at a step of 10°, and for spectral wavelengths from the UV at 340 nm to the SWIR at 1000, 1240, 1640, and 2130 nm. Figure 9 shows the histograms of the simulated TOA $L_t(\lambda)$ for wavelengths of 340, 360, 380, 412, 443, 490, 510, 555, 670, 678, 750, 865, 1000, and 1240 nm (Figure 9(a)–(c)). $L_t(\lambda)$ data for the SWIR 1640 and 2130 nm were also simulated but not shown in Figure 9. In addition, Figure 9(d) shows $L_t(\lambda)$ spectra as a function of the wavelength for a specific solar-sensor geometry, i.e., solar-zenith angle of 60°, sensor-zenith angle of 45°, and relative-azimuth angle of 90°, for aerosol models of the M80 and T80 with aerosol optical thicknesses of 0.1 and 0.2 at 865 nm, showing $L_t(\lambda)$ variations with aerosol models (M80 vs. T80) as well as aerosol optical thickness (0.1 vs. 0.2). Results in Figure 9 show the higher $L_A(\lambda)$ (thus higher $L_t(\lambda)$) contributions from the aerosol model T80 than those of the M80 (as expected), showing double mode values in histograms for some spectral bands where larger mode value corresponding to $L_t(\lambda)$ simulated from the T80 aerosols (e.g., Figure 9(b)). Therefore, the simulated TOA radiance $L_t(\lambda)$ spectra from the UV to SWIR bands include almost all practicable solar-sensor

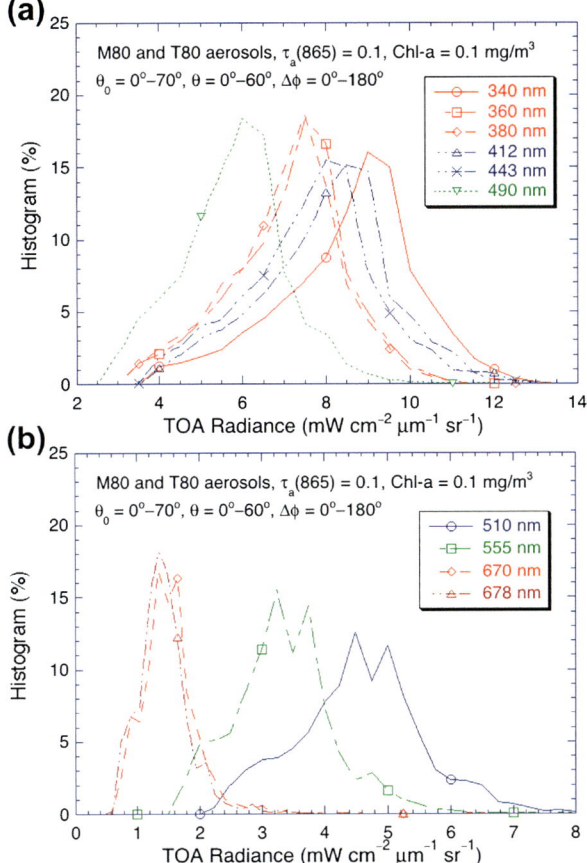

FIGURE 9 Histograms of the simulated TOA radiances at wavelengths of (a) 340, 360, 380, 412, 443, and 490 nm, (b) 510, 555, 670, and 678 nm, and (c) 750, 865, 1000, and 1240 nm for cases with solar-zenith angles of 0–70°, sensor-zenith angles of 0–60°, and relative-azimuth angles of 0–180°, and for chlorophyll-a of 0.1 mg m^{-3} and aerosol models of the M80 and T80 with aerosol optical thickness at 865 nm of 0.1. Plot (d) shows the TOA radiance spectrum as a function of the wavelength for a specific solar-sensor geometry using aerosol models of the M80 and T80 with aerosol optical thicknesses at 865 nm of 0.1 and 0.2.

geometry cases with typical aerosols and chlorophyll-a concentration for global open oceans and for clear atmosphere. The simulated mode values from histograms can be considered as typical satellite-measured TOA radiance spectra.

Table 1 provides the simulated typical TOA $L_t(\lambda)$ radiance values from the UV to SWIR wavelengths, which were obtained from histogram mode values as shown in Figure 9(a)–(c). Table 1 also provides the solar irradiance data [76] for these spectral wavelengths.

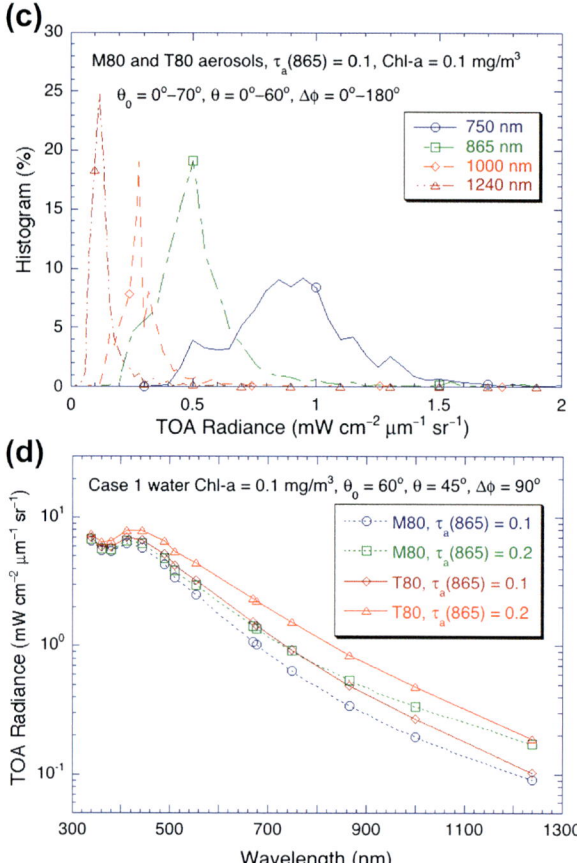

FIGURE 9 cont'd

3.4.2 Simulated Typical TOA Radiances Compared with Satellite Data

The simulated typical TOA radiance data can be compared with satellite measurements. Figure 10 shows examples of results from simulated typical TOA radiance $L_t(\lambda)$ as a function of the wavelength from 340 to 2130 nm, compared with those from SeaWiFS (412−865 nm) and VIIRS (410−2250 nm) measurements. For simulations, the typical TOA radiance spectra discussed in the previous sections are used. These data were derived from histogram distributions with assumption of no gas absorption at wavelengths of 345, 360, 380, 412, 443, 490, 510, 555, 670, 678, 750, 765, 865, 1000, 1240, 1640, and 2130 nm. SeaWiFS and VIIRS data were acquired on September 21, 2002 and January 3, 2013, respectively, from entire global data for clear atmosphere over oceans. Data from land, clouds, and sun glint were

TABLE 1 Typical TOA Radiances from the UV to SWIR Wavelengths Derived From Modes of Simulated $L_t(\lambda)$ Histogram Distributions

Wavelength λ (nm)	Solar Irradiance $F_0(\lambda)^a$	Typical TOA Radiance $L_t(\lambda)^b$
340	107.89	9.00
360	104.23	7.50
380	117.14	7.25
412	167.28	8.25
443	195.41	7.75
490	202.60	6.25
510	189.87	4.75
555	188.26	3.50
670	151.60	1.50
678	148.16	1.40
750	126.68	0.90
865	96.00	0.52
1000	72.64	0.28
1240	45.24	0.12
1640	22.70	0.05
2130	9.63	0.016

^{a}Unit of $mW\ cm^{-2}\ \mu m^{-1}$.
^{b}Unit of $mW\ cm^{-2}\ \mu m^{-1}\ sr^{-1}$.

masked out, and then mean TOA radiance values from SeaWiFS and VIIRS spectral bands were obtained. For SeaWiFS, the TOA radiances are corresponding to the SeaWiFS nominal center wavelengths at 412, 443, 490, 510, 555, 670, 765, and 865 nm, while for VIIRS, the TOA radiance data are for VIIRS nominal band center wavelengths of 410, 443, 486, 551, 671, 745, 862, 1238, 1610, and 2250 nm. It is particularly noted that the TOA radiances from VIIRS SWIR bands are also included. Thus, typical TOA radiance spectra from both SeaWiFS and VIIRS were derived from entire one-day measurements over global oceans with clear atmosphere, applicable to satellite ocean-color remote sensing. Results in Figure 10 show that simulated typical TOA radiance data are quite consistent with those from satellite measurements.

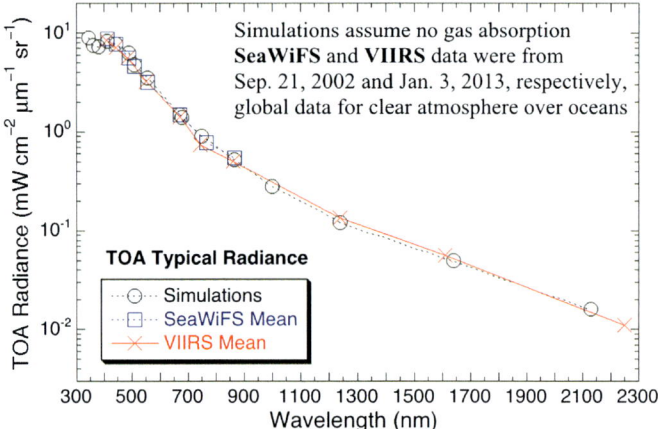

FIGURE 10 Simulated typical TOA radiance over ocean as a function of the wavelength, compared with satellite measurements from SeaWiFS and VIIRS for global daily data from clear atmosphere over oceans on September 21, 2002 and January 3, 2013, respectively.

4. SUMMARY

In this chapter, satellite sensor-measured TOA radiances over open oceans and coastal and inland waters are described with some detailed discussions for various effects from atmosphere, ocean surface, and ocean waters. It is well known that over ~90% signals of the sensor-measured TOA radiances are from atmosphere and ocean surface. For a typical open ocean case with a typical clear atmosphere, the atmospheric path radiance contributed about 94, 90, 88, 89, 95, and 99% at wavelengths of 412, 443, 490, 510, 555, and 670 nm, respectively. Thus, ocean radiance contributions are generally < ~10%. On the other hand, for the sediment-dominated waters, water-leaving radiance contributions from ocean can be quite significant (depending on sediment loading), in particular, for $nL_w(\lambda)$ at the green and red bands. For example, for the case in Figure 2(b), the TOA water-leaving reflectance $t(\lambda)t_0(\lambda)\rho_{wN}(\lambda)$ contributes about 6, 12, 26, 31, 39, 16, 8, 4, 3, and 2% for wavelengths of 412, 443, 490, 510, 555, 670, 708, 765, 779, and 865 nm, respectively. For the yellow substance (or CDOM)-dominated waters (Figure 2(b)), however, it shows that $t(\lambda)t_0(\lambda)\rho_{wN}(\lambda)$ contributions are significantly reduced to <1% for 412 and 443 nm bands and only ~1–5% for green to red wavelengths. Thus, we can do well for satellite ocean-color remote sensing for global open oceans and also for sediment-dominated waters, but it may be quite difficult to get sufficiently accurate $\rho_{wN}(\lambda)$ (particularly at the short blue bands) for yellow substance-dominated waters.

Various atmospheric and water surface effects on the TOA radiances are described and discussed in this chapter, including aerosol models, atmospheric diffuse transmittance, and ocean BRDF effects. Aerosol spectral reflectance distribution can be derived using the SSE $\varepsilon(\lambda,\lambda_0)$ parameter, which depends only on the aerosol IOPs (aerosol model) and not on aerosol optical thickness. In fact, the SSE is an important parameter in atmospheric correction for deriving corresponding aerosol models and relating it to aerosol multiple scattering effects. For ocean-color remote sensing, it is really important to accurately estimate aerosol reflectance (radiance), which is proportional to the product of aerosol single-scattering albedo, aerosol effective phase function, and aerosol optical thickness.

The typical TOA radiances for open oceans from 340 to 2130 nm are derived from histogram data analyses, including all applicable solar-sensor geometries with the M80 and T80 aerosols and for a typical aerosol optical thickness value over global open oceans. These typical TOA radiances are compared quite well with those from SeaWiFS and VIIRS measurements, including radiances at visible, NIR, and SWIR bands. Thus, these typical TOA radiance values can be used for various applications, e.g., for sensor SNR values estimations.

Finally, an extensive list of references related to various topics discussed in this chapter is provided. Interesting readers are encouraged to read proper references for further details.

DISCLAIMER

The views, opinions, and findings contained in this chapter are those of the authors and should not be construed as an official NOAA or U.S. Government position, policy, or decision.

REFERENCES

[1] M.J. Behrenfeld, J.T. Randerson, C.R. McClain, G.C. Feldman, S.O. Los, C.J. Tucker, P.G. Falkowski, C.B. Field, R. Frouin, W.E. Esaias, D.D. Kolber, N.H. Pollack, Biospheric primary production during an ENSO transition, Science 291 (2001) 2594−2597.

[2] F.P. Chavez, P.G. Strutton, C.E. Friederich, R.A. Feely, G.C. Feldman, D.C. Foley, M.J. McPhaden, Biological and chemical response of the equatorial Pacific ocean to the 1997−98 El Nino, Science 286 (1999) 2126−2131.

[3] W. Shi, M. Wang, Characterization of global ocean turbidity from moderate resolution imaging spectroradiometer ocean color observations, J. Geophys. Res. 115 (2010) C11022. http://dx.doi.org/10.1029/2010JC006160.

[4] M.J. Behrenfeld, R.T. O'Malley, D.A. Siegel, C.R. McClain, J.L. Sarmiento, G.C. Feldman, A.J. Milligan, P.G. Falkowski, R.M. Letelier, E.S. Boss, Climate-driven trends in contemporary ocean productivity, Nature 444 (2006) 752−755.

[5] W. Shi, M. Wang, Observations of a Hurricane katrina-induced phytoplankton bloom in the Gulf of Mexico, Geophy. Res. Lett. 34 (2007) L11607. http://dx.doi.org/10.1029/2007GL029724.

[6] W. Shi, M. Wang, Satellite observations of flood-driven mississippi river plume in the spring of 2008, Geophy. Res. Lett. 36 (2009) L07607. http://dx.doi.org/10.1029/2009GL037210.

[7] X. Liu, M. Wang, W. Shi, A study of a Hurricane katrina-induced phytoplankton bloom using satellite observations and model simulations, J. Geophys. Res. 114 (2009) C03023. http://dx.doi.org/10.1029/2008JC004934.

[8] N.D. Walker, R.R. Leben, S. Balasubramanian, Hurricane-forced upwelling and chlorophyll a enhancement within cold-core cyclones in the Gulf of Mexico, Geophy. Res. Lett. 32 (2005) L18610. http://dx.doi.org/10.1029/2005GL023716.

[9] W. Shi, M. Wang, Satellite observations of environmental changes from the Tonga volcano eruption in the southern tropical Pacific, Int. J. Remote Sens. 32 (2011) 5785−5796.

[10] P. Cipollini, D. Cromwell, P.G. Challenor, S. Raffaglio, Rossby waves detected in global ocean colour data, Geophys. Res. Lett. 28 (2001) 323−326.

[11] D.B. Chelton, P. Gaube, M.G. Schlax, J.J. Early, R.M. Samelson, The influence of nonlinear mesoscale eddies on near-surface oceanic chlorophyll, Science 334 (2011) 328−332.

[12] S.M. Babin, J.A. Carton, T.D. Dickey, J.D. Wiggert, Satellite evidence of hurricane-induced phytoplankton blooms in an oceanic desert, J. Geophys. Research-Oceans 109 (2004). http://dx.doi.org/10.1029/2003jc001938. Artn C03043.

[13] C. Hu, A novel ocean color index to detect floating algae in the global oceans, Remote Sens. Environ. 113 (2009) 2118−2129.

[14] W. Shi, M. Wang, Green macroalgae blooms in the yellow sea during the spring and summer of 2008, J. Geophys. Res. 114 (2009) C12010. http://dx.doi.org/10.1029/2009JC005513.

[15] S. Nakamoto, S.P. Kumar, J.M. Oberhuber, K. Muneyama, R. Frouin, Chlorophyll modulation of sea surface temperature in the arabian sea in a mixed-layer isopycnal general circulation model, Geophys. Res. Lett. 27 (2000) 747−750.

[16] S. Nakamoto, S.P. Kumar, J.M. Oberhuber, J. Ishizaka, K. Muneyama, R. Frouin, Response of the equatorial Pacific to chlorophyll pigment in a mixed layer isopycnal ocean general circulation model, Geophys. Res. Lett. 28 (2001) 2021−2024.

[17] B. Subrahmanyam, K. Ueyoshi, J.M. Morrison, Sensitivity of the Indian ocean circulation to phytoplankton forcing using an ocean model, Remote Sen. Environ. 112 (2008) 1488−1496.

[18] M.J. Behrenfeld, P.G. Falkowski, Photosynthetic rates derived from satellite-based chlorophyll concentration, Limnol. Oceanogr. 42 (1997) 1−20.

[19] L.W. Harding Jr., M.E. Mallonee, E.S. Perry, Toward a predictive understanding of primary productivity in a temperate, partially stratified estuary, Estuarine Coastal Shelf Sci. 55 (2002) 437−463.

[20] S. Son, M. Wang, L.W. Harding Jr., Satellite-measured net primary production in the Chesapeake Bay, Remote Sens. Environ. 144 (2014) 109−119.

[21] S. Son, M. Wang, J. Shon, Satellite observations of optical and biological properties in the korean dump site of the Yellow sea, Remote Sens. Environ. 115 (2011) 562−572.

[22] S. Son, M. Wang, Environmental responses to a land reclamation project in South Korea, Eos Trans. AGU 90 (2009) 398−399. http://dx.doi.org/10.1029/2009eo440002.

[23] N.P. Nezlin, P.M. DiGiacomo, D.W. Diehl, B.H. Jones, S.C. Johnson, M.J. Mengel, K.M. Reifel, J.A. Warrick, M. Wang, Stormwater plume detection by MODIS imagery in the southern California coastal ocean, Estuarine Coastal Shelf Sci. 80 (2008) 141−152. http://dx.doi.org/10.1016/j.ecss.2008.07.012.

[24] S. Son, M. Wang, Water properties in Chesapeake Bay from MODIS-Aqua measurements, Remote Sens. Environ. 123 (2012) 163−174.

[25] J.A. Warrick, D.A. Fong, Dispersal scaling from the world's rivers, Geophy. Res. Lett. 31 (2004) L04301. http://dx.doi.org/10.1029/2003GL019114.

[26] C. Hu, F.E. Muller-Karger, G.A. Vargo, M.B. Neely, E. Johns, Linkages between coastal runoff and the florida keys ecosystem: a study of a dark plume event, Geophy. Res. Lett. 31 (2004) L15307. http://dx.doi.org/10.1029/2004GL020382.

[27] L.W. Harding Jr., E.S. Perry, Long-term increase of phytoplankton biomass in Chesapeake Bay, 1950−1994, Mar. Ecol. Prog. Ser. 157 (1997) 39−52.

[28] D. Tang, H. Kawamura, H. Doan-Nhu, W. Takahashi, Remote sensing oceanography of a harmful algal bloom off the coast of southeastern Vietnam, J. Geophys. Res. 109 (2004). http://dx.doi.org/10.1029/2003JC002045.

[29] R.P. Stumpf, M.C. Tomlinson, J.A. Calkins, B. Kirkpatrick, K. Fisher, K. Nierenberg, R. Currier, T.T. Wynne, Skill assessment for an operational algal bloom forecast system, J. Marine Syst. 76 (2009) 151−161.

[30] G.A. Carvalho, P.J. Minnett, V.F. Banzon, W. Baringer, C.A. Heil, Long-term evaluation of three satellite ocean color algorithms for identifying harmful algal blooms (Karenia brevis) along the west coast of Florida: a matchup assessment, Remote Sens. Environ. 115 (2011) 1−18.

[31] M. Wang, W. Shi, J. Tang, Water property monitoring and assessment for China's inland Lake Taihu from MODIS-aqua measurements, Remote Sens. Environ. 115 (2011) 841−854.

[32] C. Hu, Z. Lee, R. Ma, K. Yu, D. Li, S. Shang, Moderate resolution imaging spectroradiometer (MODIS) observations of cyanobacteria blooms in taihu Lake, China, J. Geophys. Res. 115 (2010) C04002. http://dx.doi.org/10.1029/2009JC005511.

[33] W. Shi, M. Wang, W. Guo, "Long-term hydrological changes of the aral sea observed by satellites", J. Geophys. Res. Oceans 119 (2014) 3313−3326. http://dx.doi.org/10.1002/2014JC009988.

[34] G.K. Clarke, G.C. Ewing, C.J. Lorenzen, Spectra of backscattered light from the sea obtained from aircraft as a measure of chlorophyll, Science 167 (1970) 1119−1121.

[35] G.L. Clarke, G.C. Ewing, Remote spectroscopy of the sea for biological production studies, in: N.G. Jerlov, E. Steemann-Nielsen (Eds.), Optical Aspects of Oceanography, Academic Press, London, New York, 1974, pp. 389−413.

[36] R.W. Austin, The remote sensing of spectral radiance from below the ocean surface, in: N.G. Jerlov, E.S. Nielsen (Eds.), Optical Aspects of Oceanography, Academic, San Diego, Calif., 1974, pp. 317−344.

[37] J.E. Tyler, R.C. Smith, Measurements of Spectral Irradiance Underwater, Gordon and Breach Science Publishers, 1970.

[38] H.R. Gordon, O.B. Brown, M.M. Jacobs, Computed relationship between the inherent and apparent optical properties of a flat homogeneous ocean, Appl. Opt. 14 (1975) 417−427.

[39] A. Morel, L. Prieur, Analysis of variations in ocean color, Limnol. Oceanogr. 22 (1977) 709−722.

[40] R.C. Smith, K.S. Baker, The bio-optical state of ocean waters and remote sensing, Limnol. Oceanogr. 23 (1978) 247−259.

[41] H.R. Gordon, D.K. Clark, J.L. Mueller, W.A. Hovis, Phytoplankton pigments from the Nimbus-7 coastal zone color scanner: comparisons with surface measurements, Science 210 (1980) 63−66.

[42] W.A. Hovis, D.K. Clark, F. Anderson, R.W. Austin, W.H. Wilson, E.T. Baker, D. Ball, H.R. Gordon, J.L. Mueller, S.T.E. Sayed, B. Strum, R.C. Wrigley, C.S. Yentsch, Nimbus 7 coastal zone color scanner: system description and initial imagery, Science 210 (1980) 60–63.

[43] H.R. Gordon, A. Morel, Remote Assessment of Ocean Color for Interpretation of Satellite Visible Imagery: A Review, Springer-Verlag, New York, 1983.

[44] S.B. Hooker, W.E. Esaias, G.C. Feldman, W.W. Gregg, C.R. McClain, An overview of SeaWiFS and ocean color, in: S.B. Hooker, E.R. Firestone (Eds.), SeaWiFS Technical Report Series, Vol. 1, NASA Goddard Space Flight Center, Greenbelt, Maryland, 1992. NASA Tech. Memo. 104566.

[45] C.R. McClain, G.C. Feldman, S.B. Hooker, An overview of the SeaWiFS project and strategies for producing a climate research quality global ocean bio-optical time series, Deep-Sea Res. Part II-Topical Stud. Oceanography. 51 (2004) 5–42.

[46] C.R. McClain, A decade of satellite ocean color observations, Annu. Rev. Marine Sci. 1 (2009) 19–42.

[47] W.E. Esaias, M.R. Abbott, I. Barton, O.B. Brown, J.W. Campbell, K.L. Carder, D.K. Clark, R.L. Evans, F.E. Hodge, H.R. Gordon, W.P. Balch, R. Letelier, P.J. Minnet, An overview of MODIS capabilities for ocean science observations, IEEE Trans. Geosci. Remote Sens. 36 (1998) 1250–1265.

[48] V.V. Salomonson, W.L. Barnes, P.W. Maymon, H.E. Montgomery, H. Ostrow, MODIS: advanced facility instrument for studies of the Earth as a system, IEEE Trans. Geosci. Rem. Sens. 27 (1989) 145–153.

[49] M. Rast, J.L. Bezy, S. Bruzzi, The ESA medium resolution imaging spectrometer MERIS a review of the instrument and its mission, Int. J. Remote Sens. 20 (1999) 1681–1702.

[50] P.Y. Deschamps, F.M. Bréon, M. Leroy, A. Podaire, A. Bricaud, J.C. Buriez, G. Sèze, The POLDER mission: instrument characteristics and scientific objectives, IEEE Trans. Geosc. Rem. Sens. 32 (1994) 598–615.

[51] J. Tanii, T. Machida, H. Ayada, Y. Katsuyama, J. Ishida, N. Iwasaki, Y. Tange, Y. Miyachi, R. Sato, Ocean color and temperature scanner (OCTS) for ADEOS, SPIE 1490 (1991) 200–206.

[52] M. Wang, B.A. Franz, Comparing the ocean color measurements between MOS and SeaWiFS: a vicarious intercalibration approach for MOS, IEEE Trans. Geosci. Remote Sens. 38 (2000) 184–197.

[53] M. Wang, A. Isaacman, B.A. Franz, C.R. McClain, Ocean color optical property data derived from the Japanese ocean color and temperature scanner and the french polarization and directionality of the Earth's reflectances: a comparison study, Appl. Opt. 41 (2002) 974–990.

[54] G. Zimmermann, A. Neumann, The spaceborne imaging spectrometer MOS for ocean remote sensing, in: The 1st International Workshop on MOS-IRS and Ocean Color, 1997, pp. 1–9.

[55] C.F. Schueler, J.E. Clement, P.E. Ardanuy, C. Welsch, F. DeLuccia, H. Swenson, NPOESS VIIRS sensor design overview, in: Earth Observing Systems VI, Proc. SPIE, vol. 4483, 2002. http://dx.doi.org/10.1117/12.453451.

[56] M. Wang, X. Liu, L. Tan, L. Jiang, S. Son, W. Shi, K. Rausch, K. Voss, Impact of VIIRS SDR performance on ocean color products, J. Geophys. Res. Atmos. 118 (2013) 10347–10360. http://dx.doi.org/10.1002/jgrd.50793.

[57] F.X. Kneizys, E.P. Shettle, L.W. Abreu, J.H. Chetwynd, G.P. Anderson, W.O. Gallery, J.E.A. Selby, S.A. Clough, Users Guide to LOWTRAN-7, Air Force Geophysics Laboratory, 1988. AFGL-TR-88–0177.

[58] NOAA, NASA, USAF, U.S. Standard Atmosphere, 1976, U.S. Government Printing Office, Washington, D.C., 1976.
[59] M. Wang, W. Shi, Sensor noise effects of the SWIR bands on MODIS-derived ocean color products, IEEE Trans. Geosci. Remote Sens. 50 (2012) 3280–3292.
[60] X. Xiong, J. Sun, X. Xie, W.L. Barnes, V.V. Salomonson, On-orbit calibration and performance of aqua MODIS reflective solar bands, IEEE Trans. Geosci. Remote Sens. 48 (2010) 535–545.
[61] IOCCG, Atmospheric correction for remotely-sensed ocean-colour products, No. 10, in: M. Wang (Ed.), Reports of International Ocean-Colour Coordinating Group, IOCCG, Dartmouth, Canada, 2010.
[62] H.R. Gordon, Removal of atmospheric effects from satellite imagery of the oceans, Appl. Opt. 17 (1978) 1631–1636.
[63] H.R. Gordon, D.K. Clark, Atmospheric effects in the remote sensing of phytoplankton pigments, Boundary-Layer Meteorol. 18 (1980) 299–313.
[64] A. Morel, In-water and remote measurements of ocean color, Boundary-Layer Meteorol. 18 (1980) 177–201.
[65] H.R. Gordon, A preliminary assessment of the NIMBUS-7 CZCS atmospheric correction algorithm in a horizontally inhomogenous atmosphere, in: J.F.R. Gower (Ed.), Oceanography from Space, Plenum Press, New York and London, 1980, pp. 281–294.
[66] R.C. Smith, W.H. Wilson, Ship and satellite bio-optical research in the California Bight, in: J.F.R. Gower (Ed.), Oceanography from Space, Plenum Press, New York and London, 1981, pp. 281–294.
[67] A. Bricaud, A. Morel, Atmospheric correction and interpretation of marine radiances in CZCS imagery: use of a reflectance model, Oceanol. Acta SP (1987) 33–50.
[68] H.R. Gordon, D.K. Clark, Clear water radiances for atmospheric correction of coastal zone color scanner imagery, Appl. Opt. 20 (1981) 4175–4180.
[69] H.R. Gordon, D.J. Castaño, Coastal zone color scanner atmospheric correction algorithm: multiple scattering effects, Appl. Opt. 26 (1987) 2111–2122.
[70] H.R. Gordon, M. Wang, Retrieval of water-leaving radiance and aerosol optical thickness over the oceans with SeaWiFS: a preliminary algorithm, Appl. Opt. 33 (1994) 443–452.
[71] H.R. Gordon, J.W. Brown, R.H. Evans, Exact Rayleigh scattering calculations for use with the Nimbus-7 coastal zone color scanner, Appl. Opt. 27 (1988) 862–871.
[72] M. Wang, Aerosol polarization effects on atmospheric correction and aerosol retrievals in ocean color remote sensing, Appl. Opt. 45 (2006) 8951–8963.
[73] M. Wang, Remote sensing of the ocean contributions from ultraviolet to near-infrared using the shortwave infrared bands: simulations, Appl. Opt. 46 (2007) 1535–1547.
[74] D. Antoine, A. Morel, A multiple scattering algorithm for atmospheric correction of remotely sensed ocean colour (MERIS instrument): principle and implementation for atmospheres carrying various aerosols including absorbing ones, Int. J. Remote Sens. 20 (1999) 1875–1916.
[75] H. Fukushima, A. Higurashi, Y. Mitomi, T. Nakajima, T. Noguchi, T. Tanaka, M. Toratani, Correction of atmospheric effects on ADEOS/OCTS ocean color data: algorithm description and evaluation of its performance, J. Oceanogr. 54 (1998) 417–430.
[76] G. Thuillier, M. Herse, D. Labs, T. Foujols, W. Peetermans, D. Gillotay, P.C. Simon, H. Mandel, The solar spectral irradiance from 200 to 2400 nm as measured by the SOLSPEC spectrometer from the ATLAS and EURECA missions, Solar Phys. 214 (2003) 1–22.
[77] H.R. Gordon, Atmospheric correction of ocean color imagery in the Earth observing system era, J. Geophys. Res. 102 (1997) 17081–17106.

[78] H.R. Gordon, M. Wang, Surface roughness considerations for atmospheric correction of ocean color sensors. 1: the rayleigh scattering component, Appl. Opt. 31 (1992) 4247–4260.

[79] M. Wang, A refinement for the rayleigh radiance computation with variation of the atmospheric pressure, Int. J. Remote Sens. 26 (2005) 5651–5663.

[80] M. Wang, K.D. Knobelspiesse, C.R. McClain, Study of the sea-viewing wide field-of-view sensor (SeaWiFS) aerosol optical property data over ocean in combination with the ocean color products, J. Geophys. Res. 110 (2005) D10S06. http://dx.doi.org/10.1029/2004JD004950.

[81] P.Y. Deschamps, M. Herman, D. Tanre, Modeling of the atmospheric effects and its application to the remote sensing of ocean color, Appl. Opt. 22 (1983) 3751–3758.

[82] M. Wang, Atmospheric Correction of the Second Generation Ocean Color Sensors, University of Miami, Coral Gables, FL, 1991, p. 135.

[83] H.R. Gordon, M. Wang, Influence of oceanic whitecaps on atmospheric correction of ocean-color sensor, Appl. Opt. 33 (1994) 7754–7763.

[84] M. Wang, S. Bailey, Correction of the sun glint contamination on the SeaWiFS ocean and atmosphere products, Appl. Opt. 40 (2001) 4790–4798.

[85] H. Zhang, M. Wang, Evaluation of sun glint models using MODIS measurements, J. Quant. Spectrosc. Radiat. Transfer 111 (2010) 492–506.

[86] H. Yang, H.R. Gordon, Remote sensing of ocean color: assessment of water-leaving radiance bidirectional effects on atmospheric diffuse transmittance, Appl. Opt. 36 (1997) 7887–7897.

[87] C. Cox, W. Munk, Measurements of the roughness of the sea surface from photographs of the sun's glitter. Jour. Opt. Soc. Am. 44 (1954) 838–850.

[88] R. Frouin, M. Schwindling, P.Y. Deschamps, Spectral reflectance of sea foam in the visible and near infrared: in situ measurements and remote sensing implications, J. Geophys. Res. 101 (1996) 14361–14371.

[89] P. Koepke, Effective reflectance of oceanic whitecaps, Appl. Opt. 23 (1984) 1816–1824.

[90] E.C. Monahan, I.G. O'Muircheartaigh, Whitecaps and the passive remote sensing of the ocean surface, Int. J. Remote Sens. 7 (1986) 627–642.

[91] K.D. Moore, K.J. Voss, H.R. Gordon, Spectral reflectance of whitecaps: their contribution to water-leaving radiance, J. Geophys. Res. 105 (2000) 6493–6499.

[92] H.R. Gordon, O.B. Brown, R.H. Evans, J.W. Brown, R.C. Smith, K.S. Baker, D.K. Clark, A semianalytic radiance model of ocean color, J. Geophys. Res. 93 (1988) 10909–10924.

[93] A. Morel, S. Maritorena, Bio-optical properties of oceanic waters: a reappraisal, J. Geophys. Res. 106 (2001) 7163–7180.

[94] H.R. Gordon, Can the Lambert-Beer law be applied to the diffuse attenuation coefficient of ocean water, Limnol. Oceanogr. 34 (1989) 1389–1409.

[95] R.C. Smith, K.S. Baker, Optical properties of the clearest natural waters, Appl. Opt. 20 (1981) 177–184.

[96] D.K. Clark, Phytoplankton algorithms for the Nimbus-7 CZCS, in: J.R.F. Gower (Ed.), Oceanography from Space, Plenum, New York, 1981, pp. 227–238.

[97] A. Morel, Optical properties and radiant energy in the waters of the Guinea dome and the Mauritanian upwelling area in relation to primary production, in: Conseil International pour l'Exploration de la Mer; Rapports et Proces Verbaux, vol. 180, 1982, pp. 94–107.

[98] W. Shi, M. Wang, Satellite observations of the seasonal sediment plume in central East China Sea, J. Marine Syst. 82 (2010) 280–285.

[99] W. Shi, M. Wang, Satellite views of the Bohai sea, yellow sea, and East China sea, Prog. Oceanogr. 104 (2012) 35–45.

[100] W. Shi, M. Wang, Ocean reflectance spectra at the red, near-infrared, and shortwave infrared from highly turbid waters: a study in the Bohai sea, Yellow sea, and East China sea, Limnol. Oceanogr. 59 (2014) 427–444.

[101] M. Wang, J. Tang, W. Shi, MODIS-derived ocean color products along the China east coastal region, Geophy. Res. Lett. 34 (2007) L06611. http://dx.doi.org/10.1029/2006GL028599.

[102] J.M. Mueller, G.S. Fargion, Ocean Optics Protocols for Satellite Ocean Color Sensor Validation, Revision 3, Part I & II, NASA Goddard Space Flight Center, Greenbelt, 2002. Maryland NASA Tech. Memo. 2002–210004.

[103] M. Wang, S. Son, Y. Zhang, W. Shi, Remote sensing of water optical property for China's inland Lake Taihu using the SWIR atmospheric correction with 1640 and 2130 nm bands, IEEE J. Sel. Top. Appl. Earth Observ. Remote Sens. 6 (2013) 2505–2516.

[104] H.R. Gordon, Normalized water-leaving radiance: revisiting the influence of surface roughness, Appl. Opt. 44 (2005) 241–248.

[105] A. Morel, G. Gentili, Diffuse reflectance of oceanic waters: its dependence on sun angle as influenced by the molecular scattering contribution, Appl. Opt. 30 (1991) 4427–4438.

[106] A. Morel, G. Gentili, Diffuse reflectance of oceanic waters. II. Bidirectional aspects, Appl. Opt. 32 (1993) 6864–6879.

[107] A. Morel, G. Gentili, Diffuse reflectance of oceanic waters. III. Implication of bidirectionality for the remote-sensing problem, Appl. Opt. 35 (1996) 4850–4862.

[108] M. Wang, Effects of ocean surface reflectance variation with solar elevation on normalized water-leaving radiance, Appl. Opt. 45 (2006) 4122–4128.

[109] A. Morel, D. Antoine, B. Gentili, Bidirectional reflectance of oceanic waters: accounting for raman emission and varying particle scattering phase function, Appl. Opt. 41 (2002) 6289–6306.

[110] K.J. Voss, A. Morel, Bidirectional reflectance function for oceanic waters with varying chlorophyll concentrations: measurements versus predictions, Limnol. Oceanogr. 50 (2005) 698–705.

[111] K.J. Voss, A. Morel, D. Antoine, Detailed validation of the bidirectional effect in various case 1 waters for application to ocean color imagery, Biogeosciences 4 (2007) 781–789.

[112] A. Morel, J.L. Mueller, in: J.L. Mueller, G.S. Fargion (Eds.), Normalized Water-Leaving Radiance and Remote Sensing Reflectance: Bidirectional Reflectance and Other Factors Vol. 2, NASA Goddard Space Flight Center, Greenbelt, 2002. Maryland NASA/TM-2002–210004/Rev3.

[113] A. Morel, B. Gentili, Radiation transport within oceanic (case 1) water, J. Geophys. Res. Oceans 109 (2004). http://dx.doi.org/10.1029/2003JC002259.

[114] H. Loisel, A. Morel, Non-isotropy of the upward radiance field in typical coastal (Case 2) waters, Int. J. Remote Sens. 22 (2001) 275–295.

[115] H.C. van de Hulst, Multiple Light Scattering, Academic Press, New York, 1980.

[116] S. Chandrasekhar, Radiative Transfer, Oxford University Press, Oxford, 1950.

[117] K. Ding, H.R. Gordon, Atmospheric correction of ocean-color sensors: effects of the Earth's curvature, Appl. Opt. 33 (1994) 7096–7106.

[118] Z. Ahmad, R.S. Fraser, An iterative radiative transfer code for ocean atmosphere systems, J. Atmos. Sci. 39 (1982) 656–665.

[119] K. Ding, Radiative Transfer in Spherical Shell Atmospheres for Correction of Ocean Color Remote Sensing, University of Miami, Coral Gables, FL, 1993, p. 90.

[120] M. Wang, M.D. King, Correction of rayleigh scattering effects in cloud optical thickness retrievals, J. Geophys. Res. 102 (1997) 25915–25926.

[121] T. Tanaka, M. Wang, Solution of radiative transfer in anisotropic plane-parallel atmosphere, J. Quant. Spectrosc. Radiat. Transfer 83 (2004) 555–577.

[122] M. Wang, Light scattering from spherical-shell atmosphere: earth curvature effects measured by SeaWiFS, Eos Trans. AGU 84 (2003) 529. http://dx.doi.org/10.1029/2003EO480003.

[123] M. Wang, H.R. Gordon, Estimating aerosol optical properties over the oceans with the multiangle imaging spectroRadiometer: some preliminary studies, Appl. Opt. 33 (1994) 4042–4057.

[124] M. Wang, H.R. Gordon, Radiance reflected from the ocean-atmosphere system: synthesis from individual components of the aerosol size distribution, Appl. Opt. 33 (1994) 7088–7095.

[125] M. Wang, H.R. Gordon, Estimation of aerosol columnar size distribution and optical thickness from the angular distribution of radiance exiting the atmosphere: simulations, Appl. Opt. 34 (1995) 6989–7001.

[126] G.A. d'Almeida, P. Koepke, E.P. Shettle, Atmospheric Aerosols: Global Climatology and Radiative Characteristics, A. Deepak Publishing, Hampton, Virginia, USA, 1991.

[127] E.P. Shettle, R.W. Fenn, Models for the Aerosols of the Lower Atmosphere and the Effects of Humidity Variations on Their Optical Properties, U.S. Air Force Geophysics Laboratory, Hanscom Air Force Base, Mass, 1979. AFGL-TR-79–0214.

[128] H.C. van de Hulst, Light Scattering by Small Particles, Dover Publications, Inc., New York, 1981.

[129] O. Dubovik, B.N. Holben, T. Lapyonok, A. Sinyuk, M. Mishchenko, P. Yang, I. Slutsker, Non-spherical aerosol retrieval method employing light scattering by spheroids, Geophy. Res. Lett. 29 (2002) 1451. http://dx.doi.org/10.1029/2001GL014506.

[130] O. Dubovik, A. Sinyuk, T. Lapyonok, B.N. Holben, M. Mishchenko, P. Yang, T.F. Eck, H. Volten, O. Munoz, B. Veihelmann, W.J. van der Zande, J.-F. Leon, M. Sorokin, I. Slutsker, Application of spheroid models to account for aerosol particle nonsphericity in remote sensing of desert dust, J. Geophys. Res. 111 (2006) D11208. http://dx.doi.org/10.1029/2005JD006619.

[131] M. Wang, W. Shi, The NIR-SWIR combined atmospheric correction approach for MODIS ocean color data processing, Opt. Express 15 (2007) 15722–15733. http://dx.doi.org/10.1364/oe.15.015722.

[132] M. Wang, S. Son, W. Shi, Evaluation of MODIS SWIR and NIR-SWIR atmospheric correction algorithm using SeaBASS data, Remote Sens. Environ. 113 (2009) 635–644.

[133] M. Wang, W. Shi, L. Jiang, Atmospheric correction using near-infrared bands for satellite ocean color data processing in the turbid western Pacific region, Opt. Express 20 (2012) 741–753.

[134] M. Wang, J.H. Ahn, L. Jiang, W. Shi, S. Son, Y.J. Park, J.H. Ryu, Ocean color products from the Korean geostationary ocean color imager (GOCI), Opt. Express 21 (2013) 3835–3849.

[135] D. Doxaran, N. Lamquin, Y.J. Park, C. Mazeran, J.H. Ryu, M. Wang, A. Poteau, Retrieval of the seawater reflectance for suspended solids monitoring in the East China sea using MODIS, MERIS and GOCI satellite data, Remote Sens. Environ. 146 (2014) 36–48.

[136] Z. Ahmad, B.A. Franz, C.R. McClain, E.J. Kwiatkowska, J. Werdell, E.P. Shettle, B.N. Holben, New aerosol models for the retrieval of aerosol optical thickness and normalized water-leaving radiances from the SeaWiFS and MODIS sensors over coastal regions and open oceans, Appl. Opt. 49 (2010) 5545–5560.

[137] X. He, D. Pan, Y. Bai, Q. Zhu, F. Gong, Evaluation of the aerosol models for SeaWiFS and MODIS by AERONET data over open oceans, Appl. Opt. 50 (2011) 4353−4364.

[138] F. Melin, M. Clerici, G. Zibordi, B.N. Holben, A. Smirnov, Validation of SeaWiFS and MODIS aerosol products with globally distributed AERONET data, Remote Sens. Environ. 114 (2010) 230−250.

[139] F. Melin, G. Zibordi, J.F. Berthon, Assessment of satellite ocean color products at a coastal site, Remote Sens. Environ. 110 (2007) 192−215.

[140] G. Myhre, F. Stordal, M. Johnsrud, D.J. Diner, I.V. Geogdzhayev, J.M. Haywood, B. Holben, T. Holzer-Popp, A. Ignatov, R. Kahn, Y.J. Kaufman, N. Loeb, J. Martonchik, M.I. Mishchenko, N.R. Nalli, L.A. Remer, M. Schroedter-Homscheidt, D. Tanré, O. Torres, M. Wang, Intercomparison of satellite retrieved aerosol optical depth over ocean during the period September 1997 to December 2000, Atmos. Chem. Phys. Discuss. 4 (2004) 8201−8244.

[141] Z. Li, X. Zhao, R. Kahn, M. Mishchenko, L. Remer, K.-H. Lee, M. Wang, I. Laszlo, T. Nakajima, H. Maring, Uncertainties in satellite remote sensing of aerosols and impact on monitoring its long-term trend: a review and perspective, Ann. Geophys. 27 (2009) 2755−2770.

[142] M. Wang, Extrapolation of the aerosol reflectance from the near-infrared to the visible: the single-scattering epsilon versus multiple-scattering epsilon method, Int. J. Remote Sens. 25 (2004) 3637−3650.

[143] M. Wang, H.R. Gordon, A simple, moderately accurate, atmospheric correction algorithm for SeaWiFS, Remote Sens. Environ. 50 (1994) 231−239.

[144] H.R. Gordon, B.A. Franz, Remote sensing of ocean color: assessment of the water-leaving radiance bidirectional effects on the atmospheric diffuse transmittance for SeaWiFS and MODIS intercomparisons, Remote Sens. Environ. 112 (2008) 2677−2685.

[145] H.R. Gordon, D.K. Clark, J.W. Brown, O.B. Brown, R.H. Evans, W.W. Broenkow, Phytoplankton pigment concentrations in the middle atlantic bight: comparison of ship determinations and CZCS estimates, Appl. Opt. 22 (1983) 20−36.

[146] M. Wang, Atmospheric correction of ocean color sensors: computing atmospheric diffuse transmittance, Appl. Opt. 38 (1999) 451−455.

[147] H.R. Gordon, In-orbit calibration strategy for ocean color sensors, Rem. Sens. Environ. 63 (1998) 265−278.

[148] R.E. Eplee Jr., W.D. Robinson, S.W. Bailey, D.K. Clark, P.J. Werdell, M. Wang, R.A. Barnes, C.R. McClain, Calibration of SeaWiFS. II: Vicarious techniques, Appl. Opt. 40 (2001) 6701−6718.

[149] B.A. Franz, S.W. Bailey, P.J. Werdell, C.R. McClain, Sensor-independent approach to the vicarious calibration of satellite ocean color radiometry, Appl. Opt. 46 (2007) 5068−5082.

[150] M. Wang, H.R. Gordon, Calibration of ocean color scanners: how much error is acceptable in the near-infrared, Remote Sens. Environ. 82 (2002) 497−504.

[151] D.A. Siegel, M. Wang, S. Maritorena, W. Robinson, Atmospheric correction of satellite ocean color imagery: the black pixel assumption, Appl. Opt. 39 (2000) 3582−3591.

[152] M. Wang, W. Shi, Estimation of ocean contribution at the MODIS near-infrared wavelengths along the east coast of the U.S.: two case studies, Geophy. Res. Lett. 32 (2005) L13606. http://dx.doi.org/10.1029/2005GL022917.

[153] W. Shi, M. Wang, An assessment of the black ocean pixel assumption for MODIS SWIR bands, Remote Sens. Environ. 113 (2009) 1587−1597.

[154] K.G. Ruddick, F. Ovidio, M. Rijkeboer, Atmospheric correction of SeaWiFS imagery for turbid coastal and inland waters, Appl. Opt. 39 (2000) 897−912.

[155] S.J. Lavender, M.H. Pinkerton, G.F. Moore, J. Aiken, D. Blondeau-Patissier, Modification to the atmospheric correction of SeaWiFS ocean color images over turbid waters, Continental Shelf Res. 25 (2005) 539–555.

[156] R.P. Stumpf, R.A. Arnone, R.W. Gould, P.M. Martinolich, V. Ransibrahmanakul, A partially coupled ocean-atmosphere model for retrieval of water-leaving radiance from SeaWiFS in coastal waters, NASA Tech. Memo. 2003–206892, in: S.B. Hooker, E.R. Firestone (Eds.), SeaWiFS Postlaunch Technical Report Series, Vol. 22NASA Goddard Space Flight Center, Greenbelt, Maryland, 2003.

Chapter 4.3

Simulation and Inversion of Satellite Thermal Measurements

Christopher J. Merchant,* Owen Embury
Department of Meteorology, University of Reading, Reading, UK
*Corresponding author: E-mail: c.j.merchant@reading.ac.uk

Chapter Outline
1. Introduction — 489
2. Radiative Transfer Simulation for Thermal Remote Sensing — 490
3. Propagation of Thermal Radiation through Clear Sky — 493
4. Simulation of Interaction with Aerosol and Cloud — 500
5. Simulation of Surface Emission and Reflection — 502
6. Use of Simulations in Thermal Image Classification (Cloud Detection) — 505
7. Use of Simulations in Geophysical Inversion (Retrieval) — 509
8. Use of Simulations in Uncertainty Estimation — 516
9. Conclusion — 521
References — 523

1. INTRODUCTION

In thermal remote sensing of Earth, the aim is to infer knowledge of the state of the surface and/or atmosphere from satellite measurements of infrared radiance. This is possible because the radiance emerging from the top of the atmosphere (ToA) depends on the state in ways that we can understand and quantify by simulation. The aim of this chapter is to discuss simulation of satellite thermal measurements over the oceans, and the use of such simulations in estimating sea-surface temperature (SST) (and the associated uncertainty) when creating records of SST suitable for use in climate-change applications.

There are two aspects to "retrieval" of geophysical knowledge from remotely sensed radiances. These are often called the "forward" and "inverse" problems.

The forward problem is to quantify the dependence of measured radiances on geophysical variables of interest. Our understanding of the dependence is usually embodied in a forward model that gives a numerical simulation of radiances given the geophysical state.

The inverse problem is to infer from measured radiances the geophysical state that gave rise to them. This process is also known as "retrieval." The result of retrieval is (or should be) both an estimate of the target variable(s) and an estimate of the uncertainty in that estimate.

In thermal remote sensing of the oceans, the main retrieved quantity is generally SST. Thermal radiances do hold information relevant to other variables that may be of interest in the context of climate, such as the extent or concentration of sea ice and the locations of ocean fronts. This chapter focuses on the simulations and retrieval methods relevant specifically to SST estimation. This chapter demonstrates the key role of simulation for every step. Cloud detection, SST retrieval itself, analysis of the SST sensitivity, and SST uncertainty estimation all benefit from simulations rooted in the physics of radiative transfer (RT). Strength of approaching SST retrieval informed by RT simulations is the enhanced ability to diagnose and solve problems. A climate data record (CDR) needs to have sufficient duration, consistency, accuracy, and stability of measurement to be informative about relatively subtle long-term change in the context of variability. As will be shown, simulation helps with the design methods of retrieval, classification, and uncertainty estimation that achieve this ambitious objective.

Section 4.3.2 below discusses the role of simulation in general in thermal remote sensing of surface temperature, and introduces the variety of radiative transfer models (RTMs) that may be encountered. Sections 4.3.3–4.3.5 give qualitative overviews of the physics underlying, respectively, emission and absorption by atmospheric gases, interactions with atmospheric aerosol, and sea-surface emissions and reflection. This level of basic understanding of RTMs is the minimum required to use them appropriately. Sections 4.3.6–4.3.8 discuss, in turn, the applications of the simulations to image classification (cloud detection), retrieval, and uncertainty estimation.

2. RADIATIVE TRANSFER SIMULATION FOR THERMAL REMOTE SENSING

Top-of-atmosphere radiances at thermal wavelengths tell us about the temperature of matter observed within the field of view of the observation [1]. (This statement neglects scattering of radiance from other sources into the field of view, which is often a justified approximation in thermal remote sensing.)

Factors other than temperature also influence thermal radiances, however. For example, sea state affects surface emissivity, and concentrations of radiatively active gases affect both the absorption of radiance and the effective emissivity of the atmosphere. How thermal radiance depends on these and other factors will be reviewed briefly in Sections 4.3.3—4.3.5 of this chapter. The dependence at a given wavelength on such factors can be qualitatively understood in terms of underlying physical processes: the emission, absorption, and scattering of thermal radiation. These processes are captured in the equations of RT, which are integrals in space and in wavelength of nonlinear terms [1]. Quantitative exploration of the dependence of measured thermal radiance on the state of the surface and atmosphere requires the numerical evaluation of these integral equations in RTMs. The forward problem is therefore addressed by use of an RTM.

Any RTM of practical use for thermal remote sensing must approximate the full equations of RT. Usually the approximation involves discretization and parameterization. Discretization may mean representing the atmosphere as a number of layers and/or dividing the infrared spectrum into a number of bands. Parameterization may mean using an approximate fit to describe, for example, the pressure dependence of absorption of radiance by water vapor for a particular band of wavelengths. Any RTM simulation, therefore, has uncertainty associated with such approximations. Moreover, the RTM uses knowledge of the interaction of matter and radiation established in countless laboratory and field measurements. This knowledge consists of largely empirical data describing the complex absorption/emission features of atmospheric gases, and is assembled in impressive spectroscopic databases [2,3]. Unknown measurement errors in this spectroscopic information are propagated into RTM simulations, contributing additional uncertainty [4]. Finally, since the RTM is generally required to simulate the measurements of a particular sensor, characterization of the spectral response and calibration of the sensor inevitably add further uncertainty.

Table 1 lists and comments on some generic types of RTM, and gives examples used in thermal remote sensing.

In considering which RTM best suits a given application, questions to consider are the following:

- Is accuracy paramount? If so, this may imply use of "line-by-line, layer-by-layer" approaches.
- Is the application time or computation limited? In near-real-time processing of satellite data streams, the computational power available may necessitate use of a highly parameterized (fast) forward model.
- Are partial derivatives of radiance with respect to state parameters required? For incremental retrieval methods (see later), the answer is usually yes. In this case, does the RTM support direct output of these partial derivatives? (Brute-force perturbation methods of finding partial derivatives are generally more computationally expensive.)
- Does the RTM simulate all factors relevant to the application? Highly parameterized RTMs may neglect variability of certain atmospheric

TABLE 1 Typology of Radiative Transfer Models Relevant to Thermal Wavelengths

Type of RTM & Principle	Comments	Examples
Line by line. Calculates monochromatic optical depths from a database of spectral lines and continuum absorption, contributions from all nearby spectral lines for all gases are considered (line-by-line). The radiative transfer equation is then integrated through the atmosphere (adding surface emission and reflection if required) to give a radiance spectra.	Direct numerical solution to radiative transfer equation. In principle, most accurate, if chosen discretization is adequately fine to resolve spectral lines. Outputs must be convolved with instrument spectral response function. Models typically allow full control of viewing path. Computationally intensive.	LBLRTM [5] Reference forward model (RFM) [6]
Narrow band. Transmittance parameters are precalculated in narrow bands (typically 1 cm^{-1} bins) allowing fast calculation of band transmittance from temperature and gas concentrations. Radiative transfer equation is integrated for each band.	Packages typically include surface emissivity/reflectivity models, solar radiances, an RT solver which includes scattering, and allow full control of viewing path. Moderately computational intensity for absorption-only simulations. Intensive if full scattering calculation required.	MODTRAN [7]
Channel-integrated transmittance coefficients are precalculated for a specific instrument channel for a range of atmospheric conditions. Typical predictors are viewing angle, temperature, water vapor, and ozone concentration. Other gases assumed to have fixed concentrations. This allows very rapid calculation of channel-averaged transmittance and then integration of radiative transfer equation.	Precalculated coefficients are specific to each instrument. Accuracy limited for wide channels. Partial derivatives usually available analytically. Viewing geometry limited to surface-viewing satellites (no limb viewing etc.). Recent models may include solar radiances and simple scattering calculations. Computational demands relatively light.	RTTOV [8]
An EOF/PC analysis of precalculated spectra is used to identify a small subset of channels/wavelengths required to reconstruct the full spectrum.	Combined with channel-integrated parameterization for fast simulation of high-resolution sounders (e.g., AIRS, IASI). Computational demands relatively light.	RTTOV [9]

constituents, allowing for variable water vapor but fixed proportions of other gases, for example. A common approximation in thermal wavelengths is neglect of radiance scattering by aerosol.

As will become clear in later sections, a mix of models may be required when developing a CDR for SST. For SST retrieval, account should be taken of the changing atmospheric composition of radiatively active gases and aerosol scattering [10], requiring the use of line-by-line and scattering-capable models at some point in the design of the SST retrieval method. Line-by-line, layer-by-layer simulations of atmospheric transmittance generally run with minimal assumptions and approximations other than the discretization of the physics of RT. Their computational intensity, however, may preclude using them for some purposes (e.g., on-the-fly cloud detection, see later), and so practitioners may need to use a mix of RTMs, including faster, parameterized models. In this case, it will be useful to compare the RTMs, characterize any differences, and verify that these differences are acceptable in the context of other uncertainties (Section 4.3.8). It is useful also to have a general understanding of the workings of the RTMs being used when developing SST datasets, in order to be aware of any limitations of their applicability.

3. PROPAGATION OF THERMAL RADIATION THROUGH CLEAR SKY

Use of an RTM for simulation of thermal radiance is generally needed because the equations of RT take the form of nonlinear integrals, as we will show. Uninformed use of any simulation software is not advisable, of course: a user of an RTM needs physical insight to confidently interpret the results of their simulations. The purpose of this section is to summarize the main physics describing thermal radiative propagating through clear skies. Many texts on remote sensing provide more details, including Ref. [1].

Consider the propagation of a monochromatic thermal radiation through a clear-sky atmosphere (devoid of cloud or significant aerosol). In its passage through air, there is a probability per unit distance that a given photon of thermal radiation will be absorbed. This probability is strongly variable with wavelength.

The strong variability with wavelength arises because the energy levels of molecules are constrained to fixed values (i.e., they are quantized) and the photons absorbed must have energies corresponding to the differences in those levels. The energy of a photon is inversely proportional to wavelength: short wavelength radiation is more energetic.

The energy of a molecule is determined by the internal distribution of its electrons (electronic state), its oscillations (vibrational state), and its rotation (rotational state). When a molecule absorbs a photon, it undergoes transition to a higher energy state. Similarly, if a molecule is already in an excited state, it may go to a lower energy state by emitting a photon of appropriate energy.

Transitions between different electronic states are generally associated with the emission or absorption of an ultraviolet or visible photon; vibrational transitions are less energetic and are linked with infrared photons; finally, rotational transitions are least energetic and generally correspond to microwave photons. Subject to certain rules, changes in electronic, vibrational, and rotational states can happen together in a single transition. So, when a molecule changes vibrational energy, it may also change rotational energy.

Thus, at the finest spectral resolution, there are many wavelengths at which a radiatively active molecule present in the air absorbs. These are called spectral "lines," each corresponding to a particular transition in the energy state of the molecule. Transitions that are intrinsically more probable give rise to stronger lines. The energy of the absorbed photon supplies the energy required to raise the energy state of the molecule.

An infrared absorption feature typically comprises a central line due to a vibrational-only transition (known as the Q-branch) and two "wings" on either side. The high-energy side (known as the R-branch) includes several small, closely spaces lines where the molecule simultaneously transitions to a higher energy rotational state. The low-energy side (P-branch) corresponds to transitions resulting in a lower energy rotational state.

There is also absorption associated with water vapor (and, over a narrower range, nitrogen) that varies slowly with wavelength—i.e., is not obviously linked to particular spectral lines. This is termed "continuum absorption" [11−13].

Figure 1 (upper panel) shows the probability of absorption per unit distance (absorptivity) for a very small range of wavelength in an infrared "atmospheric window," for air under conditions typical of the troposphere (see caption for details). There are spectral lines of different intensities and also background absorption between the lines associated with water vapor continuum absorption. The absorptivity varies rapidly with wavelength, with individual features in this range as narrow as 0.002 μm. An RTM must resolve or integrate over these features adequately.

Figure 1 (lower panel) shows the spectral transmittance of the atmosphere across the main range of wavelengths associated with terrestrial thermal radiation, for a cloud-free, aerosol-free atmosphere. "Transmittance" is the fraction of vertical, surface-leaving radiance that reaches the ToA, and "spectral" means that the variations against wavelength are represented. This figure is drawn at a much coarser spectral resolution (~ 0.05 μm) than the upper panel, so that the vibration−rotation structure that is resolved in the upper panel is smoothed over in the lower panel.

The absorptivity at a given wavelength is not a constant of the clear-sky atmosphere. Most obviously, it depends on the composition of air. Other factors being equal, the absorptivity associated with a particular molecular species is essentially proportional to the number density of the molecule. The most important variable component is water vapor. Absolute humidity in the

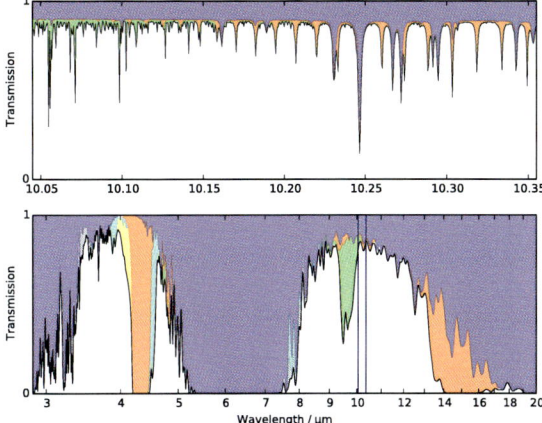

FIGURE 1 Spectrum of atmospheric vertical transmission (example). Blue: transmission accounting for water vapor (H_2O) only, showing the line features (spikes) and continuum. Pink: transmission accounting for H_2O and carbon dioxide (CO_2). Green: H_2O, CO_2, and ozone (O_3). Cyan: H_2O, CO_2, O_3, and nitrous oxide (N_2O). Gray: H_2O, CO_2, O_3, N_2O, and methane (CH_4). Yellow: H_2O, CO_2, O_3, N_2O, CH_4, and nitrogen (N_2). Upper panel is presented at high spectral resolution within the main atmospheric window, showing individual vibrational–rotational transitions. Absorption is by H_2O, CO_2, and O_3. Lower panel is presented at low spectral resolution, and shows the broad thermal atmospheric windows (3.5–4.1 μm, 8.2–9 μm, and 10–12.5 μm). Vertical lines indicate the location of the range of the upper panel. *From http://dx.doi.org/10.6084/m9.figshare.1008793, reproduced under CC-BY license.*

atmosphere varies from practically zero (cold, dry atmospheres) to ~ 20 g m^{-3} (warm, saturated air). The other radiatively active gases of concern for thermal remote sensing of the surface are listed in Table 2. The table includes information about the degree of variability in concentration of these gases, in parts per million by volume (ppmv). For a given composition and temperature, number density is proportional to the pressure. The surface pressure of Earth's atmosphere is usually within 5% of 1000 hPa, which means that the total mass of air through which vertically propagating radiation passes is variable to this extent.

As discussed further by Embury et al. [10], many of these radiatively active gases have concentrations that evolve over many years in response to human activities. (The chlorofluorocarbons are entirely anthropogenic.) For SST climate data generation over two or more decades, it is appropriate to perform simulations accounting for multiannual changes in CO_2, N_2O, CH_4, CCl_3F, and CCl_2F_2. Nitric acid (HNO_3) has a significant seasonal-latitudinal variation that it is appropriate to account for. References in Embury et al. [10] give sources of information about trace gas concentrations and profiles.

Pressure and temperature change the time interval between collisions of molecules. This has two effects of relevance.

The first effect is to change the width (in wavelength) of spectral lines, an effect called *line broadening*. The details of line broadening, line

TABLE 2 Radiative Impact of Various Gases in Thermal IR Windows. Concentrations come from a reference profile (MIPAS, 2001) except for CO_2, where a value more appropriate to present conditions has been used. BT impacts of each constituent have been calculated by comparing two runs of the RFM (Table 1). The first run has only water vapor present as a radiatively active gas, radiative impacts from all other gases having been "Turned Off." The second run is similar but with the radiative impact of the respective constituent "Turned On." The BT impact is the difference of the second from the first run, i.e., the magnitude of decrease in top of atmosphere brightness temperature caused by the gas

	Conc/ ppmv	BT Impact/mK			Spatial BT Variation/mK		
		3.7 μm	11 μm	12 μm	3.7 μm	11 μm	12 μm
CO_2	400	71	216	238			
O_3	7.5	10	6	29	2	1	6
N_2O	0.32	437	1	<1	36	<1	<1
CH_4	1.86	261	<1	<1	16	<1	0
NH_3	1.0×10^{-4}	<1	5	1			
HNO_3	8.1×10^{-3}	0	114	37	0	193	63
OCS	6.0×10^{-4}	<1	<1	4			
H_2CO	2.4×10^{-3}	1	0	0			
N_2	7.92×10^{5}	153	0	0			
C_2H_6	2.7×10^{-3}	<1	<1	6	<1	<1	6
CCl_3F	2.7×10^{-4}	0	<1	130	0	<1	21
CCl_2F_2	5.5×10^{-4}	0	231	6	0	26	1
HCFC-22	1.4×10^{-4}	0	<1	19			
CFC-113	1.9×10^{-5}	0	4	4			
CFC-114	1.2×10^{-5}	0	2	2			
CCl_4	1.1×10^{-4}	0	0	12			
HNO_4	3.0×10^{-4}	0	<1	1	0	<1	2

shapes, etc., need not be reviewed here; the purpose of the RTM is to handle these details for us. But, because temperature affects spectral absorptivity, the RTM will need to be provided with a profile of temperature versus pressure through the atmosphere in order to make calculations of absorption. (This is in addition to a profile of atmospheric composition.)

The second effect is to decouple the emission from the absorption of photons. If an isolated molecule absorbs a photon and enters an excited state, it later emits a photon of the same energy and returns to its unexcited state. If instead the molecule undergoes a collision while excited, its state may change via that collision, and any photon emitted will then be of a different wavelength (energy) to that earlier absorbed.

At pressures and temperatures representative of the troposphere, the latter case is more usual because the time between collisions is short. Collisions redistribute energy absorbed from the radiation field by individual molecules into kinetic energy of the gas: in other words, the atmosphere is heated by the radiation it absorbs.

It is sometimes stated that thermal radiation absorbed in the atmosphere is "reradiated" or "reemitted." This terminology is used in discussing the greenhouse effect, for example, but also appears in some remote-sensing texts. It is a misleading terminology since it implies equality between the absorbed and emitted radiation for a given portion of the atmosphere. There is no such equality. The quantity of radiation emitted by a portion of the atmosphere is different to the quantity absorbed, in general.

What, then, does determine the spectrum of radiation emitted by atmospheric gases? In short, the temperature and emissivity of the gases.

As for all matter close to thermodynamic equilibrium, the temperature effect is described by the "Planck function," named after physicist Max Planck. We will not reproduce here the details of this function, which are available in many texts (including Chapter 3.2 of this volume). The behavior of the Planck function is shown in Figure 2. It describes the spectral radiance emitted by an ideal emitter, as a function of wavelength, λ, and thermodynamic temperature, T, and is often written as $B(\lambda,T)$.

The Planck function describes the maximum possible thermal emission given the thermodynamic temperature. An isothermal layer of the atmosphere will emit spectral radiance (both upward and downward) equivalent to $\varepsilon B(\lambda,T)$, where σ is the emissivity of the layer along a particular direction of emission at wavelength λ, and ranges from 0 to 1. Now consider the absorption of radiance, L, that is incident on this layer along the same direction: the absorbed radiance is aL, where a is the absorptivity. Kirchhoff established that close to thermodynamic equilibrium, a and σ must equal each other. Matter that efficiently absorbs radiance also radiates efficiently. But since L and B are independent, this does not mean that the absorbed and emitted radiance equal each other.

Since the emissivity of a layer of atmosphere can be equated to the absorptivity, there is the same strong dependence of emissivity on wavelength

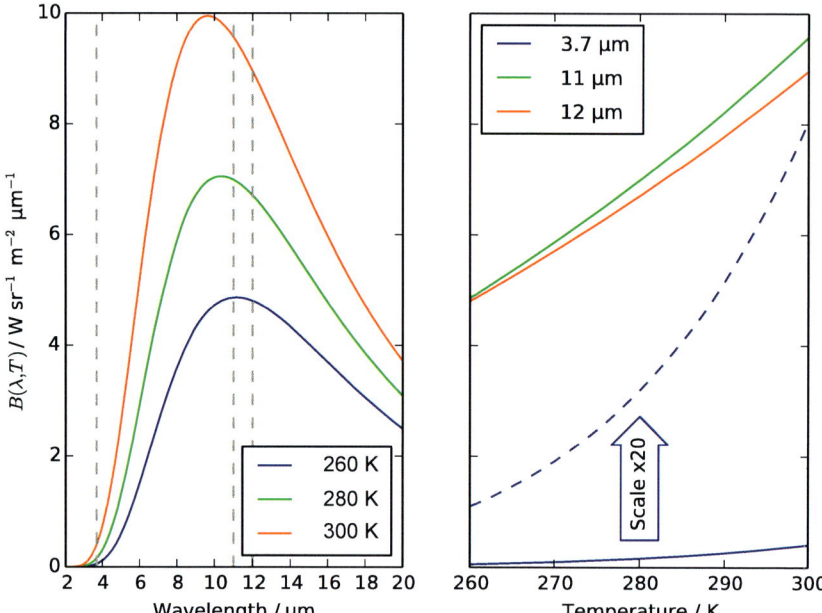

FIGURE 2 Dependence of Planck spectral radiance (*B*) on temperature and wavelength in ranges relevant to thermal remote sensing of the oceans. Left panel: *B* against wavelength for a selection of temperatures relevant to terrestrial observation. Right panel: *B* curves against temperature for a selection of thermal window wavelengths. For temperatures typical of the terrestrial surface and low clouds, the Planck function per unit wavelength interval peaks around 10 μm, i.e., in the vicinity of the main window of high atmospheric transmission. Radiance is much greater in the 10–12 μm range than around 3.7 μm, hence a version of the line multiplied by a factor of 20 is also shown, indicated by the labeled arrow. However, the latter window has the advantage for SST estimation of a much steeper dependence of radiance with temperature. *Reproduced under CC-BY license from http://dx.doi.org/10.6084/m9.figshare.1011393.*

as was discussed with regards to absorption processes. The emissivity (and absorptivity) depends also on pressure and temperature, for the reasons discussed earlier.

Now, consider the radiance at the ToA from the surface and just two isothermal layers of radiatively active atmosphere. Figure 3 shows the terms that contribute to this ToA radiance (in the absence of scattering). By following the radiative effects and their equations in Figure 3 in sequence, a clear understanding can be obtained of the nature of the numerical integration that an RTM must perform to simulate the net effect of all the physical processes discussed above. For illustration, there are only two radiative active layers in the atmosphere shown in the figure (gray shading). Each has a different spectral absorptivity (a_1 and a_2) dependent on the pressure (p), temperature (T), and composition (w). The arrows represent the radiance at different points relevant to determining the outgoing radiance at the ToA

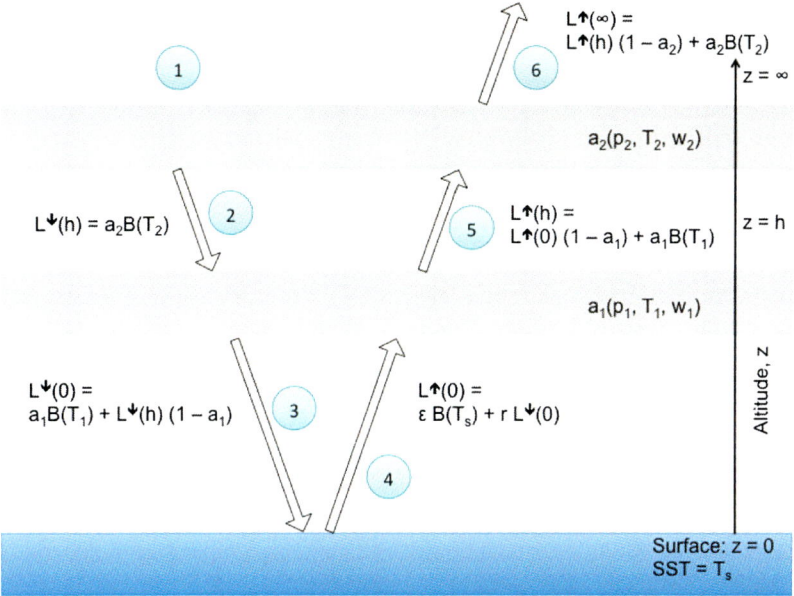

FIGURE 3 Cartoon of simulation of infrared radiative transfer for a clear-sky atmosphere over ocean of surface temperature, showing principles of calculations undertaken within an radiative transfer model (RTM). See main text for details.

(labeled 6). Working through the labeled points is in sequence: (1) Zero-incident radiance is assumed here at the ToA, although for a geometry within the solid angle of the Sun, the incoming solar radiance could be considered here, as would be necessary for the case of wavelengths around 4 μm during day time. (2) The upper layer of the atmosphere emits downward radiance according to the Planck function for the layer temperature multiplied by the absorptivity (which equals the emissivity). A significant part of the computation is the determination of the absorptivity of each layer given the atmospheric profile information. (3) The downward radiance from the lower layer comprises the transmitted fraction $(1 - a_1)$ of downward radiance incident on the layer, plus the thermal emission of the layer. (4) A fraction r of the downward irradiance is reflected at the sea surface at the specular (mirror-like) angle and adds to the thermal emission from the sea surface as determined by the SST. Since $\sigma \sim 1$ and $r \approx 1 - \sigma$, the upward radiance here is $\sim B(T_s)$, i.e., the sea-surface emission is dominant. (5) A fraction of the upward radiance incident on the lower layer is transmitted, and to this is added the thermal emission by this layer itself. Note that since the atmospheric temperature T_1 is usually less than T_s, the net effect of the layer is usually to decrease the radiance. (6) The upper layer modifies the upward radiance by transmitting only the fraction $(1 - a_2)$ of the incident radiance and by thermal emission. This evaluation completes the integration of the downward and upward radiance, yielding the radiance observable from space. Complications not represented in this cartoon are incident solar irradiance; scattering by atmospheric aerosols; effect of pressure, water vapor, and temperature variation within each radiatively active layer; how precisely r is related to σ for a rough sea surface and nonisotropic downward radiance.

For each layer in the atmosphere, the absorptivity/emissivity has to be assigned in the course of this numerical integration. For a calculation of radiance at a particular wavelength (monochromatic), this is determined by the concentration of gases that are radiatively active at that wavelength and the strength of absorption due to each gas under the particular circumstances of pressure and temperature attributed to the layer. A "line-by-line layer-by-layer" RTM consists of many discrete monochromatic radiance calculations integrated across the spectral response of the sensor. "Fast" RTMs may be built using equations similar to the monochromatic equations, under the approximation that a single vertical integration can account for many wavelengths and many relevant atmospheric constituents. Wavelength dependence within a spectral band is then addressed through parameterization of the overall absorptivity. The parameterization will include the pressure and temperature (and, for any variable constituent, concentration) of the layer in question, and may also depend on properties of other layers in the profile because these modify the spectrum incident on the layer. Numerical accuracy can be improved by careful choice of the layer temperature and channel-mean wavelength used in evaluations of the Planck emission from the layer—i.e., by using weighted averages.

While the essence of numerical evaluation of clear-sky RT is captured in Figure 3, a particular model may embody various sophisticated numerical techniques or approximations, including those referred to in Table 1.

4. SIMULATION OF INTERACTION WITH AEROSOL AND CLOUD

Solid particles and liquid droplets are often present in the atmosphere in addition to radiatively active gases. Similar to the gases, these constituents absorb and emit radiation. Because of this, they change the absorptivity/emissivity of a given atmospheric layer compared to molecular-only absorption considered in the previous sections. Unlike the gases, where absorption is determined by the quantum mechanical interaction of photons and molecular energy states, the optical properties of clouds and aerosols can be determined from classical electromagnetism—namely, Mie scattering theory.

Many aerosols and clouds are sufficiently optically thick to prevent thermal remote sensing of the ocean surface. This is the case for all but the thinnest clouds, desert dust, and ash from volcanic eruptions. When the thermal emission from the surface has been blocked or significantly reduced by cloud and aerosol, it is not possible to retrieve information about the surface using thermal radiances. While it may be possible to retrieve information about the cloud or aerosol, this is not relevant to surface retrievals, apart from the image classification step discussed in Section 4.3.6.

Surface retrievals are possible in the presence of optically thin aerosols such as marine aerosol and stratospheric aerosol. Marine aerosol comprises a mixture of water-soluble sulfates, nitrates, and particles of sea salt. Stratospheric aerosol is formed from droplets of sulfuric acid injected into the atmosphere by major volcanic eruptions and can remain for several years [14].

The optical properties (absorption coefficient, scattering coefficient, and phase function) of cloud and aerosol particles are determined from their microphysical properties (particle size distribution and refractive indices) [15,16]. The refractive index of the particles depends on their chemical composition, e.g., salts, mineral aerosol, or water.

It is often valid to use Mie scattering theory, treating the particles as spheres. Aerosol particles typically have sizes from ~ 1 nm to ~ 100 μm. Much larger particles fall out of the atmosphere relatively rapidly. The most important aerosols for RT at thermal IR wavelengths are those with diameters greater than ~ 1 μm. (For particles much smaller than the wavelength of the thermal radiation, scattering is described by the Rayleigh model. Rayleigh scattering intensity is $\propto \lambda^{-4}$, and is negligible for thermal IR wavelengths.)

Furthermore, in addition to absorption and emission, aerosols and clouds can also scatter radiance whereby the total radiance is not changed but the direction is as illustrated in Figure 4. If scattering occurs, then the RT equation becomes much harder to solve as its radiance maybe reflected back by any layer—i.e., we get a circular dependency where the radiance leaving a layer depends on the radiance incident on the layer from all directions, which depends on the radiance leaving the layer.

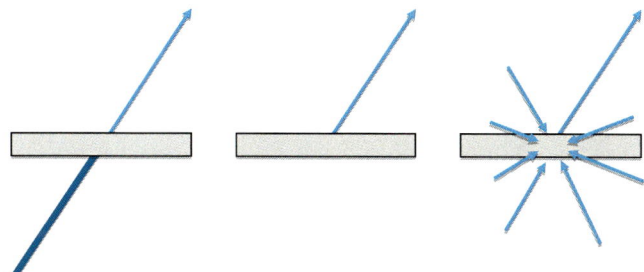

FIGURE 4 Cartoon of contributions to the outgoing radiance from a thin layer of the atmosphere. Left: radiance transmitted through the layer. Center: radiance emitted by the layer. Right: scattering by the layer. In the "absorption-only" approximation, only the first two contributions need to be considered; these can be found by adding up the contributions from each layer along a single direction. When scattering is included, it is necessary to know the complete radiance field which will depend on the outgoing radiance from the layer being considered as the outgoing radiance may be reflected back by other layers.

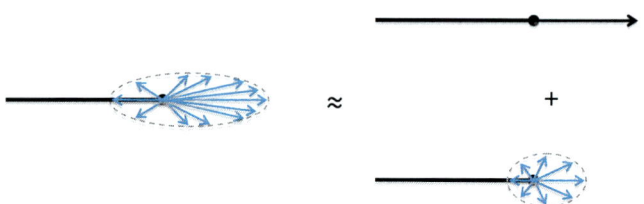

FIGURE 5 Mie scattering is strongly forward peaked, meaning that most scattered radiance continues in nearly the same direction. This may be approximated by treating the forward peak as transmitted radiance and the remaining scattering with a more isotropic distribution.

RT with scattering may be simulated with numerical methods such as ray tracing or Monte Carlo simulations. These are very computationally intensive, making them impractical for most uses. Commonly used approximation techniques include "successive orders of scattering," "doubling and adding" [17,18], and "discrete ordinates" [19]. These approximations are a tradeoff between accuracy and computational speed. Where high computational speed is essential due to the number of simulations required (e.g., real-time data assimilation), a simpler approximation may be required, which will generally be a two-stream approximation.

In a two-stream approximation, the radiance field is described by just two parameters, such as upwelling (L^+) and downwelling (L^-) radiance. Basic two-stream approximations are valid only where the solar contribution to thermal radiance is negligible—i.e., for night conditions (or to a good approximation), the 10.5–12.5 μm window. Furthermore, if the scattering phase function is highly directional (as is the case for Mie scattering), then the accuracy of two-stream approximations can be improved by treating the forward-scattered radiance as a term that "scatters" in the forward direction exactly plus an assumed distribution that is more isotropic (Figure 5) [20].

Embury et al. [10] concluded that to develop SST retrieval techniques for a CDR based on simulations, it was necessary to account for marine aerosol (sea salt) and (when relevant) stratospheric volcanic aerosol. They performed aerosol scattering calculations using a discrete ordinates solver using channel-integrated gas transmittance profiles (precalculated using a spectrally resolved RTM). For applications where greater uncertainty is tolerable (such as in cloud detection), a two-stream method for estimating the impacts of these aerosol modes is considered to be sufficient.

5. SIMULATION OF SURFACE EMISSION AND REFLECTION

The final component of the RTM setup is the sea-surface emissivity (and reflectivity), which provides the lower boundary for the RT integration. The emissivity of a flat surface may be calculated using Fresnel's equations—which give the reflection from a planar boundary between two media of different refractive indices, in this case, air and sea water. As infrared radiation is rapidly absorbed by water (penetration depth is 10–20 μm), we may consider any radiance not reflected by the surface to be absorbed. The refractive index of water [21,22] is dependent on its salinity [23,24] and temperature [25,26].

However, the sea surface is roughened by the presence of surface waves and is not flat. Surface waves are driven by the wind, with wavelengths down to ∼1 cm. This is sufficiently large compared to the wavelengths of infrared radiation that the ocean surface may be modeled as a collection of flat surfaces at different angles. The distribution of wave slope angles is dependent on the wind speed and is often modeled using the work of Cox and Munk [27]. At low wind speeds and viewing from overhead, it is sufficient to use the probability distribution of wave slopes combined with the Fresnel reflection of a particular slope to calculate the direct reflectivity and emissivity. As waves grow larger (at higher wind speeds) and at larger viewing angles, it is possible for multiple surface reflections to occur as shown in Figure 6. The top row shows the geometries of direct emission and reflection (zeroth-order interactions) which account for the majority of cases. The middle row shows the geometries which can occur with larger waves or viewing angles where an emitted or reflected ray is reflected again (first-order interaction) into the view angle. The bottom row illustrates the wave shadowing where an emitted (or reflected) ray is blocked from reaching the viewer as it intersects the water surface [28,29]. Figure 7 shows emissivity calculated including these effects [31,32] and temperature-dependent salinity from Newman et al. [26], as calculated by Filipiak [30].

An additional effect of a rough surface is that the assumption of specular reflection at the surface is invalid. Most RTMs, especially fast models, make the assumption that the surface reflection is specular such that any reflected downwelling radiance may be calculated assuming the same zenith angle as the upwelling radiance. In practice, the roughness of the surface means that downwelling radiance from a range of zenith angles is reflected into the same viewing direction; as the radiance of the sky varies with zenith angle, this

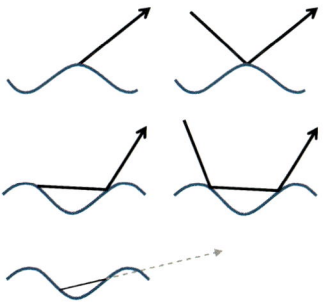

FIGURE 6 As the ocean surface is not flat, complex interactions may occur. Top left: direct emission. Top right: direct reflection. Middle left: surface emission, surface reflection (SESR). Middle right: surface reflection, surface reflection (SRSR). Bottom: shadowing.

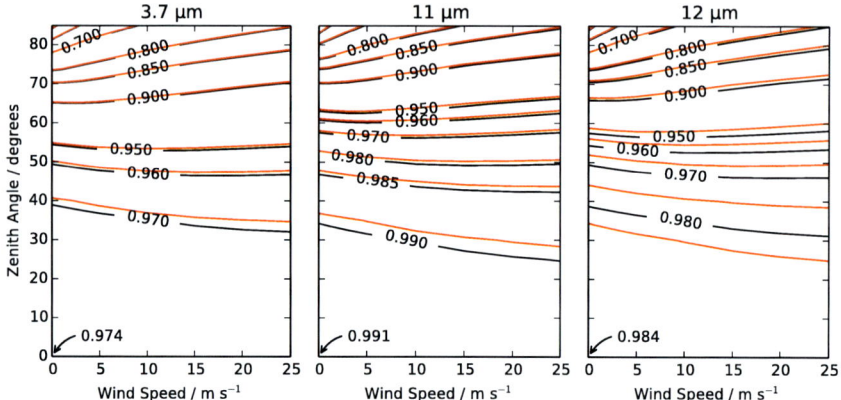

FIGURE 7 Sea-surface emissivity as a function of zenith angle and wind speed for three typical thermal channels (here, for the Advanced Along-Track Scanning Radiometer instrument). Black contours calculated for surface temperature of 280 K while red for surface temperature of 300 K. *Based on Filipiak [30].*

TABLE 3 A Sample of Published Emissivity Models Applicable to the Infrared Spectrum

Reference	Summary
Masuda et al. [67]	Direct emissivity based on Cox and Munk slope distribution.
Wu and Smith [31]	Adds to Masuda et al. first-order (SESR, SRSR) and cutoff angle to reduce shadowing effect.
Watts et al. [32]	Similar to Wu and Smith, plus adjustment for nonspecular refection for a specific satellite instrument.
Masuda [47]	Builds on Wu and Smith and Watts et al. using a probability distribution function instead of cutoff angle.
Filipiak [30]	Uses method from Wu and Smith and Watts et al. combined with temperature-dependent refractive indices from Refs [25] and [26].
Nalli et al. [33]	Adds a general adjustment for nonspecular reflection.
Yoshimori et al. [34,35]	Independent derivation using power spectrum of wave slopes rather than Cox and Munk distribution.
Bourlier [48]	Ignores multiple reflection, investigates non-Gaussian facet distribution, nonpolarized.
Bourlier [49]	Analytic calculation of zero-, first-, and second-order reflections, nonpolarized.
Caillault et al. [50]	Multiscale emissivity based on power spectrum of wave slopes.
Li et al. [36]	Based on Bourlier, adding polarization.

means the reflected sky radiance can be larger than would be the case for specular reflection [33].

Embury et al. [10] found that for defining SST retrieval for climate data generation, it was necessary to use an emissivity model accounting for the primary dependencies of emissivity on wavelength, angle, temperature, and salinity, with a model for wind-roughening effects on emissivity and effective reflectivity that accounts for surface-emission—surface reflection, double reflection, and shadowing. Table 3 lists several emissivity models of varying degrees of complexity. To our knowledge, the potential impact of recent emissivity models based on power spectrum approaches [34,35] and/or accounting for polarization [36] has not been assessed in the context of the SST CDR.

6. USE OF SIMULATIONS IN THERMAL IMAGE CLASSIFICATION (CLOUD DETECTION)

Thermal remote sensing of the ocean surface relies on cloud-free areas of imagery since all but the most tenuous clouds block or significantly attenuate the infrared radiance emitted by the sea surface. Prior to the geophysical inversion, there is a need to classify images into usable pixels ("cloud-free, acceptably low levels of aerosol, over water") and those unusable for SST estimation. Classification can include determining the nature of the pixels not to be used for SST (e.g., heavy aerosol [37], sea-ice [38], type of cloud [39]), but that is beyond the remit of this text. Hereafter, we simply refer to the usable and unusable classes as "clear" and "cloudy" for simplicity, since it is principally a matter of cloud detection.

All cloud detection schemes rely on detecting the contrasts in radiance caused by the presence clouds. Radiance contrasts are caused because clouds have reflectivity (brightness), reflectance spectrum (whiteness), temperature (infrared emission), spectral emissivity (blackness), and/or spatial coherence (image texture) that contrast with that of the surface. Note that, providing visible-wavelength and infra-red imagery is coregistered, it is not necessary to use only thermal channels for cloud detection (except at night!).

There is an extensive literature on cloud detection [40–46]. Around 20 papers per year focused on cloud detection methods are listed in major scientific citation databases. At first sight, this may be puzzling since the locations of clouds in Figure 8 seem obvious. Indeed, most cloudy pixels observed over the ocean are obvious: they are markedly more reflective, colder, or spatially contrasting than the sea. The challenge of image classification lies in the relatively small fraction of "difficult cases". Although less prevalent than "obvious" clouds, if not screened, these cloud-affected pixels return biased SSTs whose impact on SST estimation can be significant.

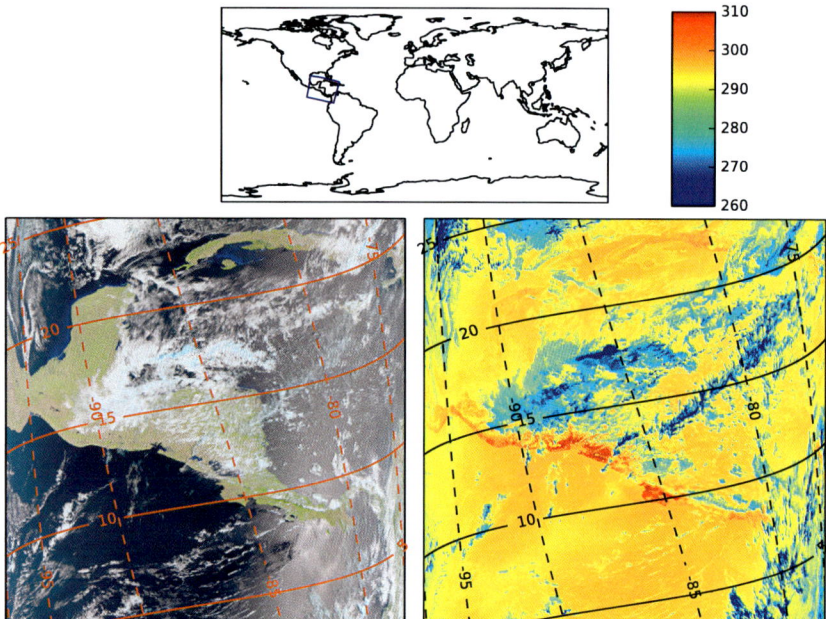

FIGURE 8 Scene observed by the Metop AVHRR instrument: lines of longitude (dashed) and latitude (solid) show how the image is distorted toward the edge of swath. Left panel: false-color RGB using 1.6 μm (red), 0.8 μm (green), and 0.6 μm (blue) channels. This choice of channels results in a "natural" looking image—oceans are blue, clouds are white, and vegetation is green. Right panel: thermal image observed at 11 μm. Clouds are colder than the surface with high clouds appearing in blue and lower clouds in green and yellow. The temperature of the land can be greater than the ocean and shows greater variability.

There are broadly three categories of approach to the cloud detection problem: predefined threshold testing, dynamic threshold testing, and probabilistic classification. Mixed approaches incorporating elements of these categories are possible.

A "threshold test" compares a metric derived from the satellite imagery to a threshold, and on this basis declares whether a pixel is plausibly clear or, on that metric, definitely cloudy. Usually, a series of several such tests are applied, and failure of any one is deemed sufficient to exclude the pixel from further SST processing. Typical tests using thermal channels are described in Table 4, based on Ref. [40] as an example.

Choice of threshold for different tests may be based on expert judgment informed by inspection of many thermal images, although it is more typical these days to select a threshold informed by simulation of the metric across a set of scenarios [39,40]. The question of how to define a set of scenarios for

TABLE 4 Typical Cloud Detection Tests Involving Thermal Channels of Sea Surface Temperature (SST) Sensors

Name of Test	Metric to Flag as Cloudy	Basis of Threshold	Comments
Gross threshold test	$BT_{12\mu m} <$ threshold	Lowest plausible clear-sky BT for location and season	Based on "worst case" simulation—e.g., climatologically cold SST and high water vapor loading
Thin cirrus test	$BT_{11\mu m} - BT_{12\mu m} >$ threshold	Largest plausible difference for clear sky, given zenith angle and $BT_{11\mu m}$	Based on simulations of the two channels across a range of representative situations
Fog-/low-stratus test	$BT_{11\mu m} - BT_{3.7\mu m} >$ threshold	Emissivity of fog/stratus at 3.7 μm is relatively low (<0.9)	Has been based on inspection of imagery, but simulations could have been used
Medium-/high-level test	$BT_{3.7\mu m} - BT_{12\mu m} >$ threshold	Largest plausible difference for clear sky, given zenith angle and $BT_{12\mu m}$	Based on simulations of the two channels across a range of representative situations

simulation is discussed in detail in the next section in the context of defining an algorithm to infer SST from BTs. Any such set of scenarios is based upon climatological knowledge of the plausible conditions of SST and atmospheric profile for a given location and season. In areas where interannual variability is great, fixed cloud detection thresholds must allow adequate margins of safety to accommodate conditions that are not well sampled in the simulation set, otherwise clear sky is likely to be wrongly flagged as cloud. In general, it is an intensive, delicate, and often subjective process to design thresholds for cloud detection that strike an adequate balance between effective detection of truly cloud pixels and avoidance of false detections.

For this reason, dynamic methods of cloud detection have been developed [39,41]. These rely on two considerations. First, RTMs exist that are sufficiently fast and accurate to simulate thermal channels for the conditions particular to specific images (i.e., for a prior estimate of the surface and atmospheric state and at the viewing geometry of the image). Second, from numerical weather prediction (NWP) at meteorological centers, estimates of the state are available that are significantly closer to the reality than is likely to be the case if working from predefined climatological information. The NWP used may be forecasts (if the cloud detection is being done in near-real time) or a retrospective "(re)analysis" for cloud detection in delayed mode. Essentially, a dynamic approach allows cloud-detection thresholds to be framed using a narrower margin of safety, informed by simulation of the expected value of the metric for the circumstances of the particular image.

The issue of how to set the margin of safety for a given threshold test still remains using dynamic simulation. A probabilistic approach that aims to address this question systematically was introduced by Merchant et al. [42]. Cloud detection fundamentally is about assessing the probability that the image pixel represents clear-sky conditions. This assessment should take account of

- the values of the reflectances and BTs actually observed;
- the values of the reflectances and BTs expected under clear-sky conditions for the time and place of observation;
- the values of the reflectances and BTs expected under cloudy conditions.

The role of simulation is to link our prior knowledge (from NWP) of plausible clear-sky conditions to the values of reflectances and BTs expected. Crucially, a probabilistic approach recognizes that both the NWP information and the simulation process involve uncertainty. The margin of safety should be based on the magnitude of these uncertainties.

Fundamentally, we want to estimate $P(\text{clear}|x, y)$, which is the probability of clear-sky conditions given the prior information about the state, x, and the observations, y. We cannot calculate this directly: instead simulation enables us to estimate the following probability distribution: $p(y|x, \text{clear})$. This is the probability of observations, given information about the state (the NWP information) for clear skies.

The steps in using an RTM to find $p(y|x, \text{clear})$ are as follows:

- simulating the best prior estimate of the satellite observations, using the NWP information;
- simulating also the most influential partial derivatives of the satellite observations with respect to the NWP information (i.e., terms such as "BT/SST");
- multiplying those partial derivatives by the degree of uncertainty of the NWP information—this gives an uncertainty in the prior estimate of the satellite observations;
- estimating the effect of uncertainty in the simulation process (e.g., from approximate RT) and from uncertainty in the sensor (e.g., radiometric noise).

This is sufficient information to estimate the probability density function, $p(y|x, \text{clear})$, that is required. A multivariate Gaussian distribution has been used for this function in Ref. [42]. Given $p(y|x, \text{clear})$, the differences between the simulation and the observation can then be assessed in terms of how probable they are to have arisen for a clear-sky condition, relative to their probability in cloudy conditions. That essentially is what Bayes' theorem does, yielding $P(\text{clear}|x, y)$. Rather than having many thresholds, a single threshold can be set that accounts for all the interchannel relationships and differences simultaneously, this being a probability threshold on $P(\text{clear}|x, y)$.

This simplified discussion does not address how $p(y|x, \text{cloudy})$ may be defined [43,51], the issues around simulation of reflectance channels, or use of spatial coherence of the image within the Bayesian framework. The aim of this short account has been to emphasize, first, that simulation of thermal radiances can improve image classification by translating prior geophysical information into the "space" of the satellite observations, and, second, that the simulation of the sensitivity of the BTs on the prior information is important. This sensitivity is embodied in the partial derivatives of the BTs.

Simulation both of BTs and of derivatives turns out also to be of use in the process of geophysical inversion which is then applied to the pixels classed as "clear sky," and in modeling the uncertainty in SST retrievals, as shown in the following two sections.

7. USE OF SIMULATIONS IN GEOPHYSICAL INVERSION (RETRIEVAL)

The process of geophysical inversion to extract SST from a number of clear-sky brightness temperature (BT) observations is often referred to as *SST retrieval*. The observations required to retrieve SST must include wavelengths that have adequately high sensitivity to SST and are differentially absorbed by atmospheric water vapor. Wavelengths between about 10 and 12.5 μm constitute the main "atmospheric window" in the thermal infrared (Figure 1), where a reasonable fraction of surface-emitted radiation reaches the ToA. The transmitted fraction varies across this window, indicating differential water vapor absorption. This is the principal window used by SST sensors, and is usually split between two channels centered around 11 and 12 μm, respectively. Channels around 3.7 μm and around 8.7 μm may also be used.

"Traditional" retrieval schemes estimate SST from a weighted combination of BTs, the weights being precalculated by some means. This is a simple, computationally efficient approach. At least two BTs are required [52], since it is necessary to infer both the SST (explicitly) and the impact of the atmosphere on BTs (implicitly).

The weights are generally termed "SST retrieval coefficients". Retrieval coefficients have often been determined empirically, using matches between in situ SST measurements and satellite observations. Various combinations of weighted BTs have been proposed and discussed in the literature, although the difference in outcome between these is often marginal [53]. For a more detailed discussion of the advantages and limitations of an empirical approach, and further references, see Merchant [54] (particularly Table 3 therein).

An alternative to empirical determination of retrieval coefficients is to base them on physics. This is achieved in practice via RT [55]. The practical steps to do this are (1) choose an RTM; (2) select/define a set of situations for which to run the RTM, and simulate the satellite observations for this set; (3) formulate the retrieval algorithm; and (4) derive the coefficients for this by a minimization procedure. Each of these steps is now discussed in turn.

Choice of RTM. The advantage of the coefficient-based approach to SST retrieval is computational, in that the coefficients are precalculated. It is therefore feasible to use a line-by-line layer-by-layer model for calculating the transmission and emission terms from the gases in the clear-sky atmosphere. Such an RTM calculates the spectral radiance at the ToA at high spectral resolution. This gives, in principle at least, the most accurate results for the RT through the gaseous atmosphere. Aerosol absorption and scattering may also need to be accounted for. To obtain the simulated channel radiance for a given sensor, the ToA spectrum must be convolved with the appropriate spectral response function (SRF). The final step in the RTM simulation is expression of the channel-integrated radiance as the equivalent BT, via a radiance-to-brightness—temperature relationship that, again, is specific to the SRF of the channel. All these steps may be available within a single RTM, or it may be preferable to combine models that have particular strengths [10].

Cases for simulation. Relevant components of the surface and atmospheric state need to be defined for each simulation. As discussed above, this usually implies specifying at least the surface (skin) SST and the profiles of atmospheric humidity and temperature (on typically ∼60 pressure levels from ∼1000 hPa to ∼1 hPa). Typically, with generic values for other state variables, this is sufficient for simulations accurate to a few tenths of Kelvin. For more accurate ($\lesssim 0.1$ K) simulations, variability in the sea-surface salinity, the wind speed, the vertical distribution of marine aerosol, and the concentrations other trace gas profiles may need to be considered, where their effects on emissivity and atmospheric RT modify the result to $\gtrsim 0.01$ K. Despite the inclusion of surface as well as atmospheric variables, the set of cases of the state variables is often called a "profile set." The profile set should encompass the full range of realistic situations to which the retrieval algorithm will be applied. A basic consideration to achieve this is geographical and seasonal distribution—all relevant regions should be sampled in all seasons of the year. However, representative sampling (in the sense of the profile set being evenly sampled over the globe in distance and time) is not required and may be disadvantageous: since variability at high latitudes exceeds variability in the tropics, it has been found to be more useful to exaggerate the proportion of high latitude profiles in the set, for example. Extreme cases should be present in the profile set, including coastal situations influenced by continental air masses, inland seas, profiles with extremely high humidity from deep convective regions, profiles with extremely low humidity from areas such as Baffin Bay, and "anomalous" profiles such as cases in the north-east subtropical Atlantic with a prominent "Saharan air layer." Radiosonde profiles have been used to form profile sets [56], as have outputs from NWP reanalysis. It is easier to construct a profile set from NWP than to collect clear-sky radiosonde profiles from round the world, but the use of NWP does raise a question that has not been fully explored. The humidity profile from an NWP model represents the mean humidity in a "grid box," which may be fully or partly cloud filled. Relative humidity in a cloud is 100%, and is <100% in clear air. The NWP humidity profile is therefore not always representative of clear-sky humidity; it may be biased toward "too wet". Nonetheless, NWP-based profile sets have been used successfully for determining SST retrieval coefficients [10].

Form of algorithm. The relationships between SST and radiances are significantly linearized by expressing the satellite observations at BTs. Retrievals have generally taken a reasonably linear form, e.g., have been variants of:

$$\widehat{x} = a + \mathbf{a}^T\mathbf{y} = a_0 + b_0 S + \sum_i (a_i + b_i S) y_i \qquad (1)$$

where $\widehat{x}$ is the retrieved SST, a is an offset coefficient, $\mathbf{a}$ is the column vector of retrieval coefficients, and $\mathbf{y}$ is the column vector of BTs. $\mathbf{a}^T\mathbf{y}$ is a way of writing the scalar product of the vectors using matrix algebra notation, and essentially gives a sum over terms as shown. The summation version also exposes an angular dependence of retrieval coefficients that has often been used; the ith element of $\mathbf{a}$ equals $a_i + b_i S$, where $S = \sec\theta - 1$, where θ is the view zenith angle of the observation. The variants of the above expression in the literature include cases where certain terms are dropped (or equivalently, have the coefficient value set to zero), or are constrained in relation to each other. It is common to vary the coefficients for different bands of latitude or other auxiliary variable. In one variant, the "nonlinear SST" (NLSST; [57]), some coefficients are made to be a weak function of a prior SST estimate (e.g., obtained from climatology). Additional terms that are quadratic in BT differences have been tried but give limited improvement in retrieval accuracy since such terms do not reflect the geographical variations in atmospheric structure that cause the errors in purely linear retrievals. The general form given in Eqn (1) is difficult to improve on significantly if one is committed to a coefficient-based approach.

Obtaining the coefficients. The coefficients are usually those that minimize $(\widehat{x} - x_p)^2$, where x_p is the profile SST used for the simulation. This is the "least squares" solution. The minimization is done across either the full profile set or, for banded coefficients, appropriate subsets. Define the data vector as $\mathbf{d}^T = [x_{p1}, x_{p2}, \ldots, x_{pj}, \ldots]$, where x_{pj} is the profile SST for the jth member of the profile set. Define the vector of coefficients (known as "model parameters" in discrete inverse theory) as $\mathbf{m}^T = [a_0, b_0, a_1, b_1, \ldots]$. Define a matrix of observations (data kernel) as

$$\mathbf{G} = \begin{bmatrix} 1 & S_1 & y_{11} & y_{11}S_1 & \cdots & y_{i1} & y_{i1}S_1 & \cdots \\ 1 & S_2 & y_{12} & y_{12}S_2 & \cdots & y_{i2} & y_{i2}S_2 & \cdots \\ \vdots & \vdots & \vdots & \vdots & \vdots & \vdots & \vdots & \vdots \\ 1 & S_j & y_{1j} & y_{1j}S_j & \cdots & y_{ij} & y_{ij}S_j & \cdots \\ \vdots & \vdots & \vdots & \vdots & \vdots & \vdots & \vdots & \vdots \end{bmatrix}$$

then the least squares solution for the coefficients is

$$\mathbf{m} = \left[\mathbf{G}^T \mathbf{G} \right]^{-1} \mathbf{G}^T \mathbf{d} \tag{2}$$

A strength of approaching SST retrieval with RT simulations is the enhanced ability to diagnose and solve problems. An example is how to adapt SST retrieval to the presence of stratospheric aerosol [58], which exploits a modification of Eqn (2), namely, least squares minimization subject to a linear constraint. Occasionally, major volcanic eruptions penetrate the stratosphere and create a haze of sulfuric acid droplets that persists at altitudes of order 20 km for a year or two [14]. This stratospheric aerosol layer has climatic impacts, and also affects remote sensing at visible and infrared wavelengths. The aerosol absorbs infrared radiation, and causes BTs to be reduced. The reduction varies between channels and with the optical depth of the aerosol. The persistent stratospheric aerosol is known to consist of sulfuric acid droplets, and some constraint can be put on the pH of the droplets (which modified their refractive index) and the droplet size distribution (which modifies their interaction with radiation as a function of wavelength). The BT impact of the stratospheric aerosol can therefore be simulated, using Mie scattering theory. The result of such a simulation is shown in Figure 9.

The impact per unit aerosol optical depth, τ, on the BTs of a typical three-channel sensor is $\frac{\partial y}{\partial \tau}$. Because the aerosol is optically thin and relatively uniform, the proportions of the impact at difference wavelengths are fairly stable, so that $\frac{\partial y}{\partial \tau} \approx \mathbf{k}$, a constant "stratospheric aerosol mode" of variability in the BTs. It is clear that the impact of the stratospheric aerosol perturbation on the retrieved SST will be $\propto \mathbf{a}^T \mathbf{k}$. Coefficients that will make the retrieved SST robust against (i.e., less sensitive to) stratospheric aerosol must therefore have the property $\mathbf{a}^T \mathbf{k} = 0$. Such an additional constraint can be

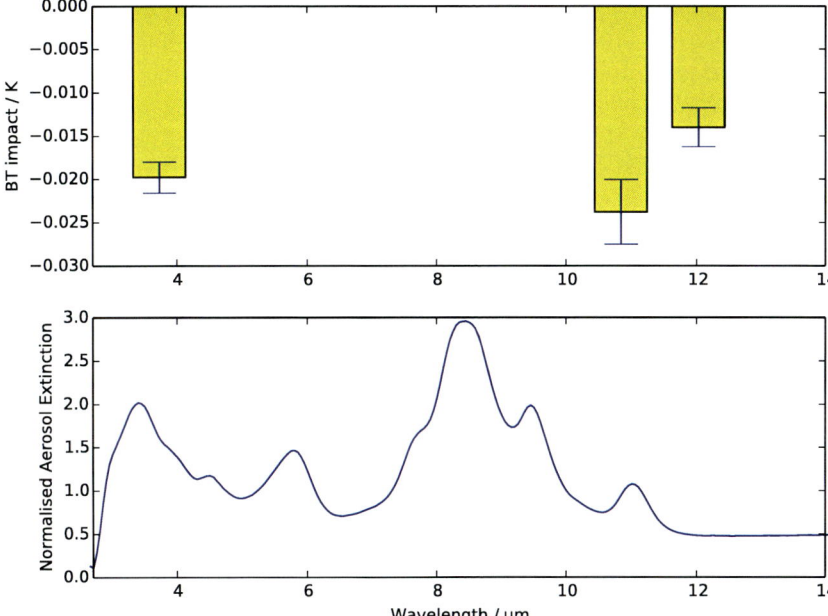

FIGURE 9 BT impact due to background stratospheric aerosol (bars) and normalized aerosol extinction coefficient (solid line). In the 11 and 12 μm channels where scattering is negligible, the ratio of the BT impacts is proportional to the aerosol extinction. However, in the 3.7 μm channel where scattering is greater, the BT impact is lower. The "error bars" on the BT impacts indicate the standard deviation of BT impact for a given stratospheric aerosol from the interaction with other aspects of atmospheric variability.

implemented by solving the following modified expression for the coefficients, **m**:

$$\begin{bmatrix} \mathbf{m} \\ \lambda \end{bmatrix} = \begin{bmatrix} \mathbf{G}^T\mathbf{G} & \mathbf{f} \\ \mathbf{f}^T & 0 \end{bmatrix}^{-1} \begin{bmatrix} \mathbf{G}^T\mathbf{d} \\ h \end{bmatrix} \quad (3)$$

where $\mathbf{f}^T = [0, 0, k_1, k_1, \ldots k_i, k_i, \ldots]$, and λ and h are auxiliary variables. (This is the solution of the least squares problem subject to a linear constraint by the method of Lagrange multipliers.) This approach has been shown to reduce the sensitivity of SSTs to stratospheric aerosol following the major eruption in 1991 of Mount Pinatubo in the Philippines [59]. This illustrates that understanding a retrieval situation by simulation can lead to useful insights when defining coefficients based on RT.

Alternatives to coefficient-based retrieval of SST involve simulation of BTs for the particular context of each observation (rather than precalculating a spatiotemporal sample, as when defining coefficients). This requires access to NWP fields and, for a practical implementation, a computationally fast simulation capability ("fast forward model"). A simulated prior estimate for the BT based on simulation is obtained for every satellite pixel. The RTM needs not necessarily be run for every pixel location; rather, the RTM can be run at the NWP resolution and interpolated. These simulated BTs can then be used in a variety of ways to give improved SST estimates.

If simulated BTs, $\mathbf{y}_b$ (where subscript b indicates BTs simulated using prior or "background" information), are used with SST retrieval coefficients, a simulated SST estimate is obtained $\widehat{x}_b = a + \mathbf{a}^T \mathbf{y}_b$. The simulation of $\mathbf{y}_b$ assumes a background SST as input to the RT model, x_b. The difference $\widehat{x}_b - x_b$ is then an estimate of a component of the retrieval error for the circumstances of the retrieval embodied in the NWP information. An SST estimate, $\widehat{x}'$, that improves upon the original estimate, $\widehat{x}$, can then be obtained:

$$\widehat{x}' = \widehat{x} - (\widehat{x}_b - x_b) = x_b + \mathbf{a}^T(\mathbf{y} - \mathbf{y}_b) \qquad (4)$$

Validation has shown that Eqn (4) reduces geographical biases and the standard deviation of discrepancies compared to using the same set of coefficients alone [60]. Success requires that simulated and observed BTs have, if necessary, been tuned to have no relative bias on average. Equation (4) shows that simulation-based bias correction (central expression) is equivalent to adjusting the background SST in the light of the discrepancy between observed and simulated BTs (rightmost expression).

This concept of retrieval by adjusting a background SST based on a simulation-minus-observation difference is common in remote sensing inverse theory, including atmospheric sounding [61]. Inverse theory gives a coherent framework for analyzing SST retrieval as an inverse problem, in terms of a fundamental understanding of how much information is truly present about SST in a given set of BTs. Different inverses for SST can be defined that optimize clearly defined aspects of the SSTs obtained. These all take the form

$$\widehat{x} - x_b = \mathbf{G}(\mathbf{y} - \mathbf{y}_b) \qquad (5)$$

with different definitions of the "gain matrix" $\mathbf{G}$. Typically, $\mathbf{G}$ includes terms that can be obtained only by simulation, namely, the derivatives of the BTs with respect to state variables. A matrix, $\mathbf{K}$, is calculated for each simulation that contains terms of the form $\frac{\partial y_i}{\partial x_k}$, where x_k represents the kth element of the state vector (i.e., of the NWP profile, which includes SST and also all the atmospheric variables on a range of altitudes). $\mathbf{K}$ may be called a "weighting function," "Jacobian," "kernel," "tangent linear matrix," or "adjoint" in different contexts in the literature. Examples of gain matrices that have been applied to the SST problem, expressed in terms of $\mathbf{K}$, are given in Table 5.

Whatever form of SST retrieval is chosen, the process can be simulated and the properties of simulation SST estimates can be explored. Expected results in terms of the SST-retrieval uncertainty will be discussed in the next section.

TABLE 5 Example Formulations of the Gain Matrix in Eqn (5)

Name of Inverse	Equation of Gain Matrix	Definitions	Comment
Modified total least squares	$(\mathbf{K}^T\mathbf{K}+\lambda\mathbf{I})^{-1}\mathbf{K}^T$	λ is a regularization parameter. $\mathbf{I}$ is the identity matrix.	Do not need to know error covariances. May vary λ to adapt the strength of regularization.
Maximum likelihood	$(\mathbf{K}^T\mathbf{S}_\varepsilon^{-1}\mathbf{K})^{-1}\mathbf{K}^T\mathbf{S}_\varepsilon^{-1}$	$\mathbf{S}_\varepsilon$ is the error covariance of the BTs.	The simulation is essentially a linearization point. Weights BTs according to BT noise.
Maximum a posteriori	$(\mathbf{K}^T\mathbf{S}_\varepsilon^{-1}\mathbf{K}+\mathbf{S}_p^{-1})^{-1}\mathbf{K}^T\mathbf{S}_\varepsilon^{-1}$	$\mathbf{S}_p$ is the error covariance of the prior NWP profile.	Gives an optimum answer if error-covariance matrices are well estimated.

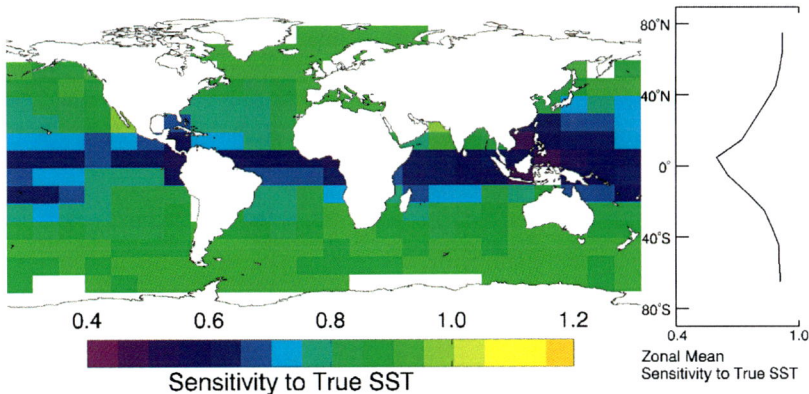

FIGURE 10 Change in retrieved SST per unit change in true SST, all other factors being held constant, for a split window SST estimate (nonlinear SST retrieval applied to the Advanced Very High Resolution Radiometer on Metop-A). *Reproduced from Merchant et al. [62].*

Another parameter of interest that is accessible via simulation is the SST sensitivity [62]. SST sensitivity is the fractional response of the retrieved SST to variation in true SST, other factors (such as the atmospheric state) being equal. Ideally, the sensitivity should be 1 K K^{-1}, so that a true change in SST causes an identical change in retrieved SST. In general, this is not the case (Figure 10).

SST sensitivity is readily calculated for a coefficient-based retrieval as

$$\frac{\partial \widehat{x}}{\partial x} = \mathbf{a}^T \frac{\partial y}{\partial x}$$

where the partial derivatives of BTs with respect to SST are calculated using RT simulation (and are row of the tangent linear matrix, $\mathbf{K}$). Where sensitivity of an SST estimate is much <1 K K^{-1}, it is expected that the strength of ocean thermal gradients is underestimated. Likewise, diurnal variations in SST are attenuated by low-sensitivity estimators [63,64].

In this section, we have made the case that simulation of thermal radiances is a major tool in understanding and improving the retrieval of SST from space. Simulation can be used to refine the design of a retrieval method, to explore the response of a retrieval method to different regimes of observations (such as when aerosol is present), and can be used directly in the retrieval process.

No retrieval scheme is perfect, of course. The following section addresses the use of simulation in characterizing the uncertainty of SST estimates.

8. USE OF SIMULATIONS IN UNCERTAINTY ESTIMATION

Uncertainty estimates should always be provided with retrieved SSTs. "Uncertainty" means "doubt," and an uncertainty estimate is a number that quantifies the degree of doubt in the measured value. A very clear way to express uncertainty is to state a confidence interval. For example, one could state that the outcome of the SST retrieval gives us 95% confidence that the SST is between 22.3 and 23.4 °C at a particular location at a given time. Our degree of doubt about the true SST lying within this range would be 5%.

Uncertainty is different from error. "Error" means "mistake," and "the measurement error" is the degree to which the measured value differs from the truth. We do not generally know the truth, and therefore do not generally know the error. One could in principle imagine making an in situ measurement of skin SST across a satellite footprint that was so accurate and precise as to be negligibly different from truth, but it has never been done.

(The above usage of the words "uncertainty" and "error" is consistent with the everyday nonscientific meanings of the words, and is in line with international standards for discussing uncertainty in measurement [65]. However, it is not uncommon among scientists to use "error" to mean both uncertainty and error, and one can sometimes find both meanings used in a single sentence.)

One reason that simulating the satellite retrieval process is valuable is that *in simulation we do know the truth* because we do each simulation for a geophysical state that we define and control. If we use a particular SST, x_b, as input to our simulation, and the (simulated) retrieved SST is $\widehat{x}$, then the (simulated) measurement error is $\widehat{x} - x_b$. As will be shown below, this capability means that aspects of SST uncertainty can be estimated using simulations.

Rather than quoting a confidence interval, a common way of presenting uncertainty information is to give a best estimate SST with its "standard uncertainty" (σ_x). The standard uncertainty is defined as the standard deviation of the error distribution. For a Gaussian error distribution, the 95% confidence interval is approximately from $\widehat{x} - 2\sigma_x$ to $\widehat{x} + 2\sigma_x$. Note that the word "standard" is often omitted, so that the use of "1σ" to characterize the uncertainty is implicit rather than explicit. This will be the case for the remainder of this chapter.

Given that we do not in general know the error in a given retrieved SST, how can the standard deviation of the error distribution be found? There are two methods: "error propagation" (which may or may not involve simulation) and simulation of the retrieval error distribution for noise-free observations.

Error propagation describes how an error in the fundamental satellite measurement (counts, radiance, or BT) changes the retrieved SST. This method is particularly appropriate to estimating the effect of sensor noise on the uncertainty of a retrieved SST. Typically a thermal imager at some point in its scan will view a calibration target, which is (ideally) a black body of known, uniform temperature. From the random fluctuations of the detector response while viewing this uniform temperature, an estimate of the sensor noise can be formed. The sensor noise is then transformed into the "noise equivalent differential temperature" (NEDT), which is the noise expressed in temperature units. In other words, NEDT is the random uncertainty in observed BT. NEDT typically has different values for different channels, depends on scene temperature (usually being greater for lower radiance scenes), and may change with operating conditions of the sensor or as the sensor ages. NEDT for all thermal channels should be routinely provided in satellite radiance (Level 1) products. Where it is not, more generic estimates of NEDT must be used, such as the design noise levels or noise characteristics established during sensor commissioning in space.

Consider a set of clear-sky BTs, y (i.e., the BTs of different channels, i, are listed in a column vector), that are the inputs to an SST retrieval algorithm R, such that $\widehat{x} = R(y)$. Let the errors due to sensor noise in the BTs be e (the channel ordering being the same as for y). The error in SST corresponding to the noise is

$$e_x = \left[\frac{\partial R}{\partial y}\right]^T e = \sum_i \frac{\partial R}{\partial y_i} e_i \qquad (6)$$

For a particular retrieval, we do not know the BT errors, and therefore cannot know e_x. We do have an estimate of the distribution from which the errors are drawn, namely, for a given channel, a Gaussian distribution with standard deviation defined by the NEDT of that channel. Let the NEDT for the

i^{th} channel be σ_i. The (standard) uncertainty in retrieval SST arising from noise is given by

$$\sigma_x = \sqrt{\sum_i \left(\frac{\partial R}{\partial y_i}\sigma_i\right)^2} \qquad (7)$$

This equation for uncertainty propagation is readily interpreted in the light of Eqn (6). The noise in each channel contributes noise in the SST equal to the NEDT multiplied by the sensitivity of the retrieved SST for that channel. Because the noise is assumed to be uncorrelated between channels, these contributions combine as the square root of the sum of the squares, according to the properties of Gaussian distributions.

In some cases, the retrieval method may be sufficiently simple that the SST noise can be estimated analytically. For example, for SST retrievals in the form of Eqn (1), we have

$$\frac{\partial R}{\partial y_i} = a_i + b_i S \qquad (8)$$

and the SST noise can be found in terms of the retrieval coefficients and the NEDT values. Different SST retrieval expressions amplify noise to different degrees. In the simple case where the NEDT is the same for all channels ($\sigma_i = \sigma_{\text{NEDT}}$), the noise amplification is

$$\frac{\sigma_x}{\sigma_{\text{NEDT}}} = \sqrt{\sum_i (a_i + b_i S)^2} \qquad (9)$$

This expression highlights that the magnitude of the retrieval coefficients directly affects the SST noise. SST noise is generally greater at the edge of the satellite swath, where S deviates most from zero (a_i and b_i are often of the same sign). Another application of Eqn (9) is to the difference in noise amplification between a dual-view radiometer and a single-view radiometer using the same channels. The dual-view radiometer, since it carries more information about the atmospheric effects from viewing a given location at two angles, is capable of less biased (more accurate) SST retrievals than from a single view. However, the noise amplification is generally greater when using two views because, with more channels, the sum of squared coefficients is greater; the SSTs are therefore likely to be more noisy (less precise) for a given pixel. SST noise suppression techniques can be applied that bring in additional information from other nearby pixels. The effect of these techniques can similarly be assessed using the principles of uncertainty propagation illustrated above.

For a retrieval of incremental form (Eqn (5)), the SST uncertainty from radiometric noise can be written as

$$\sigma_x = \sqrt{\mathbf{G}\boldsymbol{\sigma}\boldsymbol{\sigma}^T\mathbf{G}^T} = \sqrt{\mathbf{G}\mathbf{S}_\varepsilon\mathbf{G}^T} \qquad (10)$$

where the product $\boldsymbol{\sigma}\boldsymbol{\sigma}^T$ is an error-covariance matrix for the radiometric noise. If the noise is uncorrelated between channels (as we generally expect), this covariance matrix is diagonal with elements σ_i^2. However, $\boldsymbol{\sigma}\boldsymbol{\sigma}^T$ should be substituted with a nondiagonal covariance matrix, $\mathbf{S_e}$, if the BT errors are known to be correlated between channels.

If the retrieval method, $\widehat{x} = R(\mathbf{y})$, is not analytically differentiable, $\frac{\partial R}{\partial y_i}$ can be estimated numerically by perturbation. In this approach, the retrieval process is performed first for the observed BTs, $\mathbf{y}$, and then again substituting for y_i a perturbed value $y_i + \delta y$, for each channel in turn. For each perturbed retrieval, the SST obtained differs from the first result by δx_i, and $\frac{\partial R}{\partial y_i} = \frac{\delta x_i}{\delta y_i}$, and Eqn (7) can then be used. It is important to select the value of perturbation carefully if the retrieval is nonlinear, and this itself can be explored by simulation.

Noise in the BTs is not the only source of error in a retrieval of SST. There is always some degree of ambiguity in the inversion. The source of ambiguity is that many realistic surface-and-atmospheric states can give essentially the same observed radiances. These radiometrically indistinguishable states do not all have the same SST. Whatever inverse approach is used, there is an irreducible uncertainty in the retrieved SST, even if the most probable solution is correctly identified. In addition, the algorithm itself may be suboptimal, returning SSTs estimates that contain errors other than those arising from irreducible ambiguity.

Simulation of measured radiances and the SST retrieval process can be used to quantify retrieval uncertainty from algorithmic limitations. Simulations should be done for a sample of states representing the full range of application of the retrieval. The standard deviation between the simulated retrieved SST and the true SST in the simulation, $SD(R(\mathbf{y}(\mathbf{x_p})) - x_p)$, is the standard uncertainty associated with algorithmic limitations. It may be that this uncertainty varies systematically with some readily accessible factors. Obvious factors to explore include latitude, total column water vapor, satellite zenith angle, and any switches within the retrieval process (e.g., different parameters in the retrieval during day versus night). The uncertainty estimate can then be parameterized or varied as a function of such a factor. For examples, see Ref. [66] and Figure 11.

If the simulation of noise-free retrieval error is done for a particular satellite swath, it becomes clear that the error is spatially coherent (Figure 12). Given its origin (imperfect/ambiguous ability to infer the effect of the atmosphere on the radiances), this is not surprising. The atmosphere has "synoptic scales" of spatiotemporal correlation ($10^2 - 10^3$ km) that are generally much longer than the extent of a satellite pixel (1–10 km). Within those correlation scales, the noise-free retrieval error can be expected to be similar because the atmospheric profile is similar. In other words, this component of uncertainty arises from errors that are correlated on the synoptic scales of the atmosphere—they are "locally systematic."

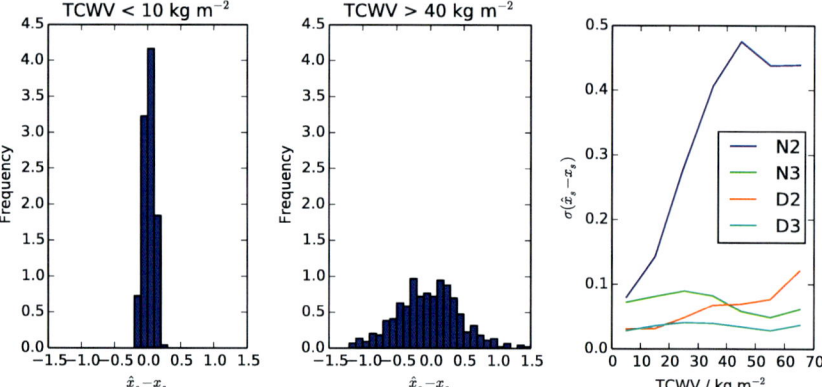

FIGURE 11 Example of algorithmic retrieval errors for a linear coefficient-based retrieval (Eqn (1)) [66]. Left panel: histogram of retrieved-true SST for dry atmospheres using 2-channel single-view algorithm. Center panel: histogram of retrieved-true SST for wet atmospheres using 2-channel single-view algorithm. Right panel: SD of retrieved-true SST as a function of TCWV for four different algorithms: nadir 2-channel (N2), nadir 3-channel (N3), dual 2-channel (D2), and dual 3-channel (D3).

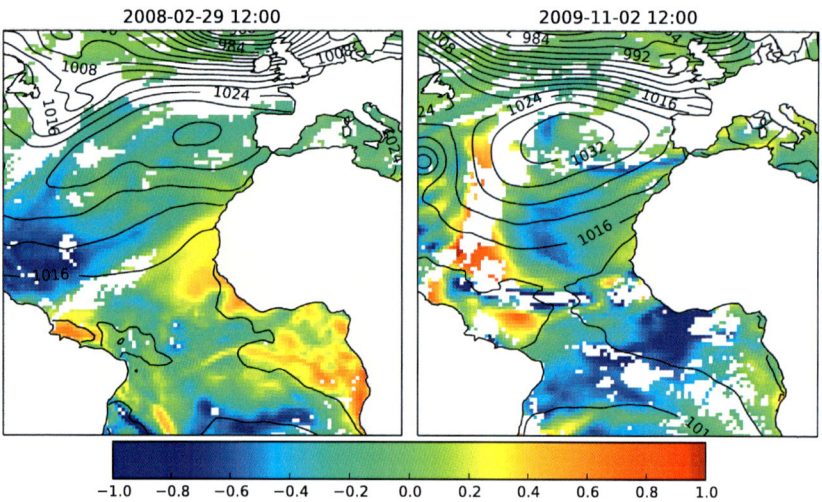

FIGURE 12 Simulated retrieval error (K) over the Atlantic on two different days with isobars of sea-level pressure overlaid. The retrieval algorithm in this example uses two channels (11 and 12 μm). Areas with 100% cloud cover in the NWP are shown in white.

It follows that this contribution to SST errors is neither fully "random" nor fully "systematic". This is probably true of all spatially extended variables estimated via Earth Observation. Most textbook discussions of errors distinguish random and systematic effects, but discussion of errors with spatiotemporal correlations over certain scales is less common outside of disciplines such as meteorology. In addition to quantifying the magnitude of this "locally systematic" component of uncertainty, simulation of a sufficient sample of realistic scenarios should in principle allow estimation of the degree of correlation as a function of spatiotemporal separation of different SSTs. In this way, the error-covariance matrix of a set of retrieved SSTs could be developed, which in turn will allow proper estimation of the uncertainty in SSTs averaged up to different space-time scales. This is an application of thermal simulation that has not been fully explored at the time of writing.

The final component of uncertainty in SST is associated with unknown systematic errors. This includes the uncertainty of calibration in each of the BTs contributing to the SST retrieval. Calibration error may change over time (e.g., as a sensor ages) and with instrumental state (e.g., with the changing instrument temperature round an orbit). Therefore, although fully systematic, these errors are not necessarily constant in space and time.

Where the SST retrieval process depends on simulation (as in "optimal estimation"), any uncorrected bias in the RTM will have an effect similar to systematic error in calibration. Because such a retrieval effectively adjusts the estimated SST on the basis of differences between simulated and observed BTs, the "calibration" of the sensor *relative to the RTM* is arguably more crucial than the absolute calibration.

The systematic error for a given SST is unknown—if one knew it, one would correct the error! So, how can the systematic uncertainty be estimated? The equations of uncertainty propagation in this case are the same as for the propagation of noise discussed above. However, the meaning is different. The uncertainty from systematic effects is not reduced in an SST averaged over many pixels, unlike the uncertainty from radiometric noise. Moreover, full characterization of the systematic uncertainty in the BTs may not be available: most sensors will have undergone detailed prelaunch calibration in controlled laboratory conditions, but the systematic uncertainty when the instrument is operated in space may be different. Absolute calibration references in space are not yet implemented for thermal Earth observations, so that developing an understanding of BT systematic errors and uncertainties usually involves considerable detective work based on understanding of the working of the individual sensors. There is a need to develop a framework for "Earth Observation metrology" to develop understanding and techniques for linking instrumental calibration to uncertainties in geophysical products. In such a framework, simulation of instruments and their observations would doubtless be central.

9. CONCLUSION

The main principles of infrared RT can be understood relatively simply in terms of the emission and absorption of monochromatic radiance in the atmosphere. Detailed calculations need spectrally resolved information about the absorption cross-section of a range of radiatively active gases as a function of the altitude-dependent atmospheric state (temperature, pressure, and concentration). Spectroscopic databases provide such information, and are exploited in RTMs.

The simulation capability provided by RTMs is central to contemporary remote sensing. Thermal remote sensing of the SST is no exception, aided by the excellent accuracy of available simulations in this wavelength domain. To obtain geophysical information from radiances, we need to know how radiances depend on the geophysical state. Simulations assist in classification of images, the inversion of radiances to geophysical estimates, the assessment of the sensitivity of estimates to geophysical variability, and in the quantification of the uncertainty in these estimates.

CDRs for SST span two or more decades of retrievals based on thermal remote sensing using atmospheric window wavelengths. The datasets need to be consistent, yet must be constructed from a series of satellite instruments that, though similar, are not identical. Simulation of clear-sky BTs for these different sensors using consistent forward modeling can give indications of any calibration inconsistency between sensors.

Over time, the concentration of radiatively active gases in the atmosphere has changed. The effects of this on satellite observations can be simulated, and this supports development of retrieval methods that account for the changing composition of the atmosphere. A more dramatic change in composition of the atmosphere, which can affect global SST retrieval adversely, is when major volcanic eruptions cause formation of a layer of aerosol in the stratosphere that persist for 1 or 2 years. Simulation of the impact of such aerosol on BTs allows retrieval methods to be devised that are relatively insensitive to those impacts (at least where BTs from four or more surface-sensitive wavelength/view-angle combinations are obtained). This is an example of the extra insight obtained by simulation being of practical help in addressing a major problem in the SST CDR.

It is important to provide useful uncertainty information in any CDR. Uncertainty information needs to discriminate more and less uncertain data. Using simulation, the magnitude of different factors that contribute to uncertainty can be evaluated, as the basis for a method for estimating uncertainty information to be associated with each SST in the CDR. Since the error-covariance structure differs between errors that arise from different factors, simulation-based insight is needed in order to develop models for uncertainty estimation across different scales of space-time averaging.

Cloud detection, retrieval of SST, estimation of uncertainty, and sensitivity: simulation capability is informative in all these steps toward creating records of SST of the quality required for applications in the area of climate.

REFERENCES

[1] R.A. Hanel, et al., Exploration of the Solar System by Infrared Remote Sensing, second ed., Cambridge University Press, Cambridge, UK, 2003.
[2] C. Hill, et al., A new relational database structure and online interface for the HITRAN database, J. Q. Spectrosc. Radiat. Transfer 130 (2013) 51−61.
[3] L.S. Rothman, I.E. Gordon, The HITRAN molecular database, in: J.D. Gillaspy, W.L. Wiese, Y.A. Podpaly (Eds.), Eighth International Conference on Atomic and Molecular Data and Their Applications, 2013, pp. 223−231.
[4] P. Lu, H. Zhang, X. Jing, The effects of different HITRAN versions on calculated long-wave radiation and uncertainty evaluation, Acta Meteorol. Sin. 26 (3) (2012) 389−398.
[5] S.A. Clough, M.W. Shephard, E.J. Mlawer, J.S. Delamere, M.J. Iacono, K. Cady-Pereira, S. Boukabara, P.D. Brown, Atmospheric radiative transfer modeling: a summary of the AER codes, J. Q. Spectrosc. Radiat. Transfer 91 (2005) 233−244.
[6] A. Dudhia, Reference Forward Model, 2014. Available from: http://www.atm.ox.ac.uk/RFM/.
[7] A. Berk, G.P. Anderson, P.K. Acharya, L.S. Bernstein, L. Muratov, J. Lee, M. Fox, S.M. Adler-Golden, J.H. Chetwynd, M.L. Hoke, R.B. Lockwood, J.A. Gardner, T.W. Cooley, C.C. Borel, P.E. Lewis, E.P. Shettle, MODTRAN5: 2006 update, in: Proceedings of SPIE, 2006, p. 6233.
[8] R. Saunders, M. Matricardi, P. Brunel, An improved fast radiative transfer model for assimilation of satellite radiance observations, Q. J. R. Meteorol. Soc. 125 (1999) 1407−1425.
[9] M. Matricardi, A principal component based version of the RTTOV fast radiative transfer model, Q. J. R. Meteorol. Soc. 136 (2010) 1823−1835.
[10] O. Embury, C.J. Merchant, M.J. Filipiak, A reprocessing for climate of sea surface temperature from the along-track scanning radiometers: basis in radiative transfer, Remote Sens. Environ. 116 (2012) 32−46.
[11] K.P. Shine, I.V. Ptashnik, G. Radel, The water vapour continuum: brief history and recent developments, Surv. Geophys. 33 (3−4) (2012) 535−555.
[12] Y.I. Baranov, W.J. Lafferty, The water vapour self- and water-nitrogen continuum absorption in the 1000 and 2500 cm^{-1} atmospheric windows, Philos. Trans. R. Soc. A 370 (1968) (2012) 2578−2589.
[13] Y.I. Baranov, et al., Water-vapor continuum absorption in the 800-1250 cm^{-1} spectral region at temperatures from 311 to 363 K, J. Q. Spectrosc. Radiat. Transfer 109 (12−13) (2008) 2291−2302.
[14] A. Lambert, R.G. Grainger, J.J. Remedios, C.D. Rodgers, M. Corney, F.W. Taylor, Measurements of the evolution of the Mt. Pinatubo aerosol cloud by ISAMS, Geophys. Res. Lett. 20 (1993) 1287−1290.
[15] M. Hess, P. Koepke, I. Schult, Optical properties of aerosols and clouds: the software package OPAC, Bull. Am. Meteorol. Soc. 79 (5) (1998) 831−844.
[16] O. Dubovik, et al., Variability of absorption and optical properties of key aerosol types observed in worldwide locations, J. Atmos. Sci. 59 (3) (2002) 590−608.
[17] G.G. Stokes, On the intensity of the light reflected from or transmitted through a pile of plates, Proc. R. Soc. Lond. 11 (1862) 545−556.
[18] J.E. Hansen, Multiple scattering of polarized light in planetary atmospheres. Part I. The doubling method, J. Atmos. Sci. 28 (1971) 120−125.
[19] K. Stamnes, et al., Numerically stable algorithm for discrete-ordinate-method radiative transfer in multiple scattering and emitting layered media, Appl. Opt. 27 (1988) 2502.

[20] J.H. Joseph, W.J. Wiscombe, J.A. Weinman, Delta-Eddington approximation for radiative flux-transfer, J. Atmos. Sci. 33 (12) (1976) 2452–2459.

[21] J.E. Bertie, Z. Lan, Infrared intensities of liquids XX: the intensity of the OH stretching band of liquid water revisited, and the best current values of the optical constants of $H_2O(l)$ at 25 °C between 15,000 and 1 cm^{-1}, Appl. Spectrosc. 50 (1996) 1047–1057.

[22] H.D. Downing, D. Williams, Optical constants of water in the infrared, J. Geophys. Res. 80 (1975) 1656–1661.

[23] D. Friedman, Infrared characteristics of ocean water (1.5–15 μm), Appl. Opt. 8 (1969) 2073.

[24] L.W. Pinkley, D. Williams, Optical properties of sea water in the infrared, J. Opt. Soc. Am. 66 (1976) 554.

[25] L.W. Pinkley, P.P. Sethna, D. Williams, Optical constants of water in the infrared: influence of temperature, J. Opt. Soc. Am. 67 (1977) 494.

[26] S.M. Newman, et al., Temperature and salinity dependence of sea surface emissivity in the thermal infrared, Q. J. R. Meteorol. Soc. 131 (2005) 2539–2557.

[27] C. Cox, W. Munk, Measurement of the roughness of the sea surface from photographs of the Sun's Glitter, J. Opt. Soc. Am. 44 (1954) 838.

[28] P.M. Saunders, Shadowing on the ocean and the existence of the horizon, J. Geophys. Res. 72 (1967) 4643–4649.

[29] P.M. Saunders, Radiance of sea and sky in the infrared window 800–1200 cm^{-1}, J. Opt. Soc. Am. 58 (1968) 645.

[30] M. Filipiak, et al., Refractive indices (500–3500 cm^{-1}) and emissivity (600–3350 cm^{-1}) of pure water and seawater [Dataset], University of Edinburgh, Edinburgh, UK, 2008.

[31] X. Wu, W.L. Smith, Emissivity of rough sea surface for 8–13 μm: modeling and verification, Appl. Opt. 36 (1997) 2609.

[32] P.D. Watts, M.R. Allen, T.J. Nightingale, Wind speed effects on sea surface emission and reflection for the along track scanning radiometer, J. Atmos. Oceanic Technol. 13 (1996) 126–141.

[33] N.R. Nalli, P.J. Minnett, P. van Delst, Emissivity and reflection model for calculating unpolarized isotropic water surface-leaving radiance in the infrared. I: theoretical development and calculations, Appl. Opt. 47 (2008) 3701.

[34] K. Yoshimori, K. Itoh, Y. Ichioka, Thermal radiative and reflective characteristics of a wind-roughened water surface, J. Opt. Soc. Am. A 11 (1994) 1886.

[35] K. Yoshimori, K. Itoh, Y. Ichioka, Optical characteristics of a wind-roughened water surface: a two-dimensional theory, Appl. Opt. 34 (1995) 6236.

[36] H. Li, N. Pinel, C. Bourlier, Polarized infrared emissivity of one-dimensional Gaussian sea surfaces with surface reflections, Appl. Opt. 50 (2011) 4611.

[37] C.J. Merchant, et al., Saharan dust in nighttime thermal imagery: detection and reduction of related biases in retrieved sea surface temperature, Remote Sens. Environ. 104 (1) (2006) 15–30.

[38] C.E. Bulgin, S. Eastwood, O. Embury, C.J. Merchant, C. Donlon, The sea surface temperature climate change initiative: Alternative image classification algorithms for sea-ice affected oceans, Remote Sens. Environ. (2014). http://dx.doi.org/10.1016/j.rse.2013.11.022.

[39] M. Derrien, H. Le Gleau, MSG/SEVIRI cloud mask and type from SAFNWC, Int. J. Remote Sens. 26 (21) (2005) 4707–4732.

[40] A.M. Závody, C.T. Mutlow, D.T. Llewellyn-Jones, Cloud clearing over the ocean in the processing of data from the along-track scanning radiometer (ATSR), J. Atmos. Oceanic Technol. 17 (2000) 595–615.

[41] A. Dybbroe, K.-G. Karlsson, A. Thoss, NWCSAF AVHRR cloud detection and analysis using dynamic thresholds and radiative Transfer modeling. Part I: algorithm description, J. Appl. Meteorol. 44 (2005) 39–54.

[42] C.J. Merchant, et al., Probabilistic physically based cloud screening of satellite infrared imagery for operational sea surface temperature retrieval, Q. J. R. Meteorol. Soc. 131 (611) (2005) 2735–2755.

[43] S. Mackie, et al., Generalized Bayesian cloud detection for satellite imagery. Part 1: technique and validation for night-time imagery over land and sea, Int. J. Remote Sens. 31 (10) (2010) 2573–2594.

[44] L.G. Istomina, et al., The detection of cloud-free snow-covered areas using AATSR measurements, Atmos. Meas. Tech. 3 (4) (2010) 1005–1017.

[45] A.A. Kokhanovsky, A semianalytical cloud retrieval algorithm using backscattered radiation in 0.4–2.4 μm spectral region, J. Geophys. Res. 108 (D1) (2003).

[46] L. Murino, et al., Cloud detection of MODIS multispectral images, J. Atmos. Oceanic Technol. 31 (2) (2014) 347–365.

[47] K. Masuda, Infrared sea surface emissivity including multiple reflection effect for isotropic Gaussian slope distribution model, Remote Sens. Environ. 103 (2006) 488–496.

[48] C. Bourlier, Unpolarized infrared emissivity with shadow from anisotropic rough sea surfaces with non-Gaussian statistics, Appl. Opt. 44 (2005) 4335.

[49] C. Bourlier, Unpolarized emissivity with shadow and multiple reflections from random rough surfaces with the geometric optics approximation: application to Gaussian sea surfaces in the infrared band, Appl. Opt. 45 (2006) 6241–6254.

[50] K. Caillault, et al., Multiresolution optical characteristics of rough sea surface in the infrared, Appl. Opt. 46 (2007) 5471.

[51] S. Mackie, et al., Generalized Bayesian cloud detection for satellite imagery. Part 2: technique and validation for daytime imagery, Int. J. Remote Sens. 31 (10) (2010) 2595–2621.

[52] D. Anding, R. Kauth, Estimation of sea surface temperature from space, Remote Sens. Environ. 1 (1970) 217–220.

[53] I.J. Barton, Satellite-derived sea surface temperatures: current status, J. Geophys. Res. 100 (1995) 8777–8790.

[54] C.J. Merchant, Thermal remote sensing of sea surface temperature, in: C. Kuenzer, S. Dech (Eds.), Thermal Infrared Remote Sensing: Sensors, Methods and Applications, Springer Netherlands, Dordrecht, 2013, pp. 287–313.

[55] C.J. Merchant, P. Le Borgne, Retrieval of sea surface temperature from space, based on modeling of infrared radiative transfer: capabilities and limitations, J. Atmos. Oceanic Technol. 21 (2004) 1734–1746.

[56] C. Francois, et al., Definition of a radiosounding database for sea surface brightness temperature simulations — application to sea surface temperature retrieval algorithm determination, Remote Sens. Environ. 81 (2–3) (2002) 309–326.

[57] K.A. Kilpatrick, G.P. Podestá, R. Evans, Overview of the NOAA/NASA advanced very high resolution radiometer pathfinder algorithm for sea surface temperature and associated matchup database, J. Geophys. Res. 106 (2001) 9179.

[58] C.J. Merchant, et al., Toward the elimination of bias in satellite retrievals of sea surface temperature 1. theory, modeling and interalgorithm comparison, J. Geophys. Res.: Oceans 104 (C10) (1999) 23565–23578.

[59] C.J. Merchant, A.R. Harris, Toward the elimination of bias in satellite retrievals of sea surface temperature: 2. Comparison with in situ measurements, J. Geophys. Res. 104 (1999) 23579.

[60] P. Le Borgne, H. Roquet, C.J. Merchant, Estimation of sea surface temperature from the spinning enhanced visible and infrared Imager, improved using numerical weather prediction, Remote Sens. Environ. 115 (1) (2011) 55–65.

[61] C.D. Rodgers, Inverse methods for atmospheric sounding – theory and practice, in: F.W. Taylor (Ed.), Series on Atmospheric Oceanic and Planetary Physics, vol. 2, World Scientific Publishing Co. Pte. Ltd, 2000.

[62] C.J. Merchant, et al., Retrieval characteristics of non-linear sea surface temperature from the advanced very high resolution radiometer, Geophys. Res. Lett. 36 (17) (2009).

[63] C.J. Merchant, et al., Extended optimal estimation techniques for sea surface temperature from the spinning enhanced visible and infra-red imager (SEVIRI), Remote Sens. Environ. 131 (2013) 287–297.

[64] C.J. Merchant, et al., Extended optimal estimation techniques for sea surface temperature from the spinning enhanced visible and infra-red imager (SEVIRI) (vol 131, pg 287, 2013), Remote Sens. Environ. 137 (2013) 331–332.

[65] [BIPM], B.I.d.P.e.M., Evaluation of Measurement Data—Guide to the Expression of Uncertainty in Measurement., 2008.

[66] O. Embury, C.J. Merchant, A reprocessing for climate of sea surface temperature from the along-track scanning radiometers: a new retrieval scheme, Remote Sens. Environ. 116 (2012) 47–61.

[67] K. Masuda, T. Takashima, Y. Takayama, Emissivity of pure and sea waters for the model sea surface in the infrared window regions, Remote Sens. Environ. 24 (2) (1988) 313–329, ISSN 0034-4257, http://dx.doi.org/10.1016/0034-4257(88)90032-6.

Chapter 5

In Situ Measurement Strategies

Giuseppe Zibordi,[1],* Craig J. Donlon[2]
[1] *European Commission, Joint Research Centre, Ispra, Italy;* [2] *European Space Agency/ESTEC, Noordwijk, The Netherlands*
*Corresponding author: E-mail: giuseppe.zibordi@jrc.ec.europa.eu

Comprehensive field measurement programs supporting satellite missions are essential for the generation of data products with the traceability, accuracy, and consistency required for climate-change investigations. Given the science requirements set for the upcoming satellite missions targeted to climate-change investigations, and taking into account the lessons learnt during previous missions, the following chapters outline the basic strategies, procedures, and protocols to be considered for the design and implementation of measurement programs delivering the necessary in situ high-quality reference measurements (Fiducial Reference Measurements, FRM) from optical radiometry.

After reviewing the strategies applied in recent decades, the chapter on ocean color missions targeted to climate-change investigations highlights the requirement for long-term FRM from:

- At least one dedicated site delivering high-quality marine radiometric data for system vicarious calibration;
- Geographically distributed sites established to maximize the number of matchups for the validation of primary satellite data products.

The validation of high-level ocean-color-derived data products together with the development and assessment of bio-optical models should be supported through globally and seasonally distributed FRM of radiometric and bio-optical matching data from

- regions representative of the variety of world aerosols and marine waters.

Field measurements should be collected and reduced through:

- measurement protocols adhering to best practice and resulting from community consensus;
- fully characterized and regularly calibrated instruments assuring the best traceability, accuracy, and stability;
- state-of-the-art processing codes relying on the most updated and community-accepted data-reduction schemes.

Ocean-color radiometric FRM should also be:

- complemented by uncertainty values quantified accounting for each potential uncertainty source;
- timely accessible at different levels of quality control.

When applicable, the former elements should benefit from:

- intercomparisons to verify each step leading to the generation of the in situ data products;
- increasing efforts on standardization and networking;
- development and implementation of advanced methods and instruments, and their gradual integration into operational field activities.

The chapter on in situ thermal infrared radiometry (TIR) focuses on the fact that ship-borne radiometers can provide FRM that are traceable to the International Temperature scale 1990. These measurements are essential to quantify uncertainties in satellite-derived sea-surface temperature (SST). This concept and the requirements to generating uncertainty budgets and estimates for ship-borne FRM TIR radiometer measurements is extensively discussed in the context of satellite SST Climate Data Records (CDR). This also includes the validation of ship-borne FRM TIR radiometer calibration performance. Emphasis is placed on the need to understand the uncertainties introduced by the validation process itself through the comparison of collocated and near-contemporaneous satellite and in situ measurements. Finally, the concept and scope of FRM ship-borne TIR radiometer network (SBRN) are defined and justified by considering requirements for a number of applications relevant to the development and traceability of SST CDR. Specifically, key requirements for an FRM TIR SBRN is the need for:

- at least one sustained, regularly repeated FRM TIR SBRN "line" (i.e., ship route) maintained for a number of defined latitudinal atmospheric regimes (LAR). It is emphasized that SRBN lines are also required between the end and the beginning of independent satellite missions where a gap may occur.

As far as possible,

- the location of FRM TIR SBRN measurements shall be continuous and optimized for maximum near-contemporaneous and colocated satellite and FRM matchups.

FRM TIR SBRN line(s) should sample

- different latitudes to resolve any thermal cycling of instruments around an orbit, and additionally in both the day- and night-time parts of the orbit.

FRM TIR SBRN measurements shall

- be collected using defined time and space matchup criteria;

- have an absolute accuracy and stability commensurate with that needed to provide the link to the satellite (for climate applications, these should be <0.05 K and <0.015 K/decade at a scale of 100–1000 km, respectively);
- be collected according to internationally agreed protocols established and maintained by scientific consensus;
- be supported by ancillary data to ensure that the environmental context of the radiometer measurements is fully quantified;
- target an absolute location accuracy of <100 m with respect to WGS-84;
- be quality controlled using outlier-tolerant (robust) statistics;
- maintain and demonstrate SI traceability with a full uncertainty analysis.

Each of the key requirements summarized above is elaborated in the following two chapters based on international consensus and agreed best practices.

Chapter 5.1

Requirements and Strategies for In situ Radiometry in Support of Satellite Ocean Color

Giuseppe Zibordi,[1,*] Kenneth J. Voss[2]
[1] *European Commission, Joint Research Centre, Ispra, Italy;* [2] *Physics Department, University of Miami, Coral Gables, FL, USA*
*Corresponding author: E-mail: giuseppe.zibordi@jrc.ec.europa.eu

Chapter Outline

1. Introduction — 532
2. Overview of Past and Current Field-Related Radiometric Activities — 533
   - 2.1 Field Measurements — 533
     - 2.1.1 Long-Term Measurements for System Vicarious Calibration — 534
     - 2.1.2 Long-Term Measurements for the Validation of Satellite Data Products — 535
     - 2.1.3 Targeted Time-Limited Comprehensive Measurements — 536
     - 2.1.4 Targeted Long-Term Comprehensive Measurements — 538
   - 2.2 Intercomparisons — 538
     - 2.2.1 Laboratory Intercomparisons — 538
     - 2.2.2 Field Intercomparisons — 540
     - 2.2.3 Data Processing — 541
   - 2.3 Data Repositories — 542
3. Requirements and Strategies for Future Satellite Ocean-Color Missions — 543
   - 3.1 Field Measurements for System Vicarious Calibration — 544
   - 3.2 Field Measurements for the Validation of Satellite Data Products — 546
   - 3.3 Field Measurements for Bio-Optical Modeling — 547
   - 3.4 Protocols Revision and Consolidation — 547
   - 3.5 Calibration and Characterization of Field Radiometers — 547
   - 3.6 Data Reduction, Quality Control, and (re) Processing — 548
   - 3.7 Accuracy Tailored to Applications — 549

3.8 Archival and Access	549	3.11 Development and Implementation	551
3.9 Intercomparisons to Secure Accuracy and Best Practice	549	**4. Summary and Way Forward**	551
3.10 Standardization and Networking	550	**References**	552

1. INTRODUCTION

Decadal time-series of Earth Observation (EO) data from multiple missions are a unique source of climate variables at different spatial scales. Nevertheless, the successful application of EO data to the detection and quantification of small trends embedded in large natural variations requires traceability to the International System of Units and, high accuracy and consistency over time of these data products. In the case of satellite ocean color, this can only be met through (1) extraordinary prelaunch sensor characterization and calibration; (2) on-orbit tracking of sensor radiometric stability (i.e., accuracy variation with time) and adjustment of responsivity changes; (3) indirect calibration of combined sensor responsivity and data reduction algorithms (i.e., system vicarious calibration); and (4) assessment (i.e., validation) of data products by quantifying statistical indices (e.g., bias and dispersion). The fulfillment of these tasks requires the definition and implementation of calibration and validation programs lasting beyond the lifetime of individual EO missions.

An essential component of ocean-color calibration and validation programs is the collection of high-quality in situ data, which are central for system vicarious calibration, validation of data products, and additionally development and assessment of bio-optical models for the generation of derived high-level products. Because of this, the collection and handling of field data have an important position in postlaunch EO strategies. For instance, in situ data supporting postlaunch system vicarious calibration must be of extremely high quality to allow accurate simulation of the radiometric signal at the EO sensor for site(s) representative of the most common marine and atmospheric conditions. These in situ data are generally produced through unique long-term deployment buoy-based systems with the aid of state-of-the-art methods for the characterization and calibration of field instruments, for data collection, and finally for data reduction and quality control. In addition to system vicarious calibration, validation and bio-optical modeling activities require in situ data representative of the multitude of observation conditions covering the variety of world marine water types. Because of this, validation data sets are commonly constructed by combining measurements performed with a number of instruments operated by independent teams on a variety of deployment platforms (i.e., ships, buoys, and offshore fixed structures). Thus, despite the effort to enforce the use of community protocols for the characterization and calibration of in situ instruments, and additionally for the collection, reduction, and quality control of data, it is unlikely that validation data sets can exhibit the same traceability, accuracy, and consistency as in situ measurements specifically produced to support system vicarious calibration.

This chapter discusses requirements for in situ optical radiometry data supporting satellite ocean-color missions targeted to climate-change applications. Emphasis is placed on basic strategies relevant to measurement programs for system vicarious calibration and validation of data products pursuing the principle that *adequately sampled, carefully calibrated, quality-controlled, and archived data for key elements of the climate system will be useful indefinitely* [1].

2. OVERVIEW OF PAST AND CURRENT FIELD-RELATED RADIOMETRIC ACTIVITIES

Since the early 1990s, during the preparatory phase for the launch of the Sea Wide Field-of-View Sensor (SeaWiFS) and Ocean Color Temperature Scanner (OCTS), a number of laboratory and field activities were established to support the forthcoming satellite ocean-color data. These activities were then continued and expanded during successive missions centered on the Moderate-Resolution Imaging Spectroradiometer (MODIS), the Global Imager (GLI), the Medium Resolution Imaging Spectrometer (MERIS), and the Visible-Infrared Imaging Radiometer (VIIRS). In addition to the planning and execution of field measurements, these activities embraced the development and assessment of measurement protocols, the design and implementation of in situ data repositories, and the execution of intercomparison activities as a way to increase traceability and accuracy of in situ data products. The following subsections summarize some of these activities with the objective of highlighting the strategies developed in the recent decades for the generation of in situ data supporting satellite ocean-color missions targeted to climate-change applications.

2.1 Field Measurements

During recent decades, in situ data for postlaunch system vicarious calibration were mostly produced with buoy-based systems. Alternately, the validation of EO data products have relied on time-series data from fixed deployment structures or moorings, and also on the episodic comprehensive data collected through oceanographic ships. The objective of this section is to provide examples of in situ measurement programs developed to specifically support satellite ocean-color activities using the capabilities offered by diverse deployment platforms such as moorings, oceanographic ships, or fixed deployment structures.

2.1.1 Long-Term Measurements for System Vicarious Calibration

Examples of in situ systems designed to support vicarious calibration of satellite ocean-color sensors are (1) the Marine Optical Buoy (MOBY) jointly developed by the National Oceanic and Atmospheric Administration and the National Aeronautics and Space Administration (NASA) for the SeaWiFS and MODIS sensors; and (2) the Buoy for the Acquisition of a Long-Term Optical Time Series (Bouée pour L'acquisition de Séries Optiques à Long Terme, BOUSSOLE) developed under the leadership of the Laboratoire d'Océanographie de Villefranche for the MERIS sensor.

Since 1997, MOBY has been deployed approximately 11 nautical miles from Lanai, Hawaii, in 1200 m water depth [2,3]. This location was chosen under the constraint that an ideal vicarious calibration site should be located in oceanic regions exhibiting spatially homogenous optical properties, a cloud-free, clear, maritime atmosphere, and an economical/convenient access to the logistical support of ships and shore facilities. Having a measurement site that is representative of the observation conditions for most of the world oceans allows the determination of system vicarious calibration coefficients applicable to a large portion of satellite ocean-color mission data. The low complexity of the marine and atmospheric optical properties, coinciding with maritime atmospheres and oligotrophic—mesotrophic waters, increases the ability to accurately simulate the radiance at the EO sensor, as required by the system-level vicarious calibration of the EO sensor and data reduction algorithms. Low cloudiness is essential to maximize the number of in situ and satellite matching observations. Homogeneity of the measurement site is required to allow comparison of in situ and satellite observations performed at very different spatial resolutions.

The MOBY system is composed of (1) a spar buoy tethered to a moored buoy to avoid drift and (2) a hyperspectral radiometer operating in the 340–955 nm spectral region with resolution better than 1 nm, coupled to a number of light collectors via fiber optics. These collectors are designed to measure in-water downward irradiance and upwelling radiance at 1, 5, and 9 m depth from three arms perpendicular to the vertical column of the buoy. Additionally, above-water downward irradiance is measured from the top of the buoy at 2.5 m high. System radiometric stability is monitored daily through internal sources. The MOBY radiometry system undergoes regular characterizations and calibrations to guarantee high accuracy and traceability to National Institute of Standards and Technology (NIST). Declared uncertainties for the upwelling radiance L_u vary from 2.4 to 3.3% in the 412–666 nm spectral interval [4]. These values were determined from the statistical composition of various contributions including the calibration source and its transfer, radiometric stability during deployment, and environmental effects.

Since 2003, BOUSSOLE has been deployed in the Ligurian Sea approximately 32 nautical miles from the coast in 2440 m water depth [5].

BOUSSOLE relies on a taut mooring applying the concept of a reversed pendulum with Archimedes buoyancy replacing gravity. Its *transparent to swell* design is optimized to minimize superstructure shading and maximize its vertical stability [5]. The instrument package includes seven-channel commercial radiometers with 10 nm bandwidth in the 412–683 nm spectral region. In-water upwelling radiance, upward irradiance, and downward irradiance are measured at 4 and 9 m. Additionally, downward irradiance is measured at 4 m above the sea surface. A spectrally independent uncertainty of 6% has been estimated for the normalized remote-sensing reflectance ρ_{WN} [6]. Propagation of the in-water measurements to the surface is performed through models, which take into account Raman effects and the resulting nonlinearity of the log-transformed radiometric measurements with depth. These corrections, determined as a function of *Chla* and sun zenith, were shown as high as a few percent at 412 nm and up to several tens percent at 670 nm [6].

Lessons learnt from the decadal deployment of optical buoys for system vicarious calibration include (1) the fundamental need for securing continuity to measurements from at least one site to safeguard consistent support to successive missions and (2) the necessity for frequent characterizations and calibrations of the in situ radiometers to ensure traceability, accuracy, and consistency with time to data products. Additional important elements include the necessity to establish robust quality assurance schemes for field measurements (e.g., to screen data affected by nonnegligible effects of buoy shading and tilt, or biofouling) and for data reduction (e.g., to extrapolate subsurface values, or minimize bidirectional effects).

2.1.2 Long-Term Measurements for the Validation of Satellite Data Products

The Ocean-Color component of the Aerosol Robotic Network (AERONET-OC) is an example of long-term measurement program supporting the validation of satellite radiometric data products [7]. AERONET-OC, resulting from the collaboration between NASA and the Joint Research Center (JRC) of the European Commission, relies on the AERONET infrastructure [8] and supports ocean-color validation activities through time-series of matching normalized water-leaving radiance L_{WN} and aerosol optical thickness τ_a from globally distributed measurement sites. Atmospheric and above-water radiometric data are collected through nine-channel multispectral radiometers with 10 nm bandwidth in the 412–1020 nm spectral region. Measurements are autonomously performed from fixed platforms generally located in coastal areas, and thus by structures characterized by high stability (i.e., not affected by perturbations like tilt and roll). The network was launched in 2002 [9] and now, through the participation of international teams establishing and managing local sites, it includes radiometers operated in the Adriatic Sea, Baltic

Sea, Black Sea, Coral Sea, Gulf of Mexico, Gulf of Thailand, North Sea, Middle Atlantic Bight, Pacific Ocean off the coast of Oregon, Persian Gulf, and Yellow Sea.

A distinctive element of the network is the standardization of measurements and data products through the use of identical instruments and protocols, the calibration of all radiometers at a single laboratory by applying a sole method and a regularly checked source, and finally, the reduction and quality control of measurements by using the same processing code. Data are available at three levels of quality. Level-1.0 comprises data products derived from any complete sequence of in situ measurements while level-1.5 indicates cloud-screened data products which also passed a series of quality control tests aiming at removing any measurement affected by significant environmental perturbations. Finally, level 2.0 includes fully quality-controlled data obtained after postdeployment calibration and extensive assessment of each individual spectrum of normalized water-leaving radiance. Another distinctive element of the network is the almost real-time accessibility to level 1.0 and level-1.5 data through a web interface. Level-2 data products are available at the end of each deployment period, approximately every 6–12 months. Uncertainties determined for L_{WN} derived from measurements performed in moderately sediment-dominated waters indicate values of 4–5% in the 412–555 nm spectral region increasing up to 8% at 670 nm due to larger perturbing effects by surface waves [7]. These uncertainty values, which may significantly vary in marine regions with bio-optical properties different from those considered for their quantification [10], are largely due to contributions from absolute calibration and environmental effects (i.e., surface perturbations and, changes in illumination conditions and water optical properties during measurement sequences).

AERONET-OC has demonstrated the strategic importance of standardization and networking for the production of cross-site consistent data from globally distributed sites. Additionally, it has highlighted the importance of repositories for (1) the archival of raw data together with information required for data processing (e.g., calibration coefficients) and (2) real-time access to data products at incremental level of quality control, and information on measurement protocols and uncertainty estimates. This allows (1) immediate and systematic reprocessing of all measurements with advances in data reduction methods or instrument characterization and (2) timely and knowledgeable use of the archived data products by the scientific community.

2.1.3 Targeted Time-Limited Comprehensive Measurements

Long-term data collection at fixed sites, even though frequently limited to a few radiometric quantities, offer the opportunity to explore the intra- and interannual uncertainties of primary EO data from which high-level products are derived. However, comprehensive in situ measurements of optical and bio-optical quantities performed in regions and seasons ideally representative of the variety of world marine water types provide the opportunity to explore the accuracy of any EO-derived data product in combination with the effects of spatial and vertical distributions of the water optically significant constituents. Oceanographic ships are excellent platforms for the execution of comprehensive field measurements along transects, even though limited in duration. Additionally, ships offer the capability of operating measurement systems presently not deployable through unmanned platforms, and thus provide the opportunity for unique investigations.

The earliest example of a measurement program devoted to support satellite ocean-color observations was that undertaken by scientists of the Coastal Zone Color Scanner (CZCS) NIMBUS Experiment Team (NET) in the late 1970s and early 1980s. This program comprised a series of oceanographic cruises to perform bio-optical measurements and in-water radiometry essential to define algorithms for the determination of phytoplankton pigments concentration [11] and the assessment of derived satellite data products [12].

More recent examples of long-term measurement programs relying on bio-optical cruises are (1) the Bermuda Bio-Optics Project (BBOP) [13] developed with the objective of adding monthly optical and bio-optical quantities to the long-term physical, biological, and biogeochemical measurements performed in the context of the Bermuda Atlantic Time Series (BATS) [14,15]; and (2) the Bio-Optical mapping of Marine Properties (BiOMaP) dedicated to the collection of comprehensive and consistent optical and bio-optical measurements in European seas to support interregional satellite ocean-color investigations [16–18]. An important feature of BiOMaP is the application of identical instrumentation, calibration schemes, measurement protocols, processing codes, and quality control criteria across the investigated regions.

Ship-based optical radiometric measurements are often performed using in-water free-fall profiling systems, which allow data collection away from the ship and thus leads to the minimization of superstructure perturbations (i.e., ship shading and reflections). As opposed to above-water or in-water fixed-depth radiometry, free-fall techniques allow for a more comprehensive characterization of the in-water vertical radiometric fields. Specifically, free-fall multispectral systems are used within BiOMaP to produce continuous depth measurements of upwelling radiance, downward irradiance, and upward irradiance in the water column, in combination with above-water downward irradiance. Uncertainties for L_{WN} measurements performed in moderately sediment-dominated waters indicate values of 4–5% in the 412–555 nm spectral range increasing to approximately 6% at 670 nm [17], with the largest contributions due to uncertainties in absolute calibration and environmental effects.

BiOMaP has shown the relevance of cross-site consistent data sets of comprehensive optical and bio-optical measurements to support satellite ocean-color validation and regional bio-optical modeling activities [17,18].

2.1.4 Targeted Long-Term Comprehensive Measurements

Long-term comprehensive bio-optical measurements are possible with the aid of moorings, which can carry a large number of instruments and offer the opportunity to collect data at high temporal resolution. Because of this, moorings are excellent platforms to investigate occasional or periodic processes in physical, bio-optical, and biogeochemical domains. One of the earliest bio-optical moorings was Biowatt established in the Sargasso Sea and supported by the U.S. Office of Naval Research [19,20] to investigate light attenuation and light production in the ocean. While this mooring had instruments to measure several bio-optical parameters like attenuation, fluorescence, and bioluminescence, it had no radiometers beyond broadband sensors to measure the photosynthetically active radiation.

An example mooring relevant for satellite ocean-color applications is the Bermuda Test Bed first deployed in 1994 in deep water off Bermuda [21]. This mooring, mainly designed to study the temporal variability of the biogeochemistry and ecology of the oligotrophic ocean in the North Atlantic, provided the capability of collecting data with a suite of optical radiometers and bio-optical instruments at different depths. Measurements of multispectral in-water upwelling radiance and downward irradiance, in combination with above-water downward irradiance, were all performed using seven-channel sensors with 10 nm bandwidth in the 412−670 nm spectral region. In the context of satellite ocean color, this was important for the development and assessment of bio-optical models, and the validation of satellite data products [22]. As with buoys specifically designed to support system vicarious calibration, biofouling was identified as a major factor limiting the quality of optical data and requiring dedicated solutions [23,24].

2.2 Intercomparisons

Verification of the correct implementation of methods, or the assessment of techniques and instruments applied in laboratory and field activities, benefits from intercomparisons. Relevant intercomparisons performed within the framework of satellite ocean color include the evaluation of different calibration techniques, measurement methods, instruments performance, and additionally processing schemes for data reduction and quality control.

This section provides an overview of sample intercomparison activities performed in the context of the main satellite ocean-color missions since the early 1990s.

2.2.1 Laboratory Intercomparisons

The SeaWiFS Round Robin Experiments (SIRREXs) were an invaluable example of laboratory intercomparisons for optical calibration standards and methods. Eight SIRREXs were performed from 1992 to 2001 with incremental objectives. The first three were held at the Center for Hydro-Optics and

Remote Sensing (CHORS) at the San Diego State University and focused on traceability of spectral irradiance and radiance sources in the visible and near-infrared spectral regions [25–27]. Results indicated differences in irradiance values of standard lamps decreasing from 8% to less than 2% from SIRREX-1 to SIRREX-2 as a result of improvements in the transfer of spectral irradiance standards. In contrast, radiance sources (both spheres and, combined lamp and reflectance plaques), exhibited differences of approximately 7% during both SIRREX-1 and SIRREX-2, only significantly decreasing below 2% during SIRREX-3. The three experiments highlighted the fundamental importance of adhering to best practices for the handling of sources, control of lamp currents, mechanical setup of lamps and radiometers, baffling of laboratory stray light, and application of bidirectional factors for reflectance plaques.

Because of this, SIRREX-4 and SIRREX-5 were structured and performed with the objective of creating a consensus on laboratory and field methods. These two SIRREXs, both held at NIST in Gaithersburg during 1994 and 1996 [28,29], pursued their objective through laboratory sessions centered on absolute calibration and monitoring of sensors stability, and additionally through the intercomparison of field measurements.

As opposed to the previous experiments, SIRREX-6 focused on the intercomparison of radiance and irradiance absolute calibration capabilities of a number of laboratories [30]. During this round-robin held in 1996, the same radiometers were calibrated at different sites using local facilities. Results indicated the ability of obtaining overall agreement generally better than 2% for both radiance and irradiance calibrations.

The following SIRREX-7, held at Satlantic Inc. in Halifax in 1999 [31], addressed the quantification of uncertainties commonly affecting the calibration of radiance and irradiance sensors in the 400–700 nm spectral interval. It specifically and extensively investigated uncertainties related to irradiance and radiance standards, power sources, plaque uniformity and bidirectionality, mechanical positioning and alignment, and polarization. Results contributed to the definition of minimum, typical, and maximum uncertainty figures for radiance and irradiance calibrations. These indicated 1.1, 2.3, and 3.4%, respectively, for irradiance calibrations and 1.5, 2.7, and 6.3%, respectively, for radiance. Even though these mean spectrally independent values were determined at a specific calibration laboratory and for a given commercial series of multispectral radiometers, results can likely be retained as uncertainty indices for calibrations performed in different laboratories and for generic optical radiometers.

SIRREX-8 was held in 2001 to primarily investigate intra- and interlaboratory uncertainties related to the determination of the immersion factor of irradiance sensors in the visible and near-infrared spectral regions [32]. The intercomparison involved three laboratories (i.e., CHORS, JRC, and Satlantic Inc.), which independently performed the characterization of a number of radiometers from the same commercial series by applying different implementations of an identical measurement method. Results obtained from individual laboratory measurements but processed with the same code indicated intralaboratory determinations of the immersion factors with a precision generally better than 0.5% and interlaboratory differences on the order of 0.6%. Results also confirmed the need for individual radiometer characterizations because of sensor-to-sensor differences of the immersion factors ranging from 1 to 5% depending on the aging of irradiance collectors or manufacturing differences. Aside the main result of assigning uncertainties to intra- and interlaboratory determinations of the immersion factor for irradiance sensors, SIRREX-8 further pointed out the importance of applying consolidated protocols and strict quality assurance schemes to laboratory measurements as a basic requirement for the accurate characterization of radiometers.

SIRREX experiments were followed by additional radiometric intercomparison activities performed within the context of the Sensor Intercomparison and Merger for Biological and Interdisciplinary Oceanic Studies (SIMBIOS) program. These experiments, performed in 2001 and 2002 [33,34] and named SIMBIOS Radiometric Intercomparison (SIMRIC), were mostly directed toward (1) ensuring a common radiometric scale among facilities committed to the calibration of in situ optical radiometers and (2) furthering improved calibration protocols. Along with the evaluation of differences in radiance measurements performed by a number of laboratories with respect to those determined by a radiance transfer radiometer having NIST absolute calibration, SIMRIC-1 and SIMRIC-2 (1) investigated the conversion factor from directional—hemispherical to directional—directional reflectance of plaques commonly used for radiance calibration and (2) stressed the importance of accounting for the accurate position of the center of lamps filament behind the posts when using calibration distances different from those applied for the absolute radiometric calibration of lamps itself.

Overall SIRREX and SIMRIC were fundamental in transferring principles of applied metrology to the ocean-color community at large and, advancing expertise on radiometers calibration and characterization.

2.2.2 Field Intercomparisons

The literature offers a number of examples of field intercomparisons of optical radiometer systems and measurement methods [35–39]. The so-called *Assessment of In Situ Radiometric Capabilities for Coastal Water Remote Sensing Applications* (ARC) performed in the context of MERIS validation activities [40] is the only one briefly presented here because of the variety of compared systems and methods, and because of the effort given in quantifying uncertainties. This intercomparison, carried out at the Acqua Alta Oceanographic Tower (AAOT) in the northern Adriatic Sea in 2010, aimed at evaluating in situ data products determined in the visible and near-infrared from simultaneous measurements performed through independent above- and in-water radiometer systems and methods. The final objective of ARC was an evaluation of the consistency of data products from various independent providers. Evaluated products were the spectral water-leaving radiance $L_w(\lambda)$, the above-water downward irradiance $E_d(0^+,\lambda)$, and the remote-sensing reflectance $R_{rs}(\lambda)$. The most important achievement of the intercomparison was the quantification of uncertainties for each independent system/method and additionally the assessment of these uncertainties with statistical results from the intercomparison.

Spectrally averaged relative differences with respect to data products from a reference system exhibited values ranging from -1 to $+6\%$, while spectrally averaged absolute differences varied from approximately 6 to 9%. These results benefitted from a laboratory intercalibration of the involved radiometers, and of almost ideal measurement conditions (i.e., relatively low sun-zenith angles, clear sky, and low sea state) in addition to the use of a stable deployment platform.

The ARC experiment illustrated the difficulty meeting the 5% uncertainty target in radiometric data products from a variety of commercial radiometers and measurement methods. In fact, conclusions pointed out the difficulty of preserving the documented level of agreement among data products from the considered systems and methods when measurements are performed in nonideal conditions (i.e., high sun zenith and sea state, illumination perturbed by clouds, and nonstable deployment platforms) and without a laboratory intercalibration of the various radiometers.

2.2.3 Data Processing

In addition to uncertainties related to instrument characterization and calibration, and the performance of measuring systems and methods, the data reduction process can affect the quality of derived radiometric products. Examples of intercomparisons focused on the processing of in-water optical radiometric profile data are the SeaWiFS Data Analysis Round Robins performed in 1994 and in 2000 (i.e., DARR-94 and DARR-00, respectively). DARR-94 [42] included comparisons of basic data products (i.e., $E_d(0^-,\lambda)$, $L_u(0^-,\lambda)$, and $K_d(\lambda)$) determined from the application of four processing methods. Results, based on the analysis of radiometric profiles representative of oceanic waters, showed coefficients of variation of approximately 3% for $L_u(0^-,\lambda)$, slightly increasing to 3–4% for $E_d(0^-,\lambda)$, and generally within 5% for $K_d(\lambda)$. DARR-00 [40] included an evaluation of three processors not assessed during DARR-94, but all referring to the same data reduction protocol [43]. The study investigated differences among independent determinations of an extended number of derived data products (i.e., $E_d(0^-,\lambda)$, $L_u(0^-,\lambda)$, $E_u(0^-,\lambda)$, $K_d(\lambda)$, $Q_n(0^-,\lambda)$, $R_{rs}(0^-,\lambda)$, and $L_{WN}(0^-,\lambda)$). The data analysis, which relied on radiometric profiles from both coastal and oceanic waters performed with free-fall and winched systems, indicated that differences were more pronounced in the red spectral region, and varied significantly from processor to processor. For instance, results from the intercomparison of $L_u(0^-,\lambda)$ determined with two processors showed differences of 2.5% in the blue-green spectral region increasing to 13% in the red. The same analysis showed convergence to within 0.5% when choosing identical processing options such as extrapolation intervals and filtering criteria for outliers.

DARR-00 unequivocally demonstrated the importance of data reduction in the generation of accurate radiometric data products and indicated that fully independent processing solutions may affect the consistency of reference data sets constructed by combining data from different sources. This shows the relevance and need for centralized processing solutions or alternatively for a shared processor supported by the scientific community. The exercise also indicated the importance of securing full reprocessing of in situ data over time to respond to advances in data reduction schemes.

2.3 Data Repositories

The SeaWiFS Bio-optical Archive and Storage System (SeaBASS) developed by NASA [44] and the MERIS Matchup In Situ Database (MERMAID) developed by ESA [45] are examples of data repositories dedicated to satellite ocean-color applications. The general objective of these databases is to support long-term archival and distribution of in situ data suitable for system vicarious calibration, validation of satellite data products, and bio-optical modeling. In both cases, in situ data are contributed by different investigators and collected using a variety of instruments, measurement methods, and platforms.

When comparing these two repositories, SeaBASS was originally restricted to radiometric and phytoplankton pigment data, and later expanded to accommodate a larger number of bio-optical and hydrographic quantities such as the inherent optical properties, concentration of optically significant constituents, salinity, and temperature. Submitted data are subject to assessment including checks on expected range of variability and, when applicable, spectral consistency. Along with regular application to the validation of data products from satellite ocean-color sensors, SeaBASS was used for the creation of the NASA bio-Optical Marine Algorithm Data (NOMAD) high-quality subset for robust bio-optical modeling [46]. This development shows the importance of labeling radiometric data as a function of their applicability driven by their uncertainties and the existence of matching bio-optical data.

As opposed to SeaBASS, MERMAID was specifically designed for the archival, evaluation, and distribution of MERIS matchups. Similar to Sea-BASS, MERMAID was initially restricted to radiometric data products complemented by quality flags, and later expanded to include a number of additional bio-optical quantities.

The previous examples indicated the importance of designing data repositories with flexible structures allowing their progressive expansion to accommodate an incremental number of variables and information for upcoming applications.

3. REQUIREMENTS AND STRATEGIES FOR FUTURE SATELLITE OCEAN-COLOR MISSIONS

As already stated, climate-change investigations require observations characterized by outstanding traceability, high accuracy, and ultimately very high consistency with time in view of detecting changes varying slowly with respect to the observation periods. This requires that continued EO observations from successive missions are tied to a unique reference, hopefully benefitting of an unbroken in situ time-series of data supporting the quantification and minimization of uncertainties.

Learning from experience gained from previous calibration and validation programs, future satellite ocean-color missions ultimately need long-term in situ radiometric measurements from:

- *At least one dedicated and sustained site delivering high-quality marine radiometric data for system vicarious calibration*; and
- *geographically distributed sites established to maximize the number of matchups for the validation of primary satellite data products.*

The validation of high-level derived data products together with the development and assessment of bio-optical models should be supported through globally and seasonally distributed in situ radiometric and bio-optical matching measurements from

- *Regions representative of the variety of world aerosols and marine waters.*

Field measurements should be collected and reduced through:

- *Measurement protocols adhering to best practice and resulting from community consensus*;
- *fully characterized and regularly calibrated instruments assuring the best traceability, accuracy, and stability*; and
- *state-of-the-art processing codes relying on the most updated and community-accepted data reduction schemes.*

In situ data products should also be:

- *complemented by uncertainty values quantified accounting for each potential uncertainty source*; and
- *timely accessible at different levels of quality control.*

When applicable, the former elements should benefit from:

- *Intercomparisons to verify each step leading to the generation of the in situ data products*;

- *increasing efforts on standardization and networking*; and
- *development and implementation of advanced methods and instruments, and their gradual integration into operational field activities.*

3.1 Field Measurements for System Vicarious Calibration

Uncertainty requirements for satellite data products generally refer to the work of Gordon and Clark [11], Gordon et al. [12], and Gordon [47], which indicate a 5% maximum uncertainty in L_W in the blue spectral wavelengths to determine chlorophyll-*a* concentration with a 35% maximum uncertainty in oligotrophic waters. This 5% uncertainty requirement, extended to all ocean-color wavelengths, was generically set as the target for the SeaWiFS primary radiometric data products [48] and kept for the successive missions.

Accuracy specifications for the calibration of satellite ocean-color sensors are thus set by the target uncertainties expected for the satellite-derived water-leaving radiance L_W [49]. Specifically, by assuming a generic uncertainty of 5% in L_W, with L_W approximately 10% of top-of-atmosphere radiance L_T, the uncertainty in L_T must be lower than approximately 0.5% (see Chapter 3.1 for details). The allowed uncertainty in L_T becomes lower than 0.3% when L_W is 5% of L_T. These estimates lead to the following general considerations: (1) the needed target uncertainties for L_T can be only met through system vicarious calibration; but (2) the achievable minimum uncertainty varies with wavelength as a function of the ratio L_W/L_T and of the uncertainty of the reference in situ L_W values applied for system vicarious calibration (this challenges the capability of meeting the target uncertainty in the red spectral region and unquestionably in the near-infrared with present in situ measurement technologies and methods).

It is also emphasized that meeting the target uncertainty requirements, and the additional needed consistency over time which is generally lower than the target uncertainty, suggests caution in the application (and interchangeability) of system vicarious calibration coefficients determined from different in situ data sets. For instance, satellite derived radiometric products resulting from the application of vicarious calibration coefficients differing by as little as 0.3% may easily exhibit a bias higher than 5% in L_W. This bias is several times higher than the 0.5% target stability value per decade required for satellite ocean-color missions devoted to climate-change investigations [50,51], and may introduce unwanted inconsistencies in long-term data records from multiple missions. This forces a careful evaluation of sites and in situ measurements supporting system vicarious calibration, which should be selected by accounting for the actual application of satellite data products and recognizing that the downstream creation of climate quality data imposes the most stringent conditions.

It is thus fundamental that future ocean-color missions supporting climate change investigations are able to rely on a main sustained long-term in situ calibration system (site and radiometry) to maximize consistency over time and thus minimize possible biases among satellite data products from different missions, which may result in artifacts-masking trends in the climate data record.

The capability of combining matchups from multiple sites is often seen as a viable solution to shorten the long time needed to accumulate the relatively large number of high-quality matchups [52] ensuring satisfactory precision to system vicarious calibration coefficients. However, even assuming equal quality for the in situ data, system vicarious calibrations relying on different sites may be affected differently by the atmospheric correction process leading to slightly diverse vicarious calibration coefficients. In fact, by recalling that system vicarious calibration is applied to compensate for errors in both the absolute radiometric calibration of the space sensor and the atmospheric correction process, different locations (due to possible differences in satellite observing geometries or marine and atmospheric optical properties) may lead to different atmospheric perturbations which would be minimized through diverse vicarious calibration coefficients. However, secondary in situ long-term calibration systems with performance equivalent to the primary one would allow the redundancy required to ensure fault-tolerance to system vicarious calibration. Additionally, these in situ secondary systems would have strategic importance for continuous verification and validation purposes to investigate effects of different observation conditions, and also to support the generation of mission-specific regional EO data products for ecological or water quality applications.

An ideal vicarious calibration site should be located far away from any land contamination at a distance from mainland that avoids potential adjacency effects in satellite observations, in regions exhibiting low cloudiness, high spatial homogeneity, and stable (within the limits of likely small seasonal variations) and accurately known (or modeled) marine and atmospheric optical properties representative for the world seas. Field radiometers must be fully characterized (in terms of linearity, temperature dependence, polarization sensitivity, and stray light perturbations) with exceptional absolute calibration traceable to National Metrological Institutes and target standard uncertainty of 2% with high stability (better than 1% per deployment with a target of 0.5%), and finally should be regularly checked and frequently swapped. Using state-of-the-art measurement technology, data reduction methods, and quality control schemes, in situ radiometric data products should aim at target standard uncertainty of 3–4% for L_w in the blue-green spectral regions and hopefully of 5% in the red, with interchannel differences well below 1%. The data rate should ensure close matchups for any satellite ocean-color mission with time differences appropriate to the site to minimize variations in bidirectional effects due to changes in sun zenith and temporal changes in the vertical distribution of phytoplankton.

Hyperspectral systems are essential to allow support for system vicarious calibration of all satellite ocean-color sensors regardless of the specific center-wavelengths and bandwidths. The additional capability of a comprehensive characterization of both atmosphere and water at the site may add benefits to system vicarious calibration of future advanced space sensors.

3.2 Field Measurements for the Validation of Satellite Data Products

Given a generic 5% uncertainty target for satellite ocean-color radiometric products in oligotrophic and mesotrophic oceanic waters, their assessment would require in situ measurements affected by much lower uncertainties. Still, the 5% value is generally considered appropriate by assuming a low contribution of any systematic component to the overall uncertainty value.

Inaccurate calibration of field instruments, irregular or regionally limited data collections, poor data reduction, inadequate archival or inconsistencies due to methodological or technological changes introduced in the course of measurement programs may diminish or even exclude the applicability of in situ data in the validation processes. This suggests that the validation of satellite data products (both primary and derived) for climate-change applications should rely on long-term measurement programs centered on consolidated technology and methods, and maintained beyond the life of any specific mission. Ideally, validation data should be globally and seasonally distributed, and ultimately represent a wide range of aerosols and water types including different trophic levels as well as waters dominated by colored dissolved organic matter or sediments.

In practical terms, specific validation strategies should be developed based on the data products to be assessed and of their target uncertainties. For instance, the validation of primary data products, such as the normalized water-leaving radiance L_{WN}, benefits from autonomous systems offering the capability of performing continuous radiometric measurements which satisfy the need for long-term measurements and maximize the number of matchups. In contrast, the validation of derived satellite data products requires comprehensive in situ measurements of bio-optical quantities (e.g., pigments concentration and absorption and backscattering coefficients), which are not always collected with the desired accuracy and geographic distribution through automated systems. Because of this, the production of these bio-optical measurements in a variety of water types should still be considered in the framework of programs relying on both dedicated moorings and oceanographic ships. Within such a general context, it is important to mention that profiling floats, drifters, and gliders are expected to expand current measurement capabilities through the regular collection of globally distributed optical and bio-optical data. Specifically, validation floats using Argo-float technology developed for physical oceanography may allow for autonomous and extensive (i.e., spatial, vertical, and temporal) measurements of optical and bio-optical quantities [53]. Because of this, these measurement systems are expected to fill the gap left by bio-optical buoys and fixed structures which provide data for near-surface water depths only at a limited number of sites, or by ships providing comprehensive vertical observations but restricted in both time and space.

3.3 Field Measurements for Bio-Optical Modeling

Bio-optical modeling, which addresses the development and assessment of algorithms for the generation of satellite products, requires in situ radiometric data and matching quantities such as seawater inherent optical properties (i.e., scattering, backscattering, and absorption coefficients) and concentration of optically significant constituents (e.g., phytoplankton pigments). As with the validation of data products, accessibility to comprehensive quality-controlled in situ data with quantified uncertainties is essential.

It is also mentioned that cross-site consistency of in situ data is essential for the development of regional bio-optical algorithms in view of minimizing geographically dependent artifacts (e.g., biases) in satellite derived products.

3.4 Protocols Revision and Consolidation

Community-shared protocols for in situ measurements, characterization and calibration of field radiometers, and data reduction are essential to create conditions for cross-mission consistency of satellite ocean-color data products. The Ocean Optics Protocols for Satellite Ocean Color Validation [54], resulting from successive revisions of the original Ocean Optics Protocols for SeaWiFS Validation [55], is definitively the most comprehensive compilation of protocols for in situ marine optical radiometry. While the original protocols focused on the basic requirements for in situ radiometric measurements in support of satellite ocean-color activities, the successive revisions incrementally included the collection and handling of the full suite of relevant radiometry and bio-optical quantities. It is expected that future protocols, designed to support ocean-color missions, would be based on these efforts, but properly updated by accounting for recent advances. However, any revision or new protocol should be supported by objective evaluation, evidence of documented results, and wide community consensus.

3.5 Calibration and Characterization of Field Radiometers

The calibration and characterization of field radiometers aim at ensuring the best traceability of measurements. When solely considering absolute radiometric calibration, the use of multiple calibration sources and methods may challenge the quantification of measurement uncertainties from different providers. Ideally, this would suggest the use of a single calibration laboratory (subject to continuous verifications). However, considering the difficulty in implementing the concept, the participation of calibration laboratories into regular intercomparisons is an alternative and feasible approach that would contribute to minimize calibration uncertainties. An additional element to consider is seeking traceability with respect to the same National Metrological Institute.

Instrument characterization must include the determination of system nonlinearity, temperature dependence, stray light perturbations, and additionally geometrical, spectral, and in-water responses (see the extensive discussion in Chapter 3.1). Missing or poor characterization of each of these terms may lead to unpredictable measurement errors.

Aside from the application of state-of-the-art methods for radiometric characterization and calibration, aging of sensors is an additional element to include in the evaluation of radiometric accuracy for field sensors. This requires regular recalibration or calibration checks of instruments, ideally performed as frequently as possible [56] to trace sensitivity changes or systems performance issues during their lifetime.

3.6 Data Reduction, Quality Control, and (re)Processing

Data reduction comprises successive numerical operations which may include the application of calibration coefficients to raw data, filtering of outliers, extrapolations for the determination of subsurface values, and additionally the application of a number of corrections such as those for the minimization of sensitivity decay with time, system nonlinearity, temperature dependence, bidirectional effects, and self-shading perturbations.

Quality control, which ensures the confident application of data in following analysis, is generally specific for the different measured quantities. It should take advantage of ancillary information provided with data themselves (e.g., cloud cover or sea state in the case of radiometric data), closure between inherent and apparent optical properties, model estimates, relative spectral consistency of data products, representativity of individual spectra in previously quality-controlled data sets, and variance of data products among sequential measurements.

Reprocessing is a pressing need for any living data set. Thus, in view of benefitting from methodological developments and advances in systems characterization, reprocessing of in situ data should be considered a necessity.

As already pointed out, centralized data processing could be an important solution to reduce inconsistencies introduced by the use of independent data reduction codes. It may additionally offer advantages such as regular systematic reprocessing of datasets and consistent quality control of data regardless of the source.

3.7 Accuracy Tailored to Applications

In situ data applicable to system vicarious calibration, validation of satellite data products, and bio-optical modeling activities should have well-defined uncertainties possibly quantified in both relative (i.e., in percent) and absolute units for each potential perturbation source [10].

Radiometric uncertainties depend on the features of the measuring system, its characterization and calibration, the measurement method, environmental conditions, and the data reduction scheme. The 1% uncertainty concept introduced during SIRREXs [57], indicating the attempt to reduce below 1% each uncertainty contributing to the overall uncertainty budget, was the rationale for dedicated studies which led to the quantification and, when possible, to the minimization of uncertainty sources. This process clearly showed that accuracy has a cost. As a result, uncertainties of in situ data products, which affect the number of measurements qualified for each specific application (e.g., bio-optical modeling, validation, and system vicarious calibration), may vary among measurement programs. This imposes the need for determining uncertainties of in situ data products for each quantity, measurement system, and likely different geographic regions characterized by diverse measurement conditions. Such a step would allow indexing in situ data products for different applications as a function of their uncertainties.

3.8 Archival and Access

Timely access to field data is ultimately a fundamental need for any calibration and validation program in view of supporting regular assessment of satellite data products. This cross-mission need suggests establishing, maintaining, and continuously expanding repositories beyond any specific mission's life. Archived data should be accessible at different quality levels and indexed as a function of their fitness for purpose.

Undoubtedly, open access to data including details on instruments, calibration history, measurement methods, data reduction algorithms, and quality control schemes allow for their independent evaluation and application. This should impose the definition of data policies facilitating access to data, but also assuring rights to data providers.

3.9 Intercomparisons to Secure Accuracy and Best Practice

In general, intercomparisons embracing calibration, measurement methods, data reduction, and quality control are effective to investigate uncertainties. In fact they are the means to evaluate the correct interpretation and implementation of measurement protocols, to identify technical issues in systems performance of instruments, to verify the applicability and accuracy of alternative methods and, additionally, to consolidate practices and disseminate new solutions.

While not representing a comprehensive list of topics for intercomparison activities relevant to in situ radiometry, the following are the areas that could be considered useful in the process.

- Methodological issues in the calibration process or changes in the performance of calibration setups can be the source of calibration errors. Cross-comparisons of calibration coefficients independently determined for the same radiometers of proven stability are a viable method to verify the performance of calibration laboratories.
- Different radiometers, often based on diverse technology, may produce different results. This may be due to a variety of characteristics such as field-of-view, acquisition rate, or spectral bands. Intercomparisons together with a detailed understanding of the instrument performance are thus essential to identify reasons for differences.
- Measurement methods generally rely on protocols sharing community consensus. Still, the understanding and implementation of measurement protocols may face objective constraints or personal interpretations. Intercomparisons of measurement methods are thus the way to verify the implementation of the protocols.
- As with measurement methods, data reduction and generation of data products are subject to the application of protocols and implementation schemes. Still, the interpretation of these protocols, their application, or adaptation to specific cases may be the source of substantial differences in derived products. Because of this, intercomparisons relying on consolidated processing codes (ideally a single reference code) are of major importance to address and solve potential sources of errors affecting data products.

An essential, but sometime overlooked, element is the need to perform intercomparisons relying on equivalent quantities. This aspect should be carefully considered each time intercomparisons refer to quantities determined from the application of diverse methods/systems or different observation conditions. For instance, if the bidirectional reflectance of sea surface and water are not taken into account, the comparison of above-water and in-water derived radiometric products may be biased due to the different viewing geometries and not by fundamental problems with method or system performance.

Finally, intercomparison results should always include uncertainties determined for each compared quantity by accounting for any significant uncertainty source [7,38].

3.10 Standardization and Networking

As already stated, uncertainties affecting field measurements from various providers are impaired by factors such as the performance of different field instruments, the use of diverse sampling methods, the application of a variety of calibration sources and protocols, and finally the adoption of assorted data reduction schemes. Standardization is a way to ensure increased consistency to measurements regardless of their origin. Networking and networks are viable solutions for the development and implementation of standard solutions [7].

It is remarked the primary importance of safeguarding long-term consistency to in situ data products for the assessment of satellite climate data products from successive missions. This need strongly suggests that once a measurement capacity of a network is established and consolidated, any change in its components (e.g., instruments or methods) should be carefully evaluated before being implemented.

3.11 Development and Implementation

Progress in technology and methods is essential to increase the accuracy of in situ measurements. However, new technology and methods require consolidation and need to be well understood prior being routinely applied. An example is offered by the frequent emphasis on the application of new technology and thus the migration from *established* to *new* systems. Reconfirming the fundamental importance of advances in technology and methods, the application of unvalidated instruments or methods may affect data products. In fact, while the initial application of new instruments or methods through occasional deployments allows advancement with limited risks, the swap of technology or methods in major programs producing time-series or geographically distributed measurements has to be handled with extreme caution to avoid affecting the consistency of data sets. Thus, close synergies should be established between development and operational programs to warrant a progressive and timely increase of in situ data quality, through advances in technology and methods, but only once these are completely validated. Best practice would also suggest a cross-over period of measurement performed with heritage and new systems/methods to document differences and provide information for future additional investigations.

4. SUMMARY AND WAY FORWARD

Comprehensive calibration and validation programs supporting satellite ocean-color missions are essential for the generation of data products with the traceability, accuracy, and consistency required for climate-change investigations. Lessons learnt indicate the fundamental need for at least one long-term reference site providing in situ optical radiometry data of exceptional quality for system vicarious calibration across successive missions. Additionally, the assessment of satellite data products and the development of bio-optical algorithms should be supported by geographically distributed radiometric measurements from regions representative of the world seas.

In all cases, data quality should be assured through the application of state-of-the-art measurement protocols, fully characterized and well-calibrated field radiometers, and finally validated processing schemes. Data, complemented by uncertainty values, should be stored in dedicated and accessible repositories.

Intercomparisons of field methods, instruments, and data reduction schemes are the ways to secure accuracy. Standardization of measurements and data reduction is an invaluable component of the overall strategy to assure high consistency to field data regardless of source and region. Additionally, development of new methods and instruments need high consideration. However, the use of newly developed methods or instruments in operational programs needs to be carefully done to avoid introducing significant discontinuities or inconsistencies in time-series or globally distributed data.

Finally, it is important to highlight that international collaboration on each element of the proposed strategies is essential in both benefiting from transnational experience and optimizing the use of resources.

REFERENCES

[1] C. Wunsch, R.W. Schmitt, D.J. Baker, Climate change as an intergenerational problem, P. Natl. Acad. Sci. 110 (2013) 4435–4436.

[2] D.K. Clark, H.R. Gordon, K.J. Voss, Y. Ge, W. Broenkow, C. Trees, Validation of atmospheric correction over the oceans, J. Geophys. Res. 102 (D14) (1997) 17209–17217.

[3] D.K. Clark, M.E. Feinholz, M. A.Yarbrough, B.C. Johnson, S.W. Brown, Y.S. Kim, R.A. Barnes, Overview of the radiometric calibration of MOBY, in: Earth Observing Systems VI, 4483, 2002, pp. 64–76.

[4] S.W. Brown, S.J. Flora, M.E. Feinholz, M.A. Yarbrough, T. Houlihan, D. Peters, K.Y.S. Kim, J.L. Mueller, B.C. Johnson, D.K. Clark, The marine optical buoY (MOBY) radiometric calibration and uncertainty budget for ocean color satellite sensor vicarious calibration, in: Remote Sensing, International Society for Optics and Photonics, 2007, pp. 67441M–67441M.

[5] D. Antoine, P. Guevel, J.F. Deste, G. Bécu, F. Louis, A.J. Scott, P. Bardey, The "BOUSSOLE" buoy-a new transparent-to-swell taut mooring dedicated to marine optics: design, tests, and performance at sea, J. Atmos. Oceanic Technol. 25 (2008) 968–989.

[6] D. Antoine, F. D'Ortenzio, S.B. Hooker, G. Bécu, B. Gentili, D. Tailliez, A.J. Scott, Assessment of uncertainty in the ocean reflectance determined by three satellite ocean color sensors (MERIS, SeaWiFS and MODIS-A) at an offshore site in the mediterranean sea (BOUSSOLE project), J. Geophys. Res. 113 (C7) (2008) C07013.

[7] G. Zibordi, B. Holben, I. Slutsker, D. Giles, D. D'Alimonte, F. Mélin, J.-F. Berthon, D. Vandemark, H. Feng, G. Schuster, B. Fabbri, S. Kaitala, J. Seppälä, AERONET-OC: a network for the validation of ocean color primary radiometric products, J. Atmos. Oceanic Technol. 26 (2009) 1634–1651.

[8] B.N. Holben, T.F. Eck, I. Slutsker, D. Tanré, J.P. Buis, A. Setzer, E. Vermote, J.A. Reagan, Y.I. Kaufman, T. Nakajima, F. Lavenu, I. Jankowiak, A. Smirnov, AERONET–A federated instrument network and data archive for aerosol characterization, Remote Sens. Environ. 66 (1998) 1–16.

[9] G. Zibordi, B.H. Holben, S.B. Hooker, F. Mélin, J.-F. Berthon, I. Slutsker, D. Giles, D. Vandemark, H. Feng, K. Rutledge, G. Schuster, A. Al Mandoos, A network for standardized ocean color validation measurements, Eos Trans. Am. Geophys. Union 87 (30) (2006) 293–297.

[10] M. Gergely, G. Zibordi, Assessment of AERONET L_{WN} uncertainties, Metrologia 51 (2014) 40–47.

[11] H.R. Gordon, D.K. Clark, Clear water radiances for atmospheric correction of coastal zone color scanner imagery, Appl. Opt. 20 (1981) 4175–4180.

[12] H.R. Gordon, D.K. Clark, J.W. Brown, O.B. Brown, R.H. Evans, W.W. Broenkow, Phytoplankton pigment concentrations in the middle atlantic bight: comparison of ship determinations and CZCS estimates, Appl. Opt. 22 (1983) 20–36.

[13] D.A. Siegel, T.K. Westberry, M.C. O'Brien, N.B. Nelson, A.F. Michaels, J.R. Morrison, A. Scott, E.A. Caporelli, J.C. Sorensen, S. Maritorena, S.A. Garver, E.A. Brody, J. Ubante, M.A. Hammer, Bio-optical modeling of primary production on regional scales: the Bermuda BioOptics project, Deep-Sea Res. II 48 (2001) 1865–1896.

[14] A.F. Michaels, A.H. Knap, R.L. Dow, K. Gundersen, R.J. Johnson, J. Sorensen, A. Close, G.A. Knauer, S.E. Lohrenz, F.A. Asper, M. Tuel, R. Bidigare, Seasonal patterns of ocean biogeochemistry at the United States JGOFS Bermuda atlantic time series study site, Deep-Sea Res. I 41 (1994) 1013–1038.

[15] D.K. Steinberg, C.A. Carlson, N.R. Bates, R.J. Johnson, A.F. Michaels, A.H. Knap, Overview of the US JGOFS Bermuda atlantic time series study (BATS): a decade-scale look at ocean biology and biogeochemistry, Deep-Sea Res. II 48 (2001) 1405–1447.

[16] J.-F. Berthon, F. Mélin, G. Zibordi, Ocean colour remote sensing of the optically complex European seas, in: Remote Sensing of the European Seas, Springer, Netherlands, 2008, pp. 35–52.

[17] G. Zibordi, J.-F. Berthon, F. Mélin, D. D'Alimonte, Cross-site consistent in situ measurements for satellite ocean color applications: the BiOMaP radiometric dataset, Remote Sens. Environ. 115 (2011) 2104–2115.

[18] D. D'Alimonte, G. Zibordi, T. Kajiyama, J.-F. Berthon, Comparison between MERIS and regional high-level products in European seas, Remote Sens. Environ. 140 (2014) 378–395.

[19] J. Marra, Eric O. Hartwig, Biowatt: a study of bioluminescence and optical variability in the sea, Eos Trans. Am. Geophys. Union 65 (1984) 732–733.

[20] T. Dickey, E. Hartwig, J. Marray, The biowatt bio-optical and physical moored program, Eos, Trans. Am. Geophys. Union 67 (1986) 650.

[21] T. Dickey, S. Zedler, X. Yu, S.C. Doney, D. Frye, H. Jannasch, D. Manov, D. Sigurdson, J.D. McNeil, L. Dobeck, T. Gilboy, C. Bravo, D.A. Siegel, N. Nelson, Physical and biogeochemical variability from hours to years at the Bermuda test bed mooring site: June 1994–March 1998, Deep-Sea Res. II 48 (2001) 2105–2140.

[22] V.S. Kuwahara, G. Chang, X. Zheng, T.D. Dickey, S. Jiang, Optical moorings-of-opportunity for validation of ocean color satellites, J. Oceanogr. 64 (2008) 691–703.

[23] F.P. Chavez, D. Wright, R. Herlien, M. Kelley, F. Shane, P.G. Strutton, A device for protecting moored spectroradiometers from biofouling, J. Atmos. Oceanic Technol. 17 (2000) 215–219.

[24] D.V. Manov, G.C. Chang, T.D. Dickey, Methods for reducing biofouling of moored optical sensors, J. Atmos. Oceanic Technol. 21 (2004) 958–968.

[25] J.L. Mueller, The first SeaWiFS Intercalibration round-robin Experiment, SIRREX-1, July 1992, in: S.B. Hooker, E.R. Firestone (Eds.), NASA Tech. Memo. 104566, vol. 14, NASA Goddard Space Flight Center, Greenbelt, Maryland, 1993, 60 pp.

[26] J.L. Mueller, B.C. Johnson, C.L. Cromer, J.W. Cooper, J.T. McLean, S.B. Hooker, T.L. Westphal, The second sea-WiFS intercalibration round-robin experiment, SIRREX-2, June 1993, in: S.B. Hooker, E.R. Firestone (Eds.), NASA Tech. Memo. 104566, vol. 16, NASA Goddard Space Flight Center, Greenbelt, Maryland, 1994, 121 pp.

[27] J.L. Mueller, B.C. Johnson, C.L. Cromer, S.B. Hooker, J.T. McLean, S.F. Biggar, The third sea-WiFS intercalibration round-robin experiment, SIRREX-3, September 1994, in: S.B. Hooker, E.R. Firestone, J.G. Acker (Eds.), NASA Tech. Memo. 104566, vol. 34, NASA Goddard Space Flight Center, Greenbelt, Maryland, 1996, 78 pp.

[28] B.C. Johnson, S.S. Bruce, E.A. Early, J.M. Houston, T.R. O'Brian, A. Thompson, S.B. Hooker, J.L. Mueller, The Fourth SeaWiFS intercalibration round-robin experiment (SIRREX-4), May 1995, in: S.B. Hooker, E.R. Firestone (Eds.), NASA Tech. Memo. 104566, vol. 37, NASA Goddard Space Flight Center, Greenbelt, Maryland, 1996, 65 pp.

[29] B.C. Johnson, H.W. Yoon, S.S. Bruce, P.-S. Shaw, A. Thompson, S.B. Hooker, R.E. Eplee Jr., R.A. Barnes, S. Maritorena, J.L. Mueller, The fifth SeaWiFS intercalibration round-robin experiment (SIRREX-5), July 1996, in: S.B. Hooker, E.R. Firestone (Eds.), NASA Tech. Memo. 1999−206892, vol. 7, NASA Goddard Space Flight Center, 1999, 75 pp.

[30] T. Riley, S. Bailey, The sixth SeaWiFS/SIMBIOS intercalibration round-robin experiment (SIRREX-6) August−December 1997, in: NASA Tech. Memo. 1998−206878, NASA Goddard Space Flight Center, Greenbelt, Maryland, 1998, 26 pp.

[31] S.B. Hooker, S. McLean, J. Sherman, M. Small, G. Lazin, G. Zibordi, J.W. Brown, The seventh SeaWiFS intercalibration round-robin experiment (SIRREX-7), March 1999, in: S.B. Hooker, E.R. Firestone (Eds.), NASA Tech. Memo. 2002−206892, vol. 17, NASA Goddard Space Flight Center, Greenbelt, Maryland, 2002, 69 pp.

[32] G. Zibordi, D. D'Alimonte, D. van der Linde, J.-F. Berthon, S.B. Hooker, J.L. Mueller, G. Lazin, S. McLean, The eighth SeaWiFS intercalibration round-robin experiment (SIRREX-8), September−December 2001, in: S.B. Hooker, E.R. Firestone (Eds.), NASA Tech. Memo. 2002−206892, vol. 21, NASA Goddard Space Flight Center, Greenbelt, Maryland, 2002, 39 pp.

[33] G. Meister, et al., The first SIMBIOS radiometric intercomparison (SIMRIC-1), April−September 2001, in: NASA/TM2002-210006, vol. 1, NASA Goddard Space Flight Center, Greenbelt, Maryland, 2002, 60 pp.

[34] G. Meister, et al., The second SIMBIOS radiometric intercomparison (SIMRIC-2), March−November 2002, in: NASA/TM-2002-210006, vol. 2, NASA Goddard Space Flight Center, Greenbelt, Maryland, 2003, 65 pp.

[35] D.A. Toole, D.A. Siegel, D.W. Menzies, M.J. Neumann, R.C. Smith, Remote-sensing reflectance determinations in the coastal ocean environment: impact of instrumental characteristics and environmental variability, Appl. Opt. 39 (2000) 456−469.

[36] S.B. Hooker, G. Lazin, G. Zibordi, S. McLean, An evaluation of above-and in-water methods for determining water-leaving radiances, J. Atmos. Oceanic Technol. 19 (2002) 486−515.

[37] S.B. Hooker, G. Zibordi, J.F. Berthon, J.W. Brown, Above−water radiometry in shallow coastal waters, Appl. Opt. 21 (2004) 4254−4268.

[38] K.J. Voss, S. McLean, M. Lewis, C. Johnson, S. Flora, M. Feinholz, M. Yarbrough, C. Trees, M. Twardowski, D. Clark, An example crossover experiment for testing new vicarious calibration techniques for satellite ocean color radiometry, J. Atmos. Oceanic Technol. 27 (2010) 1747−1759.

[39] D. Antoine, A. Morel, E. Leymarie, A. Houyou, B. Gentili, S. Victori, J.-P. Buis, S. Meunier, M. Canini, D. Crozel, B. Fougnie, P. Henry, Underwater radiance distributions measured with miniaturized multispectral radiance cameras, J. Atmos. Oceanic Technol. 30 (2013) 74−95.

[40] G. Zibordi, K. Ruddick, I. Ansko, G. Moore, S. Kratzer, J. Icely, A. Reinart, In situ determination of the remote sensing reflectance: an inter-comparison, Ocean Sci. 8 (2012) 567−586.

[41] D.A. Siegel, M.C. O'Brien, J.C. Sorensen, D.A. Konnoff, E.A. Brody, J.L. Mueller, C.O. Davis, W.J. Rhea, S.B. Hooker, Results of the SeaWiFS data analysis round-robin (DARR-94), July 1994, in: S.B. Hooker, E.R. Firestone (Eds.), NASA Tech. Memo. 104566, vol. 26, NASA Goddard Space Flight Center, Greenbelt, Maryland, 1995, 58 pp.

[42] S.B. Hooker, G. Zibordi, J.-F. Berthon, D. D'Alimonte, S. Maritorena, S. McLean, J. Sildam, Results of the second SeaWiFS data analysis round robin, March 2000 (DARR-00), in: S.B. Hooker, E.R. Firestone (Eds.), NASA Tech. Memo. 2001−206892, vol. 15, NASA Goddard Space Flight Center, Greenbelt, Maryland, 2001, 71 pp.

[43] J.L. Mueller, R.W. Austin, Ocean optics protocols for SeaWiFS validation, rev 1, in: S.B. Hooker, E.R. Firestone (Eds.), NASA Tech. Memo. 104566, vol. 25, NASA Goddard Space Flight Center, Greenbelt, Maryland, 1995, 66 pp.

[44] P.J. Werdell, S. Bailey, G. Fargion, C. Pietras, K. Knobelspiesse, G. Feldman, C.R. McClain, Unique data repository facilitates ocean color satellite validation, Eos Trans. Am. Geophys. Union 84 (2003) 377−387.

[45] K. Barker, C. Mazeran, C. Lerebourg, M. Bouvet, D. Antoine, M. Ondrusek, G. Zibordi, S. Lavender, MERMAID: the MEris MAtchup in-situ database, in: The 2nd MERIS / (A) ATSR Workshop, 22−26 September 2008, Frascati, Italy, European Space Agency, SP-666, November 2008.

[46] P.J. Werdell, S.W. Bailey, An improved in-situ bio-optical data set for ocean color algorithm development and satellite data product validation, Remote Sens. Environ. 98 (2005) 122−140.

[47] H.R. Gordon, Calibration requirements and methodology for remote sensors viewing the ocean in the visible, Remote Sens. Environ. 22 (1987) 103−126.

[48] S.B. Hooker, W.E. Esaias, G.C. Feldman, W.W. Gregg, C.R. McClain, An overview of SeaWiFS and ocean color, in: S.B. Hooker, E.R. Firestone (Eds.), NASA Tech. Memo. 1992−104566, vol. 1, NASA Goddard Space Flight Center, Greenbelt, Maryland, 1992.

[49] National Academy of Sciences, Assessing Requirements for Sustained Ocean Color Research and Operations, The National Academies Press, 2011, ISBN 978-0-309-21044-7, 126 pp.

[50] G. Ohring, B. Wielicki, R. Spencer, B. Emery, R. Datla, Satellite instrument calibration for measuring global climate change: report of a workshop, B. Am. Meteorol. Soc. 86 (2005) 1303−1313.

[51] World Meteorological Organization, Systematic Observation Requirements for Satellite-based Data Products for Climate 2011, Update Supplemental Details to the Satellite-based Component of the Implementation Plan for the Global Observing System for Climate in Support of the UNFCCC (2010 Update), World Meteorological Organization. Report GCOS − 154, 2011.

[52] B.A. Franz, S.W. Bailey, P.J. Werdell, C.R. McClain, Sensor-independent approach to the vicarious calibration of satellite ocean color radiometry, Appl. Opt. 46 (2007) 5068−5082.

[53] H. Claustre, S. Bernard, J.-F. Berthon, J. Bishop, E. Boss, C. Coatanoan, F. D'Ortenzio, K. Johnson, A. Lotliker, O. Ulloa, Bio-optical Sensors on Ar-go Floats, Reports and

Monographs of the International Ocean-Colour Coordinating Group, N. 11, IOCCG, Dartmouth, Canada, 2011.

[54] J.L. Mueller, et al., Ocean optics protocols for satellite ocean color sensor validation, revision 5, in: J.L. Mueller, G.S. Fargion, C.R. McClain (Eds.), NASA Tech. Memo. 2004-211621NASA Goddard Space Flight Center, Greenbelt, Maryland, 2004.

[55] J.L. Mueller, R.W. Austin, Ocean optics protocols for SeaWiFS validation, in: S.B. Hooker, E.R. Firestone (Eds.), NASA Tech. Memo. 104566, vol. 25, NASA Goddard Space Flight Center, Greenbelt, Maryland, 1992, 45 pp.

[56] S.B. Hooker, J. Aiken, Calibration evaluation and radiometric testing of field radiometers with the SeaWiFS quality monitor (SQM), J. Atmos. Oceanic Technol. 15 (1998) 995−1007.

[57] C.R. McClain, G.C. Feldman, S.B. Hooker, An overview of the SeaWiFS project and strategies for producing a climate research quality global ocean biooptical time-series, Deep-Sea Res. 51 (2004) 5−42.

Chapter 5.2

Strategies for the Laboratory and Field Deployment of Ship-Borne Fiducial Reference Thermal Infrared Radiometers in Support of Satellite-Derived Sea Surface Temperature Climate Data Records

Craig J. Donlon,[1,*] Peter J. Minnett,[2] Nigel Fox,[3] Werenfrid Wimmer[4]
[1] *European Space Agency/ESTEC, Noordwijk, The Netherlands;* [2] *Meteorology & Physical Oceanography, Rosenstiel School of Marine and Atmospheric Science, University of Miami, Miami, FL, USA;* [3] *National Physical Laboratory (NPL), Teddington, Middlesex, UK;* [4] *Ocean and Earth Science, University of Southampton, European Way, Southampton, UK*
*Corresponding author: E-mail: craig.donlon@esa.int

Chapter Outline

1. Introduction   558
2. Fiducial Reference Measurements for SST CDRs and Uncertainty Budgets   559
   2.1 FRM TIR Ship-Borne Radiometer Network   562
   2.2 The Importance of Uncertainty Budgets   563
     2.2.1 Uncertainty of Ship-Borne Radiometer SST_{skin} Measurements   564
     2.2.2 Uncertainties Related to the Satellite-to-SBRN Spatial and Temporal Matchup Criteria   570
     2.2.3 SBRN Requirements for Validation of SST Retrieval Algorithms   578
     2.2.4 SBRN Requirements for Validation of Satellite SST Products   581
     2.2.5 SBRN Requirements for Monitoring of Satellite Instrument Degradation   583

2.2.6 SBRN Requirements to Bridge between Different Satellite Instruments 584
3. **Laboratory Intercalibration Experiments for FRM Ship-Borne Radiometers** 585
4. **Ship-Borne Radiometer Field Intercomparison Exercises** 590
5. **Protocols to Maintain the SI Traceability of FRM Ship-Borne TIR Radiometers for Satellite SST Validation** 595
   5.1 Definition of Measurement Methodology 595
   5.2 Definition of Laboratory Calibration and Verification Methodology and Procedures 595
   5.3 Predeployment Calibration Verification 596
   5.4 Postdeployment Calibration Verification 596
   5.5 Uncertainty Budgets 596
   5.6 Improving Traceability of Calibration and Verification Measurements 596
   5.7 Accessibility to Documentation 597
   5.8 Archiving of Data 597
   5.9 Periodic Consolidation and Update of Calibration and Verification Procedures 598
6. **Summary and Future Perspectives** 598
Acknowledgments 598
References 599

1. INTRODUCTION

To retrieve sea-surface temperature (SST) retrieved from satellite top-of-atmosphere (TOA) thermal infrared (TIR) satellite measurements, an algorithm [1–11] is required to compensate for the impact of the intervening atmosphere between the surface and satellite radiometer, the nonunity value of surface emissivity [12,13] and its variability, and satellite instrument effects [14]. The Group for High-Resolution SST [15] has fostered the development of new global SST data products based on the complementary characteristics of diverse types of SST observing systems, which include uncertainty estimates derived largely from a statistical analysis of satellite data matched in space and time to global drifting buoy array SST_{depth} measurements (e.g., Ref. [16]). While these estimates of uncertainty are useful in an operational sense, the quality of drifter measurements cannot easily be verified once a drifter is deployed at sea. Furthermore, there is no internationally consistent or accepted route to establish Système International d'unités (SI) traceability for drifting buoy SST measurements. Some activities have used satellite data to identify the characteristics of overall uncertainty in ship SST, moored, and drifting buoys [17,18]. The adequacy of available ground-based reference measurements for the correction of satellite SST biases depends on the choice of uncertainty model used. Assuming that the source of uncertainty is uncorrelated can potentially lead to a false impression of adequacy and an underestimate of the derived satellite SST uncertainty. Little (if any) work has been performed to establish SI traceability to drifting buoy measurements of SST_{depth} with National Metrology Institutes (NMI).

A new approach to satellite SST validation using ship-borne TIR radiometers has been developed over the last 10 years that measure the radiometric skin temperature (SST_{skin}) of the ocean surface (as discussed in Chapter 3.2). Ship-borne radiometer measurements of SST_{skin} are complementary to and arguably more valuable than the drifting buoy subsurface SST measurements because they allow a more direct validation of SST derived from satellite infrared radiometry: SST_{skin} measured by ship-borne infrared radiometers is the same quantity that is derived from measurements of satellite infrared radiometers; ship-borne SST_{skin} measurements eliminate uncertainties related to the near-surface ocean thermal structure that complicates comparisons between subsurface thermometry and satellite measurements of SST_{skin} [15]. Furthermore, the internal calibration of ship-borne radiometers can be verified before and after each deployment in a manner that is fully traceable to the International Temperature Scale of 1990 (ITS-90) [19], i.e., the SI.

But what is the strategy for laboratory and field deployment of ship TIR radiometers in support of satellite-derived SST climate data records (CDR)? This chapter provides an overview of aspects related to answering this challenge.

2. FIDUCIAL REFERENCE MEASUREMENTS FOR SST CDRs AND UNCERTAINTY BUDGETS

Fiducial Reference Measurements (FRM) are the suite of independent ground measurements that provide the maximum scientific utility/return on investment for a satellite mission by delivering, to users, the required confidence in data products, in the form of independent validation results and satellite measurement uncertainty estimation, over the duration of the mission. The defining mandatory characteristics of an FRM are:

- FRM measurements have documented evidence of SI traceability via intercomparison of instruments under operational-like conditions.
- FRM measurements are independent from the satellite SST retrieval process.
- An uncertainty budget for all FRM instruments and derived measurements is available and maintained, traceable where appropriate to SI ideally directly through an NMI.
- FRM measurement protocols and community-wide management practices (measurement, processing, archive, documents, etc.) are defined and adhered to.

FRM are required to determine the on-orbit uncertainty characteristics of satellite measurements via independent validation activities. "The process of assessing, by independent means, the quality of the data products derived from the system outputs." Validation is a core component of a satellite mission (and should be planned for accordingly) starting at the moment satellite instrument data begin to flow until the end of the mission. Without validation, the geophysical retrieval methods and geophysical parameters derived from satellite measurements cannot be used appropriately with confidence. In the case of the SST CDR, the concept of validation is limited not only to the regular validation of a sample of SST retrievals from satellite instruments but also to the stability of the derived time series over specified time and space scales. Stability is defined as the degree to which systematic effects in SST measurements are invariant over time [20]. The Global Climate Observing System (GCOS) defines SST CDR stability requirements as 0.03 K/decade over spatial scales of 100–1000 km [21]. Furthermore, a satellite CDR shall include an uncertainty estimate for each measurement that also requires validation using independent data.

It is essential for long-term satellite SST records (e.g., Ref. [8]) that all measurements are fully anchored to SI units and that there is a direct correlation with ground reference measurements [22]. Expendable Bathy-Thermograph (XBT) measurements that profile the upper ocean temperature from ships of opportunity cannot be considered FRM suitable for validation of the SST CDR based on the fact that they cannot demonstrate SI traceability (they cannot be recovered for postdeployment calibration) and suffer from an imperfect knowledge of calibration [23]. Ship hull SST temperature measurements [24] have the potential to reach FRM status if regular SI sensor calibration traceability is maintained and a detailed knowledge of the particular ship installation is available to ascertain the depth of measurement and any potential bias associated with ship structure and activities (e.g., warming of the hull, varying load lines, ship effluent outflow). Drifting buoys [25] are used extensively for satellite validation (on the assumption that the sheer number of measurements reduces overall uncertainty by $\sqrt{n}$). Drifting buoys provide a measure of subsurface SST_{depth} over a wider geographical area, albeit with irregular spatial and temporal distributions but it cannot be shown whether their temperature measurements satisfy SST CDR stability requirements or FRM criteria because they cannot reliably maintain SI traceability (this cannot be fully assessed as the buoys are not recovered for recalibration). Deep water moorings include temperature sensors [26] with better temporal coverage but poor spatial coverage: they could reach FRM status if regular SI sensor calibration traceability is maintained. ARGO profiling floats provide vertical profiles of SST_{depth} at depths generally terminating at 5 m or more (although some high-resolution ARGO floats measure at cm vertical resolution to the ocean surface but they are limited in number) with moderate coverage but are not in general SI traceable as they are rarely recovered for recalibration [27].

The most accurate ground reference measurements and best quantified in terms of uncertainty are derived from field-deployed TIR radiometers [28,29] of varying design, operated by different teams and in different locations. These instruments are, in principle, calibrated to SI units by traceability of on-board calibration reference radiance sources and via intercomparison, using suitable reference radiometers [30,31], to precision external reference blackbody (BB) radiance targets [32−36] that themselves have been robustly traced to SI [20−23,25,37,38]. It is essential, therefore, that any potential biases between instruments used to generate and/or validate the SST CDR are understood and accounted for properly [22]. This can only be determined through formal comparison of instruments and supporting reference blackbody radiance sources, both in the laboratory and in the field to a common basis, which is traceable to the SI and for this application most commonly through the ITS-90 temperature scale [19], although linkage through radiometric quantities such as spectral radiance is also viable due to the coherence of the SI. Such comparisons also provide the opportunity to establish a fully traceable link to SI standards and a documented path to legitimately form part of the CDR [22]. To be successful, ship-borne radiometer and reference blackbody intercomparisons need to be regular, include broad international participation, and evolve with appropriate refinements to address an increasing demand for accuracy under a variety of operational conditions. It is critical that transfer reference radiometers from NMIs are also used in the comparisons. Thus, ship-borne TIR radiometers provide SST_{skin} that is potentially traceable to SI with regional coverage and excellent accuracy and stability characteristics. Data from such instruments is not used in SST retrieval methods (see Chapter 3.2), and a set of deployment protocols has been established [33]. Thus, ship-borne TIR radiometers constitute a viable FRM for satellite SST_{skin} measurements based on the criteria of SI traceability, independence, and measurement protocols with the added benefit that the measurement process can be considered equivalent to that made from a TIR satellite instrument. However, as will be discussed later, establishing an uncertainty for the end-to-end SST_{skin} measurement has yet to be fully addressed by the international community.

It is prudent to define and justify what an ideal FRM ship-borne TIR radiometer network would include. It is a challenge to establish the exact requirements for such a network based solely on the potential costs and the perceived expected benefits. Figure 1 sets out a hypothetical cost−benefit S-curve indicating the optimum cost versus benefit over time following the decision to implement a ship-borne TIR FRM radiometer network. The number of radiometers deployed at sea is determined by weighing the costs against the required temporal and spatial sampling. It is not straightforward to determine the appropriate number of radiometers, especially given the irregular distribution of ships that could host them. Figure 1 shows that costs fall as the network grows form a research and development project to a sustained operational approach that financially functions largely on the basis of maintenance costs. There is a balance between these two extremes that enables a satellite mission to deliver products with a known quality that are "Fit for

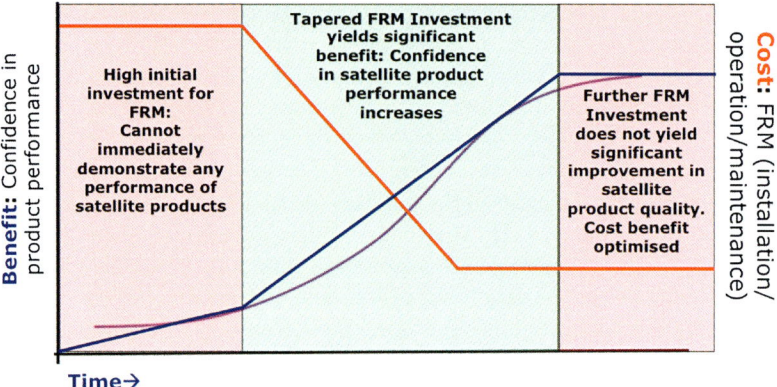

FIGURE 1 Simple cost–benefit analysis of the FRM concept. Red line indicates investment profile and cost, the blue line indicates benefits to the satellite mission (measured as "confidence in product performance"). The purple curve indicates a hypothetical optimum cost/benefit curve. FRM, fiducial reference measurements.

Purpose". This balance is also impacted by the relative accessibility of the data and the degree of international cooperation in establishing the network.

In terms of requirements for satellite SST validation, it is clear that both FRM radiometers and other SST measurements have a role to play. However, it is useful to stratify validation data into FRM and non-FRM data as follows:

- FRM TIR ship-borne radiometer network (SBRN) measurements that are regularly verified to SI standards shall be known as Class-1 (SI Traceability is maintained) FRM.
- Class-2 validation measurements are not SI traceable and include drifting and moored buoys, ARGO floats, ship thermosalinograph, and other SST measurements if appropriate. Such measurements provide a larger population than the SBRN alone although with an increased uncertainty on a case-by-case basis.

2.1 FRM TIR Ship-Borne Radiometer Network

Given the above discussion, it is clear that a sustained independent ground-based FRM network is required as a fundamental part of a satellite mission if the products from that mission are to be used with confidence as part of the SST CDR. Against this background, the aim of an SBRN is "to establish a sustained programme for acquiring measurements of SST_{skin} from FRM TIR radiometers deployed on ships of opportunity and produce long-term FRM of SST_{skin} reference data sets that can be used to tie satellite-derived SST data sets to SI standards, as a basis for the independent validation of SST data products used for climate-change monitoring" [39]. The requirements for an

FRM TIR SBRN can be defined by considering those derived from the following key applications that such a network could address:

1. To validate, using independent measurements, the performance and validity of the atmospheric compensation algorithm used in the satellite SST retrieval.
2. To monitor any degradation or change of specific satellite instrument performance across the entire mission lifetime.
3. To validate the performance of data products derived from a satellite SST mission.
4. To provide an independent reference data set that can be used to bridge between different satellite missions.
5. For satellite SST algorithm development in its own right.
6. For fundamental research and development related to air—sea interaction and the emission of TIR electromagnetic waves from the sea surface.
7. To provide an independent CDR in its own right (although limited in geographical coverage).

The following sections first consider the importance of uncertainty budgets, the criteria for matching SBRN data to satellite measurements followed by a discussion of each key application, and the requirements that it places on the design of an FRM TIR SBRN.

2.2 The Importance of Uncertainty Budgets

CDRs are required to have a comprehensive uncertainty budget [40,41]. In addition an FRM, by definition, has to be traceable to SI standards and therefore, by definition, must include an estimate of uncertainty. A reliable uncertainty estimate allows a data set to be used with confidence: without such an estimate of uncertainty, measurements cannot be compared, either among themselves or with reference standard, and in practice has little or no real meaning.

All measurements are imperfect and have uncertainties that can be of a *random* nature (e.g., detector noise) or of a *systematic* nature (e.g., an incorrectly calibrated resistance thermometer leading to a systematic offset). The effect of random uncertainty can be reduced by increasing the number of measurements and averaging those together: e.g., statistical averages to be used to reduce the random uncertainty of a blackbody measurement assuming that the source of the error can be considered random in nature. Systematic uncertainties can only be corrected using an appropriate correction factor (e.g., an offset can be applied based on laboratory (re-)calibration of a resistance thermometer). However, there will still remain an uncertainty associated with this correction. Clearly all identified and assessable systematic uncertainties must be corrected and random uncertainties reduced (to an appropriate level) to obtain the most reliable and accurate measurement.

Measurement uncertainty is defined [42] as a "nonnegative parameter characterizing the dispersion of the quantity values being attributed to a measurand (i.e., the output of a measurement model), based on the information used [to define the measurand]". The result of a measurement after correction for recognized systematic effects is still only an estimate of the true value of the measurand because uncertainties still arise due to random effects and from imperfect correction of the result for systematic effects [42]. Uncertainties arise due to many aspects that can be grouped generally into the following primary categories:

- **Instrument measurement uncertainty**: those relating to instrument hardware,
- **Retrieval/algorithm uncertainty**: those relating to derived quantities,
- **Application uncertainty**: those relating to a specific application.

For each category, good practice requires an uncertainty budget to be derived including all aspects leading to a quantification of a root-sum-square estimate of uncertainty. This is a challenging exercise but nevertheless, for climate-related applications, it is a requirement. The following section provides an example.

2.2.1 Uncertainty of Ship-Borne Radiometer SST_{skin} Measurements

The uncertainty of a ship-borne radiometer measurement is derived from a complete understanding of the uncertainty associated with all instrument hardware that may have a direct influence on the accuracy of measurement. Examples include the digitization uncertainty of A/D converters, optical misalignment, detector noise and gain and offset characteristics, signal loss due to cables, etc. Retrieval algorithm uncertainty is derived from all aspects of the geophysical algorithm used to derive a specific quantity. Examples related to ship-borne radiometric measurement of the SST_{skin} include the value of seawater emissivity and its variability with target view angle as a ship rolls and pitches, knowledge of radiometer view angle, temporal differences between sea surface and atmospheric radiance measurement, radiance-to-temperature relationships, the calibration of thermistors or at a subsystem level, the uncertainty in emissivity of a calibration blackbody radiance source, the adequacy of the number of calibration cycles used to provide a stable calibration, and the formulation of the retrieval algorithm itself. Does the retrieval algorithm account for all aspects of a particular deployment? For example, if a ship-borne radiometer is mounted high above the sea surface, does the algorithm include adequate provision to account for atmospheric attenuation and emission along the path length? Application uncertainty estimates include time and space sampling uncertainty when comparing SBRN data to satellite data, differences in the type of SST measured (e.g., SST_{depth} vs SST_{skin}), appropriate sampling of oceanographic variability in measurements or merged satellite products, knowledge of geographical position, etc. [29].

Establishing an uncertainty budget is a fundamental step that drives a better understanding of the various components of uncertainty: quite often an instrument engineer will learn much about an instrument and its fitness for purpose by attempting the derivation of a full instrument uncertainty budget—potentially leading to innovation and improvement in design. But the real driver is, if reliable and well-defined uncertainties can be provided with each ship-borne FRM, then these measurements can uniquely provide an SI traceable measurement *on a per-pixel basis when matched to satellite measurements—without the need for a large number of nominally independent observations to reduce the random uncertainty* (as in the case of drifting buoys for example). The formal NMI procedure for evaluating and expressing uncertainty includes the following eight steps [42]:

1. Define a relationship between the measurand, SST, and the most complete series of inputs X_i, (e.g., seawater emissivity, seawater radiance, sky radiance, ...) on which the measurand depends (e.g., $SST_{skin} = f(X_1, X_2, X_3, ...)$). Every quantity, including all corrections and correction factors, that can contribute a significant component of SST_{skin} uncertainty must be included.
2. Estimate the value (x_i), for each quantity X_i that defines SST_{skin} uncertainty.
3. Evaluate the standard uncertainty $u(x_i)$ for each estimate of x_i either by the statistical analysis of a series of observations (Type A uncertainty) or by other means (Type B uncertainty).
4. Evaluate the covariances associated with any input estimates that are correlated.
5. Calculate the result of the measurement, $SST(y)$, of the measurand from the f using for the input quantities X_i and the corresponding estimated values for each.
6. Determine the combined standard uncertainty $u_c(SST(y))$ of the measurement result y from the standard uncertainties and covariances associated with the input estimates.
7. If an expanded uncertainty, U, is required that specifies an interval (i.e., $y - U$ to $(SST(y) + U)$) expected to encompass most of the distribution of values that could reasonably be attributed to the measurand SST, the combined standard uncertainty $u_c(SST(y))$ is multiplied by a coverage factor k, typically in the range 2–3, to obtain $U = ku_c(y)$. k is typically selected on the basis of the level of confidence required for the chosen interval.
8. The result of the measurement $SST(y)$ is then reported together with its combined standard uncertainty $u_c(SST(y))$ or expanded uncertainty U. A full description of how $SST(y)$ and $u_c(SST(y))$ or U were obtained must be included in the report.

Steps 1 and 2 in this process lie at the core of a meaningful uncertainty analysis but, in practice, are often very challenging to achieve. Static uncertainty budgets of varying detail have been attempted for a few ship-borne radiometers including the Marine-Atmosphere Emitted Radiance Interferometer (M-AERI) [28], the Infrared Autonomous SST Radiometer (ISAR) [35], and the Calibrated Infrared in situ Measurement System (CIRIMS) [43]. However, this approach does not consider how to attach uncertainty estimates to *every* SST_{skin} measurement. A static uncertainty is sufficient to satisfy general measurement principles, but can be improved upon to meet the stringent requirements for the SST CDR. The most comprehensive uncertainty budget and analysis for a ship-borne radiometer to date was produced by Wimmer [29] for the ISAR radiometer based on a breakdown of uncertainties for the entire end-to-end instrument and data processing system. This approach was followed because the self-calibrating design of the ISAR reduces some of the instrument uncertainties and the approach enables an uncertainty value to be estimated for each SST_{skin} measurement. Wimmer [29] characterizes the uncertainties as follows:

- *Measurement uncertainty*: The uncertainty associated with the typical variability of the measured property, e.g., variability of the brightness temperature (BT) of the sea and sky view measurements, or the uncertainty of seawater emissivity, etc.
- *Instrument uncertainty*: The uncertainty related to the ISAR instrument introduces regardless of the measured property such as detector noise, thermistor calibrations, electronics digitization uncertainty, and so on.

ISAR measurement uncertainties were estimated as either Type A or Type B (Type B taking a conservative approach to values where specific robust evidence from calibration did not exist) with a full description of how the uncertainty values were established. Table 1 reproduces the instrument and measurement uncertainties from Ref. [29].

After estimating all the individual uncertainties contributing to the total ISAR instrument uncertainty (Table 1), an overall uncertainty for SST_{skin} can be estimated by propagating the individual instrument and measurement uncertainties (that must also be estimated) through the ISAR SST_{skin} data processing system shown in Figure 2. A linear approximation [44] was used in which uncertainty probability distributions are reduced to a nominal value and a standard deviation. Uncertainty related to all aspects of the ISAR processing system are included cascading from the fundamental SST_{skin} calculation. In many cases, it is not possible to calculate an exact uncertainty for each element based on statistical measurements (Type A uncertainty) and best estimates are therefore provided (Type B uncertainty) based on available literature and experience at component level. To estimate Type A, Type B, and measurement and instrument uncertainty, the ISAR processor was run four times: with one of the Type A, the Type B, or instrument uncertainty suppressed and finally with all parameters included (the latter to generate an estimate of the total uncertainty). To separate the measurement and instrument uncertainty, the detector target view (either sea or sky) uncertainty and the seawater emissivity

TABLE 1 Estimated Instrument Uncertainties for the ISAR Ship-Borne Radiometer [29] with Permission

Item (X_i)	Uncertainty ($u(x_i)$)	Unit	Uncertainty Type
Detector linearity stability	<0.01%	K/month	B
Detector noise	~0.002	Volts	A
Detector accuracy	±0.5	K	B
Analog to digital converter (ADC)	±1	LSB	B
ADC accuracy	±0.1%	Range	B
ADC zero drift	±6	µV/C	B
Reference voltage 16 bit ADC	±15	mV	B
Reference voltage 12 bit ADC	±20	mV	B
Reference resistor	1	%	B
Reference resistor temperature coefficient	±100	Ppm/C	B
Blackbody emissivity	±0.000178	Emissivity	B
Sea surface emissivity	±0.07	Emissivity	B
Steinhart–Hart approximation	±0.01	K	B
Radiative transfer approximation	±0.001	K	B
Thermistor	±0.05	K	B
Thermistor noise	~0.002	Volts	A

was set to 0.0 to generate an instrument uncertainty. The measurement uncertainty was then calculated as the difference between the instrument uncertainty and the total uncertainty.

Figure 3(a) shows the total uncertainty as calculated for ISAR data collected in July 2011 for the ship route shown in Figure 3(b). Also shown in the lower left panel of Figure 3(a) are the uncertainties split into Type A and Type B and that calculated for the instrument and in the lower right panel, the measurement uncertainty. The main contributions to the instrument uncertainty were found to be from the uncertainty of the on-board calibration blackbody thermistors and the emissivity of the calibration black bodies [29]. Figure 3(a) shows that the total SST_{skin} uncertainty is in most cases below 0.1 K, although it is not always the case. The main reason for the occasional high uncertainties (>0.1 K) is a variable target view detector

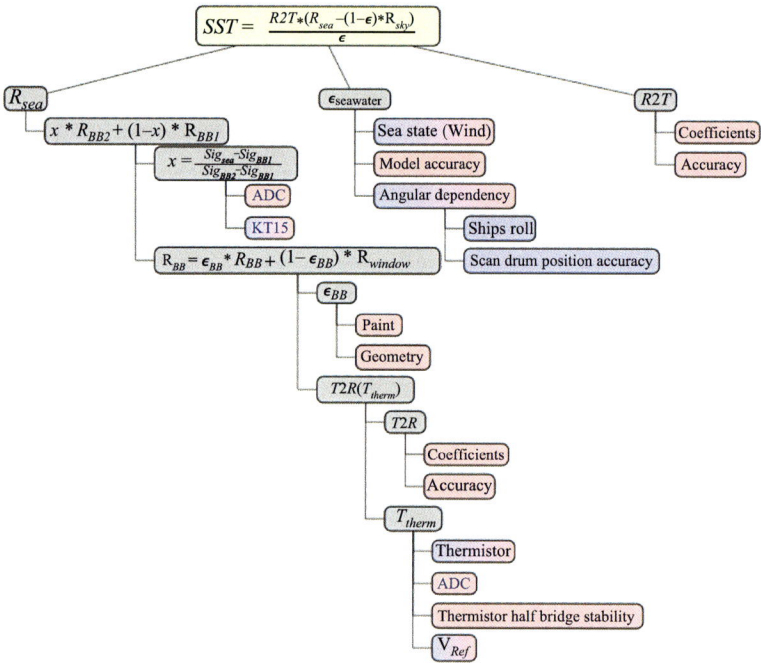

FIGURE 2 Flow chart of the ISAR SST processor. Boxes colored in blue show Type A uncertainties, boxes colored in red show Type B uncertainties, and boxes in red and blue show that the particular box has both Type A and Type B uncertainties [29]. ISAR, infrared autonomous SST radiometer; SST, sea-surface temperature; ADC, analog to digital converter.

signal and the emissivity of seawater. This mainly occurs when the ship was close to port (Figure 3(b)) or areas of mixed cloud and larger sea-state conditions. The target design uncertainty for the ISAR-derived SST_{skin} measurements is <0.1 K [45] which is largely met: the example shown in Figure 3 shows the benefit of calculating a per-measurement uncertainty estimate.

The approach of Wimmer [29] provides an excellent start but much more work on the uncertainties associated with *all* ship-borne radiometer measurements is required. Important considerations to address include the following: Are all significant terms included in the instrument and measurement uncertainty budget? How should one handle the large uncertainty associated with broken cloud conditions that result in a differently measured sky radiance compared to the sky radiance at the time of sweater radiance measurement (i.e., a cloud is present in one measurement but not the other) [46]? What is the best way to determine a geophysical component for the natural SST_{skin} variability that is expected to be more variable than the $SSTsub_{skin}$ or SST_{depth} [47]? Most importantly, there is a need to *validate* the uncertainty estimates for all ship-borne TIR FRM radiometers—this is a

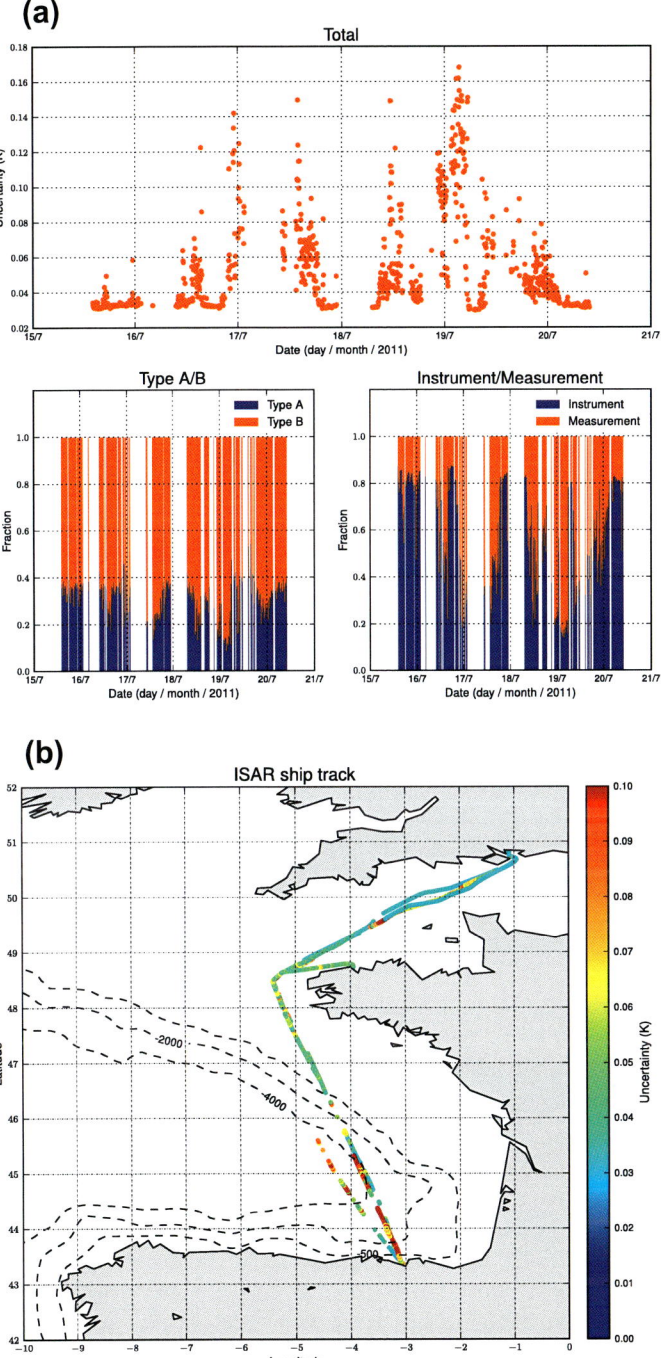

FIGURE 3 (a) ISAR radiometer uncertainties for the data collected between 15th July 2011 and 20th July 2011 along the ship-track shown in (b). The top panel shows the measurement uncertainty. The bottom left panel shows the uncertainty split into fractions of Type A and Type B uncertainty and the bottom right panel shows the fraction of instrument and measurement uncertainty [29]. ISAR, infrared autonomous SST radiometer; SST, sea-surface temperature.

clear goal for the international ship-borne radiometry community to address in the coming years.

2.2.2 Uncertainties Related to the Satellite-to-SBRN Spatial and Temporal Matchup Criteria

In principle, the concept of satellite SST product validation is straightforward: simply compare the satellite-derived SST_{skin} to the **true** SST_{skin} temperature to establish the error (if there are no errors attached to the Truth otherwise the discrepancy and its uncertainty [39] must be expressed). In practice, the "science" of validation is extremely difficult because very rarely it is possible to establish what **truth** is or to compare measurements of the same geophysical quantity, with the same spatial and temporal coverage. Consider a 1 km nadir pointing TIR satellite SST_{skin} measurement derived from spectral channels centered at 10.5 μm and 11.5 μm. This is an SST_{skin} measurement representing the SST at a depth of ~10 μm derived from the integrated radiance over a 1 km^2 area on the ocean surface obtained in ~100 μs (note some instruments such as MODIS and VIIRS have longer integration times). As such, it can be considered an instantaneous spatial average at that specific time, location, and spatial area. Ship-borne FRM TIR radiometer measurements are collected as discrete point samples (at a measurement interval) of SST_{skin}, ideally at the exact time of satellite measurement (or overpass) and, within the same spectral waveband (although the latter is not mandatory). Considering typical ship-borne radiometer configurations, each measurement takes between 1 and 7 min which, at a ship velocity of ~15 kt, equates to a sample derived along a narrow (~2−6 m) Field of View (FoV) integrated over a length of 0.46−3.26 km. This is quite different to the satellite measurement. Another related source of uncertainty is the coregistration of detectors in different bands used in the satellite SST_{skin} retrieval algorithms (see Chapter 2.4). Furthermore, if one considers typical uncertainty in satellite geo-location over the ocean (driven largely by orbital and instrument pointing knowledge: there are seldom land marks to provide geo-location reference control points) of 0.2−0.5 spatial sample distance (i.e., 2−500 m for the 1 km FoV case described above—although some instruments provide better performance [48,49]), the situation gets even worse. Further challenges are obvious for a satellite radiometer with a larger FoV.

What are the valid time-space matchup criteria required to validate the satellite-derived SST CDR? If the requirements are unattainable, it is unlikely that sufficient FRM data will be collected to provide any statistical significance in a satellite validation analysis. If the chosen matchup criteria are not severe enough, then a large number of matches will pair SST measurements from slightly different places at slightly different times. Depending on the natural variability of the SST_{skin} in space and in time, this could introduce spurious uncertainty into the validation [39]. Clearly, there is a need for a justified compromise that is both practical and affordable. Minnett [50] estimated time and space matchup criteria in the northeast Atlantic Ocean concluding that spatial separations of about 10 km and time intervals of about 2 h can introduce rms. Differences of 0.2 K into the uncertainty budget of a satellite validation data set, this being an upper limit for the meaningful validation of current infrared radiometers. Embury et al. [8] select ground reference data within 3 h of a given satellite measurement collocated in space to ±1 km. The match with the smallest time separation is retained. They note that such stringent spatial matchup criteria are, in practice, limited by the geo-location accuracy of both the satellite and ground measurement. For example, drifting buoy locations are reported at 0.1° resolution leading to spatial differences of up to ~7 km. To compensate, the assumption is made that there are no SST gradients, SST is extracted over a 5 × 5 pixel array and an average SST is computed from clear-sky pixels only: this reduces the random SST uncertainty from sensor noise by up to a factor of $5 = \sqrt{25}$.

Following Donlon et al. [47] and Wimmer et al. [39], the potential error and thus effective uncertainty in a satellite measurement of SST_{skin}, u_{sat}, is given by

$$u_{sat} = v_{sat} - v \qquad (1)$$

where v_{sat} is the satellite-estimated value of SST_{skin} and v is the true value of the SST_{skin} represented by the satellite view, i.e., its true average over the satellite pixel area at the short time interval of the overpass. In practice, v is not known precisely and instead a measured value, v_w, is used (e.g., using an FRM TIR radiometer). The uncertainty of the FRM measurement is then given by u_w, where

$$u_w = v_w - v \qquad (2)$$

Traditionally, global validation statistics (mean discrepancy between v_{sat} and v_w together with the standard deviation of the discrepancy over a specified time period, t) for satellite SST_{skin} products are based on the analysis of a well-specified matchup database (MDB) of near-colocated and near-contemporaneous satellite and FRM measurements that yields a matchup discrepancy, $u(\Delta MDB)$. Thus for a given sample pair,

$$u(\Delta MDB) = \sum_{t=0}^{t+n} v_{sat} - v_w \qquad (3)$$

This is not the same as the actual satellite measurement uncertainty u_{sat} since it includes the uncertainty associated with the FRM measurement u_w, that has been used as a proxy for the true temperature, v. In order to estimate u_{sat} and validate a satellite measurement of SST_{skin}, it is necessary to estimate u_w and, if possible, to minimize it. In the case of a ship-mounted FRM radiometer, u_w can be separated into several different types of uncertainty inherent to the validation process:

$$u_W = f(u_{Wt}, u_{Wr}, u_{Wm}, u_{Ws}, u_{Wz}, u_{Sgeoloc}) \qquad (4)$$

where u_{Wt} is the temporal mismatch uncertainty, u_{Wr} is the location mismatch uncertainty (that must include the uncertainty in satellite geo-location uncertainty $u_{Sgeoloc}$), u_{Wm} is the inherent FRM instrument measurement uncertainty, u_{Ws} is the point-in-area sampling uncertainty, and u_{Wz} is the sampling depth uncertainty intended to describe the impact of stratification or differences between SST_{depth} and SST_{skin}.

u_{Wm} must be determined independently for each instrument through careful independent sensor calibration validation. In the case of a ship-borne FRM TIR radiometer, calibration uncertainty data are obtained before and after each deployment aboard using an independent SI traceable reference radiance, i.e., a blackbody cavity (e.g., Refs [33–36,51]). This is discussed in detail in Section 3. Sometimes, independent calibration experiments require a systematic correction to be applied to the measurement that itself has an associated uncertainty [39]. Random uncertainties in u_{Wm} are more of a problem. u_{Wm} has components relating to the deployment situation for a given measurement campaign including the deployment geometry used (length of atmospheric path and atmospheric emission/absorption along that path), the impact of uncertainty related to the emissivity of seawater, sea state (ship roll and pitch), direct ship reflections at the sea surface viewed by the radiometer, ship wake contamination of the radiometer FoV, cable losses and EM interference, etc. The uncertainties associated with these aspects are some of the most difficult to establish and will vary according to instrument, deployment platform, and characteristics of the study area. Uncertainties in the specification of sea water emissivity, which depends on the geometry of deployment, can be minimized by using view angles $<40°$ from nadir view beyond which seawater emissivity in the infrared significantly diminishes (see Chapter 3.2 for detailed discussions on this aspect). The use of roll/pitch sensors is a good practice to monitor situations when ship movements are sufficient to introduce uncertainty in the SST_{skin} determination. However, the most significant measurement uncertainty (up to 0.5 K in a worst case scenario) is due to rapidly changing clouds that may invalidate the correction for the reflected sky radiance in the sea-view measurements [46]. In the case of cloud-free conditions, the uncertainty will be negligible.

u_{Ws} arises from using a point sample of v to represent the spatial average of v over the satellite FoV. Its magnitude depends on the degree of sub-FoV variability and the amplitude of spatial variability (caused by the surface expression of SST fronts, eddies, filaments, diurnal stratification, variable spatial wind speeds at local scales, etc.). Is such a linear average along a ship-track a good approximation to a 2D spatial average? If the oceanographic variability is high, does the sampling strategy employed provide an appropriate equivalence between the different measurement types? Satellite pixel assessments over, e.g., 5×5 FoV areas computed from long-term SST climatology of sufficient resolution can be used to compute useful statistical parameters describing the SST structures over the area. Figure 4 presents a

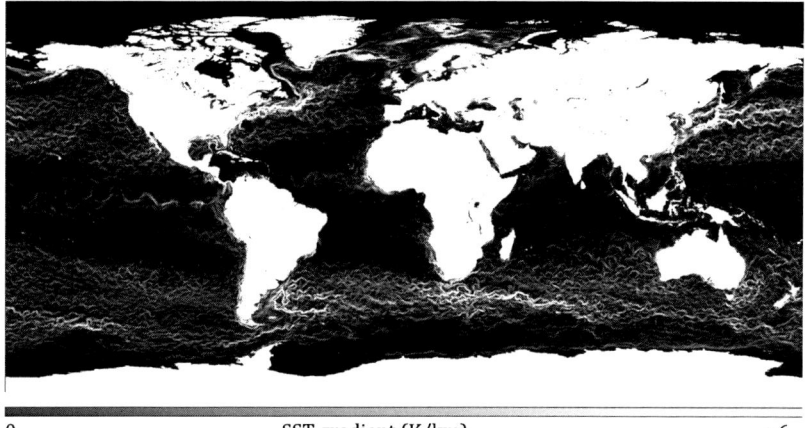

| 0 | SST gradient (K/km) | >6 |

FIGURE 4 Global SST gradients computed for 13th March 2014 using the OSTIA L4 SST analysis [52] using a 3 × 3 Sobel filter. OSTIA provides a gridded 10 km analysis with baseline correlation length scales of ~25 km. Local SST gradients can be significant and are not captured in this map. SST, sea-surface temperature.

global map of SST gradients derived from the OSTIA L4 SST analysis [52]. It is clear that in some ocean areas, SST gradients are very significant (>6 K/km) and in others they are not. Local SST gradients can be very high and are not captured in such a global analysis. Thus, the time-space matchup criteria have, in principle, a geographic variability: satellite validation matchup data obtained in regions of high SST gradients must be closely matched in space and time if significant u_{Ws} are to be avoided. Alternatively, to avoid such complexity and still be prudent, the stringent criteria appropriate to a high variability could be applied everywhere.

The inhomogeneity of atmospheric structure can also contribute to u_{Ws}. This could be investigated with atmospheric soundings of temperature and humidity and clouds estimated from the satellite data themselves, from visual inspection or all-sky camera data mounted on the ship alongside the radiometer. u_{Ws} could also be estimated from other measurements, e.g., the use of high-resolution infrared SST observations to characterize the intravariability within a microwave instrument FoV, the use of fast sampling instrumentation aboard low-flying aircraft or from ocean model outputs (if they have sufficient resolution). u_{Ws} is a challenging and highly variable uncertainty that demands careful attention when making FRM TIR radiometer measurements from ships.

u_{Wz} arises if SST_{depth} measurements are used to validate SST_{skin} (which is not the case for ship-borne TIR radiometers and is therefore conceptually set to 0). It is discussed here for completeness and because on many ships deploying FRM for SST_{skin} validation, additional SST_{depth} sensors are also deployed. Uncertainties arise due to incorrectly reported measurement depths

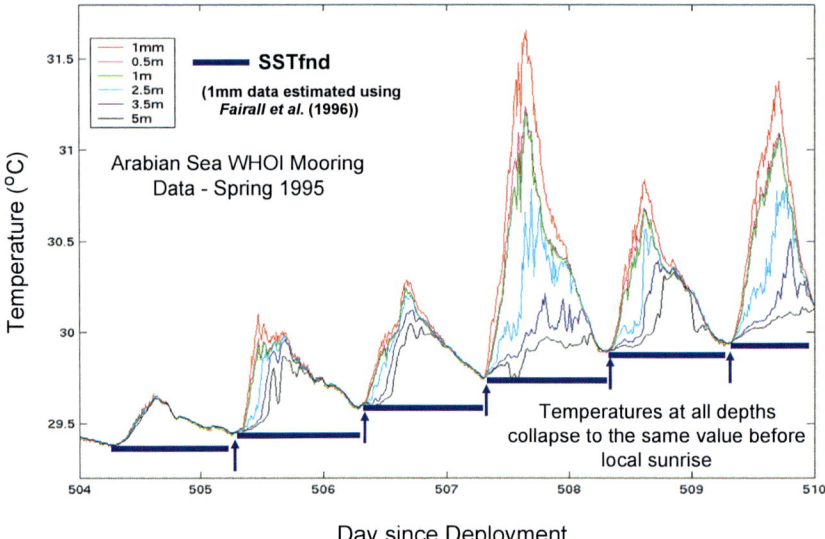

FIGURE 5 Example of diurnal stratification at the Woods Hole Arabian Sea mooring ([Trask, 1995] 15° 30′N, 61° 30′E.), Spring 1995 showing SST divergence at different depths as diurnal thermal stratification develops each day—and collapses as the surface layer nocturnally cools a and overturns in the early evening [26]. SST, sea-surface temperature.

when stratification exists at the sea (Figure 5). Problems encountered include changes in the draught of the ship—a function of the day-to-day cargo loading or, imperfect knowledge of the depth at which a trailing thermistor line will "sit" in the water as the ship moves. In practice, data likely to be affected by diurnal stratification are removed from the validation analysis based on wind speed and solar radiation measurements alone. There is scope for applying adjustments for vertical stratification and vertical displacement issues as a means to reduce u_{Wz}. Such techniques require their own uncertainty estimate to be added into the uncertainty budget for u_{Wz}.

u_{Wt} is the time-displacement uncertainty caused by the mismatch in time, t_{diff}, between the in situ sample and the satellite overpass. It can be estimated as

$$u_{Wt} = t_{diff} \frac{\partial_v}{\partial_t} \qquad (5)$$

t_{diff} could be due to timing uncertainty between a satellite measurement and an FRM measurement (normally small) but is normally a consequence of trying to maximize the number of available matchups when satellite and FRM sampling is sparse (e.g., prevented by clouds, limited FRM infrastructure, etc.). Embury et al. [8] highlight another aspect of u_{Wt} that is related to the development of a thermal stratification in the upper ocean layers when wind

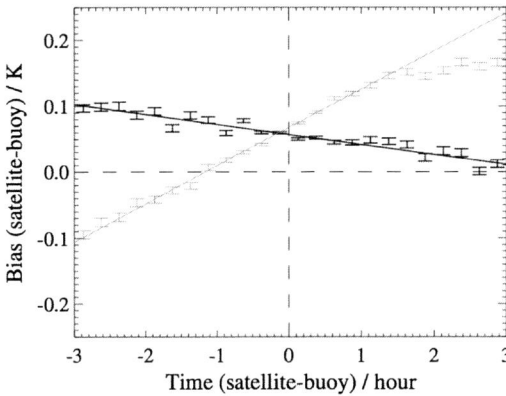

FIGURE 6 AATSR Dual-view two-channel SST 0.2 m retrieval bias as a function of satellite−buoy time difference for daytime (gray) and night-time (black) matches. Solid lines show linear best fit to data (only using time differences ±1.5 h for daytime matches). *From Ref. [8] with permission.*

speed is low and solar radiation is high [53−55] (Figure 6). A cooling trend of 0.015 K/h at night time and a warming trend of 0.058 K/h during the day over a ± 3 h matchup window is evident based on drifting buoy and AATSR satellite matchup data. The SST_{skin} layer sits on top of the thermal stratification layer meaning that u_{Wt} must also consider the effect of diurnal SST variability—highlighting the importance of closely spaced matchups in time.

u_{Wr} is the *spatial displacement uncertainty* caused when the FRM sample is displaced from the matched image pixel by a distance Δr. It can be estimated as

$$u_{Wr} = \Delta r \frac{\partial_v}{\partial_r} \qquad (6)$$

Δr could be due to spatial co-location uncertainty (depending on the severity of validation matchup criteria considered acceptable) but again, arises more significantly from trying to maximize the number of matchups with reduced space and time co-location criteria when satellite sampling is sparse (e.g., prevented by clouds, limited in situ infrastructure, etc.). u_{Wt} and u_{Wr} can be minimized by collecting more measurements but remain a significant problem for infrared sensors where cloud and aerosol contamination prevent a useful SST retrieval leading to a small number of matchup data. Given the horizontal temperature gradients between positions of the ship-borne radiometer measurement and the nearest cloud-free satellite retrieval, the size of this uncertainty can be quantified using estimates of the spatial autocorrelation of the SST [50].

u_{Ws}, u_{Wz}, u_{Wt}, and u_{Wr} are all primarily a function of vertical and horizontal ocean temperature gradients and should be much smaller in open ocean regions far from temperature fronts and tidal influences (Figure 4) and, under moderate wind speed conditions (>6 m/s), the latter which limits diurnal stratification due to wind-induced mixing [56].

Figure 7 shows a number of validation scenarios that consider the choice of matchup criteria for a ship-borne radiometer making a point sample or

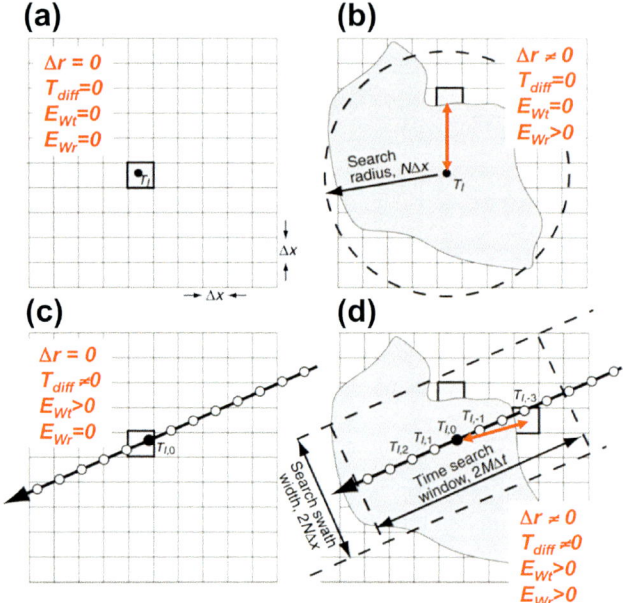

FIGURE 7 Examples of matchup situations encountered in the construction of a matchup database. (a) Point sample when there is no cloud. Match the FRM radiometer measurements closest to the time of the overpass to the satellite pixel in which it lies. Time and space uncertainty are minimal with small SST gradients, at night with a moderate wind stress. (b) Point sample obscured by cloud. Match the FRM radiometer measurements closest to the time of the satellite measurement to the closest cloud-free pixel. The search radius needs to be limited to N pixels. In this case, a search radius must be employed to overcome the problems of clouds obscuring the most appropriate pixel and establish a matchup pair. Representation uncertainties are increased. (c) FRM radiometer measurements along track in cloud-free conditions. Match the FRM radiometer measurements closest to the time of the satellite-measured pixel in which it lies. In the case of a moving ship with regular samples, time mismatch uncertainties are present, although small in this particular case. (d) FRM radiometer measurements along track in cloudy conditions. Time and space uncertainty are present in this case and displacement uncertainties are potentially large and must be constrained by appropriate boundaries. SST, sea-surface temperature; FRM, fiducial reference measurements. *From Ref. [47] with permission.*

averaging over a transect. Cases (a) and (b) consider the choice when comparing a point sample to a satellite measurement and cases (c) and (d) consider the choices when using an average along a ship transect. In this figure, T_{diff} sets the maximum allowed time discrepancy between satellite measurement and ship-borne radiometer measurement, N sets the maximum search radius such that $\Delta r < N$. Δx and Δt are the sensor sampling interval or integration time and R sets the time window of $t_{\text{diff}} = \pm R\Delta t$. The use of a search radius allows a greater sample to be obtained, minimizing the uncertainty in u_{Ws} but at the cost of increasing uncertainty in u_{Wt} and u_{Wr}. Alternatively, one could set a limit in terms of time interval, and then translate this into the number of samples of a given radiometer. In open ocean conditions, far from horizontal temperature gradients, at night (to minimize surface temperature structures due to diurnal warming and mixing) with a moderate wind speed, this may be the most appropriate validation strategy. But to determine suitable values for N and R, the MDB must contain satellite measurements over a spatial area. Furthermore, as N and R are expected to change in magnitude as the environmental conditions change (e.g., cloud, wind, diurnal variability, etc.), the MDB should also include other measurements and parameters over a similar spatial area. In terms of requirements for an SBRN,

- FRM TIR SBRN measurements shall maintain and demonstrate SI traceability with a full uncertainty analysis;
- FRM and satellite data and matchup results are quality controlled using outlier-tolerant (robust) statistics [57];
- FRM TIR SBRN measurements shall be supported by additional environmental measurements including 10 m wind speed, relative wind direction, solar radiation, air temperature subsurface SST, ship roll, pitch, velocity and location, and ship loading lines. (Note: This list shall be viewed as a minimum set of auxiliary measurements required to ensure that the environmental context of the radiometer measurements is fully quantified.)

Donlon et al. [47] and Wimmer et al. [29] discuss the process of matchups between ship-borne radiometers and satellite data. They adopted an approach in which different grades of time-space matchup severity are used to stratify a matchup data set that considers time-space mismatches that are inevitable due to cloudy conditions and imperfect positioning of a ship with respect to cloud-free pixels. The final definitions adopted [29] are the following:

- **Grade-1** defines coincidence of ship-borne FRM TIR radiometer measurement and a satellite measurement with a specific FoV (in kilometers), FoV_{sat}, within ± 2000 s time window and FoV_{sat} km search radius in space that is considered the closest coincidence considered feasible.
- **Grade-2A** maintains a temporal match within ± 2000 s but relaxes the spatial match to within ± 20 km. This grade allows a match to the nearest cloud-free pixel to a ship-track within a radius of 20 km.
- **Grade-2B** defines a temporal match within ± 2 h and spatial match within $\pm FoV_{sat}$ km. This is the criteria formally adopted by the ENVISAT AATSR Validation Protocol [58] and is similar to that proposed in Ref. [8], but somewhat greater than that in Ref. [50].
- **Grade-3** is the most relaxed validation matchup criteria considered acceptable for satellite SST validation activities and is defined as within ± 2 h and spatial match within ± 20 km.
- **Grade-4** uses matchup criteria within ± 6 h and spatial match within ± 25 km, the coarsest of the criteria used by some operational agencies for open ocean validation of satellite SST data.

These definitions provide a practical framework for the time-space criteria for validation of satellite SST measurements using a wide variety of FRM systems: they may be revised as the demonstrated accuracies of the satellite SST_{skin} retrievals improve, so that these sources of uncertainty do not become dominant.

2.2.3 SBRN Requirements for Validation of SST Retrieval Algorithms

The typical form of a satellite-derived multichannel SST estimate [1,2] is the linear (or nearly linear) combination of BTs [7] of the form

$$\widehat{s} = a_0 + a^T \widehat{y}_0$$

where $\widehat{s}$, the estimated SST, a_0, the offset coefficient, a^T, the column vector of weighting retrieval coefficients, and $\widehat{y}_0$, the BT measured in different spectral wavebands are attenuated differently by the atmosphere. Retrieval coefficients can be derived through regression of observed BTs to drifting buoy SST_{depth} measurements (e.g., Ref. [2]), or by regression against simulated satellite BTs using Radiative Transfer Modeling (RTM) (e.g., Refs [8,11,59]). A direct regression-based (DRB) approach requires uses an MDB of colocated and near-contemporaneous satellite and drifting buoy (or other SST) data. The scope of the MDB is limited in temporal and spatial extent by the number of available drifting buoy measurements and cloud-free pixel matches. This forces the recomputation of SST algorithm retrieval coefficients on a regular basis if the instrument in space is degrading [60] introducing instability and inhomogeneity. As homogeneity and stability of the SST CDR are critical for climate research, this approach is not ideal.

The use of a high-resolution line-by-line RTM (e.g., Refs [4–8,11,59]) to simulate satellite BT's that include the range of atmospheric and oceanic variability expected (e.g., changing trends of atmospheric trace gas concentrations and their seasonal and regional variability, satellite viewing geometry, the impact of atmospheric aerosol, sea water emissivity) allows algorithm retrieval coefficients to be determined for any season and any year. Output from numerical weather prediction (NWP) models (e.g., using the European Center for Medium-range Weather Forecasting (ECMWF) 40 year Reanalysis (ERA-40) [61]) can be used to condition the RTM [8]. NWP outputs provide a more representative distribution of the full variability of the atmosphere compared to radiosonde data alone [3]. The RTM approach allows a more consistent satellite SST retrieval for different satellite instrument data sets (when the RTM is configured properly for each satellite instrument) leading to a more homogeneous multisensor SST CDR [8]. In this approach, the satellite SST remain independent of ground reference SST [7,8]. This is an important consideration given the limited number of ground measurements available for validation activities—particularly in the early part of the satellite SST record. For this reason, the RTM approach to derivation of satellite SST algorithms is the basis for SST retrievals generated by the European Space Agency Climate Change Initiative [62]. Nevertheless, biases and uncertainties in retrieved SST may still occur if the forward modeling is poor, if satellite instrument artifacts are not adequately known, if clouds are undetected, when stratospheric and tropospheric aerosols are present, if an inappropriate first-guess prior is used and due to nonlinearities in the RTM at infrared wavelengths [8] (see Chapter 4.3 for more details on these aspects).

Whatever method of SST algorithm specification is used, independent validation evidence is required to continually verify the effectiveness of the SST algorithms coefficients across the duration of the satellite mission. The aim is to assess the accuracy of a specific SST retrieval scheme. This is a narrower focus than validation of an SST *product* because the quality of an SST product is determined by the end-to-end satellite processing system (i.e., includes cloud, aerosol, and sea ice detection, checks on validity of TOA radiance calibration schemes, etc.) [8]. A key requirement is that an FRM TIR SBRN used for validation of SST retrieval algorithms must adequately sample the atmospheric regimes that are both within and beyond the scope of the RTM or DRB MDB used to define the SST algorithm coefficients. Table 2 defines a set of Latitudinal Atmospheric Regimes (LAR) using an approximate latitudinal separation for this purpose (boundaries between regimes in a fluid and dynamic atmosphere are not rigid). Figure 8 shows the climatological distribution of global Total Column Water Vapor overplotted with the approximate location of each LAR regime: within each latitude band, there is a need for seasonally optimized algorithms coefficients to cater for water vapor variability. Ideally, at least one ship-borne radiometer should operate in a sustained manner for each of the LAR to provide validation of SST algorithm retrieval coefficients.

The requirements for an FRM TIR SBRN to validate satellite SST algorithm retrieval coefficients are the following:

- At least one sustained, regularly repeated FRM TIR SBRN "line" (i.e., ship route) shall be maintained for each LAR identified in Table 1.
- The location of FRM TIR SBRN measurements shall be continuous, as far as possible, and optimized for maximum near-contemporaneous and colocated satellite and FRM matchups.
- FRM TIR SBRN line(s) should sample different latitudes to resolve any thermal cycling of instruments around an orbit.
- FRM TIR SBRN line(s) shall sample in both the day- and night-time part of the orbit.
- FRM TIR SBRN measurements shall be collected using Grade 1–Grade 2 Matchup criteria.
- FRM TIR SBRN measurements shall have an absolute accuracy commensurate with that needed to provide the link to the satellite and thus for climate, this should be <0.05 K.

TABLE 2 Latitudinal Atmospheric Regimes (LAR) Requiring At Least One Radiometer Line Within an Ideal FRM TIR SBRN

SBRN LAR ID	SBRN LAR Name	Approximate LAR Latitudinal Limit	General Description of Each LAR
1-NPol 2-SPol	Polar latitude	90–75N 90–75S	High-pressure shallow atmosphere, cold and dry (cold deserts). Nearly all sea ice in the Antarctic region
3-NHL 4-SHL	High latitude	75–55N 75–55S	Includes subpolar low-pressure belt
5-NML 6-SML	Mid latitude	55–38N	Variable moist atmosphere
7-NSTHP 8-SSTHP	Subtropical high latitude	38–25N 38–25S	Sub-tropical high pressure belt, dry
9-TROP	Tropical	25–25S	Deep moist warm atmosphere. Warmest SSTs, most humid atmospheres and prone to anvil cirrus.
10-TDC	Tropical: Deep convection region	Variable	Lies within the tropical region—location of ITCZ. Mostly perennial northern hemisphere in the eastern Pacific and Atlantic Oceans, asymmetric in the Indian Ocean.
11-TMON	Monsoonal	Regional	Asian (northeast and southwest), East Asian, North American monsoon, Indo-Australian monsoon,
12-Dust	Desert dust	Variable	e.g., Atlantic Saharan outflow
13-Coast	Costal regions and marginal seas	Variable	High gradients in proximity of land
14-ISEA	Inland Seas	Mediterranean, Black Sea, Great Lakes	Atmosphere characterized by surrounding land mass and can be variable

FRM, fiducial reference measurements; TIR, thermal infrared; SBRN, ship-borne TIR radiometer network.

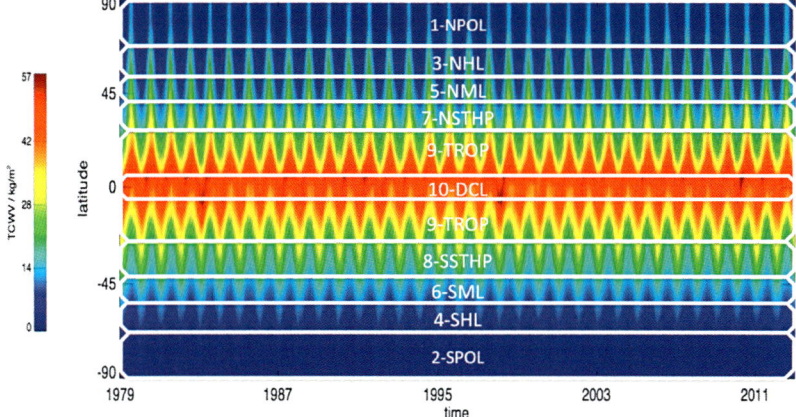

FIGURE 8 Total column water vapor from the ECMWF ERA-interim reanalysis [63] showing latitudinal atmospheric regimes defined in Table 2 requiring at least one FRM TIR SBRN line to validate satellite SST retrieval algorithms. SST, sea-surface temperature; FRM, fiducial reference measurements; TIR, thermal infrared; SBRN, ship-borne TIR radiometer network. *TCWV figure from M. Schroeder with permission.*

- FRM TIR SBRN measurements shall have an absolute stability commensurate with that needed to provide the link to the satellite and thus for climate, this should be <0.015 K/decade at a scale of 100–1000 km.
- FRM TIR SBRN measurements shall be collected according to internationally agreed protocols established and maintained by scientific consensus.
- FRM TIR SBRN measurements shall be supported by additional environmental measurements including 10 m wind speed, relative wind direction, solar radiation, air temperature subsurface SST, ship roll, pitch, velocity and location, and ship loading lines. (Note: This list shall be viewed as a minimum set of auxiliary measurements required to ensure that the environmental context of the radiometer measurements are fully quantified.)
- FRM TIR SBRN measurements shall target an absolute location accuracy of <100 m with respect to WGS-84.
- FRM TIR SBRN measurements shall be quality controlled using outlier-tolerant (robust) statistics [57].
- FRM TIR SBRN measurements shall maintain and demonstrate SI traceability with a full uncertainty analysis.

2.2.4 SBRN Requirements for Validation of Satellite SST Products

FRM TIR SBRN measurements are required to validate not only the satellite SST measurements provided in the product but also the performance and impact of cloud, sea-ice, and aerosol detection schemes used to generate the product. Large uncertainty can be introduced when such detection schemes fail [8]. In this application, validation regimes can be defined based on the ocean and atmospheric characteristics of a particular region. For example, areas with strong western boundary currents (Figure 4) will have a very dynamic ocean and atmosphere regime due to the strong gradients characteristic of those regions, whereas central ocean gyres are relatively benign. Sea ice only occurs at high latitudes. Consequently, the cloud, sea-ice, and aerosol detection schemes may have variable performance depending on the specific region. Figure 9 shows the regions used operationally by the EUMETSAT Ocean and Sea Ice Satellite Application Facilities (OSI-SAF) to validate the METOP AVHRR SST products using the global array of drifting buoys. The latitude banding is not dissimilar to that of Figure 8 but in this case, specific regions are chosen that can be used to generate regional validation statistics that may consider the regional characteristics of each area.

The requirements for an FRM TIR SBRN to address SST product validation are the following:

- At least one sustained, regularly repeated FRM TIR SBRN "line" (i.e., ship route) shall be maintained in regions that can be used to validate cloud, atmospheric aerosol, and sea ice detection schemes (bearing in mind the LAR identified in Table 1).
- The location of FRM TIR SBRN measurements shall be continuous, as far as possible, and optimized for maximum near-contemporaneous and colocated satellite and FRM matchups.
- FRM TIR SBRN line(s) should sample different latitudes to resolve any thermal cycling of instruments around an orbit.

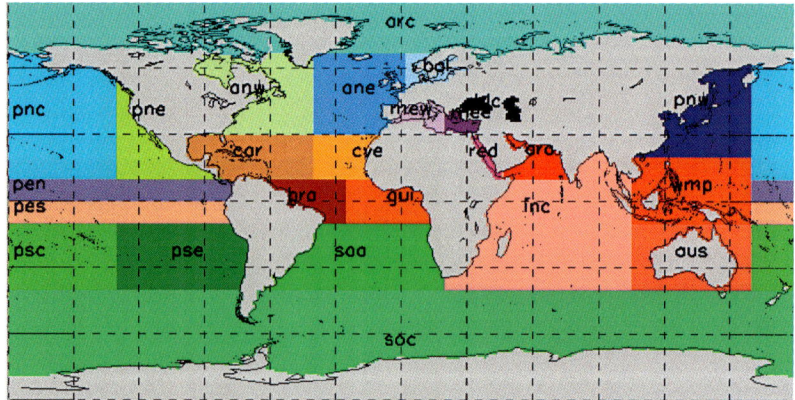

FIGURE 9 The geographical areas used for SST product validation activities using drifting buoy measurements at the EUMETSAT Ocean and Sea Ice Satellite Application Facilities. SST, sea-surface temperature. *OSI-SAF, Figure from A. Marsuin with permission.*

- FRM TIR SBRN line(s) shall sample in both the day- and night-time part of the orbit.
- FRM TIR SBRN measurements shall be collected using Grade 1–Grade 3 matchup criteria.
- FRM TIR SBRN measurements shall have an absolute accuracy commensurate with that needed to provide the link to the satellite and thus for climate, this should be <0.05 K.
- FRM TIR SBRN measurements shall have an absolute stability commensurate with that needed to provide the link to the satellite and thus for climate this should be <0.015 K/decade at a scale of 100–1000 km.
- FRM TIR SBRN measurements shall maintain and demonstrate SI traceability with a full uncertainty analysis.
- FRM and satellite data and matchup results are quality controlled using outlier-tolerant (robust) statistics [57].

2.2.5 SBRN Requirements for Monitoring of Satellite Instrument Degradation

Once a satellite instrument is launched and successfully commissioned, it may operate for many years. During the mission, lifetime degradation of the instrument is expected as optical and electronic components are exposed to the space environment (i.e., radiation, degassing of structures, orbital thermal cycling, …) and scanner and other moving parts gradually wear out (especially bearings) leading to "jitters" and noise. Furthermore, satellite and instrument subsystems degrade: cryogenic coolers become less efficient toward the end of the mission leading to a warmer instrument, satellite platform control gyroscopes, and other parts of the Attitude and Orbit Control System may fail leading to limited knowledge of instrument pointing. See Chapter 2.4 for more details.

Independent validation data is required to identify, manage, and monitor such on-orbit satellite instrument anomalies and the inevitable degradation of a satellite instrument across the entire mission lifetime. Validation is the only means to assess the impact of such events on SST product quality. The requirements for an SBRN to address satellite on-orbit instrument degradation are the following:

- At least one sustained, regularly repeated FRM TIR SBRN "line" (i.e., ship route) shall be maintained that ideally span several orbit swaths (bearing in mind the LAR identified in Table 1).
- The location of FRM TIR SBRN measurements shall be continuous, as far as possible, and optimized for maximum near-contemporaneous and colocated satellite and FRM matchups.
- FRM TIR SBRN line(s) should sample different latitudes to resolve any thermal cycling of instruments around an orbit.
- FRM TIR SBRN line(s) shall sample in both the day- and night-time part of the orbit.

- The location of FRM TIR SBRN measurements shall be optimized for maximum contemporaneous and colocated satellite and FRM matchups (i.e., care should be taken to avoid aliasing with satellite orbit progression and repeat coverage).
- FRM TIR SBRN measurements shall be collected using Grade 1–Grade 4 matchup criteria.
- FRM TIR SBRN measurements shall have an absolute accuracy commensurate with that needed to provide the link to the satellite and thus for climate, this should be <0.05 K.
- FRM TIR SBRN measurements shall have an absolute stability commensurate with that needed to provide the link to the satellite and thus for climate, this should be <0.015 K/decade at a scale of 100–1000 km.
- FRM TIR SBRN measurements shall maintain and demonstrate SI traceability with a full uncertainty analysis.
- FRM and satellite data and matchup results are quality controlled using outlier-tolerant (robust) statistics [57].

2.2.6 SBRN Requirements to Bridge between Different Satellite Instruments

The SST CDR is composed of data derived from many different polar orbiting and geostationary orbit satellite instrument series starting in the late 1970s as discussed in Chapter 2.3. Since at present no satellite instrument can reliably establish SI traceability in orbit at the necessary uncertainty levels, accurate and independent FRM TIR SBRN measurements are required to confidently relate each separate satellite instrument to another. The GCOS Climate Monitoring Principles (e.g., Ref. [21]) include a requirement for an overlap between successive instruments in a series for this reason: "A suitable period of overlap for new and old satellite systems should be ensured for a period adequate to determine inter-satellite biases and maintain the homogeneity and consistency of time-series observations".

In the case of a particular series of instruments where an interruption in data provision has occurred (due to development delays or a premature loss of mission), ground-based FRM can provide an important and essential data set to "bridge the gap" between instruments [64]. Gap-filling seeks to monitor the geophysical behavior of SST_{skin} during the gap years. The "gap" ideally also needs to be filled using an appropriate gap-filling satellite data set—ideally validated with the same FRM used to "bridge the gap". In order to study both the overall SST CDR trend and the underlying processes that contribute to climatic behavior, it is essential to adopt strategies for both gap-bridging and gap-filling. Ideally data collected just before the satellite data-gap and data collected immediately upon resumption when a new replacement satellite mission is available are, in each case, compared to FRM. In this way, the data-record is resumed with the same, totality compatible, calibration [64] within the uncertainty limitations of the FRM and its respective linkage to the Satellite.

For example, the follow-on sensor to the unique dual-view Along Track Scanning Radiometer (AATSR) series (e.g., Ref. 64) that failed in 2012 is called the Sentinel-3 Sea and Land Surface Temperature Radiometer (SLSTR [65]). Unfortunately, Sentinel-3 is not due for launch until mid-2015. The primary requirement to "bridge the gap" is to ensure that FRM are available to calibrate AATSR data products at the AATSR End of Life and SLSTR at the beginning of the Sentinel-3 mission. Experience with AATSR has demonstrated that FRM TIR ship-borne radiometers can be used (e.g., Refs [66,67]) to bridge the gap between AATSR and SLSTR [64] although to date, only a few ship-borne radiometer systems are available for this task [39].

The coverage and sampling requirements for an SBRN to bridge across satellite instrument series are the following:

- At least one sustained, regularly repeated FRM TIR SBRN "line" (i.e., ship route) shall be maintained for each LAR identified in Table 1 at the end of life of and beginning of life of satellite instruments where a mission gap occurs.
- The location of FRM TIR SBRN measurements shall be continuous, as far as possible, and optimized for maximum near contemporaneous and colocated satellite and FRM matchups.
- FRM TIR SBRN line(s) should sample different latitudes to resolve any thermal cycling of instruments around an orbit.
- FRM TIR SBRN line(s) shall sample in both the day- and night-time part of the orbit.
- FRM TIR SBRN measurements shall be collected using Grade 1–Grade 2 matchup criteria.
- FRM TIR SBRN measurements shall have an absolute accuracy commensurate with that needed to provide the link to the satellite and thus for climate, this should be <0.05 K.
- FRM TIR SBRN measurements shall have an absolute stability commensurate with that needed to provide the link to the satellite and thus for climate, this should be <0.015 K/decade at a scale of 100–1000 km.
- FRM TIR SBRN measurements shall maintain SI traceability with a full uncertainty analysis.
- FRM and satellite data and matchup results are quality controlled using outlier-tolerant (robust) statistics [57].

3. LABORATORY INTERCALIBRATION EXPERIMENTS FOR FRM SHIP-BORNE RADIOMETERS

If a ship-borne radiometer is to qualify as an FRM and satisfy the guidelines of the GCOS climate monitoring principles (e.g., Ref. [21]), each radiometer must demonstrate its traceability and thus as a minimum be validated regularly against traceable radiometric standards (e.g., Refs [22,33]). This is best achieved using an SI traceable independent reference radiance source [32–36,68] that can be viewed by all FRM TIR SBRN instruments in a controlled laboratory environment. Based on such intercomparison measurements, uncertainty estimates for individual radiometer measurements can be derived in a common manner: only by conducting detailed calibration and subsequent intercomparison experiments can field radiometers be used with any degree of confidence.

In 1996, the European Union "Concerted Action for the Study of the Ocean Thermal Skin" (CASOTS) project organized a ship-borne radiometer intercomparison experiment, held at Southampton Oceanography Center, England in June 1996 [32]. The aim of the activity was to establish the "degree of equivalence" between ship-borne SST_{skin} measurements. One of the major themes within the CASOTS project was to promote the exchange of practical experience and information on ship-borne SST_{skin} measurements, techniques, problems, and solutions. Seven groups participated in the experiment that used the AATSR flight spare calibration blackbodies [69] and specially designed CASOTS-I water-bath blackbodies as traceable reference radiance sources. The CASOTS-I radiance blackbody [32] was a low-cost, portable water bath blackbody reference source designed to perform laboratory calibrations of ship-borne radiometers before, during, and after ship deployments (Figure 10). The CASOTS-I units operated in the 278–353 K temperature range and had an emissivity of >0.998 for radiometer spot-sizes of 40 mm of less. Water bath temperatures were measured to an accuracy of 50 mK. Four units were

FIGURE 10 The CASOTS-I reference radiance blackbody [32]. The thermometer probe shown is secured to the BB for shipping purpose only. CASOTS, concerted action for the study of the ocean thermal skin.

produced and validated [51] against a National Institute of Standards and Technology (NIST) reference radiance blackbody [34] with a difference of ±20 mK [32,51]. During the CASOTS experiment, field radiometers viewed the CASOTS-I and AATSR flight spare blackbodies in standard laboratory conditions. In addition, the CASOTS-I blackbody was used in a temperature-controlled room where the temperature was reduced to a few degrees above freezing to assess the impact of radiometer internal stray radiation. While the CASOTS experiment was a starting point, it did not include an NMI or include all of the international SST_{skin} radiometers that were in use at that time as part of the experiment.

The Committee for Earth Observation Satellites (CEOS) Working Group on Calibration and Validation (WGCV) was well aware of the need and value of such comparisons and further ship-borne radiometer intercomparison activities were called for. Follow-on experiments were organized through CEOS with organizational sponsorship from space agencies. Building on the outcomes of the CASOTS experience, during March, 1998, the Rosenstiel School of Marine and Atmospheric Science (RSMAS) and CEOS hosted an international infrared validation workshop [70] to establish traceability of ship-borne TIR radiometers to SI units through the participation of an NMI. This experiment used a copy of the NIST third generation water-bath blackbody [34]. The NIST blackbody radiance sources operate from 278 to 353 K at controlled set point temperatures with water temperature combined standard uncertainties of 3.5−7.8 mK. The calculated blackbody emissivity is 0.9997 with a relative standard uncertainty of 0.0003. The emissivity increases to 0.99997 when a 50 mm limiting aperture is installed at the cavity entrance [34]. Based on this successful experiment, which laid the foundation for ship-borne radiometer SI traceability, a Second International Infrared Radiometer Calibration and Intercomparison Workshop was held in May−June, 2001 [51]. This workshop aimed to bring as many types of ship-borne radiometers used by different researchers to validate satellite SSTs together to ensure compatibility of their measurements and to determine their individual contributions to the uncertainty budgets of the satellite-derived SST fields. An ultrastable, well-characterized NIST filter transfer radiometer called the Thermal-infrared Transfer Radiometer (TXR) [30], shown in Figure 11, was used in reasonably controlled laboratory conditions to verify the performance of different reference blackbody designs used to validate the calibration of ship-borne TIR radiometers (five blackbody designs participated in the experiment). In addition, the calibration of ship-borne TIR radiometers was also validated and traced to SI. This workshop was a turning point for the ship-borne radiometer community as it set the standard and defined basic metrology protocols for the validation of the ship-borne radiometer calibration used for SST_{skin} satellite validation by establishing their traceability to SI units.

A third ship-borne radiometer intercalibration experiment was then held in 2009 at which both the National Physical Laboratory (NPL), UK and NIST

 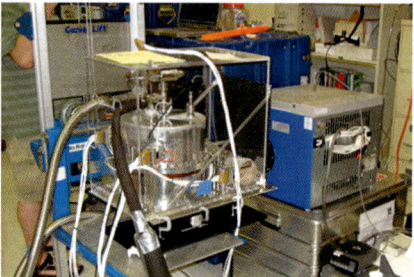

FIGURE 11 (Left) The NIST TXR transfer radiometer (right) set up to view the CASOTS-II blackbody radiance source [33] at the University of Miami second International ship-borne radiometer intercalibration experiment.

participated: NPL acted as the pilot laboratory and provided traceability to SI units during laboratory comparisons in Europe and NIST provided traceability to SI units during laboratory measurements made at RSMAS. The 2009 comparison consisted of two stages in order to allow maximum participation and enable the traceability chain to be established to both NPL and NIST. Stage 1 took place at NPL in April 2009 and involved laboratory measurements of participants' blackbodies calibrated using the NPL reference transfer radiometer (AMBER) [31,71], while participants' radiometers were calibrated using an NPL variable temperature blackbody [37]. The AMBER filter radiometer calibration is traceable to a combination of the NPL primary spectral responsivity scale and ITS-90 through a fixed point reference blackbody (Freezing point of Galium), and this calibration (and its associated uncertainties) is used to convert each blackbody temperature measurement to spectral radiance [31]. Stage 2 took place at RSMAS in May 2009 and involved laboratory measurements of participants' blackbodies calibrated using the NIST TXR, while participants' radiometers were calibrated using the RSMAS and NIST water bath blackbodies. Stage 2 also included testing of the same radiometers alongside each other, completing direct daytime and nighttime measurements of the surface temperature of the sea surface from a pier. All participants were encouraged to develop full uncertainty budgets for all measurements they reported. In order to achieve optimum comparability, lists containing the principal influence parameters for the measurements were provided to all participants. The basis for the uncertainty estimates included assessments of the following:

- Repeatability: e.g., "The typical value of the standard deviation of 180 measurements (1/s, since the radiometer response time is shorter than 1 s) at a fixed blackbody temperature without radiometer realignment."
- Reproducibility: e.g., "Typical value of difference between runs (run to run) of radiometer measurements at the same blackbody temperature including realignment."

- Linearity of radiometer: e.g., "The uncertainty of a linear regression between blackbody source X temperatures and radiometer BTs."
- Primary calibration: e.g., "Two on-board calibration blackbodies with SI traceable thermometers are used to provide a primary calibration to SI standards."

All measurements reported by the participants, along with their associated uncertainties, were analyzed by the pilot laboratory and are presented in Refs [37,71]. This latter comparison was organized to follow explicitly the principles and guidelines of the new Quality Assurance Framework for Earth Observation (QA4EO) developed by CEOS WGCV on behalf of GEO (http://qa4eo.org/docs/QA4EO_Principles_v4.0.pdf). This QA4EO framework effectively embodies the key principles of SI traceability best practice and provides translations of metrology best practice of NMIs for the Earth observation community for things like comparisons, documentary procedures, and uncertainty analysis.

One notable outcome of the 2009 experiment was the recognition that the CASOTS-I water bath blackbodies had deteriorated significantly following several years of use. Based on this result, a second-generation CASOTS-II blackbody reference radiance source was developed [33] as shown in Figure 12.

This low-cost design has proven to be extremely successful. The CASOTS-II radiance source has a 110-mm-diameter aperture cylinder-cone geometry coated with NEXTEL suede 3103 paint. Interchangeable aperture

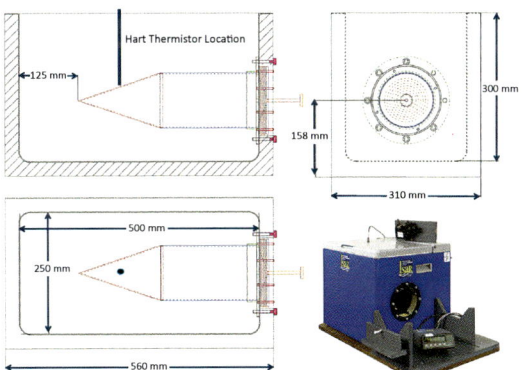

FIGURE 12 CASOTS-II blackbody system [33]. (Left) General arrangement of the CASOTS-II water bath blackbody showing the full selection of aperture-stop plates and protective cover in position. The cone (red), cylinder pot (blue), and mounting flange (green) are shown assembled inside the water bath. The nominal position of the water bath thermometer probe is shown. (Right) The ISAR radiometer-mounting jig is shown in front of the aperture with the Hart Scientific 1504 bridge. A thermometrics temperature probe 225 is shown protruding from the water bath lid and the water bath pump (fitted internally when the unit is in operation) is shown on top of the water bath lid. CASOTS, concerted action for the study of the ocean thermal skin; ISAR, infrared autonomous SST radiometer.

stops reduce the cavity aperture diameter and minimize stray radiation. Monte Carlo modeling techniques show the effective emissivity of the cavity to be 0.9999 (using a 30 mm aperture). The cavity is immersed in a water bath that is vigorously stirred using a pump that slowly heats the water bath at a mean rate of 0.6 K/h. The temperature of the water bath is measured using a thermometer traceable to SI. The worst-case radiance temperature of the CASOTS-II blackbody system is traceable to SI with an uncertainty of 58 mK [33]. When operating under typical laboratory conditions using an aperture of 40 mm, the uncertainty is 16 mK. An intercomparison with the NPL AMBER reference radiometer found no significant differences within 75 mK (110 mm aperture) or 50 mK (40 mm aperture), the combined uncertainty of the comparison [33,71].

The development of a full uncertainty analysis for each radiometer and comparison against NMI reference blackbody sources (and transfer sources) is a difficult and costly exercise. In addition, the route to SI traceability is difficult: while an SI-traceable thermometer may be used in a water-bath blackbody, it measures the temperature only at one point within the water bath. Limited measurements of temperature gradients across all surfaces of the blackbody cavity blackbody and across its wall thickness are available (e.g., Refs [33,34]) but since full characterization is not possible for every experiment, justified assumptions must be used. Furthermore, characterization measurements of blackbody cavity coating emissivity are usually performed on flat witness samples rather than on the cavity itself as it is used in practice. Assumptions must be made that imperfections accumulated during normal practical use, such as dust accumulation, residue from condensation, or sea salt spray, do not affect the measurements at the level of uncertainty required. While the assumptions are usually reasonable, an experimental verification of the resulting BT scale placed on such blackbodies is warranted in order to verify their uncertainties [51].

Regular ship-borne radiometer laboratory intercalibration experiments are a fundamental component of a credible FRM that is required to underpin not only individual satellite SST missions but also the collective SST CDR itself derived from many different missions. However, laboratory-based intercomparisons of ship-borne radiometers against NMI reference standards only provide a verification of each ship-borne radiometer *calibration* performance in terms of its ability to measure a target BT. However, such experiments do not validate the "end-to-end" SST_{skin} measurement when the radiometer is deployed on board a ship at sea. This is discussed in the next section.

4. SHIP-BORNE RADIOMETER FIELD INTERCOMPARISON EXERCISES

When ship-borne radiometers are at sea, they are subject to significant variations in atmospheric temperature, humidity, and direct warming by solar radiation that may impact the radiometer calibration. Additional calibration uncertainties can be introduced by the data logging system itself through the effects of uncompensated long cables or drifts in the electronics packages affecting analog-to-digital conversion, etc. Many instruments use slightly different spectral wave bands to determine SST_{skin}, average target measurements over different time periods, view the sea surface at different view angles, and measure sky radiance using different techniques. Critically, each ship-borne radiometer deployment aboard a ship must carefully consider the most appropriate value for seawater emissivity depending on the instrument spectral and geometric deployment characteristics of the deployment. In principle, such effects could be tested in the laboratory, or quasi-laboratory conditions on controlled water bodies but, arguably, a more robust approach to establish the degree of end-to-end equivalence between different ship-borne radiometers is to perform an intercomparison of radiometers viewing the same area of the sea surface while at sea on a ship.

An experiment of this type was performed during the second calibration and intercomparison of infrared radiometers (Miami2001) during May—June 2001 [68]. Radiometers were mounted on a research vessel, the R/V Walton Smith (Figure 13), to ensure that they can operate accurately and consistently under "at-sea" conditions. This demanded considerable planning and effort in order that all systems could view the same area of the sea surface. Over 2 days were spent working on the ship to install five radiometers and data logging systems to the port side of the ships' upper deck. This was selected as the best site from which all radiometers could view the same area of the sea surface.

FIGURE 13 (Top left) The R/V Walton Smith at sea. (Top right) Radiometers mounted on the vessel viewed from above. From left to right: (1) SISTeR, (2) ISAR-5C, (3) CIRIMS, (4) M-AERI, (5) DAR011. A JPL NNR radiometer was mounted in the bow of the ship. (Bottom left and right) Detailed views of the radiometer installations aboard the R/V Walton Smith. CIRIMS, calibrated infrared in situ measurement system.

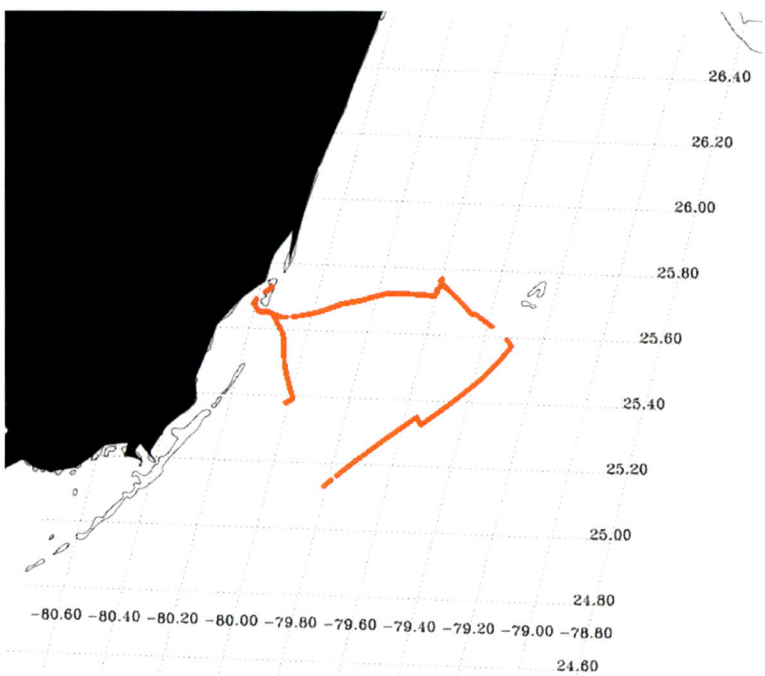

FIGURE 14 Ship-track of the R/V Walton Smith during the Second Miami ship-borne radiometer intercalibration workshop.

The R/V Walton Smith followed the track shown in Figure 14. Full use of the ships scientific facilities providing underway navigation data, thermosalinograph, ADCP, and meteorological observations was available.

In addition, several radiosonde ascents were made in order to characterize the atmospheric conditions throughout the experiment. Good atmospheric and oceanic conditions prevailed, typified by calm seas and variable cloud amount and type with only one short rain event. Initially some problems were encountered with bow wave contamination of radiometer FoV and this effect was limited by first increasing the view angle of radiometers and in addition slowing the vessel down to below 9 kt. At-sea results obtained during the experiment are shown in Figure 15. All radiometer data were averaged to 1 min mean observations (excluding the M-AERI data which provide a 10 min mean observation due to the sampling requirements of the instrument). Clearly seen in Figure 15 is the characteristic temperature difference between a conventional SST measurement made at ~ 1 m depth (shown in yellow) and the radiometer SST_{skin} measurements. It is for this reason that TIR radiometer systems must be used to accurately validate infrared satellite observations.

As a more practical and cost-effective approach compared to a dedicated at-sea multi-instrument experiment is to arrange for pairs of FRM radiometers

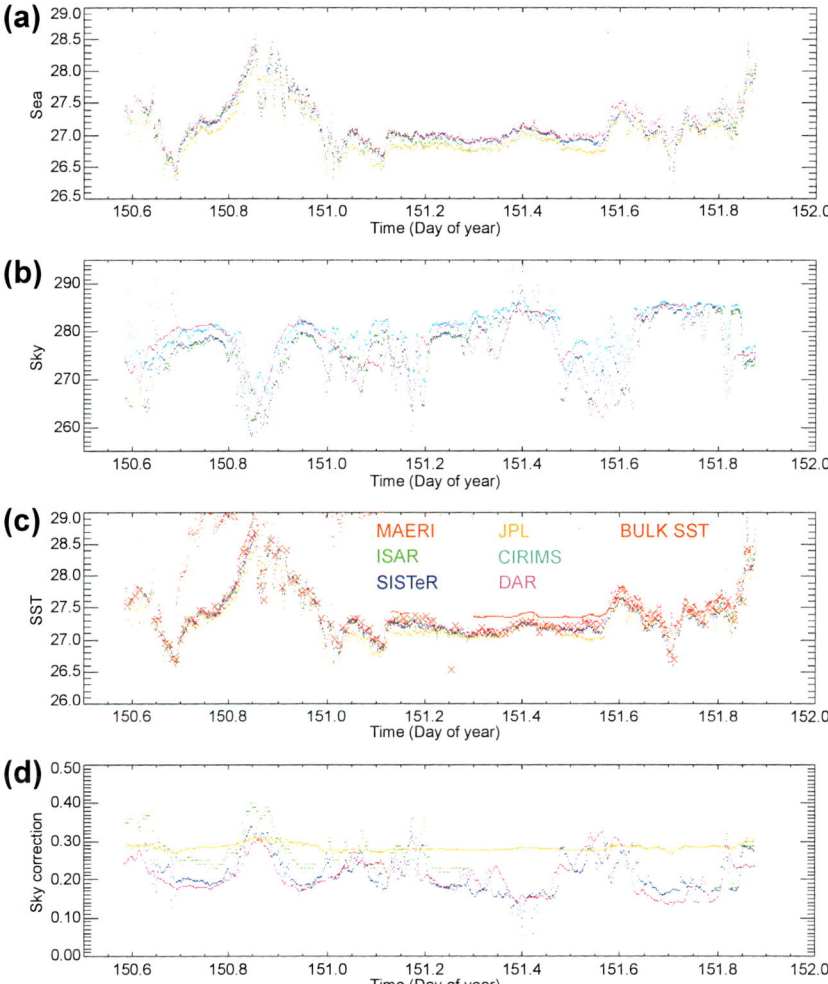

FIGURE 15 Measurements and results from all radiometers throughout the Walton Smith cruise. From the top the figure shows (a) sea brightness temperatures (BTs) as measured by the radiometers, (b) sky BTs as measured by the radiometers, (c) skin SST values derived from the sea and sky measurements, and (d) the sky correction added to the sea BT to account for reflected sky radiation.

to be mounted side-by-side in a series of "bi-lateral" experiments using sustained SBRN lines or research cruises. As an example, during the GasEx01 cruise [72], the CIRIMS [43] was operated alternately with and without an infrared transparent window to evaluate the performance of the correction developed to compensate for the effects of the window on SST_{skin} measurements (see Chapter 3.2). The rms and standard deviation were 0.05 K higher for the measurements made with the window compared to those without the window. The mean difference between contemporaneous measurements made by an M-AERI [73] spectroradiometer-derived and CIRIMS-derived SST changed from 0.04 to -0.07 K when the window was in place, indicating that the CIRIMS was measuring higher temperatures than the M-AERI. The degree to which the measurements agreed is illustrated in the time-series plot in Figure 16, which shows the M-AERI and CIRIMS data with and without the CIRIMS window in place. Based on these results, in deployments of the CIRIMS subsequent to GasEx01, the window was used only when the risk of contamination was great enough to warrant a 0.05 K increase in the standard deviation.

The type of at-sea intercomparison described above, together with regular periodic laboratory experiments, are an essential component of the verification process that is required to determine if a TIR ship-borne radiometer instrument is of sufficient quality to be used as an FRM for validation of satellite SSTs that will form part of the SST CDR.

It is clear that systematic and regular intercomparison of SBRN instruments in field conditions provide fundamental evidence that such instruments are indeed measuring the same SST_{skin} after all corrections and deployment specific factors have been accounted for. Furthermore, such deployment strategies

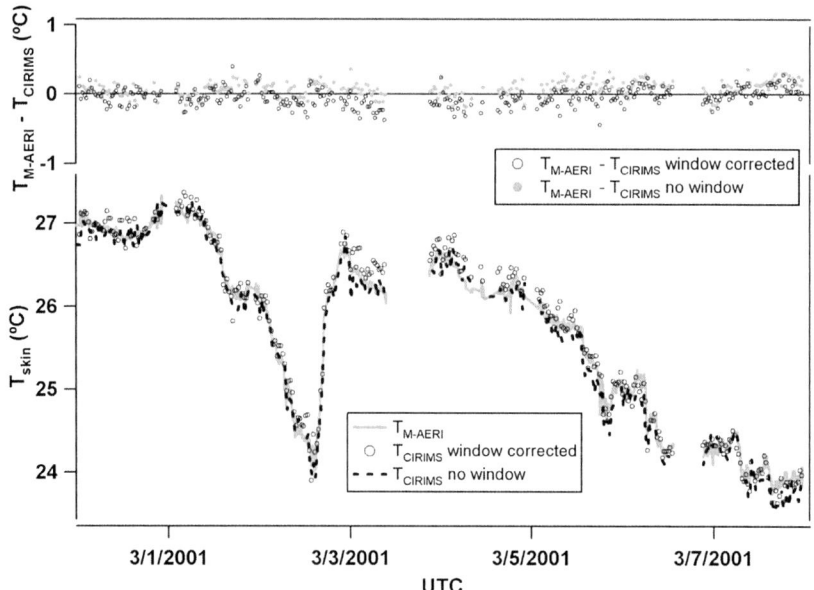

FIGURE 16 Time series of SST_{skin} measured by M-AERI and CIRIMS over an 8 day period on the GasEx01 cruise showing results with and without the IR transparent window in use. The CIRIMS measurement with the window was consistently greater than that without the window. CIRIMS, calibrated infrared in situ measurement system; SST, sea-surface temperature.

can be used to test new instrument configurations and design choices before they are used to validate satellite data contributing to the SST CDR.

5. PROTOCOLS TO MAINTAIN THE SI TRACEABILITY OF FRM SHIP-BORNE TIR RADIOMETERS FOR SATELLITE SST VALIDATION

Donlon et al. [33] define a set of nine protocols intended to guide any group collecting ship-borne infrared radiometer data for use in satellite SST validation activities toward a "common sense" best practice that will improve the quality and reduce the uncertainty in the satellite SST validation process. Each individual deployment of a ship-borne radiometer is highly specific and the protocols summarized below are considered as a minimum requirement for the FRM TIR SBRN.

5.1 Definition of Measurement Methodology

The exact methodology used to measure SST_{skin} using a ship-borne radiometer shall be fully documented. This shall include the following:

- A full technical description of the radiometer instrument (e.g., spectral characteristics, sampling characteristics, measurement technique, a description of the instrument internal calibration approach, etc.).
- The spectral characteristics of the measurement system (i.e., instrument band-pass).
- The value used for seawater emissivity.
- How the component of "sky radiance" reflected at the sea surface into the radiometer FoV is properly addressed (e.g., Ref. [46])?
- A description of the radiometer mounting arrangements and the geometric configuration of the radiometer with all measurement angles accurately documented.
- A description of steps taken to ensure that measurements are free of ship effects (e.g., ship's bow wave, significant emission from the ship superstructure, emissions from ship exhaust plumes, etc.).
- On-board instrument software used (e.g., version, release date, etc.).
- Data postprocessing software (e.g., version, release date, etc.).
- Any other aspect considered relevant to better understanding the quality of the measurements obtained.

5.2 Definition of Laboratory Calibration and Verification Methodology and Procedures

Infrared radiometers typically used for satellite validation work are calibrated using on-board calibration reference radiance sources (blackbodies). The purpose of performing pre- and postdeployment verification using external reference blackbodies is to assess the accuracy of the internal calibration system, and to provide a link in an unbroken chain of comparisons linking the ship-borne radiometer to an SI reference. The exact methodology and procedures used to perform a laboratory calibration and verification of a radiometer shall be defined and documented [37,71].

5.3 Predeployment Calibration Verification

Following the defined methodology and procedures set out under Protocol 2, the calibration performance of a ship-borne radiometer used for satellite product validation shall be verified prior to deployment using an external reference radiance source that is traceable to SI standards over the full range of SSTs expected for a deployment at sea. Ideally, the verification measurements should be repeated over a range of ambient temperatures to assess the influence of stray radiation on the radiometer measurements. The radiometer hardware, on-board configuration, on-board processing software, and data post processing software shall not be modified in any physical way between the calibration and the sea deployment (with the exception of dismounting and transporting the instrument to the calibration laboratory).

5.4 Postdeployment Calibration Verification

Following the defined methodology and procedures set out under Protocols 2 and 3, the calibration performance of a ship-borne radiometer used for satellite product validation shall be verified after deployment.

5.5 Uncertainty Budgets

Ship-borne radiometer calibration and verification data shall be linked to uncertainty budgets determined in agreement with defined NMI protocols [42,74], accounting for a comprehensive range of uncertainty sources (e.g., contributions from instruments, processing, deployment restrictions, and environmental conditions, etc.), i.e., consistent with the principles of QA4EO (http://qa4eo.org/docs/QA4EO_Principles_v4.0.pdf). An uncertainty budget for the end-to-end SST_{skin} measurement shall be provided.

5.6 Improving Traceability of Calibration and Verification Measurements

Efforts should be made where possible to define community consensus schemes and measurement protocols for calibration and verification. Well-documented data processing schemes and quality assurance criteria shall be established to ensure consistency and traceability to SI standards of in situ radiometer measurements used for satellite validation. Ship-borne radiometer users must participate regularly in intercomparison "round-robin" tests and comparison with international standards to establish SI traceability for their data. International radiometer and reference blackbody intercalibration experiments [37,51,68,70,71] are essential under this protocol and the need for regular activities of this type is obvious [22]. They promote the dissemination of state-of-art knowledge on instrument calibration, measurement methods, data processing, training opportunities, and quality assurance. In preparation for the launch of new satellite instruments and the on-going validation of currently flying satellite instruments, the CEOS community has recognized the need for a fourth FRM infrared radiometer and reference blackbody intercalibration experiment. The proposed experiment includes the following components:

- A laboratory-based comparison of the calibration processes for FRM TIR SBRN radiometers and verification of blackbody sources used to maintain calibration of FRM TIR radiometers and provide traceability to SI.
- Initiation of field intercomparisons using pairs of FRM TIR radiometers to build a database of knowledge over a several years.

The benefits of radiometer intercomparison work includes the following:

- Establish and document protocols and best practice for FRM TIR radiometer and reference blackbody intercomparisons for future use.
- Establish community best practices for FRM TIR radiometer deployments.
- Evaluate and document differences in IR radiometry primary calibrations and performances under a range of simulated environmental conditions.
- Establish and document formal SI-traceability and uncertainty budgets for participant blackbodies and radiometers.
- Evaluate and document protocols and best practice to characterize differences between FRM TIR radiometer measurements made in field (land, ocean, and ice) operational conditions.
- Follow QA4EO principles and in particular Guidelines: QA4EO-QAEO-GEN-DQK-004, version 4.0 [75].

5.7 Accessibility to Documentation

Documentation describing ship-borne radiometer calibration and verification process shall be made available to the user community to promote peer review and ensure appropriate promulgation of knowledge on ship-borne radiometer calibration and verification.

5.8 Archiving of Data

Ship-borne radiometer calibration and verification data should be archived following good data stewardship practices providing access to records by research teams on request. Laboratory calibration and verification data shall be published in a format that is freely and openly available to users of the data.

5.9 Periodic Consolidation and Update of Calibration and Verification Procedures

Ship-borne radiometer calibration and verification measurement procedures should be consolidated as a result of a critical review of those currently documented in peer-review literature or already included in compilations produced by former programs and "lessons learned" from deployments aboard ships and in the laboratory. Consolidated protocols should be maintained and published.

6. SUMMARY AND FUTURE PERSPECTIVES

This chapter has presented a review of strategies and components for the deployment of ship-borne TIR radiometers used to validate satellite infrared radiometers in support of the CDR of SST. The SST CDR is derived using combined SST measurements derived from multiple TIR satellite instruments each having specific characteristics. A variety of ground-based measurements are available, from a number of different measurement platforms, that could be used to validate and verify the satellite SST CDR. Only ship-borne TIR radiometers provide measurements that are fully traceable to SI primary standards. Such measurements are essential to quantify uncertainties in the satellite-derived SST CDR and are a fundamental component of a satellite mission. For these reasons, they are considered FRM. The requirements and approach to generating uncertainty budgets and estimates for ship-borne FRM TIR radiometers have been discussed in the context of the satellite SST CDR. This includes the validation of ship-borne radiometer calibration performance using laboratory reference radiance sources and SI transfer radiometers. To complement laboratory activities, at-sea intercomparisons of ship-borne radiometers in an end-to-end manner is also required. In addition, care is required to fully understand the uncertainties introduced by the validation process itself (i.e., the matchup of collocated and near contemporaneous satellite and FRM TIR radiometer measurements).

The scope of an FRM TIR SBRN has been defined and justified by considering requirements for a number of applications relevant to the development of and traceability, to SI, of the SST CDR. The key requirements for such a network have been presented and discussed. The challenge remains to secure the FRM TIR SBRN through international collaboration to ensure that adequate FRM are available to deliver the SST CDR that is fully traceable to SI units.

ACKNOWLEDGMENTS

The authors would like to thank Marc.Schroeder@dwd.de for the base of Figure 8 and G. Corlett for discussion on various aspects of this work.

REFERENCES

[1] D. Anding, R. Kauth, Estimation of sea surface temperature from space, Remote Sens. Environ. 1 (1970) 217–220.

[2] E.P. McClain, W.G. Pichel, C.C. Walton, Comparative performance of AVHRR-based multichannel sea surface temperatures, J. Geophys. Res. 90 (1985) 11587–11601.

[3] C.J. Merchant, A.R. Harris, Toward the elimination of bias in satellite retrievals of sea surface temperature 2. Comparison with in situ measurements, J. Geophys. Res. 104 (1999) 23579–23590. http://dx.doi.org/10.1029/1999JC900106.

[4] C.J. Merchant, P. Le Borgne, Retrieval of sea surface temperature from space, based on modeling of infrared radiative transfer: capabilities and limitations, J. Atmos. Oceanic Technol. 21 (2004) 1734–1746. http://dx.doi.org/10.1175/JTECH1667.1.

[5] C.J. Merchant, O. Embury, P. Le Borgne, B. Bellec, Saharan dust in night-time thermal imagery: detection and reduction of related biases in retrieved sea surface temperature, Remote Sens. Environ. 104 (2006) 15–30.

[6] C.J. Merchant, L. a. Horrocks, J.R. Eyre, A.G. O'Carroll, Retrievals of sea surface temperature from infrared imagery: Origin and form of systematic errors, Q. J. R. Meteorol. Soc. 132 (617) (2006) 1205–1223. http://dx.doi.org/10.1256/qj.05.143.

[7] C.J. Merchant, D. Llewellyn-Jones, R.W. Saunders, N.A. Rayner, E.C. Kent, C.P. Old, D. Berry, A.R. Birks, T. Blackmore, G.K. Corlett, O. Embury, V.L. Jay, J. Kennedy, C.T. Mutlow, T.J. Nightingale, A.G. O'Carroll, M.J. Pritchard, J.J. Remedios, S. Tett, Deriving a sea surface temperature record suitable for climate change research from the along-track scanning radiometers, Adv. Space Res. 41 (2008) 1–11. ISSN: 0273-1177, http://dx.doi.org/10.1016/j.asr.2007.07.041.

[8] O. Embury, C.J. Merchant, M.J. Filipiak, A reprocessing for climate of sea surface temperature from the along-track scanning radiometers: basis in radiative transfer, Remote Sens. Environ. 116 (2012) 32–46.

[9] J. Vazquez-Cuervo, E.M. Armstrong, A. Harris, The effect of aerosols and clouds on the retrieval of infrared sea surface temperature, J. Clim. 17 (2004) 3921–3933.

[10] Frank J. Wentz, T. Meissner, AMSR Ocean Algorithm, Version 2, Report Number 121599A-1, Remote Sensing Systems, Santa Rosa, CA, 2000, p. 66.

[11] A.M. Zàvody, C.T. Mutlow, D.T. Llewellyn-Jones, A radiative transfer model for sea surface temperature retrieval for the Along-Track Scanning Radiometer, J. Geophys. Res. 100 (1995) 937–952. http://www.agu.org/journals/jc/v100/iC01/94JC02170/.

[12] N.R. Nalli, P.J. Minnett, P. van Delst, Emissivity and reflection model for calculating unpolarized isotropic water surface-leaving radiance in the infra-red. I: theoretical development and calculations, Appl. Opt. 47 (2008) 3701–3721.

[13] N.R. Nalli, P.J. Minnett, E. Maddy, W.W. McMillan, M.D. Goldberg, Emissivity and reflection model for calculating unpolarized isotropic water surface-leaving radiance in the infrared. 2: validation using Fourier transform spectrometers, Appl. Opt. 47 (2008) 4649–4671.

[14] D. Smith, C. Mutlow, J. Delderfield, B. Watkins, G. Mason, ATSR infrared radiometric calibration and in-orbit performance, Remote Sens. Environ. 116 (2012) 4–16. http://dx.doi.org/10.1016/j.rse.2011.01.027.

[15] C.J. Donlon, N. Rayner, I. Robinson, D.J.S. Poulter, K.S. Casey, J. Vazquez-Cuervo, E. Armstrong, A. Bingham, O. Arino, C. Gentemann, D. May, P. LeBorgne, J. Piollé, I. Barton, H. Beggs, C.J. Merchant, S. Heinz, A. Harris, G. Wick, B. Emery, P. Minnett, R. Evans, D. Llewellyn-Jones, C. Mutlow, R.W. Reynolds, H. Kawamura, The global ocean

data assimilation experiment high-resolution sea surface temperature pilot project, Bull. Am. Meteorol. Soc. 88 (2007) 1197−1213. http://dx.doi.org/10.1175/BAMS-88-8-1197.

[16] D. Meldrum, E. Charpantier, M. Fedak, B. Lee, R. Lumpkin, P. Niller, H. Viola, Data buoy observations: the status quo and anticipated developments over the next decade, in: J. Hall, D.E. Harrison, D. Stammer (Eds.), Proceedings of OceanObs'09: Sustained Ocean Observations and Information for Society, vol. 2, ESA Publication WPP-306, Venice, Italy, 2000. http://dx.doi.org/10.5270/OceanObs09.cwp.62, 21−25 September 2009.

[17] J.J. Kennedy, A review of uncertainty in in situ measurements and data sets of sea surface temperature, Rev. Geophys. 52 (2014) 1−32. http://dx.doi.org/10.1002/2013RG000434.

[18] A.G. O'Carroll, R.W. Saunders, J.R. Eyre, Three-way error analysis between AATSR, AMSR-E, and in situ sea surface temperature observations, J. Atmos. Oceanic Technol. 25 (7) (2008) 1197.

[19] H. Preston-Thomas, The international temperature scale of 1990 (ITS-90), Metrologia 27 (1990) 3−10.

[20] C.J. Merchant, SST CCI Phase-II Uncertainty Characterisation Report: Sea Surface Temperature V1, SST_CCI-UCR-UOR-201, Available from the, European Space Agency, Frascati, Italy, 2014, p. 31.

[21] WMO, Systematic Observation Requirements for Satellite-based Products for Climate Supplemental Details to the Satellite-based Component of the Implementation Plan for the Global Observing System for Climate in Support of the UNFCCC - 2011 Update, 2011. GCOS-154, p. 138, Available from: http://www.wmo.int/pages/prog/gcos/Publications/gcos-154.pdf.

[22] P.J. Minnett, G.K. Corlett, A pathway to generating climate data records of sea-surface temperature from satellite measurements, Deep-sea Res. II (2012). http://dx.doi.org/10.1016/j.dsr2.2012.04.003.

[23] R. Cowley, S. Wijffels, L. Cheng, T. Boyer, S. Kizu, Biases in Expendable bathythermograph data: a new view based on Historical side-by-side comparisons, J. Atmos. Oceanic Technol. 30 (2013) 1195−1225. http://dx.doi.org/10.1175/JTECH-D-12-00127.1.

[24] W.J. Emery, K. Cherkauer, B. Shannon, R.W. Reynolds, Hull-mounted sea surface temperatures from ships of opportunity, J. Atmos. Oceanic Technol. 14 (1997) 1237−1251. http://dx.doi.org/10.1175/1520-0426(1997)014<1237:HMSSTF>2.0.CO;2.

[25] J.J. Kennedy, R.O. Smith, N.A. Rayner, Using AATSR data to assess the quality of in situ sea-surface temperature observations for climate studies, Remote Sens. Environ. 116 (2012) 79−92. http://dx.doi.org/10.1016/j.rse.2010.11.021.

[26] R.P. Trask, R.A. Weller, W. M Ostrom, Arabian Sea Mixed Layer Dynamics Experiment : Mooring Deployment Cruise Report R/V Thomas Thompson Cruise Number 46, 14 April-29 April 1995, Woods Hole Oceanographic Inst. Tech. Rep. WHOI-95-14, Woods Hole, MA, 1995, p. 88.

[27] E. Oka, K. Ando, Stability of temperature and conductivity sensors of ARGO profiling floats, J. Oceanogr. 60 (2004) 253−258.

[28] F.A. Best, H.E. Revercomb, R.O. Knuteson, D.C. Tobin, R.G. Dedecker, T.P. Dirkx, M.P. Mulligan, N.N. Ciganovich, Y. Te, Traceability of absolute radiometric calibration for the atmospheric emitted radiance interferometer (AERI), in: USU/SDL CALCON, 2003.

[29] W. Wimmer, Variability and uncertainty in measuring sea surface temperature (Ph.D. thesis), Department of Oceanography, University of Southampton, Southampton United Kingdom, 2013.

[30] J.P. Rice, B.C. Johnson, The NIST EOS thermal-infrared transfer radiometer, Metrologia 35 (1998) 505−509.

[31] E. Theocharous, N.P. Fox, V.I. Sapritsky, S.N. Mekhontsev, S.P. Morozova, Absolute measurements of black-body emitted radiance, Metrologia 35 (1998) 549.
[32] C.J. Donlon, T.J. Nightingale, L. Fielder, G. Fisher, D. Baldwin, I.S. Robinson, A low cost blackbody for the calibration of sea going infrared radiometer systems, J. Atmos. Oceanic Technol. 16 (1999) 1183−1197.
[33] C.J. Donlon, W. Wimmer, I. Robinson, G. Fisher, M. Ferlet, T. Nightingale, B. Bras, A Second-Generation Blackbody System for the Calibration and Verification of Seagoing Infrared Radiometers, J. Atmos. Oceanic Technol. 31 (2014) 1104−1127. http://dx.doi.org/10.1175/JTECH-D-13-00151.1.
[34] J.B. Fowler, A third generation water bath based blackbody source, J. Res. Natl. Inst. Stand. Technol. 100 (5) (1995) 591−599.
[35] J.B. Fowler, An oil-bath-based 293 K to 473 K blackbody source, J. Res. Natl. Inst. Stand. Technol. 101 (1996) 629.
[36] J. Geist, J.B. Fowler, A Water Bath Blackbody for the 5 to 60 Degree Temperature Range: Performance Goals, Design Concept and Test Results, 1986. U.S. National Bureau of Standards and Technology Technical Note 1228, p. 16. Available from the National Bureau of Standards and Technology, Gaithersburg, MD 20899.
[37] E. Theocharous, E. Usadi, N.P. Fox, CEOS Comparison of IR Brightness Temperature Measurements in Support of Satellite Validation. Part I: Laboratory and Ocean Surface Temperature Comparison of Radiation Thermometers, Report OP-3, National Physical Laboratory, Teddington, UK, 2010.
[38] G.C. Corlett, The ESA Climate Change Initiative SST_CCI Product Validation and Intercomparison Report (PVIR), SST_CCI-PVIR-UoL-001, available from the, European Space Agency, Frascati, Italy, 2014, p. 148.
[39] W. Wimmer, I.S. Robinson, C.J. Donlon, Long-term validation of AATSR SST data products using shipborne Radiometry in the Bay of Biscay and english channel, Remote Sens. Environ. 116 (2012) 17−31.
[40] G. Ohring, B. Wielicki, R. Spencer, W. Emery, R. Datla, Satellite instrument calibration for measuring global climate change: report of a workshop, Bull. Am. Meteorol. Soc. 86 (2005) 1303−1313.
[41] GCOS, GCOS Climate Monitoring Principles, 2003. Available from: http://www.wmo.int/pages/prog/gcos/documents/GCOS_Climate_Monitoring_Principles.pdf.
[42] JCGM, Evaluation of Measurement Data—Guide to the Expression of Uncertainty in Measurement JCGM 100:2008, GUM 1995 with Minor Corrections, First ed., September 2008. Available from: http://www.bipm.org/utils/common/documents/jcgm/JCGM_100_2008_E.pdf (2008).
[43] A.T. Jessup, R. Branch, Integrated ocean skin and bulk temperature measurements using the calibrated infrared in situ measurement system (CIRIMS) and through-hull ports, J. Atmos. Oceanic Technol. 25 (2008) 579−597.
[44] E.O. Lebigot, Uncertainties: A Python Package for Calculations with Uncertainties; Version 1.8, 2012. Available from: https://pythonhosted.org/uncertainties/index.html.
[45] C. Donlon, I.S. Robinson, M. Reynolds, W. Wimmer, G. Fisher, R. Edwards, T.J. Nightingale, An infrared sea surface temperature autonomous radiometer (ISAR) for deployment aboard volunteer observing ships (VOS), J. Atmos. Ocean. Technol. 25 (2008) 93−113.
[46] C.J. Donlon, T.J. Nightingale, The effect of atmospheric radiance errors in radiometric sea surface temperature measurements, Appl. Opt. 39 (2000) 2392−2397.
[47] C.J. Donlon, W. Wimmer, I.S. Robinson, G. Fisher, D. Poulter, G. Corlett, Validation of AATSR using in situ radiometers in the english channel and Bay of Biscay, in: ESA MERIS

and (A)ATSR Workshop 2005, ESRIN Frascati, Italy, ESA-SP-597. ISSN: 1609-042X, 2005, ISBN 92-9092-908-1. Available online at: https://earth.esa.int/workshops/meris_aatsr2005//participants/14/paper_Donlon.pdf.

[48] R.E. Wolfe, M. Nishihama, A.J. Fleig, J.A. Kuyper, D.P. Roy, J.C. Storey, F.S. Patt, Achieving sub-pixel geolocation accuracy in support of MODIS land science, Remote Sens. Environ. 83 (2002) 31−49.

[49] R.E. Wolfe, M. Nishihama, L. Guoqing, K.P. Tewari, E. Montano, MODIS and VIIRS geometric performance comparison, in: Geoscience and Remote Sensing Symposium (IGARSS), 2012 IEEE International, 2012, pp. 5017−5020.

[50] P.J. Minnett, Consequences of sea surface temperature variability on the validation and applications of satellite measurements, J. Geophys. Res. 96 (C10) (1991) 18475−18489.

[51] J. Rice, J. Butler, B. Johnson, P. Minnett, K. Maillet, T.J. Nightingale, S. Hook, A. Abtahi, C.J. Donlon, I. Barton, The Miami 2001 infrared radiometer calibration and intercomparison. Part I: laboratory characterization of blackbody targets, J. Atmos. Oceanic Technol. 21 (2004) 258−267.

[52] C.J. Donlon, M. Martin, J.D. Stark, J. Roberts-Jones, E. Fiedler, W. Wimmer, The operational sea surface temperature and sea ice analysis (OSTIA), Remote Sens. Environ. 116 (2012). http://dx.doi.org/10.1016/j.rse.2010.10.017.

[53] A.C. Stuart-Menteth, I.S. Robinson, P.G. Challenor, A global study of diurnal warming using satellite-derived sea surface temperature, J. Geophys. Res. 108 (C5) (2003) 3155. http://dx.doi.org/10.1029/2002JC001534.

[54] A.C. Stuart-Mentheth, A Global Study of Diurnal Warming (Ph.D. thesis), University of Southampton, United Kingdom, 205 pp., (2004)

[55] C.L. Gentemann, P.J. Minnett, P. Le Borgne, C.J. Merchant, Multi-satellite measurements of large diurnal warming events, Geophys. Res. Lett. 35 (2008). http://dx.doi.org/10.1029/2008GL035730.

[56] C.J. Donlon, P. Minnett, C. Gentemann, T.J. Nightingale, I.J. Barton, B. Ward, J. Murray, Towards improved validation of satellite sea surface skin temperature measurements for climate research, J. Clim. 15 (4) (2002) 353−369.

[57] P.J. Huber, Robust Statistics, Wiley, New York, 1981.

[58] I.M. Parkes, M.D. Steven, D. Llewellyn-Jones, C.T. Mutlow, C.J. Donlon, J. Foot, F. Prata, I. Grant, T. Nightingale, M.C. Edwards, AATSR Validation Measurement Protocol, May 15, 1998. PO-PL-GAD-AT-005(2), p. 18.

[59] D.T. Llewellyn-Jones, P.J. Minnett, R.W. Saunders, A.M. Zavody, Satellite multichannel infrared measurements of sea surface temperature of the N.E. Atlantic ocean using AVHRR/2, Q. J. R. Meteorol. Soc. 110 (1984) 613−631.

[60] K.S. Casey, T.B. Brandon, P. Cornillon, R. Evans, The Past, present and future of the AVHRR Pathfinder SST program, in: V. Barale, J.F.R. Gower, L. Alberotanza (Eds.), Oceanography from Space: Revisited, Springer, 2010.

[61] F. Chevallier, Sampled Database of 60-Level Atmospheric Profiles from the ECMWF Analyses, NWP SAF 4, ECMWF, Reading, UK, 2002, http://www.ecmwf.int/publications/library/do/references/show?id=83287.

[62] R. Hollmann, Coauthors, The ESA climate change initiative: satellite data records for essential climate variables, Bull. Amer. Meteor. Soc. 94 (2013) 1541−1552. http://dx.doi.org/10.1175/BAMS-D-11-00254.1.

[63] D.P. Dee, S. Uppala, Variational bias correction of satellite radiance data in the ERA-Interim reanalysis, Q. J. R. Meteorol. Soc. 135 (2009) 1830−1841.

[64] D. Llewellyn-Jones, The lessons learned from AATSR and bridging the Gap to SLSTR, in: L. Ouwehand (Ed.), Proceedings of the Sentinel-3 OLCI/SLSTR and MERIS/(A)ATSR Workshop, 15–19 October 2012, 2013, ISBN 978-92-9092-275-9. ESA SP-711.

[65] C. Donlon, B. Berruti, M-H Ferreira A Buongiorno, P. Femenias, J. Frerick, P. Goryl, U. Klein, H. Laur, C. Mavrocordatos, J. Nieke, H. Rebhan, B. Seitz, J. Stroede, R. Sciarra, The global monitoring for environment and Security (GMES) Sentinel-3 Mission, Remote Sens. Environ. 120 (2012) 27–57. http://dx.doi.org/10.1016/j.rse.2011.07.024.

[66] I.S. Robinson, W. Wimmer, C.J. Donlon, Validation of the ENVISAT advanced along track scanning radiometer in the Bay of Biscay and english channel, Remote Sens. Environ. (2011). http://dx.doi.org/10.1016/j.rse.2011.03.022.

[67] W. Wimmer, I.S. Robinson, C.J. Donlon, QA for satellite sea surface temperatures using the ISAR ship-borne radiometric system, in: Geoscience and Remote Sensing Symposium, 2009 IEEE International, IGARSS 2009, 1, no., pp. I-232,I-235, 12–17 July 2009, 2009. http://dx.doi.org/10.1109/IGARSS.2009.5416896.

[68] I.J. Barton, P.J. Minnett, C.J. Donlon, S.J. Hook, A.T. Jessup, K.A. Maillet, T.J. Nightingale, The Miami 2001 infrared radiometer calibration and inter-comparison: 2. Ship comparisons, J. Atmos. Oceanic Technol. 21 (2004) 268–283.

[69] I. Mason, P. Sheather, J. Bowles, G. Davies, Blackbody calibration sources of high accuracy for a spaceborne infrared instrument: the along track scanning radiometer, Appl. Opt. 35 (1996) 629–639.

[70] R. Kannenberg, IR instrument comparison workshop at the rosenstiel school of marine and atmospheric science (RSMAS), Earth Obs. 10 (3) (1998) 51–54.

[71] E. Theocharous, N.P. Fox, CEOS Comparison of IR Brightness Temperature Measurements in Support of Satellite Validation. Part II: Laboratory Comparison of the Brightness Temperature of Blackbodies, National Physical Laboratory, Teddington, UK, 2010. Report OP-4.

[72] D.T. Ho, C.L. Sabine, D. Hebert, D.S. Ullman, R. Wanninkhof, R.C. Hamme, P.G. Strutton, B. Hales, J.B. Edson, B.R. Hargreaves, Southern ocean gas exchange experiment: setting the stage, J. Geophys. Res. 116 (2011) C00F08. http://dx.doi.org/10.1029/2010JC006852.

[73] P.J. Minnett, R.O. Knuteson, F.A. Best, B.J. Osborne, J.A. Hanafin, O.B. Brown, The Marine-atmospheric emitted radiance interferometer (M-AERI), a high-accuracy, sea-going infrared spectroradiometer, J. Atmos. Oceanic Technol. 18 (2001) 994–1013.

[74] S. Bell, Measurement Good Practice Guide No. 11 (Issue 2), A Beginner's Guide to Uncertainty of Measurement, Available from:. ISSN: 1368-6550, National Physical Laboratory Teddington, Middlesex, United Kingdom, 1991. TW11 0LW, 41 pp.

[75] N. Fox, M.C. Greening, A guide to comparisons – organisation, operation and analysis to establish measurement equivalence to underpin the quality assurance requirements of GEO, version-4, QA4EO-QAEO-GEN-DQK-004, Available from: http://qa4eo.org/docs/QA4EO-QAEO-GEN-DQK-004_v4.0.pdf, 2010.

Chapter 6

Assessment of Satellite Products for Climate Applications

Frédéric Mélin,[1,]* Gary K. Corlett[2]

[1] European Commission, Joint Research Centre, Ispra, Italy; [2] Department of Physics and Astronomy, University of Leicester, Leicester, UK
*Corresponding author: E-mail: frederic.melin@jrc.ec.europa.eu

Climate data records (CDRs) are those data sets that are demonstrated as being suitable for climate research, and can be considered as a "time series of measurements of sufficient length, consistency, and continuity to determine climate variability and change" (see Chapter 1.1). To translate this statement into objective, quantifiable criteria in order to decide if a particular data set can be deemed a CDR is not straightforward. The character of continuity implies that any ocean CDR derived from satellite radiometers will be based on a suite of (ideally) overlapping missions, considering the expected lifetime of individual sensors (often no more than 5–10 years in the best cases). However, operating successive measurement systems applied to the same measurand is not enough to create a consistent long-term data record. Besides issues related to data documentation, management, and long-term preservation, consistency is a key feature of any potential CDR. In statistics vocabulary, it can be viewed as logical and numerical coherence. For potential CDRs, this coherence needs to be verified over time, within datasets, and across datasets.

The scientific assessment of any long-term satellite data set is the fundamental part of its consistency check. The aim of the assessment is to validate and assign confidence to the geophysical variable being measured along with its associated uncertainties, through careful consideration of the uncertainties of the reference data set and the validation process itself. The assessment can use a variety of reference data sets, including in situ or synthetic data, as well as other satellite data and/or the output of models. Ideally all measurements should be traceable to international (SI) standards.

This part of the book considers the assessment of long-term ocean data records in the domains of color and temperature, measured from radiometers operating at visible and thermal infrared wavelengths, respectively. It is shown that although the state of the art in each field is quite different, the fundamental principles of assessment in each domain are the same. Differences occur due to three main reasons. First, the availability of reference data for validation is considerably different—for ocean temperature, there are a number of nonradiometric measurement arrays covering the global oceans to supplement radiometric measurements from ships, whereas for ocean color, most radiometric data are restricted to a few localized sites and relatively scarce ship-based measurements. Second, prelaunch and in-flight calibration of radiometers is a much more tractable problem in the thermal infrared than at visible wavelengths owing to the stability and performance of on-board blackbody reference sources. Third, the ability to correct the radiances measured at the satellite for atmospheric attenuation is also less challenging in the thermal infrared domain, aided by the relative magnitude of the rate of change in geophysical variable compared to the magnitude of the atmospheric correction needed.

Chapter 6.1 focuses on the assessment of the primary ocean-color product, the spectrum of marine remote-sensing reflectance (R_{rs}). Other products such as inherent optical properties and the concentration of chlorophyll-a (Chla) will be considered as well, as both R_{rs} and Chla are listed as essential climate variables by the Global Climate Observing System. Good practices and protocols for validating satellite products with in situ data are discussed, and the current status of validation results for R_{rs} is illustrated. Model-based approaches, so far mostly relying on bio-optical models for the retrieval of in-water constituents, have shown potential for characterizing some components of the uncertainty budget. The comparison of products from different satellite missions can also contribute to that effort, and constitutes a major avenue to check consistency. Ocean-color satellite sensors have operated with different bands, and in the context of R_{rs} comparisons, differences in center wavelengths need to be accounted for. Chapter 6.1 also illustrates how time-series analysis can detect issues with the instrument radiometric characterization and calibration. The chapter ends by discussing how consistent ocean-color CDRs from different missions are expected to provide a similar picture of major oceanic phenomena, particularly in terms of seasonal and interannual signals.

Chapter 6.2 focuses on the assessment of long-term records of sea-surface temperature (SST), derived from long-term, ideally harmonized, records of brightness temperature measured top of atmosphere by a series of space sensors. The primary method of assessment of SST products and their uncertainties are comparisons to (independent) reference data measured in situ or from a ship-mounted infrared radiometer. The fundamentals of comparing a satellite field of view to a more localized reference measurement are discussed and an uncertainty budget for the validation process is defined. Qualitative and quantitative metrics are established that can be determined for any long-term SST data set, and approaches to assessing the stability of the long-term record are also presented. New methods to validate product uncertainties are discussed as well as the concept of uncertainty verification, where a confidence level is assigned to uncertainties in regions of no reference data through knowledge transfer from regions where reference data is available.

Ultimately, regardless of their amplitudes, uncertainties and inter-mission differences within CDRs have to be comprehensively documented to allow their inclusion into climate or ecosystem models through assimilation techniques, or their appropriate and informed use in statistical analyses of time series. Then, climate signals can be confidently quantified and integrated into other processes, including decision making. It is clear that creating and assessing CDRs is not work to be done in isolation in one scientific community, but benefits from interdisciplinary connections. It should be conducted as an Earth Observation community effort. The aim of the scientific assessment presented in this part of the book is to provide within each domain a consistent basis from which such higher order activities can start.

Chapter 6.1

Assessment of Satellite Ocean Colour Radiometry and Derived Geophysical Products

Frédéric Mélin,[1,]* Bryan A. Franz[2]
[1] *European Commission, Joint Research Centre, Ispra, Italy;* [2] *NASA, Goddard Space Flight Center, Greenbelt, MD, USA*
*Corresponding author: E-mail: frederic.melin@jrc.ec.europa.eu

Chapter Outline

1. Introduction 609
2. Validation of Satellite Products 610
   2.1 Validation Protocol 610
   2.2 Validation Metrics 612
   2.3 Analysis of Validation Results 614
   2.4 Model-Based Approaches to Uncertainty Analysis and Error Propagation 618
3. Comparison of Cross-Mission Data Products 621
   3.1 Band Shift Correction 622
   3.2 Point-by-Point Comparison 624
   3.3 Analysis of Time Series 626
   3.4 Climate Signal Analysis 628
4. Conclusions 631
Acknowledgments 632
References 632

1. INTRODUCTION

Standardization of methods to assess and assign quality metrics to satellite ocean color radiometry and derived geophysical products has become paramount with the inclusion of the marine reflectance and chlorophyll-a concentration (Chla) as essential climate variables (ECV; [1]) and the recognition that optical remote sensing of the oceans can only contribute to climate research if and when a continuous succession of satellite missions can be shown to collectively provide a consistent, long-term record with known uncertainties. In 20 years, the community has made significant advancements toward that objective, but providing a complete uncertainty budget for all products and for all conditions remains a daunting task. In the retrieval of marine water-leaving radiance from observed top-of-atmosphere radiance, the sources of uncertainties include those associated with propagation of sensor noise and radiometric calibration and characterization errors, as well as a multitude of uncertainties associated with the modeling and removal of effects from the atmosphere and sea surface. This chapter describes some common approaches used to assess quality and consistency of ocean color satellite products and reviews the current status of uncertainty quantification in the field. Its focus is on the primary ocean color product, the spectrum of marine reflectance R_{rs}, but uncertainties in some derived products such as the Chla or inherent optical properties (IOPs) will also be considered.

2. VALIDATION OF SATELLITE PRODUCTS

The primary method to assess satellite data is through direct comparison of a satellite product with near contemporaneous and colocated in situ measurements of the same quantity. Using the field data as a reference, such comparisons can provide estimates of the uncertainty associated with the satellite product. For derived products that rely on empirical algorithms, the in situ validation data set should be independent of any measurements used to define or tune the satellite retrieval algorithm. Unfortunately, the collection of high quality field measurements of optical radiometry for validation is challenging due to difficult environmental conditions, cloud cover and other factors that restrict remote observation, and logistical difficulties of ocean access, thus leading to a relatively limited geographic and temporal sampling of available in situ validation data. As an alternative, simulated data sets have been used for validation studies, since they can be considered error-free and can cover a large range of optical conditions [2]. They can also be produced at any desired wavelengths while validation of multispectral quantities with field data may be hampered by differences in wavelengths between the quantities to be compared. These differences need to be considered and possibly corrected in the validation exercise (see Section 3.1). This section focuses on the assessment of satellite product uncertainties through comparisons with field observations, including a description of validation protocols and metrics and a discussion on validation results. Error propagation techniques and the use of atmospheric correction and bio-optical models to assess confidence intervals are also briefly reviewed.

2.1 Validation Protocol

The validation protocol needs to be well documented to ensure consistency of approach between missions and products and reproducibility between studies. The first step is the construction of the validation, or match-up, data set. A match-up refers to the meaningful association of a satellite value with its counterpart from field observations. This entails the extraction from the overall satellite record of a subset of pixels or grid points, usually a square of $N_s \times N_s$

elements centered on the location of the field value and separated in time by less than a small interval Δt. From the extracted values, one can derive three main statistics: (1) the fraction fv of valid retrievals among the $N_s \times N_s$ potential values, (2) the average (or median) satellite value, and (3) the spatial coefficient of variation (CV_s), which is the ratio of the standard deviation within the $N_s \times N_s$ valid satellite measurements and the average value. A high CV_s means that the satellite retrievals show a large heterogeneity, which in turn suggests a reduced probability that the in situ point measurement is representative of the region observed by the satellite. In the interval $\pm \Delta t$, there might be N_t field observations collected, so that an average (or median) value and a temporal CV_t can also be calculated. In that case, a high CV_t is indicative of changing conditions at the location of the measurements. Eventually, the match-up selection protocol defines the allowed values for maximum Δt, minimum fv and maximum CV_s, as well as minimum N_t and maximum CV_t if applicable. Then, the satellite average (or median) value can be compared with the average field observation or its datum closest to the satellite overpass time.

The choice of the threshold values should allow for a sufficient number of match-ups to conduct a proper statistical analysis while maintaining the validity of the comparison. This compromise should take into account the expected environmental conditions. For instance Bailey and Werdell [3] have selected $N_s = 5$, $\Delta t = 3$-h, fv of 50% and CV_s of 15% for a global validation analysis that relies on many points in open ocean where conditions are thought more stable. Zibordi et al. [4] have used $N_s = 3$, $\Delta t = 2$-h, fv of 100% and CV_s of 20% for validation at a coastal station, the Acqua Alta Oceanographic Tower (AAOT) located in the northern Adriatic Sea. Recommended values can be given, with N_s of 3–5, Δt of 1–4-h, fv larger than 50%, and CV_s of less than 20% for some of the products being validated (typically R_{rs} at a selected wavelength). The choice of threshold values should be adapted to the conditions associated with the validation exercise, with dynamic coastal environments generally requiring more stringent criteria. It is good practice to test several thresholds to assess how validation statistics are affected. For instance, Feng et al. [5] showed how validation statistics improved with more stringent match-up selection criteria. Such an analysis can also provide insight into the degree of representativeness of the comparison, quantifying the discrepancy in scale and time of observation between the two measurement systems. For R_{rs} validation, it is also recommended to operate the selection protocol on the spectrum as a whole and not independently on separate bands; indeed the selection of varying numbers of data points for the different channels hinders a consistent assessment over the spectral domain of interest.

Ideally, a validation analysis should integrate the knowledge of the uncertainties associated with the field observations. Comprehensive validation exercises often combine in situ data collected by a variety of disparate systems and investigators using different instruments and measurement techniques. In such cases it is recommended to assess the dependence of the validation results

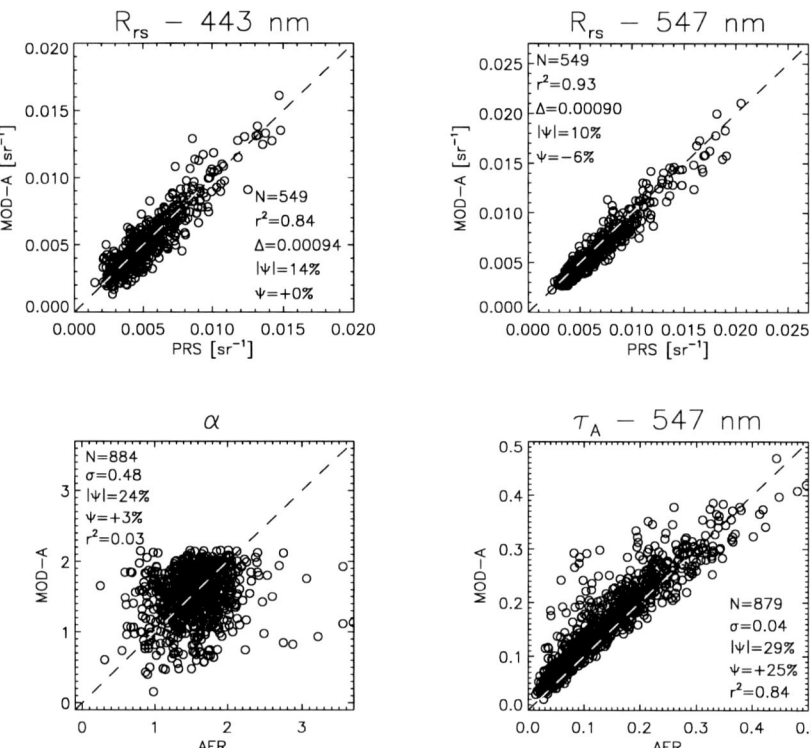

FIGURE 1 Comparison between above-water radiometry and MODIS-Aqua products for R_{rs} at 443 and 547 nm, and aerosol optical thickness τ_A at 547 nm and Ångström exponent α. Validation statistics are introduced in Section 2.2.

on the different data sets gathered for the exercise. More generally, if enough in situ data are available, the sensitivity of the validation statistics to particular sets of match-ups can be quantified by bootstrapping techniques [6].

Figure 1 shows an example of match-ups obtained at the AAOT site for the sensor MOderate Resolution Spectroradiometer (MODIS) on-board Aqua, both in terms of reflectance R_{rs} and aerosol optical thickness τ_A. Over the period 2002–2012, the number of match-ups found for R_{rs} is 549, with $N_s = 3$, $\Delta t = 1$ h, and $CV_s = 20\%$ for R_{rs} between 488 and 547 nm. More match-ups are obtained for the aerosol products, with the added conditions of $N_t = 2$ and $CV_t = 20\%$ for τ_A at 488 nm.

2.2 Validation Metrics

There are a host of statistical metrics that can be used to compare two data sets, but a minimum set for validation would include the number of match-ups (together with the number of potential match-ups) and the estimates of the scatter and systematic difference (bias) between the two distributions.

According to the range of values considered, these statistics can be expressed as absolute or relative values, with a prior log transformation typically applied for Chla or IOPs. For radiometric products, it is important to document both a measure of uncertainty in radiometric units (sr^{-1} for R_{rs}) and a measure of relative uncertainty. Indeed, relative differences tend to increase when the values of R_{rs} are small, up to tens of percent if the in situ value is near zero. In that case, the difference in radiometric units is more meaningful.

Relative differences between satellite products $(y_i)_{i=1,N}$ and field observations $(x_i)_{i=1,N}$ can be expressed in %, and computed in terms of mean absolute difference or mean difference (i.e., bias) with respect to the field observations:

$$|\psi| = 100 \cdot \frac{1}{N} \sum_{i=1}^{N} \frac{|y_i - x_i|}{x_i} \qquad (1)$$

$$\psi = 100 \cdot \frac{1}{N} \sum_{i=1}^{N} \frac{y_i - x_i}{x_i} \qquad (2)$$

while the equivalent metrics in geophysical units can be computed as:

$$|\delta| = \frac{1}{N} \sum_{i=1}^{N} |y_i - x_i| \qquad (3)$$

$$\delta = \frac{1}{N} \sum_{i=1}^{N} (y_i - x_i) = \bar{y} - \bar{x} \qquad (4)$$

where the overbar means average values. Root-mean-square (RMS) differences between the satellite and in situ measurements can be written as:

$$\Delta = \sqrt{\frac{1}{N} \sum_{i=1}^{N} (y_i - x_i)^2} \qquad (5)$$

$$\Delta_u = \sqrt{\frac{1}{N} \sum_{i=1}^{N} (y_i - \bar{y} - x_i + \bar{x})^2} = \sqrt{\Delta^2 - \delta^2} \qquad (6)$$

The total root-mean-square difference Δ can be partitioned into a part due to the bias δ and the unbiased (or centered) root-mean-square difference Δ_u quantifying non-systematic effects. In the above equation, the summation operator has been used (which means that quantities are averages), but other operators can be preferred like the median or some form of interquantile statistics. Other metrics can be included like the coefficient of determination, r^2, slope and intercept of linear regression, average ratios, etc… For spectral quantities like R_{rs}, input to bio-optical algorithms, quantifying how well the

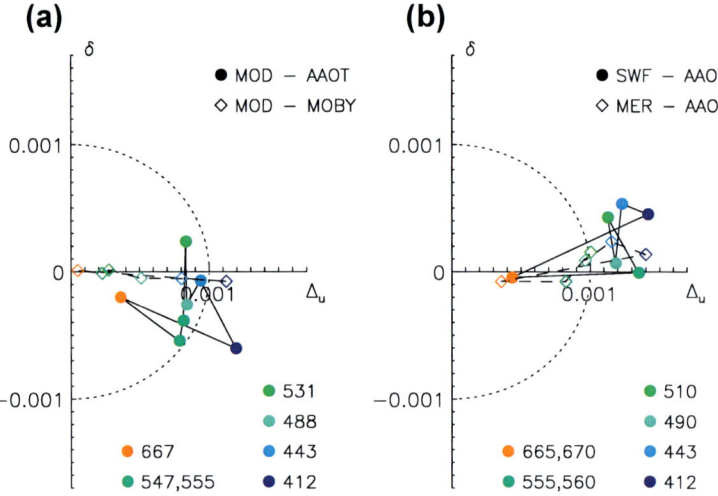

FIGURE 2 Spectral target diagram for validation results for MODIS-Aqua (a) at Marine Optical Buoy (MOBy) ($N = 229$) and Acqua Alta Oceanographic Tower (AAOT) ($N = 549$, except at 531 nm, $N = 176$) and (b) SeaWiFS ($N = 369$) and MERIS ($N = 149$) at AAOT (right). Axes are in units of sr^{-1}. See text for the definition of statistical quantities.

spectral shape is respected by the satellite products is also worthy information that can be quantified by the χ^2 distribution measuring the goodness of fit between in situ and satellite R_{rs} normalized at one wavelength of reference [7].

To document differences (including their systematic component), it is recommended to compute at least $|\psi|$, ψ, Δ (or Δ_u) and δ. As illustration, Figure 2 shows spectral target diagrams that display Δ_u and δ simultaneously for various sensors and two validation sites, AAOT and the Marine Optical Buoy (MOBy) near Hawaii [8]. By construction, Δ is the distance between a point and the origin (Equation 6). The value of Δ (or Δ_u) decreases with increasing wavelength for MODIS compared to MOBy data, with virtually no bias, which is expected considering the role of that site for vicarious calibration [9]. This is also true for the validation results obtained at AAOT, but significant values can be observed for the bias δ, generally negative for MODIS, and positive for the Medium Resolution Imaging Spectrometer (MERIS) and the Sea-viewing Wide Field-of-View Sensor (SeaWiFS) (validation results for MODIS are the same as shown on Figure 1).

2.3 Analysis of Validation Results

Performing accurate radiometric in situ oceanographic measurements to derive R_{rs} is difficult and expensive, with the implication that match-ups with satellite data are not abundant and are unevenly distributed in space and time [3]. More match-ups are available for Chla, although large expanses of ocean remain devoid of validation data [10]. Using the SeaWiFS Bio-optical Archive and

Storage System (SeaBASS) [11], a global community field data repository for marine bio-optical measurements, and following the standard protocols in [3], one obtains less than 1000 match-ups for the 13-year SeaWiFS mission (numbers vary by wavelength).

Figure 3 is a snapshot of validation results expressed as spectra of Δ between satellite and field data. MODIS-Aqua Δ values are illustrated for various data sets (Figure 3(a)), SeaBASS, BiOMaP (representative of European waters

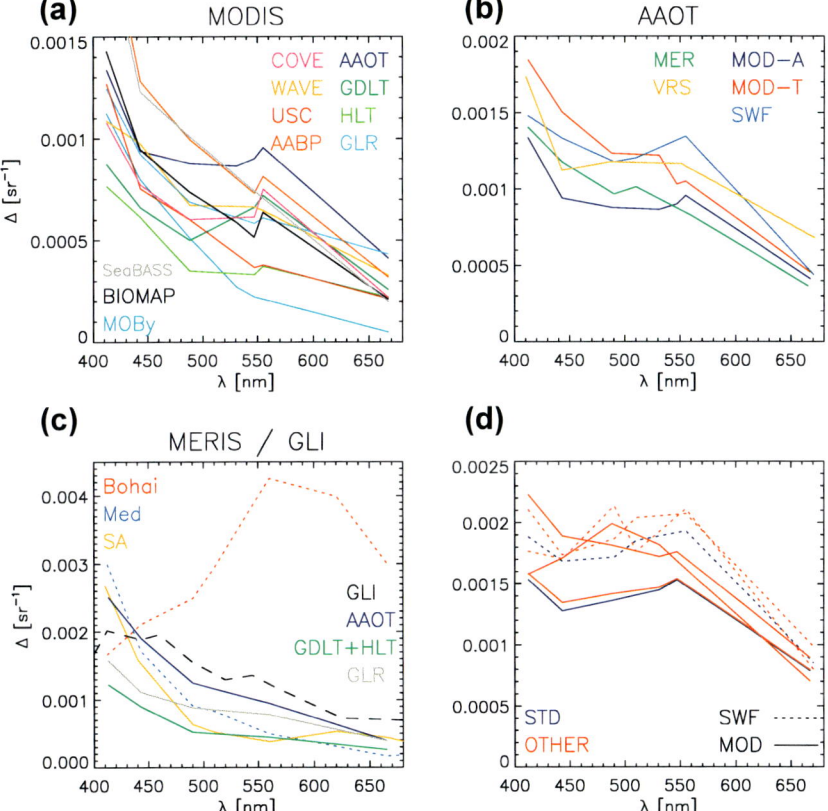

FIGURE 3 RMS differences between satellite and field data of R_{rs} (in sr^{-1}): (a) MODIS-Aqua compared to various data sets from AERONET-OC sites (see text) as well as SeaWiFS Bio-optical Archive and Storage System (SeaBASS), BiOMaP, and Marine Optical Buoy (MOBy); (b) results for MODIS-Aqua ($N = 549$) and Terra (270), MERIS ($N = 149$), SeaWiFS ($N = 369$) and VIIRS ($N = 70$) at Acqua Alta Oceanographic Tower (AAOT); (c) results for MERIS processed by ESA processor (MERIS Ground Segment, MEGS, version 7 as dotted line, version 8 otherwise) given for the Bohai Sea [12], the northwest Mediterranean Sea [13], South African (SA) coastal waters [14], and AERONET-OC sites, AAOT ($N = 86$), GDLT and HLT (Baltic Sea) and GLR (Black Sea) [15]. GLI results are represented by the black dashed line [16]. (d) Results obtained in coastal waters for SeaWiFS (dotted line) [17] and MODIS-Aqua [18] with different atmospheric correction schemes, including the standard SeaDAS (STD, in blue). When appropriate the reference source is given.

[19]), and MOBy, as well as several Aerosol Robotic Network - Ocean Color (AERONET-OC) sites [20] located in the northern Adriatic Sea (AAOT), Baltic Sea (Gustav Dalen and Helsinki Lighthouse Towers, GDLT and HLT, respectively), Black Sea (Gloria, GLR), Chesapeake Bay entrance (CERES Ocean Validation Experiment, COVE), coastal Gulf of Mexico (WAVE), coastal southern California (University of Southern California, USC), and Persian Gulf (Abu al-Bukhoosh Platform, AABP). The number of match-ups varies from 15 (AABP) to 549 (AAOT); the number of wavelengths represented is also variable (e.g., the SeaBASS validation results are shown at 412, 443, 488, and 667 nm only). Some spectra show a value at 547 and 555 nm, the latter being a band not originally intended for ocean color applications by the MODIS mission. Most Δ values are found between 0.0008 and 0.0015 sr^{-1} at 412 nm, down to between 0.0002 and 0.0004 sr^{-1} at 667 nm (with the exception of MOBy where Δ is lower). The Δ values are partly conditioned by the actual R_{rs} values; for instance, Δ is lowest at MOBy in the green bands, whereas it is lowest at the Baltic sites in the blue part of the spectrum, where R_{rs} is often very low.

There is a clearer consistency of Δ spectra when considering validation results at a single site for different missions processed with the same National Aeronautics and Space Administration (NASA)-standard algorithms (Figure 3(b) at AAOT). Results for standard European Space Agency (ESA) MERIS products are shown on Figure 3(c), for the Bohai Sea ($N = 17$) [12], the northwest Mediterranean Sea ($N = 64$ except at 412 nm) [13], South African coastal waters ($N = 14$) [14], and AERONET-OC sites in the northern Adriatic Sea (AAOT, $N = 86$), Baltic Sea (GDLT and HLT, $N = 39$) and Black Sea (GLR, $N = 12$) [15]. For completeness, Global Imager (GLI) results are also shown ($N = 435$ at 443 nm) [16]. MERIS Δ values tend to be fairly high, particularly in the blue. The case of the Bohai Sea is fairly unique and associated with highly scattering waters with R_{rs} maxima beyond 550 nm [12]. This type of Δ spectra should be confirmed with more match-ups.

Finally, interesting studies have been done to compare atmospheric correction schemes with the same validation data set [7]. Two examples of such exercises are reported on Figure 3(d), comparing SeaDAS results with other schemes [17,18]. Validation statistics appear fairly consistent for a given sensor, with the Δ values associated with the standard scheme often being the lowest. The family of Δ curves of Figure 3 could be presented for other statistical indicators. Relative differences, $|\psi|$ or ψ, would show more variations particularly between different locations, $|\psi|$ varying from 10% to tens of percent. In fact, $|\psi|$ spectra are often an inverted image of R_{rs} spectra, with $|\psi|$ values that are high in red bands in oligotrophic waters or that may exceed 100% at 412 nm in absorbing waters like in the Baltic Sea [4].

Other R_{rs} validation exercises have of course been conducted, applied to specific sensors like the Ocean Color and Temperature Scanner (OCTS) [21], Visible Infrared Imaging Radiometer Suite (VIIRS) [22], or Geostationary Ocean Color Imager (GOCI) [23,24], specific coastal regions (e.g., coastal

Chinese waters [25]), or to test alternative atmospheric corrections [26,27]. Many of these studies suffer from a limited number of match-ups often collected in restricted geographical areas, and this regional scope raises the question of the validity of validation results on larger scales. More work is needed to extend, analyze, and understand validation results across missions, atmospheric correction schemes, and field data sets or locations, with the goal to enable the extension of point validation results sparsely distributed in space to the global ocean. This issue will be further discussed in this and the following sections.

Validation analyses ideally should go beyond simply providing statistics for a given location and/or season by investigating possible dependences of the validation results on time or season, geometry of observation and illumination, atmospheric conditions, or marine properties. The benefit can be twofold, as such studies can provide insights into the reasons for discrepancies between satellite and in situ values, and also inform on other locations and times where these validation statistics may be applicable. Such analyses require a significant number of match-ups, and are therefore few in number.

The regional dependence of optical properties has been well documented. For instance, Szeto et al. [28] have related the departures from a global average relationship between R_{rs} ratios and Chla to different ocean basins, and suggested that this relationship varies across basins as a function of the relative contributions of the different optically significant constituents [28,29]. For the AAOT site, Mélin et al. [30] studied the dependence of validation results for R_{rs} on a Case-1 versus Case-2 water partition, water single scattering albedo, angles of observation and illumination, air mass, and aerosol optical thickness, using approximately 80 SeaWiFS match-ups. The only clear dependence was found for τ_A, with biases of R_{rs} significantly increasing from negative to positive with increasing τ_A. At the same site, for an updated atmospheric correction applied to SeaWiFS and MODIS, Zibordi et al. [31] highlighted an increase in bias and RMS difference for R_{rs} in winter and for high solar zenith angles. The number of match-ups was much larger for this analysis based on field observations collected by autonomous instruments. For the same match-up data set, no significant multiannual trends were found for R_{rs} validation statistics [32]. Using the large number of match-ups found at the AAOT site, D'Alimonte et al. [33] formulated a regional model of the differences between satellite and field data of R_{rs} that depended mainly on R_{rs} itself; another regional model has been defined for Baltic sites [4]. This work suggests that these differences could vary according to water optical properties.

Moore et al. [34] explored this hypothesis further in the context of optical classification applied to Chla uncertainty determination. Chla uncertainty statistics were first determined for a predefined set of optical water types (or classes), allowing the extension of these statistics to any location on the basis of the class membership of the corresponding R_{rs}. Assuming that uncertainties are indeed specific to each water type, such an approach can be used to derive

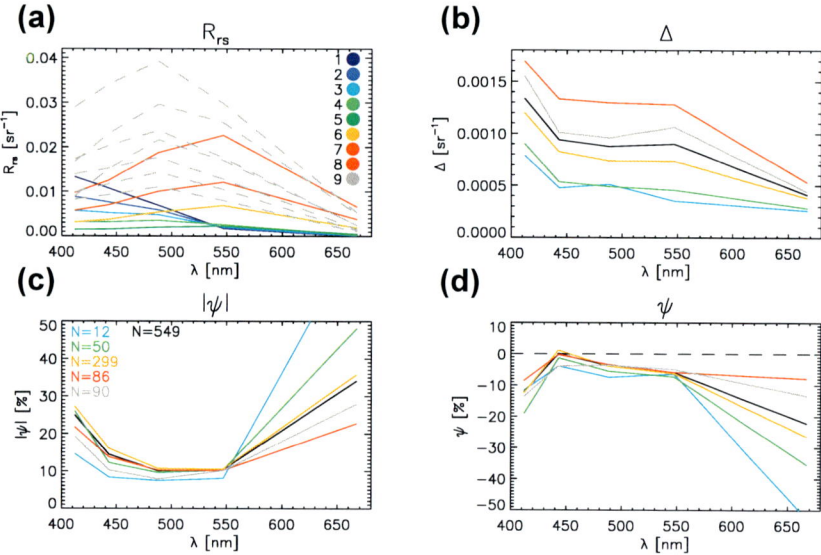

FIGURE 4 Dependence of validation statistics on optical water types at Acqua Alta Oceanographic Tower (AAOT). (a) Mean R_{rs} for the optical water types defined in [35]; type 9 includes eight subtypes. Validation results computed by types for (b) RMS difference Δ, (c) mean absolute relative difference $|\psi|$, and (d) mean relative difference ψ. Results for all match-ups combined are in black; results per optical type are shown with the same color code as for (a) only if the members number at least 10 spectra.

global maps of Chla uncertainties. Optical water types have also been used to analyze and discuss validation results for R_{rs} in coastal regions [18]. A similar exercise has been performed for the match-ups gathered at the AAOT site using a set of classes defined by Moore et al. [35]. Figure 4 shows R_{rs} associated with the considered water types (type 9 is actually an ensemble of eight subtypes originally developed to represent coccolithophore blooms) as well as validation statistics for the optical types found in the validation data set. The RMS difference Δ increases from types 3 and 4 to type 7, while that for types 6 and 9 are close to the overall average. Relative differences tend to be lower for clearer waters (type 3) in the blue part of the spectrum (for $|\psi|$), while they are higher in the red part where the signal is lower (types 3, 4, or 6). Under the assumption that validation results obtained here are inherent to each optical water type, they could be tentatively extended to similar water types in other regions. A merit of such an approach is that its uncertainty estimates remain linked to field data.

2.4 Model-Based Approaches to Uncertainty Analysis and Error Propagation

Models themselves can be used to support the assessment of satellite products, a path explored mostly for bio-optical algorithms. The retrieval of bio-optical properties from ocean color radiometry often involves the spectral matching of a bio-optical model to the spectral shape of the retrieved R_{rs}, and this inversion process can provide valuable information on the uncertainties associated with the algorithm design or sensitivity to radiometric error. The uncertainties associated with R_{rs}, if known, are propagated through the bio-optical inversion and compounded by two additional factors: (1) the approximations of the bio-optical model in its description of the relationship between inherent and apparent optical properties and its parameters describing the spectral shape of IOPs, and (2) the ambiguity of the bio-optical model, which means that the solution to the inversion is not necessarily unique [36,37].

Some studies have addressed uncertainties related to the parameters of the bio-optical model, such as the phytoplankton specific absorption, the spectral shape of the backscattering coefficient or the absorption by chromophoric dissolved organic matter and detrital particulates referred to as (CDM). Lee et al. [38] applied error propagation to an algebraic bio-optical model [39] to determine the uncertainties of the derived IOPs as a function of model parameters and the uncertainty of total absorption at a reference wavelength. The effect of the uncertainties associated with bio-optical parameters (e.g., defining the spectral shapes of phytoplankton and CDM absorption and backscattering coefficient) has been tested by running the inversion with different sets of parameters [40,41]. Wang et al. [40] studied the dispersion of the retrievals as a measure of output uncertainty. These approaches did not consider the other sources of uncertainties, including those of the input R_{rs}. Nonlinear inversion of a bio-optical model provides interesting information on the uncertainty of the output IOPs from the process of minimization of a cost function [42–45]. Additionally, cases with an unsatisfactory goodness-of-fit can be filtered out as out-of-scope conditions. The uncertainty information derived from the inversion process can accompany the retrieved IOP maps and is sometimes referred to as uncertainty maps, though inversion confidence would be more appropriate terminology. The inversion confidence can account for the variance associated with the input R_{rs}, but it is only related to how the forward model fits the input R_{rs} data, depending on the shape of the selected minimum of the cost function, and does not cope with biases affecting R_{rs} or uncertainties on model formulation and parameters. Development of quality indicator maps together with derived products has also been performed using neural networks [46].

Some atmospheric correction schemes are also based on the minimization of cost functions [47–49] and are amenable to the calculation of inversion confidence estimates as explored with bio-optical algorithms. Typically, these schemes have an embedded bio-optical model which constrains the distribution of retrieved R_{rs}. One study developed a stochastic approach to uncertainty decomposition and estimation while explicitly considering the atmospheric correction process [50]. Exercises of error propagation or accuracy analysis [51,52] can also provide valuable insight on atmospheric correction performance to support uncertainty assessments.

A model-based approach of a different kind has been proposed [53] making use of Chla algorithms applied to low-Chla waters. For these conditions, the assumption is that the difference in Chla computed with a standard band-ratio algorithm and with a three-band subtraction method [54] originates from uncertainties associated with R_{rs}. Using SeaWiFS and MODIS data, R_{rs} uncertainty estimates have been expressed as a function of Chla (see Figure 7 introduced in Section 3.2). Even though this approach does not apply to Chla values larger than 0.2 mg m^{-3}, and does not specifically account for biases, this type of technique should be further investigated.

Different methods, like those mentioned above, can inform us on various aspects of the uncertainty associated with a given retrieval, and it is desirable that their specific contributions and limitations to quantifying the overall uncertainty budget be well understood. Restricting the discussion to bio-optical algorithms producing Chla or IOPs, Table 1 is an attempt at broadly categorizing the type of uncertainty estimates obtained by various methods. The uncertainty on the derived product is assumed to stem from the uncertainty on the input R_{rs}, the potential non-uniqueness of the solution, uncertainties on the model formulation and parameters, and the uncertainties associated with the inversion process. Clearly, validation integrates all these contributions, but is affected by uncertainty in field data. Colocation techniques ([55], see Section 3.2) share this all-encompassing character but are limited in their temporal resolution and do not consider systematic effects. Uncertainty propagation techniques can potentially accommodate uncertainties on R_{rs} and model parameters, while using parameter ensembles (e.g., Ref. [40]) focuses on uncertainties on model parameters and issues of uniqueness without accounting for biases affecting R_{rs}. Finally, nonlinear inversions provide a diagnostic of product confidence given an uncertainty on R_{rs} but usually do not account for biases or parameter uncertainties. Complex approaches could combine the advantages of these various techniques.

TABLE 1 Matrix Relating Error Sources Affecting Products of Bio-Optical Algorithms and Different Methods Computing Uncertainty Terms

	R_{rs}	Uniqueness	Parameters	Inversion
Validation			x,t	
Uncertainty propagation [38]	X,T		X,T	
Parameters ensemble [40]		X,T		
Nonlinear inversion [44]	X,T			X,T
Colocation [55]			X,t	

Cells with letters indicate the contributions to the uncertainty budget that are addressed by each type of methods. Small letters refer to results obtained at selected locations x and times t, while capital letters indicate estimates potentially obtained at each pixel. The first two types of approaches can handle systematic effects (biases). References are only intended as general examples for a given approach.

3. COMPARISON OF CROSS-MISSION DATA PRODUCTS

The distribution of field observations is very uneven, with a sparse coverage of the open ocean regions, particularly for optical properties. Comparison between satellite values can build upon a much larger statistical population and can support the characterization of their uncertainties. More generally, the comparison of products from different missions over their period of temporal overlap is a key element of the consistency check of the overall data record.

The comparison of two or more data products can be conducted at several levels. First, as for validation analyses, a variety of metrics may quantify the differences between a common set of data points (e.g., average difference), which is illustrated below. In the context of earth science, it is also interesting to compare specific properties of each data record, like their spectral resolution, their spatial coverage, their inherent variability in space or time (do the data sets have the same variance, do they show the same gradients?), their seasonal cycles (do they show the same phenology?), or interannual signals. Two data series ideally should show the same behavior for these properties, but their relative importance depends on the envisioned application. Obviously, for climate research, two data sets should provide a similar picture of seasonal to interannual variations.

Even if processed consistently, which means with the same principles guiding the calibration strategy and the data processing (same algorithms and binning schemes, identical ancillary data), and compared for the same day, two ocean color data sets will show differences. These result from various elements [56] including differences in sensor design and spectral characteristics as well as their implications in the specific processing codes, such as the sensitivity to polarization, uncertainties associated with the calibration of the sensors, the sensitivity of the atmospheric correction to different aerosol types or to a different geometry of observation. Moreover, the different sensors view earth at different times of the day, generating other sources of differences. Some might be real as associated with changes in the water properties that could be more readily studied with geostationary platforms [24]. But currently and in most cases, these differences cannot be reliably distinguished from others that are occasioned by the effect on the atmospheric correction of changes in the geometry of illumination and the atmospheric content (aerosols, clouds) or simply by noise. There is also a residual spatial mismatch as a result of the remapping process or of different sizes and shapes of the pixels across the satellite track. Still an additional source of differences is introduced as time composites are created since these might be built with a different temporal sampling. Finally, different products might differ because their processing chains are not consistent, e.g., with a different calibration strategy or different algorithms. With a view to create climate data records from different missions, this should be avoided as much as possible. Such a consistency is readily achieved if the different sensors share their main characteristics, for instance being similar multispectral sensors with a wavelength range of 400–900 nm as is the case for the recent global ocean color missions SeaWiFS, MODIS, and MERIS. This consistency may be questioned as sensors are launched with new capabilities (channels in the ultraviolet or shortwave infrared, much higher spectral resolution, geostationary observations) that open up novel options for processing. A similar technical step took place from the Coastal Zone Color Scanner (CZCS) to the more recent sensors, prompting questions on how to process the data in a consistent manner [57].

This section reviews various approaches to compare satellite data sets. First, the issue of band shifting is addressed. Indeed, differences in center wavelengths need to be corrected prior to comparing spectral quantities.

3.1 Band Shift Correction

The various ocean color missions that have been in operation have a different set of bands, which is an obstacle to a straightforward comparison of their respective records of remote sensing reflectance, R_{rs}. For instance the green ocean color band, which is often used as a reference band for bio-optical algorithms, is centered at 547, 555, and 560 nm for the missions MODIS, SeaWiFS, and MERIS, respectively. In practice it is hard to know how a MODIS R_{rs} value at 547 nm compares relative to a SeaWiFS R_{rs} at 555 nm. It can be done in the framework of well-defined optical properties like a Case-1 water model [58] where a certain spectral shape for R_{rs} is expected for each value of the Chla leading to the definition of a set of consistent empirical algorithms for the different sensors [59]. Such a framework could be extended to more complex optical conditions but covering the entire natural variability does not appear realistic. A similar issue arises in validation analyses when field data that are collected as multispectral measurements are compared with satellite R_{rs}. A few studies have relied on general or regional relationships to perform an action called band shift correction, whereby the R_{rs} value is expressed at a target wavelength λ_t near an existing wavelength λ_0.

The practice of band shift correction has been developed for use in validation analyses [4,13,19,60,61], comparison between satellite products [62], and as pre-processing before merging [63], with expressions linking inherent and apparent optical properties such as:

$$R_{rs}(\lambda_t) = R_{rs}(\lambda_0) \frac{f(\lambda_t)}{Q(\lambda_t)} \frac{Q(\lambda_0)}{f(\lambda_0)} \frac{b_b(\lambda_t)}{a(\lambda_t)} \frac{a(\lambda_0)}{b_b(\lambda_0)} \qquad (7)$$

where f relates apparent optical properties (irradiance reflectance) to IOPs [64], Q is the ratio of irradiance and radiance just below the surface, and a and b_b are the total absorption and back-scattering coefficients, respectively. Equation (7) requires the value of Chla to calculate f/Q [65] through look-up tables computed in the framework of Case-1 water conditions.

A common requirement for these approaches is knowledge of inherent optical properties and/or concentrations of optically significant constituents sufficient to predict the spectral shape of R_{rs} at least within small spectral intervals. An approach recently developed makes use of the Quasi-Analytical Algorithm (QAA) [39] to compute the absorption of phytoplankton a_{ph}, that of CDM a_{cdm}, and the back-scattering coefficient associated with particles at 443 nm. Then IOPs are calculated at the target wavelength λ_t using the spectral shapes of IOPs defined in the model (with the addition of the parameterization by [66] for a_{ph} since QAA does not specify a spectral shape for that property). Finally this bio-optical model is run in forward mode to calculate R_{rs} at λ_t.

The results of this band shift correction are illustrated by Figure 5. The correction has been applied to a year (2003) of daily MODIS-Aqua R_{rs} data to express them at the SeaWiFS bands. A MODIS value is computed at 510 nm by running the conversion from 488 nm and from 531 nm to 510 nm, and then taking a weighted average. All spectra common to both sensors for a given spatial bin (of a 12th-degree grid) and day have then been accumulated (49.8 million spectra). Figure 5(a) shows the overall average over that population of the SeaWiFS R_{rs} and the MODIS-Aqua original R_{rs} as well as the R_{rs} values obtained after band shift. There is a discernible improvement in the agreement between corresponding wavelengths, at 490 and 670 nm and more clearly at 555 nm. Even the converted MODIS average value at 510 nm appears close to the SeaWiFS counterpart. Also noticeable is the fact that the use of a linear interpolation between 488 and 531 nm to compute a MODIS value at 510 nm would have resulted in a gross overestimate. Figure 5(b) is the frequency distribution of the

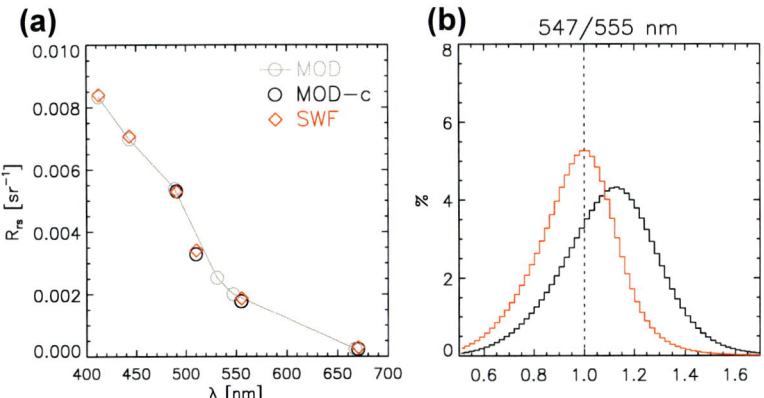

FIGURE 5 (a) Average of all daily R_{rs} coincident between MODIS-Aqua and SeaWiFS in 2003. Curve with gray circles represent MODIS values without band shifting, while black circles represent MODIS statistics computed after the band shifting correction has been applied. (b) Histogram of the ratio between MODIS (with and without band shifting, in red and black, respectively) and SeaWiFS R_{rs} in the green band.

ratio of R_{rs} at the green band before and after correction, i.e., $R_{rs,A}(547)/R_{rs,S}(555)$ and $R_{rs,A}(555)/R_{rs,S}(555)$ (where A and S denote MODIS-Aqua and SeaWiFS, respectively). The median ratio decreases from 1.10 to 0.98. Band shifting is an important tool to allow inter-mission comparison, but it also contributes its own uncertainties that should be properly estimated.

3.2 Point-by-Point Comparison

The comparison between satellite products can be conducted for each grid point in a manner similar to validation with in situ data, including with the same metrics. Here again, statistics should at least provide a measure of scatter and bias, in relative terms as well as in radiometric units for R_{rs}. In Section 2, the quantity of reference in relative differences (i.e., the denominator) was the in situ value, even though in situ observations are not error free. In the case of a comparison between satellite products, the unbiased form of the relative difference can be preferred (in %):

$$|\psi^*| = 200 \frac{1}{N} \sum_{i=1}^{N} \frac{|y_i - x_i|}{x_i + y_i} \qquad (8)$$

$$\psi^* = 200 \frac{1}{N} \sum_{i=1}^{N} \frac{y_i - x_i}{x_i + y_i} \qquad (9)$$

$|\psi^*|$ and ψ^* are referred to the average of the two products. The advantage is to avoid arbitrarily selecting one product as the value of reference, and numerically it prevents cases where only the denominator is close to zero. On the other hand, the difference cannot be easily interpreted in terms of a distance with respect to a clearly identified reference.

To compare SeaWiFS and MODIS products, the MODIS R_{rs} data were re-binned on the SeaWiFS 12th-degree grid, and then all daily values coincident on that grid were accumulated into third-degree macro cells. Figure 6(a) shows the resulting number of match-ups for the period 2003–2007. In general, the number of comparison data available for assessment decreases going poleward; on top of this, spatial patterns associated with persistent cloud or dust coverage are readily seen, e.g., along the intertropical convergence zones.

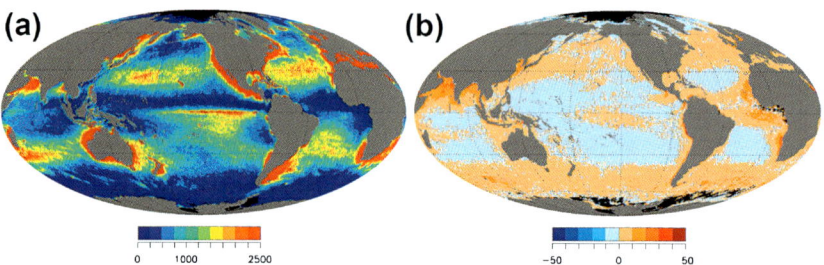

FIGURE 6 (a) Total number of match-ups between MODIS and SeaWiFS (2003–2007) on a 3rd-degree grid and (b) Mean relative difference ψ^* between SeaWiFS and MODIS (in %).

The unbiased mean relative difference ψ^* is illustrated on Figure 6(b) for R_{rs} at 443 nm. Most of the ocean is characterized by a relative bias not exceeding 5%, but larger differences can be noticed in specific coastal or tropical regions or in the northern Indian Ocean. Besides the spatial variations shown by inter-mission differences, comparison maps also show temporal variations, particularly changes associated with the seasonal cycle for apparent or inherent optical properties [62,67]. Examples of temporal analyses are provided in the next section.

If enough match-up data are available for a given grid point, more advanced statistics can be developed. Let us consider two ensembles of N coincident satellite values $(x_i)_{i=1,N}$ and $(y_i)_{i=1,N}$, each modeled as a function of a reference state r and zero-mean random errors δ and ε:

$$x_i = r_i + \delta_i \tag{10}$$

$$y_i = \alpha + \beta r_i + \varepsilon_i \tag{11}$$

with α and β additive and multiplicative biases, respectively, between x and y. Assuming that δ and ε are uncorrelated and independent of the reference state r, a mathematical development of the variance and covariance terms lead to [55]:

$$\sigma_\delta^2 = \sigma_x^2 - \frac{1}{\beta}\sigma_{xy} \tag{12}$$

$$\sigma_\varepsilon^2 = \sigma_y^2 - \beta \sigma_{xy} \tag{13}$$

which is a system of two equations with three unknowns. It can be solved with an additional assumption, for instance considering that the two satellite products, on the basis of validation analyses, have the same level of random error [55]. Solving the system may also rely on the availability of a third independent data record using a triple colocation technique. This approach is very powerful since it provides part of the uncertainty budget with the same coverage of the satellite products. Depending on the number of match-ups available, it can also be applied to separate seasons to capture variations in time.

Assuming the same level of random error for SeaWiFS and MODIS ($\sigma = \sigma_\delta = \sigma_\varepsilon$) and using the match-up data base illustrated on Figure 6(a), a global map of σ is produced, with its global average shown on Figure 7. For comparison, the average over subtropical gyre waters is also given together with the uncertainty estimates for low-Chla waters given as a function of Chla by a model-based approach [53], and the unbiased RMS difference Δ_u obtained by comparison between satellite and field data at the oligotrophic MOBy site. The spectra of σ and the results obtained by the model-based approach are fairly comparable, even though the latter are higher for the case Chla = 0.15 mg m^{-3} in South Pacific waters. The MOBy validation results Δ_u for SeaWiFS and MODIS are also comparable with σ except in blue bands where they are closer to the model-based estimate for the case

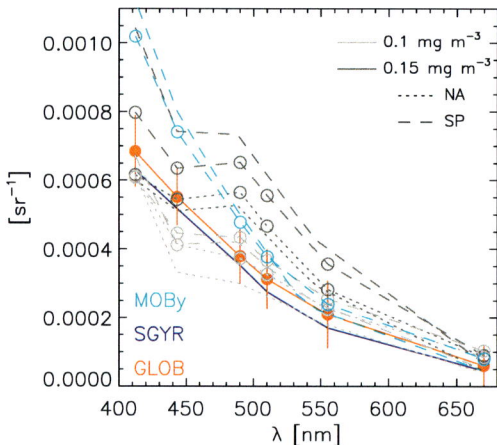

FIGURE 7 Uncertainty estimate σ obtained by colocation between SeaWiFS and MODIS, with global average (red) with standard deviation, and average over the subtropical gyres (blue). In gray are overplotted estimates of the uncertainty term proposed in [53] for SeaWiFS and MODIS as a function of Chla for the North Atlantic (NA) and South Pacific (SP) subtropical gyres. Validation results Δ_u obtained with Marine Optical Buoy (MOBy) data are shown in light blue. Curves with circles are for SeaWiFS.

Chla $= 0.15$ mg m^{-3} in South Pacific waters. Considering the diversity of methods employed (colocation, model-based, point-wise validation), the relative agreement between these curves is interesting while the sources of differences should be further investigated.

3.3 Analysis of Time Series

A primary goal in development of ocean color ECVs is to enable the assessment of long-term trends to support global climate research. This leads to stringent requirements on radiometric stability, to ensure that systematic errors such as uncorrected degradation in instrument radiometric response are not misinterpreted as geophysical change. Comparative time-series analysis of R_{rs} and derived products, either between satellite missions or relative to a historical reference, can identify issues with instrument radiometric characterization and temporal calibration stability. Analysis of the seasonal trends observed in different latitudinal zones for MODIS-Aqua R_{rs} time-series relative to SeaWiFS, for example, contributed to the discovery of an error in characterization of polarization sensitivity on MODIS [68]. Without the SeaWiFS time-series for comparison, this error may have never been identified, and seasonal cycles in the ocean color signal from MODIS in climate critical high-latitude regions would have been highly misleading. For products derived using common algorithms, relative agreement between missions also provides a measure of uncertainty for trend detection. Franz et al. [69], for example, used the average difference in regional monthly means between consistently processed SeaWiFS and MODIS data as a measure of uncertainty in the 15-year multi-mission time-series of Chla.

A typical time-series analysis starts with the data product of interest projected into a set of fixed geographic bins and averaged over specific temporal intervals. A widely used example is the SeaWiFS 9.2-km binned product: a globally distributed set of quasi-equal-area bins where the value of each bin represents the local product average over 8-day or monthly time intervals [70]. The global data set or a subset of the bins (e.g., based on geography or water-type classification) is then spatially averaged within each time interval, and the averages are trended in time. The preferred time interval for compositing is a trade-off between minimizing the geophysical variability lost to the average and maximizing the number of observed (or filled) bins. When comparing time-series between missions, it is also useful to first reduce the selected bins within each time-interval to a set of common filled bins. This is critical for the identification of anomalous sensor-calibration artifacts, as some missions show systematic geographic gaps even after 8-days of compositing, and these geographic sampling biases induce additional variability in mission-to-mission differences.

As an example, Figure 8 shows R_{rs} trends from MODIS on Terra and MODIS on Aqua, based on common bins over the overlapping missions.

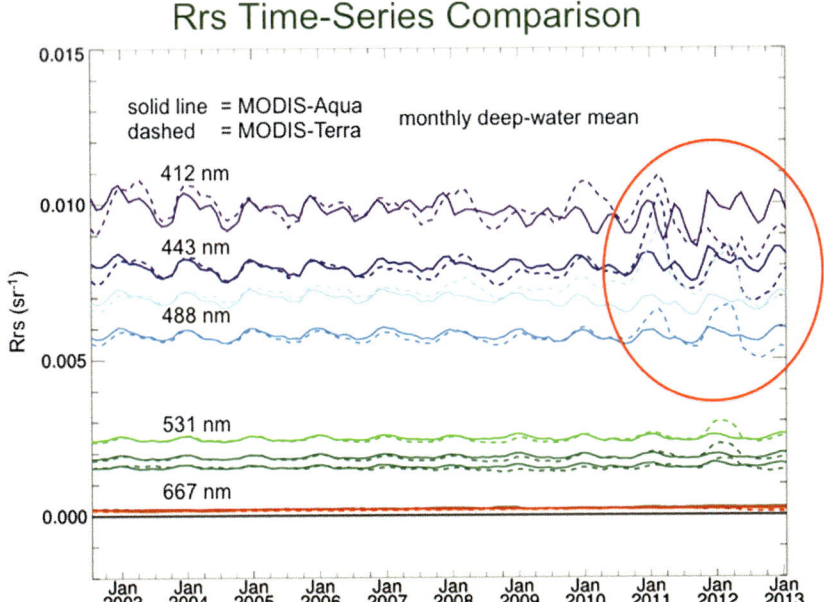

FIGURE 8 Comparative common-bin time-series of MODIS-Terra and MODIS-Aqua.

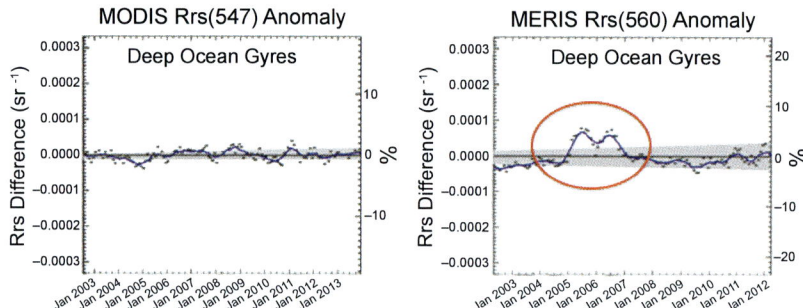

FIGURE 9 R_{rs} anomaly analysis for a global clear water region.

The measurements were restricted to include only those bins where water-depth is greater than 1000 m (to avoid the complexities and diurnal variability of the coastal regions), and an average was computed over this geographic subset for each month of the time-series. The comparison clearly demonstrates a degradation in the radiometric stability of MODIS-Terra relative to MODIS-Aqua that was traced to the MODIS-Terra instrument calibration and subsequently corrected.

For a radiometrically stable sensor, the dominant variability in the derived R_{rs} time-series for the deep oceans is a seasonal cycle associated with phytoplankton productivity. Subtraction of this mean seasonal cycle from the R_{rs} trends yields an anomaly time-series. While anomaly trends provide a mechanism for investigating long-term geophysical changes in the ocean color record, they can also serve as a powerful tool to identify sensor radiometric instabilities. Figure 9 shows the anomaly in MODIS-Aqua $R_{rs}(547)$ relative to the mean seasonal cycle for the deep ocean gyres. At this wavelength, the R_{rs} signal is relatively insensitive to small changes in Chla, and so we expect the time-series in these very low productivity regions to show little variability, as is the case for MODIS. For the 560-nm band of MERIS, however, the R_{rs} anomaly time-series shows a strong deviation in 2005–2006 suggesting a 5–10% bias that was traced to a change in the operating state of the instrument.

3.4 Climate Signal Analysis

Ocean color products are being scrutinized across a whole range of space and time scales. The various satellite data records should show the same patterns of variability, annual cycle (phenology), and trends. The global distribution of phytoplankton is well known and reproduced by all satellite products, but the advent of high resolution modeling and remote sensing is shedding new light on how phytoplankton and physics are related across spatial scales, including planetary waves [71], mesoscale, and submesoscale [72–74], or internal waves [75]. The seasonal cycle of phytoplankton is the most prominent signal in many ocean regions and phytoplankton phenology has recently been actively investigated [76]. Variations in Chla distributions have been studied at other time scales, describing intraseasonal signals such the Madden–Julian oscillation [77], tropical interannual variations like El Niño [78] or climate signals with longer time scales [79].

Temporal signals of field observations and satellite data can be compared without the need for the stringent timeliness required by match-up analyses. Some studies have checked the main temporal patterns displayed by satellite products against in situ time series, for instance comparing their respective annual cycle of Chla [80], or looking at trends in radiometric data (e.g., at AAOT [32]) or derived products at time-series stations like the Bermuda Atlantic Time series Study or the Hawaii Ocean Time series [81]. Considering the requirements for extensive in situ data sets needed for such analyses, they have been very few so far.

Notwithstanding the potential offered by satellite data for spatial and temporal analyses, how the different characteristics of each satellite product (in terms of spatial resolution, levels and structure of variance, or noise) affect analyses of climate signals or model simulations relying on data assimilation has been largely unexplored. These differences will have to be properly integrated into the long-term analysis of the biogeochemical responses of marine ecosystems to climate forcing. However, some studies have analyzed how different satellite missions represent the temporal evolution of a satellite derived product (Chla or optical properties) for specific regions [67,80,82]. Djavidnia et al. [83] have compared Chla time series averaged over the Longhurst [84] provinces, as obtained from SeaWiFS, MODIS, and MERIS. Taylor plots are useful in that regard, illustrating on the same plot the correlation between two signals, their standard deviations, and their unbiased RMS difference. From the updated results for SeaWiFS and MODIS seen on Figure 10, it appears that the correlation coefficients between monthly series are all higher than 0.8, while the variance of the MODIS time series can be lower or higher than that of SeaWiFS.

As soon as the SeaWiFS mission lifetime exceeded 5 years, investigations started studying possible trends associated with its Chla series [85,86]. The validity of these analyses was supported by the activities that ensured the characterization and stability of the instrument calibration [87]. Even if similar calibration strategies are followed for the main ocean color missions, it still appears worthwhile to compare long-term trends obtained from different missions to check that they provide the same view of interannual changes taking place in the oceans. Generally, this is unlikely to be an easy task since it requires overlaps between missions long enough for trend analysis. But the ocean color community has been fortunate to benefit from such cases with the long records of SeaWiFS, MODIS, and MERIS. For instance, the latter two missions were contemporaneously in operation for a decade. A trend analysis

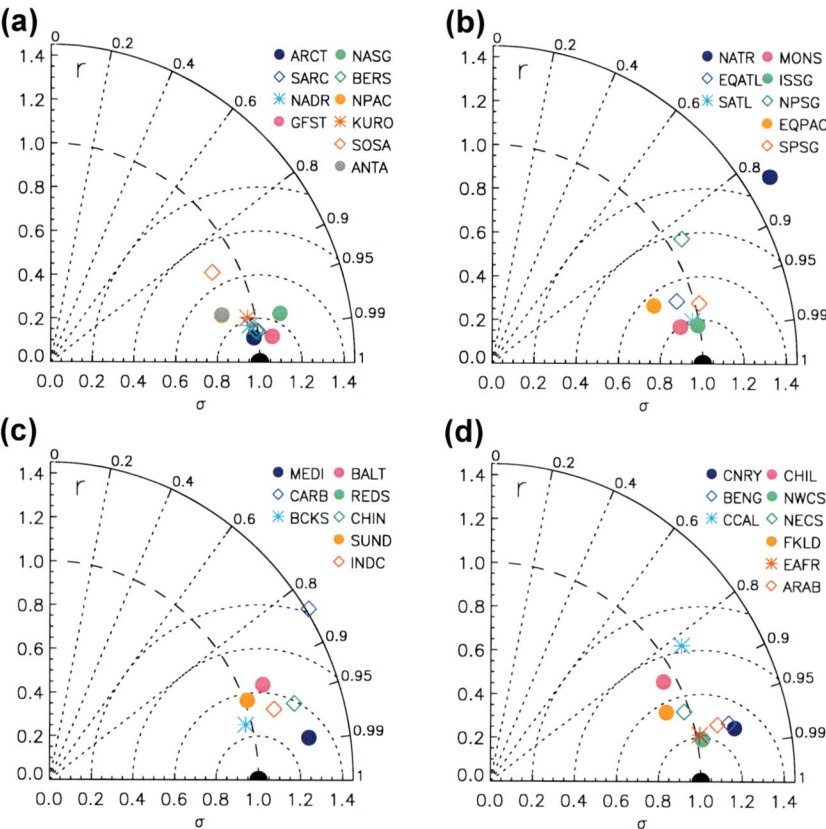

FIGURE 10 Taylor plots comparing SeaWiFS (taken as a reference) and MODIS-Aqua Chla time series averaged over biogeographic provinces associated with (a) midlatitudes, (b) subtropical regions, (c) marginal seas, and (d) shelves and upwelling regions. See Longhurst [84] for province acronyms.

was performed on the MODIS and MERIS Chla data over the period August 2002–July 2011 using a nonparametric seasonal Kendall test [79]. Figure 11 illustrates the agreement between the trend fields. The use of a contingency matrix allows the quantification of this agreement by computing the percentage of the ocean with similar or divergent behaviors (trend slopes of the same/opposite signs, significance levels, ...). For instance, 20% of the ocean is found to have a statistically significant trend ($p < 0.05$) of the same sign for both series (11% with a positive slope, 9% with a negative slope), while there is virtually no area with a statistically significant trend of opposite sign (0.005%).

Analyses checking the consistency of ocean color time series in terms of spatial distributions, phenology or trends should be seen as integral parts of an assessment strategy applied to climate data records.

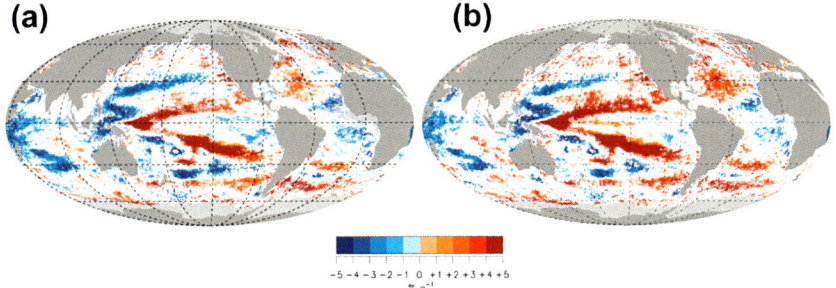

FIGURE 11 Linear trend for Chla (in % per annum) found over the period August 2002 to July 2011 for (a) MODIS-Aqua and (b) MERIS.

4. CONCLUSIONS

In the assessment of satellite products, a recurrent question is that of ranking: is a given product better than another, either because processed with different algorithms or associated with another sensor? A still larger question is: should it be deemed acceptable for climate research? Validation statistics might promote one mission for one specific product (e.g., Chla but not IOPs), using one field data set but not another, for some wavelengths only or for the spectral shape of R_{rs}. One product might also be preferred for its extensive data coverage. Ultimately, the choice of a particular product is intimately linked with the science question being addressed, and its assessment needs to be commensurate with the intended application. The stringent requirements associated with climate research call for a comprehensive approach including validation with field data, uncertainty analyses, and consistency checks like comparison between mission-specific products and time series analyses. In that regard, significant mission overlaps (at least 1 year) are an absolute prerequisite for such a strategy, besides the fact that gaps in the data records would seriously challenge our ability to use the ocean color record to detect climate signals [88].

In 1983, Gordon et al. [89] could use three spectra of water leaving radiance determined from ship-based observations to assess the atmospheric correction applied to CZCS data. Although this chapter illustrates the considerable progress made in data collection since then, the relative scarcity of high-quality R_{rs} measurements remains a limiting factor for assessing radiometric satellite ocean color products. The oceanographic community should invest in comprehensive measurement programs for validation purposes, and in the development of new technological or methodological approaches. The development of a network of automated above-water radiometers [20] represents a major progress for coastal waters. Placing bio-optical instruments onto floats [90] is also a promising avenue to increase frequency and coverage. Hyperspectral measurements are also desirable, both to fully accommodate the spectral characteristics specific to each satellite sensor, and in preparation for advanced spaceborne sensors with hyperspectral capabilities.

This chapter has also discussed approaches based on the use of models or intercomparison techniques that can profitably complement the in situ validation statistics. Their major contribution should be to allow an extension of validation results to a wider range of geometric and environmental conditions. A complete framework for error propagation is needed, but it requires a thorough and accurate characterization of the uncertainties associated with the radiance signal at top-of-atmosphere, as well as a detailed understanding of the propagation of errors through the atmospheric correction algorithm, the uncertainties associated with algorithm assumptions, and the contributions of all other ancillary inputs (e.g., meteorological conditions and atmospheric gas contributions) to the total uncertainty budget. The approaches mentioned here do not form an exhaustive list. More holistic modeling environments could be devised to support product assessment, for instance to exclude some conditions or ascertain their probability. For instance, high Chla concentrations close to river outlets are unlikely in winter, and their presence in satellite products is suspect. Ecosystem models can contribute to assess and improve satellite products, and vice-versa.

A maturity model has been proposed to assess the completeness of climate data records (CDR) [91]. This matrix model contemplates six levels of maturity for all aspects of a data set. We could argue that the ocean color products can pretend to the levels 5 or 6, synonymous of a "full operational capability", for aspects like software, documentation, metadata, public access, and utility. Understandably, product validation is lagging behind, qualifying for levels 3 or 4, "uncertainty estimated for select locations/times" or "uncertainty estimated over widely distributed times/locations by multiple investigators; differences understood". Level 5 would entail the knowledge of "consistent uncertainties estimated over most environmental conditions by multiple investigators". Progress has been made in that direction but needs to be consolidated into an operational context to allow a fully informed use of ocean color products in climate research.

ACKNOWLEDGMENTS

The authors would like to thank all the investigators contributing to the collection of in-situ data used in this chapter, particularly for the AERONET-OC and MOBy sites. Their dedication is earnestly valued.

REFERENCES

[1] GCOS-154, "Systematic Observation Requirements for Satellite-based Products for Climate," Supplemental Details to the Satellite-based Component of the "Implementation Plan for the Global Observing System for Climate in Support of the UNFCC", 2011, 138 pp.

[2] IOCCG, Remote sensing of optical properties: fundamentals, tests of algorithms, and applications, in: Z.-P. Lee (Ed.), Reports of the International Ocean-Colour Coordinating Group, No. 5, IOCCG, Dartmouth, Canada, 2006, 126 pp.

[3] S.W. Bailey, P.J. Werdell, A multi-sensor approach for the on-orbit validation of ocean color satellite data products, Remote Sens. Environ. 102 (2006) 12−23.

[4] G. Zibordi, J.-F. Berthon, F. Mélin, D. D'Alimonte, S. Kaitala, Validation of satellite ocean color primary products at optically complex coastal sites: northern Adriatic Sea, northern Baltic Proper, Gulf of Finland, Remote Sens. Environ. 113 (2009) 2574−2591.

[5] H. Feng, D. Vandemark, J.W. Campbell, B.N. Holben, Evaluation of MODIS ocean colour products at a northeast United States coast site near the Martha's Vineyard Coastal Observatory, Int. J. Remote Sens. 29 (2009) 4479−4497.

[6] R. Brewin, S. Sathyendranath, D. Mueller, C. Brockmann, P.-Y. Deschamps, E. Devred, R. Doerffer, N. Fomferra, B. Franz, M. Grant, S. Groom, A. Horseman, C. Hu, H. Krasemann, Z.-P. Lee, S. Maritorena, F. Mélin, M. Peters, T. Platt, P. Regner, T. Smyth, F. Steinmetz, J. Swinton, P.J. Werdell, G.N. White, The Ocean Colour Climate Change Initiative. III. A round-robin comparison on in-water bio-optical algorithms, Remote Sens. Environ. (2014), http://dx.doi.org/10.1016/j.rse.2013.09.016.

[7] D. Mueller, H. Krasemann, R. Brewin, C. Brockmann, P.-Y. Deschamps, R. Doerffer, N. Fomferra, B.A. Franz, M. Grant, S. Groom, F. Mélin, T. Platt, P. Regner, S. Sathyendranath, F. Steinmetz, The Ocean Colour Climate Change Initiative. II an assessment of atmospheric correction processors based on in-situ measurements, Remote Sens. Environ., in press.

[8] D.K. Clark, H.R. Gordon, K.J. Voss, Y. Ge, W. Broenkow, C. Trees, Validation of atmospheric correction over the oceans, J. Geophys. Res. 102 (1997) 17209−17217.

[9] B.A. Franz, S.W. Bailey, P.J. Werdell, C.R. McClain, Sensor-independent approach to the vicarious calibration of satellite ocean color radiometry, Appl. Opt. 46 (2007) 5068−5082.

[10] W.W. Gregg, N.W. Casey, Global and regional evaluation of the SeaWiFS chlorophyll data set, Remote Sens. Environ. 93 (2004) 463−479.

[11] P.J. Werdell, S.W. Bailey, G. Fargion, C. Pietras, K. Knobelspiesse, G.C. Feldman, C.R. McClain, Unique data repository facilitates ocean color satellite validation, EOS Trans. AGU 84 (2003) 379.

[12] T. Cui, J. Zhang, S. Groom, L. Sun, T. Smyth, S. Sathyendranath, Validation of MERIS ocean-color products in the Bohai Sea: a case study for coastal waters, Remote Sens. Environ. 114 (2010) 2326−2336.

[13] D. Antoine, F. d'Ortenzio, S.B. Hooker, G. Bécu, B. Gentili, D. Tailliez, A.J. Scott, Assessment of uncertainty in the ocean reflectance determined by three satellite ocean color sensors (MERIS, SeaWiFS and MODIS-A) at an offshore site in the Mediterranean Sea (BOUSSOLE project), J. Geophys. Res. 113 (2008) C07013. http://dx.doi.org/10.1029/2007JC004472.

[14] M.E. Smith, S. Bernard, S. O'Donoghue, The assessment of optimal MERIS ocean color products in the shelf waters of the KwaZulu-Natal Bight, South Africa, Remote Sens. Environ. 137 (2013) 124−138.

[15] G. Zibordi, F. Mélin, J.-F. Berthon, E. Canuti, Assessment of MERIS ocean color data products for European seas, Ocean Sci. 9 (2013) 521−533.

[16] H. Murakami, K. Sasaoka, K. Hosoda, H. Fukushima, M. Toratani, R. Frouin, B.G. Mitchell, M. Kahru, P.-Y. Deschamps, D. Clark, S. Flora, M. Kishino, S.-I. Saitoh, I. Asanuma, A. Tanaka, H. Sasaki, K. Yokouchi, Y. Kiyomoto, H. Saito, C. Dupouy, A. Siripong, S. Matsumura, H. Ishizaka, Validation of ADEOS-II GLI ocean color products using in-situ observations, J. Oceanogr. 62 (2006) 373−393.

[17] C. Jamet, H. Loisel, C.P. Kuchinke, K. Ruddick, G. Zibordi, H. Feng, Comparison of three SeaWiFS atmospheric correction algorithms for turbid waters using AERONET-OC measurements, Remote Sens. Environ. 115 (2011) 1955–1965.

[18] C. Goyens, C. Jamet, T. Schroeder, Evaluation of four atmospheric correction algorithms for MODIS-Aqua images over contrasted coastal waters, Remote Sens. Environ. 131 (2013) 63–75.

[19] G. Zibordi, J.-F. Berthon, F. Mélin, D. D'Alimonte, Cross-site consistent in-situ measurements for satellite ocean color applications: the BiOMaP radiometric dataset, Remote Sens. Environ. 115 (2011) 2104–2115.

[20] G. Zibordi, B.N. Holben, S.B. Hooker, F. Mélin, J.-F. Berthon, I. Slutsker, D. Giles, D. Vandemark, H. Feng, K. Rutledge, G. Schuster, A. Al Mandoos, A network for standardized ocean color validation measurements, EOS Trans. AGU 87 (2006) 293–297.

[21] M. Shimada, H. Oaku, Y. Mitomi, H. Murakami, A. Mukaida, Y. Nakamura, J. Ishisaka, H. Kawamura, T. Tanaka, M. Kishino, H. Fukushima, Calibration and validation of the ocean color version-3 product from AEOS OCTS, J. Oceanogr. 54 (1998) 401–416.

[22] S. Hlaing, T. Harmel, A. Gilerson, R. Foster, A. Weidemann, R. Arnone, M. Wang, S. Ahmed, Evaluation of the VIIRS ocean color monitoring performance in coastal regions, Remote Sens. Environ. 139 (2013) 398–414.

[23] N. Lamquin, C. Mazeran, D. Doxaran, J.-H. Ryu, Y.-J. Park, Assessment of GOCI radiometric products using MERIS, MODIS and field measurements, Ocean Sci. J. 47 (2012) 287–311.

[24] M. Wang, J.H. Ahn, L. Jiang, W. Shi, S. Son, Y.-J. Park, J.-H. Ryu, Ocean color products from the Korean Geostationary Ocean Color Imager (GOCI), Opt. Exp. 21 (2013) 3835–3849.

[25] T. Cui, J. Zhang, J. Tang, S. Sathyendranath, S. Groom, Y. Ma, W. Zhao, Q. Song, Assessment of satellite ocean color products of MERIS, MODIS and SeaWiFS along the East China Coast (in the Yellow Sea and East China Sea), ISPRS J. Photogram. Remote Sens. 87 (2014) 137–151.

[26] T. Schroeder, I. Behnert, M. Schaale, J. Fischer, R. Doerffer, Atmospheric correction algorithm for MERIS above case-2 waters, Int. J. Remote Sens. 28 (2007) 1469–1486.

[27] M. Wang, S.-H. Son, W. Shi, Evaluation of MODIS SWIR and NIR-SWIR atmospheric correction algorithms using SeaBASS data, Remote Sens. Environ. 113 (2009) 635–644.

[28] M. Szeto, P.J. Werdell, T.S. Moore, J.W. Campbell, Are the world's oceans optically different? J. Geophys. Res. 116 (2011) C00H4. http://dx.doi.org/10.1029/2011JC007230.

[29] M.J. Sauer, C.S. Roesler, P.J. Werdell, A. Barnard, Under the hood of satellite empirical chlorophyll-a algorithms: revealing the dependencies of maximum band ratio algorithms on inherent optical properties, Opt. Exp. 20 (2012) 20920–20933.

[30] F. Mélin, G. Zibordi, J.-F. Berthon, Assessment of satellite ocean color products at a coastal site, Remote Sens. Environ. 110 (2007) 192–215.

[31] G. Zibordi, F. Mélin, J.-F. Berthon, Intra-annual variations of biases in remote sensing primary ocean color products at a coastal site, Remote Sens. Environ. 124 (2012) 627–636.

[32] G. Zibordi, F. Mélin, J.-F. Berthon, Trends in the bias of primary satellite ocean color products at a coastal site, IEEE Geosci. Remote Sens. Lett. 9 (2012) 1056–1060.

[33] D. D'Alimonte, G. Zibordi, F. Mélin, A statistical method for generating cross-mission consistent normalized water-leaving radiances, IEEE Trans. Geosci. Remote Sens 46 (2008) 4075–4093.

[34] T.S. Moore, J.W. Campbell, M.D. Dowell, A class-based approach to characterizing and mapping the uncertainty of the MODIS ocean chlorophyll product, Remote Sens. Environ. 113 (2009) 2424–2430.

[35] T.S. Moore, M.D. Dowell, B.A. Franz, Detection of coccolithophore blooms in ocean color satellite imagery: a generalized approach for use with multiple sensors, Remote Sens. Environ. 117 (2012) 249–263.
[36] M. Sydor, R.W. Gould, R.A. Arnone, V.I. Haltrin, W. Goode, Uniqueness in remote sensing of the inherent optical properties of ocean water, Appl. Opt. 43 (2004) 2156–2162.
[37] M. Defoin-Platel, M. Chami, How ambiguous is the inverse problem of ocean color on coastal waters? J. Geophys. Res. 112 (2007) C03004. http://dx.doi.org/10.1029/2006JC003847.
[38] Z.-P. Lee, R.A. Arnone, C. Hu, P.J. Werdell, B. Lubac, Uncertainties of optical parameters and their propagations in an analytical ocean color inversion algorithm, Appl. Opt. 49 (2010) 369–381.
[39] Z.-P. Lee, K.L. Carder, R.A. Arnone, Deriving inherent optical properties from water color: a multiband quasi-analytical algorithm for optically deep waters, Appl. Opt. 41 (2002) 5755–5772.
[40] P. Wang, E.S. Boss, C. Roesler, Uncertainties of inherent optical properties obtained from semi-analytical inversions of ocean color, Appl. Opt. 44 (2005) 4074–4085.
[41] V. Brando, A.G. Dekker, Y.J. Park, T. Schroeder, Adaptive semi-analytical inversion of ocean color radiometry in optically complex waters, Appl. Opt. 51 (2012) 2808–2833.
[42] H.J. Van Der Woerd, R. Pasterkamp, HYDROPT: a fast and flexible method to retrieve chlorophyll-a from multispectral satellite observations of optically complex coastal waters, Remote Sens. Environ. 112 (2008) 1795–1807.
[43] M.S. Salama, A.G. Dekker, Z. Su, C.M. Mannaerts, W. Verhoef, Deriving inherent optical properties and associated inversion-uncertainties in the Dutch Lakes, Hydrol. Earth Syst. Sci. 13 (2009) 1113–1121.
[44] S. Maritorena, O.H.F. d'Andon, A. Mangin, D.A. Siegel, Merged satellite ocean color data products using a bio-optical model: characteristics, benefits and issues, Remote Sens. Environ. 114 (2010) 1791–1804.
[45] P.J. Werdell, B.A. Franz, S.W. Bailey, G.C. Feldman, E. Boss, V.E. Brando, M.D. Dowell, T. Hirata, S.J. Lavender, Z.-P. Lee, H. Loisel, S. Maritorena, F. Mélin, T.S. Moore, T.J. Smyth, D. Antoine, E. Devred, O. Fanton d'Andon, A. Mangin, A generalized ocean color inversion for retrieving marine inherent optical properties, Appl. Opt. 52 (2013) 2019–2037.
[46] H. Schiller, R. Doerffer, Improved determination of coastal water constituent concentrations from MERIS data, IEEE Trans. Geosci. Remote Sens. 43 (2005) 1585–1591.
[47] R.M. Chomko, H.R. Gordon, Atmospheric correction of ocean color imagery: test of the spectral optimization algorithm with the sea-viewing wide field-of-view sensor, Appl. Opt. 40 (2001) 2973–2984.
[48] K. Stamnes, W. Li, B. Yan, H. Eide, A. Barnard, W.S. Pegau, J.J. Stamnes, Accurate and self-consistent ocean color algorithms: simultaneous retrieval of aerosol optical properties and chlorophyll concentrations, Appl. Opt. 42 (2003) 939–951.
[49] C. Jamet, S. Thiria, C. Moulin, M. Crépon, Use of a neurovariational inversion for retrieving oceanic and atmospheric constituents from ocean color imagery: a feasibility study, J. Atmos. Ocean. Technol. 22 (2005) 460–475.
[50] M.S. Salama, A. Stein, Error decomposition and estimation of inherent optical properties, Appl. Opt. 48 (2009) 4947–4962.
[51] B. Bulgarelli, G. Zibordi, Remote sensing of ocean colour: accuracy assessment of an approximate atmospheric correction code, Int. J. Remote Sen. 24 (2003) 491–509.

[52] B. Bulgarelli, F. Mélin, G. Zibordi, SeaWiFS-derived products in the Baltic Sea: performance analysis of a simple atmospheric correction algorithm, Oceanologia 45 (2003) 655–677.
[53] C. Hu, L. Feng, Z.-P. Lee, Uncertainties of SeaWiFS and MODIS remote sensing reflectance: Implications from clear water assessments, Remote Sens. Environ. 133 (2013) 168–182.
[54] C. Hu, Z.-P. Lee, B.A. Franz, Chlorophyll a algorithms for oligotrophic oceans: a novel approach based on three-band reflectance difference, J. Geophys. Res. 117 (2012) C01011. http://dx.doi.org/10.1029/2011JC007395.
[55] F. Mélin, Global distribution of the random uncertainty associated with satellite-derived Chla, IEEE Geosci. Remote Sens. 7 (2010) 220–224.
[56] IOCCG, in: W.W. Gregg, J. Aiken, E. Kwiatkowska, S. Maritorena, F. Mélin, H. Murakami, S. Pinnock, C. Pottier (Eds.), Ocean Color Data Merging, Reports of the International Ocean-Colour Coordinating Group, No. 5, vol. 65, IOCCG, Dartmouth, Canada, 2007.
[57] D. Antoine, A. Morel, H.R. Gordon, V.F. Banzon, R.H. Evans, Bridging ocean color observations of the 1980s and 2000s in search of long-term trends, J. Geophys. Res. 110 (2005) C06009. http://dx.doi.org/10.1029/2004JC002620.
[58] A. Morel, S. Maritorena, Bio-optical properties of oceanic waters: a reappraisal, J. Geophys. Res. 106 (2001) 7163–7180.
[59] A. Morel, Y. Huot, B. Gentili, P.J. Werdell, S.B. Hooker, B.A. Franz, Examining the consistency of products derived from various ocean color sensors in open ocean (Case 1) waters in the perspective of a multi-sensor approach, Remote Sens. Environ. 111 (2007) 69–88.
[60] G. Zibordi, F. Mélin, J.-F. Berthon, Comparison of SeaWiFS, MODIS and MERIS radiometric products at a coastal site, Geophys. Res. Lett. 33 (2006) L06617. http://dx.doi.org/10.1029/2006GL025778.
[61] F. Mélin, G. Zibordi, J.-F. Berthon, S.W. Bailey, B.A. Franz, K.J. Voss, S. Flora, M. Grant, Assessment of MERIS reflectance data as processed by SeaDAS over the European seas, Opt. Exp. 19 (2011) 25657–25671.
[62] F. Mélin, G. Zibordi, S. Djavidnia, Merged series of normalized water leaving radiances obtained from multiple satellite missions for the Mediterranean Sea, Adv. Space Res. 43 (2009) 423–437.
[63] F. Mélin, V. Vantrepotte, M. Clerici, D. D'Alimonte, G. Zibordi, J.-F. Berthon, E. Canuti, Multi-sensor satellite time series of optical properties and chlorophyll a concentration in the Adriatic Sea, Prog. Oceanogr 91 (2011) 229–244.
[64] A. Morel, B. Gentili, Diffuse reflectance of oceanic waters: its dependence on sun angle as influenced by the molecular scattering contribution, Appl. Opt. 30 (1991) 4427–4438.
[65] A. Morel, D. Antoine, B. Gentili, Bidirectional reflectance of oceanic waters: accounting for Raman emission and varying particle scattering phase function, Appl. Opt. 41 (2002) 6289–6306.
[66] A. Bricaud, M. Babin, A. Morel, H. Claustre, Variability in the chlorophyll-specific absorption coefficients of natural phytoplankton: analysis and parameterization, J. Geophys. Res. 100 (1995) 13321–13332.
[67] F. Mélin, Comparison of SeaWiFS and MODIS time series of inherent optical properties for the Adriatic Sea, Ocean Sci. 7 (2011) 351–361.
[68] G. Meister, E.J. Kwiatkowska, B.A. Franz, F.S. Patt, G.C. Feldman, C.R. McClain, Moderate resolution imaging spectroradiometer ocean color polarization correction, Appl. Opt. 44 (2005) 5524–5535.

[69] B.A. Franz, D.A. Siegel, M.J. Behrenfeld, P.J. Werdell, "Global ocean phytoplankton," in State of the Climate 2012, Bull. Am. Meteorol. Soc. 94 (2013) S75–S78.
[70] J.W. Campbell, J.M. Blaisdell, M. Darzi, in: S.B. Hooker, E.R. Firestone, J.G. Acker (Eds.), Level-3 SeaWiFS Data Products: Spatial and Temporal Binning Algorithms, NASA Tech. Mem. 104566, vol. 32, NASA Goddard Space Flight Center, Greenbelt, Maryland, 1995.
[71] P. Cipollini, D. Cromwell, P.G. Challenor, S. Raffaglio, Rossby waves detected in global ocean colour data, Geophys. Res. Lett. 28 (2001) 323–326.
[72] A. Mahadevan, J.W. Campbell, Biogeochemical patchiness at the sea surface, Geophys. Res. Lett. 29 (2002) 1926. http://dx.doi.org/10.1029/2001GL014116.
[73] S.C. Doney, D.M. Glover, S.J. McCue, M. Fuentes, Mesoscale variability of sea-viewing wide field-of-view sensor (SeaWiFS) satellite ocean color: global patterns and spatial scales, J. Geophys. Res. 108 (2003) 3024. http://dx.doi.org/10.1029/2001JC000843.
[74] M. Lévy, R. Ferrari, P.J.S. Franks, A.P. Martin, P. Rivière, Bringing physics to life at the submesoscale, Geophys. Res. Lett. 39 (2012) L14062. http://dx.doi.org/10.1029/2012GL052756.
[75] J.C.B. da Silva, A.L. New, M.A. Srokosz, T.J. Smyth, On the observability of internal tidal waves in remotely-sensed ocean color data, Geophys. Lett. 29 (2002) 1569. http://dx.doi.org/10.1029/2001GL013888.
[76] M.R.P. Sapiano, C.W. Brown, S. Schollaert Uz, M. Vargas, Establishing a global climatology of marine phytoplankton phenological characteristics, J. Geophys. Res. 117 (2012) C08026. http://dx.doi.org/10.1029/2012JC007958.
[77] D. Jin, D.E. Waliser, C. Jones, R. Murtugudde, Modulation of tropical ocean surface chlorophyll by the Madden-Julian oscillation, Clim. Dyn. 40 (2013) 39–58.
[78] M.J. Behrenfeld, J.T. Randerson, C.R. McClain, G.C. Feldman, S.O. Los, C.J. Tucker, P.G. Falkowski, C.B. Field, R. Frouin, W.E. Esaias, D.D. Kolber, N.H. Pollack, Biospheric primary production during an ENSO transition, Science 291 (2001) 2594–2597.
[79] V. Vantrepotte, F. Mélin, Inter-annual variations in the SeaWiFS global chlorophyll a concentration (1997-2007), Deep-Sea Res. I 58 (2011) 429–441.
[80] P.J. Werdell, S.W. Bailey, B.A. Franz, L.W. Harding, G.C. Feldman, C.R. McClain, Regional and seasonal variability of chlorophyll-a in Chesapeake Bay as observed by SeaWiFS and MODIS-Aqua, Remote Sens. Environ. 113 (2009) 1319–1330.
[81] V.S. Saba, M.A.M. Friedrichs, M.-E. Carr, D. Antoine, R.A. Armstrong, I. Asanuma, O. Aumont, N.R. Bates, M.J. Behrenfeld, V. Bennington, L. Bopp, J. Bruggeman, E.T. Buitenhuis, M.J. Church, A.M. Ciotti, S.C. Doney, M.D. Dowell, J. Dunne, S. Dutkiewicz, W.W. Gregg, N. Hoepffner, K.J.W. Hyde, J. Ishizaka, T. Kameda, D.M. Karl, I. Lima, M.W. Lomas, J. Marra, G.A. McKinley, F. Mélin, J.K. Moore, A. Morel, J. O'Reilly, B. Salihoglu, M. Scardi, T.J. Smyth, S. Tang, J. Tjiputra, J. Uitz, M. Vichi, K. Waters, T.K. Westberry, A. Yool, Challenges of modeling depth-integrated marine primary productivity over multiple decades: a case study at BATS and HOT, Global Biogeochem. Cycles 24 (2010) GB3020. http://dx.doi.org/10.1029/2009GB003655.
[82] C. Zhang, C. Hu, S. Shang, F.E. Mueller-Karger, Y. Li, M. Dai, B. Huang, X. Ning, H. Hong, Bridging between SeaWiFS and MODIS for continuity of chlorophyll-a concentration assessments off Southeastern China, Remote Sens. Environ. 102 (2006) 250–263.
[83] S. Djavidnia, F. Mélin, N. Hoepffner, Comparison of global ocean colour data records, Ocean Sci. 6 (2010) 61–76.
[84] A. Longhurst, Ecological Geography of the Sea, Academic Press, 2006, 560 pp.

[85] C.R. McClain, S.R. Signorini, J.R. Christian, Subtropical gyre variability observed by ocean-color satellites, Deep-Sea Res. II 51 (2004) 281–301.

[86] W.W. Gregg, N.W. Casey, C.R. McClain, Recent trends in global ocean chlorophyll, Geophys. Res. Lett. 32 (2005) L03606. http://dx.doi.org/10.1029/2004GL021808.

[87] R.E. Eplee, G. Meister, F.S. Patt, R.A. Barnes, S.W. Bailey, B.A. Franz, C.R. McClain, On-orbit calibration of SeaWiFS, Appl. Opt. 51 (2012) 8702–8730.

[88] C. Beaulieu, S.A. Henson, J.L. Sarmiento, J.P. Dunne, R.R. Rykaczewski, L. Bopp, Factors challenging our ability to detect long-term trends in ocean chlorophyll, Biogeosciences 10 (2013) 2711–2724.

[89] H.R. Gordon, D.K. Clark, J.W. Brown, O.B. Brown, R.H. Evans, W.W. Broenkow, Phytoplankton pigment concentrations in the Middle Atlantic Bight: comparison between ship determinations and coastal zone color scanner estimates, Appl. Opt. 22 (1983) 20–36.

[90] IOCCG, Bio-optical sensors on Argo floats, in: H. Claustre (Ed.), Reports of the International Ocean-Colour Coordinating Group, No. 11, IOCCG, Dartmouth, Canada, 2011, 89 pp.

[91] J.J. Bates, J.L. Privette, A maturity model for assessing the completeness of climate data records, EOS Trans. Am. Geophys. Union 93 (2012) 441.

Chapter 6.2

Assessment of Long-Term Satellite Derived Sea Surface Temperature Records

Gary K. Corlett,[1,]* Christopher J. Merchant,[2] Peter J. Minnett,[3] Craig J. Donlon[4]

[1] *Department of Physics and Astronomy, University of Leicester, Leicester, UK;* [2] *Department of Meteorology, University of Reading, Reading, UK;* [3] *Meteorology & Physical Oceanography, Rosenstiel School of Marine and Atmospheric Science, University of Miami, Miami, USA;* [4] *European Space Agency/ESTEC, Noordwijk, The Netherlands*
*Corresponding author: E-mail: gkc1@leicester.ac.uk

Chapter Outline

1. Introduction 639
2. Background 640
　2.1 Assessment of Top of Atmosphere Brightness Temperatures 641
　2.2 Validation Uncertainty Budget 643
　2.3 Reference Data Sources 647
3. Assessment of Long-Term SST Datasets 649
　3.1 Example 1: Long-Term SST Data Record Assessment 652
　3.2 Example 2: Long-Term Component Assessment 654
　3.3 Quantitative Metrics 657
　3.4 Demonstrating Traceability to SI 659
　3.5 Stability 663
　3.6 Validation of Uncertainties 669
4. Summary and Recommendations 673
References 674

1. INTRODUCTION

In this chapter we consider the *assessment* of a satellite-derived ocean surface temperature climate data record (CDR) derived from thermal infrared (IR) radiances measured at the top-of-atmosphere (TOA). It is important to note, even at this stage in our discussion, any product assessment is usually done against a set of requirements, which in this case are the requirements for a satellite sea surface temperature (SST) CDR. Such requirements will usually include, among others, a specification on accuracy (e.g., require SST bias to be <0.1 K), a specification on stability (e.g., require long-term drift in SST bias to be <0.03 K/decade), and a specification on scale (e.g., accuracy to be attained on scales <1000 km). However, as noted in a recent European Space Agency (ESA) study [1] different applications of CDRs have different requirements regarding accuracy, spatial and temporal resolutions, and record length. Likewise there is no unique definition of "SST" for a CDR as different users have their own requirements on the depth at which measurements are considered valid and the space and the time scales of interest [1].

The fundamental approach to SST CDR assessment is to seek quantitative metrics that can be determined for any satellite derived SST data record. Using these metrics, a user may then make their own assessment as to the suitability of any specific data record for their particular application. Indeed, it is in the application of the data record to identify climate variability and change that truly defines if a producer has generated a CDR or not. This approach is recommended by the Group for High Resolution Sea Surface Temperature (GHRSST) CDR Technical Advisory Group (TAG) and is the basis for the GHRSST Climate Data Assessment Framework (CDAF) [2].

There are two main thermal IR CDRs we can assess: (1) the TOA radiances, which are usually expressed as brightness temperatures (BTs), and (2) the derived SSTs. In Section 2 we discuss issues encountered when comparing a satellite measurement to a measurement taken in situ or from a ship mounted IR radiometer. Then in Section 3 we look at methods for the assessment of a long-term SST data record. Finally, in Section 4, we summarize the available methods and make recommendations for future research efforts.

2. BACKGROUND

The assessment of long-term SST data records derived from IR radiometers on satellites is currently an evolving area of research. This statement may seem a little strange since an assessment of satellite SSTs was first reported in 1967 [3]. Since then there are many additional published assessments of SSTs retrieved from individual satellite sensors. However, the emphasis here is to consider long-term SST data records as a whole, i.e., data sets that span multiple decades, utilizing data from several satellites, and have (ideally) been consistently processed. We focus therefore on the assessment of long-term SST data records and will only consider individual sensors where they contribute knowledge to the assessment of an SST CDR.

The original long-term SST record from space-borne sensors is the Pathfinder data set (see Casey et al. [4] for a review of Pathfinder), which uses measurements from the Advanced Very High Resolution Radiometer (AVHRR) series of sensors (see Chapter 2.3 for further details of the AVHRRs). First published in the early 1990s, Pathfinder utilizes empirical regression algorithms

[5] that have been tuned mainly to drifting buoys to provide a self-consistent approach across the series of AVHRRs from National Oceanic and Atmospheric Administration (NOAA) 7 (1981) through to NOAA 19 for the current version 5.2 data (available through 2012). The Pathfinder data set is constructed from a single AVHRRs sensor at any one time, and the tuning to reference data is primarily to account for many instrumental issues, including errors in the calibration of the sensors [6] (Chapter 2.4), and means independence between satellite and in situ data is not maintained. More modern developments of Pathfinder will be found in the version 6.0 data set, including latitudinally banded coefficients to reduce known limitations with regression algorithms [7].

The next long-term SST record generated was that from the Along track Scanning Radiometer (ATSR) Reprocessing for Climate (ARC) project [8], which exploited benefits of the ATSR series of sensors (see Chapter 2.3 for further details of the ATSRs). The ATSRs were specifically designed to provide high quality SST suitable for climate studies unlike the AVHRR series, which were originally designed for meteorological applications. The ARC data set differs from Pathfinder in a number of ways: ARC maintains independence from in situ measurements (see Chapter 4.2), has performed a careful harmonization of the BTs from the various sensors used, and employs a physics-based retrieval of SST_{skin} using a radiative transfer model (RTM) [9,10]. This allows for independent validation of the data set using in situ measurements, accounting for skin and diurnal effects, both before [11] and after [12] harmonization of the ATSR BTs.

Although the ARC data set has been shown to be useful for quantifying interannual variability in SST and identifying major SST anomalies [8], the narrow swath of the ATSRs ($\sim$500 km) compared to the AVHRRs (where SST is usually retrieved from the central 1500 km of the swath) is a limitation. This led the ESA SST Climate Change Initiative (SST_CCI) project team to make a first attempt at using the ATSRs to reduce residual BT errors of the AVHRRs and provide an optimal data set of both ATSR and AVHRR together for the period they co-exist (1991–2012). Further details of the ESA SST_CCI data set can be found in [13] and their initial findings are reported in their Product Validation and Intercomparison Report [14] and their Climate Data Assessment Report [15]. The analysis carried out on the ESA SST_CCI data set is at the forefront of SST CDR research and assessment and we shall draw heavily on their findings throughout this chapter.

2.1 Assessment of Top of Atmosphere Brightness Temperatures

Before carrying out an assessment of the SST record it is advisable to look at the BT record from which the SST is retrieved. As shown in Embury et al. [11], differences in BT between the ATSR-2 and AATSR sensors include intersensor biases of $\sim$0.1 K beyond differences that can be accounted for by sensor characteristics. This is significant when striving for the stability levels needed for an SST CDR (<0.03 K/decade [16]). We can assess the quality of a BT CDR in three ways, (1) by simulation, (2) by comparison with other satellite data sets, and (3) by comparison to reference measurements taken on the ground adjusted by simulations. However, as for SST records, assessments of long-term harmonized BT records are not widespread and their importance is only now being recognized.

In-orbit validation of thermal infrared data acquired by satellites using surface or aircraft based reference data is a well-established practice. Such methods (see for example [17,18]) typically involve measuring the radiance emitted by the surface together with the atmospheric properties between the satellite sensor and the surface, and then using these data as input to an RTM to predict the radiance measured at the sensor. The predicted at-sensor radiance is then compared to the at-sensor radiance measured by the satellite or aircraft instrument. Similar approaches, using only satellite data, are now being exploited in real time by the Global Space Based Intercalibration System (GSICS [19]), and the Monitoring of IR Clear-sky Radiances over Oceans for SST (MICROS [20]), in order to reduce uncertainties in the TOA BTs prior to further analysis.

Long-term reference data from automated validation sites are also available: NASA JPL has been operating two sites, one at Lake Tahoe [21] and the other at Salton Sea [22]. The Lake Tahoe site has been operating since 1999 and the Salton Sea site since 2008 and together they provide a wide range of surface temperatures for calibration and validation. Measurements at Lake Tahoe and Salton Sea have been used to determine the accuracy of the mid- and thermal infrared data from several instruments including the Advanced Spaceborne Thermal Emission and Reflection Radiometer [21] on the Terra satellite, the Moderate Resolution Imaging Spectrometer [22] on the Terra and Aqua satellites, the Thematic Mapper [23], and Enhanced Thematic Mapper Plus [23] on the Landsat satellites and the Along Track Scanning Radiometer 2 [21]. See Chapter 3.2 for more details on the instruments used at these sites.

Embury and Merchant [12] have presented an assessment of a harmonized BT record by comparing clear-sky BTs measured by the ATSR series to those from an RTM. Embury and Merchant [12] used RTM simulations to show differences calculated during sensor overlap periods are only partly explained by reported sensor characteristics and that residual differences are mostly likely due to sensor calibration and/or simulation errors. By applying BT adjustments calculated from double differences between simulated and measured BTs, Embury and Merchant [12] were able to harmonize the BTs of the ATSR series. The resulting SST record was stable relative to the Global Tropical Moored Buoy Array (GTMBA) to within ±2 mK between 1995 and 2012. However, the lack of suitable reference measurements in other parts of the ocean limits this work to the region covered by the GTMBA [24], which has SI traceability [25] through regular calibration of temperature sensors in the latter part of the record; other ocean areas do not have such infrastructure available.

2.2 Validation Uncertainty Budget

In an ideal case, the assessment of a long-term satellite-derived SST data record and its associated uncertainties would require:

1. A complete characterization of the satellite instrument and SST retrieval algorithm, at all times throughout the duration of the data set, and
2. A suite of globally representative traceable reference data points that match closely the measurand (type of SST) present in the satellite data, that have accuracy and precision better than the satellite sensor, that have a known stability at least comparable to the SST stability requirement, and are provided throughout the duration of the data set.

In addition, if the sensor is part of a series spanning many years, there should be a sufficient overlap period between successive sensors to allow for a robust characterization of the intersensor period [16], using the same traceable reference data points for both sensors. As we will see later in this chapter, the reality is, unfortunately, far from ideal as the limitations in the ability to maintain traceability on-orbit (issue 1 above) and in the quality and coverage of reference data (issue 2 above). As a result of these limitations, we have two basic approaches to assess the quality of a long-term SST data record:

1. "Point": In this approach single pixel comparisons to reference data, usually measured in situ, are used.
2. "Grid": In this approach comparison to gridded products is used, which potentially improves the match-up coverage (both temporally and spatially). Also, as this type of comparison uses "averaged" data there is likely to be a lower impact from outliers on the analysis (see later discussion) but there will be additional uncertainty from the gridding process.

In addition, for reasons that will become clear later on, we must also define a third way to assess the quality of a long-term SST data record:

3. "Functional": In this approach, we transfer knowledge of an assessment against reference data in one location to another location where we have no comparison measurements.

The rest of this chapter will focus mainly on "point" comparisons as these are more widely used in satellite SST assessments. We will return briefly to "functional" assessments when we consider uncertainty validation later in the chapter in Section 3.7.

The traditional approach to satellite uncertainty estimation involves the generation of a match-up data set (MD) of coincidences between the satellite and a reference data set produced within predefined spatial and temporal limits; for example, the current GHRSST-recommended match-up limits are ± 25 km, and ± 6 h [26], although some investigators choose to use other limits. There are few assessments of the spatial and temporal limits for generating match-ups in the literature. Minnett [27] determined that limits of 10 km and 2 h would introduce an error of up to 0.2 K but this was for a very specific area of the Atlantic Ocean with relatively high temperature variability. Chapter 5.2 defines more stringent criteria for ship-borne radiometer measurements based on the work of Wimmer et al. [28]. In general, the spatial limit should be minimized by using the smallest possible spatial window. Temporal differences, no matter what limit is set, should be minimized using a diurnal variability model.

Following generation of the MD, a statistical analysis is then carried out on the MD, from which the difference and standard deviation are usually calculated. The reference data are often treated as if they are truth, i.e., they are free of error, whereas their errors are often far from negligible. Of course the difference and standard deviation calculated from such a comparison do not provide the uncertainty of each data set individually, but are simply the mean difference and combined uncertainty of a two-data set comparison. The resulting statistics may be dominated by real changes in the SST that can occur within the predefined spatial and temporal limits (geophysical effects), which can be significant, as determined by Minnett [27]. An alternative method of multi-sensor match-up processing was demonstrated by O'Carroll et al. [29], which aimed to deduce the uncertainty of an individual data set from a three-way analysis, assuming each data set to be bias free. However, this approach does not properly account for geophysical differences between the satellite and reference data sets and also assumes zero correlation between data sets.

When considering potential reference sources consideration must be given to the depth of the SST being assessed. For satellite SST retrievals produced from infrared radiances, the SST is equivalent to the temperature at a depth of ~ 10 μm and is referred to as SST_{skin}. The deviation between the skin and the sub-skin has a mean of -0.17 K when the surface wind speed is $> \sim 6$ ms^{-1} [30], and so surface wind speed measurements also form an essential component of any reference data set for satellite SST validation. Ideally, the reference source for assessing the quality of the satellite data should be a measurement at a depth that is as close as possible to that provided by the satellite, i.e., the measurands should be consistent. Indeed, where possible, it should be the same as that provided by the satellite, which is currently only possible for infrared sensors using ship-borne IR radiometers (see Chapters 3.2 and 5.2). The match-up problem is made more complex by the fact that the water surface is a dynamic system and the nature of the sea-surface is such that it is hard to ascertain a meaningful mean condition of the sea-surface over an entire satellite pixel from a point in situ measurement.

Taking all of these factors into account is essential when assessing an SST CDR. We therefore define a validation uncertainty budget (i.e. the standard deviation of a Type A statistical assessment of comparisons between satellite and reference data, σ_{Total}) as:

$$\sigma_{Total} = \sqrt{\sigma_1^2 + \sigma_2^2 + \sigma_3^2 + \sigma_4^2 + \sigma_5^2} \quad (1)$$

where the terms will be defined and discussed shortly. This approach is similar to that of Wimmer et al. [28] who defined an error model for comparisons between ship-borne radiometers and satellite SSTs (as described in Chapter 5.2). Here, we adapt the approach of Wimmer et al. [28] to cover all potential reference sources and represent the contributing terms as uncertainties rather than absolute errors.

At this point it is useful to briefly consider the difference between error and uncertainty. The Guide to the Expression of Uncertainty in Measurement (GUM [31]) defines: *measurand* as a particular quantity subject to measurement. In this chapter, the measurand is either TOA BT (Level 1b data) or SST (for Level 2 and higher data). Very few instruments directly provide a measure of the measurand, and usually detect a quantity from which the measurand is derived. As an example, in this chapter, we focus on instruments that detect infrared radiation and it is from these measurements we estimate the temperature of an object.

Any procedure to determine the value of a measurand will be inexact to some degree due to uncertainty introduced into the procedure at some point. The difference between a measurement and the true value of the measurand is called the *error*. In remote sensing, the word "error" has often been used to describe a numerical value that estimates the variability of the error if a measurement is repeated (i.e. a width of the distribution of possible errors) and this leads to confusion as to what the quoted error actually means. To avoid any misunderstanding, the GUM [31] separates these definitions and defines:

Error (of measurement): result of a measurement minus a true value of the measurand, and
Uncertainty (of measurement): is a parameter, associated with the result of a measurement that characterizes the dispersion of the values that could reasonably be attributed to the measurand.

The true value of the measurand is often not known but can be estimated using statistical methods (which the GUM refers to as a Type A assessment [31]) or empirical methods (which the GUM refers to as a Type B assessment [31]) and can be considered as having one or more random components and/or one or more systematic components. The effects causing each component of error each give rise to a component of the overall uncertainty, so that the total uncertainty arises from a mix of random and systematic effects.

Moreover, as explained further in Chapter 4.3, for satellite-retrieved SST, this neat division of error/uncertainty into random and systematic components is too simplistic. In reality, there is a spectrum of sources of error with greater or lesser degrees of spatio−temporal correlation between measurements. These correlations matter if averages or trends of satellite measurements are to be formed with appropriate attached uncertainty estimates. Also, effects can be systematic when viewed at one particular scale, while at a larger scale they can actually be random.

We now look at the terms in Eqn (1) in more detail. The main components are:

- *Satellite uncertainty* (σ_1): This is the uncertainty of the satellite measurement and should be provided with the product. It will usually vary pixel by pixel and will depend on a number of parameters affecting the retrieval (see Chapter 4.3).
- *Reference uncertainty* (σ_2): This is the uncertainty of the reference measurement and is generally unknown for individual measurements but can be estimated at the data set level (see next section).
- *Geophysical uncertainty: spatial—surface* (σ_3): This is the uncertainty arising from comparing a point measurement to a satellite footprint. For any single match-up it is systematic but can be considered to be pseudo-random for a large enough data set where the full variability of the ocean has been sampled. This term can also be reduced through averaging of the satellite data (e.g., sample 5 by 5 instead of 1 by 1) as suggested by Embury et al. [11], and by necessity includes any uncertainty in geolocation of both the satellite and the reference measurement (which for drifters can be up to 8 km when, as is common, location reports are available to only one decimal place in latitude and longitude [32]).
- *Geophysical uncertainty: spatial—depth* (σ_4): This is the uncertainty arising from comparing SSTs at different depths. Again, it is systematic for a single match-up and can be reduced by using a model of the upper ocean temperature as demonstrated by Embury et al. [11]. Likewise it can be considered pseudo-random for a large enough data set whereby the full variability of the atmosphere and ocean has been sampled.
- *Geophysical uncertainty: temporal* (σ_5): This is the uncertainty arising from comparing SSTs at different times of day (it is not always possible to get exact temporal matches between satellite and reference). This will be systematic for single match-up and may be pseudo-random for a large data set if the time differences are reduced with a model as demonstrated by Embury et al. [11]. The size of the time window is important as if it is too large then one differentially samples both heating and cooling of the ocean across its diurnal cycle [11], which will then result in a bias in the calculated statistics.

A further consideration is the actual statistical analysis itself. In a Type A statistical analysis you are fitting a model to the available data. Unfortunately this approach has poor resilience to outliers, i.e. data points that do not conform to the normal distribution. Outliers arise due to any spatial and temporal mis-match errors, and also issues with satellite or reference data quality. In particular, issues with imperfect cloud screening can lead to large errors in the retrieved SST (e.g., [33]). Consequently we expect to find a certain number of outliers in our analysis that we must (1) ensure do not overtly distort the final uncertainties, and (2) estimate in magnitude and, if possible, eliminate using clearly defined criteria.

There are several approaches to evaluate the impact of outliers on the calculated statistics. An accepted way for single sensor assessment is to first reject all match-ups outside of three-sigma from the mean (3 × the calculated standard deviation, SD) before recalculating the statistics. A three-sigma filter is somewhat arbitrary but can help to estimate the level of outliers in the data, and has been used widely in AATSR validation (for example see [34]). A critical drawback with this approach is that if there are no outliers in the actual data, then a three-sigma filter would simply throw away good data and so we will not consider the three-sigma filter any further.

An alternate way to dealing with outliers is to adopt the approach of Merchant and Harris [35] and use robust statistics. In robust statistics the influence of outliers is reduced by using a median estimator (M-est) and median absolute deviation (MAD) of the distribution, the latter being scaled to the equivalent of the normal SD and referred to as the robust standard deviation (RSD). Finally, there is the approach of fitting a Gaussian PDF to a histogram of the differences between satellite and a corresponding reference data set. This latter approach is most similar to fitting a model for a Type A uncertainty estimation.

Each of these methods can be investigated further by calculating them individually for a set of match-up data, and comparing the results to the normal standard deviation. In Figure 1 we show a typical distribution of match-up differences between the NOAA-19 AVHRR and drifting buoys (in blue) in waters around Australia. Over-plotted on Figure 1 are Gaussian PDFs from normal statistics (in green), robust statistics (in orange) and a linear least square fit of a Gaussian PDF to the data (in red).

It is clear from Figure 1 that for our example distribution (with significant outliers) that the conventional statistics (green line) significantly overestimates the standard deviation of the distribution of the majority of data. The best representations for the central part of the distribution (within one-sigma) are the linear fit (red line) and the robust statistics (orange line). From this discussion, it is clear that a distribution plot to ensure the chosen statistical model is appropriate for the data must accompany any statistical analysis.

2.3 Reference Data Sources

There are several potential reference data sources that can be used for assessing the quality of an SST CDR. Ship-borne radiometers (see Chapter 3.2) are routinely calibrated and intercalibrated and provide an estimate of the temperature at a defined depth (the skin temperature). Unfortunately their sampling is regional and sparse, which sometimes leads to difficulties in relating a single radiometer observation to the SST value from the

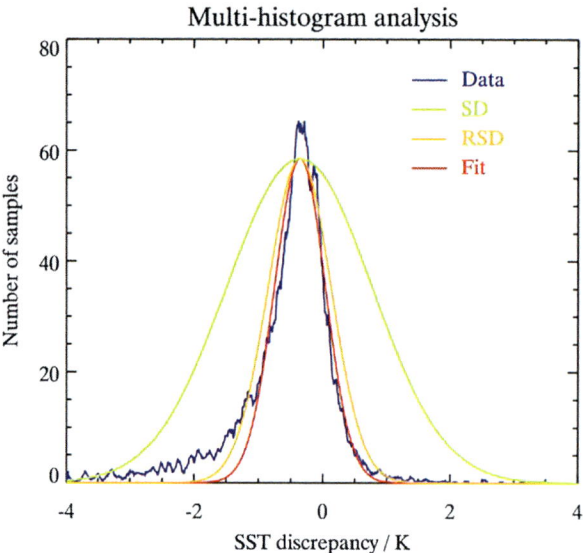

FIGURE 1 Distribution (in blue) of differences between the NOAA-19 AVHRR and drifting buoys for Australian waters. Over plotted are Gaussian PDFs from (in green) normal statistics, (in orange) robust statistics and (in red) a linear least square fit. The extreme outliers beyond the scale used in the plot force the width of the green distribution.

corresponding satellite pixel (the match-up). Although the small number of available radiometers mean that their coverage is limited compared to that offered by drifters, they are a vital source of reference data for CDR assessment as they provide the only reference data that are routinely traceable to national metrological standards (see Chapter 5.2 for further information). Our aim here is not to carry out an extensive review of other available reference sources (see Chapter 5.2) but to focus on their key aspects for SST CDR assessment.

The drifting buoy network [36] provides greatly improved coverage compared to radiometers, except in the Southern Ocean and a few other areas such as upwelling zones. The coverage of the network is good because new devices are continually deployed but, unfortunately, they are not adequately calibrated prior to deployment, and very few are recovered and recalibrated at the end of their deployment. Consequently the overall accuracy of drifters is the subject of much debate and it is currently estimated to be ~ 0.2 K [29]. Also, an obvious limitation is the fact that there is no knowledge or control over the depth at which the measurement is being made on account of the inevitable vertical motion of the buoys resulting from modulation by surface waves.

The addition of Argo profiling float [37] data to satellite SST validation is an important new development as although the floats measure SST_{depth} in a profile typically no higher than ~3 m below the ocean surface, the intrinsic high accuracy of an Argo float (~0.05 K) means the total match-up uncertainty from comparing Argo SST_{depth} to satellite SST_{skin} is comparable to that of radiometers (which is a SST_{skin} to SST_{skin} comparison) at night with a wind speed > 6 ms^{-1} and is considerably lower than both radiometers and drifters during the day (due to dynamic thermal stratification of the surface water). A new generation of Argo floats is being tested with additional high frequency sampling in the upper 4 m of the ocean surface [38]. The main issue with Argo is, like radiometers, their sampling, as each float typically only surfaces once every 10 days (although different sampling cycles can be used), and the uppermost temperature measurement has to be taken within the temporal window of the satellite overpass.

Other potential reference data include moored buoys and conventional ship measurements from ship engine room intakes or hull-mounted sensors. The GTMBA array [24] in the tropics is considered separately from other moored buoys because they are in the open ocean and far from the coastal regions which often present particular difficulties for the accurate measurements of SST from space, and where most other moored buoys are deployed. A summary of all the primary reference data sets available for SST CDR assessment, and their estimated uncertainties, is given in Table 1, [39–41, 29]

In addition to the uncertainty of each reference data set we also have to consider their temporal and spatial coverage, which has changed considerably over the years through both changes in the number and location of deployments and in sampling and reporting frequency. Figure 2 shows the temporal variation in the total number of available monthly reference data points for SST CDR assessment for the four reference data sets listed in Table 1 as well as for Voluntary Observing Ships (VOS).

In Figure 2 the variation over time is considerable, with VOS measurements staying reasonably constant but deceasing slightly over time, whereas the number of drifting buoys has increased significantly since the year 2000 but has since declined. Although small in number, but increasing in time, the high quality (see Table 1) measurements from Argo are an important reference data set, particularly moving forward in time.

3. ASSESSMENT OF LONG-TERM SST DATASETS

The importance of specific quantitative measures relevant to the climate quality aspects of SST data sets cannot be overemphasized, as these quantitative measures need to be comparable between the different sets of assessment information. In this section we mainly look at ways to assess how likely the SST at a given location may differ from the truth on average, i.e., representing the uncertainty associated with effects that are systematic.

TABLE 1 Reference Datasets Available for SST CDR Assessment and their Estimated Uncertainties

Data Type	Year	Coverage	SST[b]	Uncertainty	References
Ship-borne IR radiometers	1998	Repeated tracks in the Caribbean Sea, North Atlantic Ocean, North Pacific Ocean, and the Bay of Biscay; episodic deployments elsewhere in the world's oceans.	SSTskin	0.10 K	[39]
Argo floats	2000	Global[a] from ~2004 onwards.	SST-5 m	0.05 K	[40]
GTMBA	1979	Tropical Pacific Ocean array completed in 1998; tropical Atlantic and Indian Ocean arrays installed later.	SST-1 m	0.10 K	[41]
Drifting buoys	1981	Global[a] from ~2000 onwards.	SST-20 cm	0.20 K	[29]

[a]Data are not truly "global" but cover the majority of Earth's oceans.
[b]Depths of SST are indicative, and often the actual depth for a single measurement is unknown.

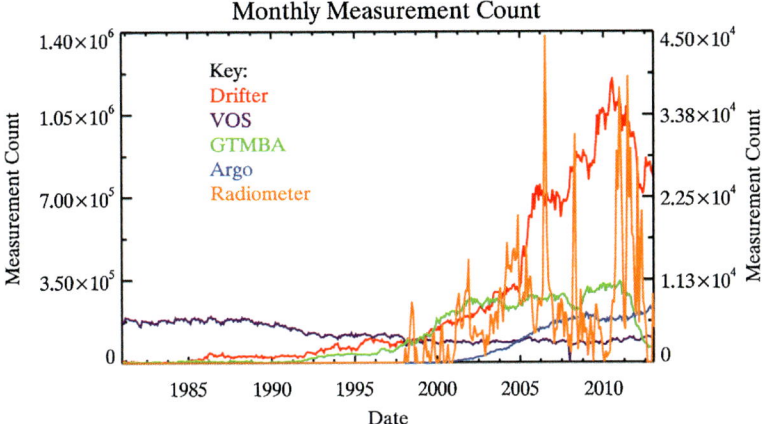

FIGURE 2 Temporal variation in total number of available monthly reference measurements for SST CDR assessment from 1981 to 2013. Counts for drifters, Voluntary Observing Ships (VOS) and GTMBA are shown on the primary y-axis and for Argo and radiometers on the secondary (right-hand) y-axis.

However, any such assessment has three important caveats that the validation scientist should be aware of:

1. As we have seen there are no globally distributed reference data with negligible errors against which satellite SST biases can be validated, and so we must use all available validation data;
2. Any averaging of SST must be done on appropriate space and time scales to avoid loss of the desired signal;
3. For comparability, strict independence of satellite and reference data is needed, which may restrict the comparisons that can be generated.

The Global Climate Observing System (GCOS) [16] statement of requirements for satellite SST is that "accuracy" should attain "0.1 K over 100 km scales", noting that "some ... datasets may approach 0.1 K accuracy on a global average basis but have biases >0.5 K for many important regions". This GCOS requirement therefore appears to be a statement about the acceptable systematic uncertainty. The relevant space-scale is 100 km (roughly 1° in latitude and longitude at the equator), according to GCOS. The present density of validation values available (mainly drifting buoys) is arguably insufficient to support assessment at this space scale, particularly prior to 2004. Merchant et al. [7] have argued that satellite SST biases can be assessed using the current array of drifting buoys on space scales of 1000 km (roughly 10° at the equator), and the real limit is likely to be somewhere in-between and geographically variable.

Assessments on coarser space scales can be made using other validation data sets. For example, the uppermost SST measurements (typically at ~4 m depth) from Argo profiling floats, which provide acceptable global coverage from roughly 2004 onwards (see Figure 2). In addition, smaller regional scale assessments can be carried out using the GTMBA [24] and the long-term shipborne radiometer deployments in the Caribbean Sea [42] and the Bay of Biscay [28]. The importance of these latter data sets should be noted as none of these data are currently used in the retrieval of satellite SSTs and are therefore remain fully independent.

3.1 Example 1: Long-Term SST Data Record Assessment

In our first example, we choose to examine the ESA SST_CCI ATSR v1.0 data set, which comprises data covering three different missions:

- Data from the AATSR on Envisat covering the period from 2002 to 2010.
- Data from the ATSR on European Remote Sensing (ERS)-2 covering the period from 1995 to 2003.
- Data from the ATSR on ERS-1 covering the period from 1991 to 1997.

A key user requirement for these data was to provide an estimate of the SST at the same depth as that measured by drifting buoys [1] so the project provides two SSTs, SST_{skin}, and $SST_{0.2m}$ in each product. It is important that both SSTs are assessed, independently using different reference data sets, which includes validating both the SST retrieval and also the model used to adjust SST_{skin} to $SST_{0.2m}$.

The spatial distribution of the temperature differences between the ESA SST_CCI ATSR V1.0 data set and drifting buoys is shown in Figure 3, which includes both the latitude/longitude variation and time/latitude variation for daytime and nighttime match-ups, respectively. The spatial resolution of the data cells is 5° in latitude and longitude, which is of order 500 km at the Equator and roughly 250 km at 60° N and S. The results are generated using a spatial limit of nearest pixel (i.e. distance <0.05°) and a time window of ±4 h. The ATSR SSTs, as retrieved from the TOA BTs, are SST_{skin} retrievals but to compare them to drifting buoys they have been adjusted to $SST_{0.2m}$ (an equivalent depth to that of the drifter measurements). In addition, an SST adjustment is also made for the difference between the drifter and satellite measurement times in order to use a larger time window (±4 h) than suggested in Chapter 5.2. To account for both of these adjustments you need to calculate two differential temperatures using a combined skin and diurnal variability model (Fairall/Kantha-Clayson, FKC [43, 44]), as demonstrated by Embury et al. [11], driven by ECMWF ERA Interim [45] fluxes interpolated to the time and place of the measurement. By making these adjustments we are minimizing σ_4 and σ_5 in Eqn (1) such that their magnitudes are <<0.1 K.

The resulting spatial maps in Figure 3 show that acceptable global coverage can be achieved for the entire ATSR mission at 5° spatial resolution (upper) and that the coverage is acceptable across all longitudes at a temporal resolution of one month from around 2002 onwards (lower). Prior to 2002 the

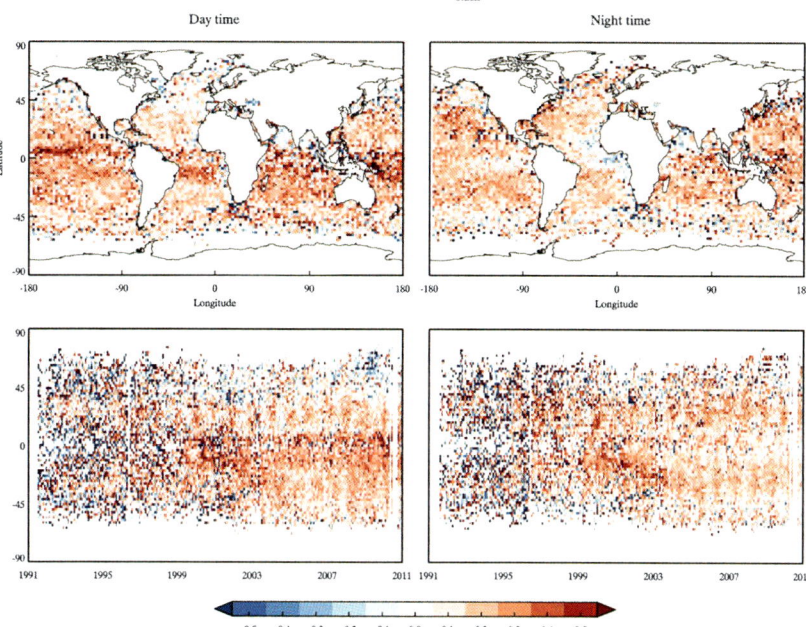

FIGURE 3 (Upper) Latitude/longitude variation of the median discrepancy for the ATSR mission compared to drifting buoys for (left) daytime and (right) nighttime, and (lower) time/latitude variation of the same statistical measure. The drifting buoy measurement has been adjusted to the satellite measurement time using the FKC model as described in the text.

data are much noisier in the time/longitude plot, which for the ATSR-2 period (1995–2002) at least is due to a reduction in match-ups from the much lower number of drifters available at this time (Figure 2). The noise in the ATSR-1 period (1991–1995) is a combination of a low number of match-ups and increased retrieval noise as a consequence of the failure of the ATSR-1 3.7 μm channel. It is clear that these data are sufficient to calculate the systematic uncertainty as defined as the RSD of the median in each cell (as proposed by the GHRSST CDAF [2]) at 1° spatial resolution.

The maps in Figure 3 show a general warm offset compared to drifters for the ESA SST_CCI ATSR V1.0 data set and have good agreement between day and night results (it is important to determine any residual impact of diurnal variability on the long-term record). There is some visible evidence of residual effects of tropospheric mineral dust in the Atlantic and NW Indian oceans (revealed as a cool bias) as well as warm offsets in regions of predominantly cirrus clouds in the daytime (across the tropics), and a cool daytime bias in high northern latitudes, particularly in the Pacific. Some of these small scale features would be masked if the resolution of the cells was too coarse and so

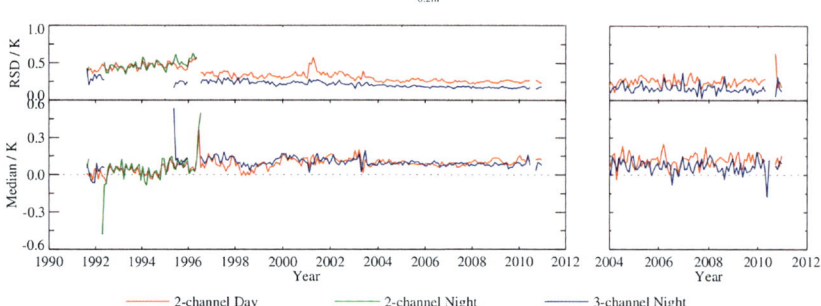

FIGURE 4 (Left) Time series of (lower) median discrepancy and (upper) robust standard deviation (RSD) for the SST_CCI ATSR mission compared to drifting buoys. Results are shown for daytime (red) and night time (blue) match-ups. Also, shown (right) is the equivalent time series for SST_CCI AVHRR compared to Argo.

one should carefully consider what spatial resolution to use, which may actually be an iterative process, taking into account the available coverage of the reference data. In any event, a standardized assessment is needed to directly compare results between data sets so such maps may have to be generated at multiple resolutions.

A time series of the ESA SST_CCI ATSR data set compared to drifting buoys can also be generated, as well as for the Argo data set when it is available, as shown in Figure 4.

The time series in Figure 4 has three regimes as identified by the changeover between ATSR-1 and ATSR-2 in 1995 and the separation between the second and third regime being identified by an occurrence towards the end of 2003 (presumed to result from a significant change in the availability of drifter measurements; Figure 2). Globally, there is good agreement between day time and night time relative biases, with a clear difference in the robust standard deviation between day time and night time: this is expected due to the increased retrieval noise of a two-channel retrieval (mainly daytime) compared to a three-channel retrieval (only nighttime), as well as differences in the cloud masking between day and night (as a result of different spectral channels being used).

3.2 Example 2: Long-Term Component Assessment

In our second example, we have chosen to examine a subset of a long-term SST data record, in this case just the AATSR sensor that was used in the full ESA SST_CCI ATSR V1.0 data set. Figure 5 shows the dependence of the median and RSD between AATSR SST_{skin} and drifter $SST_{0.2m}$ discrepancies as a function of latitude, time difference, year, total column water vapor, wind speed, solar zenith angle, across-track position, total uncertainty and

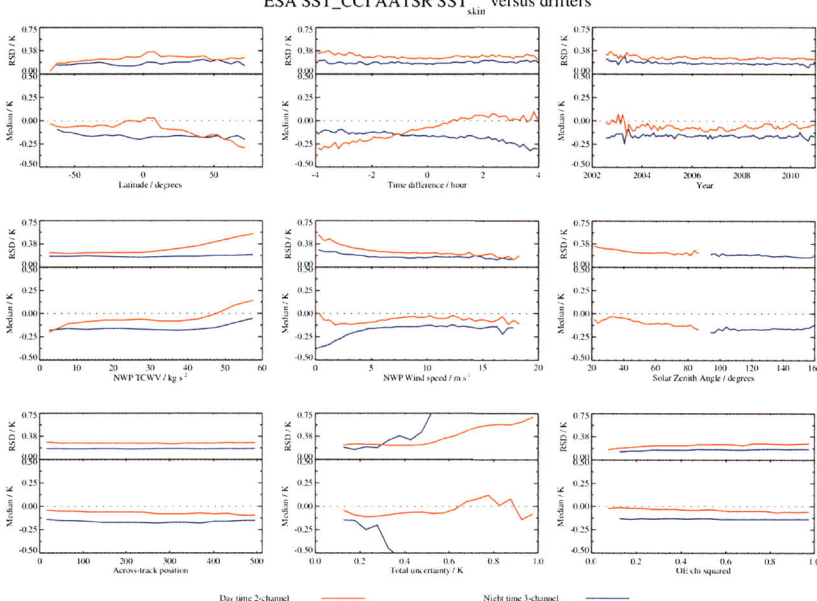

FIGURE 5 Dependence of the median and robust standard deviation between AATSR SST_{skin} and drifter $SST_{0.2m}$ discrepancies as a function of latitude, time difference, year, total column water vapor, wind speed, solar zenith angle, across-track position, total uncertainty, and retrieval chi squared function. Daytime results are shown in red and night time results are shown in blue.

the retrieval chi squared function. Daytime results are shown in red and nighttime results are shown in blue. Figure 5 shows results for the minimum set of parameters that one would wish to look at for an IR sensor assessment as they cover the main features known to affect a retrieved SST (see Chapter 4.2 for further details).

In Figure 5 we see two notable features. First, there are differences between the daytime and night time results as a function of wind speed as expected for a comparison between SST_{skin} and SST_{depth}. Second, we see a dependence on time difference as reported by Embury et al. [11] due to the average diurnal cycle sampled by the MD. These effects impact the features observed in the other dependence plots so it is not advisable to draw any conclusions from them. When looking at individual sensors of a long-term SST data record it is essential to examine both adjusted and nonadjusted comparisons of the raw satellite SSTs to the various reference data sets as one can use the suggestion of Donlon et al. [30] and simply use a 6 ms^{-1} wind speed filter to segregate the data into low and high wind speed regimes. Match-ups at wind speeds >6 ms^{-1} will be free from large diurnal signals and variability of the skin effect and one would then expect to see identical results for daytime and nighttime match-ups. The dependence on wind speed

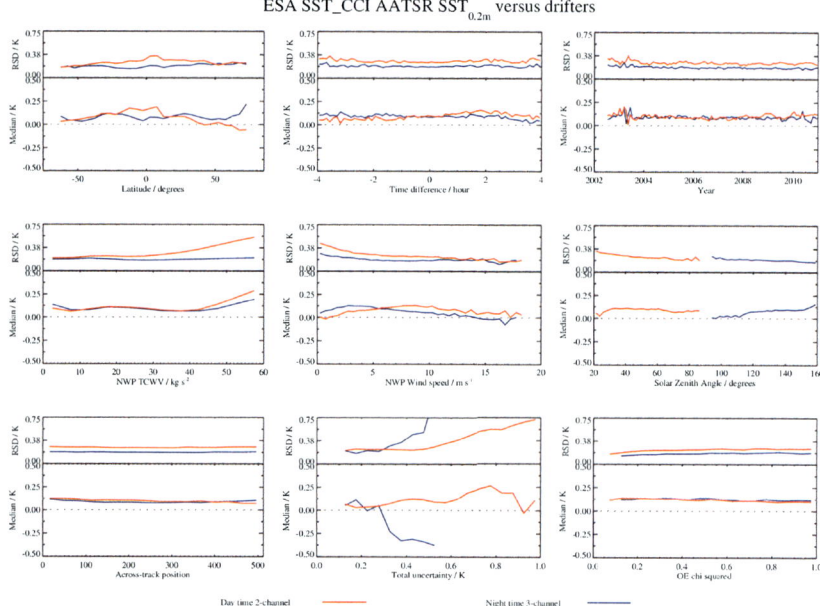

FIGURE 6 Dependence of the median and robust standard deviation between AATSR $SST_{0.2m}$ and drifter $SST_{0.2m}$ discrepancies as a function of latitude, time difference, year, total column water vapor, wind speed, solar zenith angle, across-track position, total uncertainty and retrieval chi squared function. Day time results are shown in red and night time results are shown in blue. An additional adjustment has been made using a combined diurnal variability/skin model to account for the difference in time between the satellite and drifter measurements (satellite at 10:30 am/pm local solar time).

in Figure 5 indicates an offset between the daytime and nighttime results in the region >6 ms^{-1} suggesting a bias between the 2-channel and 3-channel retrievals.

Figure 6 shows the same data as Figure 5 but after a combined skin effect/diurnal variability model (Fairall/Kantha-Clayson; [43, 44]) has been used to account for depth and time differences between the AATSR SST_{skin} and drifter $SST_{0.2m}$ (a key user requirement for SST_CCI). In Figure 6 we see that the dependence on wind speed and time difference has been reduced so we can have greater confidence in our interpretation of the other residual dependences. For this type of analysis it is vital that both the SST_{skin} and SST_{depth} comparisons are considered independently to ensure that the variability of the thermal skin over the day is not simply smoothed into the retrieval. We did not show this in Section 3.1.1 as any cool skin effect or diurnal variability effects present (due to unaccounted SST_{skin} and SST_{depth} differences) will likely hide any residual retrieval biases. We then recommend repeating the spatial analysis shown in Figure 3 for each component mission to investigate the spatial variation of any systematic differences.

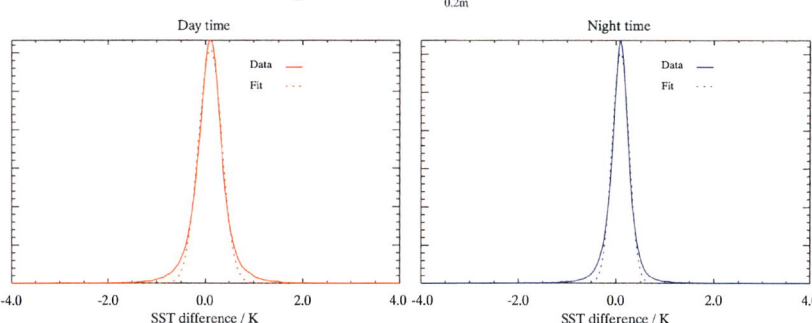

FIGURE 7 Histograms of the median discrepancy between AATSR $SST_{0.2m}$ and drifters for day time (left) and night time (right). An additional adjustment has been made using a combined diurnal variability/skin model to account for the difference in time between the satellite and drifter measurements.

As discussed in Section 2.2, it is essential that any assessment consider the shape of the distribution of the calculated differences in order to assess how well the chosen statistical model fits the data and to be able to make a quantitative statement about the impact of outliers as recommend in the GHRSST CDAF [2]. Figure 7 shows two example histograms of the median discrepancy between AATSR $SST_{0.2m}$ and drifter $SST_{0.2m}$ for daytime (left) and nighttime (right). The result of a linear least squares fit of a Gaussian PDF to the data is also shown by the dashed lines in the figure. The excellent agreement between the data and the fit implies a very low instance of outliers in this particular example.

3.3 Quantitative Metrics

Once the mainly qualitative (visual) analysis is complete (for each reference data set) several quantitative metrics can be computed. Table 2 shows example validation statistics from comparing ARC V1.0 AATSR data to various reference data sets. An adjustment has been made using a combined diurnal variability/skin model to account for differences in depth and time between the satellite and drifters, GTMBA and Argo; for radiometers only the time difference has been adjusted as they are estimates of the SST_{skin}.

In Table 2 the variation in number of match-ups between the various reference data sets is striking and also it is important to note that strictly speaking only the drifters and Argo results are directly comparable as these are "global" in nature. To correctly compare GTMBA and radiometer results, for example, one should extract "regional" drifter and Argo subsets corresponding to the spatial and temporal sampling of the GTMBA and radiometer match-ups (which is really only practical for drifters due to the low number of Argo match-ups). As well as reporting an overall systematic difference from drifting buoys,

TABLE 2 Validation Statistics from Comparing ARC AATSR $SST_{0.2m}$ to Various Reference Datasets. An additional adjustment has been made using a combined diurnal variability/skin model to account for differences in depth and time between the satellite and drifters, GTMBA and Argo; for radiometers only the time difference has been adjusted

Reference	Retrieval	Number	Median (K)	RSD (K)
Drifters	Day	670,286	+0.04	0.19
	Night	532,541	+0.02	0.17
GTMBA	Day	27,652	+0.03	0.17
	Night	20,460	+0.02	0.13
Argo	Day	7075	+0.08	0.17
	Night	3741	+0.02	0.14
Radiometers	Day	7402	+0.01	0.29
	Night	9720	+0.01	0.18

using the global median of the satellite minus drifting buoy SST difference shown in Table 2, it must be put into context regarding any geophysical difference in the type of SST. For example, if the satellite SSTs are SST_{skin} with no skin/diurnal adjustment, then a skin-effect difference of order -0.17 K is to be expected [30] in comparisons to drifting buoys. Also, only a single table covering all wind speeds is shown here, but results should be shown for both all wind speeds and for wind speeds >6 ms^{-1} separately, as discussed above.

A further measure is then to examine the geographical variation in bias (Figure 3), which can be described by the standard deviation of the satellite minus drifting buoy SST differences at a specific space scale. As discussed above, limitations in the availability of reference data greatly influences what scale this can be done over. For example, the GHRSST CDAF [2] currently recommends a scale of ~ 1000 km (averaging the data every $10°$ in latitude and longitude) for at least the period from 2005 onwards. At least for missions since around 2005, this measure should be assessable for most of the global oceans at this spatial scale.

One can then estimate the systematic difference to drifting buoys by first calculating the median ($\mu = $ median($T_{satellite}-T_{in-situ}$)) of each individual cell, and then calculating the standard deviation of μ across all cells as a measure of the systematic differences. To ensure that the results are reasonably sound statistically and to account for the uncertainty in the drifting buoys, the GHRSST CDAF [2] recommends calculating $\sigma_{cell} = (\sigma_{ref})/\sqrt{n_{cell}}$ for each cell, and then rejecting cells where $2\sigma_{cell} > 0.1$ K. This is equivalent to requiring $n_{cell} > 16$ and is expressed as such to indicate the rationale: namely that the uncertainty in the mean of the validation data should be smaller than the GCOS target with a high degree of confidence ($\sim 95\%$). It is also possible to use Argo measurements for determining the systematic difference for more recent years (2005 onwards). However, as the density of matches will be much less, the assessment has to be done on greater spatial scales.

As well as a systematic difference, the GHRSST CDAF [2] also recommends calculating a non-systematic difference, i.e. the components of uncertainty that would remain after the systematic effects have been quantified. To estimate the non-systematic uncertainty using the same data set(s) as for the systematic uncertainty, subtract the appropriate μ from each discrepancy: ($d = (x_{satellite} - x_{in-situ}) - \mu$) and then calculate the robust standard deviation of d. This process will actually overestimate the nonsystematic uncertainty, because the measure includes effects arising from the imperfect in situ observations and true geophysical variability (differences from "point to pixel" comparisons, and difference in measurement times) as discussed in Section 3.1. The overestimation is likely to be significant if the resulting value is close to about 0.2 K (for drifting buoys) or 0.1 K (for Argo).

3.4 Demonstrating Traceability to SI

It is clear that the available reference data are not ideal for assessing the quality of long-term satellite-derived SST data records as they vary considerably in availability and quality. Temperature is one of the seven base SI units and establishing traceability to SI is an essential part of any assessment. A first attempt at providing a route to SI traceability was proposed by Minnett and Corlett [46] and has subsequently evolved through a series of workshops held at the International Space Science Institute (ISSI) in Bern, Switzerland. A team of experts from the international SST community have proposed a method summarized in the flow diagram shown in Figure 8 as the main route to demonstrating SI traceability for long-term satellite-derived SST data records [47].

Figure 8 shows the key part of the process, which involves taking matchups between the satellite-derived SSTs and SI-traceable reference measurements, such as those from ship-borne IR radiometers, and comparing them to non-SI-traceable measurements such as those made by thermometers mounted on drifting buoys. The set of matchups to the ship radiometers provides traceability to SI-standards, and ideally this would be all that is required. The number of ship-borne radiometers is currently unable to provide adequate sampling of the parameter space that influences the satellite-SST retrieval and its uncertainties, except over several years' sampling, and thus the much larger data set of matchups with subsurface measurements from buoys, although with higher uncertainties and with limited traceability, has also to be used.

The key decision process hinges on whether the uncertainty budgets derived from comparisons with SI-traceable and non-SI-traceable

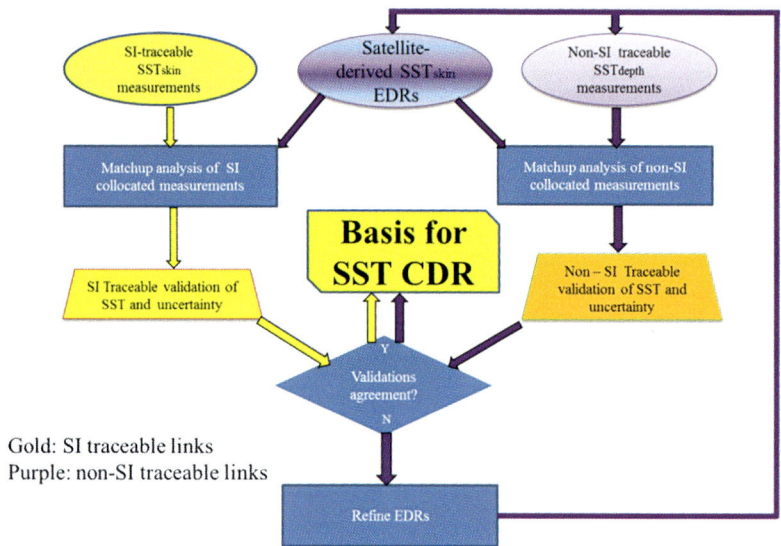

FIGURE 8 Flow diagram showing the route to demonstrating SI traceability for an SST CDR taken from an ISSI meeting report [47]. Ship-borne radiometer measurements, which are SI traceable, are included in the "match-up analysis of SI collocated measurements."

measurements are equivalent. If yes, it can be argued that the uncertainties derived from the much larger set of non-SI-traceable matchups can be used in the characterization of the long-term SST record. If not, it implies that the satellite-derived SSTs have uncertainties that are not well characterized in either sets of matchups, and that the algorithms used in deriving the satellite SSTs should be improved. Metrics for assessing the equivalence in the characteristics of the two uncertainty budgets have to be determined, taking into account the properties of the two data sets and the methods used to generate the matchups.

We can assess both SI and non-SI validation using data from the recent ESA AATSR V2.1 reprocessing, using match-ups to drifting buoys and ship-borne radiometers. A plot of the median difference between AATSR and drifting buoys (with no adjustment for depth or time of day) is shown in Figure 9 as a function of across-track position, time difference, year, latitude, wind speed and solar zenith angle. Results are shown for nadir-only (dashed) and dual-view retrievals (solid lines) and for day time two-channel (red), night time two-channel (blue) and night time 3-channel (black) cases. No adjustments for differences in depth and time are included.

In Figure 9 there is little if any dependence on across-track position or year, but a clear warming/cooling signal as a function of time difference is apparent. Also, there is some dependence on latitude (a surrogate for total column water vapor) particularly for the two-channel retrievals, and a clear

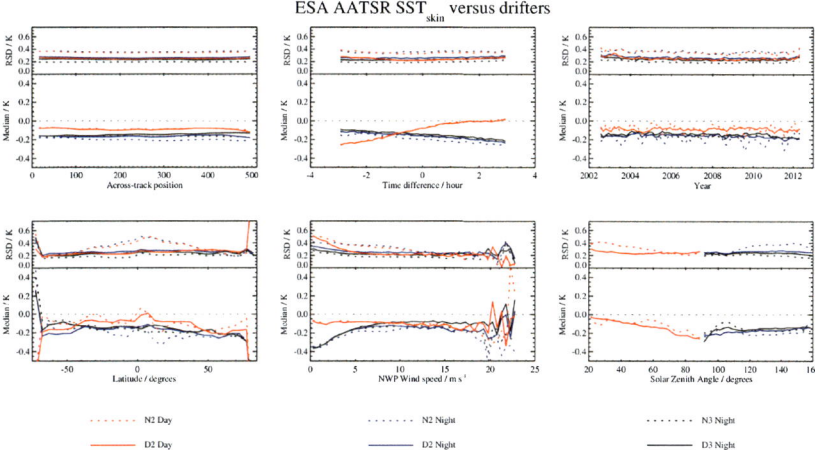

FIGURE 9 Plot of median differences between AATSR and drifting buoys for the entire AATSR mission as a function of across-track position, time difference, year, latitude, wind speed and solar zenith angle. Results are shown for nadir-only (dashed) and dual-view retrievals (solid lines) and for day time two-channel (red), night time two-channel (blue) and night time 3-channel (black) cases. No adjustments for differences in depth and time are included.

diurnal signal in the bottom three plots as one would expect for comparisons between AATSR SST_{skin} and drifter $SST_{0.2m}$. Moreover, the nighttime cool skin effect as a function of wind speed is comparable to that of Donlon et al. [30]. An equivalent plot to Figure 9 for median differences between AATSR and ship-borne radiometers (with no adjustment for time of day) is shown in Figure 10 as a function of the same six parameters. The data in Figure 10 were generated from all available radiometer data from the Calibrated InfraRed In situ Measurement System (CIRIMS), Infrared Scanning Autonomous Radiometer (ISAR), Marine-Atmospheric Emitted Radiance Interferometer (M-AERI), and Scanning Infrared Sea Surface Temperature Radiometer (SISTeR) radiometers deployed between 2002 and 2012 (see Chapter 3.2 for further information on these ship-borne radiometers).

In Figure 10 the first obvious observation is the much lower number of match-ups between AATSR and radiometers results in much noisier plots than seen in Figure 9. Indeed, one could argue that only three of the six dependences are sufficiently sampled. However, aside from the obvious dependence on time difference, one could also reasonably argue that there is no clear residual dependence on any of the other five parameters, as one would expect for comparisons between SST_{skin} and SST_{skin}. A wide time window of ±3 h has been used to generate the match-ups and so to assess properly the effect of the time difference between the satellite and ship-borne radiometer measurement, which should be minimized using a diurnal variability model (which has not been done for the example shown here).

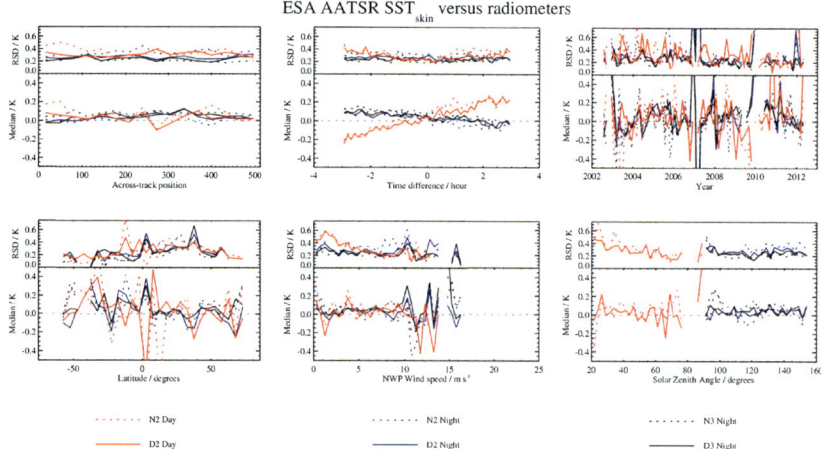

FIGURE 10 Plot of median differences between AATSR and ship-borne radiometers for the entire AATSR mission as a function of across-track position, time difference, year, latitude, wind speed and solar zenith angle. Results are shown for nadir-only (dashed) and dual-view retrievals (solid lines) and for day time two-channel (red), night time two-channel (blue) and night time 3-channel (black) cases. No adjustments for differences in depth and time are included.

The latitude dependence plot in Figure 10 is interesting in that it is least noisy in the Northern Hemisphere in the regions of the two long-term ship-borne radiometer deployments, the M-AERI deployment on the *Explorer of the Seas* in the Caribbean and the ISAR deployment on the *Pride of Bilbao* in the English Channel/Bay of Biscay. Clearly, the best possible radiometer match-ups are going to be obtained in these regions of repetitive sampling and actually these deployments offer the best possible source of all reference data for these regions as a result. However, this means it is critical that such deployments are maintained across any data gaps and instrument overlaps within a single SST CDR as otherwise these radiometric matches will not be consistent. The measurements for any given region should also ideally be undertaken with the same design of ship-borne radiometer to minimize any technological differences.

When considering what is needed for reference data it is clear that it should be sufficient to "sample" the complete measurement space and the critical factor here is what the measurement space actually is. The measurement space should cover the dependence on key parameters, such as those in Figures 5, 6, 9 and 10, and it should cover as much the entire global ocean as in the example given in Figure 3. Furthermore, it should be possible to sample the measurement space at regular intervals (ideally monthly) to identify any seasonal biases. This is challenging owing to the large variability in available reference data over time.

3.5 Stability

The calculation of the long-term stability of satellite based SST data records is a mandatory step in the assessment of the data. For SST, we define stability as the degree of invariance over time of the mean error from systematic effects. GCOS requires a stability of "<0.03 K/decade over 100 km scales" [16]. To assess stability we need a long-term reference data set of known (and ideally) higher stability than that which is being assessed.

So far only one assessment of stability capable of being informative at the level of the GCOS requirement has been published. In Merchant et al. [8] the stability of the long-term ARC record was assessed relative to components of the GTMBA [24] over a 20-year period. Merchant et al. [8] concluded that over the period 1994 to 2010 that regionally collocated ARC and GTMBA SSTs are stable, with better than 95% confidence, to within 0.005 K year^{-1}. As ARC and GTMBA are two independent data sets it is reasonable to assume that the stability of 0.005 K year^{-1} determined by Merchant et al. [8] is an upper limit on the stability of the data sets individually for at least 1994 to 2010 in that specific region. As such, we have high confidence in the stability of the GTMBA for assessment of other SST data sets across this period in the specific region sampled by the GTMBA. We attribute the high stability of the GTMBA buoys to their routine maintenance, and, crucially, their pre- and postdeployment calibration to SI [25]. However, as noted by Merchant et al. [8] stability can only be directly assessed for equatorial latitudes using the GTMBA. Other regions do not yet have adequate in situ data characterization in terms of uncertainty to establish trends although this is an on-going area for improvement and research.

Other options for stability assessment include the drifting buoy network, but it is not known to be stable to the GCOS level (at least there are no published assessments) and it does not have spatial distributions that are stable in time. As such, drifting buoys cannot be used with confidence at the level of the GCOS stability requirement, although time-series of satellite-buoy differences (as shown in Figure 4) can still be calculated in order to look at likely instability in the satellite record outside of tropical latitudes.

The Argo network of profiling floats has sensors stated to be of very high accuracy and stability [40], but as yet covers too brief a period to allow a rigorous assessment of decadal stability, and is also subject to sparse sampling and regional distributions (as the floats are often entrained into ocean dynamical structures). The earliest the network is arguably complete (i.e., capable of providing global match-ups) is 2004 and techniques needed for exploiting the Argo network for stability assessment requires further study owing to (1) its relatively short lifetime and (2) its low coverage over time at each location (so the data cannot be deseasonalized for example). Indeed, Argo is likely to provide the first global assessment of SST stability and consequently research into how to exploit the Argo data for stability is an urgent area to address.

The utility of ship-borne radiometers (1998 onwards) in areas of repeat ship tracks for stability assessment has not yet been established and is another area for on-going research. Chapter 5.2 lays out the criteria and requirements for such a network. Radiometers may have a unique advantage above other reference data sets for stability assessment as uncertainties can be provided per measurement (see Chapter 5.2), thus avoiding the need to use large numbers to reduce some of the uncertainty of the comparison between a point measurement and a satellite footprint.

The GTMBA moorings provide consistent SSTs in a defined geographical region across the whole time period of available long-term satellite based SST data sets, with the number of measurements available increasing over time due to (1) changes in reporting frequency (e.g., every hour to every minute) and (2) further deployments (e.g., addition of the Prediction and Research Moored Array in the Atlantic, PIRATA, and Research Moored Array for African-Asian-Australian Monsoon Analysis and Prediction, RAMA, arrays to the original Tropical Atmosphere Ocean, TAO, array). In Merchant et al. [8] each GTMBA location used in their stability assessment had a minimum coverage of 120 months, with five or more match-ups between ARC and buoy over the period 1991−2009. In addition, only GTMBA locations that passed strict quality control procedures were used.

The monthly mean composite time series for daytime and nighttime ARC $SST_{1.0m}$ data minus co-located GTMBA data is shown in Figure 11 (extracted from Merchant et al. [8]). The retrieved satellite SST_{skin} measurements are adjusted to $SST_{1.0m}$ to be "equivalent" to the SSTs measured by the GTMBA moorings. The 95% confidence intervals for the trends are −0.0026 to

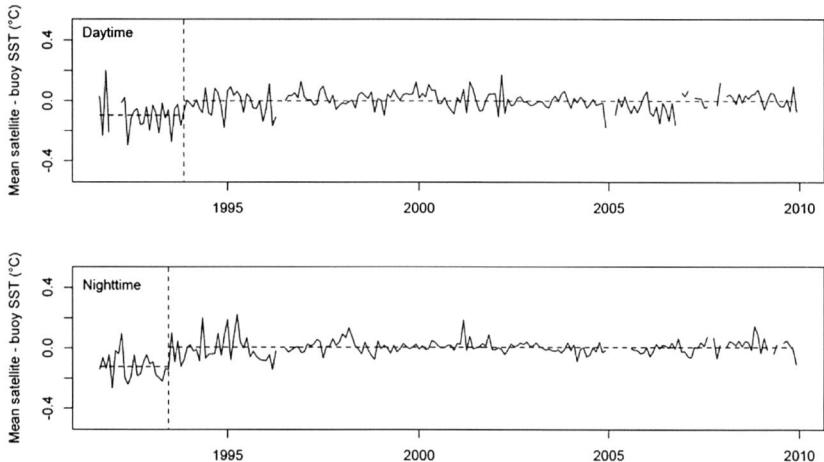

FIGURE 11 Time series of monthly mean SST differences between ARC and GTMBA SSTs. *(Taken from Merchant et al. [8])*. See text for further details and explanation of plot.

0.0015 K year^{-1} (daytime) and −0.0018−0.0019 K year^{-1} (nighttime). These results suggest that the ARC SSTs (and the GTMBA SSTs) meet the target stability defined by GCOS [16] for an area of the tropics from 1994 onwards.

The GTMBA data used by Merchant et al. [8] were extracted from the International Comprehensive Ocean-Atmosphere Data Set (ICOADS) V2.5 data set [48], which contains GTMBA data mainly obtained via the GTS system. The analysis is highly dependent on the long-term continuity of the GTMBA, particularly at sites with the longest available data records. Analysis that uses delayed mode GTMBA data is also possible and one would expect a larger range of sites to be used compared to Merchant et al. [8] as the improved reporting frequency will reduce the uncertainty at individual locations.

The use of the GTMBA for stability assessment is now an essential part of activities to analyze long-term satellite derived SST records. As such, the GHRSST CDAF [2] has such an analysis included as a core activity. The methodology for stability assessment defined in the GHRSST CDAF is simplified relative to the approach of Merchant et al. [8] and has been trialled within the ESA SST_CCI project [15]. The idea of the CDAF is to use a simplified—but scientifically robust—method that can easily be applied across multiple data sets to provide comparable metrics.

The three available ESA SST_CCI data sets (AVHRR, ATSR and analysis) have been individually matched to offline GTMBA data extracted from for the time period (1991−2010). The AVHRR data set uses data from the NOAA-12, -14, -15, -16, -17, -18, and Meteorological Operational (MetOp) A platforms. The high-temporal resolution GTMBA data have a sampling resolution of 5, 10 or 60 min and the highest available time-resolution is always used if multiple resolutions are available. The GTMBA data are matched to the nearest SST_CCI pixel centre and a maximum time difference of 30 min is used as a threshold (afforded by using the offline GTMBA data).

Following the initial match-up process the monthly median ESA SST_CCI minus GTMBA difference for each GTMBA location was calculated. Then for each month of the year and location, the multiyear average of the monthly median ESA SST_CCI minus GTMBA differences was calculated. For each month the data were then deseasonalized by subtracting the multiyear average for the appropriate month of the year from each month of the time series. For the AVHRR and ATSR data sets separate multiyear averages were used for day and night. The data are deseasonalized to minimize any potential aliasing of any annual cycle in residual time series following the approach of Merchant et al. [8]. Retaining at this point only locations where buoy data were available for >15 years within the 1991−2010 period, the monthly mean difference across all locations was determined to end up with a single ESA SST_CCI minus GTMBA SST time series for each ESA SST_CCI data set (and for day and night for AVHRR and ATSR). A least squares linear fit to each time series of monthly mean differences was calculated and 95% confidence intervals were determined.

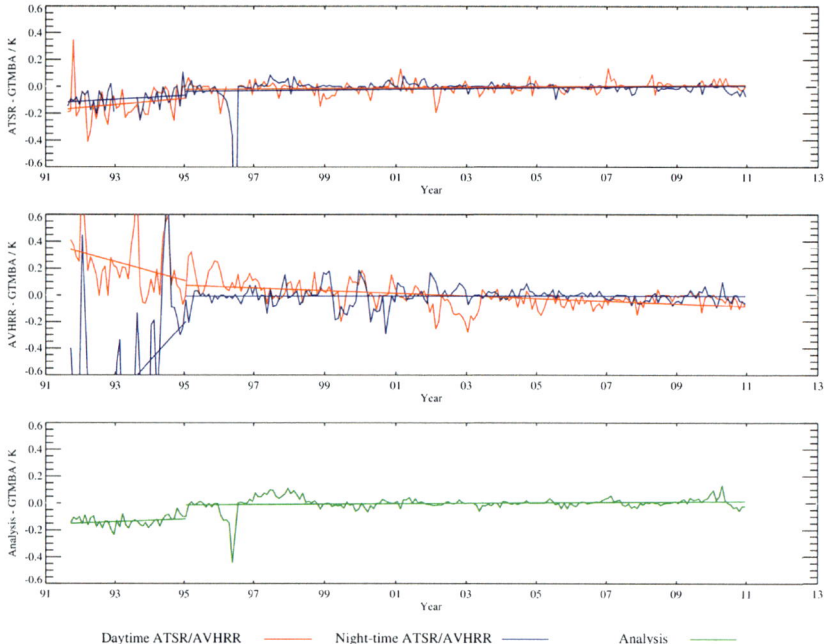

FIGURE 12 Time series of deseasonalized composite monthly mean differences (K) between the SST_CCI products and the GTMBA. Separate day and night time series are provided for the L2P AVHRR and L3U ATSR data sets. Also, plotted are the results of a least squares linear fit for the 1991 to May 1995 and June 1995 to 2010 periods (see text for further discussion).

The results from the SST_CCI stability assessment are shown in Figure 12. A step-change is apparent from 1995 onwards, which is most likely due to the switch between ATSR-1 and ATSR-2. As the ATSRs were used to bias correct the radiances for the AVHRRs the feature is apparent in all time series. Consequently, the 95% confidence interval on the slope of the fit was calculated for two separate periods, 1991 to May 1995 covering the ATSR-1 period and June 1995 to 2010 covering the ATSR-2/AATSR period. The resulting 95% confidence intervals for the least squares linear fits to the time series of ESA SST_CCI minus GTMBA differences are summarized in Table 3.

For the ATSR product, the night time trend in the differences to the GTMBA measurements for the 1995–2010 period is comparable to that calculated by Merchant et al. [8]. However, the daytime stability confidence interval does not include zero, and relative to [8] is somewhat less stable; nonetheless, the true stability is still likely to be within the GCOS requirements [16]. For the ATSR-1 period (1991–1994), both the day and night trends calculated for the ATSR product have improved stability (based on the most likely relative trend)

TABLE 3 Summary of 95% Confidence Intervals for Least Squares Linear Fits to ESA SST_CCI Minus GTMBA Monthly Mean Difference Time Series for 1991 to May 1995 and June 1995 to 2010

ESA SST_CCI 95% Confidence Interval (mK year^{-1}) for 1991–1995

	Day	Night	Both
AVHRR	−137.9 < Trend < −2.4	105.9 < Trend < 462.3	
ATSR	−13.6 < Trend < 60.1	−7.4 < Trend < 36.8	
Analysis			−1.8 < Trend < 22.1

ESA SST_CCI 95% confidence interval (mK year^{-1}) for 1995–2010

	Day	Night	Both
AVHRR	−12.3 < Trend < −7.4	−2.0 < Trend < 2.0	
ATSR	0.7 < Trend < 3.2	−1.4 < Trend < 6.4	
Analysis			0.1 < Trend < 3.2

compared to that reported in Merchant et al. [8], although there is nonetheless likely to be a positive trend artifact that is outside the GCOS target.

Regarding the AVHRR product there is no comparable analysis in the literature for precursor data sets. We note that, as for the ATSR product, the daytime stability is poorer than for nighttime. This may reflect the greater amplification of error in two-channel relative to three-channel SST retrieval that is common to all IR sensors and retrieval methods. The AVHRR time series in Figure 12 appears to have a step improvement in interannual stability from 2003/4. The cause of this is not clear, as this does not correspond to a changeover between sensors within the time series. Being tied to the ATSR calibration, the analysis product has a stability over the period 1995–2010 that likely meets the GCOS requirement (the confidence interval is mostly within the interval −3 to +3 mK/year).

Unlike those applied by in Merchant et al. [8], step-detection techniques are not specified in the GHRSST CDAF [2], as these require significant resources and expertise to implement. A consequence, of not using step-detection techniques is that step changes (such as the change between ATSR-1 and ATSR-2 evident in Figure 13) have to be identified visually/subjectively, with a corresponding chance of both steps being missed and steps being falsely imputed. Development of step-detection methods, ideally applicable across CDR domains to identify correlations between geophysical variables, is very much needed.

668 Optical Radiometry for Ocean Climate Measurements

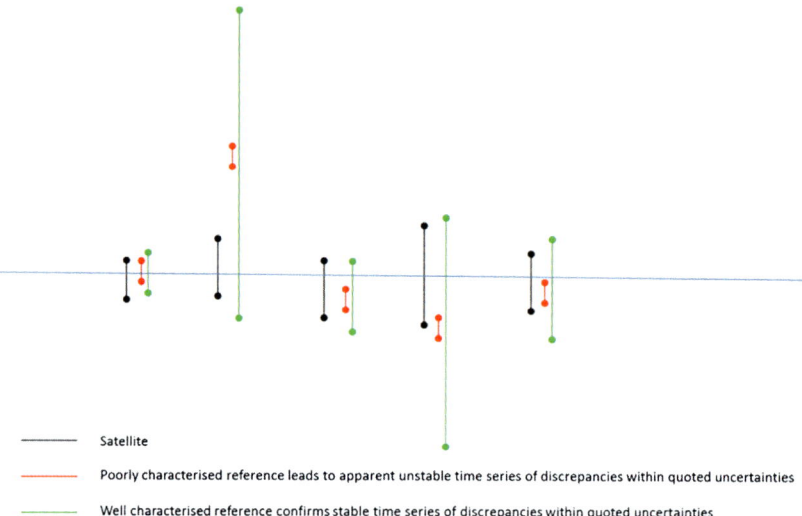

FIGURE 13 Schematic example of three time series of match-ups between satellite and reference data shown as uncertainties (the dispersion of the values that could reasonably be attributed to the measurand). The satellite data is shown in black and two different instances of the reference data are shown in red, for a poorly characterized reference data set whereby a mean uncertainty is used for each measurement, and in green, for a well-characterized reference data set in which each measurement has its own uncertainty.

The approaches presented so far for stability assessment rely on (1) estimating the stability of the reference data set and (2) use the rule of large numbers to reduce the uncertainty due to random effects. If we had a reference data set with known uncertainties per measurement then we would not have to use the rule of large match-ups for assessment, but instead could use individual measurements with their uncertainties and assess their degree of equivalence as shown schematically in Figure 13. In Figure 13 we show three hypothetical data sets (1) in black, the satellite data, (2) in red, a poorly characterized reference data set whereby a mean uncertainty for the entire data set has to be used for each measurement, and (3) in green, a well characterized reference data set in which each measurement has its own uncertainty. For our idealized case one can see that any stability assessment carried out between the black and red points would lead one to conclude the satellite time series has a very low stability, whereas a stability assessment carried out between the black and green points would lead one to conclude the satellite time series has a very good stability when the measurement uncertainties are considered.

So far none of the available reference data sets can provide individual measurement uncertainties. However, efforts are well underway to quantify them for ship-borne radiometers, as summarized in Chapter 5.2.

3.6 Validation of Uncertainties

Given the significant variations in available reference data (in space, time and accuracy) the only pragmatic approach to providing uncertainties for a satellite SST CDR is to generate the uncertainties using an uncertainty model, which itself needs to be validated. A key step in evaluating CDRs of SST is therefore to assess both the SST and its quoted uncertainty. The principal approach to validation of uncertainties is to examine the distribution of satellite-reference SST differences as a function of uncertainty. This method has been trialed by Lean and Saunders [49] in their assessment of the ARC data set.

In an ideal case, the standard deviation of the differences between the satellite SST and a reference SST would equal the satellite uncertainty,

$$\sigma_{sat-ref} = \sigma_{sat} \qquad (2)$$

However, the reference data has its own uncertainties to consider, as discussed in Section 2.3. Consequently the standard deviation of the differences between the satellite SST and a reference SST is really a combination of both the uncertainty in the satellite SST and the uncertainty in the reference SST,

$$\sigma_{sat-ref} = \sqrt{\sigma_{sat}^2 + \sigma_{ref}^2} \qquad (3)$$

As we discussed in Section 2.2, there are of course the other terms to consider relating to:

- The difference in spatial sampling (a point reference measurement versus a satellite pixel);
- The difference in depth of the measurements;
- The difference in time of the measurements.

Such an approach naturally considers the uncertainty due to environmental effects related to the homogeneity of a region/process. For example, validation in a region dominated by strong SST fronts at low wind speed will mean the uncertainty due to spatial sampling will be systematic for any one single match-up. However, as the number of match-ups increases the uncertainty will reduce by $1/\sqrt{N}$ as you sample the variability at multiple locations (unless you always sample on one side of a front, say). Consequently, the effect is considered to be a pseudo-random term across a set of validation data and not systematic. Likewise, in an area of strong solar radiation and low wind speed, the uncertainty due to difference in depth would be systematic for any one match-up. Therefore, these latter terms can be reduced to being <<0.1 K in the mean through the use of a depth/time adjustment, a large number of match-ups (to reduce pseudo-random terms), and through like versus like (SST_{skin} versus SST_{skin} or SST_{depth} versus SST_{depth}) comparisons.

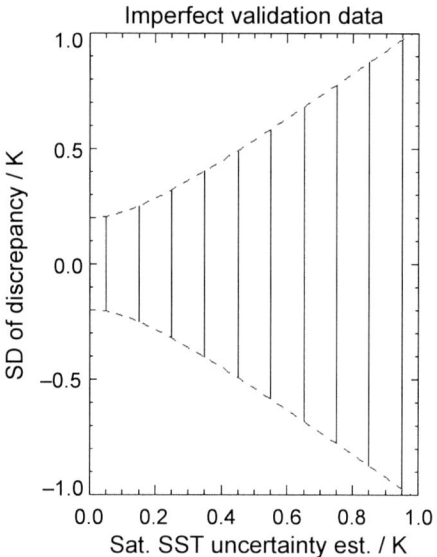

FIGURE 14 Idealized Uncertainty Validation plot, assuming validation against data with Gaussian errors and an SD of 0.2 K. Vertical lines span -1 to $+1$ times the SD of discrepancy, for data binned into 0.1 K bins of estimated satellite SST uncertainty. When the satellite SST uncertainty is small, the SD of discrepancy is dominated by the in situ uncertainty. For large satellite SST uncertainty, the SD of discrepancy approaches the estimated uncertainty of the satellite. The Dotted line gives the locus of the results if the satellite SST uncertainty is perfectly estimated. Deviations from the dotted line indicate biases in uncertainty estimation.

An idealized Uncertainty Validation plot, assuming validation against data with Gaussian errors and an SD of 0.2 K, is shown in Figure 14. Vertical lines span -1 to $+1$ times the SD of discrepancy, for data binned into 0.1 K bins of the estimated satellite SST uncertainty. When the satellite SST uncertainty is small, the SD of discrepancy is dominated by the in situ uncertainty. For large satellite SST uncertainty, the SD of discrepancy approaches the estimated uncertainty of the satellite. The Dotted line gives the locus of the results if the satellite SST uncertainty is perfectly estimated. Deviations from the dotted line indicate biases in uncertainty estimation.

One can easily see in Figure 14 that at low satellite uncertainties the SD of the differences is dominated by the uncertainty in the reference data and as satellite uncertainties grow the satellite uncertainty dominates the statistics, as the reference uncertainty becomes a less significant contribution to the total uncertainty. In fact, the uncertainty of the reference data can dominate the statistics at low satellite uncertainties meaning it may not be possible to validate the uncertainty model once this limit has been reached. Also, it is clear that as uncertainties are added in quadrature, the geophysical terms assumed to be small will be more significant at lower satellite uncertainties and a "geophysical limit" will be present even for reference data with uncertainties <<0.1 K.

To demonstrate this approach with a practical example, we show results for the uncertainty validation for the ESA SST_CCI analysis data set in Figure 15. The spread of uncertainties in the products is from ~0.05 to 1.5 K and the agreement between the theoretical and measured RSD values is excellent across the full range of uncertainties. Some divergence is seen for uncertainties above 1.2 K but the increase in spread of the standard error indicates a low number of match-ups at these levels. The standard error is reasonably consistent although there is a slight non-linearity at low uncertainties of 0.05 K and a small cold bias at uncertainties of 0.6 K and above.

The provision and validation of product uncertainties derived from a physical-based uncertainty model is deemed essential for the assessment of any long-term SST record owing to the variations in reference data over time; arguably it is essential for any long-term geophysical measurement record. However, we still have to find a way to provide confidence in the measurand and its associated uncertainties even when no reference data are available. This is the idea behind "functional" validation, which uses knowledge gained from comparisons to reference data to provide an assessment in the regions with no reference data.

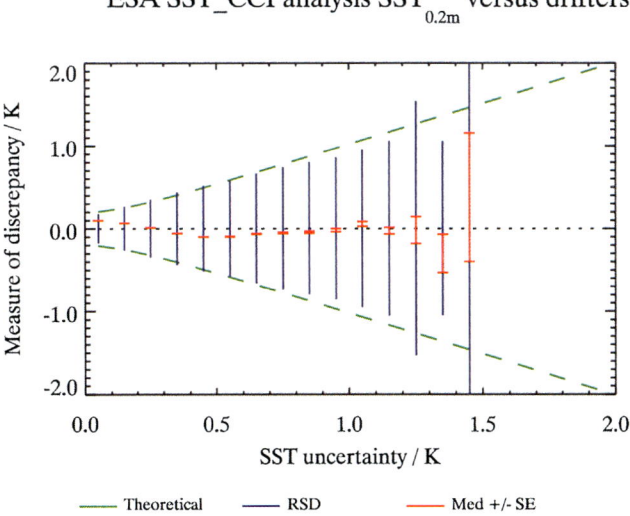

FIGURE 15 Plot of ESA SST_CCI analysis product uncertainty against the RSD of the discrepancies between the analysis and drifting buoys. The green lines indicated the theoretical dispersion of uncertainties assuming an average drifter buoy measurement uncertainty of 0.2 K. The blue lines indicated the measured dispersion for each uncertainty level. The red lines indicate the standard error for each uncertainty level and also provide an indication of the number of match-ups.

Within the ESA SST_CCI project [14] an attempt was made to provide verification maps to indicate where independent verification of product uncertainties could provide confidence that they were of the right order of magnitude. The verification maps were generated by computing the percentage difference between the calculated and theoretical RSD of the differences between the various ESA SST_CCI data sets and drifting buoys. The comparisons were carried out across the full range of uncertainties at each 15 degrees of latitude and longitude (taking into account uncertainties in the drifting buoy data). The median % difference for each latitude/longitude cell is then scaled to give an indication of verification according to:

- Very high—uncertainties are confirmed to be within 20% of their quoted values
- High—uncertainties are confirmed to be within 20–40% of their quoted values
- Medium—uncertainties are confirmed to be within 40–60% of their quoted values
- Low—uncertainties are confirmed to be within 60–80% of their quoted values
- Very low—uncertainties are confirmed to be within 80–100% of their quoted values

A "not verifiable" category was also included where it was not possible to independently verify the product uncertainties using the reference data set. It is important to recognize that this does not mean that the product uncertainties should not be used in these cases, just that they could not be confirmed independently.

An example uncertainty verification map for the ESA SST_CCI analysis data set is shown in Figure 16. The coverage is very good with very few unverified regions. On average the uncertainties are of high quality compared to the reference data set, and in general, regions of medium and low quality occur in areas that contain low numbers of drifting buoys.

The initial attempt at uncertainty verification from the ESA SST_CCI project has its limitations in that the distribution of sea ice has not been factored in, so verification results in Polar Regions are going to be lower than they actually are. Also, the maps do not distinguish between cases where (1) the degree of verification is low due to a low number of match-ups from cases where (2) the degree of the verification is low due to the measured uncertainties being overestimated or underestimated.

Nevertheless these maps can provide users with a unique assessment of the product uncertainty quality and will allow them to scale the product uncertainties should they wish to do so in regions of low degree of verification. A method for implementing "functional" verification can then be added using match-ups with similar dependence on Total Column Water Vapour, Solar Zenith Angle, or Aerosol Optical Depth, etc. that can be identified

ESA SST_CCI analysis SST$_{0.2m}$ versus drifters

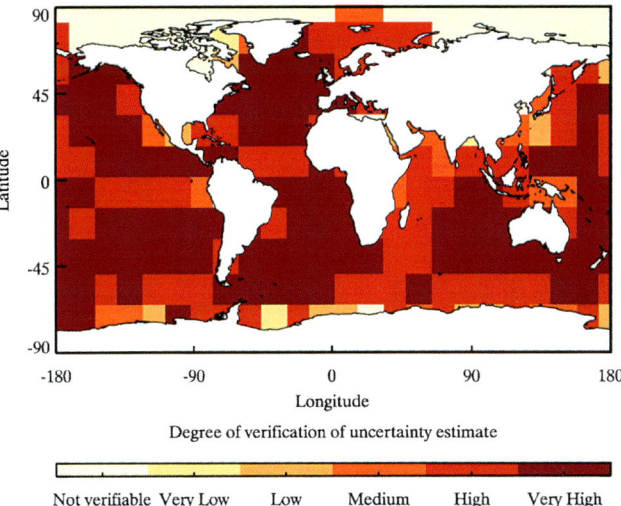

FIGURE 16 Verification maps for OSTIA SST$_{depth}$ uncertainties assessed using drifting buoy SST$_{depth}$. This plot shows the degree to which the SST CCI product uncertainties can be verified using independent reference data. It should not be taken as an indication of SST CCI product data quality and is intended to help the user interpret their own results from applying product uncertainties in their analysis.

using dummy locations (i.e. no reference data). It is reasonable to assume the uncertainty model is correlated between such locations and therefore the uncertainties are of the right order of magnitude even though this cannot be independently verified.

4. SUMMARY AND RECOMMENDATIONS

In this chapter we have shown that an ocean thermal infrared SST CDR is a challenge in many regards, with multiple solutions that vary in space, time and quality, and on the application. The assessment of thermal infrared ocean CDRs is an ongoing area of active research and the development of quantitative metrics that allow a user to decide which data are right for them is recommended. We have presented a review of the problems encountered when carrying out a CDR assessment and have presented current methodologies for the assessment itself. Careful consideration of both the primary measurand and its associated uncertainty is essential and harmonization of satellite records through a considered analysis of overlaps between constituent sensors is mandatory. Significant variations in the quality and coverage of available reference data mean approaches that use uncertainty models, which are validated, are recommended. Assessment methods should "follow the physics" (e.g., should compare like with like) and a validation uncertainty budget should be defined covering all known effects. Determining the stability of the CDR is a fundamental requirement and the development of practical step-detection methods that are applicable to a range of CDRs is a recommended research priority. New methods to establish SI traceability are needed, especially for non-radiometric reference data, and a better understanding of the uncertainties of the various reference data is required.

REFERENCES

[1] S. Good, N. Rayner, SST CCI User Requirements Document, 2010. Project Document SST_CCI-URD-UKMO-001 Issue 2, Available from: www.esa-sst-cci.org.
[2] C.J. Merchant, J. Mittaz, G.K. Corlett, Group for High Resolution Sea Surface Temperature (GHRSST) Climate Data Assessment Framework (CDAF), 2014. Project Document CDR-TAG_CDAF Version 1.0.4, Available from: www.ghrsst.org.
[3] L.J. Allison, J. Kennedy, An Evaluation of Sea Surface Temperature as Measured by the Nimbus I High Resolution Infrared Radiometer, NASA Technical Note No. D-4078, National Aeronautics and Space Administration, Washington, D.C, 1967.
[4] K.S. Casey, T.B. Brandon, P. Cornillon, R. Evans, The past, present, and future of the AVHRR pathfinder SST program, in: V. Barale, J. Gower, L. Alberotanza (Eds.), Oceanography from Space, Springer, Netherlands, Dordrecht, 2010, pp. 273–287. ISBN 978-90-481-8681-5.
[5] K.A. Kilpatrick, G.P. Podestá, R. Evans, Overview of the NOAA/NASA advanced very high resolution radiometer pathfinder algorithm for sea surface temperature and associated matchup database, J. Geophys. Res. 106 (2001) 9179–9197.
[6] J.P.D. Mittaz, A.R. Harris, J.T. Sullivan, A physical method for the calibration of the AVHRR/3 thermal IR channels 1: the prelaunch calibration data, J. Atmos. Oceanic Technol. 26 (2009) 996–1019.
[7] C.J. Merchant, A.R. Harris, H. Roquet, P. Le Borgne, Retrieval characteristics of non-linear sea surface temperature from the advanced very high resolution radiometer, Geophys. Res. Lett. 36 (2009) L17604.
[8] C.J. Merchant, O. Embury, N.A. Rayner, D.I. Berry, G.K. Corlett, K. Lean, K.L. Veal, E.C. Kent, D.T. Llewellyn-Jones, J.J. Remedios, R. Saunders, A 20 year independent record of sea surface temperature for climate from along-track scanning radiometers, J. Geophys. Res. 117 (2012) C12013.
[9] O. Embury, C.J. Merchant, M.J. Filipiak, A reprocessing for climate of sea surface temperature from the along-track scanning radiometers: basis in radiative transfer, Remote Sens. Environ. 116 (2012) 32–46.
[10] O. Embury, C. J.Merchant, A reprocessing for climate of sea surface temperature from the along-track scanning radiometers: a new retrieval scheme, Remote Sens. Environ. 116 (2012) 47–61.
[11] O. Embury, C. J.Merchant, G.K. Corlett, A reprocessing for climate of sea surface temperature from the along-track scanning radiometers: initial validation, accounting for skin and diurnal variability effects, Remote Sens. Environ. 116 (2012) 62–78.
[12] O. Embury, C.J. Merchant, A reprocessing for climate of sea surface temperature from the along-track scanning radiometers: harmonisation of satellite datasets into a single record, Remote Sens. Environ. (2014). Submitted for publication.

[13] C.J. Merchant, O. Embury, J. Roberts-Jones, E. Fiedler, C.E. Bulgin, G.K. Corlett, S. Good, A. McLaren, N. Rayner, S. Morak-Bozzio, C. Donlon, Sea surface temperature datasets for climate applications from phase 1 of the European Space Agency Climate Change Initiative (SST CCI), Geosci. Data J. (2014). http://dx.doi.org/10.1002/gdj3.20

[14] G. Corlett, C. Atkinson, N. Rayner, S. Good, E. Fiedler, A. McLaren, J. Hoeyer, C. Bulgin, SST CCI Product Validation and Intercomparison Report, 2014. Project Document SST_CCI-PVIR-UoL-001 Issue 1, Available from: www.esa-sst-cci.org.

[15] N. Rayner, J. Kennedy, C. Atkinson, T. Graham, E. Fiedler, A. McLaren, G. Corlett, SST CCI Climate Assessment Report, 2014. Project Document SST_CCI-CAR-UKMO-001 Issue 1, Available from: www.esa-sst-cci.org.

[16] GCOS-154, Systematic Observation Requirements for Satellite-Based Products for Climate, supplemental details to the satellite-based component of the "Implementation plan for the Global Observing System for Climate in Support of the UNFCC, 2011, 138 pp.

[17] J.R. Schott, W.J. Volchok, Thematic mapper thermal infrared calibration, Photogramm. Eng. Rem. S. 51 (1985) 1351−1357.

[18] S.J. Hook, K. Okada, In-flight wavelength correction of thermal infrared multispectral scanner (TIMS) data acquired from the ER-2, IEEE T. Geosci. Remote 34 (1996) 179−188.

[19] M. Goldberg, G. Ohring, J. Butler, C. Cao, R. Datla, D. Doelling, V. Gärtner, T. Hewison, B. Iacovazzi, D. Kim, T. Kurino, J. Lafeuille, P. Minnis, D. Renaut, J. Schmetz, D. Tobin, L. Wang, F. Weng, X. Wu, F. Yu, P. Zhang, T. Zhu, The global space-based inter-calibration system, B. Am. Meteorol. Soc. 92 (4) (2011).

[20] X. Liang, A. Ignatov, Monitoring of IR clear-sky radiances over oceans for SST (MICROS), J. Atmos. Ocean. Tech. 28 (10) (2011).

[21] S.J. Hook, A.J. Prata, R.E. Alley, A. Abtahi, R.C. Richards, S.G. Schladow, S.Ó. Pálmarsson, Retrieval of lake bulk-and skin-temperatures using along track scanning radiometer (ATSR) data: a case study using Lake Tahoe, CA, J. Atmos. Ocean. Tech. 20 (2003) 534−548.

[22] S.J. Hook, R.G. Vaughan, H. Tonooka, S.G. Schladow, Absolute radiometric in-flight validation of mid infrared and thermal infrared data from ASTER and MODIS on the terra spacecraft using the Lake Tahoe, CA/NV, USA, automated validation site, IEEE T. Geosci. Remote 45 (2007) 1798−1807.

[23] S.J. Hook, G. Chander, J.A. Barsi, R.E. Alley, A. Abtahi, F.D. Palluconi, B.L. Markham, R.C. Richards, S.G. Schladow, D.L. Helder, In-flight validation and recovery of water surface temperature with landsat 5 thermal infrared data using an automated high altitude lake validation site at Lake Tahoe CA/NV, USA, IEEE T. Geosci. Remote 42 (2004) 2767−2776.

[24] M.J. McPhaden, K. Ando, B. Bourlès, H.P. Freitag, R. Lumpkin, Y. Masumoto, V.S.N. Murty, P. Nobre, M. Ravichandran, J. Vialard, D. Vousden, W. Yu, The global tropical moored buoy array, in: Proceedings of OceanObs 9, 2010.

[25] H.P. Freitag, T.A. Sawatzky, K.B. Ronnholm, M.J. McPhaden, Calibration Procedures and Instrumental Accuracy Estimates of Next Generation Atlas Water Temperature and Pressure Measurements, 2005. NOAA Technical Memorandum OAR PMEL-128.

[26] GHRSST Science Team, The Recommended GHRSST Data Specification (GDS) 2.0 Document Revision 5, 2012. Available from: http://www.ghrsst.org.

[27] P.J. Minnett, Consequences of sea surface temperature variability on the validation and applications of satellite measurements, J. Geophys. Res. 96 (1991) 18,475−18,489.

[28] W. Wimmer, I.S. Robinson, C.J. Donlon, Long-term validation of AATSR SST data products using shipborne radiometry in the Bay of Biscay and English channel, Remote Sens. Environ. 116 (2012) 17−31.

[29] A.G. O'Carroll, J.R. Eyre, R.W. Saunders, Three-way error analysis between AATSR, AMSR-E, and in situ sea surface temperature observations, J. Atmos. Ocean. Tech. 25 (2008) 1197−1207. http://dx.doi.org/10.1175/2007JTECHO542.1.

[30] C.J. Donlon, P.J. Minnett, C. Gentemann, T.J. Nightingale, I.J. Barton, B. Ward, M.J. Murray, Toward improved validation of satellite sea surface skin temperature measurements for climate research, J. Climate 15 (2002) 353−369.

[31] Bureau International des Poids et Mesures, Guide to the Expression of Uncertainty in Measurement (GUM), JCGM 100:2008, 2008. Available online at: http://www.bipm.org/en/publications/guides/gum.html.

[32] W.J. Emery, D.J. Baldwin, P. Schlüssel, R.W. Reynolds, Accuracy of in situ sea surface temperatures used to calibrate infrared satellite measurements, J. Geophys. Res. 106 (2001) 2387−2405.

[33] C.J. Merchant, A.R. Harris, E. Maturi, S. Maccallum, Probabilistic physically based cloud screening of satellite infrared imagery for operational sea surface temperature retrieval, Q. J. R. Meteorol. Soc. 131 (2005) 2735−2755.

[34] A.G. O'Carroll, J.G. Watts, L.A. Horrocks, R.W. Saunders, N.A. Rayner, Validation of the AATSR meteo product sea-surface temperature, J. Atmos. Ocean. Technol. 23 (2005) 711−726.

[35] C.J. Merchant, A.R. Harris, Toward the elimination of bias in satellite retrievals of sea surface temperature: 2. Comparison with in situ measurements, J. Geophys. Res.: Oceans 104 (C10) (1999) 23579−23590.

[36] R. Lumpkin, M. Pazos, Measuring surface currents with surface velocity program drifters: the instrument, its data, and some recent results, in A. Griffa, A.D. Kirwan, A.J. Mariano, T. Ozgokmen, T. Rossby (Eds.), Lagrangian Analysis and Prediction of Coastal and Ocean Dynamics (LAPCOS), 2007, p. 500.

[37] D. Roemmich, G.C. Johnson, S. Riser, R. Davis, J. Gilson, W.B. Owens, S.L. Garzoli, C. Schmid, M. Ignaszewski, The argo program: observing the global ocean with profiling floats, Oceanography 22 (2009) 34−43.

[38] J.E. Anderson, S. Riser, Near-surface variability of temperature and salinity in the near tropical ocean: observations from profiling floats, Journal of Geophysical Research, 2014. Submitted for publication.

[39] I.J. Barton, P.J. Minnett, K.A. Maillet, C.J. Donlon, S.J. Hook, A.T. Jessup, T.J. Nightingale, The Miami2001 infrared radiometer calibration and intercomparison. Part II: shipboard results, J. Atmos. Ocean. Technol. 21 (2004) 268−283.

[40] E. Oka, K. Ando, Stability of temperature and conductivity sensors of argo profiling floats, J. Oceanogr. 60 (2) (2004) 253−258.

[41] J.J. Kennedy, R.O. Smith, N.A. Rayner, Using AATSR data to assess the quality of in situ sea-surface temperature observations for climate studies, Remote Sens. Environ. 116 (2012) 79−92.

[42] E.J. Noyes, P.J. Minnett, J.J. Remedios, G.K. Corlett, S.A. Good, D.T. Llewellyn-Jones, The accuracy of the AATSR sea surface temperatures in the caribbean, Remote Sens. Environ. 101 (2006) 38−51.

[43] C. Fairall, E. Bradley, J. Godfrey, G. Wick, J. Edson, G. Young, Cool-skin and warm-layer effects on sea surface temperature, J. Geophys. Res. 101 (C1) (1996) 1295−1308.

[44] L.H. Kantha, C.A. Clayson, An improved mixed layer model for geophysical applications, J. Geophys. Res. 99 (C12) (1994) 25235−25266.

[45] A. Simmons, S. Uppala, D. Dee, S. Kobayashi, ERA-Interim: new ECMWF reanalysis products from 1989 onwards, ECMWF Newslett. 110 (2006) 26–35.
[46] P.J. Minnett, G.K. Corlett, A pathway to generating climate data records of sea-surface temperature from satellite measurements, Deep-sea Res. Part II 77–80 (2012) 44–51.
[47] P.J. Minnett, G.K. Corlett, International Teams in Space Science, Generation of Climate Data Records of Sea-surface Temperature from Current and Future Satellite Radiometers, 2012. Report of the Second Workshop, available from, http://www.issibern.ch/teams/satradio/.
[48] S.D. Woodruff, S.J. Worley, S.J. Lubker, Z. Ji, J.E. Freeman, D.I. Berry, P. Brohan, E.C. Kent, R.W. Reynolds, S.R. Smith, C. Wilkinson, ICOADS release 2.5: extensions and enhancements to the surface marine meteorological archive, Int. J. Climatol. 31 (2011) 951–967.
[49] K. Lean, R.W. Saunders, Validation of the ATSR reprocessing for climate (ARC) dataset using data from drifting buoys and a three-way error analysis, J. Climate 26 (2013) 13.

Index

Note: Page numbers followed by "f" and "t" indicate figures and tables respectively

A

A-Si. *See* Amorphous silicon
AAOT. *See* Acqua Alta Oceanographic Tower
AATSR. *See* Advanced Along Track Scanning Radiometer
Above water methods, 281
  See also In water methods
  modeling, 282
  sky-light polarization, 281–282
  superstructure perturbations, 282
  wave perturbations, 281
Above water systems, 267–268
  See also In water systems
Absolute response, 263–264
Accuracy, 43, 549
Acqua Alta Oceanographic Tower (AAOT), 279, 611
ADC. *See* Analog-to-digital converters
Adding–doubling method, 464
ADF. *See* Angular distribution function
Advanced Along Track Scanning Radiometer (AATSR), 155, 585
  CASOTS-I radiance blackbody, 586–587
  dual-view two-channel SST, 575
  radiometric calibration, 204
    blackbody cross-over test, 204
    blackbody temperatures and IR channel blackbody signals, 208f
    blackbody thermometer readings, 206t
    NEΔT, 209, 211t
    orbital variation, 205f
    preflight calibration activities, 204
    results for, 204–206
    temperature uncertainties, 208f
    TIR channels, 209
    trends for, 207f
Advanced Spaceborne Thermal Emission and Reflection Radiometer (ASTER), 225–227, 368
Advanced Very High Resolution Radiometer (AVHRR), 69, 165–166, 209–211, 640–641
  calibration uncertainties, 212
  changes in temperature, 211f
  magnitudes of temperature variations, 211–212, 212f
  NOAA polar-orbiting satellites, 212–213
  prelaunch calibration and characterization, 213
  product, 667
AERONET-OC. *See* Ocean-Color component of the Aerosol Robotic Network
Aerosol
  models, 464–466
  spectral reflectance, 466–467
AIRS. *See* Atmospheric Infrared Sounder
Along track scanning radiometer (ATSR), 161, 203, 641
  AATSR optical configuration, 163
  ATSR-2, 368
  AVHRR bands, 161
  calibration configuration, 203
  conical scanning geometry, 163
  design feature, 162–163
  detector signals, 163–164
  IR channels, 163
  scanning geometry, 163f
  spectral bands, 162t
Aluminum (Al), 343
Amorphous silicon (A-Si), 335
Analog-to-digital converters (ADC), 89–90, 353–354
Angular distribution function (ADF), 421
Anti-reflection (AR), 336
Antimony (Sb), 332–333
ANX. *See* Ascending Node Crossing
Application uncertainty, 564
AR. *See* Anti-reflection
ARC project. *See* ATSR Reprocessing for Climate project
Argo
  network, 663
  program, 236
Ascending Node Crossing (ANX), 165–166
ASTER. *See* Advanced Spaceborne Thermal Emission and Reflection Radiometer

679

680 Index

Atmosphere
  brightness temperature assessment, 641–642
  correction, 227
  diffuse transmittance, 470–471
  path radiance contributions
    aerosol models, 464–466
    aerosol spectral reflectance, 466–467
    radiative transfer simulation, 464
    single-scattering epsilon, 467f
  path reflectance, 468–470
Atmospheric Infrared Sounder (AIRS), 223
Atmospheric transmission window. *See* Split window
ATSR. *See* Along Track Scanning Radiometer
ATSR Reprocessing for Climate project (ARC project), 641
Auto-self protection system, 353–354
AVHRR. *See* Advanced Very High Resolution Radiometer

B

Bad pixel replacement (BPR), 391–393
Ball Experimental SST radiometer (BESST radiometer), 322–324, 342–343
Band pass filters, 338
Band shift correction, 622–624
Bayesian classifier, 230
BBAR coatings. *See* Broadband antireflective coatings
Beam positioning, 345–350
Beam shaping, 341–345
BESST radiometer. *See* Ball Experimental SST radiometer
Biconical reflectance, 25–26
Bidirectional reflectance distribution function (BRDF), 16, 36, 86–87, 122–123, 457
  absolute, 28
  bidirectional reflectance, 24–25
  diffuse, 26
  ocean BRDF effects
    in-water BRDF effects, 463–464
    ocean surface, 463
  reflection properties, 24
  of seawater, 28
Bidirectional reflectance factor, 27, 37–38
Bio-optical
  buoys, 265
  modeling, 7, 289–291, 547
Bio-Optical mapping of Marine Properties (BiOMaP), 537–538

Biofouling, 279–280
BiOMaP. *See* Bio-Optical mapping of Marine Properties
Bismuth (Bi), 332–333
Blackbody calibration, 182
  emissivity, 183–184, 185t
  thermometry, 182–183
Blackbody
  cross-over test, 204
  sources, 30–32
Blooming, 99
Bouée pour l'acquisition de Séries Optiques à Long Terme (BOUSSOLE), 57, 534–535
BPR. *See* Bad pixel replacement
BRDF. *See* Bidirectional reflectance distribution function
Brewster's angle, 40
"Bright target recovery" problem, 99
Brightness temperature (BT), 509, 593f, 640
Broadband antireflective coatings (BBAR coatings), 337
BT. *See* Brightness temperature

C

Calibrated Infrared in situ Measurement System (CIRIMS), 322–324, 371–374, 371f, 565–566
Calibration model, 172
  nonlinearity, 174–176
  offset variations, 176
  radiometric
    errors, 173–174
    noise, 174
  two-point calibration scheme, 172–173
  uncertainty
    budget, 174, 174t
    combination, 173
Calibration system, 324, 354–355
  external stirred water bath calibration, 355–358
  NNR calibration, 361
  self calibrating radiometers, 358–361
CASOTS. *See* Concerted Action for the Study of Ocean Thermal Skin
CASOTS-I radiance blackbody, 586–587, 586f
CASOTS-II blackbody system, 589f
CDAF. *See* Climate Data Assessment Framework

Index

CDM. *See* Chromophoric dissolved organic matter and detrital particulates
CDOM. *See* Colored dissolved organic matter
CDR. *See* Climate data record
CDRP. *See* Climate Data Record program
Center for Hydro-Optics and Remote Sensing (CHORS), 539
Center wavelength (CWL), 338
CEOS. *See* Committee on Earth Observing Satellites
Check standard (CS), 50–51
Chlorophyll-a (Chla), 74, 606, 609–610
Chopping, 333–334
CHORS. *See* Center for Hydro-Optics and Remote Sensing
Chromophoric dissolved organic matter and detrital particulates (CDM), 619, 623
CIRIMS. *See* Calibrated Infrared in situ Measurement System
Clear-sky ocean pixels, 220–222
Climate Data Assessment Framework (CDAF), 640
Climate Data Record (CDR), 4, 5t, 490, 605
  assessment, 639–640
  essential climate variables to, 10–11
  for ocean color data, 121–122
  uncertainties and intermission differences, 607
  uncertainty requirements for SST, 307
Climate Data Record program (CDRP), 9
Climate signal analysis, 628–630
  See also Time series analysis
Climate-observing system
  CDRs, 6–10
  characteristics, 4–5
  ECVs, 6–10
Cloud
  cloud-free conditions, 306–307
  detection, 504–509, 507t
  screening, 227, 229–230
Coastal typical case-2 waters, 458–463
Coastal Zone Color Scanner (CZCS), 74, 248, 452–453, 621–622
  atmospheric contribution, 455
Code transition noise. *See* Input-referred noise
Coefficient of variation (CV), 439, 610–611
Colored dissolved organic matter (CDOM), 75–76, 458
Commercial radiometer "head" detectors, 335–336

Committee on Earth Observing Satellites (CEOS), 6, 587
Concerted Action for the Study of Ocean Thermal Skin (CASOTS), 586–587
Configuration factor, 21–22
Cooled long-wave cameras, 391–393
Correction schemes, 420, 426
Correlations, 58–59
Cosine error, 258
  effects, 276
Cross-mission data products comparison, 621
  band shift correction, 622–624
  climate signal analysis, 628–630
  common-bin time-series, 627f
  point-by-point comparison, 624–626
  satellite data sets comparison, 622
  SeaWiFS and MODIS-Aqua time series, 630f
  time series analysis, 626–628
CS. *See* Check standard
CV. *See* Coefficient of variation
CWL. *See* Center wavelength
CZCS. *See* Coastal Zone Color Scanner

D

Data
  acquisition, 353–354
  processing, 541–542
  reduction, 548–549
  repositories, 542–543
Degree of linear polarization (DoLP), 288
Degree of polarization (DOP), 77–78, 78f
Depolarizers, 72
Detector noise, 101
Deuterated L-alanine doped triglycine sulfate (DLATGS), 365
Diffuse attenuation coefficients, 268
Diffuse reflectance standards, 36
  BRDF, 36–37
  diffuse directional-hemispherical reflectance, 36, 38
  incident radiation, 38
  lamp-to-plaque distances, 36
Diffusers, 251
Digital object identifier (DOI), 9
Digitization noise, 102
Direct regression-based approach (DRB approach), 578
Directional-hemispherical reflectance, 25–26

Disk-integrated lunar irradiances, 129
DLATGS. *See* Deuterated L-alanine doped triglycine sulfate
DOI. *See* Digital object identifier
DoLP. *See* Degree of linear polarization
DOP. *See* Degree of polarization
Double differences approach, 220–222
DRB approach. *See* Direct regression-based approach
Drifting buoy network, 648
Dynamic two-point calibration, 372

E

Earth Observation (EO), 532
  metrology, 521
  satellites, 4, 308
Earth Observing System (EOS), 37–38
Earth Research Satellite (ERS), 652
Earth-shine plate (ESP), 186
ECMWF. *See* European Center for Medium-range Weather Forecasting
ECV. *See* Essential Climate Variable
Effective focal length (EFL), 96–98
EFL. *See* Effective focal length
Electrical substitution radiometer (ESR), 39–40
Electro-optics cameras, 266–267
Electromagnetic spectrum (EM spectrum), 306
Electrostatic discharge (ESD), 106
Ellipsoid reflectors, 344
EM spectrum. *See* Electromagnetic spectrum
End-of-life performance (EOL performance), 103–104
Environmental system, 351–353
EO. *See* Earth Observation
EOL performance. *See* End-of-life performance
EOS. *See* Earth Observing System
Error, 645
Error propagation, 517, 618–620
ERS. *See* Earth Research Satellite
ERS-1. *See* European Remote-sensing Satellite-1
ESA. *See* European Space Agency
ESA SST_CCI project, 671f, 672
ESD. *See* Electrostatic discharge
ESP. *See* Earth-shine plate
ESR. *See* Electrical substitution radiometer
Essential Climate Variable (ECV), 4, 5t, 609–610

European Center for Medium-range Weather Forecasting (ECMWF), 578–579
European Remote-sensing Satellite-1 (ERS-1), 161
European Space Agency (ESA), 9, 452–453, 639–640
External stirred water bath calibration, 355–358

F

FCDR. *See* Fundamental Climate Data Record
Felyx project, 9
Fiducial reference measurements (FRM), 245, 308, 527, 559
  See also Ship-borne radiometer network (SBRN)
  cost–benefit analysis, 562f
  for SST CDR, 559
    long-term satellite SST records, 560
    TIR SBRN, 561–563
    uncertainty budgets, 563–585
Field intercomparisons, 540–541
Field measurements, 533–534
  long-term
    comprehensive measurements, 538
    measurements, 534–536
  for system vicarious calibration, 544–546
  time-limited comprehensive measurements, 536–538
Field of view (FOV), 417–418, 570
Field radiometer design, TIR, 321–322
  beam positioning, 345–350
  beam shaping, 341–345
  calibration system, 354–361
  data acquisition, 353–354
  detector system, 328
    commercial radiometer "head" detectors, 335–336
    IR detectors, 330t
    microbolometer, 334–335
    pyroelectric detectors, 333–334
    quantum detectors, 329–332
    spectral response characteristics, 329f
    thermopile detectors, 332–333
  environmental system, 351–353
  FTM TIR SST$_{skin}$ radiometers, 361–362
  GPS, 363
  instrument control, 353–354
  scan drum/mirror arrangement, 348f
  ship borne TIR radiometers, 322–324
  spectral definition, 336–339
    TASCO THI500 radiometer, 339f

transmission characteristics, 337f
subsystems, 324
thermal control system, 350−351
Field radiometers, 548
  Hyperspectral radiometer, 249−250
  irradiance sensors, 250−252
  multispectral radiometer, 249−250
  radiance sensors, 252−254
  systems, 249
Filter radiometers, 1−2
  and HTBB for realization, 50
    irradiance measurement, 51−52
    receiving aperture, 50−51
    uncertainty evaluations, 52−57
  for validation or measurement, 46−47
    ancillary calibration measurements, 47
    effective spectral width, 49
    filter detector system, 46f
    Gershun tube radiometer, 47
    measurement equations, 50
    self-consistent approach, 49
    VXR, 47, 48f
FLIP. *See* Floating Instrument Platform
"Flip-flop" mechanism, 171
Floating Instrument Platform (FLIP), 391
Flux transfer equation, 22
Focal plane assembly (FPA), 159, 167
Forward problem, 490−491
Fourier transform interferometer (FTIR), 335−336
Fournier−Forand analytical formulation, 413, 428
FOV. *See* Field of view
FPA. *See* Focal plane assembly
FRM. *See* Fiducial reference measurements
FRM ship borne TIR radiometer
  air temperature derivation, 387−389
  CIRIMS, 371−374, 371f, 374t
  DAR-011 filter radiometer, 363−364
  ISAR, 375−380
  M-AERI interferometer, 384f
  NASA/JPL NNR radiometer, 368−371
  SISTeR filter radiometer, 364−368
  spectroradiometers, 382−387
  TIR cameras, 389−393
  UAV using BESST radiometer, 380−382
FTIR. *See* Fourier transform interferometer
Full width at half maximum (FWHM), 338
"Functional" approach, 643
Fundamental Climate Data Record (FCDR), 5t
FWHM. *See* Full width at half maximum

G

Gaussian−Poisson statistical model (GP statistical model), 434
GCOS. *See* Global Climate Observing system
Geophysical inversion process, 509
  cases for simulation, 510−511
  form of algorithm, 511
  gain matrix, 515t
  retrieval coefficients, 509−510
  SST coefficient-based retrieval, 513−514
  SST sensitivity, 516
  "traditional" retrieval, 509
Geophysical retrieval validation, 225
  argo program, 236
  atmospheric correction, 227, 230−232
  band-to-band registration, 228f
  brightness temperatures, 226f
  cloud screening, 229−230
  GTMBA, 235
  independent temperature measurements, 232
  MODIS, 227
  moored buoys, 236
  satellite SST accuracies, 233−234
  ship-board radiometers, 236−237
  sources of uncertainties, 228f
  stability of skin SSTs, 235
  standard deviation of buoy measurements, 232−233
  surface temperatures, 234
Geophysical uncertainty, 646
Geostationary orbits, 79
Germanium (Ge), 336
Gershun tube, 46−47
GHRSST. *See* Group for High Resolution Sea Surface Temperature
GLI. *See* Global Imager
Global Climate Observing System (GCOS), 6, 307, 559−560, 651
  climate-monitoring principles, 6
  requirements for ECVs and CDRs, 6, 7t
  SOCR, 7−8
  SST, 8−10
Global Imager (GLI), 74−75
Global Positioning System (GPS), 363
Global Space Based Intercalibration System (GSICS), 642
Global Tropical Moored Buoy Array (GTMBA), 235, 642
GOCI. *See* Korean Geostationary Ocean Color Imager

Gold (Au), 343
GP statistical model. *See* Gaussian—Poisson statistical model
GPS. *See* Global Positioning System
"Grid" approach, 643
Group for High Resolution Sea Surface Temperature (GHRSST), 8—9, 640
GSICS. *See* Global Space Based Intercalibration System
GTMBA. *See* Global Tropical Moored Buoy Array
Guide to expression of uncertainty in measurement (GUM), 42, 645
GUM. *See* Guide to expression of uncertainty in measurement

H

Hemispherical-directional reflectance, 25—26
Henyey and Greenstein (HG), 412
HG. *See* Henyey and Greenstein
High-temperature blackbody (HTBB), 31—32
Historical sensors, 109
    CZCS and OCTS, 110—111
    MERIS, 115—116
    MODIS, 113—115
    SeaWiFS, 111—113
HMOMC scheme. *See* Hybrid Matrix Operational-Monte Carlo scheme
HNO_3. *See* Nitric acid
Hot junctions, 332—333
HTBB. *See* High-temperature blackbody
Hybrid Matrix Operational-Monte Carlo scheme (HMOMC scheme), 436—437
    See also Monte Carlo scheme (MC scheme)
Hyperspectral
    radiometer, 249—250
    systems, 546

I

IASI. *See* Infrared Atmospheric Sounding Interferometer
ICOADS. *See* International Comprehensive Ocean-Atmosphere Data Set
ICT. *See* Internal Calibration Target
IFOV. *See* Instantaneous field of view
Imaging radiometer, 253f
Immersion
    effects, 260
        irradiance sensors, 260—261
        radiance sensors, 262—263

factor, 277
IMU. *See* Inertial measurement unit
In situ measurement strategies
    FRM, 527
        ocean-color radiometric, 528
        ship-borne FRM TIR radiometer, 528
        TIR, 528
In situ optical radiometry, 245, 248, 533
    See also Satellite ocean-color missions
    applications, 285
        bio-optical models, 289—291
        in-water light field polarization, 287—289
        satellite adiometric products validation, 291—293
        sky and sea radiance distribution, 285—287
        system vicarious calibration, 293—294
    errors and uncertainty estimation, 273—285
    field radiometer systems, 249—254
    measurement methods, 264—272
    system calibration, 254—264
In situ radiometric measurement simulation, 413—414
    overstructure perturbations, 414—429
    perturbations inducing, 429—441
In water methods, 277
    biofouling, 279—280
    inelastic scattering, 280—281
    modeling, 280
    self-shading, 278—279
    superstructure perturbations, 279
    tilt effects, 279
    wave focusing and defocusing, 277—278
In water systems, 265
    fixed-depths, 265
    profiling, 266
    radiance distribution systems, 266—267
In-water light field polarization, 287—289
Index of refraction, 17
Indium antimonide (InSb), 329—331, 383
    detectors, 157
Inelastic scattering, 280—281
Inertial measurement unit (IMU), 391
Infrared (IR), 389—391
    absorption, 494
    radiation, 306, 311
Infrared Atmospheric Sounding Interferometer (IASI), 223
Infrared Autonomous SST Radiometer (ISAR), 322—324, 565—566
    instrument, 375—380

Inherent optical properties (IOPs), 80, 265, 457–458, 609–610
Inland typical case-2 waters, 458–463
Input-referred noise, 102
Instantaneous field of view (IFOV), 78–79, 129
Instrument
 characterization, 60
 instrument-mounting bracket, 380–382
 measurement uncertainty, 564
 radiometric calibration, 184
  blackbody sources, 186–188
  data analysis, 191–192
  facility, 186
  instrument level test campaign, 184–186
  test procedures, 189–191
  test results, 192–196
 uncertainty, 566, 567t
Integrating sphere receiver (ISR), 51–52
Integrating spheres, 275
Intensity, 15
Internal Calibration Target (ICT), 166
International Comprehensive Ocean-Atmosphere Data Set (ICOADS), 665
International Ocean Colour Coordinating Group (IOCCG), 8, 452–453
International Space Science Institute (ISSI), 659
International System of Units (SI), 4–5, 17, 245
International-Temperature-Scale of 1990 (ITS-90), 182–183, 559
Inverse problem, 490
Inverse theory, 514
IOCCG. See International Ocean Colour Coordinating Group
IOPs. See Inherent optical properties
IR. See Infrared
Irradiance, 16, 21
 See also Radiance
 configuration factor, 21–23
 flux transfer equation, 22
 measurement, 51–52
  analysis, 431–432
 propagation of optical flux, 21
 responsivity, 41–42
 sensors, 250–252, 260–261
  angular response, 258–260
 spectral flux, 21
 Taylor series expansion, 22–23
 uniform and Lambertian source, 22

ISAR. See Infrared Autonomous SST Radiometer
ISR. See Integrating sphere receiver
ISSI. See International Space Science Institute
ITS-90. See International-Temperature-Scale of 1990

K

Korean Geostationary Ocean Color Imager (GOCI), 79

L

Laboratory intercomparisons, 539–540
Lambertian emitter, 15
Lamp sources, 32–33
Lamp-illuminated integrating spheres, 33–35
Lamps-under-test (LUTs), 50–51
LAR. See Latitudinal atmospheric regimes
Laser-illuminated integrating spheres, 35
Laser-illuminated sphere approach, 41–42
Latitudinal atmospheric regimes (LAR), 528, 579, 580t
Lead zirconate titanate (PZT), 333–334
LEO. See Low earth orbits
Light grasp, 345
Liquid nitrogen, 331
Lithium tantalate (LiTaO$_3$), 333–334
Long-term measurements
 comprehensive measurements, 538
 for satellite data products validation, 535–536
 for system vicarious calibration, 534–535
Long-term satellite data set, 605
Long-term SST data record assessment, 640, 649–651
  See also Satellite products assessment
 ARC project, 641
 assessments on coarser space scales, 651–652
 atmosphere brightness temperature assessment, 641–642
 demonstrating traceability to SI, 659–662
 ESA SST_CCI ATSR v1.0 data set, 652–654
 GCOS, 651
 long-term component assessment, 654–657
 quantitative metrics, 657–659
 reference data sources, 647–649
 stability, 663–668

Long-term SST data record assessment (*Continued*)
  validation
    of uncertainties, 669—673
    uncertainty budget, 643—647
Look-up table (LUT), 410
Low earth orbits (LEO), 78—79
  missions, 69—71
Lunar calibration, 123—124, 128, 134f
  See also Solar calibration
  instrument relative calibration biases, 135t
  intercomparisons, 133—135
  lunar radiometric response trending, 130
  residuals *vs.* phase angle, 136f
  ROLO photometric model of Moon, 129—130
  uncertainties, 131—132
Lunar radiometric response trending, 130
LUT. *See* Look-up table
LUTs. *See* Lamps-under-test

M

M-AERI. *See* Marine Atmospheric Emitted Radiance Interferometer; Marine-Atmosphere Emitted Radiance Interferometer
M-est. *See* Median estimator
MAD. *See* Median absolute deviation
MAERI, 332, 358—359, 387
Marine Atmospheric Emitted Radiance Interferometer (M-AERI), 322—324
Marine Optical Buoy (MOBy), 141, 534, 614
  validation, 625—626
Marine reflectance, 609—610
Marine-Atmosphere Emitted Radiance Interferometer (M-AERI), 565—566
Match-up data set (MD), 643—644
Match-up process, 665
Matchup database (MDB), 571
MC scheme. *See* Monte Carlo scheme
MCST. *See* MODIS Characterization Support Team
MD. *See* Match-up data set
MDB. *See* Matchup database
Mean wavelength, 49
Measurement equation, 2, 42
  examples, 46—57
  GUM, 42
  sensitivity coefficients, 45
  uncertainty in ocean color measurements, 57
  comparisons and reproducibility, 60—61
  correlations, 58—59
  uncertainty values, 44
Measurement methods, 264—265, 595
  radiometric data products, 268—272
  in water systems, 265—267
  above water systems, 267—268
Measurement uncertainty, 564, 566
Median absolute deviation (MAD), 647
Median estimator (M-est), 647
Medium Resolution Imaging Spectrometer (MERIS), 69—71, 74—75, 86, 115—116, 452—453, 533
  degradation of solar diffuser, 92
  reflective solar bands, 123t
MEM technology. *See* Microelectromechanical technology
Mercury cadmium telluride (HgCdTe), 329—331, 383
  detectors, 157
MERIS. *See* Medium Resolution Imaging Spectrometer
MERIS Matchup In Situ Database (MERMAID), 542
Meteorological Operational (MetOp), 665
Metrological traceability, 30
Microbolometer, 334—335
Microelectromechanical technology (MEM technology), 335
Mid-infrared (MIR), 306—307
Mie scattering theory, 501—502
MIR. *See* Mid-infrared
MISTRC. *See* Multi-channel Infrared Sea Truth Radiometric Calibrator
MOBy. *See* Marine Optical Buoy
Model parameters. *See* Vector of coefficients
Model-based approach, 618—620
Moderate Resolution Imaging Spectroradiometer (MODIS), 36, 69, 74—75, 166—167, 368, 452—453, 533
  design, 113—115
  Google Earth image, 382f
  MODIS-Aqua true color, 461f
  MODIS-Aqua-derived climatology, 462f
  radiometric calibration, 213—214
  reflective solar bands, 123t
  spectral bands, 168t—169t
  SRCA, 214—216
MODIS Characterization Support Team (MCST), 126—128
Modular Optoelectronic Scanner (MOS), 452—453

Monte Carlo scheme (MC scheme), 408
  case studies, 434–441
  RTE solutions, 410–413
  simulations, 183, 440–441
MOS. *See* Modular Optoelectronic Scanner
Mueller matrix, 256
Multi-channel Infrared Sea Truth Radiometric Calibrator (MISTRC), 322–324, 356–357
Multispectral radiometer, 249–250

N

Narrow spectral band pass filters, 338
NASA. *See* National Aeronautics and Space Administration
NASA/JPL NNR radiometer, 368, 369t
  calibration, 369–371
  design philosophy, 368
  operation, 368–369
National Aeronautics and Space Administration (NASA), 37–38, 534
  Ocean Biology Processing Group, 465–466
National Institute of Standards and Technology (NIST), 370–371, 534, 586–587
National Marine Electronics Association (NEMA), 354
National Measurement Institute. *See* National Metrological Institute (NMI)
National Metrological Institute (NMI), 30, 137, 558–559
National Oceanic and Atmospheric Administration (NOAA), 640–641
National Physical Laboratory (NPL), 39–40, 587–589
National Polar-orbiting Operational Environmental Satellite System (NPOESS), 224–225
NEΔT. *See* Noise equivalent differential temperature (NEDT)
Near infrared (NIR), 74–75, 306–307, 405, 453–454
  channels, 161
Near Nulling Radiometer (NNR), 322–324, 369t
NEDT. *See* Noise equivalent differential temperature
NEMA. *See* National Marine Electronics Association
NEP. *See* Noise equivalent power
Networking, 551
NIR. *See* Near infrared

NIST. *See* National Institute of Standards and Technology
NIST Spectral Tri-function Automated Reference Reflectometer (STARR), 36–37
Nitric acid (HNO_3), 495
NLSST. *See* Nonlinear sea surface temperature
NMI. *See* National Metrological Institute
NNR. *See* Near Nulling Radiometer
NOAA. *See* National Oceanic and Atmospheric Administration
Noise, 100
  detector, 101
  digitization, 102
  equivalent radiance, 103
  noise-free retrieval error simulation, 519
  read, 101–102
  SNR and noise equivalent radiance, 103
  total system noise reduction, 102–103
Noise equivalent differential temperature (NEDT), 209, 210f, 211t, 517
Noise equivalent power (NEP), 328
Non-uniformity correction (NUC), 391–393
Nonlinear sea surface temperature (NLSST), 511
Nonlinearity, 174–176
NOP. *See* Numerical ocean prediction
NPL. *See* National Physical Laboratory
NPOESS. *See* National Polar-orbiting Operational Environmental Satellite System
NUC. *See* Non-uniformity correction
Numerical approaches, 408
Numerical ocean prediction (NOP), 306
Numerical weather prediction (NWP), 306, 507–508, 578–579

O

OBPG. *See* Ocean Biology Processing Group
Ocean and Sea Ice Satellite Application Facilities (OSI-SAF), 581–582
Ocean Biology Processing Group (OBPG), 8, 126–128
Ocean climate measurements, 1
Ocean color
  applications, 437–438
  data, 121–122
  products, 628–629
  remote sensing, 414
  satellite sensors, 606
  uncertainty in measurements, 57–61

Ocean Color and Temperature Sensor (OCTS), 69, 111, 452–453
Ocean Color Imager (OCI), 74–75
Ocean color radiometry (OCR), 4, 7–8
Ocean Color Radiometry Virtual Constellation (OCR-VC), 8
Ocean radiance contributions, 457
  coastal typical case-2 waters, 458–463
  inland typical case-2 waters, 458–463
  MODIS-Aqua true color, 461f
  MODIS-Aqua-derived climatology, 462f
  ocean BRDF effects, 463–464
  open ocean case-1 waters, 457–458
Ocean-Color component of the Aerosol Robotic Network (AERONET-OC), 535–536
Ocean–atmospheric system, 455–457
  radiative transfer simulation for, 464
Oceanic radiometric measurements, 1
OCI. *See* Ocean Color Imager
OCR. *See* Ocean color radiometry
OCR-VC. *See* Ocean Color Radiometry Virtual Constellation
OCTS. *See* Ocean Color and Temperature Sensor
On-board calibration, 92–93, 176–177, 203
  AATSR
    blackbody calibration sources, 182f
    radiometric calibration, 204–209
  ATSR
    on-board blackbody cavity, 181f
    series, 203
  AVHRR, 209–213
  black plates, 179–180
  blackbody source uncertainty budget, 177–178, 177t
  calibration equation, 177
  cavity blackbodies, 180–181
  deep space view, 178–179
  MODIS
    mirror response *vs.* scan angle, 216–218
    radiometric calibration, 213–214
    SRCA, 214–216
  radiance, 177
  radiometric errors, 178f
  sources of uncertainties, 203
  VIIRS radiometric calibration, 213–214
On-orbit calibration uncertainties, 142–143, 149–150
  accuracy, 143
  assessment combination, 144
  on-orbit characterization, 87
  TOA radiances
    long-term stability of, 143–144
    precision, 144
OOB response. *See* Out-of-band response
Open ocean case-1 waters, 457–458
Optical flux, 19
Optical region, 14
OSI-SAF. *See* Ocean and Sea Ice Satellite Application Facilities
Out-of-band response (OOB response), 84–86
Overstructure perturbations, 414
  modeling features, 414–415
  self-shading effects, 423–429
  ship-shading effect, 415–420
  theoretical investigations, 414–415
  tower-shading effects, 420–423

P

Parabolic mirror, 343–344, 344f
Parts per million by volume (ppmv), 494–495
Path radiance, 405–406
Pathfinder data set, 640–641
Payload Electronics Module simulator, 186
PCBs. *See* Printed circuit boards
PDF. *See* Probability Density Function
Peltier devices, 331
PHO-TRAN backward MC code, 427–428
Photoconductive detectors, 329–331
Photon noise, 160
Phytoplankton, 628–629
Pixel aggregation, 72
Planck function, 50–51, 155, 497
  spectral radiance dependency, 498f
  thermal emission, 497
Planck's radiation law, 154
Plane-parallel atmosphere (PPA), 464
Platinum resistance thermometer (PRT), 181, 204
Point spread function (PSF), 257
Point-by-point comparison, 624–626
"Point" approach, 643
Polarization, 90–91
  sensitivity, 256
Polarization and directionality of the Earth's reflectances (POLDER), 452–453
Polarizers, 254

Index   **689**

POLDER. *See* Polarization and directionality of the Earth's reflectances
Polytetrafluoroethylene (PTFE), 35−36
PPA. *See* Plane-parallel atmosphere
ppmv. *See* Parts per million by volume
Preamplifier noise. *See* Read noise
Precision, 43
Predeployment calibration verification, 596
Prelaunch absolute
  radiance-based radiometric calibration, 88
  reflectance-based radiometric calibration, 88−89
Prelaunch characterization, 87
Printed circuit boards (PCBs), 106
Probability Density Function (PDF), 646−647
PRT. *See* Platinum resistance thermometer
PSF. *See* Point spread function
PTFE. *See* Polytetrafluoroethylene
Pyroelectric detectors, 333−334
PZT. *See* Lead zirconate titanate

Q

Q-branch. *See* Vibrational-only transition
QA4EO. *See* Quality Assurance Framework for Earth Observation
QAA. *See* Quasi-Analytical Algorithm
QE. *See* Quantum efficiency
Qualitative analysis, 657
Qualitative metrics, 606−607
Quality Assurance Framework for Earth Observation (QA4EO), 589
Quality control, 548−549
Quantitative metrics, 606−607, 657−659
Quantum detectors, 329−332
  See also Thermopile detectors
Quantum efficiency (QE), 98−99
Quasi operational ocean field radiometers, 375−380
Quasi-Analytical Algorithm (QAA), 623

R

R-branch, 494
R/V. *See* Research Vessel
Radiance, 15−17
  See also Irradiance
  conservation, 19
  differential element, 20
  distribution measurement system, 253
  invariance, 20
  optical flux, 19

  optical radiation, 18f
  polarization measurement system, 253−254
  projected solid angle, 20−21
  propagation of fluxes, 18−19
  reduced, 19−20
  responsivity, 41−42
  sensors, 252, 262−263
    Gershun tube, 252−253
    imaging radiometer, 253f
  Snell's law application, 19−20
Radiance Distribution System (RADS), 266−267, 427−428
Radiative transfer (RT), 490
  simulation, 490−493
Radiative transfer equation (RTE), 407−410
  deterministic solutions, 410
  MC methods, 410−413
  numerical approaches, 408
  solution methods, 408
Radiative transfer model (RTM), 490, 578, 641
  typology of, 492t
Radiometers, 14−15, 38−39
  ESR, 39−40
  radiance and irradiance responsivity, 41−42
Radiometric uncertainty, 87, 549
  prelaunch absolute
    radiance-based radiometric calibration, 88
    reflectance-based radiometric calibration, 88−89
  relative radiometric calibration, 89
Radiometry, 14
  calibration testing, 184
  data products, 268
    above-water measurements, 269−270
    in-water measurements, 268−269
    water-leaving radiance normalization, 270−272
  detectors, 17
  distance and aperture areas in, 28−30
  index of refraction, 17
  irradiance, 21−23
  noise, 174
  radiance, 17−21
  radiometric quantities, 15t
  reflectance, 23−28
  source of radiation, 16f
  standards and scale realizations, 30
    radiometers, 38−42

Radiometry (*Continued*)
   sources, 30–38
   wavelength regions, 14t
RADS. *See* Radiance Distribution System
RAL. *See* Rutherford Appleton Laboratory
RAL/SIL radiometer, 345–346
Random error, 43
Random Telegraph Signal (RTS), 329–331
Rayleigh scattering, 455
Read noise, 101–102
Reduced radiance, 19–20
Reference data sources, 647–649, 650t
Reference satellite sensor comparisons, 218
   instruments on same satellite, 223–225
   SNO, 222–223
   spatial comparisons, 219–220, 219f
   temporal comparisons, 220–222
Reference uncertainty, 646
Reflectance, 16, 23–25
   biconical reflectance, 25–26
   BRDF, 24, 26
   conical-directional, 24–25
   diffuse directional-hemispherical, 26–27
   downwelling spectral irradiance, 27–28
   factors, 16, 27
   reflectance factor, 27
   spectral irradiance, 28
Reflectivity versus scan angle (rvs), 216–218
Regular ship-borne radiometer, 590
Relative humidity (RH), 465
Relative radiometric calibration, 89
Relative spectral response (RSR), 84, 222–223, 223f
Remote sensing systems
   See also Thermal remote sensing
   ATSR, 161–164
   AVHRR, 165–166
   MODIS, 166–167
   reflectance, 270–271
   SEVIRI, 171
   SLSTR, 164–165
   VIIRS, 167–171
Reprocessing, 548–549
Reproducibility, 10–11
Research Vessel (R/V), 356
Retrieval, 509–516
Retrieval problem. *See* Inverse problem
Retrieval/algorithm uncertainty, 564
RH. *See* Relative humidity
Rhodium (Rh), 343
RMS. *See* Root mean square

Robust standard deviation (RSD), 647
Rolloff of imaging systems, 260
ROLO photometric model of Moon, 129–130
Root mean square (RMS), 100–101
Root sum square (RSS), 100–101
Rosenstiel School of Marine and Atmospheric Science (RSMAS), 587
"Round-robin" tests, 596–597
RSD. *See* Robust standard deviation
RSMAS. *See* Rosenstiel School of Marine and Atmospheric Science
RSR. *See* Relative spectral response
RSS. *See* Root sum square
RT. *See* Radiative transfer
RTE. *See* Radiative transfer equation
RTM. *See* Radiative transfer model
RTS. *See* Random Telegraph Signal
Rutherford Appleton Laboratory (RAL), 322–324
rvs. *See* Reflectivity versus scan angle

S

S-NPP satellite. *See* Suomi National Polar-orbiting Partnership satellite
Santa Barbara Research Center (SBRC), 111–112
Satellite adiometric product validation, 291–293
Satellite data product validation, 546–547
Satellite ocean color radiometer (SOCR), 7
Satellite ocean color sensor
   chronological sequence, 81t–83t
   evolution of science objectives, 80–84
   measurement fundamentals and science objectives, 75–79
   performance parameters and specifications, 84
      coverage and spatial resolution, 86–87
      on-board calibration systems, 92–93
      polarization, 90–91
      radiometric uncertainty, 87–89
      SNR and quantization, 89–90
      spectral coverage and dynamic range, 84–86
      straylight, 91
   remote-sensing products, 452
   requirements, 80–84
Satellite ocean-color missions, 533
   data repositories, 542–543
   field measurements, 533–534
      for bio-optical modeling, 547

Index 691

long-term comprehensive
measurements, 538
long-term measurements, 534–536
for satellite data products validation,
546–547
for system vicarious calibration,
544–546
time-limited comprehensive
measurements, 536–538
field-related radiometric activities, 533
intercomparisons, 538
data processing, 541–542
field intercomparisons, 540–541
laboratory intercomparisons, 539–540
requirements and strategies, 543
accuracy, 549
archival and access, 549
consolidation, 547
data reduction, 548–549
development and implementation, 551
field radiometers, 548
networking, 551
protocols revision, 547
quality control, 548–549
reprocessing, 548–549
secure accuracy, 550–551
standardization, 551
Satellite products
See also Long-term SST data record
assessment
assessment, 639–640
CDR, 605
long-term satellite data set, 605
primary ocean-color product, 606
validation, 610
analysis of validation results, 614–618
matrix relating error sources, 620t
metrics, 612–614
model-based approach, 618–620
protocol, 610–612
statistics dependency, 618f
Satellite radiometers, 201–202
Satellite radiometry, 69
assessment, 609–610
cross-mission data products comparison,
621–630
Satellite retrieval process, 516
Satellite thermal measurements, 489
emissivity models, 505t
geophysical inversion process, 509–516
metop AVHRR instrument, 506f
Mie scattering, 502f

radiative transfer simulation, 490–493
sea-surface emissivity, 504f
simulation
of aerosol *vs.* cloud, 500–502
of surface emission *vs.* reflection,
502–504
in thermal image classification,
504–509
thermal radiation propagation, 493–500
uncertainty estimation, 516–521
Satellite uncertainty, 646
Satellite-measured radiance spectra, 455
Satellites International Radiometer (SIL),
322–324
SBRC. *See* Santa Barbara Research Center
SBRN. *See* Ship-borne radiometer network
Scanning Infrared Sea surface Temperature
Radiometer (SISTeR), 322–324,
364–365, 365f
mounting and support, 366–368
operation, 366
self-calibrating design, 365
skin SST measurements, 366
"Scanning mirror" approach, 158, 346–347
Scattering phase functions, 412
Scene mirror, 203
Science traceability matrix (STM), 80
SD. *See* Standard deviation
SDs. *See* Solar diffusers
SDSM. *See* Solar diffuser stability monitor
Sea and Land Surface Temperature Radiometer
(SLSTR), 164–165, 165t, 585
Sea radiance distribution, 285–287
Sea surface, 503
Sea surface depth temperature (SST_{depth}), 310
Sea surface skin temperature (SST_{skin}), 309
determination using spectroradiometers,
382–387
ship borne radiometer measurement,
317–320
practical measurement, 320–321
Sea surface subskin temperature ($SST_{subskin}$),
309–310
Sea surface temperature (SST), 4, 8–10, 69,
202, 308–309, 309f, 452, 528,
558–559, 606–607, 639–640
See also Geophysical inversion process
ARC data set, 641
atmosphere brightness temperatures
assessment, 641–642
CDR assessment, 640
ESA SST_CCI analysis, 671f

692 Index

Sea surface temperature (SST) (*Continued*)
  FRM for SST CDR, 559
  long-term satellite SST records, 560
  TIR SBRN, 561−563
  global SST gradients, 573f
  ISAR SST processor, 568f
  long-term assessment, 640−641, 649−673
  product validation activities, 582f
  quantitative metrics, 657−659
  reference data sources, 647−649, 650t
  retrieval, 509
    for climate data generation, 504
  SI traceability, 659−662, 660f
  stability, 663−668
  TIR used for, 323f
  uncertainty budgets, 563−585
  validation, 595
    accessibility to documentation, 597
    archiving of data, 597
    calibration measurements, 596−598
    laboratory calibration, 595−596
    measurement methodology, 595
    periodic consolidation, 598
    postdeployment calibration verification, 596
    predeployment calibration verification, 596
    procedures, 595−596
    statistics, 658t
    of uncertainties, 669−673, 670f
    uncertainty budgets, 596, 643−647
    verification measurements, 596−598
    verification methodology, 595−596
Sea-surface waves, perturbations inducing by, 429−430
  experimental findings, 431−433
  MC case studies, 434−441
  statistical modeling, 433−434
  wind-generated sea, 430−431
Sea-viewing Wide Field-of-view Sensor (SeaWiFS), 7−8, 69, 70f, 74−75, 452−453
  lunar calibration time series, 131f
  mission, 629−630
  on-orbit precision estimation, 145t
  reflective solar bands, 123t
  TOA uncertainty assessment, 146t
  vicarious calibration stability estimation, 145t
SeaBASS. *See* SeaWiFS Bio-optical Archive and Storage System

Seawater
  analysis of light penetration in, 416
  IOPs, 433−434
  phase function, 413
SeaWiFS. *See* Sea-viewing Wide Field-of-view Sensor
SeaWiFS Bio-optical Archive and Storage System (SeaBASS), 8, 542, 614−615
SeaWiFS Intercalibration Round-Robin Experiments (SIRREXs), 37−38, 539−540
Self calibrating radiometers, 358−361
Self-shading, 278−279
  effects, 423−429, 424t
Sensitivity coefficients, 45
Sensor engineering, 93−95
  design fundamentals and radiometric equations, 96−99
  orbit geometry terms, 98f
  performance considerations, 99
    dynamic range and sensitivity, 99−100
    EOL performance, 103−104
    noise, 100−103
  sensor implementation, 104−107
  whiskbroom and pushbroom designs, 94f−95f, 95−96
Sensor implementation, 104
  design controls and margins, 104−105
  electronic parts selection, 105
  environmental test and performance verification, 106−107
  life test and component screening, 105
  materials selection and control, 105
  process controls, 106
  reviews and schedule, 107
Sensor Intercomparison and Merger for Biological and Interdisciplinary Oceanic Studies (SIMBIOS), 540
Sensor noise, 517
SESR. *See* Surface emission, surface reflection
SEVIRI. *See* Spinning Enhanced Visible and Infrared Imager
Ship borne radiometer, 350−351
  FRM ship borne TIR radiometer, 363−393
  Ship borne TIR radiometers, 311, 322
    practical measurement, 320−321
  SST_{skin} ship borne radiometer measurement, 317−320
  TIR field radiometer design, 321−363
Ship optical radiometric measurements, 537
Ship-board radiometers, 236−237

Ship-borne radiometer network (SBRN), 562–563, 590–591
  at-sea intercomparison, 594
  different satellite instruments, 584–585
  laboratory intercalibration experiments for FRM, 585–586
    CASOTS-II blackbody system, 589f
    CEOS, 587
    NIST TXR transfer radiometer, 588f
    QA4EO framework, 589
    regular ship-borne radiometer, 590
    uncertainty analysis, 590
  R/V Walton Smith, 591f–592f
  for satellite instrument degradation monitoring, 583–584
  for satellite SST products validation, 581–583
  satellite-to-SBRN spatial, 570–578
  for SST retrieval algorithms validation, 578–581
  SST_{skin} measurements, 564–570, 594f
Ship-shading effect, 415–420, 417t, 418f
Shortwave infrared (SWIR), 74–75, 405, 453–454
SI. *See* International System of Units; Système International d'unitès
SI traceability, 659–662, 660f
Signal to noise ratio (SNR), 69–71, 74–75, 124–125, 160–161, 209, 357, 453–454
  noise equivalent radiance, 103
  on orbit, 128
  and quantization, 89–90
SIL. *See* Satellites International Radiometer
Silicon (Si), 337–338
Silver (Ag), 343
SIMBIOS. *See* Sensor Intercomparison and Merger for Biological and Interdisciplinary Oceanic Studies
SIMBIOS Radiometric Intercomparison (SIMRIC), 540
Simulated typical TOA radiances, 471–472
  examples of, 472–474
  open oceans, 474–475
  with satellite data, 476–477
Simultaneous Nadir Overpass (SNO), 222–223
Single-scattering epsilon (SSE), 466
SIRCUS. *See* Spectral Irradiance and Radiance responsivity Calibrations using Uniform Sources

SIRREXs. *See* SeaWiFS Intercalibration Round-Robin Experiments
SIS. *See* Spherical integrating spheres
SISTeR. *See* Scanning Infrared Sea surface Temperature Radiometer
Sky radiance distribution, 285–287
Sky-light polarization, 281–282
SLSTR. *See* Sea and Land Surface Temperature Radiometer
SNO. *See* Simultaneous Nadir Overpass
SNR. *See* Signal to noise ratio
SOCR. *See* Satellite ocean color radiometer
Solar calibration, 122–125
  See also Vicarious calibration
  geometric corrections, 125
  SNR on orbit, 128
  uncertainties, 128
Solar diffuser stability monitor (SDSM), 92–93, 122–123
Solar diffusers (SDs), 92–93, 122
  degradation, 125
  radiometric response trends, 126–128
Solar-reflected wavelengths (SR wavelengths), 17
Soldering, 106
SOS. *See* Successive order of scattering
Southampton Underwater Multiparameter Optical Spectrograph System (SUMOSS), 427
Space flight mission, 104
Spectral
  calibration of grating instruments, 135
  flux, 21
  irradiance reflectance, 28
  response, 257–258
Spectral Irradiance and Radiance responsivity Calibrations using Uniform Sources (SIRCUS), 35
Spectral response function (SRF), 510
Spectralon SD, 135
Spectralon®, 35
Spectro-radiometric calibration assembly (SRCA), 214–216
Spectroradiometer, 14–15
  air temperature derivation, 387–389
  M-AERI interferometer, 384f
  SST_{skin} determination, 382–387
Spheres integration, 33–36
Spherical integrating spheres (SIS), 88
Spinning Enhanced Visible and Infrared Imager (SEVIRI), 171, 231
  spectral bands, 173t

Split window, 230
SPRTs. *See* Standard platinum resistance thermometers
SR wavelengths. *See* Solar-reflected wavelengths
SRCA. *See* Spectro-radiometric calibration assembly
SRF. *See* Spectral response function
SSE. *See* Single-scattering epsilon
SST. *See* Sea surface temperature
Standard deviation (SD), 647
Standard platinum resistance thermometers (SPRTs), 182—183
Standard uncertainty, 517
Standardization, 551
STARR. *See* NIST Spectral Tri-function Automated Reference Reflectometer
Statistical modeling, 433—434
Stirling cycle cooler, 331—332
Stirred tank method. *See* External stirred water bath calibration
STM. *See* Science traceability matrix
Stochastic approach, 408
Stray light, 91
  perturbations, 257
Strip lamps. *See* Tungsten ribbon filaments
Successive order of scattering (SOS), 464
SUMOSS. *See* Southampton Underwater Multiparameter Optical Spectrograph System
Sun, 14
Suomi National Polar-orbiting Partnership satellite (S-NPP satellite), 72, 453
Superstructure perturbations, 279, 282
Surface emission, surface reflection (SESR), 504
SWIR. *See* Shortwave infrared
Synoptic scales, 519
System calibration, 254—255
  absolute response, 263—264
  angular response of irradiance sensors, 258—260
  immersion effects, 260—263
  linearity response, 255
  polarization sensitivity, 256
  rolloff of imaging systems, 260
  spectral response, 257—258
  stray light perturbations, 257
  temperature response, 255—256
  vicarious calibration, 293—294
    field measurements, 544—546
    long-term measurements, 534—535

Systematic error, 43
Système International d'unités (SI), 558—559

T

TAG. *See* Technical Advisory Group
TCDR. *See* Thematic climate data record
TDI. *See* Time-delay-integration
TEC. *See* Thermoelectric heater/cooler unit
Technical Advisory Group (TAG), 640
Television Infrared Observation Satellite-N (TIROS-N), 165—166
Temporal comparisons, 220—222
Thematic climate data record (TCDR), 5t
Theoretical investigations
  simulation capabilities, 405
  TOA radiance spectra, 405—406
Thermal control system, 324, 350—351
Thermal image classification, 504—509
Thermal infrared (TIR), 306, 405, 558—559
  See also Field radiometer design; TIR; FRM ship borne TIR radiometer
  cameras, 389—393
  general considerations, 311—317
  IRemission spectra, 313f
  pure water emissivity, 316f
  radiometry, 246, 528
  region, 14
  ship borne TIR radiometer system, 311
    design characteristics, 325t—326t
    "embedded" relationship, 327f
    signal path and components, 327f
    SST_{skin} ship borne radiometer measurement, 317—321
Thermal infrared satellite radiometers
  calibration model, 172—176
  design principles, 155—156
    ATSR TIR channel optical chain, 158f
    band average radiance, 157
    earth-scene radiances, 156
    IR instruments, 157—158
    performance model, 159—160
    scanning mirror, 158
    signal to noise, 160—161
    SLSTR performance requirements, 156f
  geophysical retrievals validation, 225—237
  on-board calibration, 176—181, 203—218
  Planck's radiation law, 154
  pre-launch characterization and calibration
    blackbody calibration, 182—184
    instrument radiometric calibration, 184—196

reference satellite sensor comparisons, 218–225
remote sensing systems, 161–171
SST, 154
Thermal radiation propagation, 493
 absorptivity/emissivity, 500
 atmospheric vertical transmission spectrum, 495f
 clear-sky RT, 500
 infrared absorption feature, 494
 molecule energy, 493–494
 Planck function, 497
 Planck spectral radiance dependency, 498f
 radiative impact, 496t
Thermal remote sensing
 of Earth, 489
 ocean surface, 504
 radiative transfer simulation, 490–493
 of SST, 490, 522
Thermal-infrared Transfer Radiometer (TXR), 587
Thermoelectric heater/cooler unit (TEC), 372
Thermopile detectors, 332–333
3D backward MC PHO-TRAN code, 421
Threshold test, 506
Throughput of optical arrangement, 21–22
Tilt effects, 279
Time series analysis, 626–628
Time-delay-integration (TDI), 78–79
Time-limited comprehensive measurements, 536–538
Time-space matchup criteria, 570–571
TIR. See Thermal infrared
TIROS-N. See Television Infrared Observation Satellite-N
Top of atmosphere (TOA), 7, 75–76, 121–122, 306, 405, 454, 457, 489, 558–559, 639–640
 atmospheric diffuse transmittance, 470–471
 atmospheric path radiance contributions
  aerosol models, 464–466
  aerosol spectral reflectance, 466–467
  atmospheric path reflectance, 468–470
  radiative transfer simulation, 464
  single-scattering epsilon, 467f
 ocean radiance contributions, 457–464
 simulated typical TOA radiances, 471–472
  examples, 472–474
  open oceans, 474–475
  with satellite data, 476–477
Total solar irradiance (TSI), 29
Total system noise reduction, 102–103

Tower-shading effects, 420–423, 423t
Transmittance, 494
True value, 43
TSI. See Total solar irradiance
TTHG. See Two-terms Henyey and Greenstein
Tungsten ribbon filaments, 32
Two free parameters, 413
Two-parametric expression, 413
Two-terms Henyey and Greenstein (TTHG), 412
TXR. See Thermal-infrared Transfer Radiometer
Type A uncertainties, 44
Type B uncertainties, 44

U

UAS. See Unmanned airborne systems
UAV. See Unmanned autonomous vehicles
Ultraviolet (UV), 14, 453–454
Uncertainty
 analysis, 618–620
 calibration specific sources, 274
  irradiance standards, 274–275
  radiance standards, 275
 comparison, 145–148
 estimation, 516
  algorithmic retrieval error, 520f
  Earth Observation metrology, 521
  noise-free retrieval error simulation, 519
  retrieval method, 518–519
  satellite retrieval process, 516
  simulated retrieval error, 520f
  SST retrieval process, 521
  standard uncertainty, 517
 instrument specific sources, 276
  cosine error effects, 276
  immersion factor, 277
 maps, 619
 methods and field specific source, 277
  uncertainty budget for radiometric products, 282–285
  in water methods, 277–281
  above water methods, 281–282
 in ocean color measurements, 57–61
Uncertainty budgets, 563
 See also Validation—uncertainty budget
 ISAR SST processor, 568f
 measurement uncertainty, 564
 satellite-to-SBRN spatial, 570–578
 SBRN requirements

696 Index

Uncertainty budgets (*Continued*)
　different satellite instruments, 584–585
　for satellite instrument degradation monitoring, 583–584
　for satellite SST products validation, 581–583
　for SST retrieval algorithms validation, 578–581
　SBRN SST$_{skin}$ measurements, 564–570
　temporal matchup criteria, 570–578
　total column water vapor, 581f
Uninterruptable power supplies (UPS), 353
Unmanned airborne systems (UAS), 389–391
Unmanned airborne vehicles BESST radiometer, 380
　components and modular design, 381f
　Google Earth image with overlays of MODIS image, 382f
　instrument-mounting bracket, 380–382
Unmanned autonomous vehicles (UAV), 322–324
UPS. *See* Uninterruptable power supplies
UV. *See* Ultraviolet

V

Validation, 559–560
　analysis, 611–612
　metrics, 612–614
　protocol, 610–612
　statistics dependency, 618f
　uncertainty budget, 643
　　components, 646
　　conventional statistics, 647
　　long-term SST data record, 643
　　robust statistics, 647
　　SST$_{skin}$, 644
　　statistical analysis, 644
　　traditional approach, 643–644
Vanadium oxide (VOx), 335
Variable temperature blackbody (VTBB), 54–55
Vector of coefficients, 511–512
Vibrational-only transition, 494
Vicarious calibration, 137
　advantage of, 138
　alternative approaches, 142
　coefficient, 138
　NIR/SWIR bands, 138–140
　system, 293–294
　　field measurements, 544–546
　　long-term measurements, 534–535

　　uses of, 137
　　visible band calibration, 140–142
VIIRS. *See* Visible Infrared Imaging Radiometer Suite
Virtual optical profiles, 438, 440
VIS channels. *See* Visible channels
Visible and near-infrared (VNIR), 7
Visible band calibration, 140
　calibration coefficients, 142f
　gains, 141–142
　hyperspectral instrument, 140
　MOBy, 141
　NMI-traceable instrumentation, 140
　operational data processing, 142
　satellite-based remote sensor, 141
　surface reflectances, 141
　TOA signal, 141
Visible channels (VIS channels), 161
Visible Infrared Imaging Radiometer Suite (VIIRS), 72, 89–90, 167–171, 453
　lunar calibration time series, 133f
　radiometric calibration, 213–214
　reflective solar bands, 123t
　solar calibration time series, 127f
　spectral bands, 170t–171t
Visible Transfer Radiometer (VXR), 47
Visual analysis. *See* Qualitative analysis
VNIR. *See* Visible and near-infrared
Volume Scattering Function (VSF), 408–409
Voluntary Observing Ships (VOS), 649
VOx. *See* Vanadium oxide
VSF. *See* Volume Scattering Function
VTBB. *See* Variable temperature blackbody
VXR. *See* Visible Transfer Radiometer

W

"Warm-up and cool-down" sequence (WUCD sequence), 214
Water bath blackbody (WBBB), 31
Water-leaving radiance normalization, 270–272
Wave perturbations, 281
WBBB. *See* Water bath blackbody
Wein's displacement law, 312–313
WGCV. *See* Working Group on Calibration and Validation
Wind-generated sea, 430–431
Wire-Stabilized Profiling Environmental Radiometer system (WiSPER), 420
WiSPER. *See* Wire-Stabilized Profiling Environmental Radiometer system

Working Group on Calibration and Validation (WGCV), 587
Working standard (WS), 50–51
WS. *See* Working standard
WUCD sequence. *See* "Warm-up and cool-down" sequence

Z

Zinc selenide (ZnSe), 337–338

Edwards Brothers Malloy
Thorofare, NJ USA
November 18, 2014